T0236964

# Electromagnetics Made Easy

S. Balaji

# Electromagnetics Made Easy

Springer

S. Balaji
Indira Gandhi Centre for Atomic Research
Kalpakkam, Tamil Nadu, India

ISBN 978-981-15-2660-2 ISBN 978-981-15-2658-9 (eBook)
https://doi.org/10.1007/978-981-15-2658-9

© Springer Nature Singapore Pte Ltd. 2020, corrected publication 2020
This work is subject to copyright. All rights are reserved by the Publisher, whether the whole or part of the material is concerned, specifically the rights of translation, reprinting, reuse of illustrations, recitation, broadcasting, reproduction on microfilms or in any other physical way, and transmission or information storage and retrieval, electronic adaptation, computer software, or by similar or dissimilar methodology now known or hereafter developed.
The use of general descriptive names, registered names, trademarks, service marks, etc. in this publication does not imply, even in the absence of a specific statement, that such names are exempt from the relevant protective laws and regulations and therefore free for general use.
The publisher, the authors and the editors are safe to assume that the advice and information in this book are believed to be true and accurate at the date of publication. Neither the publisher nor the authors or the editors give a warranty, expressed or implied, with respect to the material contained herein or for any errors or omissions that may have been made. The publisher remains neutral with regard to jurisdictional claims in published maps and institutional affiliations.

This Springer imprint is published by the registered company Springer Nature Singapore Pte Ltd.
The registered company address is: 152 Beach Road, #21-01/04 Gateway East, Singapore 189721, Singapore

# Preface

Engineering electromagnetics is an interesting subject in the electrical/electronics engineering curriculum. However, students find it difficult to grasp the concepts and usually complain the paper is tedious. The common conception about electromagnetics is that the subject is hard to understand and is not as beautiful as other papers. As an example, mechanics appears to be friendlier to students. Concrete concepts and dealing with tangible objects make the learning of mechanics simple for students. Studies on projectile thrown in space, an object sliding on a floor, can be clearly visualized. On the contrary, electromagnetics is characterized by abstract concepts and intangible fields. Along with complex mathematics, imperceptible concepts and invisible fields confuse the students, and understanding electromagnetics becomes a dream for the reader.

An attempt has been made in this book to change the above scenario. Real-life examples have been used throughout the book for easy grasping of the abstract concepts. This book is a result of a decade of teaching electromagnetics for electrical/electronics engineering students and physics postgraduates. Motivated by the positive feedback from students for the simple and lucid form of presentation of the subject, a need was felt to turn the presentation in the form of a textbook for a wider audience.

The book contains ten chapters. The backbone of electromagnetics is vector analysis, and to make sure the reader becomes more familiar with vector calculus, Chap. 1 has been devoted to the study of vectors. Various theorems in vector calculus and coordinate systems, which are the fundamental concepts required to understand electromagnetics, are discussed in detail in the first chapter.

Chapters 2–10 can be grouped under four major sections—electrostatics, magnetostatics, time-varying fields and applications of electromagnetics. Chapters 2 and 3 focus on electrostatics. Chapter 2 commences with an introduction to Coulomb's law and proceeds to calculate electric field for various charge distributions. Chapter 3 continues with the calculation of electric field and ends up with a detailed discussion on dielectrics and capacitors. Chapters 4 and 5 concentrate on magnetostatics. Chapter 4 introduces Biot–Savart law and continues with the calculation of magnetic flux density using various methods, and the methods are compared with their electrostatic counterpart. Chapter 5 elaborates about magnetization and magnetic materials. Chapter 6 introduces time-varying fields by discussing Faraday's law and induction, followed by a detailed account on Maxwell's equation. The last sections of Chap. 6 describe the gauge transformations, retarded potentials and time-harmonic fields. Chapter 7 gives a detailed account on the flow of energy in electromagnetic wave—the Poynting theorem, wave polarization and reflections and transmission of electromagnetic waves. Chapters 8–10 describe the application part of electromagnetics. The theory so developed in Chaps. 1–7 is applied in the development of transmission lines, waveguides and antennas in Chaps. 8–10.

Sequential development of the subject and utilizing concrete examples to explain abstract theory are the distinct features of the book. There are a number of topics in this book making it unique in the presentation of the concepts. Few such topics are mentioned below:

1. Real-life examples have been used to explain
   (a) that "Vectors are difficult to deal with as compared to scalars" in Sect. 1.1
   (b) Flux in Sect. 1.12
   (c) Existence of source and sink as related to the divergence of a vector in Sect. 1.14
   (d) Non-existence of magnetic monopoles in Sect. 4.12
   (e) Deficiency as related to magnetic scalar potential in the calculation of **B** in Sect. 4.15.
2. In spherical polar coordinates, $\theta$ is allowed to vary from 0 to $\pi$ and not up to $2\pi$. The reason for such a variation is explained with suitable figures in Sect. 1.19a.
3. For a given charge distribution all the three methods (Coulomb's law, Gauss's law and potential formulation) have been utilized to calculate the electric field.
4. In Sect. 2.18, the three methods (Coulomb's law, Gauss's law and potential formulation) utilized for the calculation of the electric field for a given charge distribution have been compared.
5. A detailed discussion on reference point R in the calculation of potential is elaborated in Sect. 2.22.

Much deeper explanations are put forward for a number of concepts, which makes the subject easy and understandable. Here are few examples.

When the instructor states "Vectors are difficult to deal with as compared to scalars," at once students question—Why is it so? Section 1.1 of this book explains why.

When the students ask—why magnetic scalar potential $V_m$ cannot be used in the calculation, the reply given will be $V_m$ is a multi-valued function. At once the question to shoot up in the students' mind is—let $V_m$ be a multi-valued function, then how does it affect the calculation? How the multi-valuedness of $V_m$ affects the calculation is explained in Sect. 4.15.

Likewise, there are a number of questions in the students' mind for which they get partial answers which do not satisfy the students' quest to learn, leading to students not understanding the subject, and finally feel that electromagnetics is difficult and can never be understood. So, as much as possible effort has been made in this book to make sure that all the questions are answered and students can grasp the subject easily.

Understanding the concepts is a different skill, while applying the concepts to solve the problems is a different skill. To make sure that the students apply whatever they have learnt, in solving problems, a number of examples have been included throughout the text. A list of exercises is given at the end of each chapter, and it is emphasized that the students should attempt those exercises on their own to further strengthen their problem-solving skill.

Finally, I should thank my wife E. Sumathi and my daughter B. S. Nakshathra for their extreme patience they had during the development of this book.

Kalpakkam, India

S. Balaji

---

The original version of the book was published without incorporating the belated corrections provided by the author. The correction to the book is available at https://doi.org/10.1007/978-981-15-2658-9_11

# Contents

# About the Author

**Dr. S. Balaji** is a Scientific Officer at the Indira Gandhi Center for Atomic Research (IGCAR), Kalpakkam, India. He obtained his Ph.D. in physics from the University of Madras. His research primarily focuses on radiation damage, application of ion beams in material science, and ion beam analysis. He is also actively involved in the development of experimental facilities related to ion beam irradiation. He has published 20 articles in international peer-reviewed journals and conference proceedings. Having a passion for teaching, he has taught engineering electromagnetics/electromagnetic theory to engineering undergraduates and physics postgraduates for over a decade. Currently, he teaches the same topic to the junior research fellows at IGCAR.

# Chapter 1
# Vector Analysis

## 1.1 Introduction

(a)

Several complex problems in electromagnetism involve many physical quantities having both magnitude and direction. These quantities called vectors play an important role in understanding the multitude of phenomena in electrostatics, magnetostatics, etc.

A scalar is a quantity which involves magnitude alone. A vector is a quantity which involves both magnitude and direction. Examples of scalar quantities are mass, length, temperature, etc., and examples of vector quantities are displacement, velocity, force, etc.

The following example will make the difference between vector and scalar much clearer. Suppose, assume that there are two persons, person A and person B, trying to add rice in an empty box as shown in Fig. 1.1a, b. Person A is having 7 kg of rice and person B is having 2 kg of rice.

In Fig. 1.1a person A and person B are standing in same direction and are adding the rice in the empty box. After adding the rice in the box, the box would contain 9 kg of rice. In Fig. 1.1b person A and person B are standing opposite to each other and are adding the rice in the empty box. Once again the box would then contain 9 kg of rice after the rice has been added in the box. So, the conclusion is, whatever direction the mass is added the net effect is same, we get 9 kg of rice in total. The total (9 kg in our example) simply depends on the initial ***magnitude*** (7 and 2 kg in our example) of the masses and not on the direction in which the masses are added.

Now consider Fig. 1.2a, b. In Fig. 1.2a person A and person B are kicking a football with a force of 7 N and 2 N (Numbers are not to be taken seriously) in the same direction. That is person A and person B are standing adjacent to each other and kicking the football. The total effect or the total force is 7 N + 2 N = 9 N and the ball will move in the direction of the kick. In Fig. 1.2b person A and person B are standing opposite to each other and kicking the football with a force of 7 N and

© Springer Nature Singapore Pte Ltd. 2020

S. Balaji, *Electromagnetics Made Easy*,
https://doi.org/10.1007/978-981-15-2658-9_1

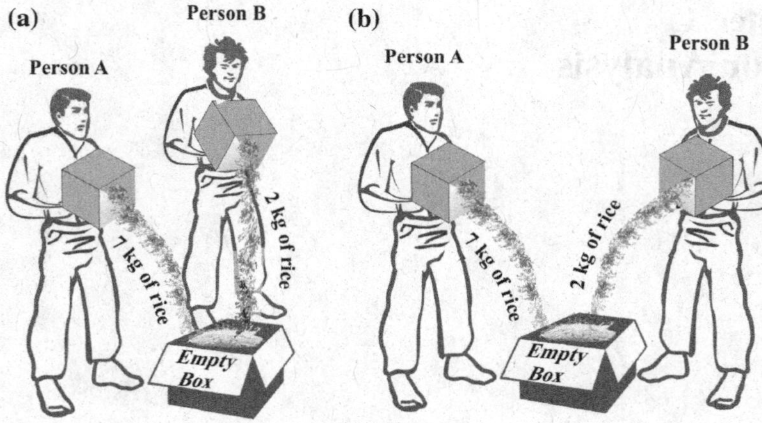


**Fig. 1.1** **a** Person A and Person B standing in same direction and adding rice in the empty box. **b** Person A and Person B standing in the opposite direction and adding rice in the empty box

2 N, respectively. The net or total force is 7 N − 2 N = 5 N acting in the direction of kick of person A. Now, the conclusion is the total force (9 N and 5 N in our example) depends on, not only on the ***magnitude*** (7 N and 2 N in our example) but also on the ***direction*** of the initial forces. Person A and person B can stand in any possible direction and kick the football. Few such possible directions are shown in Fig. 1.2c. For such situations shown in Fig. 1.2c the net or total force and its direction must be obtained by using laws of vector addition.

Thus, irrespective of whatever direction the initial masses are added we get the same total mass. However when forces are added, depending upon the direction of the initial forces, we get a different net or total force.

Quantities which behave like mass are called scalars and quantities which behave like force are called vectors. Thus, the above example clearly distinguishes between a vector and scalar.

We have another important information from the above example. During addition of the scalars one need not bother about their direction. One must take into account their magnitude alone. Because irrespective of whatever direction the scalars (masses in our example) are added we get the same total or net scalar (mass in our example) we need not bother about their direction.

However, during addition of vectors we need to keep track of not only their magnitude, but also their direction. In our force example in Fig. 1.2a–c the initial direction of 7 N and 2 N must always be tracked if we want to get the total or net force.

As we will see shortly the direction of the total or net force must also be tracked. Because the direction of the initial vector (7 N and 2 N in our force example in Fig. 1.2a–c) must always be tracked in order to obtain the value of the direction and magnitudes of the total or net vector ***vector addition becomes complicated than scalar addition***. Scalar addition is simple. One needs to bother about their

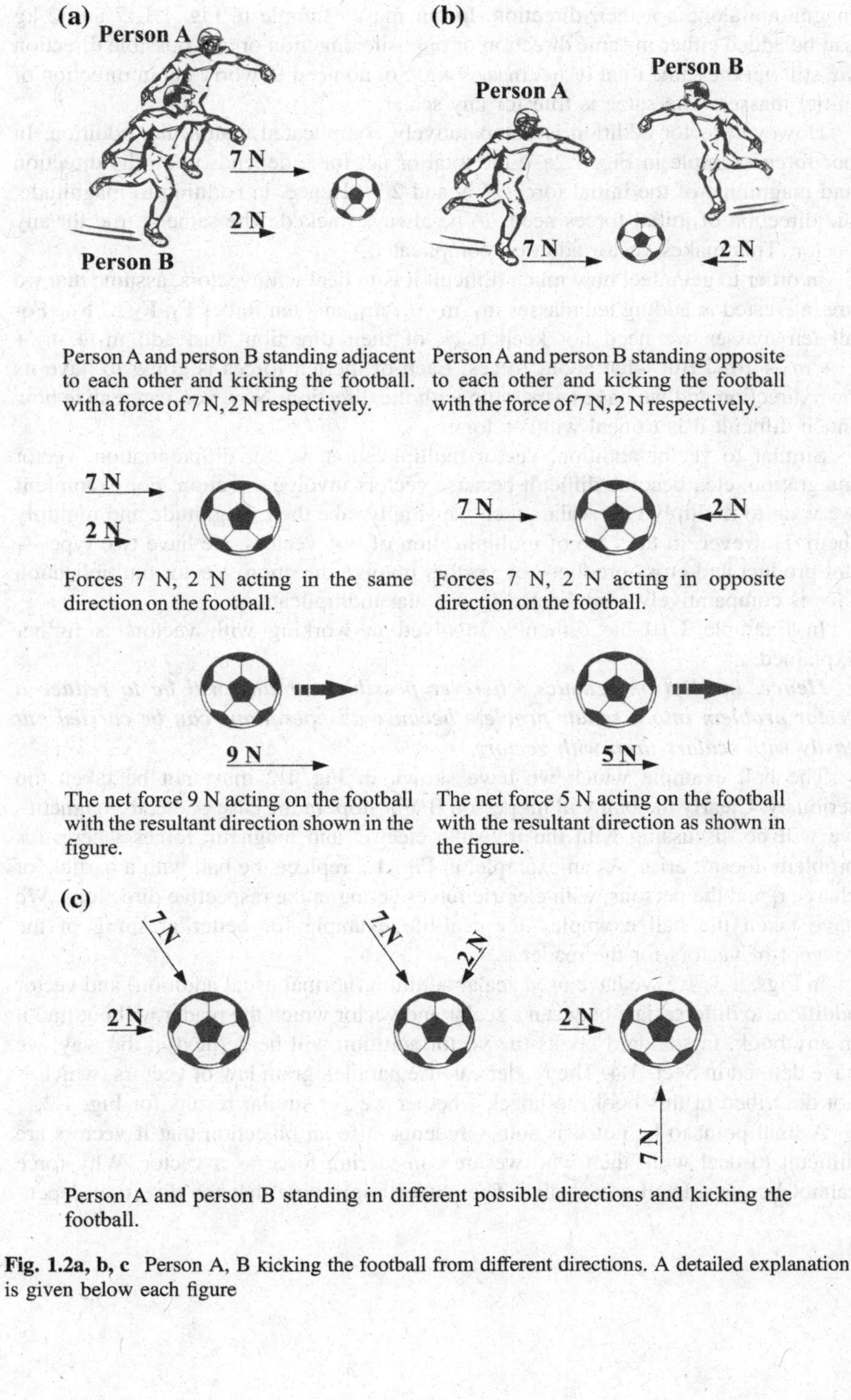
(a)
Person A
7 N
2 N
Person B
(b)
Person B
Person A
7 N
2 N
Person A and person B standing adjacent to each other and kicking the football with a force of 7 N, 2 N respectively.
Person A and person B standing opposite to each other and kicking the football with the force of 7 N, 2 N respectively.
7 N
2 N
7 N
2 N
Forces 7 N, 2 N acting in the same direction on the football.
Forces 7 N, 2 N acting in opposite direction on the football.
9 N
5 N
The net force 9 N acting on the football with the resultant direction shown in the figure.
The net force 5 N acting on the football with the resultant direction as shown in the figure.
(c)
7N
2 N
7N
2N
2 N
7N
Person A and person B standing in different possible directions and kicking the football.

**Fig. 1.2a, b, c** Person A, B kicking the football from different directions. A detailed explanation is given below each figure

magnitude alone not their direction. In our mass example in Fig. 1.1, 7 and 2 kg can be added either in same direction or opposite direction or any possible direction we still get the same total or net mass 9 kg. So, no need to worry about direction of initial masses. The same is true for any scalar.

However vector addition is comparatively complicated than scalar addition. In our force example in Fig. 1.2a–c the total or net force depends upon the direction and magnitude of the initial forces 7 N and 2 N. Hence, in addition to magnitude, the direction of initial forces needs to be always tracked. The same is true for any vector. This makes vector addition complicated.

In order to get a feel how much difficult it is to deal with vectors, assume that we are interested in adding ten masses $m_1$, $m_2$, … $m_{10}$ and ten forces $\mathbf{F_1}$, $\mathbf{F_2}$ … $\mathbf{F_{10}}$. For all ten masses we need not keep track of their direction. Just add $m_1 + m_2 + \cdots + m_9 + m_{10}$. But what about forces. Each of the ten forces is going to have its own direction and we have to mess up with the direction. Now you can realize how much difficult it is to deal with vectors.

Similar to vector addition, vector multiplication, vector differentiation, vector integration, etc., become difficult because vectors involve direction. For example if we want to multiply two scalars then we simply take their magnitude and multiply them. However, in the case of multiplication of two vectors, we have two types—dot product and cross product—as vectors involve direction. Vector multiplication thus is comparatively complicated than scalar multiplication.

In Example 1.10 the difficulty involved in working with vectors is further explained.

***Hence, in electromagnetics wherever possible our aim will be to reduce a vector problem into a scalar problem because all operations can be carried out easily with scalars than with vectors.***

The ball example which we have shown in Fig. 1.2 must not be taken too seriously. Clearly the ball will hit person B and stop. In the case of electromagnetics we will be discussing with the invisible electric and magnetic forces where this problem doesn't arise. As an example, in Fig. 1.2 replace the ball with a particle of charge q and the persons, with electric forces acting in the respective directions. We have taken the ball example—the real-life example for better grasping of the concept of vectors for the reader.

In Figs. 1.1, 1.2 we have used scalar addition (normal usual addition) and vector addition, to differentiate between a scalar and vector which the reader will not find it in any book. In standard books the vector addition will be defined in the way, we have defined in Sect. 1.4. The reader can use parallelogram law of vectors (which is not described in this book) to check whether we get similar results for Fig. 1.2.

A final point to be noted is some students raise an objection that if vectors are difficult to deal with, then why we are considering force as a vector. Why force cannot be considered as a scalar. The reader might note that the direction dependency of the force shown in Fig. 1.2a, b is not our wish. We are not forcing the force to act in such a manner; adding and cancelling depending upon direction. Direction dependency of force (vectors) is naturally and intrinsically present in it and direction independency of mass (scalars) is naturally and intrinsically present in it.

(b)
Previously we noted that not only the direction of the initial forces must be tracked but the direction of the net or total force must also be tracked. In order to understand the above point consider Fig. 1.2d. In Fig. 1.2d person A and person B are kicking the football exactly opposite to the manner shown in Fig. 1.2b. Clearly, in Fig. 1.2d the resultant force is also 5 N but now it is opposite to the resultant force 5 N in Fig. 1.2b. Thus, the resultant force in Fig. 1.2b, d is 5 N but acting completely in opposite direction. The result is football in Fig. 1.2d moves completely in opposite direction as compared to Fig. 1.2b. Thus although the total force in Fig. 1.2b, d is 5 N the result they produce is different (football moving in opposite direction) because the direction in which the resultant force 5 N acting in Fig. 1.2b, d is exactly opposite.

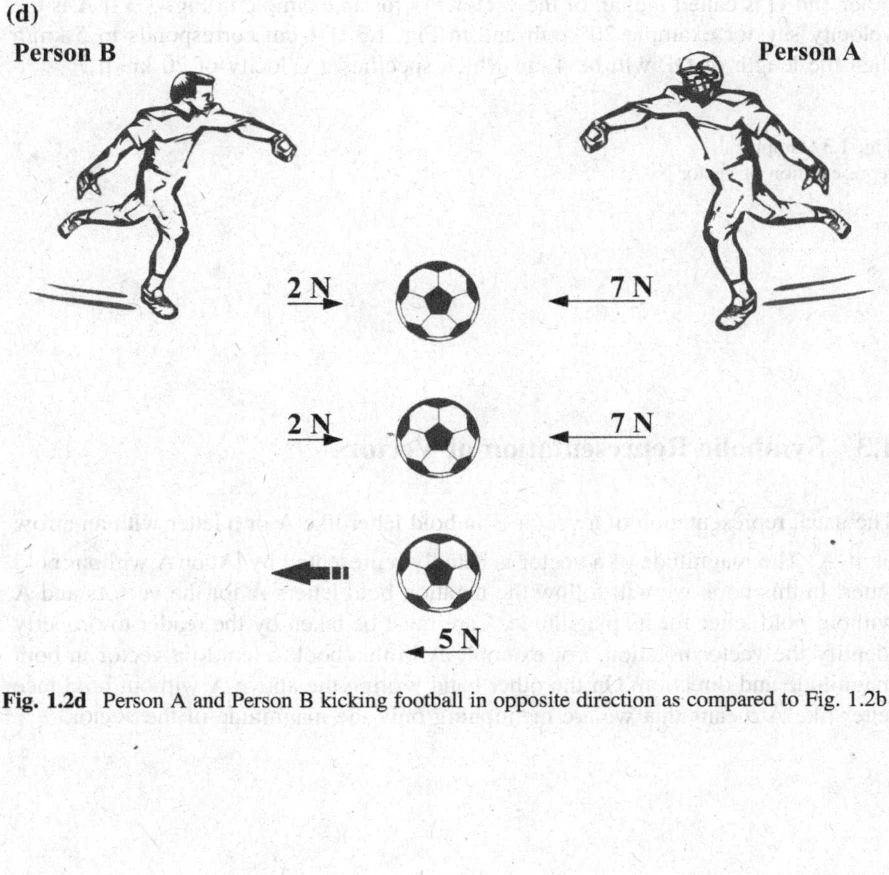


**Fig. 1.2d** Person A and Person B kicking football in opposite direction as compared to Fig. 1.2b

***Thus the conclusion is although the magnitude of two forces are same if their directions are different they are completely different forces. The same is true for any vector.***

However, we need not track the direction of total or net mass 9 kgs in Fig. 1.1a, b because mass doesn't involve direction. The same is true for any scalar.

In your lower classes you would have learnt how to add, subtract, multiply differentiate and integrate scalars. It's very easy. Now in the coming chapters we will learn how add, subtract, multiply, differentiate and integrate vectors.

## 1.2 Graphical Representation of Vectors

A vector is graphically represented by an arrow. The length of the arrow represents the magnitude of the vector with respect to a pre-chosen scale. The direction of the vector is given by the arrowhead. For example Fig. 1.3 shows the graphical representation of a vector **A**. The arrow pointing from O to P represents the direction of the vector and the length represents the magnitude (to a suitable scale) of the vector. The end P containing the arrowhead is called the head of the vector whereas the other end O is called the tail of the vector. As for an example in Fig. 1.3 if **A** is the velocity say for example 20 km/h and in Fig. 1.3 if 1 cm corresponds to 5 km/h then the length of OP will be 4 cm which specifies a velocity of 20 km/h.

**Fig. 1.3** Graphical representation of vector

## 1.3 Symbolic Representation of Vectors

The usual representation of a vector is in bold letter like **A** or a letter with an arrow on it $\overrightarrow{A}$. The magnitude of a vector is usually represented by |**A**| or A without bold letter. In this book we will follow the notation bold letters **A** for the vectors and A without bold letter for its magnitude. Care must be taken by the reader to properly identify the vector notation. For example **A** in this book refers to a vector in both magnitude and direction. On the other hand writing the above **A** without bold face letter like A means that we are mentioning only the magnitude of the vector.

## 1.4 Vector Addition

Figure 1.2 illustrates addition of vectors is not as simple as addition of scalars. In Fig. 1.4a the addition of two vectors **A** and **B** is shown. The tail of the **B** is placed on the head of **A**. The resultant **R** is given by the vector drawn from the tail of the vector **A** to the head of the vector **B**.

In Fig. 1.4a we add vector **B** to **A**. In Fig. 1.4b we add vector **A** to **B**. In both the cases the resultant is same. Hence vector addition is commutative.

$$\mathbf{A}+\mathbf{B}=\mathbf{B}+\mathbf{A} \tag{1.1}$$

Also vector addition is associative

$$(\mathbf{A}+\mathbf{B})+\mathbf{C}=\mathbf{A}+(\mathbf{B}+\mathbf{C}) \tag{1.2}$$

## 1.5 Subtraction of Vectors

Figure 1.5 shows two vectors **A** and **B**. For subtracting **B** from **A** reverse the direction of **B** (without changing the magnitude) as shown in Fig. 1.6 and then add **B** to **A**.

**Fig. 1.5** Subtraction of vectors

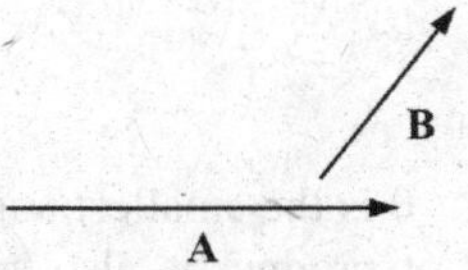


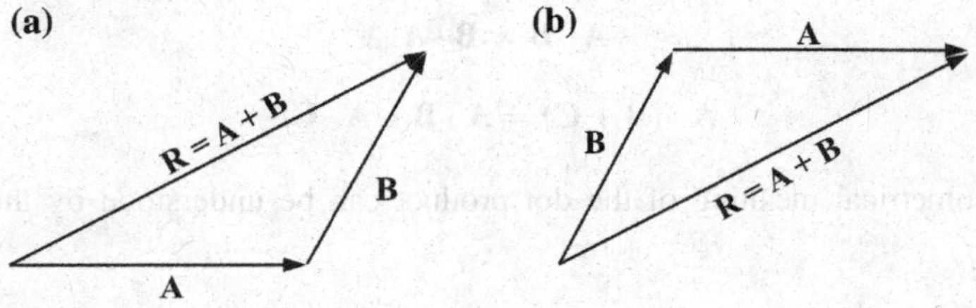


**Fig. 1.4** Addition of vectors

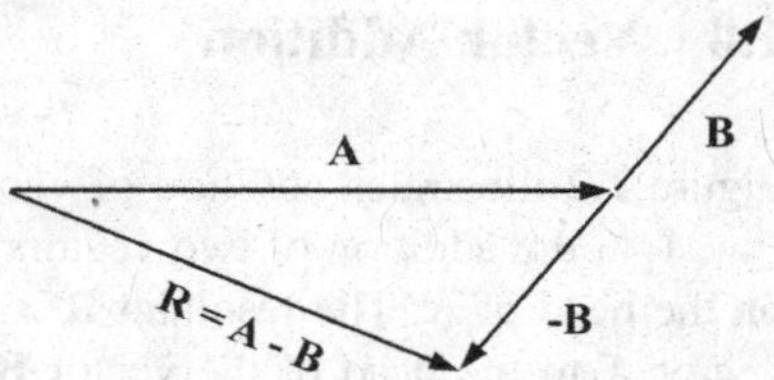


**Fig. 1.6** Subtraction of vectors

## 1.6 Multiplication of a Vector by a Scalar

If a vector is multiplied by a scalar u then the magnitude of the vector is increased by u times and the direction of the vector remains unchanged. If u = 3 then the magnitude of vector is increased by three times as shown in Fig. 1.7. On the other hand if u is negative then the direction of the vector is reversed.

Multiplication of vector by a scalar is distributive.

$$u(\mathbf{A}+\mathbf{B}) = u\mathbf{A} + u\mathbf{B} \tag{1.3}$$

## 1.7 Multiplication of Vectors: Dot Product of Two Vectors

The dot product of two vectors **A**, **B** is defined by

$$\mathbf{A}\cdot\mathbf{B} = AB\cos\theta \tag{1.4}$$

θ is the smallest angle between the two vectors when they are placed tail to tail. Dot product is also sometimes called scalar product because the result of a dot product is a scalar. Dot product obeys commutative and distributive laws.

$$\mathbf{A}\cdot\mathbf{B} = \mathbf{B}\cdot\mathbf{A} \tag{1.5}$$

$$\mathbf{A}\cdot(\mathbf{B}+\mathbf{C}) = \mathbf{A}\cdot\mathbf{B} + \mathbf{A}\cdot\mathbf{C} \tag{1.6}$$

The geometrical meaning of the dot product can be understood by the aid of Fig. 1.8.

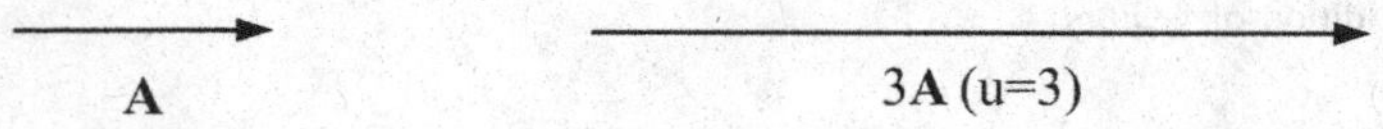


**Fig. 1.7** Multiplication of a vector by a scalar

Fig. 1.8 Dot product of vectors

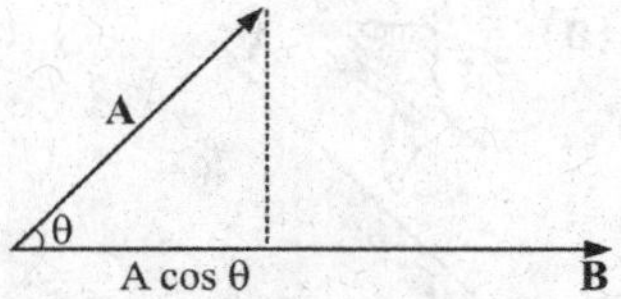


As can be seen from Fig. 1.8 the projection of **A** on **B** is A cosθ. Geometrically the dot product $\mathbf{A} \cdot \mathbf{B}$ is the product of B and the projection of **A** on **B**.

If two vectors are parallel then $\mathbf{A} \cdot \mathbf{B} = \mathrm{AB} \cos 0 = \mathrm{AB}$ and if they are perpendicular then $\mathbf{A} \cdot \mathbf{B} = \mathrm{AB} \cos 90 = 0$.

## 1.8 Multiplication of Vectors—Cross—Product of Two Vectors

The cross product of two vectors is denoted by $\mathbf{A} \times \mathbf{B}$ and is defined by

$$\mathbf{A} \times \mathbf{B} = \mathrm{AB} \sin \theta \, \hat{\mathbf{n}} \tag{1.7}$$

Here $\hat{\mathbf{n}}$ is the unit vector in the direction perpendicular to thè plane containing **A**, **B**. However there are two perpendicular directions to the plane as shown in Fig. 1.9. The proper direction of $\hat{\mathbf{n}}$ is given by right-hand rule. Let all the fingers of the right hand except the thumb point in the direction of **A** and let the fingers curl towards **B** through the small angle between **A** and **B**. Then the thumb gives the direction of $\hat{\mathbf{n}}$ or the direction of $\mathbf{A} \times \mathbf{B}$. The entire sequence is shown in Fig. 1.10a. Similar sequence for obtaining the direction of $\mathbf{B} \times \mathbf{A}$ is shown in Fig. 1.10b. The direction of $\mathbf{A} \times \mathbf{B}$ and $\mathbf{B} \times \mathbf{A}$ as per right-hand rule is indicated in Fig. 1.9. From the figure it is clear that

$$\mathbf{A} \times \mathbf{B} = -(\mathbf{B} \times \mathbf{A}) \tag{1.8}$$

Fig. 1.9 Cross product of vectors

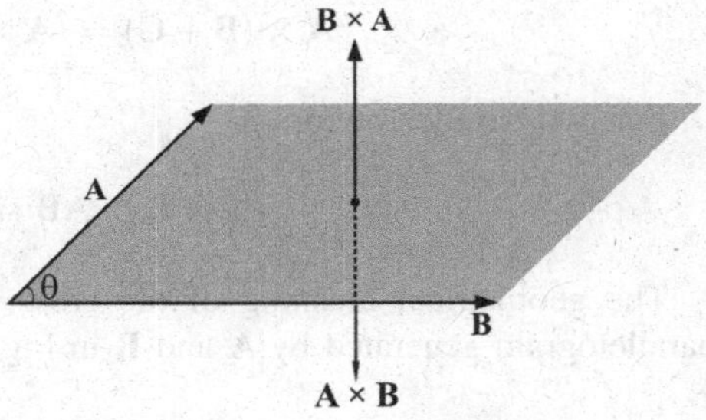

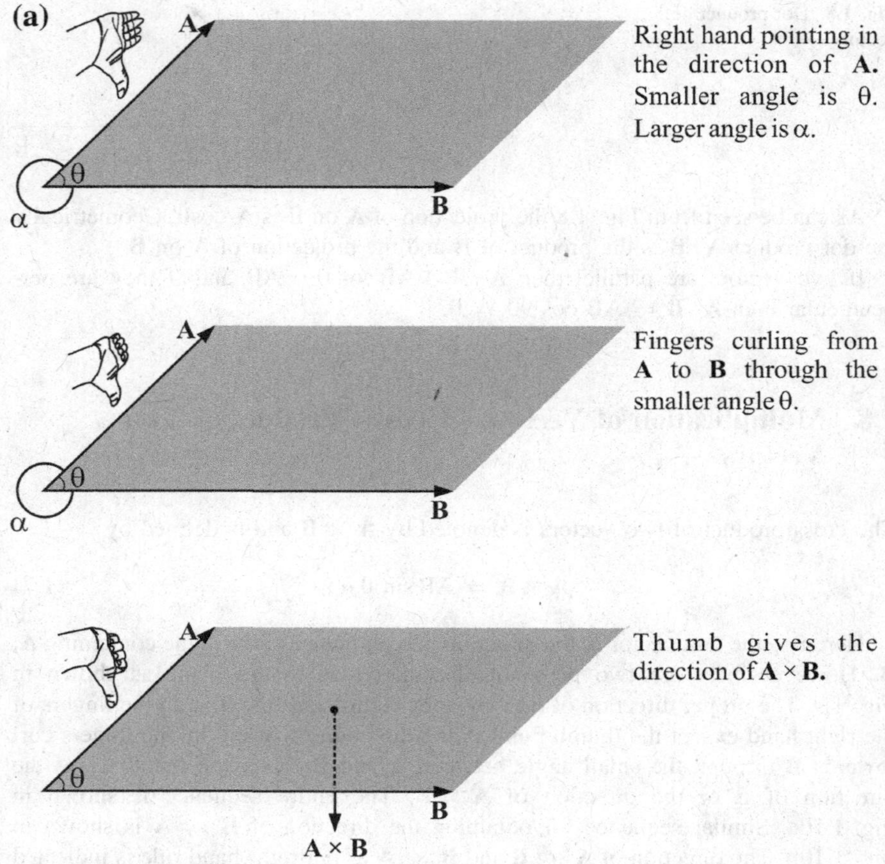
(a)
A
Right hand pointing in the direction of A. Smaller angle is θ. Larger angle is α.
θ
α
B
A
Fingers curling from A to B through the smaller angle θ.
θ
α
B
A
Thumb gives the direction of A × B.
θ
B
A × B

**Fig. 1.10a** Fixing the direction for the cross product $\mathbf{A} \times \mathbf{B}$

Hence the cross product is not commutative. However the cross product is distributive if the order of vectors is not changed.

$$\mathbf{A} \times (\mathbf{B} + \mathbf{C}) = (\mathbf{A} \times \mathbf{B}) + (\mathbf{A} \times \mathbf{C}) \tag{1.9}$$

For two parallel vectors

$$\mathbf{A} \times \mathbf{B} = \text{AB} \sin 0 \; \hat{\mathbf{n}} = 0$$

The geometrical meaning of the cross product is $|\mathbf{A} \times \mathbf{B}|$ is the area of the parallelogram generated by **A** and **B** in Fig. 1.9.

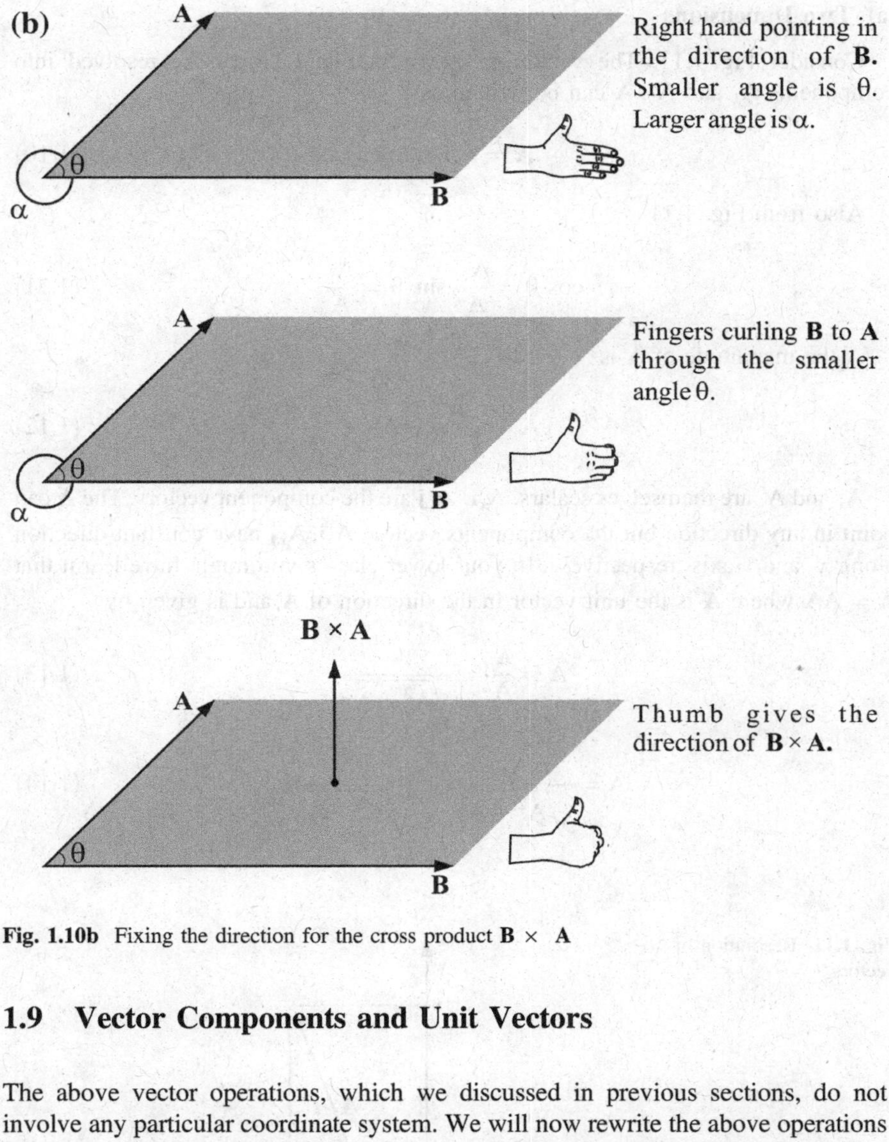
(b)
A
θ
α
B
Right hand pointing in the direction of **B**. Smaller angle is θ. Larger angle is α.
A
θ
α
B
Fingers curling **B** to **A** through the smaller angle θ.
B × A
A
θ
B
Thumb gives the direction of **B** × **A**.

**Fig. 1.10b** Fixing the direction for the cross product **B** × **A**

## 1.9 Vector Components and Unit Vectors

The above vector operations, which we discussed in previous sections, do not involve any particular coordinate system. We will now rewrite the above operations in terms of components in the most widely used coordinate system—Cartesian coordinate system. The unit vectors in the direction of *x*-, *y*- and *z*-axis in the Cartesian coordinate system are $\hat{\mathbf{i}}, \hat{\mathbf{j}}, \hat{\mathbf{k}}$. We will discuss vector components in two dimensions and then extend the results to three dimensions.

(a) **Two Dimensions**

Consider Fig. 1.11. The vector **A** shown in Fig. 1.11 can be resolved into components $A_x$ and $A_y$. **A** can be written as

$$\mathbf{A} = A_x\hat{\mathbf{i}} + A_y\hat{\mathbf{j}} \tag{1.10}$$

Also from Fig. 1.11

$$\cos\theta = \frac{A_x}{A}, \sin\theta = \frac{A_y}{A} \tag{1.11}$$

A the magnitude of **A** is

$$A = \sqrt{A_x^2 + A_y^2} \tag{1.12}$$

$A_x$ and $A_y$ are themselves scalars. $A_x\hat{\mathbf{i}}$, $A_y\hat{\mathbf{j}}$ are the component vectors. The **A** can point in any direction but the components vectors $A_x\hat{\mathbf{i}}$, $A_y\hat{\mathbf{j}}$ have constant direction along *x*- and *y*-axis, respectively. In your lower classes you might have learnt that $\mathbf{A} = A\hat{\mathbf{A}}$ where $\hat{\mathbf{A}}$ is the unit vector in the direction of **A** and is given by

$$\hat{\mathbf{A}} = \frac{\mathbf{A}}{A} = \frac{\mathbf{A}}{\sqrt{A_x^2 + A_y^2}} \tag{1.13}$$

$$\hat{\mathbf{A}} = \frac{A_x}{\sqrt{A_x^2 + A_y^2}}\hat{\mathbf{i}} + \frac{A_y}{\sqrt{A_x^2 + A_y^2}}\hat{\mathbf{j}} \tag{1.14}$$

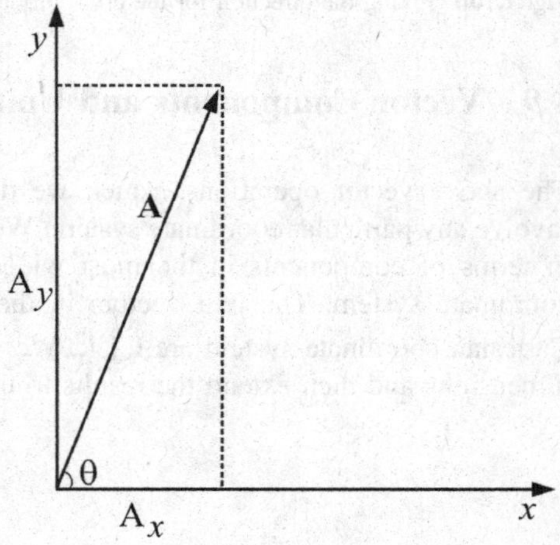


**Fig. 1.11** Resolution of vectors

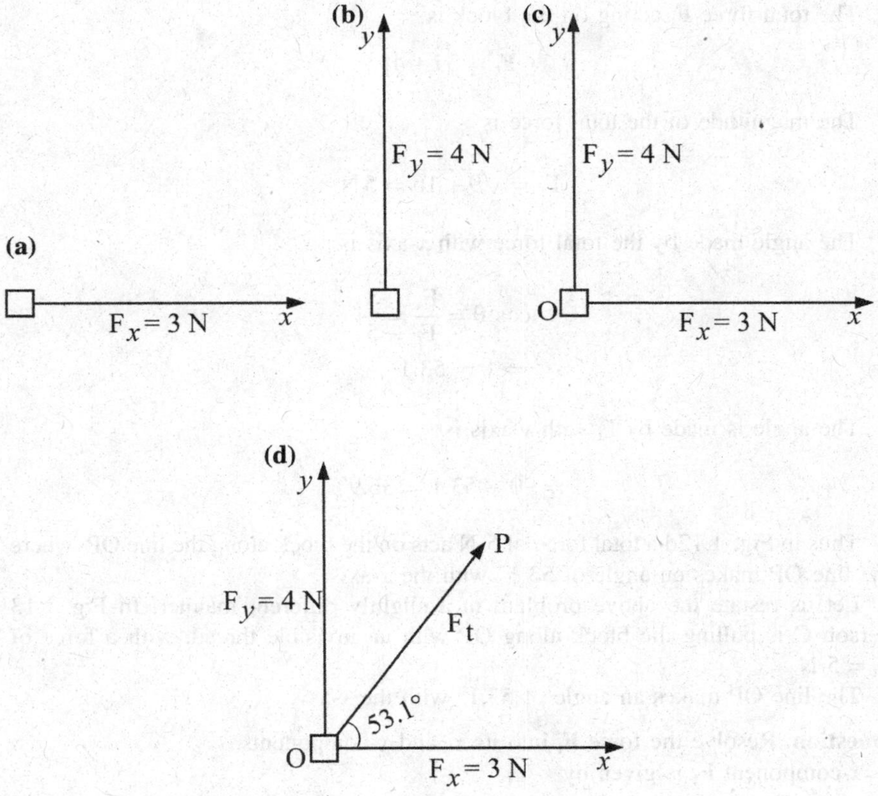


**Fig. 1.12** Example illustrating the resolution of forces

To make the above points clearer let us consider an example. In Fig. 1.12a person A is pulling a block of iron along *x*-axis with a force of 3 N using a thread tied to the block. The block will move along *x*-axis. In Fig. 1.12b person B is pulling the same block with a force of 4 N along *y*-axis using a thread tied to the block. Now the block will move along *y*-axis (In the above cases numbers are not to be taken seriously). Now consider Fig. 1.12c. Person A is pulling the block along *x*-axis with a force of 3 N while person B is pulling the block along *y*-axis with a force of 4 N using threads tied to the block. Assume that the threads are "invisible"—in the sense person A and person B are able to exert forces on the block (or) pull the block using the thread but the block is not constrained to move along *x*- or *y*-axis because of the thread or in other words the block can move in the *x*–*y* plane because of the forces acting on it without getting constrained by the threads.

**Question**: What is the total force $\mathbf{F}_t$ acting on the block. What is the direction of $\mathbf{F}_t$ or in which direction the block will move?

The total force $\mathbf{F}_t$ acting on the block is

$$\mathbf{F}_t = 3\hat{\mathbf{i}} + 4\hat{\mathbf{j}}$$

The magnitude of the total force is

$$F_t = \sqrt{9+16} = 5\,N$$

The angle made by the total force with $x$-axis is

$$\cos\theta = \frac{F_x}{F_t} = \frac{3}{5}$$
$$\Rightarrow \theta = 53.1^\circ$$

The angle is made by $F_t$ with $y$-axis is

$$= 90 - 53.1 = 36.9^\circ$$

Thus in Fig. 1.12d a total force of 5 N acts on the block along the line OP, where the line OP makes an angle of 53.1° with the $x$-axis.

Let us restate the above problem in a slightly different manner. In Fig. 1.13 person C is pulling the block along OP with an invisible thread, with a force of $F_t = 5$ N.

The line OP makes an angle of 53.1° with the $x$-axis.

**Question**: Resolve the force $\mathbf{F}_t$ into its $x$- and $y$-components.

$x$-component $F_x$ is given by

$$F_x = F_t \cos\theta = 5\cos 53.1 = 3\,N$$

$y$-component $F_y$ is given by

$$F_y = F_t \sin\theta = 5\ \sin 53.1$$
$$F_y = 4\,N.$$

Thus person C pulling the block along line OP in Fig. 1.13 is equivalent to person A and B pulling the same block with a force of 3 N and 4 N along $x$- and $y$-axis, respectively, as in Fig. 1.12d. Or in other words, we have resolved $\mathbf{F}_t$ in

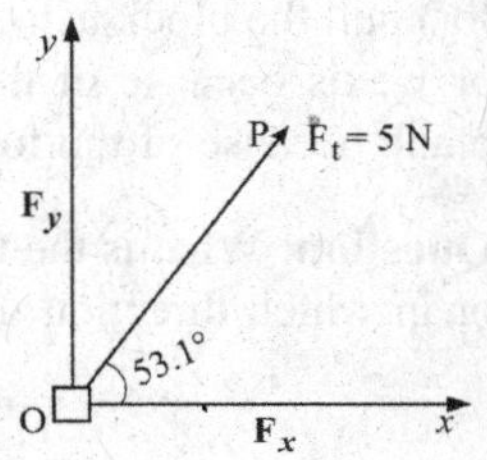


**Fig. 1.13** Example illustrating the resolution of forcesExample illustrating the resolution of forces

**Fig. 1.14** Figure illustrating that mass cannot be resolved into components

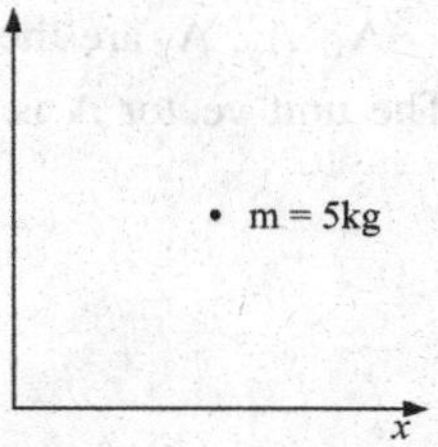


Fig. 1.13 into equivalent $x$- and $y$-components $\mathbf{F}_x$ and $\mathbf{F}_y$, respectively. The same is true for any vector. That is, any vector can be resolved into its components.

What about resolving scalars. To find out, consider a particle of mass 5 kg located as shown in Fig. 1.14. Can you resolve the mass of 5 kg into its components along $x$- and $y$-axis. Mass being a scalar quantity is independent of direction and cannot be resolved into components. The same is true for any scalar.

(b) **Three Dimensions**

Now let us generalize the above results for three dimensions. In Fig. 1.15a we show $x$, $y$, $z$ along with their unit vectors $\hat{\mathbf{i}}$, $\hat{\mathbf{j}}$, $\hat{\mathbf{k}}$. The vector $\mathbf{A}$ shown in Fig. 1.15b can be written in terms of $\hat{\mathbf{i}}$, $\hat{\mathbf{j}}$, $\hat{\mathbf{k}}$ as

$$\mathbf{A} = A_x\hat{\mathbf{i}} + A_y\hat{\mathbf{j}} + A_z\hat{\mathbf{k}} \tag{1.15}$$

and

$$\cos\alpha = \frac{A_x}{A}, \cos\beta = \frac{A_y}{A}, \cos\Gamma = \frac{A_z}{A} \tag{1.16}$$

A the magnitude of $\mathbf{A}$ is

$$A = \sqrt{A_x^2 + A_y^2 + A_z^2} \tag{1.17}$$

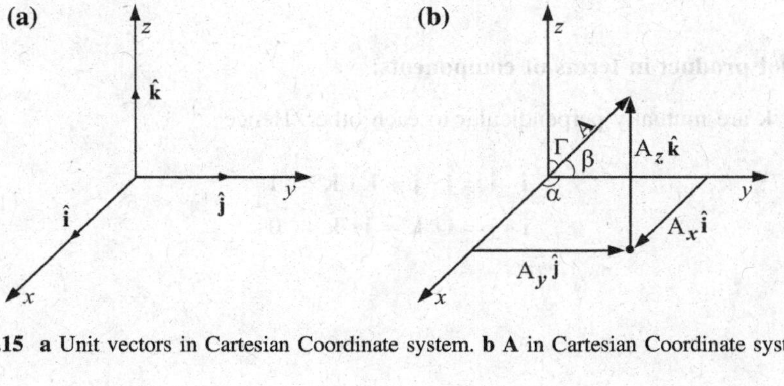


**Fig. 1.15** **a** Unit vectors in Cartesian Coordinate system. **b A** in Cartesian Coordinate system

$A_x$, $A_y$, $A_z$ are themselves scalars. $A_x\,\hat{\mathbf{i}}$, $A_y\,\hat{\mathbf{j}}$, $A_z\,\hat{\mathbf{k}}$ are the component vectors. The unit vector $\hat{\mathbf{A}}$ is given by

$$\hat{\mathbf{A}} = \frac{\mathbf{A}}{A} = \frac{A_x\hat{\mathbf{i}} + A_y\hat{\mathbf{j}} + A_z\hat{\mathbf{k}}}{\sqrt{A_x^2 + A_y^2 + A_z^2}} \tag{1.18}$$

$$\hat{\mathbf{A}} = \frac{A_x}{\sqrt{A_x^2 + A_y^2 + A_z^2}}\hat{\mathbf{i}} + \frac{A_y}{\sqrt{A_x^2 + A_y^2 + A_z^2}}\hat{\mathbf{j}} + \frac{A_z}{\sqrt{A_x^2 + A_y^2 + A_z^2}}\hat{\mathbf{k}} \tag{1.19}$$

Writing any vector in terms of its components has number of advantages.

(c) **Vector addition and subtraction in terms of components**:

To add any two vectors **A** and **B** just add the components

$$\begin{aligned} \mathbf{A} + \mathbf{B} &= (A_x\hat{\mathbf{i}} + A_y\hat{\mathbf{j}} + A_z\hat{\mathbf{k}}) + (B_x\hat{\mathbf{i}} + B_y\hat{\mathbf{j}} + B_z\hat{\mathbf{k}}) \\ \mathbf{A} + \mathbf{B} &= (A_x + B_x)\hat{\mathbf{i}} + (A_y + B_y)\hat{\mathbf{j}} + (A_z + B_z)\hat{\mathbf{k}} \end{aligned} \tag{1.20}$$

To subtract two vectors **A** and **B** just subtract the components

$$\begin{aligned} \mathbf{A} - \mathbf{B} &= (A_x\hat{\mathbf{i}} + A_y\hat{\mathbf{j}} + A_z\hat{\mathbf{k}}) - (B_x\hat{\mathbf{i}} + B_y\hat{\mathbf{j}} + B_z\hat{\mathbf{k}}) \\ \mathbf{A} - \mathbf{B} &= (A_x - B_x)\hat{\mathbf{i}} + (A_y - B_y)\hat{\mathbf{j}} + (A_z - B_z)\hat{\mathbf{k}} \end{aligned} \tag{1.21}$$

Addition and subtraction of components are similar to that of scalars.

(d) **Multiplication of a vector by a scalar**:

To multiply a vector by a scalar say u, multiply each component by a respective scalar

$$\begin{aligned} u\mathbf{A} &= u(A_x\hat{\mathbf{i}} + A_y\hat{\mathbf{j}} + A_z\hat{\mathbf{k}}) \\ u\mathbf{A} &= uA_x\hat{\mathbf{i}} + uA_y\hat{\mathbf{j}} + uA_z\hat{\mathbf{k}} \end{aligned} \tag{1.22}$$

(e) **Dot product in terms of components:**

$\hat{\mathbf{i}}, \hat{\mathbf{j}}, \hat{\mathbf{k}}$ are mutually perpendicular to each other. Hence

$$\begin{aligned} \hat{\mathbf{i}}\cdot\hat{\mathbf{i}} = \hat{\mathbf{j}}\cdot\hat{\mathbf{j}} = \hat{\mathbf{k}}\cdot\hat{\mathbf{k}} &= 1 \\ \hat{\mathbf{i}}\cdot\hat{\mathbf{j}} = \hat{\mathbf{i}}\cdot\hat{\mathbf{k}} = \hat{\mathbf{j}}\cdot\hat{\mathbf{k}} &= 0 \end{aligned} \tag{1.23}$$

Thus the dot product can be written as

$$\begin{aligned} \mathbf{A}\cdot\mathbf{B} &= (A_x\hat{\mathbf{i}} + A_y\hat{\mathbf{j}} + A_z\hat{\mathbf{k}})\cdot(B_x\hat{\mathbf{i}} + B_y\hat{\mathbf{j}} + B_z\hat{\mathbf{k}}) \\ \mathbf{A}\cdot\mathbf{B} &= A_xB_x + A_yB_y + A_zB_z \end{aligned} \quad (1.24)$$

(f) **Cross product in terms of components**:

$$\left.\begin{aligned} \hat{\mathbf{i}}\times\hat{\mathbf{i}} &= \hat{\mathbf{j}}\times\hat{\mathbf{j}} = \hat{\mathbf{k}}\times\hat{\mathbf{k}} = 0 \\ \hat{\mathbf{i}}\times\hat{\mathbf{j}} &= -\hat{\mathbf{j}}\times\hat{\mathbf{i}} = \hat{\mathbf{k}} \\ \hat{\mathbf{j}}\times\hat{\mathbf{k}} &= -\hat{\mathbf{k}}\times\hat{\mathbf{j}} = \hat{\mathbf{i}} \\ \hat{\mathbf{k}}\times\hat{\mathbf{i}} &= -\hat{\mathbf{i}}\times\hat{\mathbf{k}} = \hat{\mathbf{j}} \end{aligned}\right\} \quad (1.25)$$

Therefore

$$\mathbf{A}\times\mathbf{B} = (A_x\hat{\mathbf{i}} + A_y\hat{\mathbf{j}} + A_z\hat{\mathbf{k}})\times(B_x\hat{\mathbf{i}} + B_y\hat{\mathbf{j}} + B_z\hat{\mathbf{k}}) \quad (1.26)$$

$$= (A_yB_z - A_zB_y)\hat{\mathbf{i}} + (A_zB_x - A_xB_z)\hat{\mathbf{j}} + (A_xB_y - A_yB_x)\hat{\mathbf{k}} \quad (1.27)$$

which can be written in much more familiar form as

$$\mathbf{A}\times\mathbf{B} = \begin{vmatrix} \hat{\mathbf{i}} & \hat{\mathbf{j}} & \hat{\mathbf{k}} \\ A_x & A_y & A_z \\ B_x & B_y & B_z \end{vmatrix} \quad (1.28)$$

## 1.10 Triple Products

Scalar triple product is defined as $\mathbf{A}\cdot(\mathbf{B}\times\mathbf{C})$. The geometrical meaning of $\mathbf{A}\cdot(\mathbf{B}\times\mathbf{C})$ can be understood with the aid of Fig. 1.16.

$$\mathbf{A}\cdot(\mathbf{B}\times\mathbf{C}) = \mathbf{A}\cos\theta\,(\mathbf{B}\times\mathbf{C}) \quad (1.29)$$

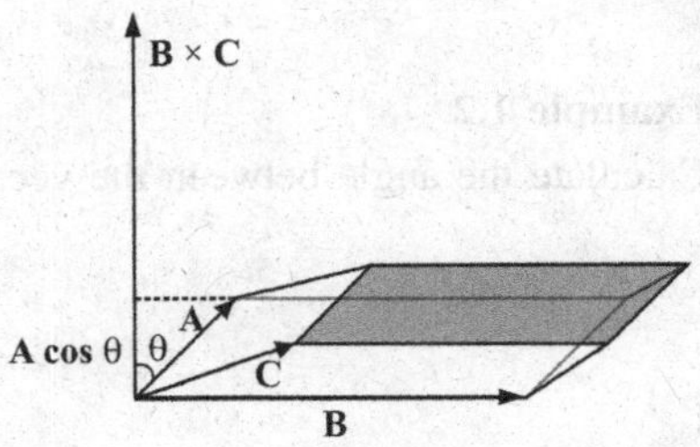


**Fig. 1.16** Geometrical meaning of scalar triple product

We have already seen that cross product $\mathbf{B} \times \mathbf{C}$ is the area of the parallelogram (Area of the base in Fig. 1.16). Also from Fig. 1.16 A cos θ is the vertical height of the parallelepiped. Hence,

$$\mathbf{A} \cdot (\mathbf{B} \times \mathbf{C}) = (\text{Vertical height of parallelepiped}) \times (\text{Area of base})$$
$$= \text{Volume of the parallelepiped.}$$

Thus $\mathbf{A} \cdot (\mathbf{B} \times \mathbf{C})$ is the volume of the parallelepiped, whose sides are **A**, **B** and **C**. As any face of the parallelepiped can be taken as the base

$$\mathbf{A} \cdot (\mathbf{B} \times \mathbf{C}) = \mathbf{B} \cdot (\mathbf{C} \times \mathbf{A}) = \mathbf{C} \cdot (\mathbf{A} \times \mathbf{B})$$

Note that in the above expression cyclic alphabetical order **A**, **B**, **C** is maintained. In component form

$$\mathbf{A} \cdot (\mathbf{B} \times \mathbf{C}) = \begin{vmatrix} A_x & A_y & A_z \\ B_x & B_y & B_z \\ C_x & C_y & C_z \end{vmatrix} \tag{1.30}$$

(b) **Vector triple product**:

The vector triple product is defined as

$$\mathbf{A} \times (\mathbf{B} \times \mathbf{C}) = \mathbf{B}(\mathbf{A} \cdot \mathbf{C}) - \mathbf{C}(\mathbf{A} \cdot \mathbf{B}) \tag{1.31}$$

The reader should note that we have written $\mathbf{B}(\mathbf{A} \cdot \mathbf{C})$. The statement $\mathbf{B} \cdot (\mathbf{A} \cdot \mathbf{C})$ has no meaning. Similarly the statement $\mathbf{B} \times (\mathbf{A} \cdot \mathbf{C})$ also has no meaning.

**Example 1.1**

Find the value of t if the vectors $\mathbf{A} = 2\hat{\mathbf{i}} + \hat{\mathbf{j}} + \hat{\mathbf{k}}$ and $\mathbf{B} = 5\hat{\mathbf{i}} + t\hat{\mathbf{j}} + 6\hat{\mathbf{k}}$ are perpendicular to each other.

**Solution**

If **A** and **B** are perpendicular to each other then $\mathbf{A} \cdot \mathbf{B} = 0$

$$\Rightarrow (2\hat{\mathbf{i}} + \hat{\mathbf{j}} + \hat{\mathbf{k}}) \cdot (5\hat{\mathbf{i}} + t\hat{\mathbf{j}} + 6\hat{\mathbf{k}}) = 0$$
$$\Rightarrow 10 + t + 6 = 0$$
$$\Rightarrow t = -16$$

**Example 1.2**

Calculate the angle between the vectors $\mathbf{A} = 3\hat{\mathbf{i}} + \hat{\mathbf{j}} + 2\hat{\mathbf{k}}$, $\mathbf{B} = 3\hat{\mathbf{i}} - 6\hat{\mathbf{j}} + 9\hat{\mathbf{k}}$.

**Solution**
We have

$$\mathbf{A}\cdot\mathbf{B}=\left(3\hat{\mathbf{i}}+\hat{\mathbf{j}}+2\hat{\mathbf{k}}\right)\cdot\left(3\hat{\mathbf{i}}-6\hat{\mathbf{j}}+9\hat{\mathbf{k}}\right)$$
$$=9-6+18=21.$$
$$A=\sqrt{9+1+4}=\sqrt{14}$$
$$B=\sqrt{9+36+81}=\sqrt{126}$$

We know that

$$\mathbf{A}\cdot\mathbf{B}=AB\cos\theta$$

$$\cos\theta=\frac{\mathbf{A}\cdot\mathbf{B}}{AB}=\frac{21}{\sqrt{14}\sqrt{126}}$$
$$\theta=\cos^{-1}\left(\frac{21}{\sqrt{14}\sqrt{126}}\right)$$
$$\theta=60^\circ$$

**Example 1.3**
Check whether the vectors $\mathbf{A}=6\hat{\mathbf{i}}-9\hat{\mathbf{j}}-3\hat{\mathbf{k}}$ and $\mathbf{B}=-2\hat{\mathbf{i}}+3\hat{\mathbf{j}}+\hat{\mathbf{k}}$ are parallel to each other.

**Solution**
We have

$$\mathbf{A}\times\mathbf{B}=\begin{vmatrix}\hat{\mathbf{i}} & \hat{\mathbf{j}} & \hat{\mathbf{k}}\\ 6 & -9 & -3\\ -2 & 3 & 1\end{vmatrix}$$
$$=\hat{\mathbf{i}}[-9+9]-\hat{\mathbf{j}}[6-6]+\hat{\mathbf{k}}[18-18]=0$$
$$=0$$
$$\mathbf{A}\times\mathbf{B}=0$$
$$\Rightarrow A\,B\ \ \sin\theta=0$$

As A, B the magnitude of **A**, **B** are not zero

$$\sin\theta=0\ \Rightarrow\theta=0.$$

Hence **A** and **B** are parallel to each other.

**Example 1.4**

Check whether the vectors $\mathbf{A} = 2\hat{\mathbf{i}} + 4\hat{\mathbf{j}} + 18\hat{\mathbf{k}}$, $\mathbf{B} = 9\hat{\mathbf{i}} + 3\hat{\mathbf{j}} + 6\hat{\mathbf{k}}$ and $\mathbf{C} = 4\hat{\mathbf{i}} + 2\hat{\mathbf{j}} + 6\hat{\mathbf{k}}$ are coplanar.

**Solution**

The condition that the three vectors are coplanar is that their scalar triple product is zero

$$\mathbf{A} \cdot (\mathbf{B} \times \mathbf{C}) = 0$$

we have

$$\begin{aligned}
\mathbf{A} \cdot (\mathbf{B} \times \mathbf{C}) &= \begin{vmatrix} 2 & 4 & 18 \\ 9 & 3 & 6 \\ 4 & 2 & 6 \end{vmatrix} \\
&= 2(18 - 12) - 4(54 - 24) + 18(18 - 12) \\
&= 2(6) - 4(30) + 18(6) \\
&= 0.
\end{aligned}$$

Hence the vectors are coplanar.

## 1.11 Line, Surface and Volume Integration

In this section we will learn about the integration of vectors. There are three different types of vector integration, namely line, surface and volume integrations.

(a) **Line Integration**

Let us assume a scalar function u and a vector function **A**. There are three important line integrations

$$\int_P^Q u\, d\boldsymbol{l}, \int_P^Q \mathbf{A} \cdot d\boldsymbol{l}, \int_P^Q \mathbf{A} \times d\boldsymbol{l},$$

where in general the line integral will depend on the path. For example the above integrations will yield different results for path 1 and path 2 as shown in Fig. 1.17.

Out of the above three integrals $\int_P^Q \mathbf{A} \cdot d\boldsymbol{l}$ will often appear in electromagnetics. If the path over which the line integral is carried out is closed we shall put a circle over the integration indicating that the path is closed

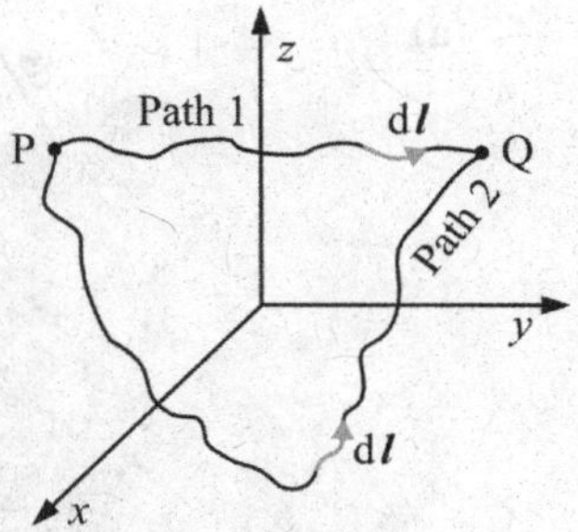


**Fig. 1.17** Illustration to explain line integration

$$\oint \mathbf{A} \cdot \mathrm{d}\boldsymbol{l}$$

Although in general the line integration depends on path, for certain vectors the line integration doesn't depend on path, but only on end points. For example for these certain vectors the line integration over the path 1 and path 2 in Fig. 1.17 will yield same result and will depend on only the end points P and Q.

(b) **Surface Integration**

Consider a surface **S**. Let the surface **S** be divided into small elements ds as shown in Fig. 1.18. There are three different surface integrals.

$$\iint_S \mathrm{u}\ \mathrm{ds}, \iint_S \mathbf{A} \cdot \mathbf{ds}, \iint_S \mathbf{A} \times \mathbf{ds}.$$

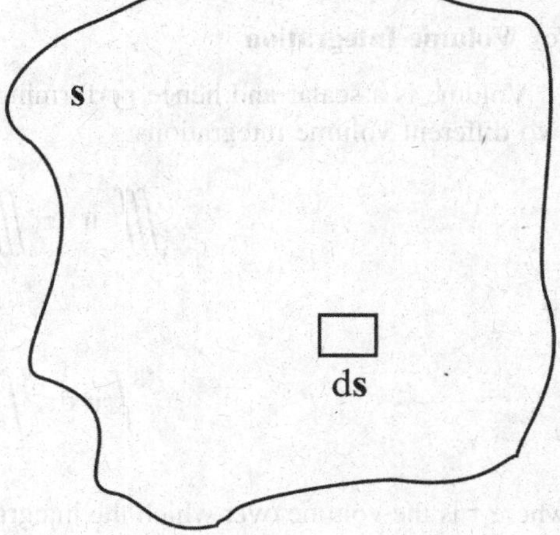


**Fig. 1.18** Figure showing Surface **S**

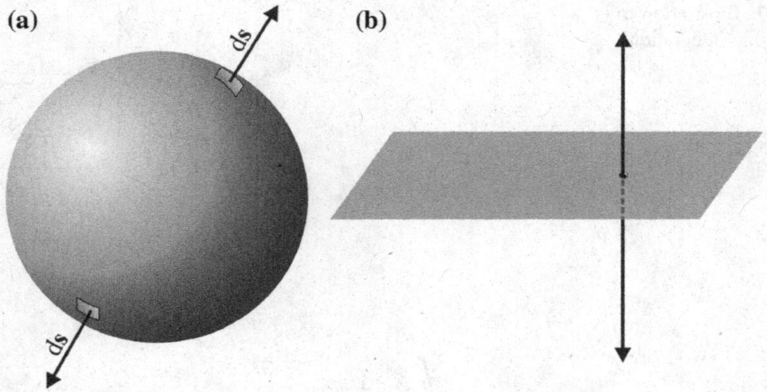


**Fig. 1.19** *a* Closed surface showing direction of the surface. **b** Open surface showing direction of the surface

where u is a scalar. Out of the three integrals the surface integral $\iint_S \mathbf{A} \cdot \mathbf{ds}$ will often appear in electromagnetics and it is customary to write the above integral for open and closed surface as

$$\int_S \mathbf{A} \cdot \mathbf{ds} \text{ and } \oint_S \mathbf{A} \cdot \mathbf{ds}$$

Surface is a vector and has direction. For a closed surface (assume the surface of a tennis ball, which is a closed surface) the outward drawn normal is the direction of the surface at that point as shown in Fig. 1.19a.

For an open surface the direction of the surface is the perpendicular direction to the surface at that point (assume a piece of paper which is an open surface). However there are two perpendicular direction to the open surface and hence the direction is arbitrary as shown in Fig. 1.19b.

(c) **Volume Integration**

Volume is a scalar and hence performing volume integration is easy. There are two different volume integrations

$$\iiint_\tau \mathrm{u}\, d\tau, \iiint_\tau \mathbf{A}\, d\tau$$

or

$$\int_\tau \mathrm{u}\, d\tau, \int_\tau \mathbf{A}\, d\tau$$

where $\tau$ is the volume over which the integration is performed and $d\tau$ is the volume element.

## 1.12 Flux

Flux is an important quantity which we will encounter often in electromagnetics. In this section we will learn how to define flux.

Assume a rectangular cardboard abcd as shown in Fig. 1.20, whose surface area is S. Suppose the cardboard is exposed to sun rays and is held perpendicular to the sun rays as shown in Fig. 1.20a. A shadow forms below as shown in the figure. Because the cardboard is held perpendicular to the sun rays the area of the shadow formed will be exactly S.

However now if the cardboard is held at an angle θ to the sun rays that is, instead of being perpendicular to the sun rays, the cardboard is inclined with respect to the direction of the sun rays at an angle θ as shown in Fig. 1.20b.

Now the area of the shadow will be lesser. This is because the effective area shown by the cardboard to the incoming sun rays is lesser than S, when the cardboard is inclined at an angle θ to the direction of sun rays. When the cardboard is perpendicular to the sun rays all the area S of the cardboard is shown to the sun rays. Hence the area of the shadow formed is S. But when the cardboard is inclined at angle θ to the sun rays the effective area of the cardboard shown to the sun rays is $S \cos \theta$ and hence the area of the shadow formed is $S \cos \theta$. In other words the projection of surface S is $S \cos \theta$. Of course when the cardboard is held perpendicular to the sun rays $\theta = 0$ and hence the effective area is $S \cos 0 = S$.

Now suppose in the place of cardboard assume a surface abcd whose area is S and in the place of sun rays assume an electric field **E** vector. The effective area shown by the surface abcd to the incoming vector **E** will be $S \cos \theta$.

Flux is defined as the number of lines passing through the given area. Hence the total flux passing through the surface abcd in Fig. 1.20b is

$$= \mathbf{E}(S \cos \theta) = \mathbf{E} \cdot \mathbf{S}$$

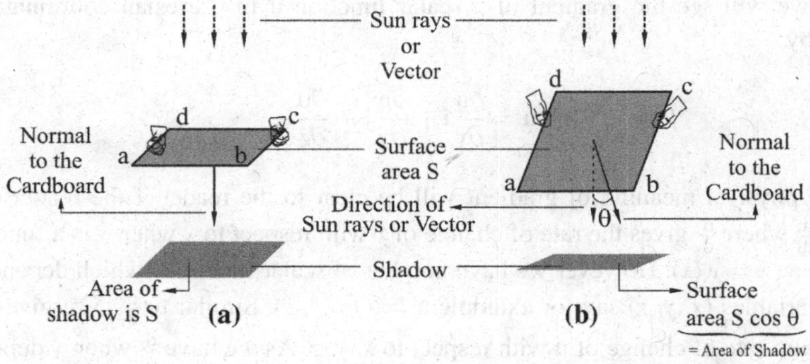


**Fig. 1.20** **a** Cardboard held perpendicular to the sun rays. **b** Cardboard held at an inclined angle to the sun rays

However if we divide the surface into small elements ds then the total flux through abcd will be

$$\text{Flux} = \iint_s \mathbf{E} \cdot \mathrm{ds}$$

In Sect. 1.11b we have treated area as a vector. Throughout electromagnetics we will be treating area as a vector. Now a general question is whether area is vector or a scalar. In Sect. 1.1 and in Fig. 1.1 masses sum up and give the same result in whichever direction they are added. Quantities which behave like mass are called scalars. In Fig. 1.2 forces sum up or cancel out depending upon the direction in which they act. Quantities which behave like force are called vectors. If you add area of two plots (empty land) they sum up irrespective of whatever direction they are added and hence area must be a scalar. But in electromagnetics we treat area as vector. How is this possible? The answer lies in Fig. 1.20. The effective area shown by area abcd to the incoming vector depends on the "DIRECTION" of orientation of normal of the area abcd to the incoming vector **E**. If the "DIRECTION" of the normal of the area abcd is along the direction of the vector then the effective area shown to the incoming vector is S cos 0 = S. If the "DIRECTION" of the normal of the area abcd is oriented at an angle θ to the incoming vector the effective area shown to the incoming vector is S cos θ.

**Because the amount of effective area shown by the area abcd to the incoming vector depends on the angle made by 'DIRECTION' of the normal of the area abcd to the incoming vector, area must be treated as vector.**

## 1.13 Vector Differentiation: Gradient of a Scalar Function

There are three important types of vector differentiation—gradient, divergence and curl. In this section we will discuss gradient. In Sects. 1.14, 1.15 we will discuss divergence and curl.

As we will see the gradient of a scalar function u in Cartesian coordinates is given by

$$\nabla \mathrm{u} = \frac{\partial \mathrm{u}}{\partial x}\hat{\mathbf{i}} + \frac{\partial \mathrm{u}}{\partial y}\hat{\mathbf{j}} + \frac{\partial \mathrm{u}}{\partial z}\hat{\mathbf{k}}$$

The physical meaning of gradient will be clear to the reader if the reader considers $\frac{dy}{dx}$ where $\frac{dy}{dx}$ gives the rate of change of *y* with respect to *x* when *y* is a function of *x* alone *y* = *y* (*x*). However we have number of scalar functions which depend on three variables (*x, y, z*) say for example u = u (*x, y, z*). Similar to $\frac{dy}{dx}$, ∇u gives the maximum rate of change of u with respect to *x, y, z*. As we have $\frac{dy}{dx}$ when *y* depends

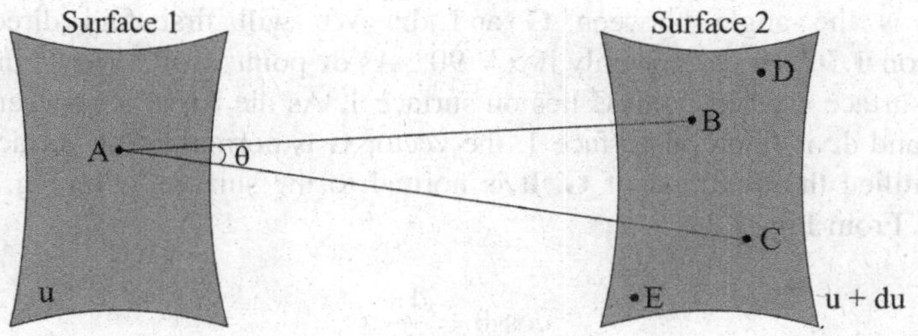


**Fig. 1.21** Figure showing Gradient of a Scalar

on single variable $y = y(x)$, we have $\nabla$ u when the given scalar function u depends on three variables $(x, y, z)$.

Consider a scalar function u $(x, y, z)$ which is constant over the surface 1 as shown in Fig. 1.21. Consider another surface, surface 2 on which the scalar function has a value u + du where du is small change in u. Point A lies on surface 1 and points B and C lie on surface 2. Consider an arbitrary origin and with respect to the origin let the position vectors of A and C be **r** and **r** + d**r**. We have AC = d**r**.

In Cartesian coordinates

$$\mathbf{dr} = dx\,\hat{\mathbf{i}} + dy\,\hat{\mathbf{j}} + dz\,\hat{\mathbf{k}} \tag{1.32}$$

From calculus

$$du = \frac{\partial u}{\partial x}dx + \frac{\partial u}{\partial y}dy + \frac{\partial u}{\partial z}dz \tag{1.33}$$

The above expression can be written as

$$du = \left(\frac{\partial u}{\partial x}\hat{\mathbf{i}} + \frac{\partial u}{\partial y}\hat{\mathbf{j}} + \frac{\partial u}{\partial z}\hat{\mathbf{k}}\right)\cdot\left(dx\,\hat{\mathbf{i}} + dy\,\hat{\mathbf{j}} + dz\,\hat{\mathbf{k}}\right) \tag{1.34}$$

Let us denote the vector $\left(\frac{\partial u}{\partial x}\hat{\mathbf{i}} + \frac{\partial u}{\partial y}\hat{\mathbf{j}} + \frac{\partial u}{\partial x}\hat{\mathbf{k}}\right)$ as **G**.

Hence

$$du = \mathbf{G}\cdot\mathbf{dr} \tag{1.35}$$

Now let us assume point C lies on surface 1. In that case du = 0.

Hence Eq. 1.35 is

$$\begin{aligned}\mathbf{G}\cdot\mathbf{dr} &= 0.\\ G\,dr\,\cos\alpha &= 0\end{aligned} \tag{1.36}$$

where $\alpha$ is the angle between **G** and **dr**. We will first find direction of **G**. Equation 1.36 can be zero only if $\alpha = 90°$. As **dr** points from A to C, **dr** will be lying on surface 1 when point C lies on surface 1. As the angle $\alpha$ between **dr** and **G** is 90° and **dr** is lying on surface 1, the vector **G** is normal to the surface 1. We have identified the direction of **G**. It is normal to the surface 1. In Fig. 1.21 let AB = d**n**. From Fig. 1.21

$$\cos\theta = \frac{dn}{dr} \tag{1.37}$$

$$\Rightarrow dn = dr\cos\theta \tag{1.38}$$

Let $\hat{\mathbf{n}}$ be the unit vector in the direction of AB. Then Eq. 1.38 can be written as

$$dn = \hat{\mathbf{n}} \cdot \mathbf{dr} \tag{1.39}$$

As we move from point A to B u increases by du. Hence

$$\begin{aligned} u &= u(n) \\ du &= \frac{\partial u}{\partial n} dn \end{aligned} \tag{1.40}$$

Substituting (1.39) in (1.40)

$$du = \frac{\partial u}{\partial n}\hat{\mathbf{n}} \cdot \mathbf{dr} \tag{1.41}$$

Comparing (1.41) and (1.35)

$$\mathbf{G} = \frac{\partial u}{\partial n}\hat{\mathbf{n}} \tag{1.42}$$

Thus the magnitude of **G** is $\frac{\partial u}{\partial n}$ and its direction is normal to the surface 1. Consider $\frac{\partial u}{\partial n}, \frac{\partial u}{\partial r}$ in Fig. 1.21. We see that out of the two derivatives $\frac{\partial u}{\partial n}, \frac{\partial u}{\partial r}, \frac{\partial u}{\partial n}$ is greater than $\frac{\partial u}{\partial r}$ because dn in Fig. 1.21 is lesser than dr. If we consider D, E or any other point on surface 2 always AB = dn will be minimum as compared to AD, AE, etc. Hence $\frac{\partial u}{\partial n}$ will be the maximum value of the derivative pointing normal to surface 1.

Hence from Eqs. 1.34, 1.35, 1.42

$$\mathbf{G} = \frac{\partial u}{\partial x}\hat{\mathbf{i}} + \frac{\partial u}{\partial y}\hat{\mathbf{j}} + \frac{\partial u}{\partial z}\hat{\mathbf{k}} = \frac{\partial u}{\partial n}\hat{\mathbf{n}} \tag{1.43}$$

"Thus the gradient of the scalar function u is a vector. Its magnitude is equal to the maximum rate of change of scalar function u with respect to scalar variables. The direction of this vector is along this maximum rate of change of scalar function".

We denote the gradient of a scalar function as grad u or ∇u.

$$\mathbf{G} = \text{grad}\,\mathrm{u} = \nabla \mathrm{u} = \frac{\partial \mathrm{u}}{\partial x}\hat{\mathbf{i}} + \frac{\partial \mathrm{u}}{\partial y}\hat{\mathbf{j}} + \frac{\partial \mathrm{u}}{\partial z}\hat{\mathbf{k}} \tag{1.44}$$

The above equation can be written as

$$\nabla \mathrm{u} = \left[\hat{\mathbf{i}}\frac{\partial}{\partial x} + \hat{\mathbf{j}}\frac{\partial}{\partial y} + \hat{\mathbf{k}}\frac{\partial}{\partial z}\right]\mathrm{u} \tag{1.45}$$

Here $\nabla = \hat{\mathbf{i}}\frac{\partial}{\partial x} + \hat{\mathbf{j}}\frac{\partial}{\partial y} + \hat{\mathbf{k}}\frac{\partial}{\partial z}$ is called "del" operator.

∇u serves the same purpose in three dimensions as $\frac{\mathrm{d}y}{\mathrm{d}x}$ in one dimension. That is in Fig. 1.21 if we represent AD = $\mathbf{dr_1}$, AE = $\mathbf{dr_2}$, etc. We have number of derivations $\frac{\partial \mathrm{u}}{\partial \mathrm{n}}, \frac{\partial \mathrm{u}}{\partial \mathrm{r}}, \frac{\partial \mathrm{u}}{\partial r_1}, \frac{\partial \mathrm{u}}{\partial r_2} \ldots$ etc. Out of the above available derivatives $\frac{\partial \mathrm{u}}{\partial \mathrm{n}}$ is the derivative selected to represent the rate of change of the scalar function with space variables (*x, y, z*) because $\frac{\partial \mathrm{u}}{\partial \mathrm{n}}$ gives the maximum rate of change of the scalar function with respect to space variables.

As $\frac{\mathrm{d}y}{\mathrm{d}x}$ gives the rate of change of *y* with respect to *x* when *y* depends on *x* only that is *y* = *y* (*x*), ∇u gives the information how fast u varies with *x, y, z* when u depends on (*x, y, z*) that is u = u (*x, y, z*). Some of the examples in which the given scalar function depends on three variables (*x, y, z*) are pressure of a gas in a cubical box P = P (*x, y, z*), temperature inside a room T = T (*x, y, z*). The other derivatives like $\frac{\mathrm{du}}{\mathrm{dr}}$ are related to gradient as follows. From Eqs. 1.32, 1.34 and 1.44 we can write

$$\mathrm{du} = \nabla \mathrm{u}.\mathbf{dr}$$

$$\frac{\mathrm{du}}{\mathrm{dr}} = \nabla \mathrm{u} \cdot \hat{\mathbf{r}}$$

The magnitude of $\frac{\mathrm{du}}{\mathrm{dr}}$ depends on direction of **dr** and hence it is called as directional derivative.

## 1.14 Vector Differentiation: Divergence of a Vector

The word divergence reminds us the word "diverge" which means to "spread out". Divergence is the measure of how much the given vector diverges or spreads out. For example for the vector shown in Fig. 1.22a the vector spreads out and is having large positive divergence. The vector shown in Fig. 1.22b is having large negative divergence. The vector shown in Fig. 1.22c is having zero divergence.

Suppose we put a hole (marked p in Fig. 1.23a) in a floor and fix a pipe in it and allow water to come out as shown in Fig. 1.23a, water spreads out; hence water is

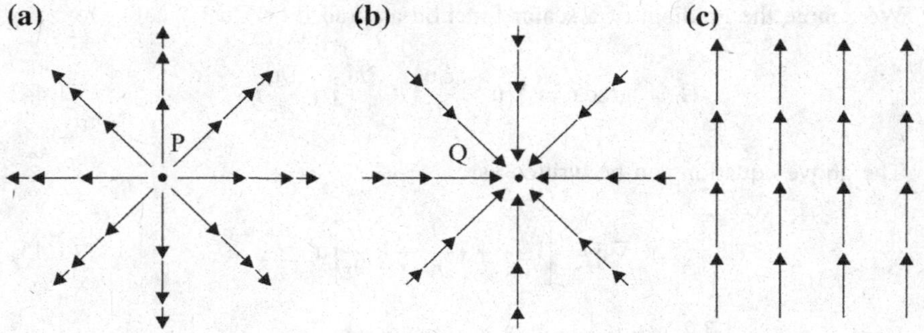


**Fig. 1.22** Divergence of vector fields

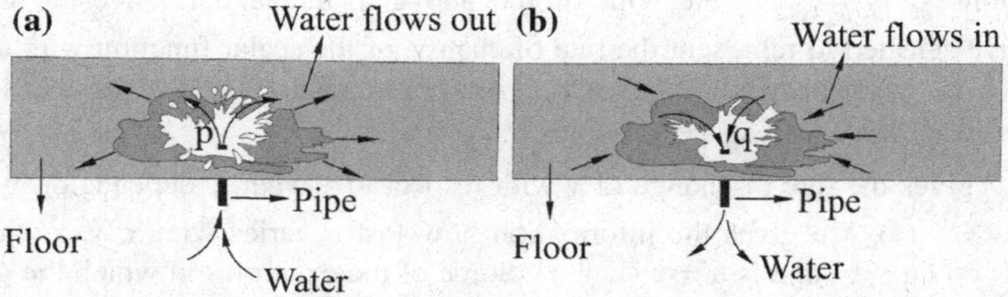


**Fig. 1.23** **a** Water flowing outside the pipe. **b** Water flowing inside the pipe

diverging and is having positive divergence (in vectorial terms the velocity of water is diverging). On the other hand, we put a hole (marked q in Fig. 1.23b) in a cone-like structure and allowed water to flow inside, from the edges of the cone, then the water is having negative divergence (in vectorial terms it is the velocity of water having negative divergence).

We can get another important physical information from Fig. 1.23. In Fig. 1.23a point p is acting like a source from which water emanates from and diverges. Hence presence of positive divergence implies implicitly that there is a source for the associated vector from which the vector emanates from. In Fig. 1.23b point q is acting like a sink into which water flows in. Hence presence of negative divergence implies that there is a sink for the associated vector to which the vector converges to.

In Cartesian coordinates the divergence is definedas (for any vector **A**)

$$\nabla \cdot \mathbf{A} = \left[\hat{\mathbf{i}}\frac{\partial}{\partial x} + \hat{\mathbf{j}}\frac{\partial}{\partial y} + \hat{\mathbf{k}}\frac{\partial}{\partial z}\right] \cdot \left[\mathrm{A}_x\hat{\mathbf{i}} + \mathrm{A}_y\hat{\mathbf{j}} + \mathrm{A}_z\hat{\mathbf{k}}\right] \tag{1.46}$$

$$\nabla \cdot \mathbf{A} = \frac{\partial \mathrm{A}_x}{\partial x} + \frac{\partial \mathrm{A}_y}{\partial y} + \frac{\partial \mathrm{A}_z}{\partial z} \tag{1.47}$$

(1) In Fig. 1.22a if the diverging vector is **A** then $\nabla \cdot \mathbf{A} > 0$ (This implies that there is a source for **A**. Point P in Fig. 1.22a is acting like a source for **A**).
(2) In Fig. 1.22b if the vector depicted is **A** having negative divergence then $\nabla \cdot \mathbf{A} < 0$ (This implies that there is a sink for **A**. Point Q in Fig. 1.22b is acting like a sink for **A**).
(3) In Fig. 1.22c if the depicted vector is **A** then $\nabla \cdot \mathbf{A} = 0$ (There is no source or sink for **A**).

As we will see in Chap. 4 the above example will be used to explain magnetic monopoles do not exist.

A vector alone can have divergence and a scalar cannot have divergence. For example if the mass of an object is 200 kg, it is 200 kg as itself and it cannot diverge or spread out. (If the mass of 200 kg explodes and spreads out, it will be the velocity of each piece which will be diverging, not the mass of the individual piece).

We will now see the mathematical description of divergence. As shown in Fig. 1.22 the divergence of any vector **A** represents flux. In Fig. 1.22a there is outward flow of flux from the source at point P. In Fig. 1.22b there is inward flow of flux to the sink at point Q. In Fig. 1.22c there is an equal amount of inward and outward flux going through any volume resulting in zero net flux.

"Thus the divergence of a vector function **A** is defined as the flux per unit volume in the limit of volume enclosed by the closed surface tends to zero".

Consider a point P′ whose coordinates are $x', y', z'$ with respect to some arbitrary origin. Consider a small closed surface S surrounding the point P′. Let $\tau$ be the small volume enclosed by the surface S. Then the divergence of **A** at point P′ is defined as

$$\text{div}\,\mathbf{A} = \lim_{\tau \to 0} \frac{\oint \mathbf{A} \cdot \mathbf{ds}}{\tau} \tag{1.48}$$

where in the above equation $\oint \mathbf{A} \cdot \mathbf{ds}$ represents flux (see Sect. 1.12).

Consider an infinitesimal rectangular box whose sides are $\Delta x$, $\Delta y$, $\Delta z$ as shown in Fig. 1.24. The box has six faces—ABQP, CRSD, ABCD, PQRS, BCRQ, ADSP. From Fig. 1.24

$$\text{Area } (\Delta\,\mathbf{s})_{\text{ABQP}} = \Delta y\,\Delta z\,\hat{\mathbf{i}}$$

$$\text{Area } (\Delta\,\mathbf{s})_{\text{CRSD}} = -\Delta y\,\Delta z\,\hat{\mathbf{i}}$$

$$\text{Area } (\Delta\,\mathbf{s})_{\text{ABCD}} = -\Delta x \Delta z\,\hat{\mathbf{j}}$$

$$\text{Area } (\Delta\,\mathbf{s})_{\text{PQRS}} = \Delta x \Delta z\,\hat{\mathbf{j}}$$

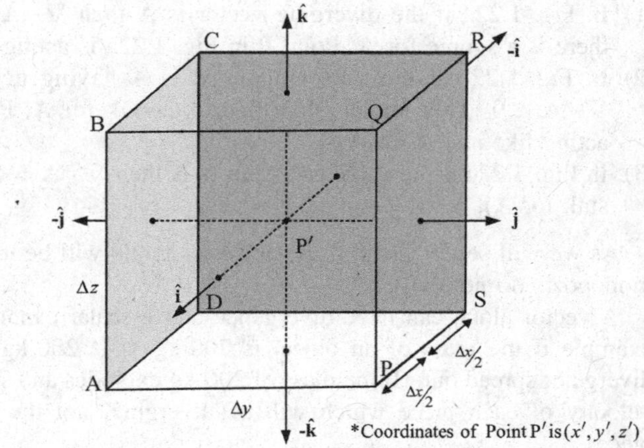


**Fig. 1.24** Differential volume in Cartesian coordinate system

$$\text{Area}\,(\Delta\,\mathbf{s})_{BCRQ} = \Delta\,x\Delta y\hat{\mathbf{k}}$$

$$\text{Area}\,(\Delta\,\mathbf{s})_{ADSP} = -\Delta\,x\Delta y\hat{\mathbf{k}}$$

The vector **A** in Cartesian coordinates can be written as

$$\mathbf{A} = \mathrm{A}_x\hat{\mathbf{i}} + \mathrm{A}_y\hat{\mathbf{j}} + \mathrm{A}_z\hat{\mathbf{k}}$$

For the box shown in Fig. 1.24 the integral $\oint \mathbf{A}.\mathbf{ds}$ can be written as

$$\oint_S \mathbf{A}\cdot\mathbf{ds} = \int_{ABQP} \mathbf{A}\cdot\mathbf{ds} + \int_{CRSD} \mathbf{A}\cdot\mathbf{ds} + \int_{ABCD} \mathbf{A}\cdot\mathbf{ds} + \int_{PQRS} \mathbf{A}\cdot\mathbf{ds} + \int_{BCRQ} \mathbf{A}\cdot\mathbf{ds} + \int_{ADSP} \mathbf{A}\cdot\mathbf{ds} \tag{1.49}$$

Now consider the integral $\int_{ABQP} \mathbf{A}\cdot\mathbf{ds}$

$$\int_{ABQP} \mathbf{A}\cdot\mathbf{ds} = (\mathbf{A})_{ABQP}\cdot(\Delta\mathbf{s})_{ABQP} \tag{1.50}$$

$$= \left[\mathrm{A}_x\hat{\mathbf{i}} + \mathrm{A}_y\hat{\mathbf{j}} + \mathrm{A}_z\hat{\mathbf{k}}\right]_{ABQP}\cdot\left(\Delta y\Delta z\hat{\mathbf{i}}\right) \tag{1.51}$$

$$\int_{ABQP} \mathbf{A}\cdot d\,\mathbf{s} = [\mathrm{A}_x]_{ABQP}\Delta y\Delta z \tag{1.52}$$

Because the box is infinitesimal the value of $A_x$ at the centre of the ABQP face can be taken to be average value of $A_x$ over the entire face. Hence Eq. 1.52 is

$$\int_{ABQP} \mathbf{A} \cdot \mathbf{ds} = [A_x \Delta y \Delta z]_{\text{at}\left(x' + \frac{\Delta x}{2}, y', z'\right)} \tag{1.53}$$

Applying Taylor's series expansion and neglecting higher order terms as the box is infinitesimal

$$[A_x]_{\text{at}\left(x' + \frac{\Delta x}{2}, y', z'\right)} = A_x(x', y', z') + \frac{\Delta x}{2}\left(\frac{\partial A_x}{\partial x}\right)_{(x', y', z')}$$

Substituting the above equation in 1.53

$$\int_{ABQP} \mathbf{A} . \mathbf{ds} = \left[A_x(x', y', z') + \frac{\Delta x}{2}\left(\frac{\partial A_x}{\partial x}\right)_{(x', y', z')}\right] \Delta y \Delta z \tag{1.54}$$

Similarly for the face CRSD

$$\int_{CRSD} \mathbf{A} \cdot \mathbf{ds} = (\mathbf{A})_{CRSD} \cdot \Delta \mathbf{s}_{CRSD} \tag{1.55}$$

$$= \left[A_x \hat{\mathbf{i}} + A_y \hat{\mathbf{j}} + A_z \hat{\mathbf{k}}\right]_{CRSD} \cdot \left(\Delta y \Delta z \left(-\hat{\mathbf{i}}\right)\right) \tag{1.56}$$

$$= -[A_x \Delta y \Delta z]_{\text{at}\left(x' - \frac{\Delta x}{2}, y', z'\right)} \tag{1.57}$$

Using Taylor series expansion and neglecting higher order terms

$$[A_x]_{\text{at}\left(x' + \frac{\Delta x}{2}, y', z'\right)} = A_x(x', y', z') - \frac{\Delta x}{2}\left(\frac{\partial A_x}{\partial x}\right)_{(x', y', z')}$$

Substituting the above equation in 1.57

$$\int_{CRSD} \mathbf{A} \cdot \mathbf{ds} = -\left[A_x(x', y', z') - \frac{\Delta x}{2}\left(\frac{\partial A_x}{\partial x}\right)_{(x', y', z')}\right] \Delta y \, \Delta z \tag{1.58}$$

Adding Eqs. 1.54, 1.58 we get

$$\int_{ABQP} \mathbf{A} \cdot \mathbf{ds} + \int_{CRSD} \mathbf{A} \cdot \mathbf{ds} = \left(\frac{\partial A_x}{\partial x}\right)_{(x', y', z')} \Delta x \, \Delta y \, \Delta z \tag{1.59}$$

Applying the same procedure to ABCD and PQRS we get

$$\int_{\text{ABCD}} \mathbf{A}\cdot \mathbf{ds}+\int_{\text{PQRS}} \mathbf{A}\cdot \mathbf{ds}=\left(\frac{\partial \text{A}_y}{\partial \text{y}}\right)_{(x',y',z')}\Delta x\,\Delta y\,\Delta z \tag{1.60}$$

Similarly for the faces BCRQ and ADSP

$$\int_{\text{BCRQ}} \mathbf{A}\cdot \mathbf{ds}+\int_{\text{ADSP}} \mathbf{A}\cdot \mathbf{ds}=\left(\frac{\partial \text{A}_z}{\partial z}\right)_{(x',y',z')}\Delta x\,\Delta y\,\Delta z \tag{1.61}$$

Substituting Eqs. 1.59, 1.60, 1.61 in Eq. 1.49

$$\oint_{\text{S}} \mathbf{A}\cdot \mathbf{ds}=\left(\frac{\partial \text{A}_x}{\partial x}+\frac{\partial \text{A}_y}{\partial y}+\frac{\partial \text{A}_z}{\partial \text{z}}\right)_{(x',y',z')}\Delta x\,\Delta y\,\Delta z \tag{1.62}$$

With $\tau=\Delta x\,\Delta y\,\Delta z$, the volume of the box, the above equation becomes

$$\oint_{\text{S}} \mathbf{A}\cdot \mathbf{ds}=\left(\frac{\partial \text{A}_x}{\partial x}+\frac{\partial \text{A}_y}{\partial y}+\frac{\partial \text{A}_z}{\partial z}\right)_{(x',y',z')}\tau \tag{1.63}$$

Substituting Eq. 1.63 in Eq. 1.48

$$\text{div}\,\mathbf{A}=\frac{\partial \text{A}_x}{\partial x}+\frac{\partial \text{A}_y}{\partial y}+\frac{\partial \text{A}_z}{\partial z} \tag{1.64}$$

The above equation applies for any point $(x',y',z')$. Equation 1.64 can be written as

$$\text{div}\,\mathbf{A}=\left(\hat{\mathbf{i}}\frac{\partial}{\partial x}+\hat{\mathbf{j}}\frac{\partial}{\partial y}+\hat{\mathbf{k}}\frac{\partial}{\partial y}\right)\cdot\left(\text{A}_x\hat{\mathbf{i}}+\text{A}_y\hat{\mathbf{j}}+\text{A}_z\hat{\mathbf{k}}\right) \tag{1.65}$$

$$\text{div}\,\mathbf{A}=\nabla\cdot\mathbf{A} \tag{1.66}$$

## 1.15 Vector Differentiation: Curl of a Vector

In your daily life you could have come across the word "curly hair". The meaning of word curl is circulation. Curl of the vector is the measure of how much the given vector circulates about the given point in question.

For example, the vector shown in Fig. 1.25 is having high curl as it is circulating about a given point. A vector alone can curl, scalar cannot curl. For example if we consider a block of mass 200 kg, it is 200 kg as itself and the value 200 kg doesn't

Fig. 1.25 Curl of a vector field

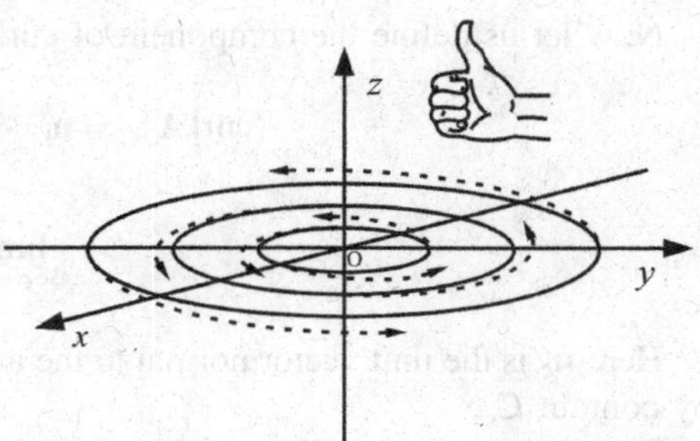


Fig. 1.26 A point of high curl

circulate. On the other hand consider a swirl as shown in Fig. 1.26 which is having high curl or circulation. The velocity of water at each point is different and circulates about the central point. We will now see the mathematical description of curl.

The curl of the vector field at any point is a vector quantity, the magnitude of the vector quantity being given by maximum line integral per unit area, the line integral being carried out along the boundary of infinitesimal test area at that point and the direction of the vector is perpendicular to the plane of the test area.

As per the above definition the curl of any vector **A** is defined as

$$\text{curl}\,\mathbf{A} = \lim_{\Delta s \to 0} \frac{\left[\hat{\mathbf{n}} \oint_C \mathbf{A} \cdot d\boldsymbol{l}\right]_{max}}{\Delta s} \tag{1.67}$$

Here $\Delta s$ is the test area which is enclosed by contour C as shown in Fig. 1.27. Here $\hat{\mathbf{n}}$ is the unit vector normal to the test area. We will see how to fix the direction of curl or $\hat{\mathbf{n}}$ later on.

Fig. 1.27 Figure showing relation between different unit vectors

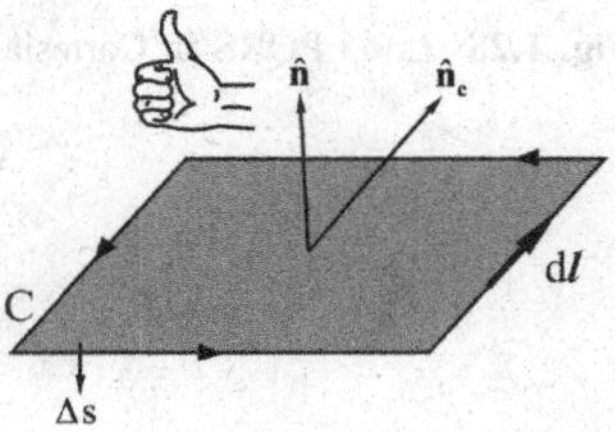

Now let us define the component of curl

$$
\begin{aligned}
(\text{curl}\,\mathbf{A})_{C} &= \hat{\mathbf{n}}_{C}\cdot(\text{curl}\,\mathbf{A}) \\
&= \lim_{\Delta s_C \to 0} \frac{\oint_{C_C} \mathbf{A}\cdot d\boldsymbol{l}}{\Delta s_C}
\end{aligned}
\tag{1.68}
$$

Here $\hat{\mathbf{n}}_{\mathbf{c}}$ is the unit vector normal to the area $\Delta s_C$, where $\Delta s_C$ is the area bounded by contour $C_c$.

The direction of $\hat{\mathbf{n}}$ in Fig. 1.27 is given by right-hand rule. When the fingers curl around C the thumb gives the direction of $\hat{\mathbf{n}}$ which is perpendicular to the area $\Delta s$ . The same right-hand rule is true for $\hat{\mathbf{n}}_{\mathbf{c}}$ which is perpendicular to area $\Delta s_C$, area $\Delta s_C$ being bounded by contour $C_C$. Contour $C_C$ is not shown in Fig. 1.27.

Consider an infinitesimal loop PQRS in the $z-y$ plane surrounding the point P′ in the region of vector function **A**, as shown in Fig. 1.28. The positive direction around the loop is assigned by right-hand rule. When the fingers of the right hand point in the direction of the arrows in the contour PQRS in Fig. 1.28 then the thumb gives the direction of the positive normal to the surface. This positive normal for the loop PQRS is along positive $x$-axis.

Now let us calculate the line integral of **A** along PQRS.

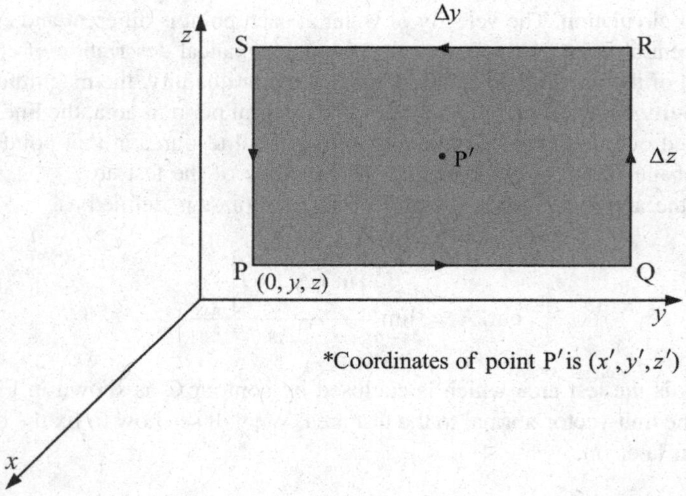


**Fig. 1.28** Loop PQRS in Cartesian coordinate system

The line integral of **A** along PQ is

$$\int_{PQ} \mathbf{A} \cdot d\boldsymbol{l} = \left[A_x\hat{\mathbf{i}} + A_y\hat{\mathbf{j}} + A_z\hat{\mathbf{k}}\right] \cdot \Delta y\hat{\mathbf{j}}$$

$$= \left[A_y \Delta y\right]_{\text{at PQ}}$$

Because the loop is infinitesimal the value of $A_x$ at the centre of the line PQ can be taken to be average value of $A_x$ over the entire line. Hence the above equation becomes

$$\int_{PQ} \mathbf{A} \cdot d\boldsymbol{l} = \left[A_y \Delta y\right]_{\text{at}\left(x',y',z'-\frac{\Delta z}{z}\right)} \tag{1.69}$$

The line integral of **A** along RS will be

$$\begin{aligned}\int_{RS} \mathbf{A} \cdot d\boldsymbol{l} &= \left[A_x\hat{\mathbf{i}} + A_y\hat{\mathbf{j}} + A_z\hat{\mathbf{k}}\right] \cdot \left(-\Delta y\hat{\mathbf{j}}\right) \\ &= -\left[A_y \Delta y\right]_{\text{at}\left(x',y',z'+\frac{\Delta z}{z}\right)}\end{aligned} \tag{1.70}$$

The line integral of **A** along SP is

$$\begin{aligned}\int_{SP} \mathbf{A} \cdot d\boldsymbol{l} &= \left[A_x\hat{\mathbf{i}} + A_y\hat{\mathbf{j}} + A_z\hat{\mathbf{k}}\right] \cdot \left(-\Delta z\hat{\mathbf{k}}\right) \\ &= -\left[A_z \Delta z\right]_{\text{at}\left(x',y'-\frac{\Delta y}{z},z'\right)}\end{aligned} \tag{1.71}$$

The line integral of **A** along QR is

$$\begin{aligned}\int_{QR} \mathbf{A} \cdot d\boldsymbol{l} &= \left[A_x\hat{\mathbf{i}} + A_y\hat{\mathbf{j}} + A_z\hat{\mathbf{k}}\right] \cdot \left(\Delta z\hat{\mathbf{k}}\right) \\ &= \left[A_z \Delta z\right]_{\text{at}\left(x',y'+\frac{\Delta y}{z},z'\right)}\end{aligned} \tag{1.72}$$

The line integral of $\oint_{PQRS} \mathbf{A} \cdot d\boldsymbol{l}$ can be written as

$$\oint_{PQRS} \mathbf{A} \cdot d\boldsymbol{l} = \oint_{PQ} \mathbf{A} \cdot d\boldsymbol{l} + \oint_{RS} \mathbf{A} \cdot d\boldsymbol{l} + \oint_{SP} \mathbf{A} \cdot d\boldsymbol{l} + \oint_{QR} \mathbf{A} \cdot d\boldsymbol{l} \tag{1.73}$$

Substituting Eqs. 1.69, 1.70, 1.71, 1.72 in Eq. 1.73 we get

$$\oint_{PQRS} \mathbf{A} \cdot \mathrm{d}\boldsymbol{l} = \left[(\mathrm{A}_y)_{(x',y',z'-\frac{\Delta z}{z},)} - (\mathrm{A}_y)_{(x',y',z'+\frac{\Delta z}{z},)}\right]\Delta y + \left[(\mathrm{A}_z)_{(x',y'+\frac{\Delta y}{z},z')} - (\mathrm{A}_z)_{(x',y'-\frac{\Delta y}{z},z')}\right]\Delta z \tag{1.74}$$

Using Taylor's series expansion, neglecting higher order terms we get

$$\left[(\mathrm{A}_y)_{(x',y',z'-\frac{\Delta z}{z},)} - (\mathrm{A}_y)_{(x',y',z'+\frac{\Delta z}{z},)}\right] = -\frac{\partial\, \mathrm{A}_y}{\partial z}\Delta z \tag{1.75}$$

and

$$\left[(\mathrm{A}_z)_{(x',y'+\frac{\Delta y}{z},z')} - (\mathrm{A}_z)_{(x',y'-\frac{\Delta y}{z},z')}\right] = \frac{\partial\, \mathrm{A}_z}{\partial_y}\Delta y \tag{1.76}$$

Substituting Eqs. 1.75 and 1.76 in Eq. 1.74

$$\oint_{PQRS} \mathbf{A} \cdot \mathrm{d}\boldsymbol{l} = \frac{\partial\, \mathrm{A}_z}{\partial y}\Delta y \Delta z - \frac{\partial\, \mathrm{A}_y}{\partial z}\Delta z \Delta y \tag{1.77}$$

With $\Delta \mathrm{s} = \Delta y \Delta z$ we get

$$\oint_{PQRS} \mathbf{A} \cdot \mathrm{d}\boldsymbol{l} = \left[\frac{\partial\, \mathrm{A}_z}{\partial y} - \frac{\partial\, \mathrm{A}_y}{\partial z}\right]\Delta \mathrm{s} \tag{1.78}$$

Substituting Eq. 1.78 in Eq. 1.68

$$\begin{aligned}(\mathrm{curl}\,\mathbf{A})_x &= \hat{\mathbf{i}} \cdot (\mathrm{curl}\,\mathbf{A}) \\ &= \lim_{\Delta \mathrm{s}\to 0} \frac{\oint_{PQRS} \mathbf{A} \cdot \mathrm{d}\boldsymbol{l}}{\Delta \mathrm{s}} = \left[\frac{\partial\, \mathrm{A}_z}{\partial y} - \frac{\partial\, \mathrm{A}_y}{\partial z}\right]\end{aligned} \tag{1.79}$$

Similarly

$$(\mathrm{curl}\,\mathbf{A})_y = \frac{\partial\, \mathrm{A}_x}{\partial z} - \frac{\partial\, \mathrm{A}_z}{\partial x} \tag{1.80}$$

and

$$(\text{curl}\,\mathbf{A})_z = \frac{\partial\,\mathrm{A}_y}{\partial x} - \frac{\partial\,\mathrm{A}_x}{\partial y} \tag{1.81}$$

We know that

$$(\text{curl}\,\mathbf{A}) = (\text{curl}\,\mathbf{A})_x\hat{\mathbf{i}} + (\text{curl}\,\mathbf{A})_y\hat{\mathbf{j}} + (\text{curl}\,\mathbf{A})_z\hat{\mathbf{k}} \tag{1.82}$$

Substituting Eqs. 1.79, 1.80, 1.81 in Eq. 1.82

$$\text{curl}\,\mathbf{A} = \left[\frac{\partial \mathrm{A}_z}{\partial y} - \frac{\partial \mathrm{A}_y}{\partial z}\right]\hat{\mathbf{i}} + \left[\frac{\partial \mathrm{A}_x}{\partial z} - \frac{\partial \mathrm{A}_z}{\partial x}\right]\hat{\mathbf{j}} + \left[\frac{\partial \mathrm{A}_y}{\partial x} - \frac{\partial \mathrm{A}_x}{\partial y}\right]\hat{\mathbf{k}} \tag{1.83}$$

The above equation can be written as

$$\text{curl}\,\mathbf{A} = \left[\hat{\mathbf{i}}\frac{\partial}{\partial x} + \hat{\mathbf{j}}\frac{\partial}{\partial y} + \hat{\mathbf{k}}\frac{\partial}{\partial z}\right] \times \left[\mathrm{A}_x\hat{\mathbf{i}} + \mathrm{A}_y\hat{\mathbf{j}} + \mathrm{A}_z\hat{\mathbf{k}}\right] \tag{1.84}$$

$$\text{curl}\,\mathbf{A} = \nabla \times \mathbf{A} \tag{1.85}$$

In determinant form $\nabla \times \mathbf{A}$ is

$$\nabla \times \mathbf{A} = \begin{vmatrix} \hat{\mathbf{i}} & \hat{\mathbf{j}} & \hat{\mathbf{k}} \\ \frac{\partial}{\partial x} & \frac{\partial}{\partial y} & \frac{\partial}{\partial z} \\ \mathrm{A}_x & \mathrm{A}_y & \mathrm{A}_z \end{vmatrix} \tag{1.86}$$

We have seen three different types of vector differentiation.

1. Gradient: The operator $\nabla$ operates on an scalar function. In Cartesian coordinates

$$\nabla \mathrm{u} = \frac{\partial \mathrm{u}}{\partial x}\hat{\mathbf{i}} + \frac{\partial \mathrm{u}}{\partial y}\hat{\mathbf{j}} + \frac{\partial \mathrm{u}}{\partial z}\hat{\mathbf{k}}$$

As seen above gradient of a scalar function is a vector.
Divergence: The operator $\nabla$ operates on a vector by means of dot product. In Cartesian coordinates

$$\nabla \cdot \mathbf{A} = \frac{\partial \mathrm{A}_x}{\partial x} + \frac{\partial \mathrm{A}_y}{\partial y} + \frac{\partial \mathrm{A}_z}{\partial z}$$

As seen above divergence of a vector function is a scalar.

Curl: The operator $\nabla$ operates on a vector by means of cross product. In Cartesian coordinates

$$\nabla \times \mathbf{A} = \left[\frac{\partial A_z}{\partial y} - \frac{\partial A_y}{\partial z}\right]\hat{\mathbf{i}} + \left[\frac{\partial A_x}{\partial z} - \frac{\partial A_z}{\partial x}\right]\hat{\mathbf{j}} + \left[\frac{\partial A_y}{\partial x} - \frac{\partial A_x}{\partial y}\right]\hat{\mathbf{k}}$$

As seen above the curl of a vector function is a vector. Because curl of any vector is a vector it must have both magnitude and direction. Let us see an example.

In Fig. 1.25 let the velocity vector $\mathbf{v}$ given by $v_x\hat{\mathbf{i}} + v_y\hat{\mathbf{j}} + v_z\hat{\mathbf{k}}$ be curling around with O as centre. Let us define curl of $\mathbf{v}$ as

$$\mathbf{C} = \nabla \times \mathbf{v} = \left[\frac{\partial v_z}{\partial y} - \frac{\partial v_y}{\partial z}\right]\hat{\mathbf{i}} + \left[\frac{\partial v_x}{\partial z} - \frac{\partial v_z}{\partial x}\right]\hat{\mathbf{j}} + \left[\frac{\partial v_y}{\partial x} - \frac{\partial v_x}{\partial y}\right]\hat{\mathbf{k}}$$

Then the direction of $\mathbf{C}$ and the curl of $\mathbf{v}$ are given by right-hand rule. When the fingers of right hand point in the direction of circulating $\mathbf{v}$ as shown in Fig. 1.25 then the thumb gives the direction of $\mathbf{C}$ which is along $z$-axis. Thus there are two vectors $\mathbf{v}$, $\mathbf{C}$ – .$\mathbf{v}$ the velocity which is circulating about and $\mathbf{C}$, the curl of v pointing along $z$-direction.

2. We have seen that divergence of any vector means how much the vector is spreading out. Curl of any vector means how much the vector is circulating about.
   If $\nabla \cdot \mathbf{A} = 0$ then $\mathbf{A}$ is called solenoidal vector
   If $\nabla \times \mathbf{A} = 0$ then $\mathbf{A}$ is known as irrotational vector.
   For all the vectors shown in Fig. 1.22 curl of these vectors is zero because nothing is circulating. For the vector shown in Fig. 1.25 divergence is zero because nothing is spreading out.
3. Divergence and curl operate on a vector. As we have already seen divergence and curl of a scalar don't have a meaning.

**Example 1.5**

Find the gradient of u at the point (1, 2, 3) if $u = x^2 - yz$.

**Solution**

$$\nabla u = \left(\hat{\mathbf{i}}\frac{\partial}{\partial x} + \hat{\mathbf{j}}\frac{\partial}{\partial y} + \hat{\mathbf{k}}\frac{\partial}{\partial z}\right)(x^2 - yz)$$

$$\Rightarrow \nabla u = 2x\hat{\mathbf{i}} + z\hat{\mathbf{j}} + y\hat{\mathbf{k}}$$

Gradient of a scalar function u, $\nabla u$ is a vector as we discussed previously.

$\nabla$u at the point (1, 2, 3) is then

$$\nabla u = 2\hat{\mathbf{i}} + 3\hat{\mathbf{j}} + 2\hat{\mathbf{k}}$$

**Example 1.6**

Find the divergence of the vector $\mathbf{A} = x\hat{\mathbf{i}} + y\hat{\mathbf{j}} + z\hat{\mathbf{k}}, \mathbf{B} = x^2 y\hat{\mathbf{i}} + z^3 y\hat{\mathbf{j}}$.

**Solution**

$$\nabla \cdot \mathbf{A} = \left(\hat{\mathbf{i}}\frac{\partial}{\partial x} + \hat{\mathbf{j}}\frac{\partial}{\partial y} + \hat{\mathbf{k}}\frac{\partial}{\partial z}\right).\left(x\hat{\mathbf{i}} + y\hat{\mathbf{j}} + z\hat{\mathbf{k}}\right)$$
$$\nabla \cdot \mathbf{A} = 3$$
$$\nabla \cdot \mathbf{B} = \left(\hat{\mathbf{i}}\frac{\partial}{\partial x} + \hat{\mathbf{j}}\frac{\partial}{\partial y} + \hat{\mathbf{k}}\frac{\partial}{\partial z}\right) \cdot \left(x^2 y\hat{\mathbf{i}} + z^3 y\hat{\mathbf{k}}\right)$$
$$\nabla \cdot \mathbf{B} = 2xy + z^3$$

Divergence of **A** is 2, divergence of **B** is $2xy + z^3$.

**Example 1.7**

Calculate curl of vector $\mathbf{A} = y\hat{\mathbf{k}}$.

**Solution**

$$\nabla \times \mathbf{A} = \begin{vmatrix} \hat{\mathbf{i}} & \hat{\mathbf{j}} & \hat{\mathbf{k}} \\ \frac{\partial}{\partial x} & \frac{\partial}{\partial y} & \frac{\partial}{\partial z} \\ 0 & 0 & y \end{vmatrix}$$

$$\nabla \times \mathbf{A} = \hat{\mathbf{i}}\left[\frac{\partial y}{\partial y} - 0\right] - \hat{\mathbf{j}}[0 - 0] + \hat{\mathbf{k}}[0 - 0]$$

$$\nabla \times \mathbf{A} = \hat{\mathbf{i}}$$

**Example 1.8**

Prove that for any vector **A**, div curl **A** = 0. That is prove that $\nabla \cdot (\nabla \times \mathbf{A}) = 0$.

$$\nabla \cdot (\nabla \times \mathbf{A}) = \nabla \cdot \begin{vmatrix} \hat{\mathbf{i}} & \hat{\mathbf{j}} & \hat{\mathbf{k}} \\ \frac{\partial}{\partial x} & \frac{\partial}{\partial y} & \frac{\partial}{\partial z} \\ A_x & A_y & A_z \end{vmatrix}$$

$$= \left[\hat{\mathbf{i}}\frac{\partial}{\partial x} + \hat{\mathbf{j}}\frac{\partial}{\partial y} + \hat{\mathbf{k}}\frac{\partial}{\partial z}\right] \cdot \left[\left(\frac{\partial A_z}{\partial y} - \frac{\partial A_y}{\partial z}\right)\hat{\mathbf{i}} + \left(\frac{\partial A_x}{\partial z} - \frac{\partial A_z}{\partial x}\right)\hat{\mathbf{j}} + \left(\frac{\partial A_y}{\partial x} - \frac{\partial A_x}{\partial y}\right)\hat{\mathbf{k}}\right]$$

$$= \frac{\partial}{\partial x}\left(\frac{\partial A_z}{\partial y} - \frac{\partial A_y}{\partial z}\right) + \frac{\partial}{\partial y}\left(\frac{\partial A_x}{\partial z} - \frac{\partial A_z}{\partial x}\right) + \frac{\partial}{\partial z}\left(\frac{\partial A_y}{\partial x} - \frac{\partial A_x}{\partial y}\right)$$

$$= \frac{\partial^2 A_z}{\partial x\, \partial y} - \frac{\partial^2 A_y}{\partial x\, \partial z} + \frac{\partial^2 A_x}{\partial y\, \partial z} - \frac{\partial^2 A_z}{\partial y\, \partial x} + \frac{\partial^2 A_y}{\partial z\, \partial x} - \frac{\partial^2 A_x}{\partial z\, \partial y} = 0$$

**Example 1.9**

Prove that curl grad u = 0. That is prove that $\nabla \times \nabla u = 0$ where u is a scalar function.

**Solution**

$$\nabla \times \nabla u = \begin{vmatrix} \hat{\mathbf{i}} & \hat{\mathbf{j}} & \hat{\mathbf{k}} \\ \frac{\partial}{\partial x} & \frac{\partial}{\partial y} & \frac{\partial}{\partial z} \\ \frac{\partial u}{\partial x} & \frac{\partial u}{\partial y} & \frac{\partial u}{\partial z} \end{vmatrix}$$

$$= \left(\frac{\partial^2 u}{\partial y\, \partial z} - \frac{\partial^2 u}{\partial z\, \partial y}\right)\hat{\mathbf{i}} + \left(\frac{\partial^2 u}{\partial z\, \partial x} - \frac{\partial^2 u}{\partial x\, \partial z}\right)\hat{\mathbf{j}} + \left(\frac{\partial^2 u}{\partial x\, \partial y} - \frac{\partial^2 u}{\partial y\, \partial x}\right)\hat{\mathbf{k}} = 0$$

## 1.16 Divergence Theorem

Divergence theorem is helpful in transforming a surface integral of a vector (flux of the vector) into a volume integral of the divergence of the vector and vice versa. The theorem states that the volume integral of the divergence of the vector **A** is equal to the flux of the vector over the closed surface S enclosing the volume $\tau$ over which the volume integral is calculated, i.e.

$$\int_\tau \nabla \cdot \mathbf{A} d\tau = \oint_S \mathbf{A} \cdot d\mathbf{s} \tag{1.87}$$

Divergence theorem is also known as Gauss's divergence theorem.

We have already seen that divergence of a vector is a scalar quantity (see Sect. 1.15).

The volume $\tau$ is also a scalar quantity. Hence when dealing with integral $\int_\tau \nabla \cdot \mathbf{A} d\tau$ we will be working with scalars, on the other hand in the integral $\oint_S \mathbf{A} \cdot d\mathbf{s}$,

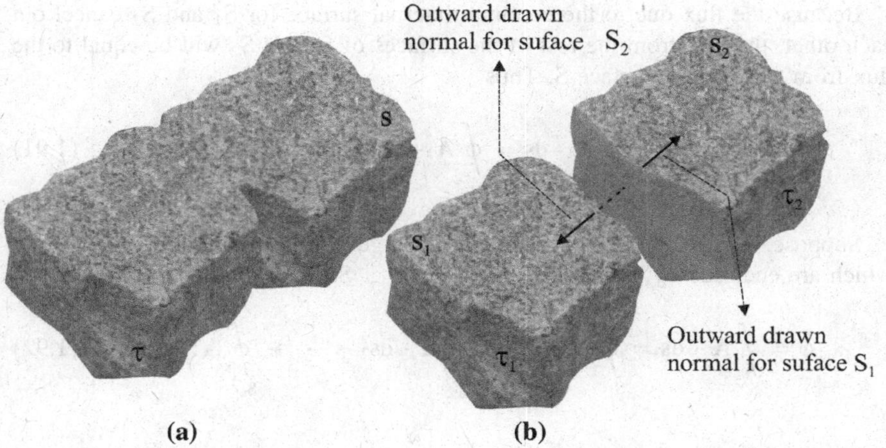


**Fig. 1.29** **a** Arbitrary volume bounded by a surface. **b** Arbitrary volume divided into two small volumes

surface d**s** is a vector quantity. Hence when working with integral $\oint_S \mathbf{A} \cdot d\mathbf{s}$ we will be working with vectors. This point will be more clear in Example 1.10.

Now we shall see the proof for divergence theorem.

Let us consider some arbitrary volume $\tau$ which is enclosed by surface S as shown in Fig. 1.29a. The arbitrary volume is present in a region of vector function **A**. The flux diverging from the surface S of volume $\tau$ is

$$\phi = \oint_S \mathbf{A} \cdot d\mathbf{s} \tag{1.88}$$

Now let us divide the volume $\tau$ into two parts with volume $\tau_1$ and $\tau_2$ which are enclosed by surfaces $S_1$ and $S_2$, respectively, as shown in Fig. 1.29b.

We see that in Fig. 1.29b at the common internal surface for $S_1$ and $S_2$ the outward drawn normals of the two parts point in opposite direction. Hence the contribution of the internal common surface to the flux of the two parts will cancel each other.

$$\text{The flux emerging out of surface } S_1 = \oint_{S_1} \mathbf{A} \cdot d\mathbf{s}_1 \tag{1.89}$$

$$\text{The flux emerging out of surface } S_2 = \oint_{S_2} \mathbf{A} \cdot d\mathbf{s}_2 \tag{1.90}$$

Because the flux due to the common internal surface for $S_1$ and $S_2$ cancel out each other, the flux from the rest of the surfaces of $S_1$ and $S_2$ will be equal to the flux from the original surface S. Thus

$$\phi = \oint_S \mathbf{A} \cdot \mathbf{ds} = \oint_{S_1} \mathbf{A} \cdot \mathbf{ds_1} + \oint_{S_2} \mathbf{A} \cdot \mathbf{ds_2} \tag{1.91}$$

Suppose we divide the volume $\tau$ into a large number of volumes $\tau_1, \tau_2 \ldots \tau_n$ which are enclosed by the surfaces $S_1$, $S_2$, $S_3$, ... $S_n$, respectively, then

$$\phi = \oint_S \mathbf{A} \cdot \mathbf{ds} = \oint_{S_1} \mathbf{A} \cdot \mathbf{ds_1} + \oint_{S_2} \mathbf{A} \cdot \mathbf{ds_2} + \cdots + \oint_{S_n} \mathbf{A} . \mathbf{ds_n} \tag{1.92}$$

$$\phi = \oint_S \mathbf{A} \cdot \mathbf{ds} = \sum_{j=1}^{N} \oint_{S_j} \mathbf{A} \cdot \mathbf{ds_j} \tag{1.93}$$

The above equation can be written as

$$\oint_S \mathbf{A} \cdot \mathbf{ds} = \sum_{j=1}^{N} \tau_j \frac{\oint_{S_j} \mathbf{A} \cdot \mathbf{ds_j}}{\tau_j} \tag{1.94}$$

The volume $\tau_j$ is infinitely small if N is very large, i.e. if $N \to \infty, \tau_j \to 0$. In this limit using Eqs. 1.48, equation 1.94 can be written as

$$\oint_S \mathbf{A} \cdot \mathbf{ds} = \sum_{j=1}^{N} \Delta\tau_j \nabla \cdot \mathbf{A} \tag{1.95}$$

We write $\Delta\tau_j$ instead of $\tau_j$ in the above equation because volume $\tau_j$ is infinitely small.

Converting the summation symbol into integration and expressing $\Delta\tau_j$ as $d\tau$ we can write the above equation as

$$\oint_S \mathbf{A} \cdot \mathbf{ds} = \int_\tau \nabla \cdot \mathbf{A} d\tau \tag{1.96}$$

Hence divergence theorem is proved. Divergence theorem is an important identity in vector analysis. We will use the theorem to establish other theorems in electromagnetics.

Now we will see the physical interpretation of divergence theorem. In Fig. 1.23a we have shown a pipe fitted in a floor and water spreads out from the floor. We used

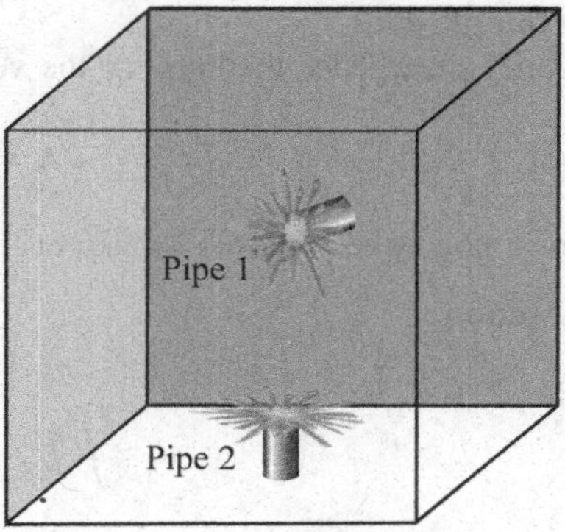


**Fig. 1.30** Pipes fixed inside a cubical structure

this example to explain that whenever there is divergence it implicitly means that there is a source present. Let us restate the above example in a slightly different manner. Instead of fixing the pipe in the floor, let us fix two pipes inside a cubical structure which will allow water to pass through (see Fig. 1.30).

Assuming that water is incompressible whatever water comes out of the pipes flows out through the six faces of the cube.

Now let us generalize the above example to any given vector. Consider Fig. 1.31 (Fig. 1.31 can be of any irregular shape). There are n numbers of sources giving rise to divergence. The n numbers of sources are contained in volume $\tau$ which is bounded by surface S. In Sect. 1.12 we saw that $\oint \mathbf{A} \cdot d\mathbf{s}$ is the flux of the vector. Just as in Fig. 1.30 whatever water comes out of the pipe flows out of the surfaces of the cube, similarly in Fig. 1.31 whatever diverges from the n number of sources $\left(= \int_\tau \nabla \cdot \mathbf{A} d\tau\right)$ flows out of surface in the form of flux $\left[= \oint \mathbf{A} \cdot d\mathbf{s}\right]$.

Hence $\int_\tau \nabla \cdot \mathbf{A} d\tau = \oint_S \mathbf{A} \cdot d\mathbf{s}$.

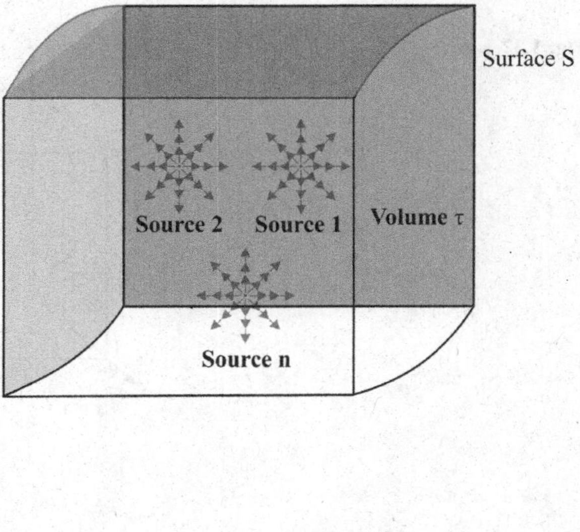


**Fig. 1.31** Vectors diverging inside a irregular structure

**Example 1.10**
Verify divergence theorem for the vector

$$\mathbf{A} = x\hat{\mathbf{i}} + y\hat{\mathbf{j}} + z\hat{\mathbf{k}}$$

for a unit cube which is located at the origin.

**Solution**

$$\int_{\tau} \nabla \cdot \mathbf{A} d\tau = \oint_{S} \mathbf{A} \cdot \mathbf{ds}$$

First let us evaluate the integral $\int_{\tau} \nabla \cdot \mathbf{A} d\tau$

$$\nabla \cdot \mathbf{A} = \left(\hat{\mathbf{i}}\frac{\partial}{\partial x} + \hat{\mathbf{j}}\frac{\partial}{\partial y} + \hat{\mathbf{k}}\frac{\partial}{\partial z}\right) \cdot \left(x\hat{\mathbf{i}} + y\hat{\mathbf{j}} + z\hat{\mathbf{k}}\right)$$
$$= 1 + 1 + 1 + = 3$$

Now

$$\int_{\tau} \nabla \cdot \mathbf{A} d\tau = \int_{x=0}^{1}\int_{y=0}^{1}\int_{z=0}^{1} 3 dx\, dy\, dz$$
$$= 3[x]_0^1[y]_0^1[z]_0^1 = 3$$

Now let us evaluate the surface integral. The cube has six faces. So, the surface integral has to be evaluated at all the six faces as shown in Fig. 1.32.

**Fig. 1.32**

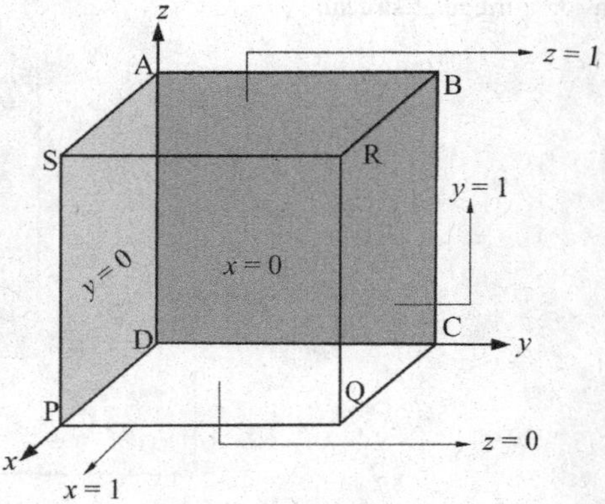

(i) For face ABCD, $x = 0$ $d\mathbf{s} = -dy\,dz\,\hat{\mathbf{i}}$ as outward drawn normal to ABCD is along $x$-axis

$$\iint_{\substack{ABCD \\ x=0}} \mathbf{A} \cdot dy\,dz(-\hat{\mathbf{i}}) = \int_{\substack{ABCD \\ x=0}} \left(x\hat{\mathbf{i}} + y\hat{\mathbf{j}} + z\hat{\mathbf{k}}\right) \cdot dy\,dz(-\hat{\mathbf{i}})$$

(ii) For the face PQRS, $x = 1$ and $d\mathbf{s} = dy\,dz\,\hat{\mathbf{i}}$ . Hence

$$\iint_{\substack{PQRS \\ x=1}} \mathbf{A} \cdot dy\,dz\,\hat{\mathbf{i}} = \int_{\substack{PQRS \\ x=1}} \left(x\hat{\mathbf{i}} + y\hat{\mathbf{j}} + z\hat{\mathbf{k}}\right) \cdot dy\,dz\,\hat{\mathbf{i}}$$

$$\int_{\substack{PQRS \\ x=1}} x\,dy\,dz = 1$$

(iii) For the face SADP, $y = 0$ and $d\mathbf{s} = -dx\,dz\,\hat{\mathbf{j}}$ as the outward drown normal to SADP is along $y$-axis

$$= \int_{\substack{SADP \\ y=0}} \mathbf{A} \cdot dx\,dz(-\hat{\mathbf{j}})$$

$$=$$

(iv) For the face RBCQ, $y = 1$ and $d\mathbf{s} = dx\,dz\,\hat{\mathbf{j}}$

$$= \iint_{\substack{RBCQ \\ y=1}} \mathbf{A} \cdot (dx\,dz)\hat{\mathbf{j}}$$

$$= \iint_{\substack{RBCQ \\ y=1}} \left(x\hat{\mathbf{i}} + y\hat{\mathbf{j}} + z\hat{\mathbf{k}}\right) \cdot (dx\,dz)\hat{\mathbf{j}}$$

$$= \iint_{\substack{RBCQ \\ y=1}} y\,dx\,dz = 1$$

(v) For the face PQCD, $z = 0$ and $\mathbf{ds} = \mathrm{d}x\,\mathrm{d}y\left(-\hat{\mathbf{k}}\right)$ as the outward drawn normal is along $z$-axis

$$= \iint_{\substack{PQCD\\ z=0}} \mathbf{A}\cdot(\mathrm{d}x\,\mathrm{d}y)(-\hat{\mathbf{k}}) =$$

(vi) For the face SRBA, $z = 1$ and $\mathbf{ds} = \mathrm{d}x\,\mathrm{d}y\,\hat{\mathbf{k}}$

$$= \iint_{\substack{SRBA\\ z=1}} \mathbf{A}\cdot \mathrm{d}x\,\mathrm{d}y\,\hat{\mathbf{k}} =$$

Thus

$$\oint \mathbf{A}\cdot\mathbf{ds} = \int_{ABCD} \mathbf{A}\cdot\mathbf{ds} + \int_{PQRS} \mathbf{A}\cdot\mathbf{ds} + \int_{SADP} \mathbf{A}\cdot\mathbf{ds} + \int_{RBCQ} \mathbf{A}\cdot\mathbf{ds} + \int_{PQCD} \mathbf{A}\cdot\mathbf{ds} + \int_{SRBA} \mathbf{A}\cdot\mathbf{ds} = 3$$

Hence divergence theorem is verified. We evaluated two integrals $\int_\tau \nabla\cdot\mathbf{A}\mathrm{d}\tau$ and $\oint \mathbf{A}\cdot\mathbf{ds}$. To evaluate $\int_\tau \nabla\cdot\mathbf{A}\mathrm{d}\tau$ we did single integration. However to evaluate $\oint \mathbf{A}\cdot\mathbf{ds}$ we performed six integrations. $\nabla\cdot\mathbf{A}$ and volume are scalar quantities. In $\int_\tau \nabla\cdot\mathbf{A}\mathrm{d}\tau$ thus only scalar quantities are involved which doesn't have direction. Hence single integration is sufficient to arrive at result. However in the integral $\oint \mathbf{A}\cdot\mathbf{ds}$ surface is a vector quantity which involves direction. For the cube shown in Fig. 1.31 there are six different faces or surfaces each having its own direction. Hence six different integrations for six different directions. Thus evaluation of $\oint \mathbf{A}\cdot\mathbf{ds}$ becomes lengthy and laborious. $\int_\tau \nabla\cdot\mathbf{A}\mathrm{d}\tau$ is easy as it involves single integration because the integration involves scalars. $\oint \mathbf{A}\cdot\mathbf{ds}$ is lengthy as it involves vectors. The above example apart from verifying divergence theorem proves the advantage of working with scalars as compared to working with vectors. Scalars are easy to work with as compared to vectors, as scalars don't involve direction.

## 1.17 Stoke's Theorem

Stoke's theorem is helpful in transforming a surface integral of a vector into a line integral around the boundary C, where C is the boundary of the respective surface. The theorem states that

$$\int_S (\nabla \times \mathbf{A}) \cdot \mathbf{ds} = \oint_C \mathbf{A} \cdot \mathbf{d}\boldsymbol{l} \tag{1.97}$$

Here C is the boundary of the surface S. $\mathbf{ds}$, $\mathbf{d}\boldsymbol{l}$ are the small elements of S and C, respectively.

Both line and surface integrals involve vectors. But however as will see in Example 1.11 line integrals are easy to evaluate as compared to surface integrals.

Now let us see the proof for Stoke's theorem.

Consider $\oint_C \mathbf{A} \cdot \mathbf{d}\boldsymbol{l}$ where C is the boundary of surface S as shown in Fig. 1.33a in the region of vector **A**. Let us divide the surface S in two parts $S_1$ and $S_2$ having boundaries $C_1$ and $C_2$, respectively, as shown in Fig. 1.33b. The line integral of the vector **A** over the boundaries $C_1$ and $C_2$ can be written as $\oint_{C_1} \mathbf{A} \cdot \mathbf{d}\boldsymbol{l}_1$ and $\oint_{C_2} \mathbf{A} \cdot \mathbf{d}\boldsymbol{l}_2$. From Fig. 1.33b as we traverse the line PQ along boundary $C_1$ and line PQ along boundary $C_2$, the line integral of **A** is equal and opposite and cancels out each other, the remaining boundaries being same as that of C. Hence

$$\oint_C \mathbf{A} \cdot \mathbf{d}\boldsymbol{l} = \oint_{C_1} \mathbf{A} \cdot \mathbf{d}\boldsymbol{l}_1 + \oint_{C_2} \mathbf{A} \cdot \mathbf{d}\boldsymbol{l}_2 \tag{1.98}$$

In Fig. 1.33c we divide the surface S into number of surface $s_1, s_2, s_3, \ldots s_N$ whose boundaries are $C_1, C_2, C_3, \ldots C_N$. A small part of Fig. 1.33c is shown in Fig. 1.34. Clearly at the common interface the line integral of **A** cancels out. Using the same argument as we did for Fig. 1.33b to arrive at Eq. 1.98, we can write for Fig. 1.33c

$$\oint_C \mathbf{A} \cdot \mathbf{d}\boldsymbol{l} = \oint_{C_1} \mathbf{A} \cdot \mathbf{d}\boldsymbol{l}_1 + \oint_{C_2} \mathbf{A} \cdot \mathbf{d}\boldsymbol{l}_2 + \cdots + \oint_{C_N} \mathbf{A} \cdot \mathbf{d}\boldsymbol{l}_N \tag{1.99}$$

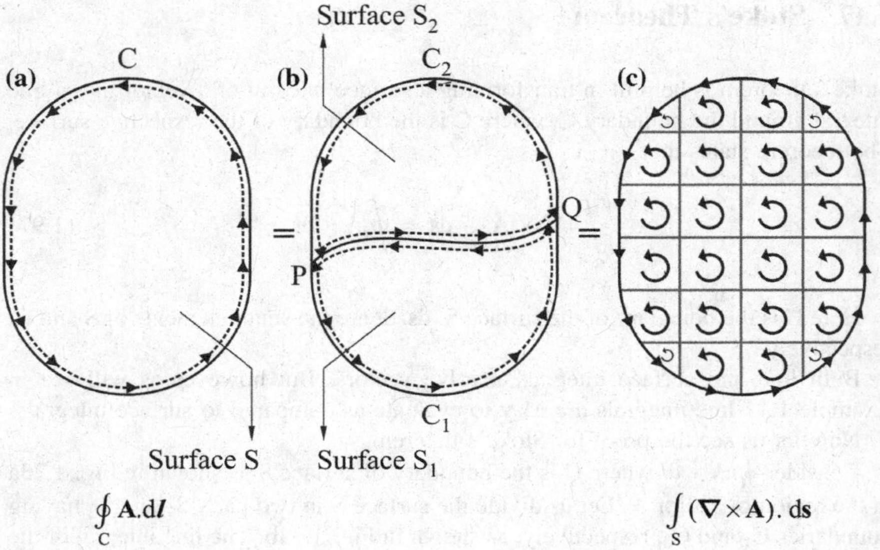


**Fig. 1.33** **a** Surface S with boundary C. **b** Surface S divided into two equal parts. **c** Surface S divided into N number of parts

**Fig. 1.34** .

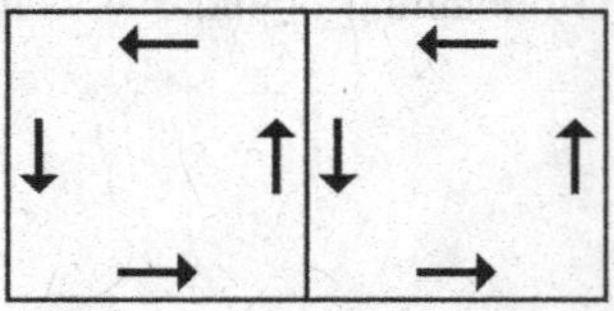

$$\Rightarrow \oint_C \mathbf{A}\cdot d\boldsymbol{l} = \sum_{j=1}^{N} \oint_{Cj} \mathbf{A}\cdot d\boldsymbol{l}_j \tag{1.100}$$

The above equation can be written as

$$\oint_C \mathbf{A}\cdot d\boldsymbol{l} = \sum_{j=1}^{N} s_j \frac{\oint_{Cj} \mathbf{A}\cdot d\boldsymbol{l}_j}{s_j} \tag{1.101}$$

Suppose if N is very large then the surface $s_j$ becomes very small, i.e. if $N \to \infty$, $s_j \to 0$ so that we can write Eq. 1.68

$$\lim_{\Delta s_j \to 0} \frac{\oint_{Cj} \mathbf{A}\cdot d\boldsymbol{l}_j}{\Delta s_j} = \text{curl}\,\mathbf{A}\cdot\hat{\mathbf{n}}_j \tag{1.102}$$

where $\hat{\mathbf{n}}_j$ is the unit vector normal to surface $\Delta s_j$.

In the limit $N \to \infty$, $s_j \to 0$ Eq. 1.101 can be written as

$$\oint_C \mathbf{A} \cdot d\boldsymbol{l} = \sum_{j=1}^{N} \Delta s_j \frac{\oint_{C_j} \mathbf{A} \cdot d\boldsymbol{l}_j}{\Delta s_j} \tag{1.103}$$

Here we have used $\Delta s_j$ instead of $s_j$ to note that $s_j \to 0$ in the limit .
Using Eq. 1.102 in Eq. 1.103

$$\oint_C \mathbf{A} \cdot d\boldsymbol{l} = \sum_{j=1}^{N} \Delta s_j \text{ curl } \mathbf{A} \cdot \hat{\mathbf{n}}_j \tag{1.104}$$

Replacing the summation by integration and $\Delta s_j$ by ds, $\hat{\mathbf{n}}_j$ by $\hat{\mathbf{n}}$

$$\oint_C \mathbf{A} \cdot d\boldsymbol{l} = \int_S (\nabla \times \mathbf{A}) \cdot \hat{\mathbf{n}} ds \tag{1.105}$$

$$\Rightarrow \oint_C \mathbf{A} \cdot d\boldsymbol{l} = \int_S (\nabla \times \mathbf{A}) \cdot d\mathbf{s} \tag{1.106}$$

Hence Stoke's theorem is proved.

The physical interpretation of Stoke's theorem is as follows. We know that $\nabla \times \mathbf{A}$ is the measure of how much the vector **A** circulates about. As shown in Fig. 1.33c there is large circulation over surface S. Hence $\int_S (\nabla \times \mathbf{A}) \cdot d\mathbf{s}$ is the measure of total circulation over the surface S. However as shown in Fig. 1.34 at the common interface the line integral of **A** cancels out. So in Fig. 1.33c at all common interfaces the circulations are in opposite direction and cancel out. Hence finally we arrive at Fig. 1.33a. In Fig. 1.33a we calculate the line integral $\oint_C \mathbf{A} \cdot d\boldsymbol{l}$ which is equivalent to $\int_S (\nabla \times \mathbf{A}) \cdot d\mathbf{s}$ as the above argument shows.

Thus to calculate the curl over the surface S in Fig. 1.33c you can go over the entire surface S and calculate total curl using the relation $\int_S (\nabla \times \mathbf{A}) \cdot d\mathbf{s}$ (or) go over the entire boundary C and calculate the circulation along the boundary C in Fig. 1.33a using the relation $\oint_C \mathbf{A} \cdot d\boldsymbol{l}$.

However while integrating the line integral which way to go around either clockwise or anticlockwise, along the boundary C of Fig. 1.33a. The direction over which the boundary C is to be traversed is fixed by the direction of **ds** in Fig. 1.33. Suppose assume that **ds** in Fig. 1.33 is outward perpendicular to the plane of the paper. Let the thumb of your right hand point along **ds**, i.e. along the outward perpendicular to the plane of the paper. Then the fingers of right hand will be along the arrow marks along the boundary shown in Fig. 1.33a which is the direction in which the integration of line integral $\oint_C \mathbf{A} \cdot d\boldsymbol{l}$ is to be carried out.

**Example 1.11**

Verify Stoke's theorem for the surface shown in Fig. 1.35 if $\mathbf{A} = 4\hat{\mathbf{i}} + z^2y^3\hat{\mathbf{j}}$. The square surface is of unit dimension.

**Solution**
First let us calculate the line integral

(a) For the line a shown in Fig. 1.35 $x = 0, y = 0 \to 1, z = 0$ Hence $\mathbf{A} = 4\hat{\mathbf{i}}$

$$\mathbf{A} \cdot \mathrm{d}\boldsymbol{l} = 4\mathrm{d}x \Rightarrow \int \mathbf{A} \cdot \mathrm{d}\boldsymbol{l} = \int 4\mathrm{d}x = 0 \text{ as } \mathrm{d}x = 0$$

(b) For the line b shown in Fig. 1.35 $x = 0, y = 1, z = 0 \to 1$ Hence $\mathbf{A} = 4\hat{\mathbf{i}} + z^2\hat{\mathbf{j}}$

$$\int \mathbf{A} \cdot \mathrm{d}\boldsymbol{l} = \int 4\mathrm{d}x + z^2\mathrm{d}y = 0 \text{ as } \mathrm{d}x,\ \mathrm{d}y = 0$$

(c) For the line c shown in Fig. 1.35 $x = 0, y = 1 \to 0, z = 1$ Hence $\mathbf{A} = 4\hat{\mathbf{i}} + y^3\hat{\mathbf{j}}$

$$\mathbf{A} \cdot \mathrm{d}\boldsymbol{l} = 4\mathrm{d}x + y^3\mathrm{d}y = y^3\mathrm{d}y \text{ as } \mathrm{d}x = 0$$

$$\mathbf{A} \cdot \mathrm{d}\boldsymbol{l} = \int_1^0 y^3\mathrm{d}y = \left.\frac{y^4}{4}\right|_1^0 = \frac{-1}{4}$$

(d) For the line d shown in Fig. 1.35 $x = 0, y = 0, z = 1 \to 0$ $\mathbf{A} = 4\hat{\mathbf{i}}$

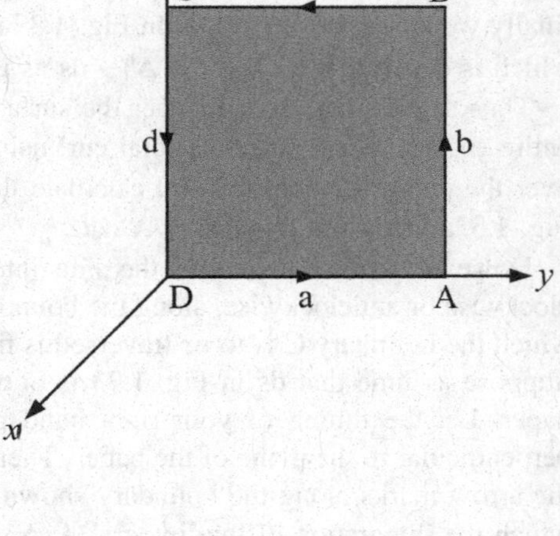


**Fig. 1.35** A square surface ABCD in Cartesian coordinate system

$$\int \mathbf{A} \cdot \mathrm{d}\boldsymbol{l} = 0 \text{ as } \mathrm{d}x = 0$$

Hence the total sum $\oint \mathbf{A} \cdot \mathrm{d}l = 0 + 0 - \frac{1}{4} + 0 = -\frac{1}{4}$
We have $\nabla \times \mathbf{A} = -2zy^3\hat{\mathbf{i}}$
By right-hand rule $\mathbf{ds} = \mathrm{d}y\ \mathrm{d}z\ \hat{\mathbf{i}}$
Thus $\int_S (\nabla \times \mathbf{A}) \cdot \mathbf{ds} = \int_0^1 \int_0^1 (-2zy^3)\mathrm{d}y\mathrm{d}z = -\frac{1}{4}$
Hence Stoke's theorem is verified.

In Fig. 1.35 there is only one surface ABCD to work with. Suppose assume that ABCD in Fig. 1.35 is an elastic membrane. We have redrawn ABCD elastic membrane in Fig. 1.36a. Let us stretch the elastic membrane in the form of a cube as shown in Fig. 1.36b. As ABCD the elastic membrane is stretched into a cube, face ABCD is now open. Thus there are five surfaces of cube to work with. Hence for the given loop ABCD in Fig. 1.36a there is only one surface, while in Fig. 1.36b there are five different surfaces. For the same loop ABCD if the elastic membrane in Fig. 1.36a is blown up in the form of balloon as shown in Fig. 1.36c surface integration becomes even more complicated. In all the three figures the loop ABCD is two dimensional, just one need $z$- and $y$-axis to describe loop ABCD. But the corresponding surface associated with ABCD can be three dimensional as in Fig. 1.36b, c and hence surface integration might become difficult.

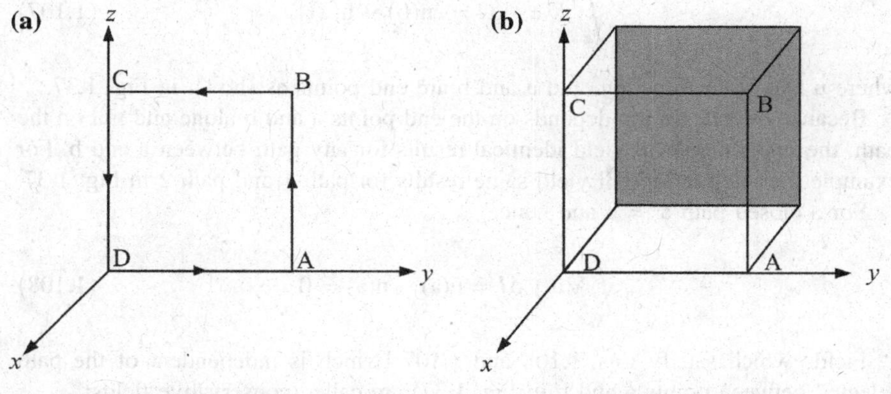


**Fig. 1.36a, b** **a** Elastic membrane ABCD. **b** Elastic membrane stretched in the form of cube

**Fig. 1.36c** Elastic membrane blown in the form of balloon

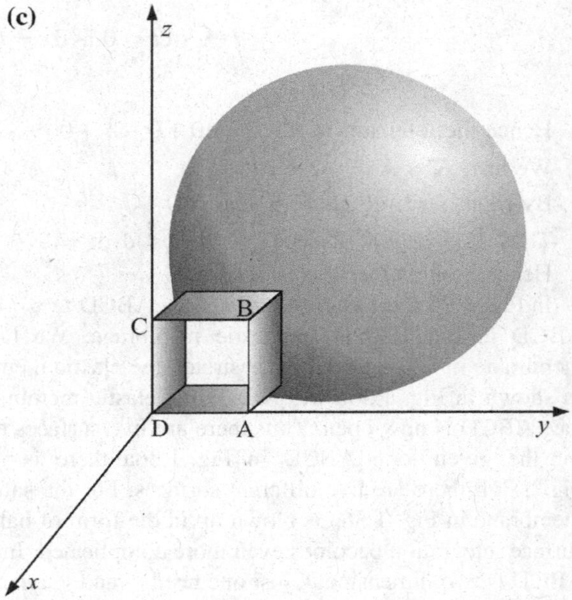


## 1.18 The Gradient Theorem

The Gradient theorem states that

$$\int_a^b \nabla u \cdot d\boldsymbol{l} = u(b) - u(a) \tag{1.107}$$

where u is a scalar function, and a and b are end points as shown in Fig. 1.37.

Because the integration depends on the end points a and b alone and not on the path, the integration will yield identical results for any path between a and b. For example the integration will yield same results for path 1 and path 2 in Fig. 1.37.

For a closed path a = b and hence

$$\oint \nabla u \cdot d\boldsymbol{l} = u(a) - u(a) = 0 \tag{1.108}$$

Fields which satisfy Eqs. 1.108 and 1.107 (which is independent of the path selected between points a and b in Fig. 1.37) are called conservative fields.

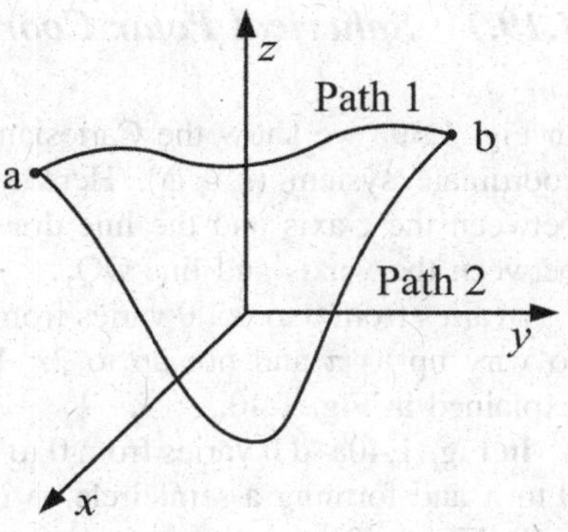


**Fig. 1.37** End points a, b along with Path 1 and Path 2 in Cartesian coordinate system

## 1.19 Others Coordinate Systems

In two dimensions the most familiar coordinate system is Cartesian coordinate system with $x$- and $y$-axis as shown in Fig. 1.38. Another coordinate system in two dimensions is polar coordinate system (r, θ). The Cartesian coordinates and polar coordinates are related to each other by

$$x = \mathrm{r}\cos\theta, y = \mathrm{r}\sin\theta$$

The point P is Fig. 1.38 can be represented by either Cartesian coordinates $x$, $y$ or polar coordinates (r, θ). If we are interested in calculating electric (or) magnetic fields at point P due to some charge or current distributions then these physical quantities can be expressed in terms of Cartesian coordinates ($x$, $y$) or polar coordinates (r, θ).

Similarly in three dimensions in addition to Cartesian coordinates ($x$, $y$, $z$) there are number of other coordinate systems, out of which two are very important spherical coordinate system and cylindrical coordinate system.

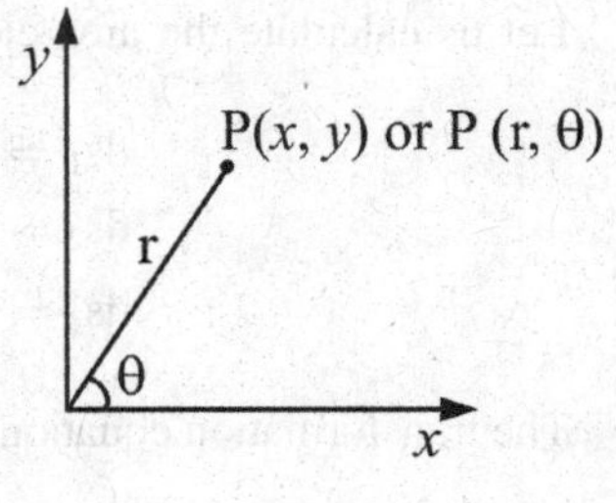


**Fig. 1.38** Cartesian and polar coordinate system

### 1.19.1 Spherical Polar Coordinates

In Fig. 1.39a we show the Cartesian coordinate system (*x*, *y*, *z*) and the spherical coordinate system $(r, \theta, \phi)$. Here r is the distance from origin. θ is the angle between the *z*-axis and the line drawn from the origin to point P. ϕ is the angle between the *x*-axis and line OQ.

r varies from 0 to $\infty$, θ varies from 0 to $\pi$ and ϕ varies from 0 to $2\pi$. θ is allowed to vary up to $\pi$ and not up to $2\pi$ The reason for such limited variation of θ in explained in Fig. 1.40.

In Fig. 1.40a–d θ varies from 0 to $\pi$ and forms a semicircle. After varying θ from 0 to $\pi$ and forming a semicircle, ϕ is varied in Fig. 1.40e–g.

In Fig. 1.40e we vary ϕ from 0 to 90°. In Fig. 1.40f ϕ is varied to 270° and in Fig. 1.40g ϕ completes $2\pi$ degrees which finally forms sphere. Thus variation of from 0 to $\pi$ forms a semicircle and when the semicircle is rotated through $2\pi$ degrees (i.e. we vary ϕ) we finally form a sphere.

To form a hemisphere we vary θ from 0 to $\pi/2$ as shown in Fig. 1.40h–j. Then we rotate ϕ from 0 to $2\pi$ in Fig. 1.40k–m which finally forms a hemisphere.

An elemental length d***l*** can be expressed in spherical polar coordinates as

$$\mathrm{d}\boldsymbol{l} = \mathrm{d}l_1\hat{\mathbf{r}} + \mathrm{d}l_2\hat{\boldsymbol{\theta}} + \mathrm{d}l_3\hat{\boldsymbol{\phi}} \tag{1.109}$$

Using the result from Fig. 1.39b–d

$$\mathrm{d}\boldsymbol{l} = \mathrm{d}r\hat{\mathbf{r}} + r\mathrm{d}\theta\,\hat{\boldsymbol{\theta}} + r\sin\theta\,\mathrm{d}\phi\hat{\boldsymbol{\phi}} \tag{1.110}$$

The volume element $\mathrm{d}\tau$ shown in Fig. 1.39e, Fig. 1.41 can be expressed as

$$\begin{aligned} \mathrm{d}\tau &= (\mathrm{d}l_1)\times(\mathrm{d}l_2)\times(\mathrm{d}l_3) \\ \mathrm{d}\tau &= (\mathrm{d}r)(r\mathrm{d}\theta)(r\sin\theta\,\mathrm{d}\phi) \end{aligned} \tag{1.111}$$

$$\mathrm{d}\tau = r^2\sin\theta\,\mathrm{d}r\,\mathrm{d}\theta\,\mathrm{d}\phi \tag{1.112}$$

Let us calculate the area elements in Fig. 1.41

$$\left.\begin{aligned} \mathrm{d}s_1 &= \mathrm{d}l_1\times\mathrm{d}l_3 = r\sin\theta\,\mathrm{d}r\,\mathrm{d}\phi \\ \mathrm{d}s_2 &= \mathrm{d}l_2\times\mathrm{d}l_1 = r\,\mathrm{d}r\,\mathrm{d}\theta \\ \mathrm{d}s_3 &= \mathrm{d}l_2\times\mathrm{d}l_3 = r^2\sin\theta\,\mathrm{d}\theta\,\mathrm{d}\phi \end{aligned}\right\} \tag{1.113}$$

The transformation equation from Cartesian coordinates to spherical coordinates is

$$\left.\begin{aligned} x &= r\sin\theta\cos\phi \\ y &= r\sin\theta\sin\phi \\ z &= r\cos\theta \end{aligned}\right\} \tag{1.114}$$

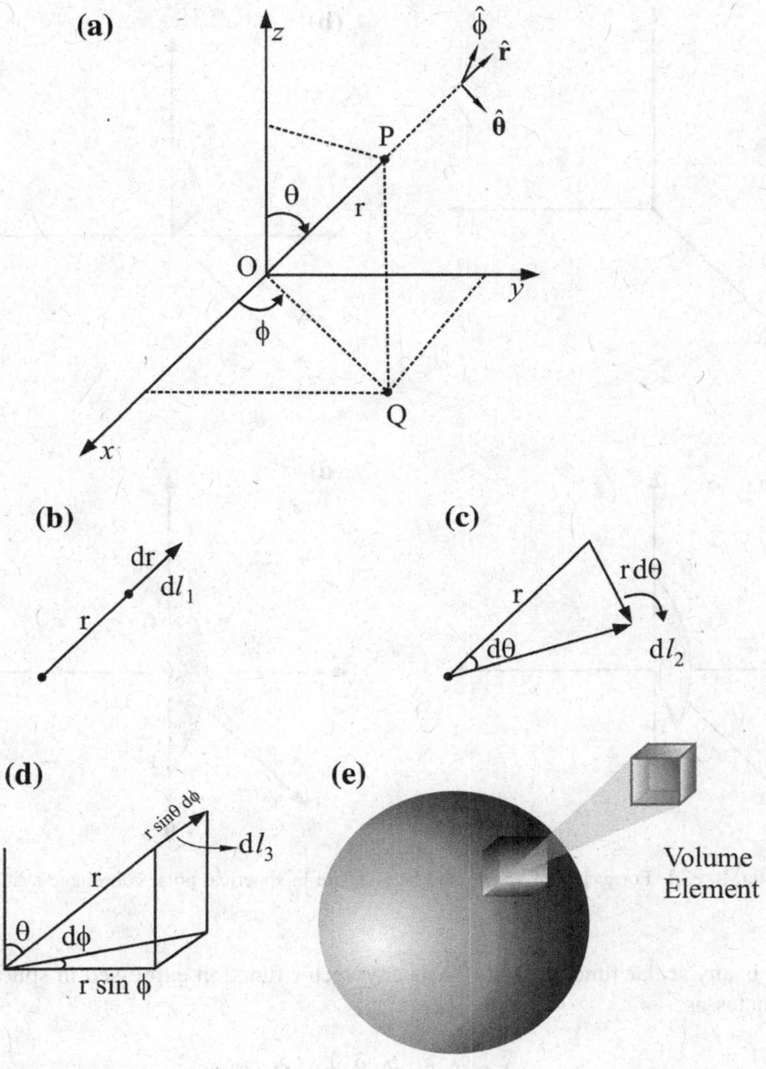


**Fig. 1.39a, b, c, d, e** Spherical polar coordinates along with line and volume elements

and

$$\left.\begin{aligned} r &= \sqrt{x^2 + y^2 + z^2} \\ \theta &= \tan^{-1}\frac{\sqrt{x^2 + y^2}}{z} \\ \phi &= \tan^{-1}\frac{y}{x} \end{aligned}\right\} \tag{1.115}$$

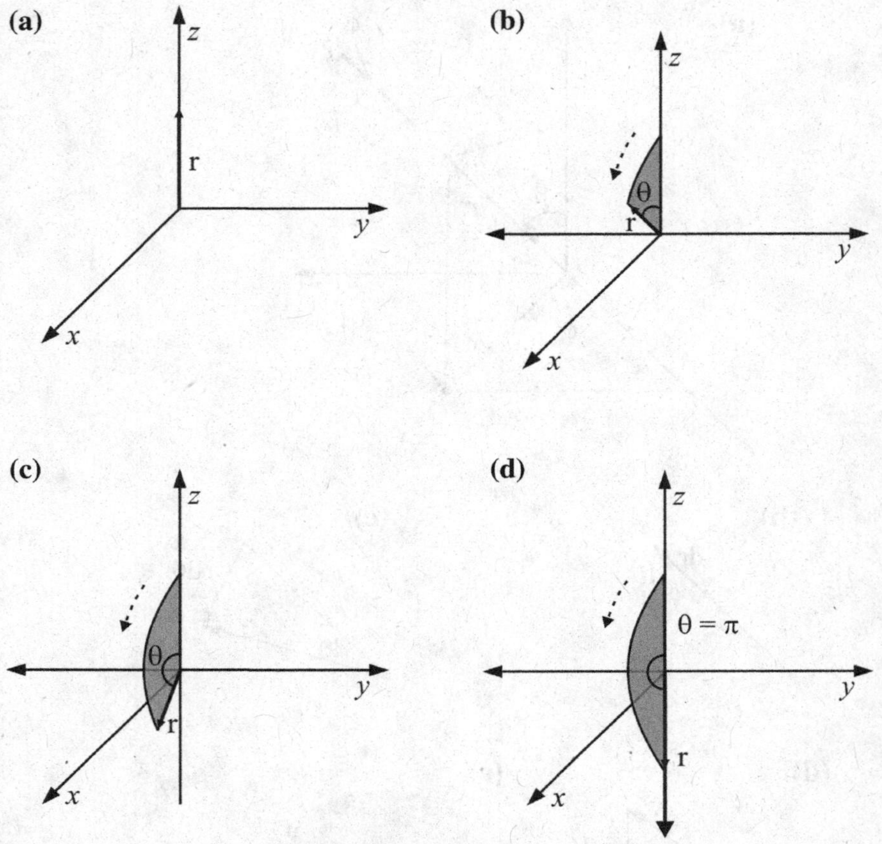


**Fig. 1.40a, b, c, d** Formation of sphere and hemisphere in spherical polar coordinate system

If u is any scalar function and if **A** is any vector function expressed in spherical coordinates as

$$\mathbf{A} = A_r\hat{\mathbf{r}} + A_\theta\hat{\boldsymbol{\theta}} + A_\phi\hat{\boldsymbol{\phi}}$$

then gradient, divergence and curl can be expressed in spherical coordinates as

$$\nabla u = \frac{\partial u}{\partial r}\hat{\mathbf{r}} + \frac{1}{r}\frac{\partial u}{\partial \theta}\hat{\boldsymbol{\theta}} + \frac{1}{r\sin\theta}\frac{\partial u}{\partial \phi}\hat{\boldsymbol{\phi}} \tag{1.116}$$

$$\nabla \cdot \mathbf{A} = \frac{1}{r^2}\frac{\partial}{\partial r}\left[r^2 A_r\right] + \frac{1}{r\sin\theta}\frac{\partial}{\partial\theta}[\sin\theta\, A_\theta] + \frac{1}{r\sin\theta}\frac{\partial A_\phi}{\partial\phi} \tag{1.117}$$

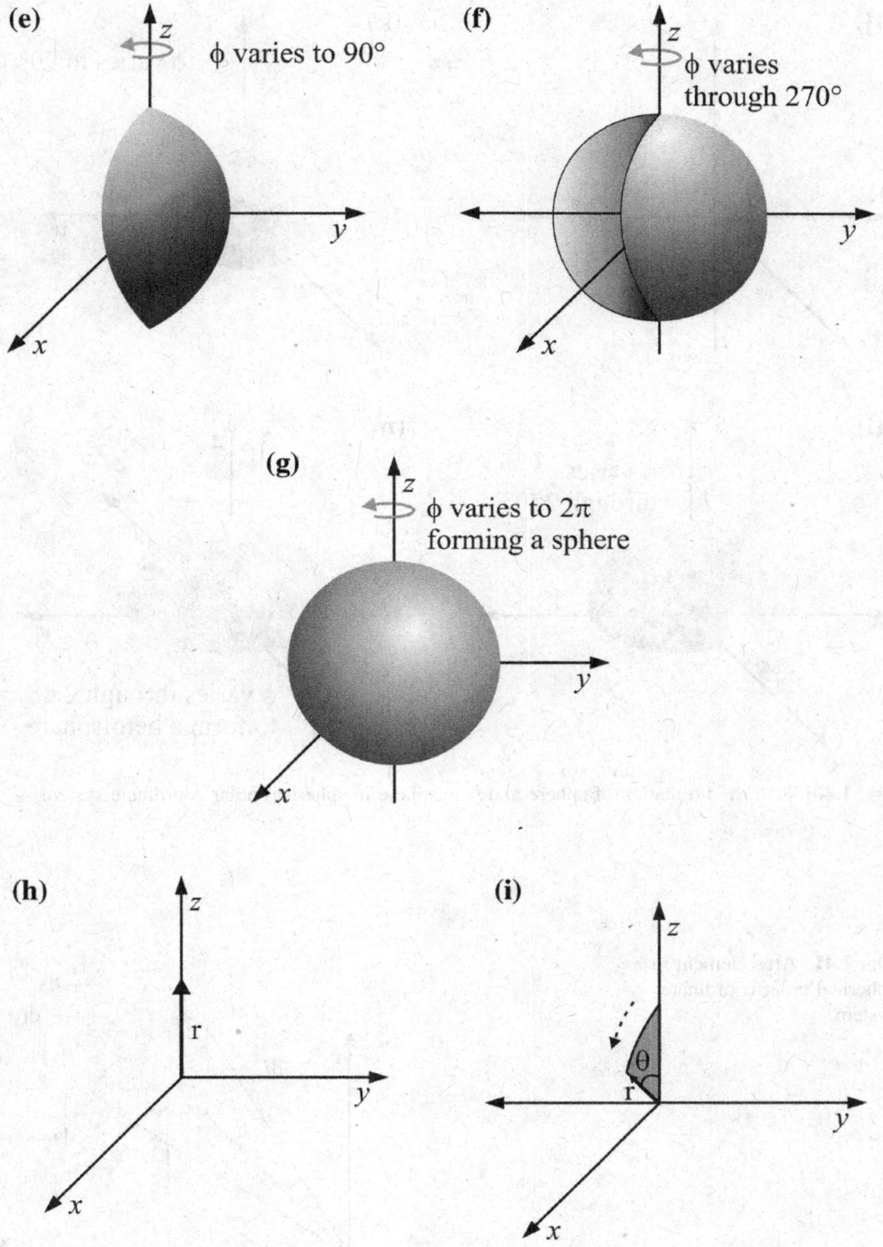


**Fig. 1.40e, f, g, h, i** Formation of sphere and hemisphere in spherical polar coordinate system

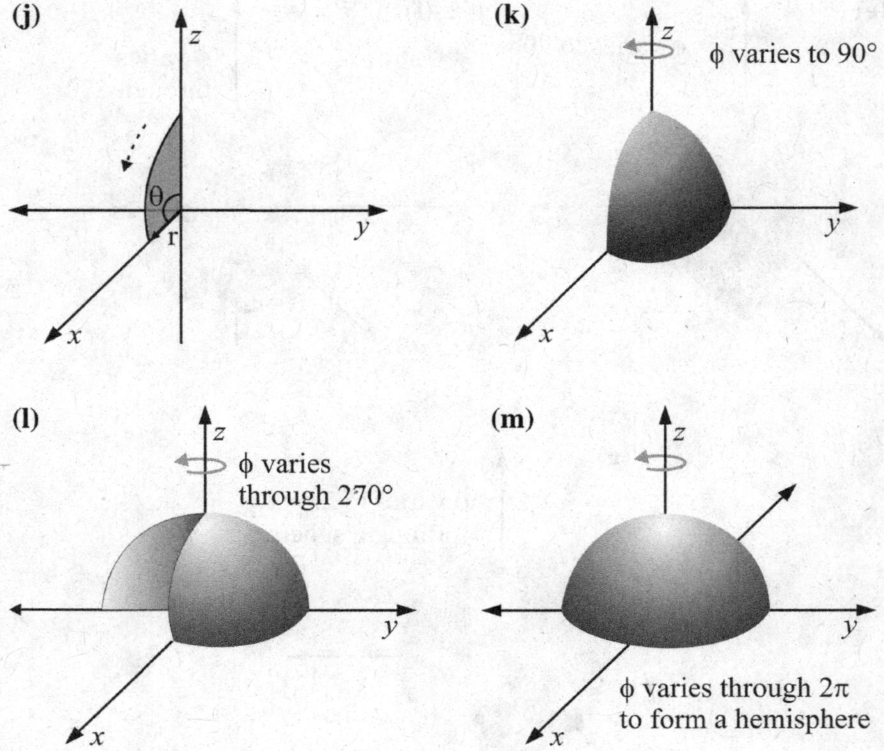


**Fig. 1.40j, k, l, m** Formation of sphere and hemisphere in spherical polar coordinate system

**Fig. 1.41** Area element in spherical polar coordinate system

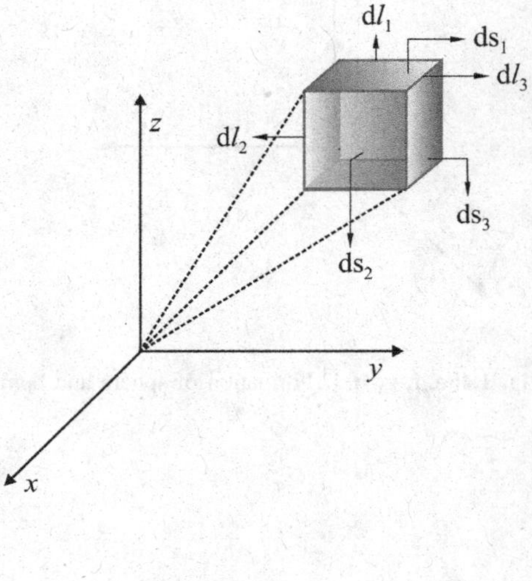

$$\nabla \times \mathbf{A} = \frac{1}{r \sin\theta}\left[\frac{\partial}{\partial \theta}(\sin\theta\, A_{\phi}) - \frac{\partial A_{\theta}}{\partial \phi}\right]\hat{\mathbf{r}} + \frac{1}{r}\left[\frac{1}{\sin\theta}\frac{\partial A_r}{\partial \phi} - \frac{\partial}{\partial r}(rA_{\phi})\right]\hat{\boldsymbol{\theta}} + \frac{1}{r}\left[\frac{\partial}{\partial r}(rA_{\theta}) - \frac{\partial A_r}{\partial \theta}\right]\hat{\boldsymbol{\phi}} \tag{1.118}$$

The $\nabla^2$ operator called Laplacian is given by

$$\nabla^2 u = \frac{1}{r^2}\frac{\partial}{\partial r}\left[r^2 \frac{\partial u}{\partial r}\right] + \frac{1}{r^2 \sin\theta}\frac{\partial}{\partial \theta}\left[\sin\theta \frac{\partial u}{\partial \theta}\right] + \frac{1}{r^2 \sin^2\theta}\frac{\partial^2 u}{\partial \phi^2} \tag{1.119}$$

### 1.19.2 Cylindrical Coordinates

In Fig. 1.42 Cartesian coordinate system along with cylindrical coordinate system $(r_C, \phi, z)$ is shown. Here $r_C$ is the distance from point P to the $z$-axis, $\phi$ is the angle between $x$-axis and line O Q and $z$ is the usual Cartesian coordinate.

Here $r_C$ varies from 0 to $\infty$, $\phi$ varies from 0 to $2\pi$ and $z$ varies from $-\infty$ to $\infty$. An element $d\boldsymbol{l}$ in cylindrical coordinates can be expressed as

$$d\boldsymbol{l} = d l_1 \hat{\mathbf{r}}_C + d l_2 \hat{\boldsymbol{\phi}} + d l_3 \hat{\mathbf{z}} \tag{1.120}$$

Using the results from Fig. 1.43 we can write

$$d\boldsymbol{l} = dr_C \hat{\mathbf{r}}_C + r_C d\phi \hat{\boldsymbol{\phi}} + dz \hat{\mathbf{z}} \tag{1.121}$$

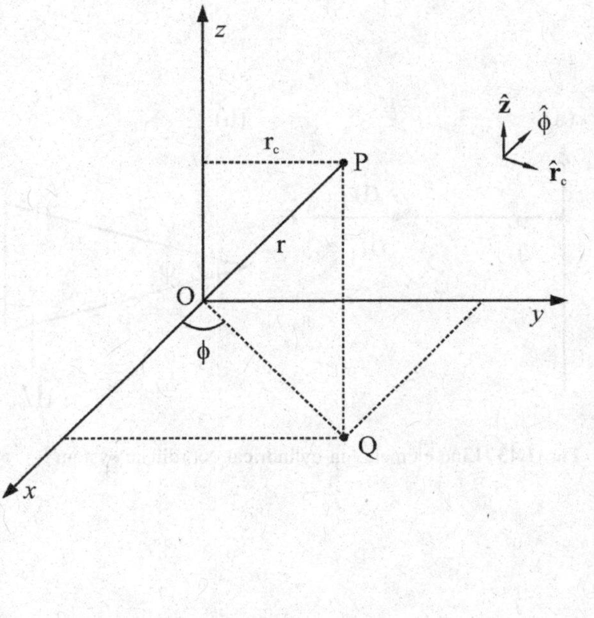


**Fig. 1.42** Cartesian and cylindrical coordinate system

The volume element shown in Fig. 1.44 can be written as

$$d\tau = dl_1\, dl_2\, dl_3 \tag{1.122}$$

$$d\tau = r_C\, dr_C\, d\phi\, dz \tag{1.123}$$

Now let us calculate the area element in Fig. 1.44

$$\left.\begin{aligned} ds_1 &= dl_1 dl_3 = dr_C\, dz \\ ds_2 &= dl_1 dl_2 = r_C\, dr_C d\phi \\ ds_3 &= dl_2 dl_3 = r_C dr_C\, dz \end{aligned}\right\} \tag{1.124}$$

The transformation equation from Cartesian to cylindrical coordinate system is given by

$$x = r_C \cos\phi, y = r_C \sin\phi, z = z \tag{1.125}$$

Also

$$\left.\begin{aligned} r_C &= \sqrt{x^2 + y^2} \\ \phi &= \tan^{-1}\frac{y}{x} \\ z &= z \end{aligned}\right\} \tag{1.126}$$

If u is any scalar function and if any vector **A** is expressed in cylindrical coordinates as

$$\mathbf{A} = A_{r_C}\hat{\mathbf{r}}_C + A_\phi\hat{\boldsymbol{\phi}} + A_z\hat{\mathbf{z}}$$

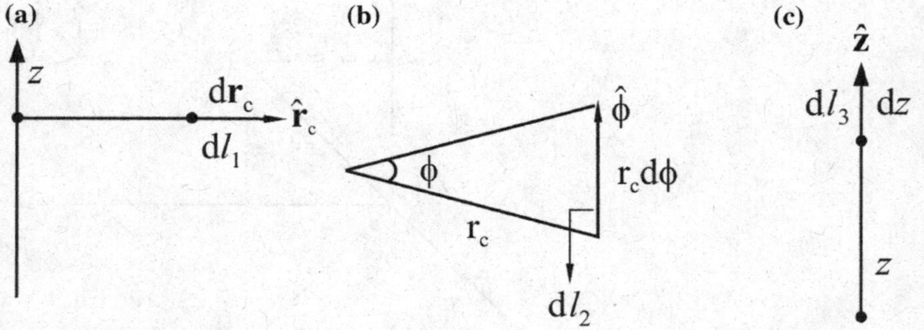


**Fig. 1.43** Line elements in cylindrical coordinate system

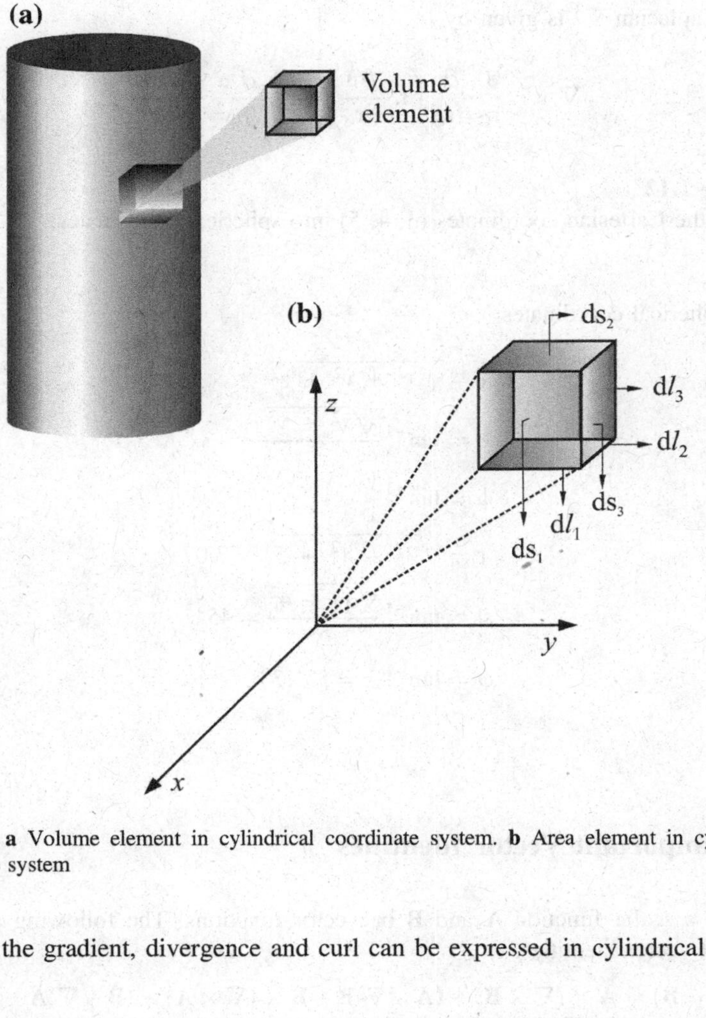


**Fig. 1.44** **a** Volume element in cylindrical coordinate system. **b** Area element in cylindrical coordinate system

Then the gradient, divergence and curl can be expressed in cylindrical coordinates as

$$\nabla \mathrm{u} = \frac{\partial \mathrm{u}}{\partial \mathrm{r_C}}\hat{\mathbf{r}}_\mathrm{C} + \frac{1}{\mathrm{r_C}}\frac{\partial \mathrm{u}}{\partial \phi}\hat{\boldsymbol{\phi}} + \frac{\partial \mathrm{u}}{\partial z}\hat{\mathbf{z}} \tag{1.127}$$

$$\nabla \cdot \mathbf{A} = \frac{1}{\mathrm{r_C}}\frac{\partial}{\partial \mathrm{r_C}}(\mathrm{r_C}\,\mathrm{A}_{\mathrm{r_c}}) + \frac{1}{\mathrm{r_C}}\frac{\partial \mathrm{A}_\phi}{\partial \phi} + \frac{\partial \mathrm{A}_z}{\partial z} \tag{1.128}$$

$$\begin{aligned}\nabla \times \mathbf{A} = &\left(\frac{1}{\mathrm{r_C}}\frac{\partial \mathrm{A}_z}{\partial \phi} - \frac{\partial \mathrm{A}_\phi}{\partial z}\right)\hat{\mathbf{r}}_\mathrm{C} + \left(\frac{\partial \mathrm{A}_{\mathrm{r_C}}}{\partial z} - \frac{\partial \mathrm{A}_z}{\partial \mathrm{r_C}}\right)\hat{\phi} \\ &+ \frac{1}{\mathrm{r_C}}\left(\frac{\partial \mathrm{c}}{\partial \mathrm{r}}(\mathrm{r_C}\mathrm{A}_\phi) - \frac{\partial \mathrm{A}_{\mathrm{r_C}}}{\partial \phi}\right)\hat{\mathbf{z}}\end{aligned} \tag{1.129}$$

The Laplacian $\nabla^2$ is given by

$$\nabla^2 u = \frac{1}{r_C}\frac{\partial}{\partial r_C}\left(r_C\frac{\partial u}{\partial r_C}\right) + \frac{1}{r_C^2}\frac{\partial^2 u}{\partial \phi^2} + \frac{\partial^2 u}{\partial z^2} \tag{1.130}$$

**Example 1.12**
Convert the Cartesian coordinates (3, 4, 5) into spherical coordinates.

**Solution**

(a) In spherical coordinates

$$r = \sqrt{x^2 + y^2 + z^2}$$
$$\theta = \tan^{-1}\frac{\sqrt{x^2 + y^2}}{z}$$
$$\phi = \tan^{-1}\frac{y}{x}$$
$$r = \sqrt{3^2 + 4^2 + 5^2} = 7.07$$
$$\theta = \tan^{-1}\frac{\sqrt{3^2 + 4^2}}{5} = 45^\circ$$
$$\phi = \tan^{-1}\frac{4}{3} = 53.13^\circ$$

## 1.20 Important Vector Identities

Let u be a scalar function **A** and **B** be vector functions. The following are few important vector identities.

1. $\nabla(\mathbf{A}\cdot\mathbf{B}) = \mathbf{A}\times(\nabla\times\mathbf{B}) + (\mathbf{A}\cdot\nabla)\mathbf{B} + \mathbf{B}\times(\nabla\times\mathbf{A}) + (\mathbf{B}\cdot\nabla)\mathbf{A}$
2. $\nabla\times(\mathbf{A}+\mathbf{B}) = \nabla\times\mathbf{A} + \nabla\times\mathbf{B}$
3. $\nabla\cdot(\nabla\times\mathbf{A}) = 0$
4. $\nabla\times\nabla u = 0$
5. $\nabla\cdot(\mathbf{A}+\mathbf{B}) = \nabla.\mathbf{A} + \nabla.\mathbf{B}$
6. $\nabla\times(u\mathbf{A}) = u\nabla\times\mathbf{A} + \nabla u\times\mathbf{A}$
7. $\nabla\times(\nabla\times\mathbf{A}) = \nabla(\nabla\cdot\mathbf{A}) - \nabla^2\mathbf{A}$
8. $\nabla\cdot(\mathbf{A}\times\mathbf{B}) = \mathbf{B}\cdot(\nabla\times\mathbf{A}) - \mathbf{A}\cdot(\nabla\times\mathbf{B})$
9. $\nabla\times(\mathbf{A}\times\mathbf{B}) = (\mathbf{B}\cdot\nabla)\mathbf{A} - (\mathbf{A}\cdot\nabla)\mathbf{B} + \mathbf{A}(\nabla\cdot\mathbf{B}) - \mathbf{B}(\nabla\cdot\mathbf{A})$

10. $\nabla \cdot (\mathrm{u}\mathbf{A}) = \mathrm{u}\nabla \cdot \mathbf{A} + \mathbf{A} \cdot \nabla \mathrm{u}$

11. $\nabla \cdot \nabla = \left(\hat{\mathbf{i}}\frac{\partial}{\partial x} + \hat{\mathbf{j}}\frac{\partial}{\partial y} + \hat{\mathbf{k}}\frac{\partial}{\partial z}\right) . \left(\hat{\mathbf{i}}\frac{\partial}{\partial x} + \hat{\mathbf{j}}\frac{\partial}{\partial y} + \hat{\mathbf{k}}\frac{\partial}{\partial z}\right)$

$\Rightarrow \nabla^2 = \frac{\partial^2}{\partial x^2} + \frac{\partial^2}{\partial y^2} + \frac{\partial^2}{\partial z^2}$ is called Laplacian operator.

The above form of Laplacian operator is in Cartesian coordinates. We have seen the form of Laplacian in spherical coordinates in Sect. 1.19a and in cylindrical coordinates in Sect. 1.19b.

**Example 1.13**

Calculate $\nabla \mathrm{r}$ if $\mathrm{r} = (x^2 + y^2 + z^2)^{1/2}$ where $\hat{\mathbf{r}} = x\hat{\mathbf{i}} + y\hat{\mathbf{j}} + z\hat{\mathbf{k}}$

**Solution**

$$\nabla \mathrm{r} = \frac{\partial \mathrm{r}}{\partial x}\hat{\mathbf{i}} + \frac{\partial \mathrm{r}}{\partial y}\hat{\mathbf{j}} + \frac{\partial \mathrm{r}}{\partial z}\hat{\mathbf{k}}$$

$$\begin{aligned}
\nabla \mathrm{r} &= \frac{\partial}{\partial x}(x^2 + y^2 + z^2)^{1/2}\hat{\mathbf{i}} + \frac{\partial}{\partial y}(x^2 + y^2 + z^2)^{1/2}\hat{\mathbf{j}} \\
&\quad + \frac{\partial}{\partial z}(x^2 + y^2 + z^2)^{1/2}\hat{\mathbf{k}} \\
&= \frac{1}{2}(x^2 + y^2 + z^2)^{-1/2} \cdot 2x\hat{\mathbf{i}} \\
&\quad + \frac{1}{2}(x^2 + y^2 + z^2)^{-1/2}(2y)\hat{\mathbf{j}} \\
&\quad + \frac{1}{2}(x^2 + y^2 + z^2)^{-1/2}(2z)\hat{\mathbf{k}} \\
&= (x^2 + y^2 + z^2)^{-1/2}(x\hat{\mathbf{i}} + y\hat{\mathbf{j}} + z\hat{\mathbf{k}}) \\
&= \frac{\mathbf{r}}{\mathrm{r}} = \frac{\mathrm{r}\hat{\mathbf{r}}}{\mathrm{r}} = \hat{\mathbf{r}}
\end{aligned}$$

**Example 1.14**

Show that $\nabla\left(\frac{1}{\mathrm{r}}\right) = \frac{-\mathbf{r}}{\mathrm{r}^3} = \frac{-\hat{\mathbf{r}}}{\mathrm{r}^2}$

where $\mathbf{r}$ is the position vector of a point $\mathbf{r} = x\hat{\mathbf{i}} + y\hat{\mathbf{j}} + z\hat{\mathbf{k}}$

**Solution**

$$\nabla\frac{1}{\mathrm{r}} = \frac{\partial}{\partial x}\left(\frac{1}{\mathrm{r}}\right)\hat{\mathbf{i}} + \frac{\partial}{\partial y}\left(\frac{1}{\mathrm{r}}\right)\hat{\mathbf{j}} + \frac{\partial}{\partial z}\left(\frac{1}{\mathrm{r}}\right)\hat{\mathbf{k}}$$

with $r^2 = x^2 + y^2 + z^2$

$$\begin{aligned}
\frac{\partial}{\partial x}\left(\frac{1}{r}\right) &= \frac{\partial}{\partial x}\left(\frac{1}{(x^2+y^2+z^2)^{1/2}}\right) \\
&= \frac{\partial}{\partial x}\left[(x^2+y^2+z^2)^{-1/2}\right] \\
&= \frac{-1}{2}\frac{2x}{(x^2+y^2+z^2)^{3/2}} \\
&= \frac{-x}{\left((x^2+y^2+z^2)^{\frac{3}{2}}\right)} = \frac{-x}{r^3}
\end{aligned}$$

Similarly

$$\frac{\partial}{\partial y}\left(\frac{1}{r}\right) = \frac{-y}{r^3}$$
$$\frac{\partial}{\partial y}\left(\frac{1}{r}\right) = \frac{-z}{r^3}$$

Thus

$$\nabla\left(\frac{1}{r}\right) = \left(\frac{-x}{r^3}\right)\hat{\mathbf{i}} + \left(\frac{-y}{r^3}\right)\hat{\mathbf{j}} + \left(\frac{-z}{r^3}\right)\hat{\mathbf{k}}$$
$$\nabla\left(\frac{1}{r}\right) = \frac{-(x\hat{\mathbf{i}} + y\hat{\mathbf{j}} + z\hat{\mathbf{k}})}{r^3} = \frac{-\mathbf{r}}{r^3}$$
$$\Rightarrow \nabla\left(\frac{1}{r}\right) = \frac{-\mathbf{r}}{r^3} = \frac{-r\hat{\mathbf{r}}}{r^3} = \frac{-\hat{\mathbf{r}}}{r^2}$$

## 1.21 Two and Three Dimensions

Throughout electromagnetics we will be discussing two- and three-dimensional problems. If the reader has difficulty in visualizing two and three dimensions go to a corner of the room as shown in Fig. 1.45. In that figure, O is the corner of the room. Fix O, the corner of the room as the origin. Two walls along with the floor are shown in the same figure. Take the line joining the walls and floor as *x*-, *y*-, *z*-axis. Floor is the *x*–*y* plane. Wall 1 is the *x*–*z* plane; wall 2 is the *z*–*y* plane. We have shown an ant moving on *x*–*z* plane, that is a wall 1. If we want to describe the motion of the ant then we need *x*- and *z*-axis alone. No need for *y*-axis as long as the ant moves in *x*–*z* plane. Thus the problem is two dimensional. On the other hand if a bee is flying inside the room as shown in the same figure then we need all the three *x*-, *y*-, *z*-axis to describe the motion of bee then the problem is three dimensional.

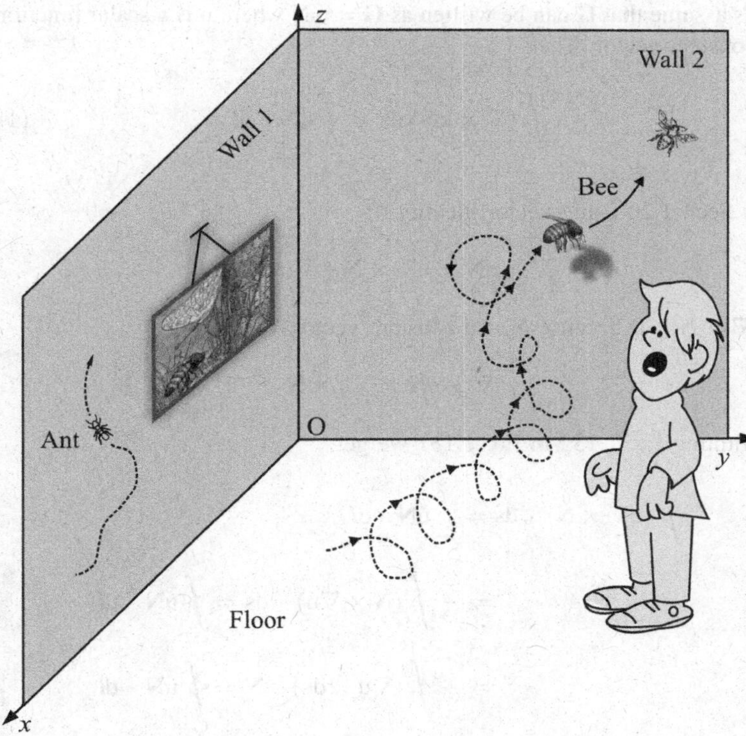


**Fig. 1.45** Two and three dimensions

**Example 1.15**
Prove that

$$-\int_{S} \nabla u \times \mathbf{ds} = \oint_{C} u d\boldsymbol{l}$$

where u is a scalar.

**Solution**
Consider any vector **G**, from Stoke's theorem

$$\int_{S} (\nabla \times \mathbf{G}) \cdot \mathbf{ds} = \oint_{C} \mathbf{G} \cdot \mathbf{d}\boldsymbol{l}$$

Let us assume that **G** can be written as **G** = u **N** where u is a scalar function and **N** is a constant vector. Then

$$\int (\nabla \times u\mathbf{N}).\mathbf{ds} = \oint_C u\mathbf{N} \cdot \mathbf{d}\boldsymbol{l} \tag{1.131}$$

From Sect. 1.20 using vector identity 6

$$\nabla \times u\mathbf{N} = u\nabla \times \mathbf{N} + \nabla u \times \mathbf{N}$$

But $\nabla \times \mathbf{N} = 0$ because **N** is a constant vector. Hence

$$\nabla \times u\mathbf{N} = \nabla u \times \mathbf{N} \tag{1.132}$$

Substituting Eq. 1.132 in Eq. 1.131 we get

$$\int (u\nabla \times \mathbf{N}) \cdot \mathbf{ds} = \oint_C u\mathbf{N} \cdot \mathbf{d}\boldsymbol{l}$$

$$\Rightarrow -\int (\mathbf{N} \times \nabla u) \cdot \mathbf{ds} = \oint_C u\mathbf{N} \cdot \mathbf{d}\boldsymbol{l}$$

$$\Rightarrow -\int (\nabla u \times \mathbf{ds}) \cdot \mathbf{N} = \oint_C u\mathbf{N} \cdot \mathbf{d}\boldsymbol{l}$$

As **N** is a constant vector, pulling **N** out of integral and cancelling both sides.

$$\Rightarrow -\int_S \nabla u \times \mathbf{ds} = \oint_C u\,\mathbf{d}\boldsymbol{l} \tag{1.133}$$

Hence proved.

**Example 1.16**
Prove that for any two vector **Q**, **R** where **Q** is a constant vector

$$\nabla \times (\mathbf{Q} \times \mathbf{R}) = \mathbf{Q}(\nabla \cdot \mathbf{R}) - (\mathbf{Q} \cdot \nabla)\mathbf{R}$$

**Solution**
From Sect. 1.20 using vector identity 9 we get

$$\nabla \times (\mathbf{Q} \times \mathbf{R}) = (\mathbf{R} \cdot \nabla)\mathbf{Q} - (\mathbf{Q} \cdot \nabla)\mathbf{R} + \mathbf{Q}(\nabla \cdot \mathbf{R}) - \mathbf{R}(\nabla \cdot \mathbf{Q})$$

As **Q** is a constant vector

$$\nabla \cdot \mathbf{Q} = 0, (\mathbf{R} \cdot \nabla)\mathbf{Q} = 0$$

Hence

$$\nabla \times (\mathbf{Q} \times \mathbf{R}) = \mathbf{Q}(\nabla \cdot \mathbf{R}) - (\mathbf{Q}.\nabla)\mathbf{R} \qquad (1.134)$$

**Example 1.17**
For a constant vector **Q** and scalar function u prove that

$$\mathbf{S} \times \mathbf{Q} = \oint_C \mathbf{Q} \cdot \mathbf{r} dl$$

where S is the surface bounded by loop C and $\mathbf{r} = x\hat{\mathbf{i}} + y\hat{\mathbf{j}} + z\hat{\mathbf{k}}$.

**Solution**
Let u $=$ $\mathbf{Q} \cdot \mathbf{r}$ where **Q** is a constant vector. From Sect. 1.20 using vector identity 1

$$\nabla \text{u} = \nabla(\mathbf{Q} \cdot \mathbf{r}) = \mathbf{Q} \times (\nabla \times \mathbf{r}) + (\mathbf{Q} \cdot \nabla)\mathbf{r} + \mathbf{r} \times (\nabla \times \mathbf{Q}) + (\mathbf{r} \cdot \nabla)\mathbf{Q}$$

Because **Q** is a constant vector

$$\nabla \times \mathbf{Q} = 0 \text{ and } (\mathbf{r} \cdot \nabla)\mathbf{Q} = 0$$

Hence

$$\nabla \text{u} = \nabla(\mathbf{Q} \cdot \mathbf{r}) = \mathbf{Q} \times (\nabla \times \mathbf{r}) + (\mathbf{Q} \cdot \nabla)\mathbf{r}$$

But $\nabla \times \mathbf{r} = 0$. Hence

$$\begin{aligned}\nabla \text{u} &= \nabla(\mathbf{Q} \cdot \mathbf{r}) = (\mathbf{Q} \cdot \nabla)\mathbf{r} \\ &= \left(\text{Q}_x \frac{\partial}{\partial x} + \text{Q}_y \frac{\partial}{\partial y} + \text{Q}_z \frac{\partial}{\partial z}\right) \cdot \left(x\hat{\mathbf{i}} + y\hat{\mathbf{j}} + z\hat{\mathbf{k}}\right) \\ &= \text{Q}_x\hat{\mathbf{i}} + \text{Q}_y\hat{\mathbf{j}} + \text{Q}_z\hat{\mathbf{k}} = \mathbf{Q}\end{aligned}$$

Subsituting $\nabla \text{u} = \mathbf{Q}$ and u = $(\mathbf{Q} \cdot \mathbf{r})$ in Eq. 1.133

$$-\int_S \mathbf{Q} \times \text{d}\mathbf{s} = \oint_C (\mathbf{Q} \cdot \mathbf{r}) dl$$

Because **Q** is a constant vector

$$
\begin{aligned}
-\mathbf{Q}\times\int_S \mathrm{d}\mathbf{s} &= \oint_C (\mathbf{Q}\cdot\mathbf{r})\mathrm{d}\boldsymbol{l}\\
\Rightarrow -\mathbf{Q}\times\mathbf{S} &= \oint_C (\mathbf{Q}\cdot\mathbf{r})\mathrm{d}\boldsymbol{l}
\end{aligned}
\tag{1.135}
$$

$$
\Rightarrow \mathbf{S}\times\mathbf{Q} = \oint_C (\mathbf{Q}\cdot\mathbf{r})\mathrm{d}\boldsymbol{l} \tag{1.136}
$$

## Exercises

**Problem 1.1**
Check whether the two vectors $\mathbf{P} = 12\,\hat{\mathbf{i}} - 6\,\hat{\mathbf{j}} + 3\,\hat{\mathbf{k}}$ and $\mathbf{Q} = \hat{\mathbf{i}} - \hat{\mathbf{j}} - z\,\hat{\mathbf{k}}$ are perpendicular to each other.

**Problem 1.2**
If $\mathbf{P} = 10\hat{\mathbf{i}} - 4\hat{\mathbf{j}} + 2\hat{\mathbf{k}}$ find **Q** if **Q** is parallel to **P**.

**Problem 1.3**
Check whether the vectors $\mathbf{P} = 2\hat{\mathbf{i}} + 4\hat{\mathbf{j}} + 18\hat{\mathbf{k}}$,
$\mathbf{Q} = 9\hat{\mathbf{i}} + 3\hat{\mathbf{j}} + 6\hat{\mathbf{k}}$ and $\mathbf{R} = 10\hat{\mathbf{i}} + 5\hat{\mathbf{j}} + 15\hat{\mathbf{k}}$ are coplanar.

**Problem 1.4**
If $\mathbf{P} = 3\hat{\mathbf{i}} + 2\hat{\mathbf{j}} + 5\hat{\mathbf{k}}$ find the unit vector in the direction of **P**.

**Problem 1.5**
Check whether the three vectors $\mathbf{A} = 2\hat{\mathbf{i}} + 3\hat{\mathbf{j}} + 5\hat{\mathbf{k}}$,
and $\mathbf{B} = \hat{\mathbf{i}} + \hat{\mathbf{j}} - \hat{\mathbf{k}}$, $\mathbf{C} = 3\hat{\mathbf{i}} + 4\hat{\mathbf{j}} + 4\hat{\mathbf{k}}$ form a right angled triangle.

**Problem 1.6**
A vector **P** is directed from (3, 4, 5) to (1, 3, 2). Determine $|\mathbf{P}|$ and unit vector in the direction of **P**.

**Problem 1.7**
Find the angle between the two vectors $\mathbf{P} = \hat{\mathbf{i}} - \hat{\mathbf{j}} + \hat{\mathbf{k}}$ and $\mathbf{Q} = 3\hat{\mathbf{i}} - 2\hat{\mathbf{j}} - \hat{\mathbf{k}}$ using dot product and cross product.

**Problem 1.8**
Find the value of a such that the three vectors $10\hat{\mathbf{i}} - 5\hat{\mathbf{j}} + 5\hat{\mathbf{k}}$,
$2\hat{\mathbf{i}} + 4\hat{\mathbf{j}} - 6\hat{\mathbf{k}}$ and $3\hat{\mathbf{i}} + 2\hat{\mathbf{j}} + a\hat{\mathbf{k}}$ are coplanar.

**Problem 1.9**

Find the projection of vector $\mathbf{P} = \hat{\mathbf{i}} + 3\hat{\mathbf{j}} + 2\hat{\mathbf{k}}$ on vector $\mathbf{Q} = 2\hat{\mathbf{i}} + \hat{\mathbf{j}} - 3\hat{\mathbf{k}}$

**Problem 1.10**

If $\mathbf{P} + \mathbf{Q} + \mathbf{R} = 0$ then show that $\mathbf{P} \times \mathbf{Q} = \mathbf{Q} \times \mathbf{R} = \mathbf{R} \times \mathbf{P}$.

**Problem 1.11**

Calculate $(\mathbf{P} \times \mathbf{Q}) \cdot (\mathbf{R} \times \mathbf{S})$

**Problem 1.12**

Show that $\nabla \log r = \frac{\mathbf{r}}{r^2}$ where **r** is the position vector.

**Problem 1.13**

If **r** is the position vector show that

(i) $\nabla r^n = rn^{n-2}\mathbf{r}$
(ii) $\text{div}\,\mathbf{r} = 3$
(iii) $\text{div}\, r^n\mathbf{r} = (n + 3)r^n$
(iv) $\text{curl}(r^n\mathbf{r}) = 0$

**Problem 1.14**

Calculate $\nabla u$ at $(1, -5, 3)$ if $u = x^2z + y^3z^2$.

**Problem 1.15**

Show that $\nabla \cdot (\mathbf{P} + \mathbf{Q}) = \nabla \cdot \mathbf{P} + \nabla \cdot \mathbf{Q}$

**Problem 1.16**

Calculate a if $\mathbf{A} = (3x + y)\hat{\mathbf{i}} + (z - x)\hat{\mathbf{j}} + (ay + z)\hat{\mathbf{k}}$ is solenoidal.

**Problem 1.17**

If $\mathbf{P} = 3x^2\hat{\mathbf{i}} - y^2z\hat{\mathbf{j}} + 2x^3z\hat{\mathbf{k}}$ find

(a) $\nabla \times \mathbf{P}$
(b) $\nabla \times (\nabla \times \mathbf{P})$

**Problem 1.18**

If $\mathbf{P} = (2x^2 + y)\hat{\mathbf{i}} - 10yz\hat{\mathbf{j}} + 12xz^2\hat{\mathbf{k}}$

Calculate $\int \mathbf{P} \cdot d\mathbf{r}$ along the straight line from (0, 0, 0) to (1, 1, 0) and then to (1, 1, 1).

**Problem 1.19**

Calculate the gradients of

(a) $u(r, \theta, \phi) = 3r \sin\theta - 4\phi + 1$
(b) $u(r, \theta, z) = 3\cos\phi - rz$

**Problem 1.20**

Calculate div **P** if

(a) $\mathbf{P} = 3\hat{\mathbf{r}} + r\sin\theta\,\hat{\boldsymbol{\theta}} + r\hat{\boldsymbol{\phi}}$

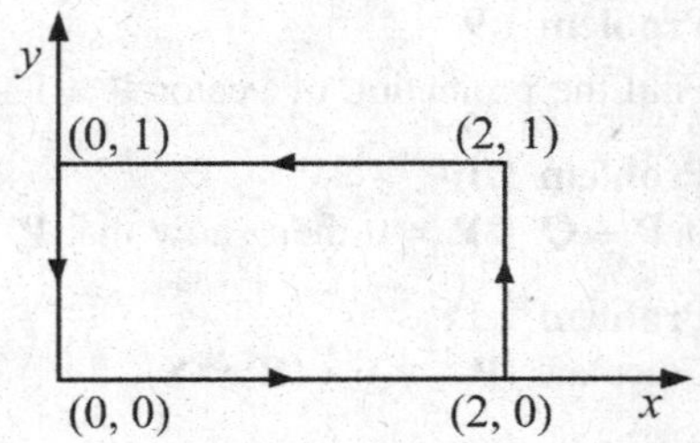


**Fig. 1.46** A rectangular area in Cartesian coordinates (Problem 1.24)

(b) $\mathbf{P} = r\hat{\mathbf{r}} + z\cos\phi\,\hat{\boldsymbol{\phi}} + 3\hat{\mathbf{z}}$

**Problem 1.21**

Calculate the net flux of the vector field $\mathbf{P}(x, y, z) = 3xy^2\hat{\mathbf{i}} + z^2\hat{\mathbf{j}} + y^3\hat{\mathbf{k}}$ emerging from a cube of dimensions $0 \le x, y, z \le 1$

**Problem 1.22**

Solve Problem 1.21 using Gauss's divergence theorem.

**Problem 1.23**

Verity divergence theorem for the vector $\mathbf{P} = x^3\hat{\mathbf{i}} + y^3\hat{\mathbf{j}} + z^3\hat{\mathbf{k}}$ for the cube $0 \le x, y, z \le 1$ shown in Fig. 1.25.

**Problem 1.24**

Verify Stoke's theorem for the vector $\mathbf{P} = (x + y)\hat{\mathbf{i}} + (y + z)\hat{\mathbf{j}} + (x + z)\hat{\mathbf{k}}$ for a plane rectangular area with vertices (0, 0), (2, 0), (2, 1), (0, 1) as shown in Fig. 1.46.

**Problem 1.25**

Find the value of a, b and c if

$\mathbf{P} = (4x + 5y + \mathrm{a}\,z)\hat{\mathbf{i}} + (3x + \mathrm{b}\,y + z)\hat{\mathbf{j}} + (\mathrm{c}\,x + 2y - 3z)\,\hat{\mathbf{k}}$ is irrotational.

**Problem 1.26**

Calculate the surface area and volume of a sphere by integrating surface and volume elements in spherical polar coordinates.

**Problem 1.27**

Calculate the surface area and volume of a cylinder by integrating surface and volume elements in cylindrical coordinates.

**Problem 1.28**

Prove that for any vector $\mathbf{G}$ $\int_{\tau} \nabla \times \mathbf{G}\, d\tau = -\oint_{S} \mathbf{G} \times d\mathbf{s}$.

# Chapter 2
# Electric Charges at Rest: Part I

In this chapter we will learn how to calculate electric field for static charge distributions. Greeks were the first to observe electrical phenomena. They observed that a piece of amber when rubbed with fur or cat skin acquired the property of attracting small bits of paper and straw. Later on, Coulomb French Scientist conducted experiments using a torsion balance set-up to study forces between charged objects.

In the following sections we will describe Coulomb's law, discuss Gauss's law and then see how potential is used to calculate electric field. For complete grasp of the above topics reader is advised to read Sects. 2.1–2.18 continuously without breaking any section or example in between. After reading up to Sect. 2.18 the reader will be able to appreciate why there are three different methods, Coulomb's law, Gauss's law and potential to calculate electric field for a given charge distribution.

## 2.1 Coulomb's Law

Charles-Augustin de Coulomb, French Scientist, using his torsion balance set-up performed a series of experiments to study the nature of forces between two charged objects. The results obtained by Coulomb using his torsion balance set-up are called Coulomb's law. As per Coulomb's law the force between two charges

(i) is directly proportional to the product of their charges.
(ii) is inversely proportional to the square of the distance between the two charges and acts along the line joining the two charges.
(iii) depends on the nature of the medium in which the charged objects are situated.

© Springer Nature Singapore Pte Ltd. 2020
S. Balaji, *Electromagnetics Made Easy*,
https://doi.org/10.1007/978-981-15-2658-9_2

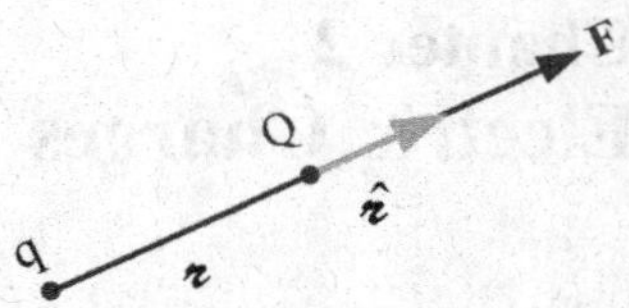


**Fig. 2.1** A source charge q separated from a test charge Q

A source charge q is separated from the test charge Q by a distance $\mathbf{r}$ as shown in Fig. 2.1.

The force exerted by q on Q as per Coulomb's law is then

$$\mathbf{F} = k\frac{qQ}{r^2}\hat{\mathbf{r}} \tag{2.1}$$

Here $\hat{\mathbf{r}}$ is the unit vector in the direction of **F**. k is the proportionality constant whose value depends on the medium in which the charges are situated, and units in which force, charge and distance are expressed. Because in Fig. 2.1 the charges are like charges, the forces are repulsive in nature. If the charges are unlike charges, the forces are attractive in nature.

In SI units force is measured in Newton, distance in metre and charge in Coulomb. The proportionality constant k is written as

$$k = \frac{1}{4\pi\varepsilon_0}$$

So that Eq. 2.1 is

$$\mathbf{F} = \frac{1}{4\pi\varepsilon_0}\frac{qQ}{r^2}\hat{\mathbf{r}} \tag{2.2}$$

Here $\varepsilon_0$ is the absolute permittivity of free space whose value is given by

$$\varepsilon_0 = 8.85 \times 10^{-12} C^2 N^{-1} m^{-2}$$

If the charges are not situated in vacuum but in a medium of permittivity $\varepsilon$, then $\varepsilon = \varepsilon_0 \varepsilon_r$. The force between the two charges is

$$\mathbf{F} = \frac{1}{4\pi\varepsilon}\frac{qQ}{r^2}\hat{\mathbf{r}} \tag{2.3}$$

$$\mathbf{F} = \frac{1}{4\pi\varepsilon_0\varepsilon_r}\frac{qQ}{r^2}\hat{\mathbf{r}} \tag{2.4}$$

Here $\varepsilon_r$ is the dielectric constant of the medium and is a dimensionless quantity. For air and vacuum $\varepsilon_r = 1$.

## 2.2 Electric Field Intensity

The electric field intensity **E** is defined as the force exerted on a unit positive charge placed at that point in an electrostatic field. Here electrostatic field refers to the space surrounding an electrostatic charge where the electrostatic forces can be detected.

The electric field produced by charge q in Fig. 2.1 is

$$\mathbf{E} = \frac{\mathbf{F}}{Q} \tag{2.5}$$

Substituting Eq. 2.2 in 2.5

$$\mathbf{E} = \frac{1}{4\pi\varepsilon_0}\frac{q}{\boldsymbol{\mathscr{r}}^2}\hat{\boldsymbol{\mathscr{r}}} \tag{2.6}$$

Consider Fig. 2.2. Point P is situated at a distance of r from the origin, and charge q is situated at a distance of r′ from origin. The distance between q and point P is $\boldsymbol{\mathscr{r}}$. The electric field at point P is then

$$\mathbf{E}(\mathbf{r}) = \frac{1}{4\pi\varepsilon_0}\frac{q}{\boldsymbol{\mathscr{r}}^2}\hat{\boldsymbol{\mathscr{r}}} \tag{2.7}$$

where $\boldsymbol{\mathscr{r}} = \mathbf{r} - \mathbf{r}'$.

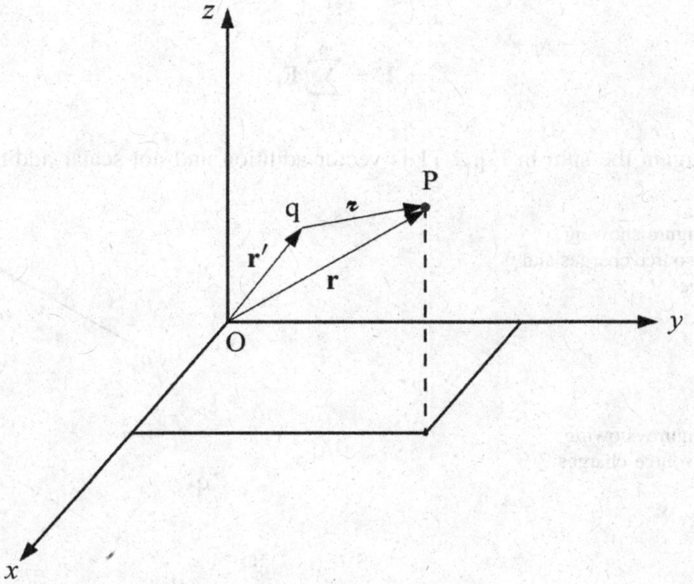


**Fig. 2.2** Charge q in Cartesian coordinate system

## 2.3 Electric Field Intensity Due to a Group of Discrete Point Charges

Consider a group of discrete point charges $q_1$, $q_2$, $q_3 \ldots q_n$ as shown in Fig. 2.3 exerting force on test charge Q. We are interested in calculating the force exerted by the charges $q_1$, $q_2$, $q_3 \ldots q_n$ on the test charge Q. The net force acting on charge Q is

$$\mathbf{F} = \mathbf{F_1} + \mathbf{F_2} + \mathbf{F_3} + \cdots + \mathbf{F_n} \tag{2.8}$$

$$\mathbf{F} = \sum_{i=1}^{n} \mathbf{F_i} \tag{2.9}$$

where the force $\mathbf{F_i}$ in the above equation is given by Eq. 2.3 replacing q by $q_i$, $\boldsymbol{\mathscr{r}}$ by $\boldsymbol{\mathscr{r}}_i$ and $\hat{\boldsymbol{\mathscr{r}}}$ by $\hat{\boldsymbol{\mathscr{r}}}_i$.

Thus the net force acting on the test charge Q is found by calculating the force due to individual charges separately and then summing all the individual forces. However the addition is vector addition and not a scalar addition. That is forces in Eq. 2.8 must be vectorically added (see Sect. 1.1).

Now consider Fig. 2.4. We are interested in calculating the net electric field at point P.

From Eq. 2.9

$$\frac{\mathbf{F}}{Q} = \sum_{i=1}^{n} \frac{\mathbf{F_i}}{Q} \tag{2.10}$$

$$\Rightarrow \mathbf{E} = \sum_{i=1}^{n} \mathbf{E_i} \tag{2.11}$$

Once again the sum in Eq. 2.11 is vector addition and not scalar addition.

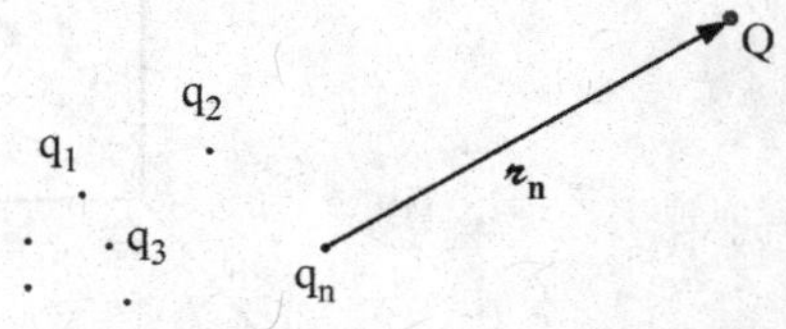


**Fig. 2.3** Figure showing number of source charges and a test charge

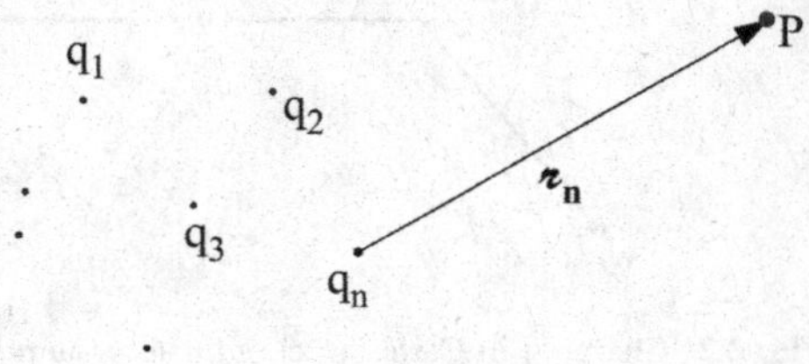


**Fig. 2.4** Figure showing number of source charges

With the aid of Eqs. 2.7, 2.11 can be written as

$$\mathbf{E} = \frac{1}{4\pi\varepsilon_o} \sum_{i=1}^{n} \frac{q_i}{\mathscr{r}_i^2} \hat{\mathscr{r}}_i \tag{2.12}$$

## 2.4 Continuous Charge Distributions

In the previous section we defined the electric field intensity due to a group of discrete point charges. Now let us consider charges that are spread over space. There are three different continuous charge distributions as shown in Fig. 2.5.

(1) *Line Charge*: Charge is distributed over a line as shown in Fig. 2.5a. Mathematically we describe the line charge with symbol $\lambda$

$$\lambda = \frac{dq}{dl'} \tag{2.13}$$

where $\lambda$ is charge per unit length.

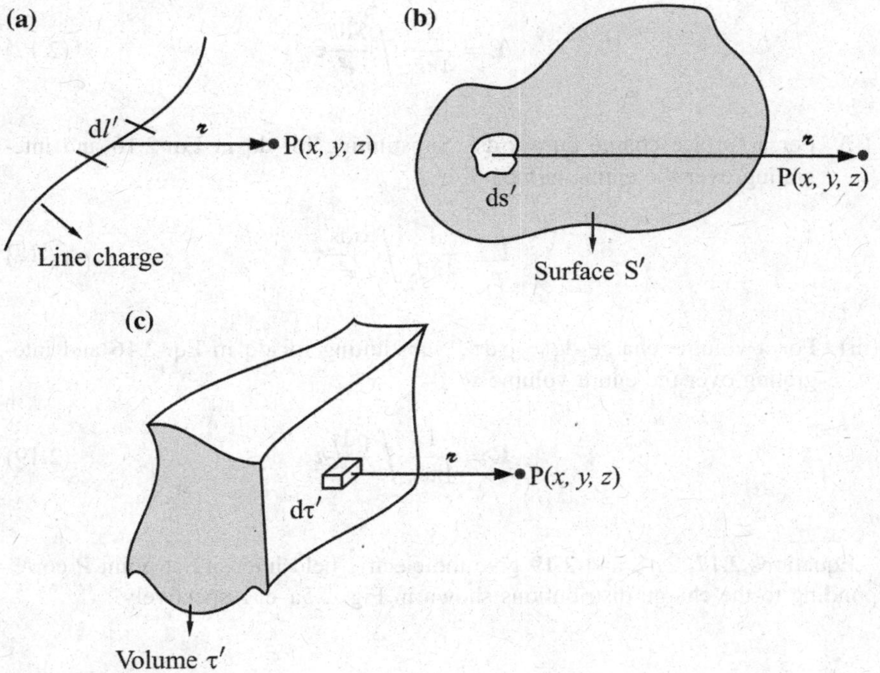


**Fig. 2.5** Continuous charge distributions, a. Line charge distribution, b. Surface charge distribution, c. Volume charge distribution

(2) *Surface Charge*: Charge is distributed over a surface as shown in Fig. 2.5b. Mathematically we describe the surface charge with symbol $\sigma$

$$\sigma = \frac{dq}{ds'} \tag{2.14}$$

where $\sigma$ is charge per unit area.

(3) *Volume Charge*: Charge is distributed over a volume as shown in Fig. 2.5c. Mathematically we describe volume charge with symbol $\rho$

$$\rho = \frac{dq}{d\tau'} \tag{2.15}$$

where $\rho$ is charge per unit volume.

For a small charge dq Eq. 2.7 can be written as

$$\mathbf{E}_{dq} = \frac{1}{4\pi\varepsilon_0}\frac{dq}{\mathscr{r}^2}\hat{\mathscr{r}} \tag{2.16}$$

(i) For a line charge $dq = \lambda dl'$. Substituting for dq in Eq. 2.16 and integrating over the entire line $l'$

$$\mathbf{E} = \frac{1}{4\pi\varepsilon_0}\int_{l'}\frac{\lambda dl'}{\mathscr{r}^2}\hat{\mathscr{r}} \tag{2.17}$$

(ii) For a surface charge $dq = \sigma ds'$. Substituting for dq in Eq. 2.16 and integrating over the entire surface $S'$

$$\mathbf{E} = \frac{1}{4\pi\varepsilon_0}\int_{S'}\frac{\sigma ds'}{\mathscr{r}^2}\hat{\mathscr{r}} \tag{2.18}$$

(iii) For a volume charge $dq = \rho\, d\tau'$. Substituting for dq in Eq. 2.16 and integrating over the entire volume $\tau'$

$$\mathbf{E} = \frac{1}{4\pi\varepsilon_0}\int_{\tau'}\frac{\rho d\tau'}{\mathscr{r}^2}\hat{\mathscr{r}} \tag{2.19}$$

Equations 2.17, 2.18 and 2.19 give the electric field intensity at point P corresponding to the charge distributions shown in Fig. 2.5a–c, respectively.

Finally the total charge enclosed for the line charge is

$$q_{tot} = \int_{l'} \lambda dl' \tag{2.20}$$

surface charge is

$$q_{tot} = \int_{S'} \sigma ds' \tag{2.21}$$

and volume charge is

$$q_{tot} = \int_{\tau'} \rho\, d\tau' \tag{2.22}$$

In Fig. 2.5 we have used primed coordinates $dl', ds', d\tau'$ instead of $dl, ds, d\tau$. The above difference has been made in order to distinguish between source points where charge is located and field points where the field is calculated. For example in Fig. 2.5c within the volume all the points will be denoted with prime coordinates, i.e. $(x', y', z')$ because within the volume $\tau'$ all the charge is located. However for points like P in Fig. 2.5c we use unprimed coordinates $(x, y, z)$ to denote that it is the point where field is calculated. Thus we use primed symbols for source points and unprimed symbols for field points. One must be careful enough to integrate with respect to either source points or field points wherever necessary.

## 2.5 A Note about Coulomb's Law

Equations (2.12), (2.17), (2.18) and (2.19) calculate electric field vector **E** directly in terms of vector. The reader can observe that $\hat{\boldsymbol{\mathscr{r}}}$ is present in the equations mentioned above signifying that **E** is directly calculated in terms of a vector. We have already seen in Sect. 1.1 and Example 1.10 that vectors are very difficult to deal with. Hence, calculation of **E** using the above-mentioned equations becomes complicated even for simple charge distributions. To circumvent this problem much easier methods to calculate **E** have been developed—Gauss's law and potential formulation. As we will see, in Gauss's law we use symmetry of the problem to calculate **E** easily, while in potential formulation we reduce the vector problem into a scalar problem so that **E** is calculated easily. This point will be more clear to the reader when the reader completes Sect. 2.18.

## 2.6 Calculating Electric Field E Using Coulomb's Law

Initially we will provide a note on calculating electric field **E** in two dimensions in Cartesian coordinate system. The points discussed here are valid in three dimensions and other coordinate systems. The electric field **E** in two dimensions is given by

$$\mathbf{E} = E_x\hat{\mathbf{i}} + E_y\hat{\mathbf{j}}$$

While solving problems using Coulomb's law it will be sufficient to calculate $E_x$ and $E_y$ for the given charge distribution and substitute the values in the above equation to get **E**. However in calculating **E** for certain charge distributions like the ones solved in Examples 2.4, 2.5 and 2.7 either $E_x$ or $E_y$ might cancel out each other and the remaining component alone is to be calculated. But for certain charge distributions as shown in Example 2.9, Sect. 2.14, neither component will cancel out and hence $E_x$ and $E_y$ need to be calculated separately. Whether the components will cancel out (or) get added up depends on the charge distribution and the point where the electric field is calculated. The above conclusions hold good for three dimensions and other coordinate systems.

**Example 2.1**
Two charges of magnitude $5 \times 10^{-9}$ C and $7 \times 10^{-9}$ C are separated by 25 cm in vacuum. Find the force of interaction between them. If the same charges are separated by the same distance in mica ($\varepsilon_r = 6$), what is the force of interaction?

**Solution**
We know that

$$F = \frac{1}{4\pi\varepsilon_0}\frac{qQ}{r^2}$$

$$F = \frac{1}{4\pi\varepsilon_0}\frac{(5 \times 10^{-9})(7 \times 10^{-9})}{(25 \times 10^{-2})^2}$$

$$\text{But } \frac{1}{4\pi\varepsilon_0} = \frac{1}{4\pi(8.85 \times 10^{-12})} = 9 \times 10^9.$$

$$\text{Hence } F = (9 \times 10^9)\frac{(5 \times 10^{-9})(7 \times 10^{-5})}{(25)^2}$$

$$F = 0.504 \times 10^{-5}\ \text{N}$$

For mica $\varepsilon_r = 6$

$$\text{Hence } F_{mica} = \frac{F}{\varepsilon_r} = \frac{0.504 \times 10^{-5}}{6}$$

$$F_{mica} = 8.4 \times 10^{-7}\ \text{N}.$$

**Example 2.2**

A line charge $\lambda = 5(y')^4$ nC b/m is situated along *y*-axis from 0 to 7 cm. Calculate the total charge.

**Solution**

From Eq. 2.20

$$q_{tot} = \int \lambda dl'$$

$$\Rightarrow q_{tot} = \int_0^7 5(y')^4 dy' = 5\left[\frac{(y')^5}{5}\right]_0^7$$

$$\Rightarrow q_{tot} = 7^5\ \text{nC}$$

$$\Rightarrow q_{tot} = 16.807\ \mu\text{C}$$

**Example 2.3**

In a circular disc of radius R charge density is given by $\sigma = k(r')^2$. Calculate the total charge.

**Solution**

From Eq. 2.21

$$q_{tot} = \int \sigma ds'$$

From Fig. 2.6

$$ds' = r'\,dr'\,d\phi'$$

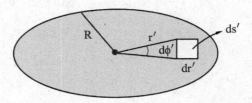

**Fig. 2.6** Charge distributed over a circular disc

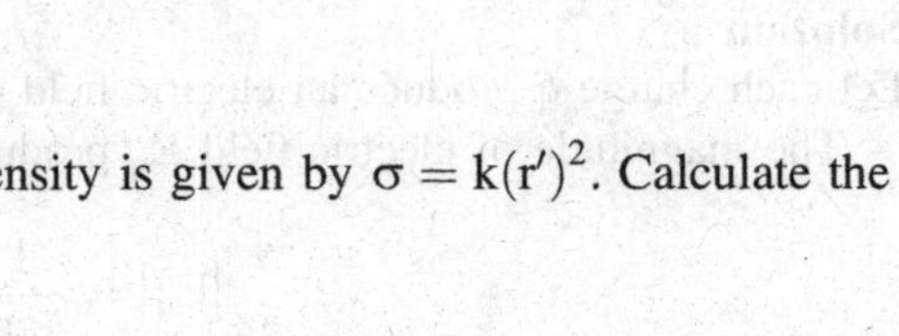

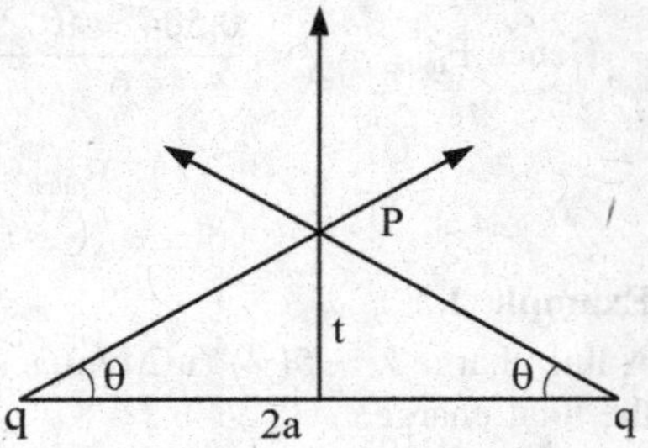


**Fig. 2.7** Two point charges q separated by distance 2a

$$\text{Hence } q_{tot} = \int_0^R \int_0^{2\pi} k(r')^2 r' \, dr' \, d\phi'$$

$$\Rightarrow q_{tot} = 2\pi k \frac{R^4}{4}$$

$$\Rightarrow q_{tot} = \frac{\pi k R^4}{2}$$

**Example 2.4**

Two charges each having magnitude q are separated by a distance 2a as shown in Fig. 2.7. Find the net electric field at point P.

**Solution**

Let each charge q produce an electric field of $\mathbf{E}_q$ at point P as shown in Fig. 2.8a.

The magnitude of electric field $\mathbf{E_q}$ produced by q at point P is

$$|\mathbf{E_q}| = \frac{1}{4\pi\varepsilon_o} \frac{q}{PB^2}$$

The line CPD shown in Fig. 2.8a is shown separately in Fig. 2.8b. Resolving $\mathbf{E_q}$ into $\mathbf{E}_{qx}$ and $\mathbf{E}_{qy}$ in Fig. 2.8b we see that the *x* components $\mathbf{E}_{qx}$ cancel out each other while $\mathbf{E}_{qy}$ components add up. Hence, the net electric field at point P by the two charges each of magnitude q is given by

$$\mathbf{E} = 2E_{qy}\hat{\mathbf{j}}$$

From Fig. 2.8c

$$\cos(\pi/2 - \theta) = \frac{E_{qy}}{E_q}$$

$$\Rightarrow E_{qy} = E_q \sin\theta$$

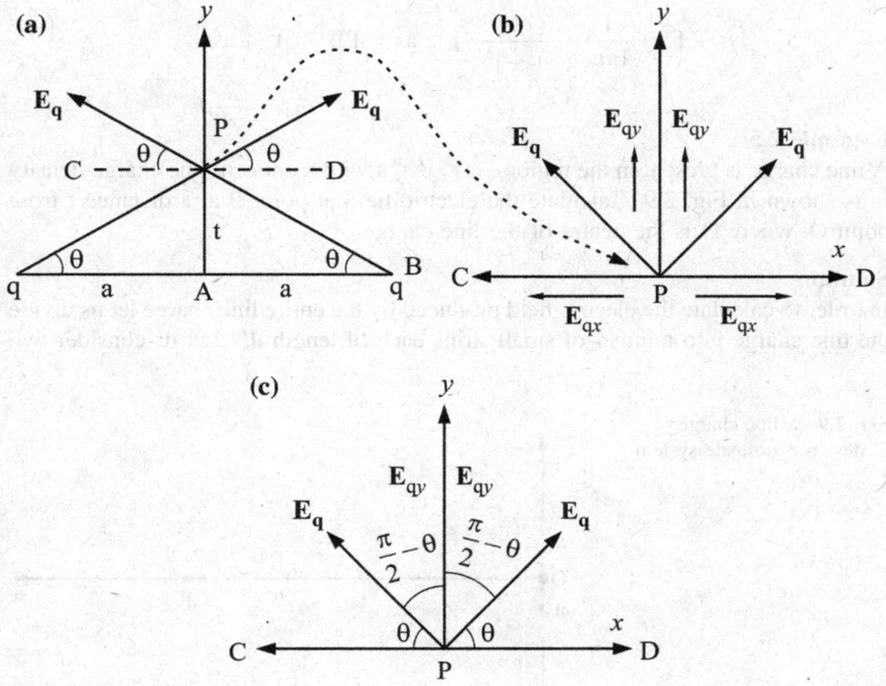


**Fig. 2.8** Electric field produced by both the charges q and resolution of electric field into components is shown in the figure

Substituting for $E_{qy}$ in **E**

$$\mathbf{E} = 2E_q \sin\theta\,\hat{\mathbf{j}}$$

From Fig. 2.8a

$$\sin\theta = \frac{AP}{PB} = \frac{t}{\sqrt{t^2 + a^2}}$$

Substituting for sin θ in **E**

$$\mathbf{E} = 2\,E_q \frac{t}{\sqrt{t^2 + a^2}}\hat{\mathbf{j}}$$

Substituting for $E_q$

$$\mathbf{E} = 2\frac{1}{4\pi\varepsilon_o}\frac{q}{PB^2}\frac{t}{\sqrt{t^2 + a^2}}\hat{\mathbf{j}}$$

$$\mathbf{E} = \frac{1}{4\pi\varepsilon_o}\frac{qt}{[t^2 + a^2]^{3/2}}\hat{\mathbf{j}} \quad \text{as} \quad PB^2 = t^2 + a^2$$

**Example 2.5**
A line charge is present in the region $-a \leq y' \leq a$ with a uniform line charge density $\lambda$ as shown in Fig. 2.9. Calculate the electric field at point P at a distance t from point O, where O is the centre of the line charge.

**Solution**
In order to calculate the electric field produced by the entire line charge let us divide the line charge into number of small strips each of length $dl'$. Let us consider two

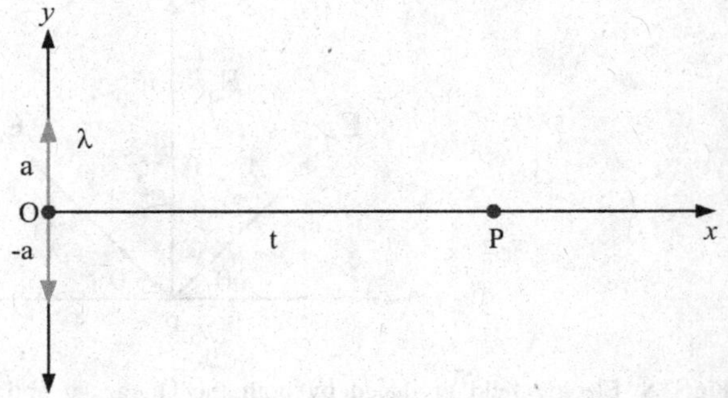


**Fig. 2.9** A line charge in Cartesian coordinate system

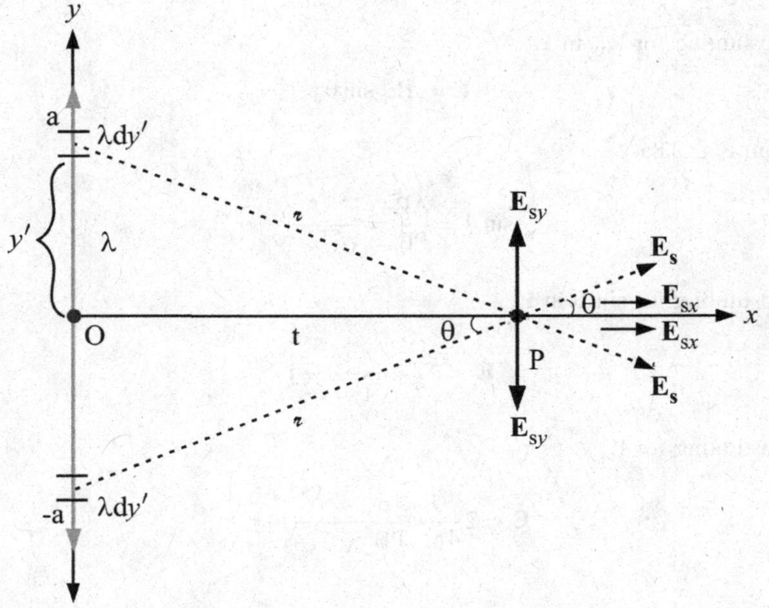


**Fig. 2.10** The line charge divided into number of small strips

strips $dy'$ each holding a charge $\lambda dy'$ as shown in Fig. 2.10. Let us calculate the electric field by these two strips initially and finally integrate between 0 and a in order to calculate the net electric field produced by the line charge lying between –a and a.

The magnitude of electric field produced by single strip $\lambda dy'$ at point P as shown in Fig. 2.10 is

$$|\mathbf{E_s}| = \frac{1}{4\pi\varepsilon_o}\frac{\lambda dy'}{r^2}$$

$$|\mathbf{E_s}| = \frac{1}{4\pi\varepsilon_o}\frac{\lambda dy'}{\left[(y')^2 + t^2\right]}$$

Resolving the electric field $\mathbf{E_s}$ due to the two strips into $\mathbf{E}_{sx}$ and $\mathbf{E}_{sy}$ components as shown in Fig. 2.10 we see that $\mathbf{E}_{sy}$ components cancel out each other while $\mathbf{E}_{sx}$ components add up. Let us denote the net electric field at point P due to the two strips as $\mathbf{dE}$ and $\mathbf{dE}$ is given by

$$\mathbf{dE} = 2\,E_{sx}\hat{\mathbf{i}}$$

From Fig. 2.10 we can write

$$\cos\theta = \frac{E_{sx}}{E_s} = \frac{t}{r}$$

$$\Rightarrow E_{sx} = \frac{t}{r}E_s$$

$$E_{sx} = \frac{t}{\sqrt{(y')^2 + t^2}}E_s$$

Substituting the value of $E_{sx}$ in $\mathbf{dE}$

$$\mathbf{dE} = 2\frac{t}{\sqrt{(y')^2 + t^2}}E_s\hat{\mathbf{i}}$$

Substituting the value of $E_s$ in $\mathbf{dE}$

$$\mathbf{dE} = 2\frac{t}{\sqrt{(y')^2 + t^2}}\left(\frac{1}{4\pi\varepsilon_o}\right)\frac{\lambda dy'}{\left[(y')^2 + t^2\right]^2}\hat{\mathbf{i}}$$

$$\mathbf{dE} = \frac{1}{2\pi\varepsilon_o}\frac{\lambda\, t\, dy'}{\left[(y')^2 + t^2\right]^{3/2}}\hat{\mathbf{i}}$$

This is the field produced by two line elements $\lambda dy'$ at point P. The field produced by the entire line charge at point P is

$$\mathbf{E} = \int_0^a d\mathbf{E} = \frac{\lambda t}{2\pi\varepsilon_o}\int_0^a \frac{dy'}{\left[(y')^2 + t^2\right]^{3/2}}\hat{\mathbf{i}}$$

$$= \frac{\lambda t}{4\pi\varepsilon_o}\left[\frac{y'}{t^2\sqrt{(y')^2 + t^2}}\right]_0^a \hat{\mathbf{i}}$$

$$\mathbf{E} = \frac{\lambda}{2\pi\varepsilon_o}\left[\frac{a}{t\sqrt{a^2 + t^2}}\right]\hat{\mathbf{i}}$$

**Example 2.6**
Two point charges $q_1 = 10$ nC and $q_2 = 20$ nC are located at (1, 1, 3) m and at (2, 3, 2) m, respectively. Calculate the magnitude of force on $q_1$ by $q_2$.

**Solution**
The distance between the two charges is

$$\mathscr{r} = \sqrt{(2-1)^2 + (3-1)^2 + (2-3)^3} = \sqrt{6}\, m$$

The force exerted by $q_2$ on$q_1$ is given by

$$\mathbf{F} = \frac{1}{4\pi\varepsilon_o}\frac{(10 \times 10^{-9})(20 \times 10^{-9})}{(\sqrt{6})^2}\hat{\boldsymbol{\mathscr{r}}}$$

Here $\hat{\boldsymbol{\mathscr{r}}}$ is the unit vector pointing from $q_2$ to $q_1$

$$\mathbf{F} = 300 \times 10^{-9}\,\hat{\boldsymbol{\mathscr{r}}}\,\mathrm{N}$$

In order to find $\hat{\boldsymbol{\mathscr{r}}}$ let us calculate the vector $\boldsymbol{\mathscr{r}}$ pointing from $q_2$ to $q_1$

$$\boldsymbol{\mathscr{r}} = (2-1)\hat{\mathbf{i}} + (3-1)\hat{\mathbf{j}} + (2-3)\hat{\mathbf{k}}$$

$$\boldsymbol{\mathscr{r}} = \hat{\mathbf{i}} + 2\hat{\mathbf{j}} - \hat{\mathbf{k}}$$

and $|\boldsymbol{\mathscr{r}}| = \sqrt{1+4+1} = \sqrt{6}$

$$\hat{\boldsymbol{\mathscr{r}}} = \frac{\boldsymbol{\mathscr{r}}}{|\boldsymbol{\mathscr{r}}|} = \frac{\hat{\mathbf{i}} + 2\hat{\mathbf{j}} - \hat{\mathbf{k}}}{\sqrt{6}}$$

$$\mathbf{F} = \frac{300}{\sqrt{6}} \times 10^{-9}(\hat{\mathbf{i}} + 2\hat{\mathbf{j}} - \hat{\mathbf{k}})\,\mathrm{N}$$

$$\text{Now } |\mathbf{F}| = \frac{300 \times 10^{-9}}{\sqrt{6}}\sqrt{6} = 300 \times 10^{-9}\,\mathrm{N}.$$

**Example 2.7**

Charge is uniformly distributed on a circular disc of radius R. Calculate the electric field at point P at distance $y$ from the centre of the disc. Charge density σ on the surface of the disc is uniform.

**Solution**

In order to calculate the electric field produced by the entire surface charge σ divide the surface of the entire disc into number of small surface elements ds′. Let us consider two surface elements ds′ each holding a charge σds′ as shown in Fig. 2.11a, b. Let us calculate the electric field due to these two surface elements initially and finally integrate between 0 and π, 0 to R in order to calculate net electric field produced by the entire disc.

The electric field produced by the single surface element ds′ at point P is

$$|\mathbf{E_e}| = \frac{1}{4\pi\varepsilon_o}\frac{\sigma ds'}{\left[y^2 + (r')^2\right]}$$

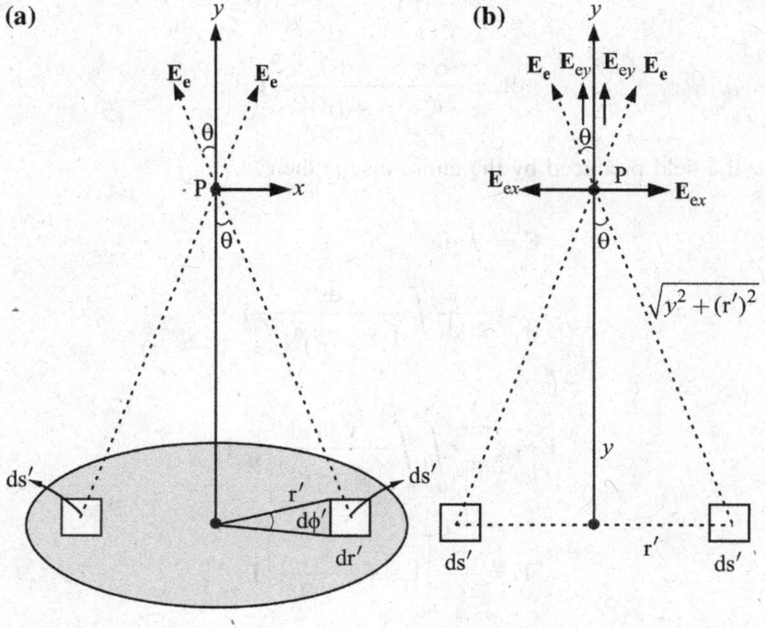


**Fig. 2.11** **a** Chargeuniformly distributed over a circular disc. **b** Resolution of electric field vectors due to two surface elements

Resolve $|\mathbf{E_e}|$ into $\mathbf{E}_{ex}$ and $\mathbf{E}_{ey}$ components as shown in Fig. 2.11b. Clearly $\mathbf{E}_{ex}$ components cancel out, while $\mathbf{E}_{ey}$ components add up. Let $\mathbf{dE}$ be the electric field produced by the two elements at point P. Then as per the above discussion

$$\mathbf{dE} = 2\,\mathrm{E}_{ey}\,\hat{\mathbf{j}}$$

From Fig. 2.11b

$$\cos\theta = \frac{\mathrm{E}_{ey}}{\mathrm{E_e}} = \frac{y}{\sqrt{y^2 + (\mathrm{r}')^2}}$$

$$\Rightarrow \mathrm{E}_{ey} = \frac{y}{\sqrt{y^2 + (\mathrm{r}')^2}}\mathrm{E_e}$$

Substituting the value of $\mathrm{E}_{ey}$ in $\mathbf{dE}$

$$\mathbf{dE} = 2\frac{y}{\sqrt{y^2 + (\mathrm{r}')^2}}\mathrm{E_e}$$

Substituting for $\mathrm{E_e}$ in $\mathbf{dE}$

$$\mathbf{dE} = 2\frac{y}{\sqrt{y^2 + (\mathrm{r}')^2}}\frac{1}{4\pi\varepsilon_o}\frac{\sigma \mathrm{ds}'}{\left[y^2 + (\mathrm{r}')^2\right]}\hat{\mathbf{j}}$$

$$\mathbf{dE} = \frac{\sigma\, y}{2\pi\varepsilon_o}\frac{\mathrm{ds}'}{[y^2 + (\mathrm{r}')]^{3/2}}\hat{\mathbf{j}}$$

Now the field produced by the entire disc is then

$$\mathbf{E} = \int \mathbf{dE}$$

$$= \frac{\sigma y}{2\pi\varepsilon_o}\int \frac{\mathrm{ds}'}{\left[y^2 + (\mathrm{r}')^2\right]^{3/2}}\hat{\mathbf{j}}$$

$$\mathbf{E} = \frac{\sigma y}{2\pi\varepsilon_o}\int_0^{\pi}\int_0^{R}\frac{\mathrm{r}'\,\mathrm{dr}'\,\mathrm{d}\phi'}{\left[y^2 + (\mathrm{r}')^2\right]^{3/2}}\hat{\mathbf{j}}$$

$$\mathbf{E} = \frac{\sigma}{2\varepsilon_o}\left[1 - \frac{y}{[y^2 + \mathrm{R}^2]}\right]\hat{\mathbf{j}}$$

**Example 2.8**
A 1 nC charge is situated at (2, 0, 0) m, and 2 nC charge is situated at (0, 1, 0) m. Find the electric field at point P whose coordinates are (0, 0, 4) m.

**Solution**
The position vector of 1 nC charge with respect to point P is

$$\boldsymbol{\mathscr{r}}_1 = -2\hat{\mathbf{i}} + 4\hat{\mathbf{k}}$$

The position vector of 2 nC charge with respect to point P is

$$\boldsymbol{\mathscr{r}}_1 = -1\hat{\mathbf{j}} + 4\hat{\mathbf{k}}$$

The electric field of 1 nC charge at point P is

$$\mathbf{E}_1 = \frac{1}{4\pi\varepsilon_0}\frac{1 \times 10^{-9}}{\mathscr{r}_1^2}\hat{\boldsymbol{\mathscr{r}}}_1$$

Now $|\boldsymbol{\mathscr{r}}_1| = \sqrt{4+16} = \sqrt{20}$

$$\hat{\boldsymbol{\mathscr{r}}}_1 = \frac{\boldsymbol{\mathscr{r}}_1}{|\boldsymbol{\mathscr{r}}_1|} = \frac{-2\hat{\mathbf{i}} + 4\hat{\mathbf{k}}}{\sqrt{20}}$$

Hence, $\mathbf{E}_1 = \dfrac{1}{4\pi\varepsilon_0}\dfrac{1 \times 10^{-9}}{20}\dfrac{(-2\hat{\mathbf{i}} + 4\hat{\mathbf{k}})}{\sqrt{20}}$

$$\Rightarrow \mathbf{E}_1 = -0.20125\,\hat{\mathbf{i}} + 0.4025\,\hat{\mathbf{k}}$$

Now $|\boldsymbol{\mathscr{r}}_2| = \sqrt{1+16} = \sqrt{17}$

$$\hat{\boldsymbol{\mathscr{r}}}_1 = \frac{\boldsymbol{\mathscr{r}}_2}{|\boldsymbol{\mathscr{r}}_2|} = \frac{-\hat{\mathbf{j}} + 4\hat{\mathbf{k}}}{\sqrt{17}}$$

$$\Rightarrow \mathbf{E}_2 = (9 \times 10^9)\frac{(2 \times 10^{-9})}{(17)}\frac{(-\hat{\mathbf{j}} + 4\hat{\mathbf{k}})}{\sqrt{17}}$$

$$\mathbf{E}_2 = -0.2568\,\hat{\mathbf{j}} + 1.027\,\hat{\mathbf{k}}$$

The net electric field at point P is

$$\mathbf{E} = \mathbf{E}_1 + \mathbf{E}_2 = (-0.20125\,\hat{\mathbf{i}} + 0.4025\,\hat{\mathbf{k}}) + (-0.2568\,\hat{\mathbf{j}} + 1.027\,\hat{\mathbf{k}})$$
$$\mathbf{E} = -0.20125\,\hat{\mathbf{i}} - 0.2568\,\hat{\mathbf{j}} + 1.4295\,\hat{\mathbf{k}}$$

**Example 2.9**
Three charges of magnitude $q_1$, $q_2$ and $q_3$ are situated at the three corners of a square of side b. Calculate the net electric field produced at the fourth corner of the square.

**Solution**
Consider Fig. 2.12. As shown the electric field produced by $q_1$ is acting along *x*-axis and is given by

$$\mathbf{E}_1 = |\mathbf{E}_1|\hat{\mathbf{i}}$$

The electric field produced by $q_3$ is acting along *y*-axis and is given by

$$\mathbf{E}_3 = |\mathbf{E}_3|\hat{\mathbf{j}}$$

The electric field produced by $q_2$ is along the diagonal and is denoted as $\mathbf{E}_2$. The net electric field at point O is given by

$$\mathbf{E} = \mathbf{E}_1 + \mathbf{E}_2 + \mathbf{E}_3$$

To find **E** we need to resolve $\mathbf{E}_2$ into its *x* and *y* components as shown in Fig. 2.12b

$$\mathbf{E}_2 = \mathrm{E}_{2x}\hat{\mathbf{i}} + \mathrm{E}_{2y}\hat{\mathbf{j}}$$

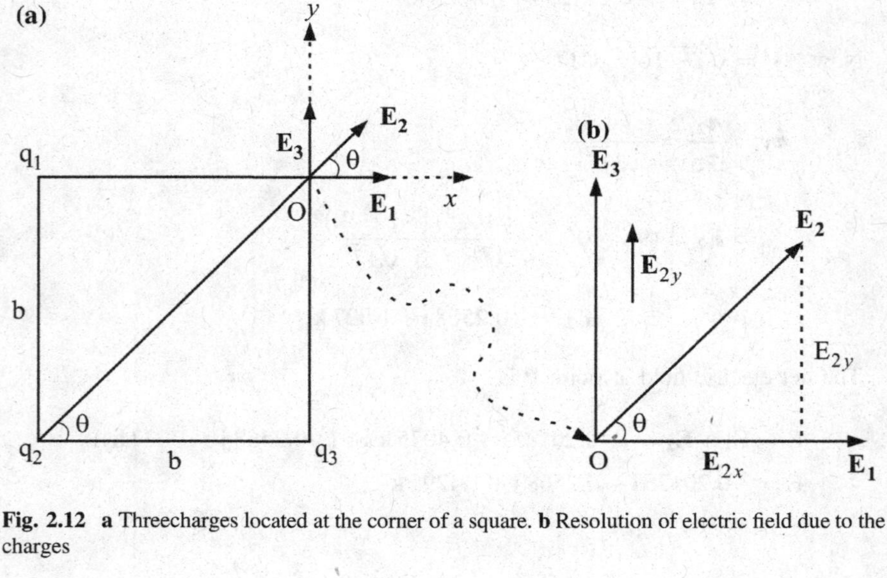


**Fig. 2.12** **a** Threecharges located at the corner of a square. **b** Resolution of electric field due to the charges

Thus

$$\mathbf{E} = \mathrm{E}_1\hat{\mathbf{i}} + \mathrm{E}_{2x}\hat{\mathbf{i}} + \mathrm{E}_{2y}\hat{\mathbf{j}} + \mathrm{E}_3\hat{\mathbf{j}}$$
$$\mathbf{E} = (\mathrm{E}_1 + \mathrm{E}_{2x})\hat{\mathbf{i}} + (\mathrm{E}_{2y} + \mathrm{E}_3)\hat{\mathbf{j}}$$

**Example 2.10**
Three charges shown in Fig. 2.13 are in equilibrium. Find the value of a.

**Solution**
For the charges to be in equilibrium

$$\frac{(27)(12)}{a^2} = \frac{(27)(102)}{20} = \frac{(12)(102)}{(20-a)^2}$$

$$\Rightarrow a^2 = \frac{(27)(12)(20)}{(27)(102)}$$

$$\Rightarrow a = 1.5339\,\text{mm}$$

## 2.7 Solid Angle

In the lower classes the reader would have learnt about ordinary angle which is angle in two dimensions. The ordinary angle θ is shown in Fig. 2.14a. It is the angle between two lines. In Fig. 2.14a it is the angle between lines OA and OB.

When θ is in radians

$$\theta = \frac{l}{r}$$

where $l$ is the arc length and OA = OB = r in Fig. 2.14a. As we increase OA, OB in Fig. 2.14b to $OA_1$, $OB_1$, θ remains the same.

That is $\theta = \frac{l}{r} = \frac{l_1}{r_1}\,[OA_1 = OB_1 = r_1]$.

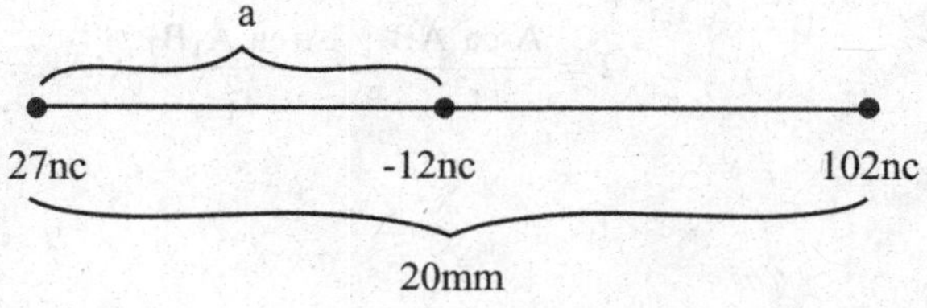


**Fig. 2.13** Three charges separated by specified distance

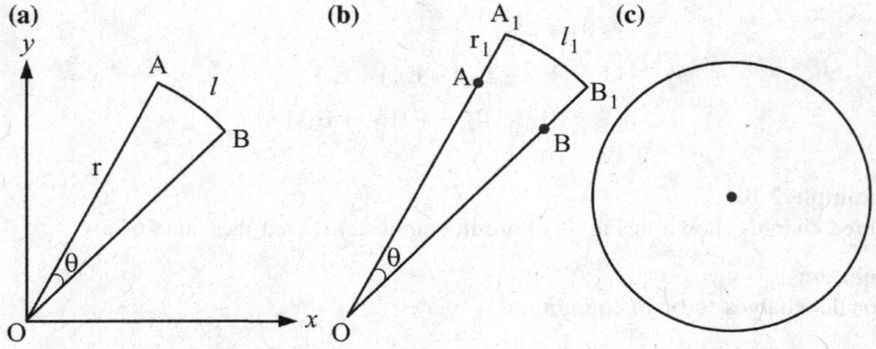


**Fig. 2.14** Angle in two dimensions

As we increase $l$ it finally forms a circle as shown in Fig. 2.14c. For a circle $l = 2\pi r$.

$$\theta = \frac{l}{r} = \frac{2\pi r}{r} = 2\pi$$

Similarly for a semicircle $l = \pi r$

$$\theta = \frac{l}{r} = \frac{\pi r}{r} = \pi$$

Thus the ordinary angle is the angle in two dimensions.

Solid angle is the angle in three dimensions. Take a plain sheet of paper, and rotate it in the form of cone as shown in Fig. 2.15. The angle subtended is the solid angle.

The solid angle is denoted by $\Omega$. Now consider Fig. 2.16a. We have shown solid circle $\Omega$ along with $x$-, $y$-, $z$-axes. Now in Fig. 2.16a AB is area. The solid angle is given by

$$\Omega = \frac{\text{Area AB}}{r^2}$$

From Fig. 2.16a as r is increased to $r_1$ in Fig. 2.16b still $\Omega$ remains constant

$$\Omega = \frac{\text{Area AB}}{r^2} = \frac{\text{Area } A_1B_1}{r_1^2} [OA_1 = OB_1 = r_1]$$

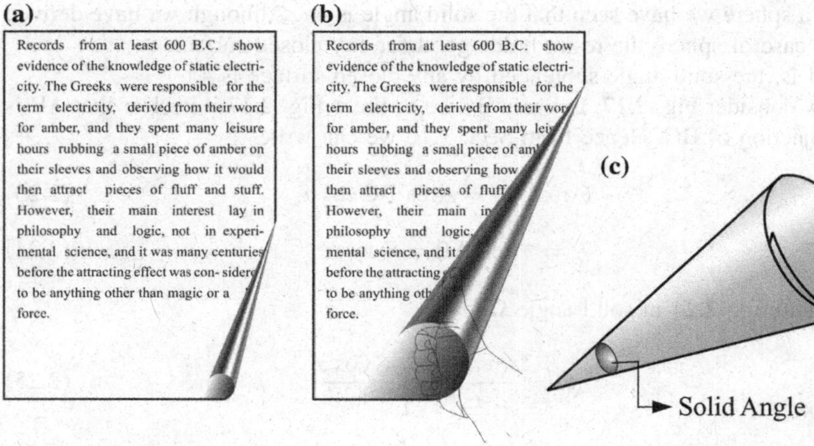


**Fig. 2.15** Solid Angle—Angle in three dimensions

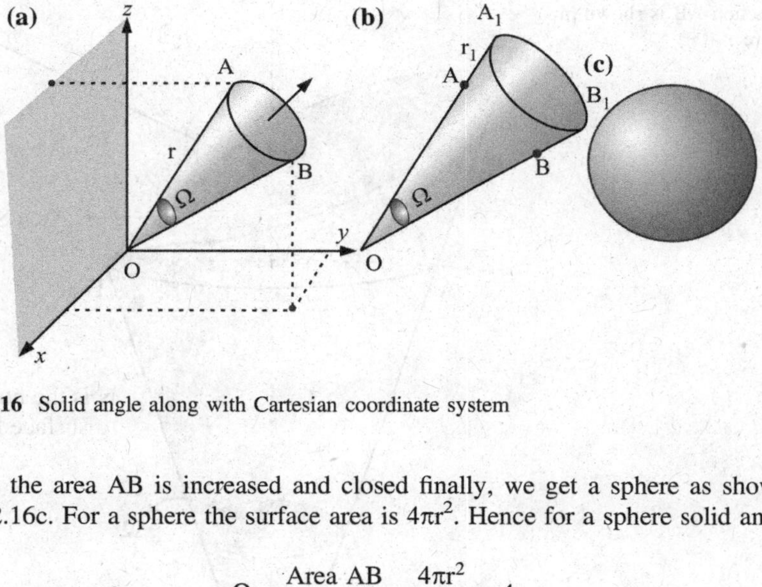


**Fig. 2.16** Solid angle along with Cartesian coordinate system

As the area AB is increased and closed finally, we get a sphere as shown in Fig. 2.16c. For a sphere the surface area is $4\pi r^2$. Hence for a sphere solid angle is

$$\Omega = \frac{\text{Area AB}}{r^2} = \frac{4\pi r^2}{r^2} = 4\pi$$

For a hemisphere surface area is $2\pi r^2$. Thus solid angle for hemisphere is

$$\Omega = \frac{2\pi r^2}{r^2} = 2\pi$$

For a sphere we have seen that the solid angle is $4\pi$. Although we have derived for the case of sphere the result holds good for any closed surface.

That is, the solid angle subtended by any closed surface is $4\pi$.

Now consider Fig. 2.17. Let area BC be S. From Fig. 2.17 it is clear that AB is the projection of BC. Hence from Sect. 1.12 we can write

$$\text{Area AB} = \text{Area BC}\cos\alpha \tag{2.23}$$

$$\Rightarrow \text{Area AB} = \text{S}\cos\alpha \tag{2.24}$$

Substituting 2.24 in solid angle $\Omega$

$$\Omega = \frac{\text{Area AB}}{\text{r}^2} = \frac{\text{S}\cos\alpha}{\text{r}^2} \tag{2.25}$$

**Fig. 2.17** Area S along with its projection AB is shown in the figure

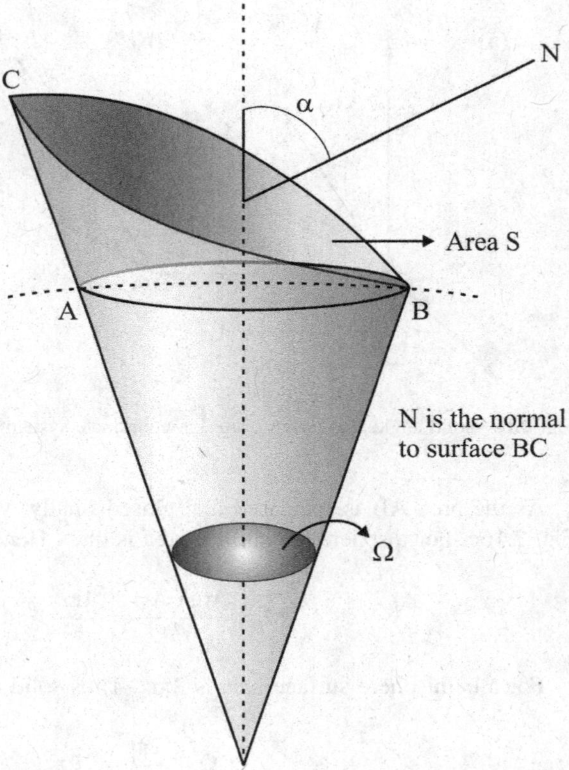

## 2.8 Gauss's Law

In this section we will prove Gauss's law and later we will see how Gauss's law simplifies the calculation of electric field as compared to Coulomb's law.

Consider an arbitrary surface S which is enclosinga charge q as shown in Fig. 2.18a. Let **E** be the electric field produced by q as shown in Fig. 2.18a. $\hat{\boldsymbol{\imath}}$ is the unit vector in the direction of **E**. Consider a small strip of area **ds**, and $\hat{\mathbf{n}}$ is the unit vector normal to ds. The angle between $\hat{\mathbf{n}}$ and $\hat{\boldsymbol{\imath}}$ is $\alpha$.

The flux passing through the small surfaceds is

$$= \mathbf{E} \cdot \mathbf{ds} \tag{2.26}$$

The total flux passing through the closed surface S is

$$= \oint_S \mathbf{E} \cdot \mathbf{ds} \tag{2.27}$$

(Reader should note that we are using unprimed symbols instead of primed symbols because we are dealing with field points).

But $\mathbf{E} = \frac{q}{4\pi\varepsilon_0 \imath^2}\hat{\boldsymbol{\imath}}$, $\mathbf{ds} = \mathrm{ds}\,\hat{\mathbf{n}}$.

Substituting the above equations in Eq. 2.27

$$\oint \mathbf{E} \cdot \mathbf{ds} = \oint \frac{q}{4\pi\varepsilon_0} \frac{1}{\imath^2} \hat{\boldsymbol{\imath}} \cdot \hat{\mathbf{n}}\, \mathrm{ds}$$

Because $\hat{\boldsymbol{\imath}}, \hat{\mathbf{n}}$ are unit vectors, the dot product $\hat{\boldsymbol{\imath}} \cdot \hat{\mathbf{n}} = \cos\alpha$. Thus

$$\text{Total flux} = \frac{q}{4\pi\varepsilon_0} \oint \frac{\mathrm{ds}\cos\alpha}{\imath^2} \tag{2.28}$$

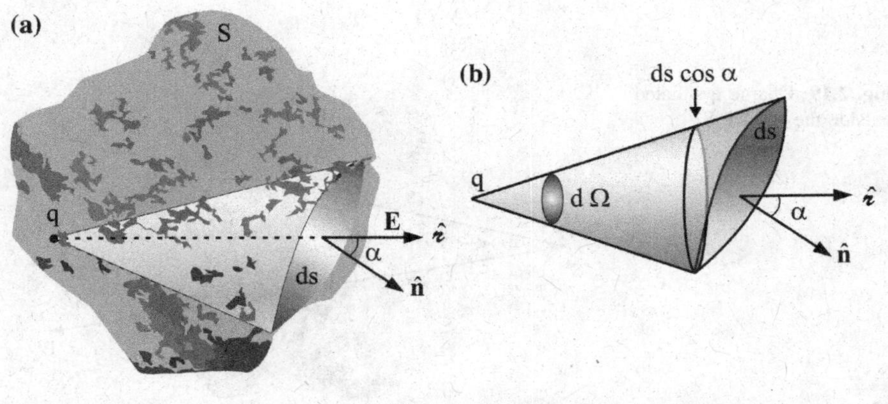


**Fig. 2.18** **a**Arbitrary surface S enclosing charge q. **b** Charge q and small strip of area ds.

But from Fig. 2.18b and using Eq. 2.25

$$\text{Total flux} = \frac{q}{4\pi\varepsilon_o}\oint d\Omega \tag{2.29}$$

where $d\Omega$ is the solid angle as shown in Fig. 2.18b. But $\oint d\Omega = 4\pi$ because the solid angle subtended by any closed surface is $4\pi$. Hence,

$$\oint \mathbf{E}\cdot \mathbf{ds} = \frac{q}{4\pi\varepsilon_o}4\pi = \frac{q}{\varepsilon_o} \tag{2.30}$$

Suppose the charge q is situated outside the surface S as shown in Fig. 2.19. Consider two small strips $ds_1$ and $ds_2$. At surface $ds_1$ the electric field is pointing opposite to $\hat{\mathbf{n}}_1$, the unit vector normal to $ds_1$. Hence flux at $ds_1$ is (using Eq. 2.30)

$$= \frac{-q d\Omega}{4\pi\varepsilon_0}$$

(Here we need not integrate because we are considering only small strip $ds_1$ and not the entire surface S).

At $ds_2$ the **E** is pointing in the same direction of $\hat{\mathbf{n}}_2$; hence the flux at $ds_2$ is

$$= \frac{q d\Omega}{4\pi\varepsilon_0}$$

Hence the net flux through the elements $ds_1$ and $ds_2$ is

$$= -\frac{q d\Omega}{4\pi\varepsilon_o} + \frac{q d\Omega}{4\pi\varepsilon_o} = 0$$

The above fact is true for any cone drawn through the surface S in Fig. 2.19. Thus if the charge is lying outside the surface the net flux enclosed by the surface due to the charge is zero.

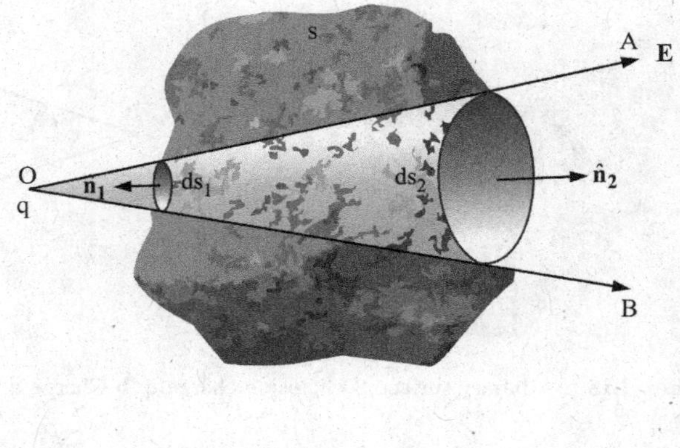


**Fig. 2.19** Charge q situated outside the surface S

In Fig. 2.18 if the charge is situated in a medium of dielectric constant $\varepsilon_1$ then the flux is

$$\oint \mathbf{E} \cdot \mathbf{ds} = \frac{q}{\varepsilon_0 \varepsilon_r} \tag{2.31}$$

In Fig. 2.18 instead of one charge if the surface S encloses n number of charges $q_1, q_2, q_3, \ldots q_n$, then told flux enclosed by the surface S is

$$\text{Total flux} = (\text{flux})_1 + (\text{flux})_2 + \cdots + (\text{flux})_n \tag{2.32}$$

$$\oint \mathbf{E} \cdot \mathbf{ds} = \frac{q_1}{\varepsilon_0} + \frac{q_2}{\varepsilon_0} + \cdots + \frac{q_n}{\varepsilon_0} \tag{2.33}$$

Here **E** is the "net" electric field due to all charges

$$\oint_S \mathbf{E} \cdot \mathbf{ds} = \frac{q_1 + q_2 + q_3 + \cdots + q_n}{\varepsilon_0} \tag{2.34}$$

$$\oint_S \mathbf{E} \cdot \mathbf{ds} = \frac{q_{enc}}{\varepsilon_0} \tag{2.35}$$

where $q_1 + q_2 + \ldots + q_n = q_{enc}$ is the total charge enclosed by the surface S and $(\text{Flux})_1$, $(\text{Flux})_2$, … $(\text{Flux})_n$ are the fluxes through the surface S due to charges $q_1$, $q_2 \ldots q_n$.

From Eq. 2.22

$$q_{enc} = \oint_\tau \rho\, d\tau \tag{2.36}$$

assuming that a volume charge density ρ is enclosed in surface S in Fig. 2.18. τ is the volume enclosed by surface S in Fig. 2.18. In Eq. 2.36 we use unprimed symbol instead of primed symbol because we are dealing with field points.

With aid of Gauss's divergence theorem

$$\oint_S \mathbf{E} \cdot \mathbf{ds} = \int_\tau \nabla \cdot \mathbf{E} d\tau \tag{2.37}$$

Substituting Eqs. 2.37 and 2.36 in Eq. 2.35

$$\int_\tau \nabla \cdot \mathbf{E}\, d\tau = \frac{\int_\tau \rho\, d\tau}{\varepsilon_0} \tag{2.38}$$

The above integration considers a small volume $d\tau$ and then integrates over the entire volume $\tau$. Equation 2.38 is valid even for the small volume $d\tau$. In that case we need not integrate. Hence,

$$\nabla \cdot \mathbf{E} d\tau = \frac{\rho \, d\tau}{\varepsilon_o} \tag{2.39}$$

$$\Rightarrow \nabla \cdot \mathbf{E} = \frac{\rho}{\varepsilon_o} \tag{2.40}$$

The above form of Gauss's law is called differential or point form of Gauss's law.

## 2.9 Sketches of Field Lines

In Chap. 1 we discussed vectors in detail and the concept of flux was introduced in Sect. 1.12. Flux is defined as the number of lines passing through the given area. Or in other words we have used "flux lines" to define flux. In Fig. 1.20 we have shown lines of the vector passing through the area abcd which are the flux lines.

We have read about Coulomb's law from which we observe that the field of the point charge falls off as $1/r^2$. Also Eq. 2.40 tells us the fact that divergence of the electric field produced by a static charge distribution is nonzero. Recall that divergence means spreading out of vectors. Equation 2.40 says that the electric field of a static charge distribution spreads out. For example we sketch the electric field of a point charge in Fig. 2.20a and see that electric field spreads out. In Fig. 2.20a we have sketched the electric field in terms of vectors. Because the field falls off as $1/r^2$ as per Coulomb's law, the length of the vectors becomes shorter as we move away from the charge. Alternatively we can sketch the electric field using "flux lines" or "field lines" as shown in Fig. 2.20b. The flux lines or field lines are

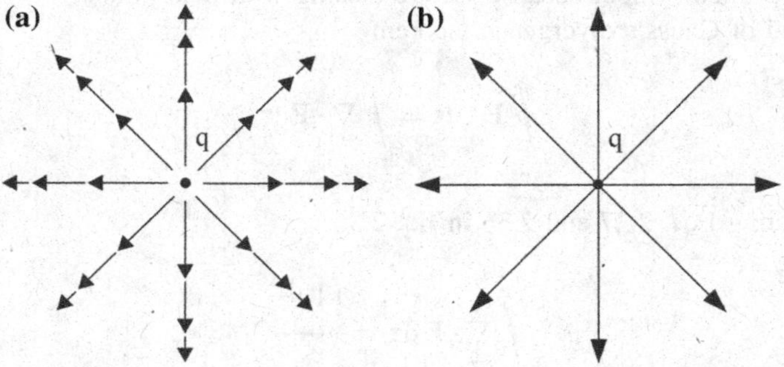


**Fig. 2.20** **a**Electric field of a point charge. **b** Field lines of a point charge

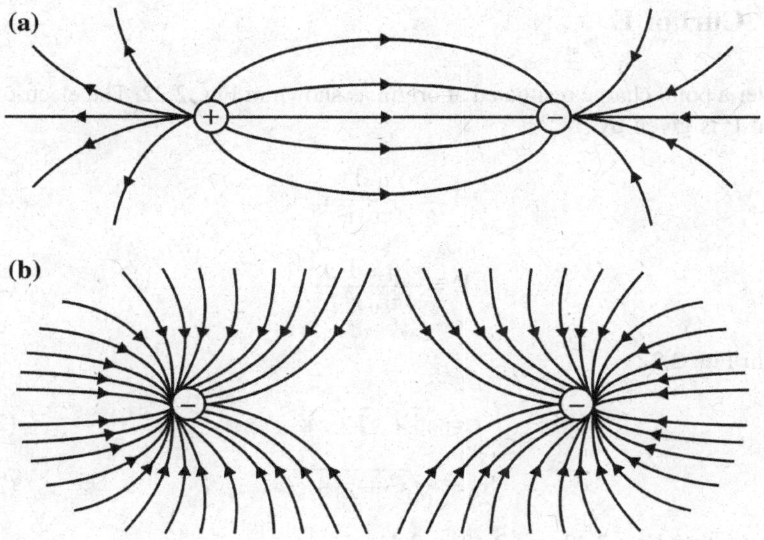


**Fig. 2.21** **a** Fieldlines of a positive and negative charge. **b** Field lines of two negative charges

directed lines. At each point surrounding the charge the direction of the line indicates the direction of the vector. The magnitude of the electric field is indicated by the "density" of the field lines. The density of the lines is stronger near the charge indicating that the field at those points is stronger. Far away from the charge the density of the lines is less indicating the field is weaker at those points.

The electric field pattern due to a positive and negative charge and two negative charges has been represented in Fig. 2.21 with the help of field line concept. The field lines have no real existence, but however as shown in Fig. 2.21 the field lines are helpful in visualizing the fields. The following points should be noted about the field lines.

(a) Field lines cannot cross each other—otherwise field would have two different directions at that point.
(b) Field lines start on positive charge and end on negative charge as shown in Fig. 2.21a—otherwise, they extend to infinity as shown in Fig. 2.21b.
(c) Assuming a sphere of radius r the centre of which contains a point charge q the field lines emanate outwards perpendicularly and uniformly through the surface of the sphere. As the surface area of the sphere is $4\pi r^2$ the density of the lines—that is the flux lines per unit area is inversely proportional to $r^2$.

## 2.10 Curl of E

Consider a point charge q situated at origin as shown in Fig. 2.22. The electric field at point P is given by

$$\mathbf{E} = \frac{q}{4\pi\varepsilon_0}\frac{1}{r^2}\hat{r}$$

$$\mathbf{E} = \frac{q}{4\pi\varepsilon_0}\frac{1}{\mathbf{r}^2}\frac{\mathbf{r}}{|\mathbf{r}|} \tag{2.41}$$

From Fig. 2.22

$$\mathbf{r} = x\hat{\mathbf{i}} + y\hat{\mathbf{j}} + z\hat{\mathbf{k}} \tag{2.42}$$

$$|\mathbf{r}| = \sqrt{x^2 + y^2 + z^2} \tag{2.43}$$

Substituting Eqs. 2.43 and 2.42 in 2.41

$$\mathbf{E} = \frac{q}{4\pi\varepsilon_0}\frac{x\hat{\mathbf{i}} + y\hat{\mathbf{j}} + z\hat{\mathbf{k}}}{[x^2 + y^2 + z^2]^{3/2}} \tag{2.44}$$

$$\begin{aligned}\Rightarrow \mathbf{E} = &\frac{q}{4\pi\varepsilon_o}\frac{x}{[x^2 + y^2 + z^2]^{3/2}}\hat{\mathbf{i}}\\ &+ \frac{q}{4\pi\varepsilon_o}\frac{y}{[x^2 + y^2 + z^2]^{3/2}}\hat{\mathbf{j}}\\ &+ \frac{q}{4\pi\varepsilon_o}\frac{z}{[x^2 + y^2 + z^2]^{3/2}}\hat{\mathbf{k}}\end{aligned} \tag{2.45}$$

$$\Rightarrow \mathbf{E} = \mathrm{E}_x\hat{\mathbf{i}} + \mathrm{E}_y\hat{\mathbf{j}} + \mathrm{E}_z\hat{\mathbf{k}} \tag{2.46}$$

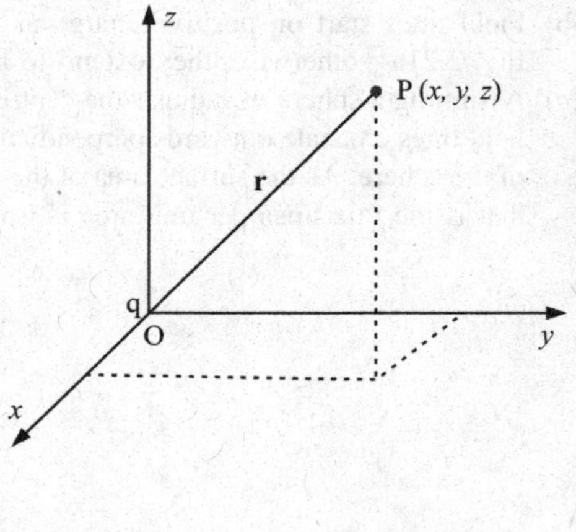


**Fig. 2.22** Point charge q situated in origin

where

$$E_x = \frac{q}{4\pi\varepsilon_o}\frac{x}{[x^2+y^2+z^2]^{3/2}} \tag{2.47}$$

$$E_y = \frac{q}{4\pi\varepsilon_o}\frac{y}{[x^2+y^2+z^2]^{3/2}} \tag{2.48}$$

$$E_z = \frac{q}{4\pi\varepsilon_o}\frac{z}{[x^2+y^2+z^2]^{3/2}} \tag{2.49}$$

We know that

$$\begin{aligned}\nabla\times\mathbf{E} &= \begin{vmatrix}\hat{\mathbf{i}} & \hat{\mathbf{j}} & \hat{\mathbf{k}}\\ \frac{\partial}{\partial x} & \frac{\partial}{\partial y} & \frac{\partial}{\partial z}\\ E_x & E_y & E_z\end{vmatrix}\\ &= \left(\frac{\partial E_z}{\partial y}-\frac{\partial E_y}{\partial z}\right)\hat{\mathbf{i}}+\left(\frac{\partial E_x}{\partial z}-\frac{\partial E_z}{\partial x}\right)\hat{\mathbf{j}}+\left(\frac{\partial E_y}{\partial x}-\frac{\partial E_x}{\partial y}\right)\hat{\mathbf{k}}\end{aligned} \tag{2.50}$$

Let us calculate

$$\left(\frac{\partial E_z}{\partial y}-\frac{\partial E_y}{\partial z}\right) \tag{2.51}$$

$$\frac{\partial E_z}{\partial y} = \frac{q}{4\pi\varepsilon_o}\frac{\partial}{\partial y}\left[\frac{z}{(x^2+y^2+z^2)^{3/2}}\right]$$

$$\frac{\partial E_z}{\partial y} = \frac{-3q}{4\pi\varepsilon_o}\left[\frac{zy}{(x^2+y^2+z^2)^{5/2}}\right] \tag{2.52}$$

$$\begin{aligned}\frac{\partial E_y}{\partial z} &= \frac{q}{4\pi\varepsilon_o}\frac{\partial}{\partial z}\left[\frac{y}{(x^2+y^2+z^2)^{3/2}}\right]\\ &= \frac{-3q}{4\pi\varepsilon_o}\left[\frac{yz}{(x^2+y^2+z^2)^{5/2}}\right]\end{aligned} \tag{2.53}$$

Now

$$\left(\frac{\partial E_z}{\partial y} - \frac{\partial E_y}{\partial z}\right) = \frac{-3q}{4\pi\varepsilon_0}\left[\frac{zy}{(x^2+y^2+z^2)^{5/2}}\right] + \frac{3q}{4\pi\varepsilon_0}\left[\frac{yz}{(x^2+y^2+z^2)^{5/2}}\right]$$

$$\Rightarrow \left(\frac{\partial E_z}{\partial y} - \frac{\partial E_y}{\partial z}\right) = 0 \tag{2.54}$$

Similarly

$$\frac{\partial E_x}{\partial z} - \frac{\partial E_z}{\partial x} = 0 \tag{2.55}$$

$$\frac{\partial E_y}{\partial x} - \frac{\partial E_x}{\partial y} = 0 \tag{2.56}$$

Substituting Eqs. 2.54, 2.55 and 2.56 in Eq. 2.50

$$\nabla \times \mathbf{E} = 0 \tag{2.57}$$

Thus curl of **E** is zero. Suppose there are a number of charges each producing the field $\mathbf{E_1}$, $\mathbf{E_3}$, …$\mathbf{E_n}$, then from Eq. 2.11

$$\mathbf{E} = \mathbf{E_1} + \mathbf{E_2} + \cdots + \mathbf{E_n} \tag{2.58}$$

With the help of Eq. 2.57 we see that

$$\left.\begin{array}{l} \nabla \times \mathbf{E_1} = 0 \\ \nabla \times \mathbf{E_2} = 0 \\ \cdots \\ \nabla \times \mathbf{E_n} = 0 \end{array}\right\} \tag{2.59}$$

for individual charges.

Thus for the net electric field **E** due to all charges given in Eq. 2.58

$$\nabla \times \mathbf{E} = \nabla \times \mathbf{E_1} + \nabla \times \mathbf{E_2} + \cdots + \nabla \times \mathbf{E_n} \tag{2.60}$$

From Eq. 2.59 then

$$\nabla \times \mathbf{E} = 0 \tag{2.61}$$

The above equation is true for any static chargedistribution. Recall that curl is a measure of circulation or rotation. $\nabla \times \mathbf{E} = 0$ says that electric fields due to static charges do not circulate about. Electric fields due to static charges are irrotational fields.

## 2.11 Potential of Discrete and Continuous Charge Distributions

In this section we will derive expression for potential. In the later sections we will see how potential simplifies the calculation of electric field as compared to Coulomb's law.

From Eq. 2.61

$$\nabla \times \mathbf{E} = 0 \tag{2.62}$$

From Example 1.9, curl gradient of any scalar is zero. Hence,

$$\nabla \times (-\nabla \mathrm{V}) = 0 \tag{2.63}$$

Here V is a scalar quantity and is known as electric potential. The inclusion of the negative sign in the above equation will be explained later.

Comparing Eqs. 2.62 and 2.63

$$\mathbf{E} = -\nabla \mathrm{V} \tag{2.64}$$

Equation 2.64 can be written as

$$\int_{\mathrm{R}}^{\mathrm{P}} \mathbf{E} \cdot \mathrm{d}\boldsymbol{l} = -\int_{\mathrm{R}}^{\mathrm{P}} \nabla \mathrm{V} \cdot \mathrm{d}\boldsymbol{l}$$

Applying gradient theorem (Eq. 1.107)

$$-\int_{\mathrm{R}}^{\mathrm{P}} \mathbf{E} \cdot \mathrm{d}\boldsymbol{l} = \mathrm{V(P)} - \mathrm{V(R)}$$

where R the reference point is the "zero of the potential", i.e. where we fix the potential to be zero, and P is the point where the potential V is calculated. We will learn more about point R in future sections. For the time being V(R) = 0.

Hence

$$\mathrm{V(P)} = -\int_{\mathrm{R}}^{\mathrm{P}} \mathbf{E} \cdot \mathrm{d}\boldsymbol{l} \tag{2.65}$$

Let us fix the "zero of the potential" at ∞.

$$V = -\int_{\infty}^{P} \mathbf{E} \cdot d\boldsymbol{l} \tag{2.66}$$

Consider Fig. 2.23. The potential at ∞ is zero. Charge q is located at origin. The point P lies at a distance r from charge q.

In Fig. 2.23

$$\mathbf{E} = \frac{1}{4\pi\varepsilon_0}\frac{q}{r^2}\hat{\mathbf{r}} \tag{2.67}$$

From Fig. 2.23

$$d\boldsymbol{l} = dr\,\hat{\mathbf{r}} \tag{2.68}$$

From Eqs. 2.67 and 2.68

$$\mathbf{E} \cdot d\boldsymbol{l} = \frac{q}{4\pi\varepsilon_0}\left(\frac{1}{r^2}\right)\hat{\mathbf{r}} \cdot (dr\,\hat{\mathbf{r}}) \tag{2.69}$$

$$\mathbf{E} \cdot d\boldsymbol{l} = \frac{q}{4\pi\varepsilon_0}\left(\frac{1}{r^2}\right)dr \tag{2.70}$$

Substituting Eq. 2.70 in Eq. 2.66

$$V = -\int_{\infty}^{r} \frac{q}{4\pi\varepsilon_0}\left(\frac{1}{r^2}\right)dr \tag{2.71}$$

$$V = -\frac{q}{4\pi\varepsilon_0}\left[\frac{-1}{r}\right]_{\infty}^{r} \tag{2.72}$$

$$V = \frac{q}{4\pi\varepsilon_0}\frac{1}{r} \tag{2.73}$$

The negative sign in Eq. 2.63 was selected to make the potential of the positive charge come out positive.

Equation 2.73 holds good for Fig. 2.23 where the point charge q lies at origin. In general the point charge q need not lie in origin as shown in Fig. 2.24. For such situations the potential is given by

**Fig. 2.23** Point charge located at the origin

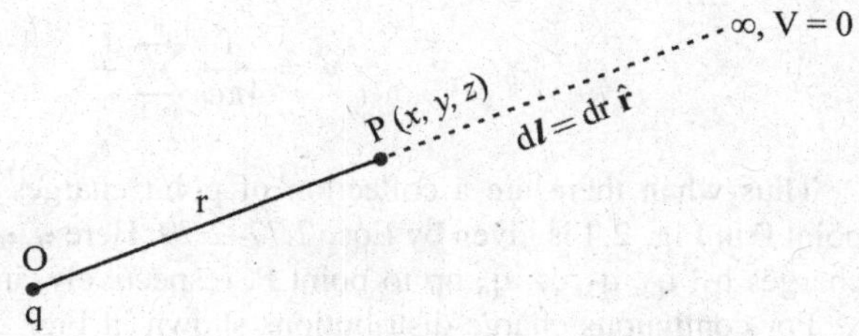


**Fig. 2.24** Point charge q in Cartesian coordinate system

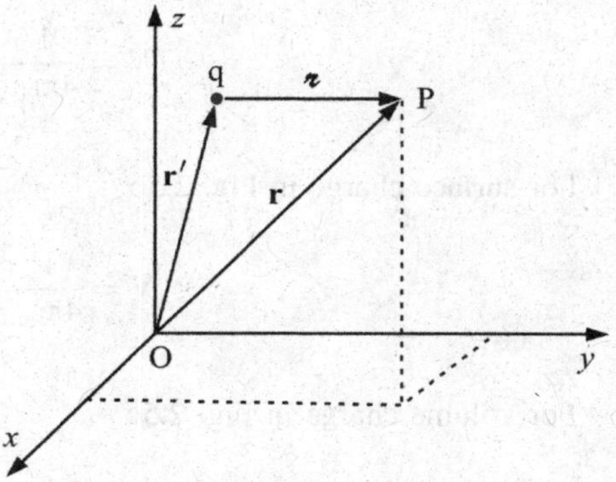


$$V = \frac{q}{4\pi\varepsilon_o}\frac{1}{\mathscr{r}} \tag{2.74}$$

If there is a collection of n number of point charges as shown in Fig. 2.4, then the total potential at point P can be calculated as follows. From Eq. 2.11 the net electric field at point P in Fig. 2.4 is

$$\mathbf{E} = \mathbf{E_1} + \mathbf{E_2} + \cdots + \mathbf{E_n} \tag{2.75}$$

Integrating

$$\int_{\infty}^{P} \mathbf{E}\cdot d\boldsymbol{l} = \int_{\infty}^{P} \mathbf{E_1}\cdot d\boldsymbol{l_1} + \int_{\infty}^{P} \mathbf{E_2}\cdot d\boldsymbol{l_2} + \cdots + \int_{\infty}^{P} \mathbf{E_n}\cdot d\boldsymbol{l_n} \tag{2.76}$$

$$\Rightarrow V = V_1 + V_2 + \cdots + V_n \tag{2.77}$$

Using Eq. 2.74 we get

$$V = \frac{1}{4\pi\varepsilon_o}\frac{q_1}{\mathscr{r}_1} + \frac{1}{4\pi\varepsilon_o}\frac{q_2}{\mathscr{r}_2} + \cdots + \frac{1}{4\pi\varepsilon_o}\frac{q_n}{\mathscr{r}_n} \tag{2.78}$$

$$V = \frac{1}{4\pi\varepsilon_0} \sum_{i=1}^{n} \frac{q_n}{\mathscr{r}_n} \tag{2.79}$$

Thus when there are a collection of point charges in Fig. 2.4 the potential at point P in Fig. 2.4 is given by Eqs. 2.77–2.79. Here $\mathscr{r}_1, \mathscr{r}_2, \mathscr{r}_3 \ldots \mathscr{r}_n$ are the distances of charges $q_1$, $q_2$, $q_3$, … $q_n$ up to point P, respectively, in Fig. 2.4.

For continuous charge distributions shown in Fig. 2.5a–c the potential is:

For line charge in Fig. 2.5a

$$V = \frac{1}{4\pi\varepsilon_0} \int_{l'} \frac{\lambda dl'}{\mathscr{r}} \tag{2.80}$$

For surface charge in Fig. 2.5b

$$V = \frac{1}{4\pi\varepsilon_0} \int_{S'} \frac{\sigma ds'}{\mathscr{r}} \tag{2.81}$$

For volume charge in Fig. 2.5c

$$V = \frac{1}{4\pi\varepsilon_0} \int_{\tau'} \frac{\rho\, d\tau'}{\mathscr{r}} \tag{2.82}$$

For a point charge situated at origin in Fig. 2.22 **E** from Eq. 2.46 is

$$\mathbf{E} = E_x \hat{\mathbf{i}} + E_y \hat{\mathbf{j}} + E_z \hat{\mathbf{k}} \tag{2.83}$$

where the components $E_x$, $E_y$, $E_z$ are given by Eqs. 2.47–2.49.

For a point charge situated in origin in Fig. 2.23 the potential is given by Eq. 2.73

$$V = \frac{1}{4\pi\varepsilon_0} \frac{q}{r} = \frac{1}{4\pi\varepsilon_0} \frac{q}{\sqrt{x^2 + y^2 + z^2}}$$

As we see there are no components for V because V is a scalar. Vectors alone have components. Scalars don't have components. This point has been discussed in detail in Sect. 1.9.

Electric field **E**is a vector and hence can be resolved into components as shown inFig. 2.25a. However thepotential at point P in Fig. 2.25b is V and does not have components. That is along *x*-, *y*-, *z*-axes the potential is V, not $V_x$, $V_y$, $V_z$. To make this point more clear consider an object of mass m in Fig. 2.25c. The mass of the object along *x*-, *y*-, *z*-axes is m, not $m_x$, $m_y$, $m_z$, because mass is a scalar quantity. If the mass of the object is 50 kg, it will be always 50 kg along *x*-, *y*-, *z*-axes.

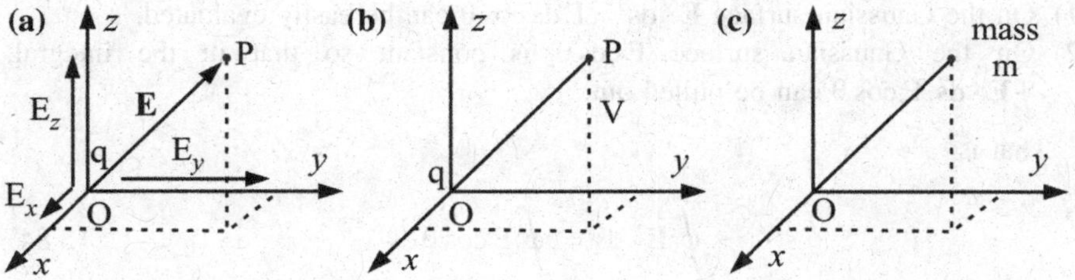


**Fig. 2.25** **a**Resolution of electric field into components. **b** Charge q and potential V in Cartesian coordinate system. **c** Mass m in Cartesian coordinate system

Similarly the potential V is a scalar and hence along *x*-, *y*-, *z*-axes its value is V and not $V_x$, $V_y$, $V_z$. However the electric field **E** is a vector, and hence it can be split up into three components $E_x$, $E_y$ and $E_z$.

Hence if you are using Coulomb's law directly, to calculate electric field then you have to work with three components $E_x$, $E_y$ and $E_z$. Hence you have to fuss with direction. However if you are working with potential, you are working with only one component V alone and need not bother about direction because along *x*-, *y*-, *z*-directions the value is V alone. This makes working with potential V easy as compared to working with **E** directly. Once V has been found, then **E** can be calculated using the relation

$$\mathbf{E} = -\nabla V$$

The way in which potential simplifies calculation of electric field as compared to Coulomb's law will be more clear to the reader when the reader goes through Sect. 2.18.

A final point. The potentials calculated in Eqs. 2.80, 2.81 and 2.82 will tend to infinity if the charge distribution itself extends to infinity. In that case we select a suitable different reference point.

## 2.12 Calculating Electric Field Using Gauss's Law and Potential

We know that Gauss's law is

$$\oint \mathbf{E} \cdot \mathbf{ds} = \frac{q_{enc}}{\varepsilon_0}$$

Our aim is to calculate **E** using the above equation. However **E** is inside the integral. If we want to calculate **E** using the above equation, then **E** must be taken out of the integral. For this, for a given charge distribution we construct an imaginary Gaussian surface such that the following conditions are satisfied.

(1) On the Gaussian surface $\mathbf{E} \cdot \mathbf{ds} = \mathrm{E\,ds}\cos\theta$ can be easily evaluated.
(2) On the Gaussian surface $\mathrm{E}\cos\theta$ is constant so that in the integral, $\oint \mathbf{E} \cdot \mathbf{ds}$, $\mathrm{E}\cos\theta$ can be pulled out.

That is

$$\oint \mathbf{E} \cdot \mathbf{ds} = \oint \mathrm{E}\cos\theta\,\mathrm{ds} \tag{2.84}$$

Because $\mathrm{E}\cos\theta$ is constant over Gaussian surface

$$\oint \mathrm{E}\cos\theta\,\mathrm{ds} = \mathrm{E}\cos\theta \oint \mathrm{ds} \tag{2.85}$$

Now, $\oint \mathrm{ds}$, the calculation of area of the Gaussian surface is simple mathematics. Thus from Gauss's law

$$\oint \mathbf{E} \cdot \mathbf{ds} = \frac{q_{enc}}{\varepsilon_0}$$

$$\Rightarrow \oint \mathrm{E\,ds}\cos\theta = \frac{q_{enc}}{\varepsilon_0} \tag{2.86}$$

Because E cos θ is constant over Gaussian surface

$$\mathrm{E}\cos\theta \oint \mathrm{ds} = \frac{q_{enc}}{\varepsilon_0} \tag{2.87}$$

$$\Rightarrow \mathrm{E} = \frac{q_{enc}}{\varepsilon_0} \frac{1}{\cos\theta \oint \mathrm{ds}} \tag{2.88}$$

Hence from the above equation E can be easily calculated.

If for a given charge distribution, a Gaussian surface cannot be constructed such that $\mathbf{E} \cdot \mathbf{ds}$ cannot be easily evaluated (or) if E cos θ is not constant over Gaussian surface, then **E** cannot be pulled out of the integral. In such situations calculation of **E** using Gauss's law becomes difficult.

Then the alternative method to calculate **E** is by calculating potential initially and then determining **E** using the relation

$$\mathbf{E} = -\nabla \mathrm{V}$$

In the following sections we will calculate the electric field by using Coulomb's law, Gauss's law and potential for different charge distributions and compare the results and methods.

## 2.13 Electric Field Due to an Infinite Line Charge

Consider an infinite uniform line charge present along $y$-axis as shown in Fig. 2.26. The charge per unit length of the line charge is $\lambda$. Let us calculate the electric field at point P which is situated at a distance of r from point O as shown in Fig. 2.26. Now let us calculate the electric field at point P using Coulomb's law, Gauss's law and potential formulation.

(a) **Using Coulomb's Law**

Let us divide the entire line charge into small strips $dy'$ and consider only two strips as shown in Fig. 2.26. Each strip holds a charge of $\lambda\, dy'$. Let us initially calculate the electric field due to these two strips. Finally we integrate to calculate the net electric field at point P due to the entire line charge.

Resolving the electric field $\mathbf{E_s}$ due to each strip into $x$ and $y$ components $\mathbf{E}_{sx}$ and $\mathbf{E}_{sy}$ we see that $\mathbf{E}_{sy}$ components cancel out each other while $\mathbf{E}_{sx}$ components add up. Let the net electric field at point P due to the two strips be $\mathbf{dE}$. The $\mathbf{dE}$ is given by

$$\mathbf{dE} = 2\,E_{sx}\,\hat{\mathbf{i}} \tag{2.89}$$

From Fig. 2.26

$$\cos\theta = \frac{E_{sx}}{E_s} \tag{2.90}$$

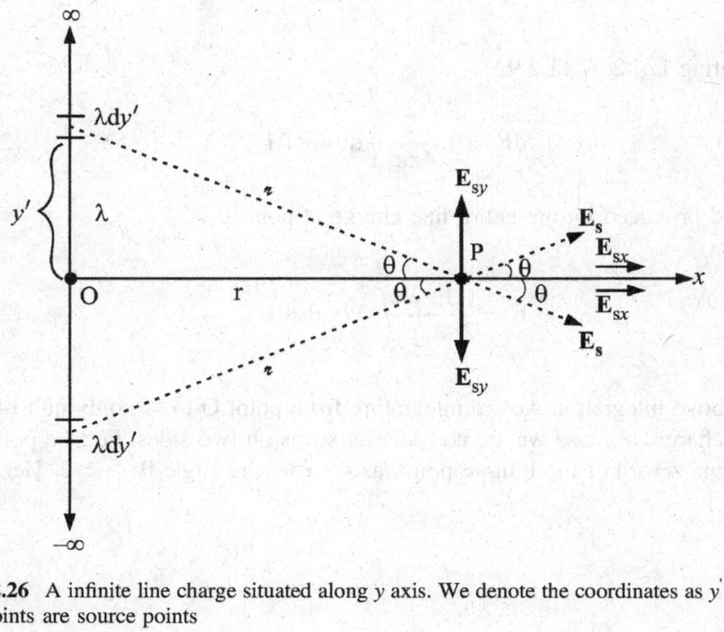


**Fig. 2.26** A infinite line charge situated along $y$ axis. We denote the coordinates as $y'$ to note that the points are source points

$$E_{sx} = E_s \cos\theta \tag{2.91}$$

Substituting 2.91 in 2.89

$$d\mathbf{E} = 2\,E_s \cos\theta\,\hat{\mathbf{i}} \tag{2.92}$$

The electric field $\mathbf{E}_s$ produced by the single element is given by

$$E_s = \frac{1}{4\pi\varepsilon_0}\frac{\lambda dy'}{\imath^2} \tag{2.93}$$

From Fig. 2.26

$$y' = r\tan\theta \Rightarrow dy' = r\sec^2\theta\,d\theta \tag{2.94}$$

and

$$\imath = \frac{\imath}{\cos\theta} \tag{2.95}$$

Substituting Eqs. 2.95 and 2.94 in 2.93

$$E_s = \frac{1}{4\pi\varepsilon_0}\frac{\lambda r\sec^2\theta d\,\theta}{r^2}\cos^2\theta \tag{2.96}$$

$$E_s = \frac{1}{4\pi\varepsilon_0}\frac{\lambda}{r}d\theta \tag{2.97}$$

Substituting Eq. 2.97 in 2.92

$$d\mathbf{E} = 2.\frac{1}{4\pi\varepsilon_0}\frac{\lambda}{r}d\theta\cos\theta\,\hat{\mathbf{i}} \tag{2.98}$$

The field produced by the entire line charge at point P is

$$\mathbf{E} = \frac{1}{2\pi\varepsilon_0}\frac{\lambda}{r}\int_0^{\pi/2}\cos\theta\,d\theta\,\hat{\mathbf{i}} \tag{2.99}$$

In the above integration we are integrating from point O to $\infty$, only half of the entire line charge, because we are considering strips on two sides. For the point O the angle $\theta$ is zero. For the infinite point, as $y \to \infty$ the angle $\theta \to \pi/2$. Hence

$$\mathbf{E} = \frac{1}{2\pi\varepsilon_o}\frac{\lambda}{r}[\sin\theta]_0^{\pi/2}\,\hat{\mathbf{i}} \tag{2.100}$$

$$\mathbf{E} = \frac{\lambda}{2\pi\varepsilon_o r}\hat{\mathbf{i}} \tag{2.101}$$

(b) **Using Gauss's Law**

In Fig. 2.27 we show the infinite line charge. We have to calculate field due to infinite Line Charge by applying Gauss's law.

The first step is to get "the feel of the field". In Fig. 2.26 we see that only $x$ components survive, while $y$ components cancel out. Hence the field of the infinite line charge "looks" radial as shown in Fig. 2.27a. Next step is to construct a Gaussian surface, keeping in mind that evaluation of $\mathbf{E} \cdot \mathbf{ds}$ must be easy and $E \cos\theta$ must be constant over the Gaussian surface. A little thought shows that the cylindrical Gaussian surface satisfies the above requirement. So we construct the cylindrical Gaussian surface as shown in Fig. 2.27b. The Gaussian surface is an imaginary surface. Let r and L be the radius and length of the cylindrical Gaussian surface. Applying Gauss's law to the cylindrical Gaussian surface

$$\oint \mathbf{E} \cdot \mathbf{ds} = \frac{q_{enc}}{\varepsilon_o}$$

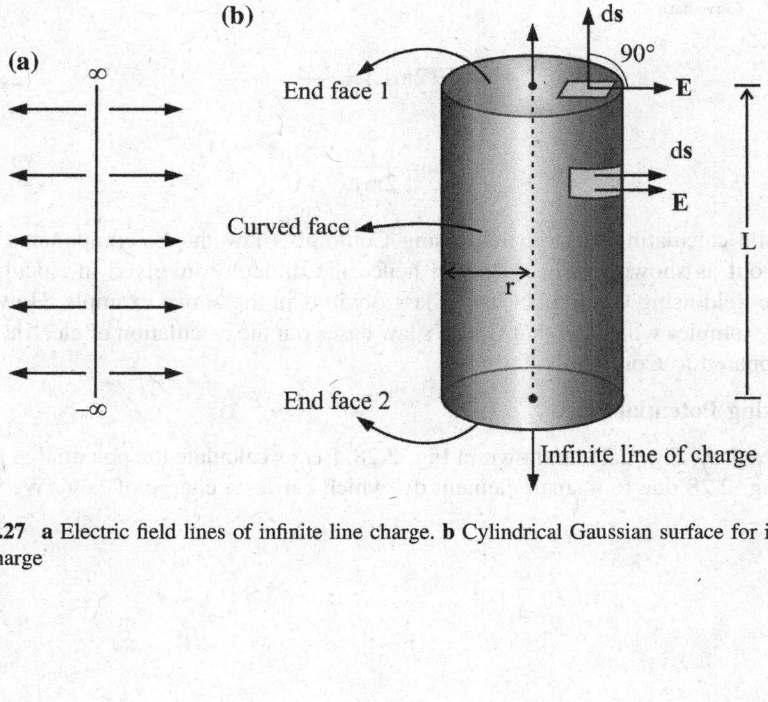


**Fig. 2.27** **a** Electric field lines of infinite line charge. **b** Cylindrical Gaussian surface for infinite line charge

$$\Rightarrow \int_{\text{Endface 1}} \mathbf{E}\cdot \mathbf{ds} + \int_{\text{Curvedface}} \mathbf{E}\cdot \mathbf{ds} + \int_{\text{Endface 2}} \mathbf{E}\cdot \mathbf{ds} = \frac{\lambda L}{\varepsilon_o} \tag{2.102}$$

where $q_{enc} = \lambda L$ is the total charge enclosed by the Gaussian surface. End face 1, 2 and curved face of the cylindrical Gaussian surface are shown in Fig. 2.27b.

At the end faces the angle between **E** and **ds** is 90° and at the curved surface the angle between **E** and **ds** is zero degrees

Hence Eq. 2.102 is

$$\int_{\text{Endface1}} \mathrm{E\,ds} \cos 90^\circ + \int_{\text{Curvedface}} \mathrm{E\,ds} \cos 0^\circ + \int_{\text{Endface2}} \mathrm{E\,ds} \cos 90^\circ = \frac{\lambda L}{\varepsilon_o} \tag{2.103}$$

As $\cos 90^\circ = 0$, $\cos 0^\circ = 1$ Eq. 2.103 is

$$\int_{\text{Curvedface}} \mathrm{E\,ds} = \frac{\lambda L}{\varepsilon_o} \tag{2.104}$$

From Fig. 2.27b it is clear that E is constant over the curved surface and hence

$$\mathrm{E} \int_{\text{Curvedface}} \mathrm{ds} = \frac{\lambda L}{\varepsilon_o} \tag{2.105}$$

Now $\int_{\text{Curvedface}}$ ds is the area of the curved face which is equal to $2\pi rL$. Hence

$$\Rightarrow \mathrm{E}(2\pi rL) = \frac{\lambda L}{\varepsilon_o} \tag{2.106}$$

$$\mathrm{E} = \frac{\lambda}{2\pi r\varepsilon_o} \tag{2.107}$$

While calculating electric field using Coulomb's law the $E_y$ components cancelled out as shown in Fig. 2.26 and hence the difficulty involved in calculating electric field using Coulomb's law is less obvious in the above example. However future examples will show that Gauss's law eases out the calculation of electric field as compared to Coulomb's law.

(c) **Using Potential**

Let us redraw Fig. 2.26 as shown in Fig. 2.28. Let us calculate the potential at point P in Fig. 2.28 due to a small element $dy'$ which carries a charge of $\lambda dy'$. We have

**Fig. 2.28** .

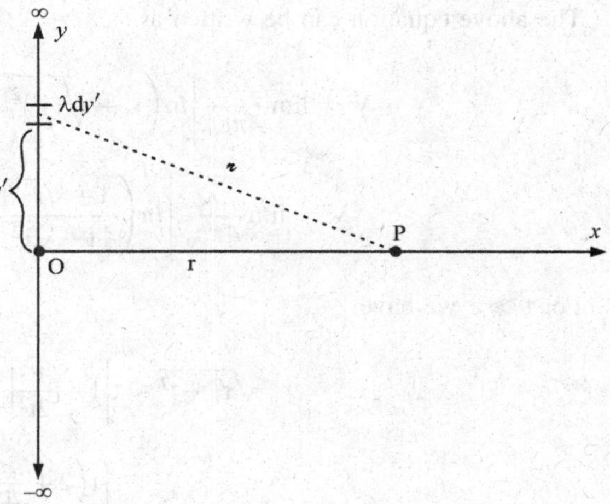


$$V = \frac{1}{4\pi\varepsilon_o}\frac{\lambda dy'}{\mathscr{r}} \tag{2.108}$$

The total potential due to the entire line charge is

$$V = \frac{1}{4\pi\varepsilon_o}\int_{-\infty}^{\infty}\frac{\lambda dy'}{\mathscr{r}} \tag{2.109}$$

From Fig. 2.28 $\mathscr{r}^2 = (y')^2 + r^2$. Hence

$$V = \frac{1}{4\pi\varepsilon_o}\int_{-\infty}^{\infty}\frac{\lambda dy'}{\sqrt{(y')^2 + r^2}} \tag{2.110}$$

$$V = \frac{\lambda}{4\pi\varepsilon_o}\left[\ln\left(y' + \sqrt{(y')^2 + r^2}\right)\right]_{-\infty}^{\infty} \tag{2.111}$$

$$V = \infty$$

As previously mentioned if the charge distribution extends to infinity, then the potential calculated using Eq. 2.80 goes to infinity. However we can calculate the potential difference between two points. We have

$$V = \frac{\lambda}{4\pi\varepsilon_o}\left[\ln\left(y' + \sqrt{(y')^2 + r^2}\right)\right]_{-\infty}^{\infty} \tag{2.112}$$

The above equation can be written as

$$V = \lim_{t\to\infty} \frac{\lambda}{4\pi\varepsilon_o} \left[ ln\left( y' + \sqrt{(y')^2 + r^2} \right) \right]_{-t}^{t} \tag{2.113}$$

$$V = \lim_{t\to\infty} \frac{\lambda}{4\pi\varepsilon_o} \left[ ln\left( \frac{t + \sqrt{r^2 + t^2}}{-t + \sqrt{r^2 + t^2}} \right) \right] \tag{2.114}$$

For $t \gg r$ we have

$$\sqrt{r^2 + t^2} \approx t\left|1 + \frac{r^2}{2t^2}\right|$$

$$V = \lim_{t\to\infty} \frac{\lambda}{4\pi\varepsilon_o} ln\left[ \frac{t\left(2 + \frac{r^2}{2t^2}\right)}{r^2/2t^2} \right] \tag{2.115}$$

Consider a reference point $r_o$. The potential at the reference point $r_o$ is

$$V_o = \lim_{t\to\infty} \frac{\lambda}{4\pi\varepsilon_o} ln\left[ \frac{t\left(2 + \frac{r_o^2}{2t^2}\right)}{r_0^2/2t^2} \right] \tag{2.116}$$

$$V - V_o = \lim_{t\to\infty} \frac{\lambda}{4\pi\varepsilon_o} ln\left[ \frac{t\left(2 + \frac{r^2}{2t^2}\right)}{r^2/2t^2} \cdot \frac{r_0^2/2t^2}{t\left(2 + \frac{r_o^2}{2t^2}\right)} \right] \tag{2.117}$$

$$V - V_o = \lim_{t\to\infty} ln\left[ \frac{r_o^2}{r^2} \cdot \frac{\left(2 + \frac{r^2}{2t^2}\right)}{\left(2 + \frac{r_o^2}{2t^2}\right)} \right] \tag{2.118}$$

Applying the limit

$$V - V_o = \frac{\lambda}{4\pi\varepsilon_o} ln \frac{r_o^2}{r^2} \tag{2.119}$$

$$V - V_o = \frac{\lambda}{2\pi\varepsilon_o} ln\left(\frac{r}{r_o}\right) \tag{2.120}$$

Now the electric field is

$$\mathbf{E} = -\nabla(V - V_o) = -\frac{\partial(V - V_o)}{\partial r} \tag{2.121}$$

$$E = \frac{\partial}{\partial r}\left[\frac{\lambda}{4\pi\varepsilon_o} ln\left(\frac{r}{r_o}\right)\right] \tag{2.122}$$

$$E = \frac{\lambda}{2\pi\varepsilon_o}\frac{1}{r/r_o}\frac{1}{r_o} \tag{2.123}$$

$$\Rightarrow E = \frac{\lambda}{2\pi\varepsilon_o r} \tag{2.124}$$

The reason for using potential difference instead of potential to calculate **E** and the physical significance of the reference point will be clear to the reader in future sections.

## 2.14 Electric Field Due to the Finite Line Charge

(a) **Using Coulomb's Law**

Consider a uniformly charged wire of length L, the charge per unit length being λ. Let us calculate the electric field at point P as shown in Fig. 2.29. The coordinates of point P are $(x, y)$. Let us divide the entire line charge into small elements $dx'$ each carrying a charge of $\lambda dx'$. We will calculate the field produced by the small element $\lambda dx'$ and then integrate over the entire length to get the field of the entire line charge. The point of intersection of point P on the line charge is at a distance of $x$ from one end of the line charge, and the element $dx'$ is situated at a distance of $x'$ from origin O as shown in Fig. 2.29. The electric field produced by the element $dx'$ is $d\mathbf{E}$. Let us resolve $d\mathbf{E}$ into $x$ and $y$ components $d\mathbf{E}_x$ and $d\mathbf{E}_y$, respectively. From Fig. 2.29 we can write

$$dE_x = \frac{1}{4\pi\varepsilon_o}\frac{\lambda dx'}{\mathscr{r}^2}\cos\theta \tag{2.125}$$

But from Fig. 2.29

$$\mathscr{r}^2 = y^2 + (x - x')^2 \tag{2.126}$$

$$\cos\theta = \frac{x - x'}{\sqrt{y^2 + (x - x')^2}} \tag{2.127}$$

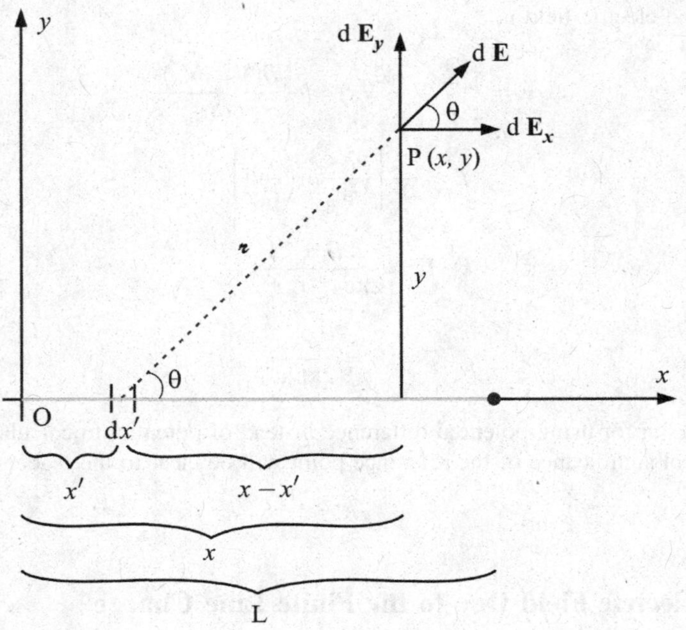


**Fig. 2.29** Finite line of charge

Substituting Eqs. 2.126 and 2.127 in 2.125

$$dE_x = \frac{1}{4\pi\varepsilon_o}\frac{\lambda dx'}{\left[y^2+(x-x')^2\right]}\frac{(x-x')}{\sqrt{y^2+(x-x')^2}} \tag{2.128}$$

$$dE_x = \frac{1}{4\pi\varepsilon_o}\frac{\lambda(x-x')dx'}{\left[y^2+(x-x')^2\right]^{3/2}} \tag{2.129}$$

The $x$ component of the electric field due to the entire line charge of length L is

$$E_x = \frac{\lambda}{4\pi\varepsilon_o}\int_0^L \frac{(x-x')dx'}{\left[y^2+(x-x')^2\right]^{3/2}} \tag{2.130}$$

$$E_x = \frac{\lambda}{4\pi\varepsilon_o}\left[\frac{1}{\sqrt{y^2+(x-L)^2}} - \frac{1}{\sqrt{x^2+y^2}}\right] \tag{2.131}$$

Now let us calculate the $y$ component

$$dE_y = \frac{1}{4\pi\varepsilon_0}\frac{\lambda dx'}{r^2}\sin\theta \tag{2.132}$$

From Fig. 2.29

$$\sin\theta = \frac{y}{\sqrt{y^2 + (x - x')^2}} \tag{2.133}$$

Substituting Eqs. 2.126 and 2.133 in Eq. 2.132

$$dE_y = \frac{1}{4\pi\varepsilon_0}\frac{\lambda dx'}{\left[y^2 + (x - x')^2\right]}\cdot\frac{y}{\sqrt{y^2 + (x - x')^2}} \tag{2.134}$$

$$dE_y = \frac{1}{4\pi\varepsilon_0}\frac{\lambda y dx'}{\left[y^2 + (x - x')^2\right]^{3/2}} \tag{2.135}$$

The $y$ component of the electric field due to the entire line charge of length L is

$$E_y = \frac{\lambda y}{4\pi\varepsilon_0}\int_0^L \frac{dx'}{\left[y^2 + (x - x')^2\right]^{3/2}} \tag{2.136}$$

$$E_y = \frac{\lambda}{4\pi\varepsilon_0 y}\left[\frac{L - x}{\sqrt{y^2 + (x - L)^2}} + \frac{x}{\sqrt{y^2 + L^2}}\right] \tag{2.137}$$

The net electric field at point P due to the entire line charge is then given by

$$\mathbf{E} = E_x\hat{\mathbf{i}} + E_y\hat{\mathbf{j}} \tag{2.138}$$

where $E_x$ and $E_y$ are given by Eqs. 2.131 and 2.137, respectively.

(b) **Using Gauss's Law**

Step one is to plot the electric field or "to get a feel of the field". For comparison purposes let us sketch the field due to the finite and infinite line charge.

Initially we show a line charge with two elements λdx' in Fig. 2.30a. At point P the $\mathbf{E}_y$ components cancel out each other leaving only $\mathbf{E}_x$ components.

Now let us consider the infinite line charge as shown in Fig. 2.30b. In this case because the line charge is infinite, for each line element $\lambda dx'$ there will be always

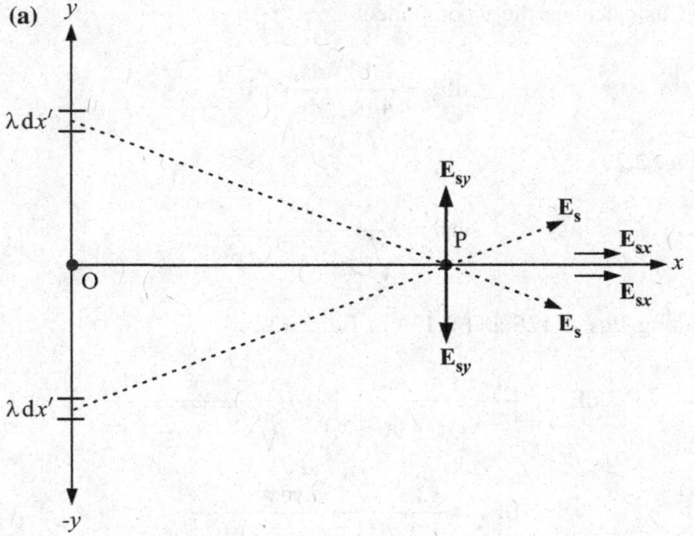


**Fig. 2.30a** Line charge with two line elements.

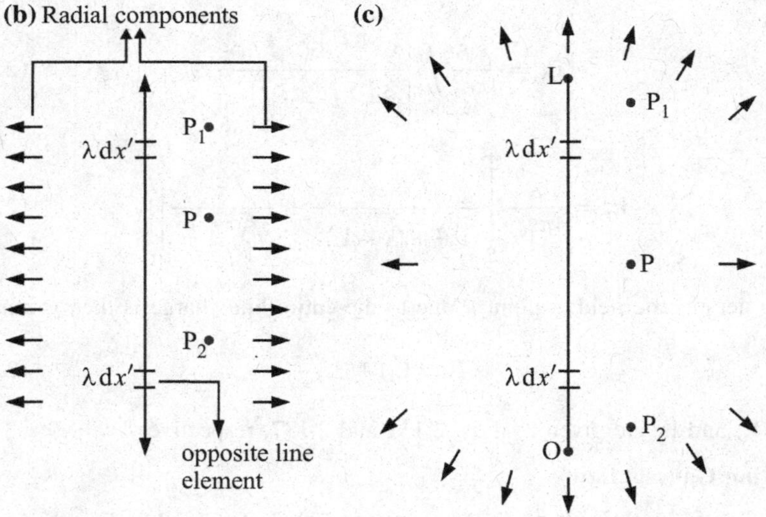


**Fig. 2.30b, c** **b** Electric field of a infinite line charge. **c** Electric field of a finite line charge

opposite line element $\lambda dx'$ such that $\mathbf{E}_y$ components cancel out and $\mathbf{E}_x$ components alone add. Hence it is either point P or $P_1$ or $P_2$ or any other point in Fig. 2.30b and the field is always radial as shown in Fig. 2.30b because at every point in Fig. 2.30b the $\mathbf{E}_y$ components will cancel out and $\mathbf{E}_x$ components will add up. Hence radial components are alone present in Fig. 2.30b. Therefore we were able to construct a cylindrical Gaussian surface in Fig. 2.27b such that the electric field **E**·**ds** can be easily evaluated for the cylindrical Gaussian surface and also **E** is constant over the cylindrical Gaussian surface. Thus using Gauss's law we calculated **E** for infinite line charge in Sect. 2.13b.

Now let us consider finite line charge. In Fig. 2.30c we see that at point P the $E_y$ components cancel out and $E_x$ components are alone present. Thus field at point P in Fig. 2.30c is radial. However at points like $P_1$ and $P_2$ the $E_y$ components will not cancel out. Hence the field will not be radial at points like $P_1$ and $P_2$ as shown in Fig. 2.30c. Now let us construct a cylindrical Gaussian surface for the finite line charge as shown in Fig. 2.31. Clearly at point like P, **E** and **ds** are in the same direction. But at points like $P_1$, $P_2$, **E** makes an angle with **ds** as shown in Fig. 2.31. Thus at edges of the cylindrical Gaussian surface **E** and **ds** have different directions.

Hence at the edges or at any other point where **E** and **d*s*** are at some angle to each other evaluation of **E**·**d*s*** is difficult over the entire Gaussian surface. Even if we somehow manage to calculate **E**·**d*s***, $E\cos\theta$ is not constant over the Gaussian surface and hence taking $E\cos\theta$ out of the integral $\oint \mathbf{E}\cdot\mathbf{ds} = \oint E ds\cos\theta$ becomes difficult. It is still true that over the cylindrical Gaussian surface

$$\oint \mathbf{E}\cdot\mathbf{ds} = \frac{q_{enc}}{\varepsilon_0}$$

But however evaluation of $\mathbf{E}\cdot\mathbf{ds}$ and taking $E\cos\theta$ out of the integral $\oint \mathbf{E}\cdot\mathbf{ds}$ are difficult. Hence using Gauss's law to calculate electric field in the above case, finite line charge, becomes tedious. Instead of selecting a cylindrical Gaussian

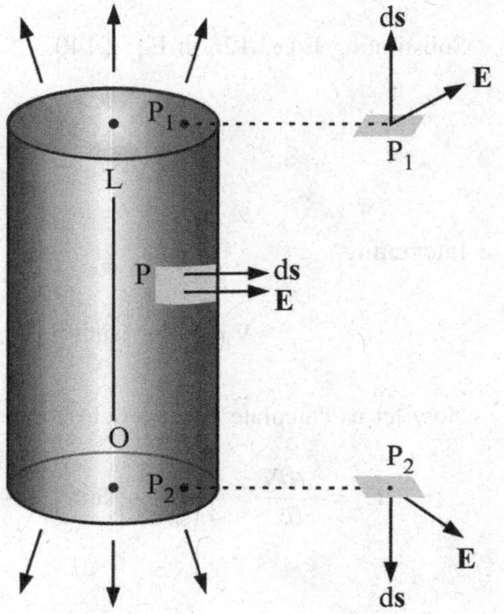


**Fig. 2.31** Cylindrical Gaussian surface for finite line of charge

surface we can select a Gaussian surface of different shapes. However until $\mathbf{E} \cdot \mathbf{ds}$ can be easily evaluated and E cos θ can be taken out of the integral $\oint \mathbf{E} \cdot \mathbf{ds}$, easy evaluation of **E** using Gauss's law is not possible.

Thus we have seen using Gauss's law evaluation of **E** for a finite line charge is very difficult.

"The same is true for any charge distribution. For a given charge distribution if we are interested in using Gauss's law to calculate **E**, then we must be able to construct a Gaussian surface such that $\mathbf{E} \cdot \mathbf{ds}$ must be evaluated easily and E cos θ must be constant over Gaussian surface so that E cos θ can be pulled out of $\oint \mathbf{E} \cdot \mathbf{ds}$. If we are unable to construct a Gaussian surface which satisfies the above requirement, then Gauss's law cannot be easily applied to calculate **E**. It is still true that $\oint \mathbf{E} \cdot \mathbf{ds} = \frac{q_{enc}}{\varepsilon_o}$ over the Gaussian surface, but Gauss's law is not useful in calculating **E**.

(c) **Using Potential**

The potential at point P in Fig. 2.29 due to the element $dx'$ which carries a charge of $\lambda dx'$ is

$$dV = \frac{1}{4\pi\varepsilon_o}\frac{\lambda dx'}{\mathscr{r}} \tag{2.139}$$

The potential at point P due to the entire line charge in Fig. 2.29 is

$$V = \frac{1}{4\pi\varepsilon_o}\int_0^L \frac{\lambda dx'}{\mathscr{r}} \tag{2.140}$$

Substituting Eq. 2.126 in Eq. 2.140

$$V = \frac{\lambda}{4\pi\varepsilon_o}\int_0^L \frac{dx'}{\sqrt{y^2 + (x - x')^2}} \tag{2.141}$$

Integrating

$$V = \frac{\lambda}{4\pi\varepsilon_o}\left[\sinh^{-1}\left(\frac{x - L}{y}\right) - \sinh^{-1}\frac{x}{y}\right] \tag{2.142}$$

Now let us calculate electric field components

$$E_x = \frac{-\partial V}{\partial x} = \frac{\lambda}{4\pi\varepsilon_0}\frac{\partial}{\partial x}\left[\sinh^{-1}\left(\frac{x - L}{y}\right) - \sinh^{-1}\left(\frac{x}{y}\right)\right] \tag{2.143}$$

$$E_x = \frac{\lambda}{4\pi\varepsilon_o}\left[\frac{1}{\sqrt{\left(\frac{x-L}{y}\right)^2+1}}\frac{1}{y} - \frac{1}{\sqrt{\left(\frac{x}{y}\right)^2+1}}\frac{1}{y}\right] \tag{2.144}$$

$$E_x = \frac{\lambda}{4\pi\varepsilon_o}\left[\frac{1}{\sqrt{(x-L)^2+y^2}} - \frac{1}{\sqrt{x^2+y^2}}\right] \tag{2.145}$$

Now the $E_y$ component is

$$E_y = \frac{-\partial V}{\partial y} = \frac{\lambda}{4\pi\varepsilon_o}\frac{\partial}{\partial y}\left[\sin h^{-1}\left(\frac{x-L}{y}\right) - \sin h^{-1}\left(\frac{x}{y}\right)\right] \tag{2.146}$$

$$E_y = \frac{\lambda}{4\pi\varepsilon_o}\left[\frac{-1}{\sqrt{\frac{(x-L)^2}{y^2}+1}}\frac{x-L}{y^2} - \frac{1}{\sqrt{\frac{x^2}{y^2}+1}}\left(\frac{-x}{y}\right)\right] \tag{2.147}$$

$$E_y = \frac{\lambda}{4\pi\varepsilon_o y}\left[\frac{L-x}{\sqrt{(x-L)^2+y^2}} + \frac{x}{\sqrt{x^2+y^2}}\right] \tag{2.148}$$

Now **E** is given by

$$\mathbf{E} = E_x\hat{\mathbf{i}} + E_y\hat{\mathbf{j}} \tag{2.149}$$

where $E_x$, $E_y$ the components are given by Eqs. 2.145, 2.148.

## 2.15 Electric Field Along the Axis of a Uniformly Charged Circular Disc

Consider a circular disc of radius R on which charge is uniformly distributed with surface charge density $\sigma$. Let us calculate the electric field at point P at a distance $y$ from the centre of the disc as shown in Fig. 2.32.

(a) **Using Coulomb's Law**

We have calculated the electric field for the above problem in Example 2.7 using Coulomb's law.

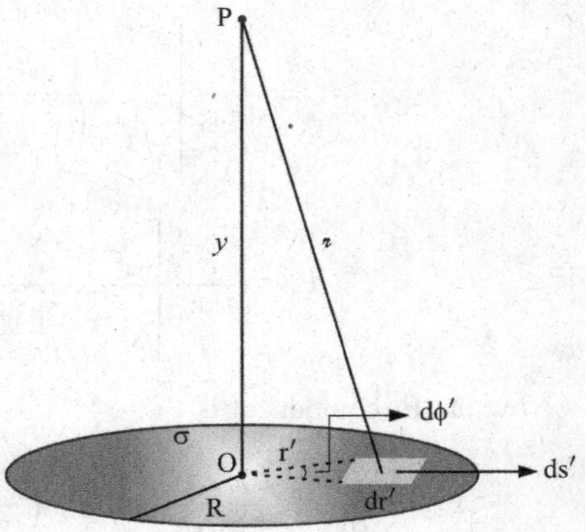


**Fig. 2.32** Circular disc carrying uniform surface charge density

(b) **Using Gauss's Law**

In Fig. 2.33 field distribution for a uniformly charged disc at few points is shown. Clearly construction of a Gaussian surface (over which $\mathbf{E}\cdot\mathbf{ds}$ can be easily evaluated) is not that easy. Hence for similar reasons mentioned in Sect. 2.14b Gauss's law is not useful here in calculating **E**.

(c) **Using Potential**

Let us calculate the electric field at point P in Fig. 2.32 using potential. Consider a surface ds′ element situated at a distance of $\mathscr{r}$ from point P. The potential due to the surface element at point P is given by

$$V = \frac{1}{4\pi\varepsilon_0}\int\limits_{S'}\frac{\sigma ds'}{\mathscr{r}} \tag{2.150}$$

From Fig. 2.32

$$ds' = r'\,dr'\,d\phi' \tag{2.151}$$

$$\mathscr{r} = \sqrt{y^2 + (r')^2} \tag{2.152}$$

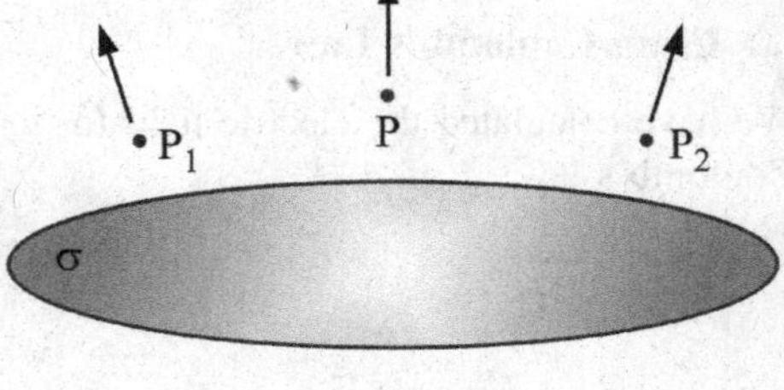


**Fig. 2.33** Electric field distribution at few points for a circular disc carrying a uniform surface charge density

Substituting Eqs. 2.151 and 2.152 in Eq. 2.150

$$V = \frac{1}{4\pi\varepsilon_0}\int_0^{2\pi}\int_0^{R}\frac{\sigma r'\,dr'\,d\phi'}{\sqrt{y^2+(r')^2}} \tag{2.153}$$

$$V = \frac{2\pi\sigma}{4\pi\varepsilon_0}\int_0^{R}\frac{r'\,dr'}{\sqrt{y^2+(r')^2}} \tag{2.154}$$

$$V = \frac{\sigma}{2\varepsilon_0}\left[\sqrt{y^2+R^2}-y\right] \tag{2.155}$$

$$\mathbf{E} = -\nabla V = -\left[\hat{\mathbf{i}}\frac{\partial}{\partial x}+\hat{\mathbf{j}}\frac{\partial}{\partial y}\right]\left[\frac{\sigma}{2\varepsilon_0}\left(\sqrt{y^2+R^2}-y\right)\right] \tag{2.156}$$

As $\dfrac{\partial V}{\partial x} = 0$

$$\mathbf{E} = -\frac{\partial}{\partial y}\left[\frac{\sigma}{2\varepsilon_0}\left(\sqrt{y^2+R^2}-y\right)\right]\hat{\mathbf{j}} \tag{2.157}$$

$$\mathbf{E} = \frac{\sigma}{2\varepsilon_0}\left[1-\frac{y}{\sqrt{y^2+R^2}}\right]\hat{\mathbf{j}} \tag{2.158}$$

**Note** Integration here is carried from 0 to $2\pi$, that is over the entire disc, because we consider only one surface element $ds'$. In Example 2.7 we carry out integration from 0 to $\pi$ for $\phi'$, that is over only half of the disc, because we consider two surface elements.

## 2.16 Electric Field Due to an Infinite Plane Sheet of Charge

Consider an infinite plane sheet of charge carrying a uniform surface charge density $\sigma$ as shown in Fig. 2.34. We are interested in calculating the electric field at point P. We will use cylindrical coordinates (r, $\phi$, $z$). Consider two strips situated opposite to each other. The electric field $\mathbf{E_s}$ due to each strip is shown in the figure. Resolving the $\mathbf{E_s}$ into components $\mathbf{E_{sr}}$, $\mathbf{E_{sz}}$ we see that $\mathbf{E_{sr}}$ components cancel out while $\mathbf{E_{sz}}$ components add up.

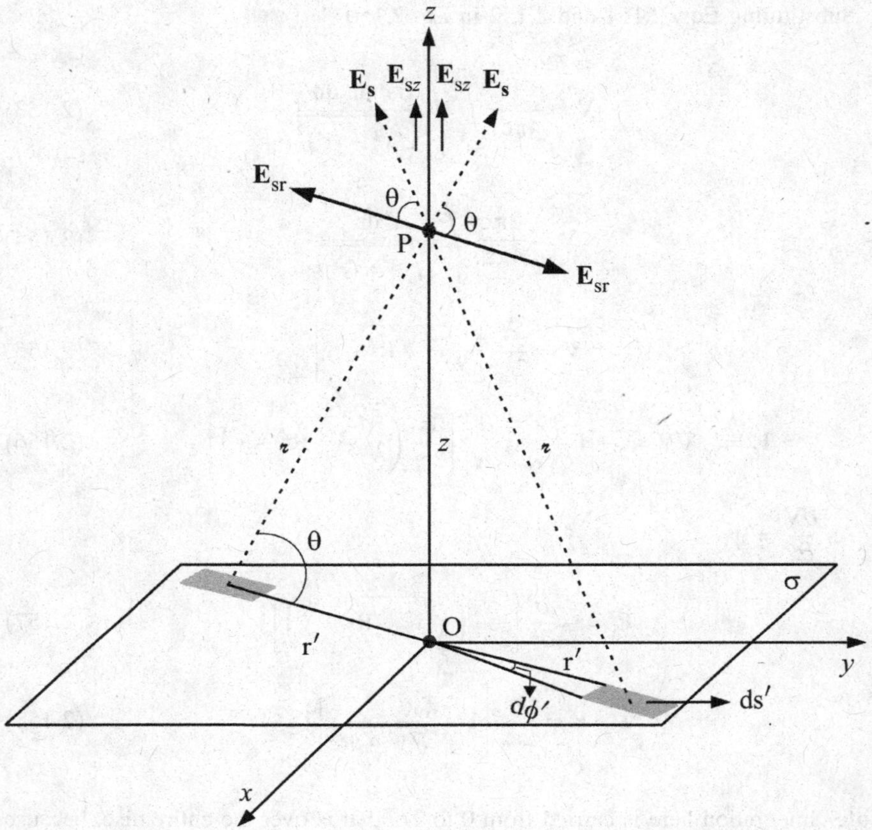


**Fig. 2.34** Infinite plane sheet carrying a uniform surface charge density

Hence the net field due to the two strips at point P is

$$\mathbf{dE} = 2E_{sz}\,\hat{\mathbf{z}} \tag{2.159}$$

From Fig. 2.34

$$\sin\theta = \frac{E_{sz}}{E_s}$$

$$\Rightarrow E_{sz} = E_s \sin\theta \tag{2.160}$$

The field due to single strip ds′ at point P is

$$E_s = \frac{1}{4\pi\varepsilon_0}\frac{\sigma ds'}{\mathscr{r}^2} \tag{2.161}$$

From Fig. 2.34

$$\sin\theta = \frac{z}{\mathscr{r}}$$

But

$$\mathscr{r} = \sqrt{z^2 + (r')^2} \tag{2.162}$$

Hence

$$\sin\theta = \frac{z}{\sqrt{z^2 + (r')^2}} \tag{2.163}$$

Substituting Eqs. 2.161–2.163 in Eq. 2.159

$$\mathbf{dE} = 2E_s \sin\theta\, \hat{\mathbf{z}}$$

$$\mathbf{dE} = 2\frac{1}{4\pi\varepsilon_0}\frac{\sigma ds'}{\mathscr{r}^2}\sin\theta\,\hat{\mathbf{z}} \tag{2.164}$$

$$\mathbf{dE} = 2\frac{1}{4\pi\varepsilon_0}\frac{\sigma ds'}{\left[z^2 + (r')^2\right]}\frac{z}{\sqrt{z^2 + (r')^2}}\hat{\mathbf{z}} \tag{2.165}$$

$$\mathbf{dE} = \frac{\sigma z}{2\pi\varepsilon_0}\frac{ds'}{\left[z^2 + (r')^2\right]^{3/2}}\hat{\mathbf{z}} \tag{2.166}$$

The field due to the entire plane sheet of charge is then

$$\mathbf{E} = \frac{\sigma z}{2\pi\varepsilon_0}\int\limits_{S'}\frac{ds'}{\left[z^2 + (r')^2\right]^{3/2}}\hat{\mathbf{z}} \tag{2.167}$$

$$\mathbf{E} = \frac{\sigma z}{2\pi\varepsilon_0}\int\limits_0^{\pi}\int\limits_0^{\infty}\frac{r'\,dr'\,d\phi'}{\left[(r')^2 + z^2\right]^{3/2}}\hat{\mathbf{z}} \tag{2.168}$$

$$\mathbf{E} = \frac{\sigma z}{2\varepsilon_0} \int_0^\infty \frac{r'\,dr'}{\left[(r')^2 + z^2\right]^{3/2}} \hat{\mathbf{z}} \tag{2.169}$$

$$\mathbf{E} = \frac{\sigma z}{2\varepsilon_0} \left[ \frac{-1}{\left((r')^2 + z^2\right)^{1/2}} \right]_0^\infty \hat{\mathbf{z}} \tag{2.170}$$

$$\mathbf{E} = \frac{\sigma z}{2\varepsilon_0} \frac{1}{z} \hat{\mathbf{z}} \tag{2.171}$$

$$\mathbf{E} = \frac{\sigma}{2\varepsilon_0} \hat{\mathbf{z}} \tag{2.172}$$

(a) **Using Gauss's Law**

Now let us calculate the electric field of an infinite plane sheet of charge using Gauss's law. Step one is to get feel of the field. As shown in Figs. 2.34 and 2.35a the field due to a infinite plane sheet of charge points in $z$ direction.

Hence as shown in Fig. 2.35b we construct a Gaussian surface in the shape of pillbox with surface area S. Half of the pillbox lies above the plane sheet of charge, and half of the pillbox lies below the plane sheet of charge. Bottom half of the pillbox is not shown in the figure.

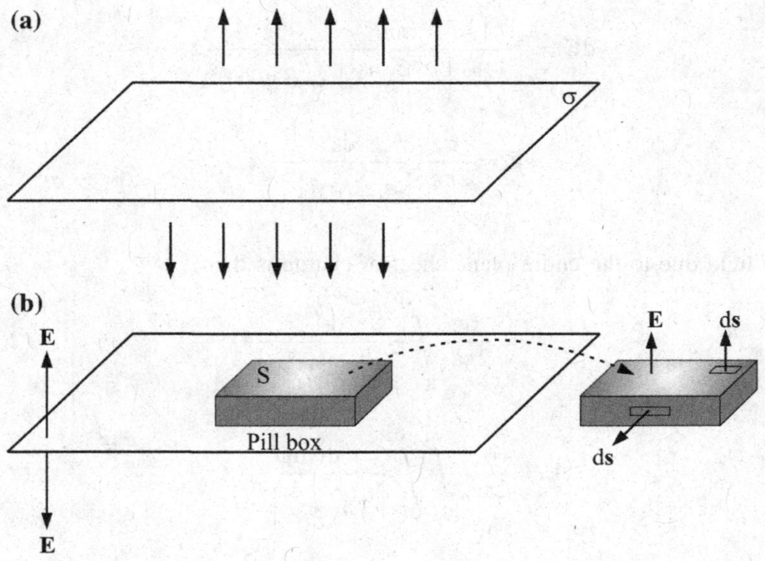


**Fig. 2.35** **a** Electric field lines for infinite plane sheet carrying uniform surface charge density. **b** Gaussian Surface for infinite plane sheet carrying uniform surface charge density

As per Gauss's law

$$\oint \mathbf{E}\cdot d\mathbf{s} = \frac{q_{enc}}{\varepsilon_o}$$

The above integral can be written for the pillbox as

$$\left|\int \mathbf{E}\cdot d\mathbf{s}\right|_{top} + \left|\int \mathbf{E}\cdot d\mathbf{s}\right|_{bottom} + \left|\int \mathbf{E}\cdot d\mathbf{s}\right|_{sides} = \frac{q_{enc}}{\varepsilon_o} \tag{2.173}$$

For the sides of two pillboxes as shown in Fig. 2.35b the angle between **E** and ds is 90°. Hence **E** · ds = 0 for sides of the pillbox. Equation 2.173 is then

$$\left|\int \mathbf{E}\cdot d\mathbf{s}\right|_{top} + \left|\int \mathbf{E}\cdot d\mathbf{s}\right|_{bottom} = \frac{q_{enc}}{\varepsilon_o} \tag{2.174}$$

For the top and bottom of pillbox the angle between **E** and **ds** is zero. Hence

$$\left|\int_S |\mathbf{E}||d\mathbf{s}|\cos 0\right|_{top} + \left|\int_S |\mathbf{E}||d\mathbf{s}|\cos 0\right|_{bottom} = \frac{q_{enc}}{\varepsilon_o} \tag{2.175}$$

$$\Rightarrow \int_S |\mathbf{E}||d\mathbf{s}| + \int_S |\mathbf{E}||d\mathbf{s}| = \frac{q_{enc}}{\varepsilon_o} \tag{2.176}$$

Because |**E**| is constant over the top and bottom faces of the pillbox

$$|\mathbf{E}|\int_S |d\mathbf{s}| + |\mathbf{E}|\int_S |d\mathbf{s}| = \frac{q_{enc}}{\varepsilon_o} \tag{2.177}$$

$$ES + ES = \frac{q_{enc}}{\varepsilon_o} \tag{2.178}$$

From Fig. 2.35 $q_{enc} = \sigma S$ hence

$$2ES = \frac{\sigma S}{\varepsilon_o} \tag{2.179}$$

$$\Rightarrow E = \frac{\sigma}{2\varepsilon_o} \tag{2.180}$$

$$\Rightarrow \mathbf{E} = \frac{\sigma}{2\varepsilon_o}\hat{\mathbf{z}} \tag{2.181}$$

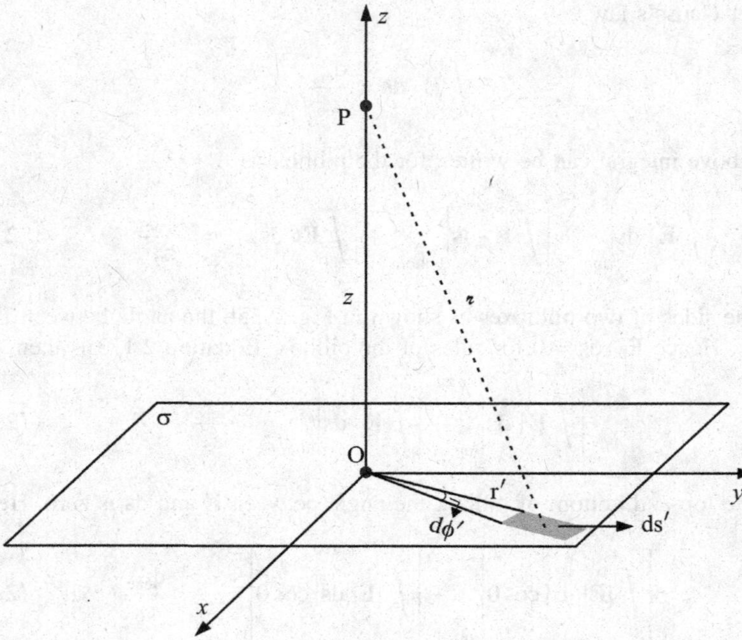


**Fig. 2.36** Infinite plane sheet carrying a uniform surface charge density

(b) **Using Potential**

The potential at point P due to entire plane in Fig. 2.36 is

$$V = \frac{1}{4\pi\varepsilon_0} \int_{S'} \frac{\sigma ds'}{ᴚ} \tag{2.182}$$

$$V = \frac{\sigma}{4\pi\varepsilon_0} \int_0^{2\pi} \int_0^{\infty} \frac{r' \, dr' \, d\phi'}{\sqrt{(r')^2 + z^2}} \tag{2.183}$$

by substituting Eq. 2.162.

$$V = \frac{\sigma}{2\varepsilon_0} \int_0^{\infty} \frac{r' \, dr'}{\sqrt{(r')^2 + z^2}} \tag{2.184}$$

The above equation can be written as

$$V = \lim_{t\to\infty} \frac{\sigma}{2\varepsilon_o} \int_0^t \frac{r'\,dr'}{\sqrt{(r')^2 + z^2}} \tag{2.185}$$

$$V = \lim_{t\to\infty} \frac{\sigma}{2\varepsilon_0} \left[\sqrt{(r')^2 + z^2}\right]_0^t \tag{2.186}$$

$$V = \lim_{t\to\infty} \frac{\sigma}{2\varepsilon_0} \left[\sqrt{t^2 + z^2} - z\right] \tag{2.187}$$

As $t \gg z$,

$$\sqrt{t^2 + z^2} \approx t\left[1 + \frac{z^2}{2t^2}\right]$$

Hence

$$V = \lim_{t\to\infty} \frac{\sigma}{2\varepsilon_0} \left[t\left[1 + \frac{z^2}{2t^2}\right] - z\right] \tag{2.188}$$

$$V = \lim_{t\to\infty} \frac{\sigma}{2\varepsilon_0} \left[t + \frac{z^2}{2t} - z\right] \tag{2.189}$$

Consider a reference point $z_0$. The potential at reference point $z_0$ is

$$V_o = \lim_{t\to\infty} \frac{\sigma}{2\varepsilon_0} \left[t + \frac{z_o^2}{2t} - z_o\right] \tag{2.190}$$

The potential difference between points $z$ and $z_0$ is

$$V - V_o = \frac{\sigma}{2\varepsilon_0} \lim_{t\to\infty} \left[\frac{z^2}{2t} - \frac{z_o^2}{2t} + z_o - z\right] \tag{2.191}$$

Applying the limit

$$V - V_o = \frac{\sigma}{2\varepsilon_0} [z_o - z] \tag{2.192}$$

Now the electric field is

$$\mathbf{E} = -\nabla(V - V_o) \tag{2.193}$$

In cylindrical coordinates

$$\mathbf{E} = -\left[\frac{\partial(V - V_o)}{\partial r}\hat{\boldsymbol{r}} + \frac{1}{r}\frac{\partial(V - V_o)}{\partial \phi}\hat{\boldsymbol{\phi}} + \frac{\partial(V - V_o)}{\partial z}\hat{\mathbf{z}}\right] \tag{2.194}$$

$$E = \frac{-\partial(V - V_o)}{\partial z}\hat{\mathbf{z}} \tag{2.195}$$

as $V - V_o$ in Eq. 2.192 doesn't contain terms with r, $\phi$.

Substituting Eq. 2.192 in 2.195

$$E = \frac{\sigma}{2\varepsilon_o}\hat{\mathbf{z}} \tag{2.196}$$

The reason for using potential difference instead of potential to calculate electric field and the physical significance of reference point $z_o$ will be clear to the reader in future sections.

## 2.17 Electric Field of a Uniformly Charged Spherical Shell

Consider a spherical shell of radius R as shown in Fig. 2.37. Charge q is distributed uniformly on the shell with surface charge density $\sigma$.

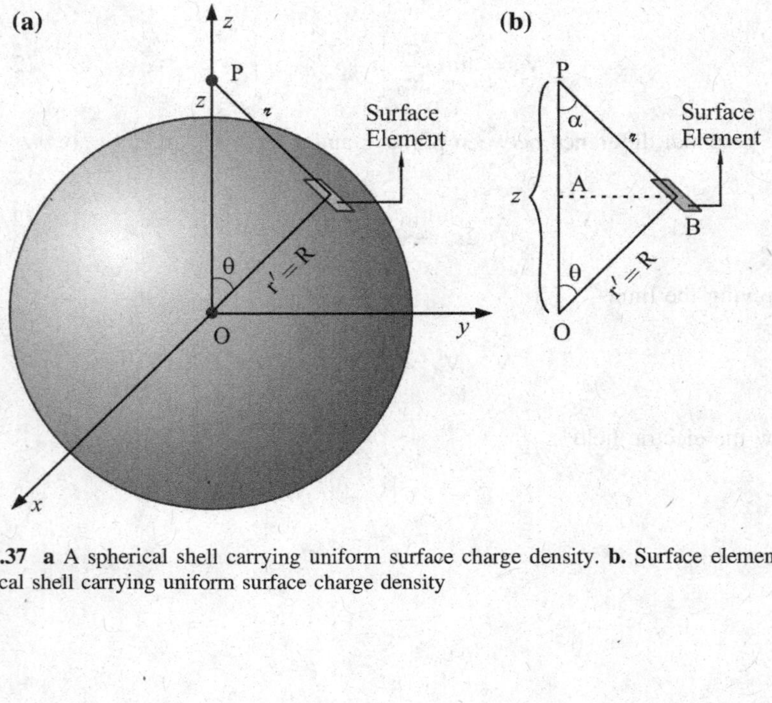


**Fig. 2.37** **a** A spherical shell carrying uniform surface charge density. **b.** Surface element of a spherical shell carrying uniform surface charge density

**Fig. 2.38** Two opposite surface elements of a spherical shell carrying uniform surface charge density

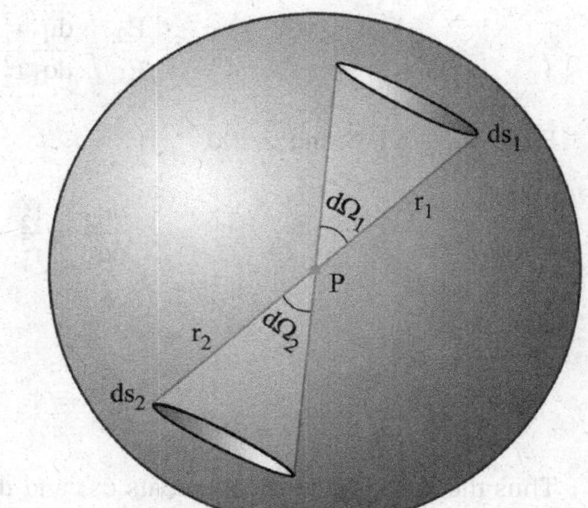


(a) **Using Coulomb's Law**

(i) ***At a point interior to the spherical shell***

Consider point P in the interior of the spherical shell as shown in Fig. 2.38. Consider two cones with point P as their apex extending up to the surface of the sphere where they cut the sphere. Cone 1 and cone 2 subtend a solid angle of $d\Omega_1$ and $d\Omega_2$, respectively. Both the solid angles are equal

$$d\Omega_1 = d\Omega_2 \tag{2.197}$$

$$\Rightarrow \frac{ds_1}{r_1^2} = \frac{ds_2}{r_2^2} \tag{2.198}$$

$$\Rightarrow \frac{ds_2}{ds_1} = \frac{r_2^2}{r_1^2} \tag{2.199}$$

Because the spherical shell is uniformly charged, the charge on each element is proportional to the area. Hence,

$$\frac{dq_2}{dq_1} = \frac{ds_2}{ds_1} \tag{2.200}$$

The ratio between electric fields at point P due to the elements $ds_1$ and $ds_2$ is

$$\frac{E_2}{E_1} = \frac{\frac{1}{4\pi\varepsilon_0}\frac{dq_2}{r_2^2}}{\frac{1}{4\pi\varepsilon_0}\frac{dq_1}{r_1^2}} \tag{2.201}$$

$$\frac{E_2}{E_1} = \frac{dq_2\, r_1^2}{dq_1\, r_2^2} \tag{2.202}$$

From Eqs. 2.199 and 2.200

$$\frac{dq_2}{dq_1} = \frac{r_2^2}{r_1^2} \tag{2.203}$$

Hence

$$\frac{E_2}{E_1} = 1 \tag{2.204}$$

Thus the field due to the elements $ds_1$ and $ds_2$ at point P is equal in magnitude. From Fig. 2.38 the field due to $ds_1$ and $ds_2$ at point P is oppositely directed. Hence the field at point P is zero. This is true for any point in the interior of the sphere.

(ii) ***At a point exterior to the spherical shell***

Consider Fig. 2.37a. Clearly the *x*, *y* components of electric field at point P due to the symmetrically placed surface elements cancel out and only the *z* components add up. Thus we have to integrate over the *z* components, $dE_s \cos\alpha$ of the elements, and here $dE_s$ is field produced by the surface element at P given by

$$dE_s = \frac{1}{4\pi\varepsilon_0}\frac{dq}{\mathscr{r}^2} \tag{2.205}$$

We have

$$dq = \sigma\, ds' = \sigma R^2 \sin\theta'\, d\theta'\, d\phi' \tag{2.206}$$

in spherical polar coordinates.

From Fig. 2.37b

$$\mathscr{r}^2 = \mathscr{R}^2 + \mathscr{z}^2 - 2\mathscr{R}\mathscr{z}\cos\theta' \tag{2.207}$$

($\theta$ in Fig. 2.37b is denoted as $\theta'$ in Eq. 2.207 to note that it is a source point.) Also,

$$\cos\alpha = \frac{z - R\cos\theta'}{\mathscr{r}} \tag{2.208}$$

Now the electric field at point P due to the entire shell is

$$E_z = \int dE_s \cos\alpha \tag{2.209}$$

Substituting Eq. 2.205 in Eq. 2.209

$$E_z = \int \frac{1}{4\pi\varepsilon_o} \frac{dq}{r^2} \cos\alpha \tag{2.210}$$

Substituting Eqs. 2.206, 2.207 and 2.208 in Eq. 2.210

$$E_z = \frac{\sigma R^2}{4\pi\varepsilon_o} \int_0^{2\pi} d\phi' \int_0^{\pi} \frac{(z - R\cos\theta')\sin\theta' \, d\theta'}{(R^2 + z^2 - 2Rz\cos\theta')} \tag{2.211}$$

$$E_z = \frac{2\pi\,\sigma\,R^2}{4\pi\varepsilon_o} \int_0^{\pi} \frac{(z - R\cos\theta')\sin\theta' \, d\theta'}{(R^2 + z^2 - 2Rz\cos\theta')} \tag{2.212}$$

Let $t = \cos\theta'$ $\quad dt = -\sin\theta' \, d\theta'$

$$\theta' = 0 \Rightarrow t = 1$$
$$\theta' = \pi \Rightarrow t = -1$$

Hence

$$E_z = \frac{2\pi\sigma R^2}{4\pi\varepsilon_o} \left[ \int_{-1}^{1} \frac{(z - Rt)dt}{(R^2 + z^2 - 2R\,zt)^{1/2}} \right] \tag{2.213}$$

$$E_z = \frac{2\pi\,\sigma\,R^2}{4\pi\varepsilon_o} \left[ \frac{1}{z^2} \frac{(zt - R)}{\sqrt{R^2 + z^2 - 2R\,zt}} \right]_{-1}^{1} \tag{2.214}$$

$$E_z = \frac{2R^2}{4_o} \frac{1}{z^2} \left[ \left( \frac{z - R}{\sqrt{(z - R)^2}} + \frac{z + R}{\sqrt{(z + R)^2}} \right) \right] \tag{2.215}$$

$$E_z = \frac{2\pi\,\sigma\,R^2}{4\pi\varepsilon_o} \left[ \frac{2}{z^2} \right] = \frac{4\pi\,\sigma\,R^2}{4\pi\varepsilon_o} \frac{1}{z^2} \tag{2.216}$$

But $4\pi\,R^2\,\sigma = q$ the total charge enclosed by the spherical surface. Hence

$$\mathbf{E}_z = \frac{1}{4\pi\varepsilon_o} \frac{q}{z^2} \hat{\mathbf{k}} \tag{2.217}$$

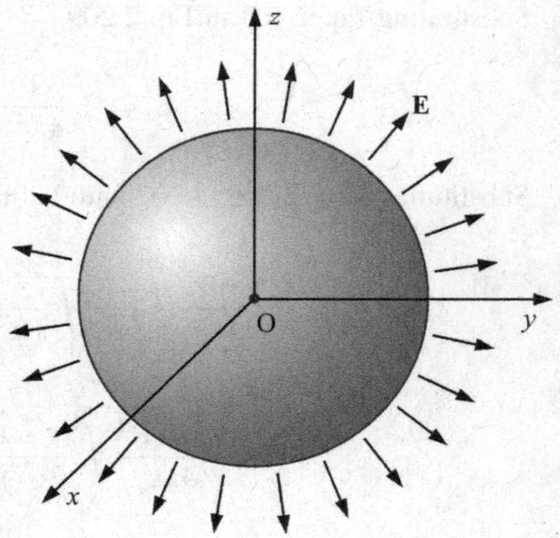


**Fig. 2.39** Electric field lines of a spherical shell carrying uniform surface charge density

Now the total electric field at point P due to the entire spherical shell is

$$\mathbf{E} = E_x\hat{\mathbf{i}} + E_y\hat{\mathbf{j}} + E_z\hat{\mathbf{k}} = \frac{1}{4\pi\varepsilon_0}\frac{q}{z^2}\hat{\mathbf{k}}$$

as $E_x$ and $E_y$ are zero.

(b) **Using Gauss's Law**

First step is to plot the electric field of the uniformly charged spherical shell as shown in Fig. 2.39. Clearly if we want to evaluate $\mathbf{E} \cdot \mathbf{ds}$ easily, then we have to construct the Gaussian surface to be a sphere.

(i) ***At a point interior to the spherical shell***

We construct a Gaussian surface as shown in Fig. 2.40 for an interior point P. Gauss's law states

$$\oint \mathbf{E} \cdot \mathbf{ds} = \frac{q_{enc}}{\varepsilon_0}$$

From Fig. 2.40 the angle between **E** and **ds** is zero and the charge $q_{enc}$ enclosed by the Gaussian surface is zero. Hence from Gauss's law

$$\begin{aligned} &\int |\mathbf{E}||\mathbf{ds}|\cos 0 = 0 \\ &\Rightarrow |\mathbf{E}| = 0 \end{aligned} \qquad (2.218)$$

Hence electric field at the interior point of the charged spherical shell is zero.

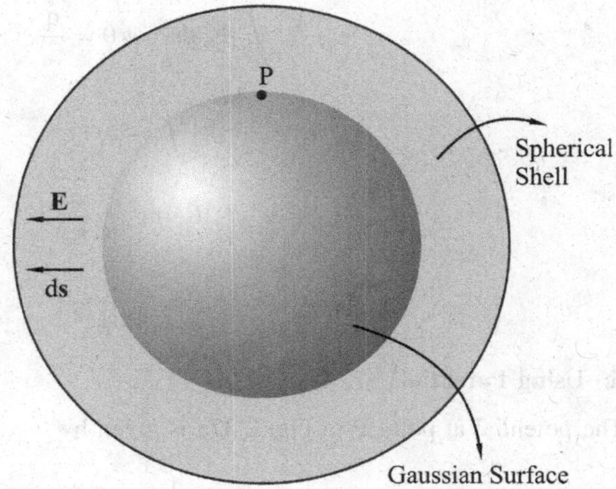


**Fig. 2.40** Gaussian surface for an interior point of a spherical shell carrying uniform surface charge density

(ii) ***At a point exterior to the spherical shell***

We construct a Gaussian surface for an exterior point P as shown in Fig. 2.41. The angle between **E** and d**s** is zero, and the total charge enclosed by the spherical shell is q. Applying Gauss's law

$$\oint \mathbf{E} \cdot \mathbf{ds} = \frac{q_{enc}}{\varepsilon_o} \tag{2.219}$$

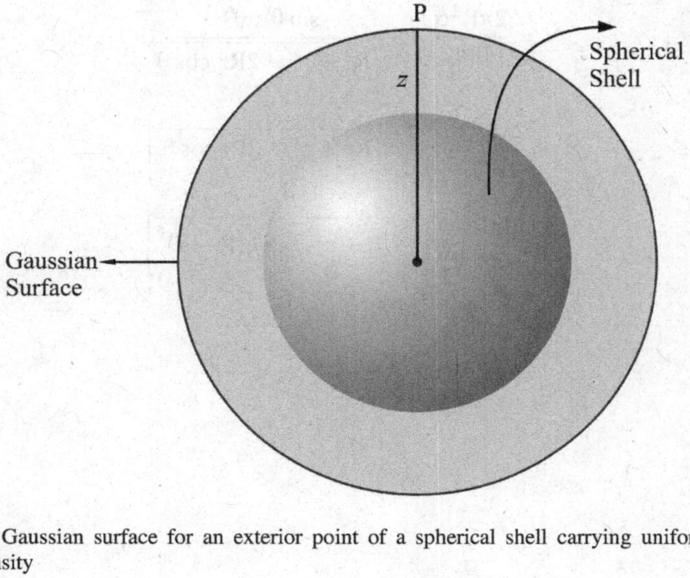


**Fig. 2.41** Gaussian surface for an exterior point of a spherical shell carrying uniform surface charge density

$$\oint |\mathbf{E}||\mathbf{ds}| \cos 0 = \frac{q}{\varepsilon_0} \tag{2.220}$$

$$\Rightarrow |\mathbf{E}| \oint |\mathbf{ds}| = \frac{q}{\varepsilon_0} \tag{2.221}$$

$$\Rightarrow |\mathbf{E}| 4\pi z^2 = \frac{q}{\varepsilon_0} \tag{2.222}$$

$$\Rightarrow E = \frac{q}{4\pi\varepsilon_0 z^2} \tag{2.223}$$

(c) **Using Potential**

The potential at point P in Fig. 2.37a is given by

$$V = \frac{1}{4\pi\varepsilon_0} \int \frac{\sigma\, ds'}{\mathscr{r}} \tag{2.224}$$

Substituting Eq. 2.207 and equation for $ds'$ we get

$$V = \frac{1}{4\pi\varepsilon_0} \sigma \int \frac{R^2 \sin\theta'\, d\theta'\, d\phi'}{\sqrt{R^2 + z^2 - 2Rz \cos\theta'}} \tag{2.225}$$

$$V = \frac{1}{4\pi\varepsilon_0} \int_0^{2\pi} d\phi' \int_0^{\pi} \frac{R^2 \sin\theta'\, d\theta'}{\sqrt{R^2 + z^2 - 2Rz \cos\theta'}} \tag{2.226}$$

$$V = \frac{2\pi R^2 \sigma}{4\pi\varepsilon_0} \int_0^{\pi} \frac{\sin\theta'\, d\theta'}{\sqrt{R^2 + z^2 - 2Rz \cos\theta'}} \tag{2.227}$$

$$V = \frac{2\pi R^2\, \sigma}{4\pi\varepsilon_0} \left[\frac{1}{Rz} \sqrt{R^2 + z^2 - 2Rz \cos\theta'}\right]_0^{\pi} \tag{2.228}$$

$$V = \frac{2\pi R^2\, \sigma}{4\pi\varepsilon_0 Rz} \left[\sqrt{(R+z)^2} - \sqrt{(R-z)^2}\right] \tag{2.229}$$

(i) ***Inside the spherical shell***

Inside the spherical shell $z < R$.
Hence Eq. 2.229 is

$$V = \frac{2\pi R^2 \sigma}{4\pi\varepsilon_0 R z}\left[\sqrt{(R+z)^2} - \sqrt{(R-z)^2}\right] \tag{2.230}$$

$$V = \frac{4\pi R^2 \sigma}{4\pi\varepsilon_0 R} \tag{2.231}$$

$$V = \frac{q}{4\pi\varepsilon_0 R} \tag{2.232}$$

Now

$$\begin{aligned} \mathbf{E} &= -\nabla V \\ &= -\left[\hat{\mathbf{i}}\frac{\partial}{\partial x} + \hat{\mathbf{j}}\frac{\partial}{\partial y} + \hat{\mathbf{k}}\frac{\partial}{\partial z}\right]\frac{q}{4\pi\varepsilon_0 R} \\ \mathbf{E} &= 0 \end{aligned}$$

(ii) ***Outside the spherical shell***

Outside the spherical shell $z > R$.
Hence Eq. 2.229 is

$$V = \frac{2\pi R^2 \sigma}{4\pi\varepsilon_0 R z}\left[\sqrt{(R+z)^2} - \sqrt{(z-R)^2}\right] \tag{2.233}$$

$$V = \frac{2\pi R^2 \sigma}{4\pi\varepsilon_0 R z}[(R+z) - (z-R)] \tag{2.234}$$

$$V = \frac{4\pi R^2 \sigma}{4\pi\varepsilon_0 z} = \frac{q}{4\pi\varepsilon_0 z} \tag{2.235}$$

Now

$$\begin{aligned} \mathbf{E} &= -\nabla V \\ &= -\left[\hat{\mathbf{i}}\frac{\partial}{\partial x} + \hat{\mathbf{j}}\frac{\partial}{\partial y} + \hat{\mathbf{k}}\frac{\partial}{\partial z}\right]\frac{q}{4\pi\varepsilon_0 z} \end{aligned} \tag{2.236}$$

$$\mathbf{E} = \frac{q}{4\pi\varepsilon_0 z^2}\hat{\mathbf{k}} \tag{2.237}$$

Of the three methods discussed here Gauss's law is the easiest as the reader may observe.

The next two methods are Coulomb's law and potential formulation. Luckily in this problem $E_x$, $E_y$ components cancelled out and $E_z$ component alone exists, and the reader can see how much steps are involved in calculating $E_z$ component using Coulomb's law. Suppose for a given charge distribution in addition to $E_z$ component if $E_x$ and $E_y$ components do not cancel out then $E_x$ and $E_y$ must be calculated separately. That is, as we calculated $E_z$ through Eqs. 2.205–2.217 we have to repeat similar procedure for $E_x$, $E_y$ each separately. The reader can see how much tedious procedure it would be. The problem is Coulomb's law calculates electric field directly in terms of vector. There are three different directions $x$, $y$ and $z$ and hence three components $E_x$, $E_y$ and $E_z$. However the potential V is a scalar quantity and we have already seen that it doesn't have components. So we perform only single integration as we have done in Eqs. 2.224–2.229 and finally calculate E as

$$\mathbf{E} = -\nabla V = -\left[\frac{\partial V}{\partial x}\hat{\mathbf{i}} + \frac{\partial V}{\partial y}\hat{\mathbf{j}} + \frac{\partial V}{\partial z}\hat{\mathbf{k}}\right]$$

$$\mathbf{E} = E_x\hat{\mathbf{i}} + E_y\hat{\mathbf{j}} + E_z\hat{\mathbf{k}}$$

which is easy as compared to Coulomb's law.

## 2.18 Comparison of Coulomb's Law, Gauss's Law and Potential Formulation

Let us now compare the methods we used to calculate the electric field for a given charge distribution—Coulomb's law, Gauss's law and potential formulation.

Out of the three methods we see that Gauss's law is the easiest method. From sections 2.13, 2.16 and 2.17 we observe Gauss's law is able to calculate the electric field intensities for the respective charge distributions very easily as compared to Coulomb's law and potential formulation.

Successful application of Gauss's law requires construction of Gaussian surface for a given charge distribution over which $\mathbf{E} \cdot \mathbf{ds}$ can be easily evaluated and over the Gaussian surface $|\mathbf{E}| \cos\theta$ must be constant so that $|\mathbf{E}| \cos\theta$ can be pulled out of the integral

$$\oint \mathbf{E} \cdot \mathbf{ds} = \oint |\mathbf{E}| \cos\theta \, ds = |\mathbf{E}| \cos\theta \oint ds$$

Evaluation of the area $\oint ds$ is simple mathematics. E can thus be calculated using Gauss's law

$$\oint \mathbf{E} \cdot \mathbf{ds} = |\mathbf{E}| \cos\theta \oint ds = \frac{q_{enc}}{\varepsilon_o}$$

$$|\mathbf{E}| = \frac{1}{\cos\theta \oint ds} \frac{q_{enc}}{\varepsilon_o}$$

The above conditions are easily satisfied when symmetry exists in the given charge distributions. In Sect. 2.13 cylindrical symmetry exists. In Sect. 2.16 planar symmetry exists. In Sect. 2.17 spherical symmetry exists. Existence of symmetries in the respective charge distributions facilitated easy calculation of $\oint \mathbf{E} \cdot \mathbf{ds}$.

If symmetry doesn't exist in the given problem, then easy calculation of $\mathbf{E} \cdot \mathbf{ds}$ is not possible using Gauss's law. In such situations Gauss's law is not useful in calculation of **E** although it is still true that $\oint \mathbf{E} \cdot \mathbf{ds} = \frac{q_{enc}}{\varepsilon_o}$.

This point has been made clear in Sects. 2.14 and 2.15.

When Gauss' law is not useful, we are left out with two options to calculate the electric field—Coulomb's law and potential formulation.

As we have previously mentioned Coulomb's law calculates **E** directly in terms of vector. This makes the calculation of **E** using Coulomb's law difficult. Section 2.17 illustrates this point. If we are interested in calculating **E** using Coulomb's law, then we have to work with three components—$E_x$, $E_y$ and $E_z$. However working with potential is easy. Potential is a scalar quantity and doesn't have to be split into components.

Working with vectors is comparatively difficult than working with scalars. This fact has been made clear in Sect. 1.1 and Example 1.10. This point has also been made clear in Sect. 2.17.

We will see one more example below to show working with vectors is difficult as compared to working with scalars.

We redraw Fig. 2.11a, 2.32 in Fig. 2.42a, b.

In Fig. 2.42a the electric fields due to two surface elements $ds_1'$ and $ds_2'$ are shown. The electric field produced by the surface elements at A, B acts along AC and BD, respectively. These fields are denoted as $\mathbf{E_1}$ and $\mathbf{E_2}$, respectively. Resolving $\mathbf{E_1}$ and $\mathbf{E_2}$ as shown in Fig. 2.43 into $x$ and $y$ components we see that $x$ components cancel out while $y$ components add up.

Then we need to integrate the $E_y$ components to calculate the net electric at point P due to entire disc.

Now consider Fig. 2.44a.

We are interested in calculating the electric field at point $P_1$ due to the charge density $\sigma$ over the entire disc, and we have resolved the electric fields, $\mathbf{E_1}$, $\mathbf{E_2}$, due to the elements $\mathbf{ds_1}$, $\mathbf{ds_2}$ into $x$, $y$ components in Fig. 2.44b. Clearly both $x$ and $y$ components are adding up. $E_{1x}$ and $E_{2x}$ and $E_{1y}$ and $E_{2y}$ are in same direction.

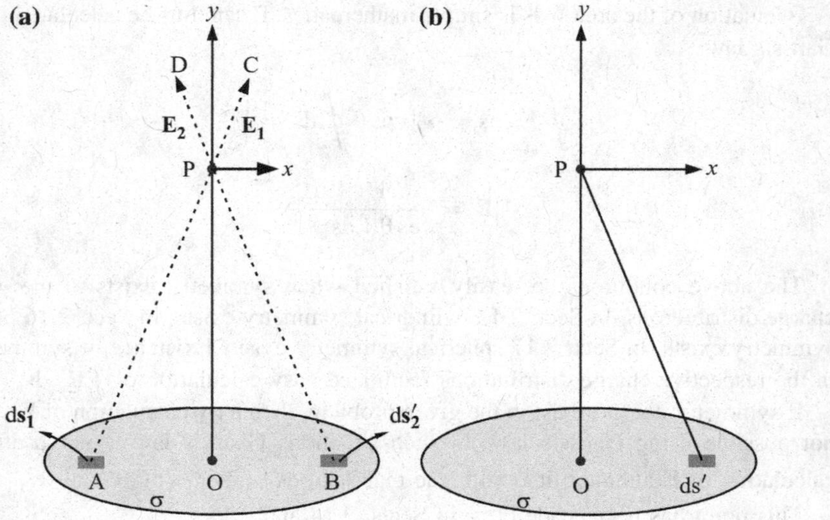


**Fig. 2.42** **a** Charge uniformly distributed over a circular disc along with two surface elements and its field vectors at point P. **b** Charge uniformly distributed over a circular disc along with one surface element.

**Fig. 2.43** Resolution of electric field at point P due to two strips of surface elements shown in Fig. 2.42a.

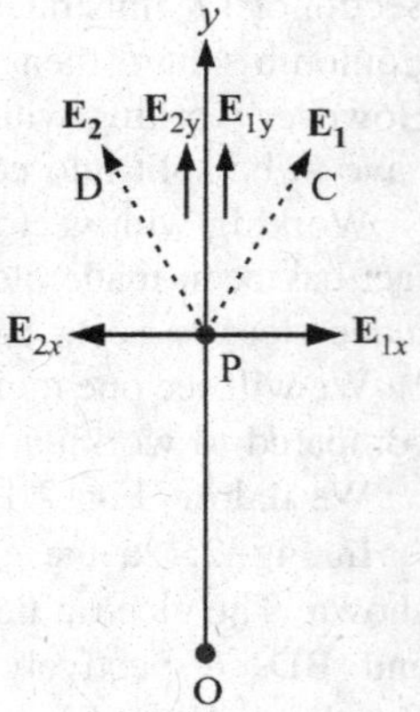


Summing up we get $x$ and $y$ components due to the two elements. Then we integrate over the entire disc to get $E_x$ and $E_y$ and hence the electric field.

If the problem happens in three dimensions, then we need to keep track of three components $E_x$, $E_y$ and $E_z$ as we have mentioned at the end of Sect. 2.17.

The trouble with above method is electric field being a vector has a direction. For example in Fig. 2.44a $\mathbf{E_1}$ is in the direction from $P_1$ to $C_1$ while $\mathbf{E_2}$ is in the direction from $P_1$ to $D_1$. Hence you have to resolve $\mathbf{E_1}$, $\mathbf{E_2}$ into components, then add the $x$, $y$ components separately and then finally integrate it to find the field at $P_1$

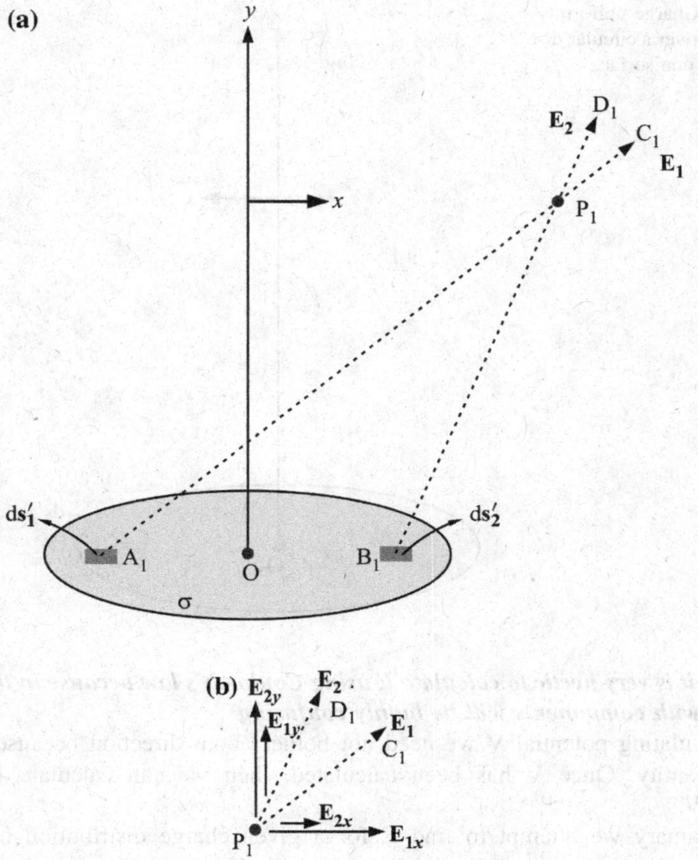


**Fig. 2.44** **a** Charge uniformly distributed over a circular disc along with two surface elements and its field vectors at point $P_1$. **b** Resolution of electric field at point P1 due to two strips of surface elements shown in Fig. 2.44a

due to the entire disc. In Fig. 2.44a, the problem is two-dimensional - What about three dimensions?

The problem becomes even more complicated in three dimensions because there are now three components to work with $E_x$, $E_y$ and $E_z$.

Now consider calculating the electric field **E** using potential formulation. As shown in Fig. 2.45 we need to calculate the potential at point $P_1$ without worrying about "***what is the direction of the potential***" because potential being a scalar doesn't have direction. One need not get confused with components whenever working with V because potential being a scalar doesn't have "***direction and hence components***".

Thus if we are going to calculate **E** using Coulomb's law, then we have to keep track of the direction and mess up with components. "***If the charge distribution is***

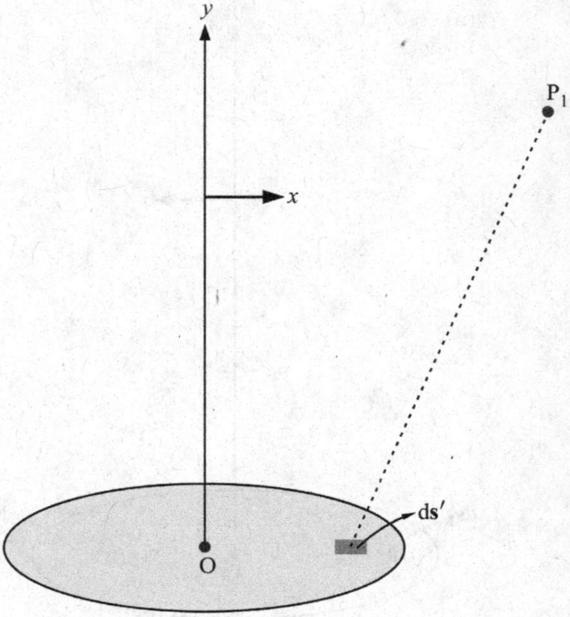


**Fig. 2.45** Charge uniformly distributed over a circular disc along with one surface element

***complex, it is very hectic to calculate E using Coulomb's law because in that case working with components will be highly confusing***".

In calculating potential V we need not bother about direction because V is a scalar quantity. Once V has been calculated, then we can calculate **E** using $\mathbf{E} = -\nabla V$.

In summary we attempt to find **E** for a given charge distribution by using Gauss's law. However when Gauss's law cannot be used, then we calculate **E**, first by calculating potential V. Coulomb's law is rarely used.

Up to this point we have discussed how to calculate electric field for various charge distributions. However based on symmetry and charge distribution we can find with which coordinate the field does not vary. For example consider the infinite line charge distribution shown in Fig. 2.26. Considering cylindrical coordinates ($r_c$, $\phi$, $z$) when we keep $r_c$ and $\phi$ constant and vary $z$ the line charge appears to be same for every point of $z$. Therefore the field doesn't vary with $z$. Similarly holding $r_c$ and $z$ constant if we vary $\phi$ once again the line charge appears to be same at every point of $\phi$. Hence the field doesn't vary with $\phi$ also. On the other hand if we hold $\phi$ and $z$ constant and vary $r_c$, then there is a change in the charge distribution as seen by every point corresponding to given $r_c$ and hence field changes with $r_c$.

Considering the infinite plane sheet of charge, the field doesn't vary with $r_c$ also (Eq. 2.172) and is constant to a near point or a very far away point from the plane. As we move away from the plane sheet of charge, more and more charge comes into the field of view and hence the field doesn't decrease.

Finally we have developed three different methods to calculate the electric field of the given charge distribution. Given the initial charge distribution we plug in charge distribution into the respective equations and finally find the electric field. However in most practical cases the "initial charge distribution" is not known. Without even knowing the initial charge distribution we are supposed to find the electric field.

Two methods have been developed to calculate electric field in such situations: method of images and Laplace's equation method. Before discussing these methods we will discuss some intermediate topics like uniqueness theorem, boundary conditions, etc., which we will be using in the above methods.

## 2.19 Electric Field of a Dipole

We will calculate the electric field of the dipole in this section. We will be using the concept of dipole in future sections.

Two charges of equal magnitude and opposite sign separated by a small distance form a dipole.

Figure 2.46a shows a dipole in which there are two charges q and –q separated by a small distance d.

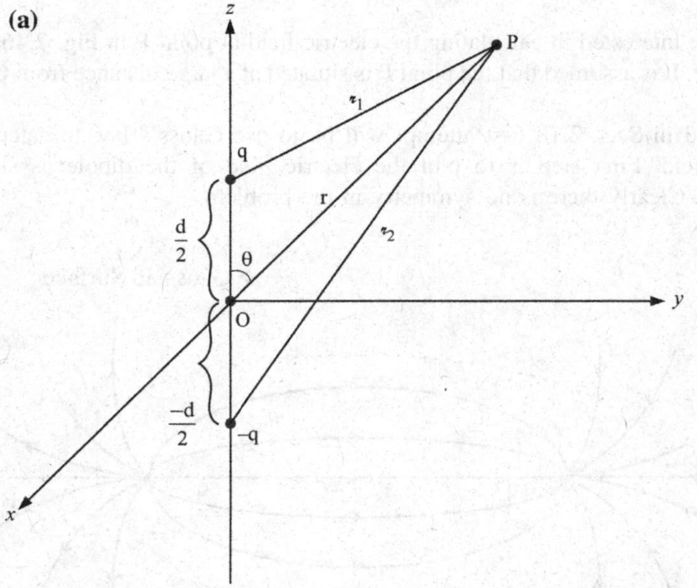


**Fig. 2.46a** **a** A dipole - two equal and opposite charges separated by a small distance d

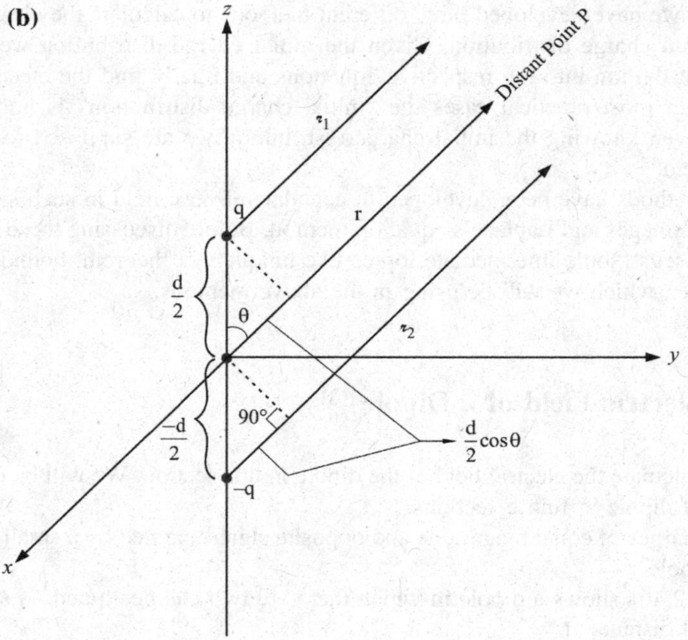


**Fig. 2.46b** **a** The figure illustrates that at a distant point the lines from q and -q become almost parallel

We are interested in calculating the electric field at point P in Fig. 2.46a due to the dipole. It is assumed that the point P is situated at a large distance from O so that $d \ll r$.

As said in Sect. 2.18 first attempt will be to use Gauss's law to calculate the electric field. First step is to plot the electric field of the dipole as shown in Fig. 2.47. Clearly there is no symmetry in the problem.

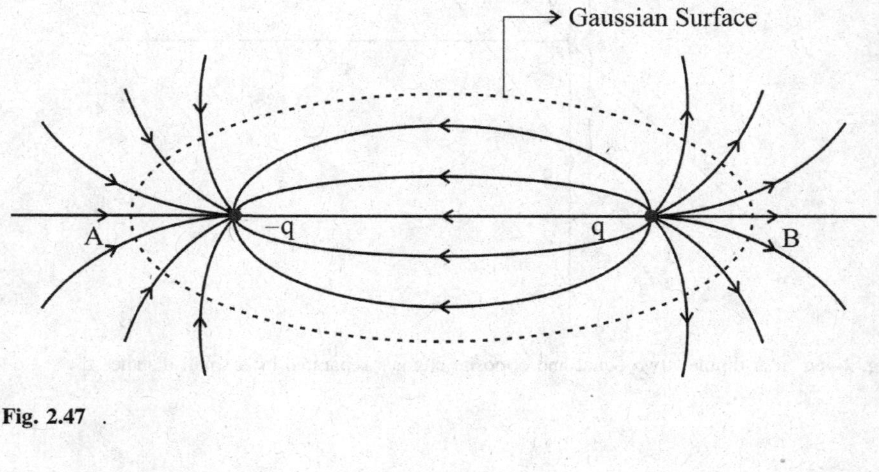


**Fig. 2.47** .

Construction of a Gaussian surface over which $\oint \mathbf{E} \cdot \mathbf{ds}$ can be easily evaluated is not possible. Hence Gauss's law is not useful here. Let us construct a Gaussian surface as shown in Fig. 2.47. Let us apply Gauss's law to the Gaussian surface.

$$\oint \mathbf{E} \cdot \mathbf{ds} = \frac{q_{enc}}{\varepsilon_o}$$

Here $q_{enc} = +q - q = 0$. Hence

$$\oint \mathbf{E} \cdot \mathbf{ds} = 0$$

Based on the above equation can we conclude **E** is zero everywhere due to the dipole? Clearly **E** is not zero everywhere. As we know $\oint \mathbf{E}.\mathbf{ds}$ gives the flux due to **E**. In Fig. 2.47 at point A field lines are entering the Gaussian surface. At point B field lines are leaving the Gaussian surface. Thus over the entire Gaussian surface $\oint \mathbf{E} \cdot \mathbf{ds} = 0$ which means the net flux is zero. "Thus $\oint \mathbf{E} \cdot \mathbf{ds} = 0$ might mean that either **E** can be zero as it happens in Eq. 2.218 or the net flux can be zero as it happens in this dipole case. General thinking by plotting the electric field will show, why $\oint \mathbf{E} \cdot \mathbf{ds} = 0$ for the given charge distribution".

With Gauss's law not helpful here we now turn to potential to calculate the electric field.

The potential at point P due to +q charge is

$$= \frac{1}{4\pi\varepsilon_o}\frac{q}{\mathscr{r}_1} \tag{2.238}$$

The potential at point P due to –q charge is

$$= \frac{-1}{4\pi\varepsilon_o}\frac{q}{\mathscr{r}_2} \tag{2.239}$$

The potential at point P due to both the charges +q and –q is then

$$V = \frac{1}{4\pi\varepsilon_o}\left[\frac{q}{\mathscr{r}_1} - \frac{q}{\mathscr{r}_2}\right] \tag{2.240}$$

$$V = \frac{q}{4\pi\varepsilon_o}\left[\frac{\mathscr{r}_2 - \mathscr{r}_1}{\mathscr{r}_1\mathscr{r}_2}\right] \tag{2.241}$$

From Fig. 2.46b we see that as $d \ll r$

$$\mathscr{r}_1 \approx \mathscr{r} - \frac{1}{2}d\cos\theta \tag{2.242}$$

$$r_2 \approx r + \frac{1}{2} d \cos\theta \tag{2.243}$$

and

$$r_1 r_2 \approx r^2 - \left(\frac{1}{2} d \cos\theta\right)^2 \approx r^2 \tag{2.244}$$

Substituting Eqs. 2.242–2.244 in Eq. 2.241.

$$V = \frac{q}{4\pi\varepsilon_0} \frac{d \cos\theta}{r^2} \tag{2.245}$$

$$\text{If } \theta = 90^\circ \text{ then } V = 0 \tag{2.246}$$

That is the plane at $y = 0$ is at zero potential.
We know that in spherical polar coordinates

$$\mathbf{E} = -\nabla V$$

$$\mathbf{E} = -\left(\frac{\partial V}{\partial r}\hat{\mathbf{r}} + \frac{1}{r}\frac{\partial V}{\partial \theta}\hat{\boldsymbol{\theta}} + \frac{1}{r \sin\theta}\frac{\partial V}{\partial \phi}\hat{\boldsymbol{\phi}}\right) \tag{2.247}$$

Differentiating Eq. 2.245

$$\frac{\partial V}{\partial r} = \frac{-qd \cos\theta}{2\pi\varepsilon_0 r^3} \tag{2.248}$$

$$\frac{\partial V}{\partial \theta} = \frac{-qd \sin\theta}{2\pi\varepsilon_0 r^2} \tag{2.249}$$

$$\frac{\partial V}{\partial \phi} = 0 \tag{2.250}$$

Substituting Eqs. 2.248–2.250 in Eq. 2.247

$$\mathbf{E} = \frac{qd}{4\pi\varepsilon_0 r^3}\left[2\cos\theta\,\hat{r} + \sin\theta\,\hat{\boldsymbol{\theta}}\right] \tag{2.251}$$

The above equation can be written as

$$\mathbf{E} = \mathbf{E}_r + \mathbf{E}_\theta = E_r\,\hat{\mathbf{r}} + E_\theta\,\hat{\boldsymbol{\theta}} \tag{2.252}$$

where

$$E_r = \frac{qd\cos\theta}{2\pi\varepsilon_o r^3}, \quad E_\theta = \frac{qd\sin\theta}{4\pi\varepsilon_o r^3}$$

The magnitude of the resultant dipole field is given by

$$E = \sqrt{E_r^2 + E_\theta^2}$$

$$E = \sqrt{\left(\frac{qd\cos\theta}{2\pi\varepsilon_o r^3}\right)^2 + \left(\frac{qd\sin\theta}{4\pi\varepsilon_o r^2}\right)^2} \tag{2.253}$$

$$E = \frac{qd}{4\pi\varepsilon_o r^3}\sqrt{3\cos^2\theta + 1} \tag{2.254}$$

Now let us define the electric dipole moment. The electric dipole moment **p** is defined as **p** = q**d** [**d** points from –q to +q].

Then we can write Eq. 2.245 as

$$V = \frac{1}{4\pi\varepsilon_o}\frac{qd\cos\theta}{r^2}$$

$$V = \frac{1}{4\pi\varepsilon_o}\frac{q\mathbf{d}.\hat{\mathbf{r}}}{r^2}$$

$$\Rightarrow V = \frac{1}{4\pi\varepsilon_o}\frac{\mathbf{p}\cdot\hat{\mathbf{r}}}{r^2} \tag{2.256}$$

From Fig. 2.48 we can write

$$p_r = p\cos\theta$$
$$p_\theta = -p\sin\theta$$

We have

$$\mathbf{p} = p_r\,\hat{\mathbf{r}} + p_\theta\,\hat{\boldsymbol{\theta}} \tag{2.257}$$

$$\Rightarrow \mathbf{p} = p\cos\theta\,\hat{\mathbf{r}} - p\sin\theta\,\hat{\boldsymbol{\theta}} \tag{2.258}$$

Now using Eq. 2.258 we can write

$$\begin{aligned} 3(\mathbf{p}.\hat{\mathbf{r}})\hat{\mathbf{r}} - \mathbf{p} &= 3p\cos\theta\,\hat{\mathbf{r}} - [p\cos\theta\,\hat{\mathbf{r}} - p\sin\theta\,\hat{\boldsymbol{\theta}}] \\ &= 2p\cos\theta\,\hat{\mathbf{r}} + p\sin\theta\,\hat{\boldsymbol{\theta}} \end{aligned} \tag{2.259}$$

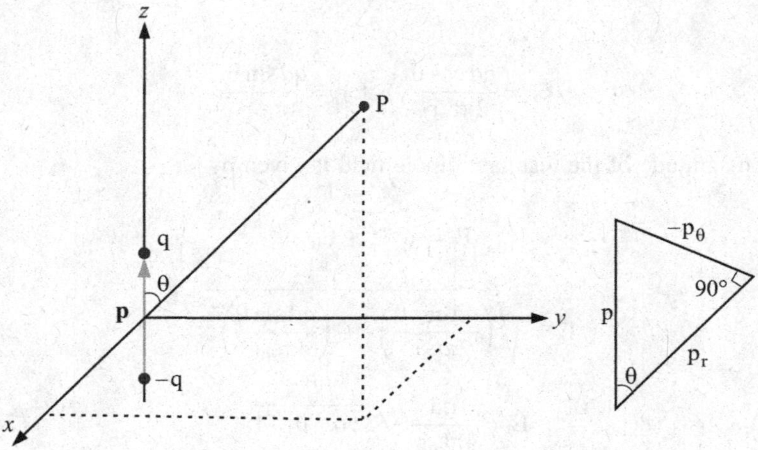


**Fig. 2.48** Figure shows electric dipole moment of a dipole

Equation 2.251 is

$$\mathbf{E} = \frac{1}{4\pi\varepsilon_0 r^3}[2p\cos\theta\,\hat{\mathbf{r}} + p\sin\theta\,\hat{\boldsymbol{\theta}}] \tag{2.260}$$

Substituting Eq. 2.259 in 2.260

$$\mathbf{E} = \frac{1}{4\pi\varepsilon_0 r^3}[3(\mathbf{p}\cdot\hat{\mathbf{r}})\hat{\mathbf{r}} - \mathbf{p}] \tag{2.261}$$

Equation 2.261 gives an interesting result. The electric field of a dipole falls off like $1/r^3$. In comparison the electric field of a point charge falls off like $1/r^2$. Thus electric field of a dipole falls off more rapidly than that of the point charge. This is physically expected because at far away distances the opposite charges +q and −q appear to be closely spaced and hence the fields cancel out more rapidly than that of the point charge.

In Fig. 2.49 we show the field of a point dipole. Point dipole is one in which we allow d to approach zero while q approaches infinity keeping the product p = qd constant.

## 2.20 Calculation of Potential Using $V = \int \mathbf{E}\cdot d\mathit{l}$

In the previous sections we saw how to calculate the electric field **E** once the potential V is calculated. However if we are interested in calculating potential or potential difference between two points, instead of calculating electric field, then

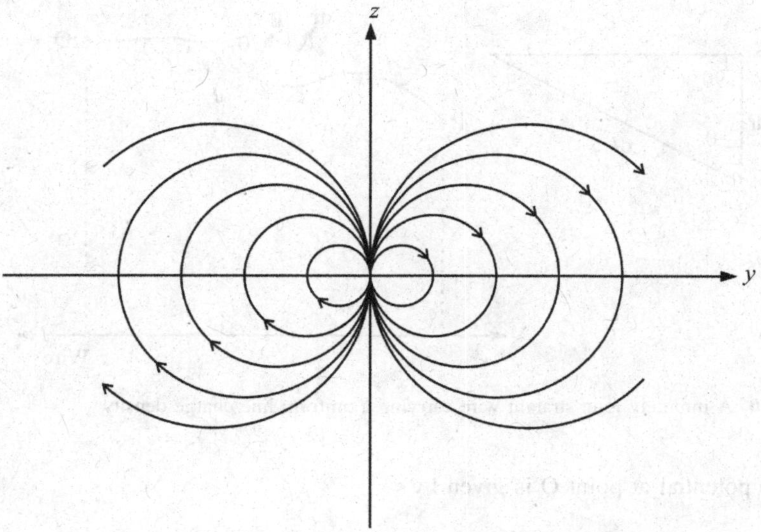


**Fig. 2.49** Electric field pattern of a point dipole

there are two methods to calculate it. The first method is to calculate potential directly using Eqs. 2.80–2.82. Another method is to calculate **E** using Gauss's law; if symmetry exists for the given charge distribution, then calculate potential using the relation $\mathbf{V} = \int \mathbf{E} \cdot d\boldsymbol{l}$. We will discuss the second method here.

(a) ***Potential due to infinitely long charged wire***

In Sect. 2.13 we calculated the electric field due to infinitely long straight wire by applying Gauss's law

$$\mathbf{E} = \frac{\lambda}{2\pi\varepsilon_0 r}$$

In Fig. 2.50 we show the infinitely long straight wire carrying a uniform line charge density $\lambda$. Consider two points P and Q situated at a distance $r_P$ and $r_Q$ from the wire, respectively. We are interested in calculating the potential difference between the two points, P and Q. The potential at point P is given by

$$V_P = -\int_{\infty}^{P} \mathbf{E} \cdot d\boldsymbol{l} \qquad (2.262)$$

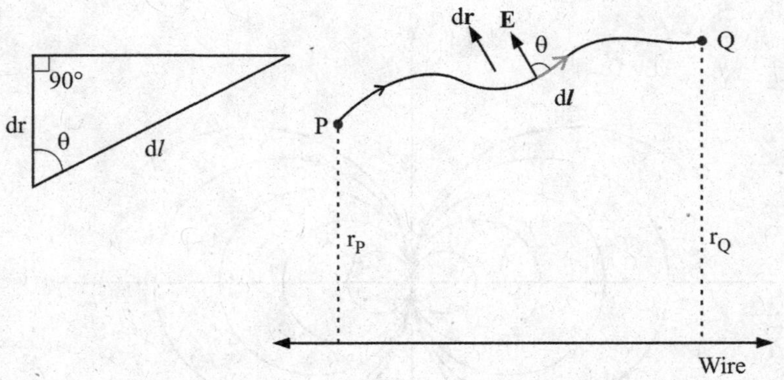


**Fig. 2.50** A infinitely long straight wire carrying a uniform line charge density

The potential at point Q is given by

$$V_Q = -\int_{\infty}^{Q} \mathbf{E} \cdot d\boldsymbol{l} \tag{2.263}$$

Now,

$$V_P - V_Q = -\int_{\infty}^{P} \mathbf{E} \cdot d\boldsymbol{l} + \int_{\infty}^{Q} \mathbf{E} \cdot d\boldsymbol{l} \tag{2.264}$$

$$V_P - V_Q = -\int_{\infty}^{P} \mathbf{E} \cdot d\boldsymbol{l} - \int_{Q}^{\infty} \mathbf{E} \cdot d\boldsymbol{l} \tag{2.265}$$

$$V_P - V_Q = -\int_{Q}^{P} \mathbf{E} \cdot d\boldsymbol{l} \tag{2.266}$$

$$V_P - V_Q = -\int_{\infty}^{P} E\, dl \cos\theta \tag{2.267}$$

$$V_P - V_Q = -\int_{Q}^{P} \frac{\lambda}{2\pi\varepsilon_0 r} dl \cos\theta \tag{2.268}$$

From Fig. 2.50

$$dl\cos\theta = dr \tag{2.269}$$

Substituting Eq. 2.269 in Eq. 2.268

$$V_p - V_Q = -\frac{-\lambda}{2\pi\varepsilon_o}\int_{r_Q}^{r_P}\frac{dr}{r} \tag{2.270}$$

$$V_p - V_Q = \frac{\lambda}{2\pi\omega_o}\,ln\frac{r_Q}{r_P} \tag{2.271}$$

(b) **Potential of a Spherical Shell**

The electric field of spherical shell has been calculated by Gauss's law in Sect. 2.17.

$$\mathbf{E} = 0 \quad \text{if} \quad r<R$$

$$\mathbf{E} = \frac{q}{4\pi\varepsilon_o r^2}\hat{\boldsymbol{r}} \quad \text{if} \quad r > R$$

where R is radius of the spherical shell.

For a point interior to the spherical shell as shown in Fig. 2.51a the potential is

$$V = -\int_{\infty}^{r}\mathbf{E}\cdot d\boldsymbol{l}$$

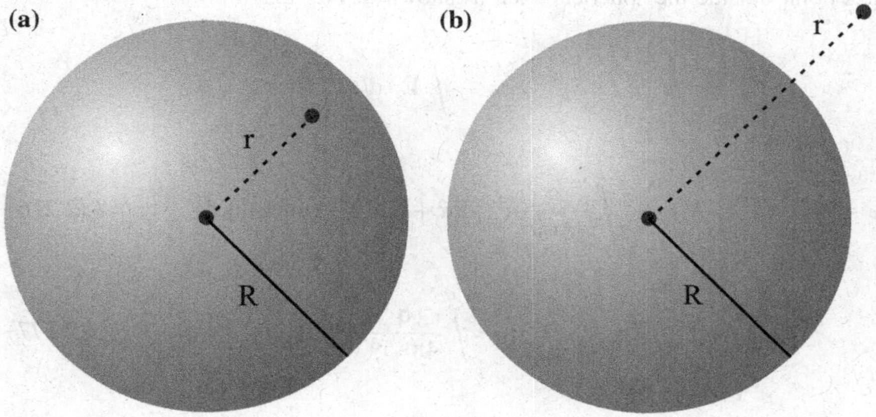


**Fig. 2.51** **a** An interior point in a spherical shell carrying a uniform surface charge density. **b** An exterior point in a spherical shell carrying a uniform surface charge density

In spherical polar coordinates

$$dl = dr\,\hat{\mathbf{r}} + r\,d\theta\hat{\boldsymbol{\theta}} + r\sin\theta\,d\phi\,\hat{\boldsymbol{\phi}}$$

(i) ***Inside the Spherical Shell***

For a point inside the spherical shell as shown in Fig. 2.51a

$$V = -\int_{\infty}^{r} \mathbf{E}\cdot d\boldsymbol{l} = -\int_{\infty}^{R} \mathbf{E}\cdot d\boldsymbol{l} - \int_{R}^{r} \mathbf{E}\cdot d\boldsymbol{l} \tag{2.272}$$

$$V = -\int_{\infty}^{R} \frac{q}{4\pi\varepsilon_0 r^2}\hat{\mathbf{r}}.(dr\,\hat{\mathbf{r}} + rd\theta\,\hat{\boldsymbol{\theta}} + r\sin\theta\,d\phi\,\hat{\boldsymbol{\phi}}) - \int_{R}^{r} 0\cdot d\boldsymbol{l} \tag{2.273}$$

$$V = -\int_{\infty}^{R} \frac{q}{4\pi\varepsilon_0 r^2}dr$$

$$V = \frac{q}{4\pi\varepsilon_0}\left[\frac{1}{r}\right]_{\infty}^{R} \tag{2.274}$$

$$V = \frac{q}{4\pi\varepsilon_0 R} \tag{2.275}$$

Thus inside the spherical shell the potential is constant while the field is zero.

(ii) ***Outside the Spherical Shell***

At a point outside the spherical shell as shown in Fig. 2.51b

$$V = -\int_{\infty}^{r} \mathbf{E}\cdot d\boldsymbol{l}$$

$$V = -\int_{\infty}^{r} \frac{q}{4\pi\varepsilon_0 r^2}\hat{\mathbf{r}}\cdot[dr\,\hat{\mathbf{r}} + r\,d\theta\,\hat{\boldsymbol{\theta}} + r\sin\theta\,d\phi\,\hat{\boldsymbol{\phi}}] \tag{2.276}$$

$$V = -\int_{\infty}^{r} \frac{q}{4\pi\varepsilon_0 r^2}dr \tag{2.277}$$

$$V = \frac{q}{4\pi\varepsilon_0 r} \tag{2.278}$$

## 2.21 The Conservative Nature of Electric Field

Consider a point charge situated in origin as shown in Fig. 2.52. The electric field produced by the point charge at any point situated at a distance of r from origin is

$$\mathbf{E} = \frac{1}{4\pi\varepsilon_o}\frac{q}{r^2}\hat{\boldsymbol{r}}$$

Let us calculate the line integral of $\mathbf{E}$ from point A to B in Fig. 2.52. In spherical polar coordinates

$$d\boldsymbol{l} = dr\,\hat{\boldsymbol{r}} + rd\theta\,\hat{\boldsymbol{\theta}} + r\sin\theta\, d\phi\,\hat{\boldsymbol{\phi}}$$

Hence

$$\mathbf{E}\cdot d\boldsymbol{l} = \frac{1}{4\pi\varepsilon_o}\frac{q}{r^2}dr$$

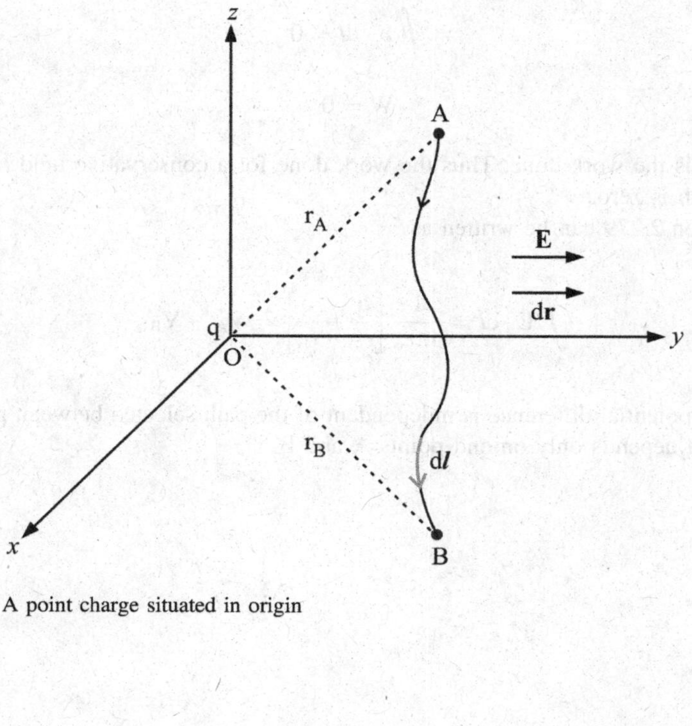


**Fig. 2.52** A point charge situated in origin

$$\int_A^B \mathbf{E} \cdot d\boldsymbol{l} = \frac{q}{4\pi\varepsilon_0} \int_A^B \frac{dr}{r^2} = \frac{q}{4\pi\varepsilon_0} \left[\frac{1}{r_A} - \frac{1}{r_B}\right] \tag{2.279}$$

Thus $\int_A^B \mathbf{E} \cdot d\boldsymbol{l}$ depends only on the end points $r_A$ and $r_B$ and is a constant. The integral $\int_A^B \mathbf{E} \cdot d\boldsymbol{l}$ is independent of path.

One can go from A to B in any path in evaluating the integral $\int_A^B \mathbf{E} \cdot d\boldsymbol{l}$, but still whatever path selected the result of integration $\int_A^B \mathbf{E} \cdot d\boldsymbol{l}$ is always $\frac{1}{4\pi\varepsilon_0}\left[\frac{q}{r_A} - \frac{q}{r_B}\right]$.

If the path AB is a closed path, then $r_A = r_B$ and hence

$$\oint \mathbf{E} \cdot d\boldsymbol{l} = 0 \tag{2.280}$$

If the line integral of a vector field around a closed path is zero, such a field is known as conservative field.

From Eq. 2.280

$$\oint Q\mathbf{E} \cdot d\boldsymbol{l} = 0 \tag{2.281}$$

$$\oint \mathbf{F} \cdot d\boldsymbol{l} = 0 \tag{2.282}$$

$$W = 0 \tag{2.283}$$

where W is the work done. Thus the work done for a conservative field around a closed path is zero.

Equation 2.279 can be written as

$$\int_A^B \mathbf{E} \cdot d\boldsymbol{l} = \frac{1}{4\pi\varepsilon_0}\left[\frac{q}{r_A} - \frac{q}{r_B}\right] = V_A - V_B \tag{2.284}$$

Hence potential difference is independent of the path selected between points A and B and depends only on end points A and B.

## 2.22 The Reference Point R in the Equation $V = -\int_R^P \mathbf{E} \cdot d\mathbf{l}$

In Eq. 2.65 we have defined the potential as

$$V = -\int_R^P \mathbf{E} \cdot d\mathbf{l}$$

and noted down that R is the reference point where we set the potential to be zero. In this section we will discuss about the reference point.

If we want to measure the length of an object using a metre scale, then we fix zero marking on the metre scale as zero length and with respect to the zero marking we measure the length of any object. If we want to measure the temperature of a body using mercury thermometer, the zero marking on the mercury thermometer is taken to be zero and with respect to the zero reference point we measure the temperature of the body. This is shown in Fig. 2.53. Similarly for the case of potential we require a reference point where the potential is set to be zero and with respect to the zero reference point we measure the potential of any other point.

The normal procedure is that the infinite point is taken as the reference point where we set the potential to be zero. Hence

$$V = -\int_\infty^P \mathbf{E} \cdot d\mathbf{l}$$

This is shown in Fig. 2.54. ∞ is the point where we fix the potential to be zero and from infinity we measure the potential V at point P.

We fix the infinity as reference point for free charges situated in space. But in actual practical problems ground is the reference point where we fix the zero reference point for potential. That is we change the reference point from infinity to ground. In order to understand this point consider Fig. 2.55 where we change the reference point from R to $R_1$. It can be shown that

$$V = \kappa + V_1 \tag{2.285}$$

where

$V = -\int_R^P \mathbf{E} \cdot d\mathbf{l}$ is the potential at point P with respect to reference point R in Fig. 2.55.

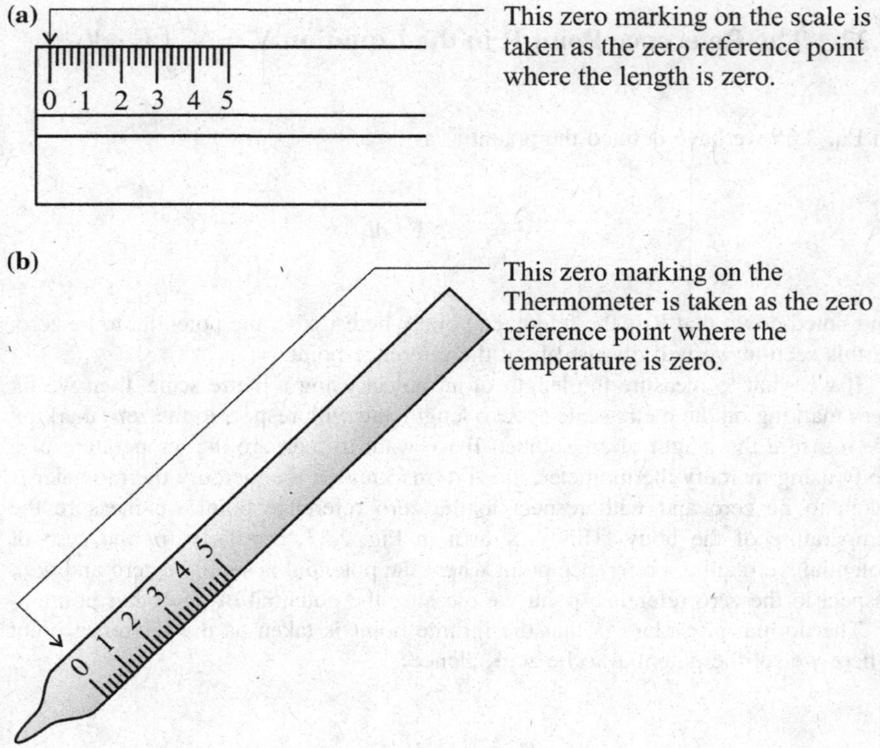
(a)
0 1 2 3 4 5
This zero marking on the scale is taken as the zero reference point where the length is zero.
(b)
0 1 2 3 4 5
This zero marking on the Thermometer is taken as the zero reference point where the temperature is zero.

**Fig. 2.53** Figure showing zero reference point for a) Scale b) Thermometer

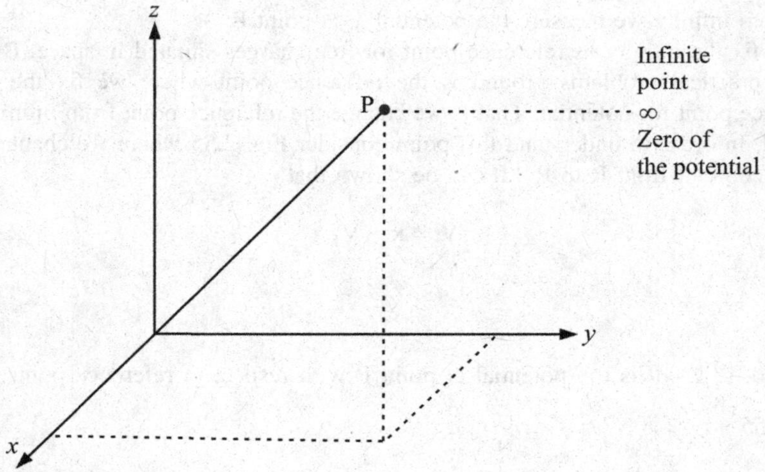
z
y
x
P
Infinite point
∞
Zero of the potential

**Fig. 2.54** Figure showing zero reference point for potential

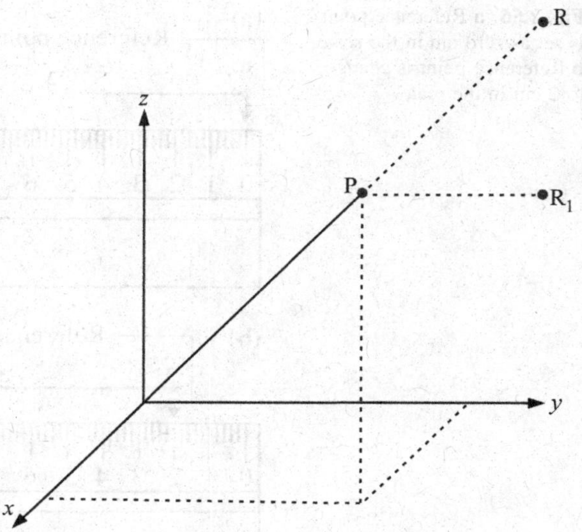


**Fig. 2.55** Figure showing change of reference point from R to $R_1$ to measure potential at point P

$V_1 = -\int_{R_1}^{P} \mathbf{E} \cdot d\boldsymbol{l}$ is the potential at point P with respect to reference point $R_1$ in Fig. 2.55.

$\kappa = -\int_{R}^{R_1} \mathbf{E} \cdot d\boldsymbol{l}$ is a constant as the line integral depends only on the end points (see Sect. 2.21).

Taking gradient of Eq. 2.285

$$-\nabla V = -\nabla\kappa - \nabla V_1 \Rightarrow -\nabla V = -\nabla V_1 \Rightarrow \mathbf{E} = \mathbf{E}_1 \qquad (2.286)$$

Here $\nabla\kappa = 0$ because $\kappa$ is a constant.

That is changing the reference point doesn't alter the actual electric field.

If you are bit confused, see Fig. 2.56. In Fig. 2.56a we are interested in measuring the length of the thread $A_1$ $A_2$ whose length is 5 cm. We see that the length of the thread $A_1$ $A_2$ with reference point set at zero cm is 5 cm. In Fig. 2.56b we have changed the zero of the reference point to 2 cm. We see that the length of the thread $A_1$ $A_2$ is still 7 cm − 2 cm = 5 cm.

Hence the conclusion is if we change the reference point the final reading gets changed (in Fig. 2.56b we see final value 7 cm). But the actual quantity the length of the thread $A_1$ $A_2$ remains unchanged. Similarly when we change the reference point in our potential, the final reading of potential changes because some constant $\kappa$ is added to existing potential. But the actual quantity the electric field **E** remains unchanged (Eq. 2.286).

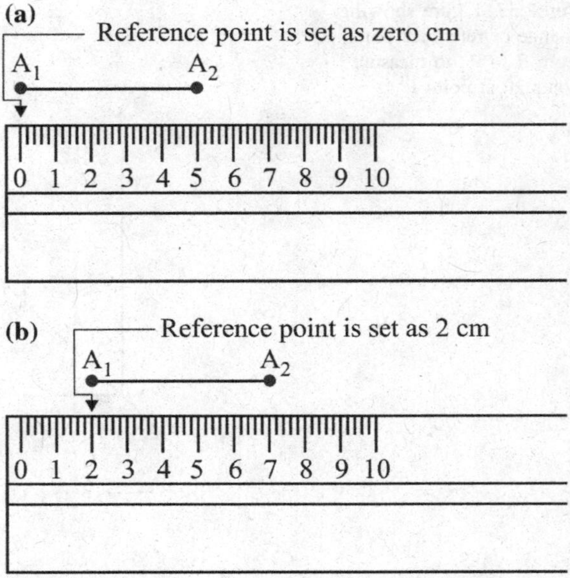


**Fig. 2.56** **a** Reference point is set as zero cm in the scale. **b** Reference point is changed to 2 cm in the scale

As the electric field doesn't change when we change the reference point, in Sects. 2.13c and 2.16c we calculate potential at some reference point $r_o$ to determine the electric field whenever the charge distribution tends to infinity.

Another important point to be noted is it is the potential difference between two points which has physical significance and not the potential itself. As an example if you are asked to measure the distance between earth and sun, then you measure the distance of sun from earth and report the value. In the above statement it is implicitly assumed earth is the zero reference point for distance. Fixing earth as zero reference point for distance, **from earth** the distance of sun is measured. Suppose assume that in this universe only sun is present no earth, no planets, no stars, nothing else. Then in that case distance between what and what? When sun alone is present in empty space, then distance has no physical meaning in the above example. Or in other words it is the difference in distance between two points that has physical significance and not the distance itself. Similarly, it is the potential difference between two points, which has physical significance and not the potential itself.

Fixing earth as zero reference point for distance, from earth the distance of sun is measured in the above example. Similarly in Eq. 2.66 we measure the potential at point P from infinity where we have fixed the zero reference point of the potential. Some students raise objection by looking at Eq. 2.140 - that in Eq. 2.140 we are integrating from 0 to L and not from $\infty$ to L. But there is a difference between integrations in Eq. 2.66 and Eq. 2.140. Can the reader say what is the difference?

**Example 2.11**

Consider two concentric spherical shells of radii $R_1$ and $R_2$ as shown in Fig. 2.57. The inside shell with radius $R_1$ carries a charge $q_1$, and the outside shell with radius $R_2$ carries a charge $q_2$ with the charges $q_1$ and $q_2$ uniformly distributed over the shells, respectively.

Find the electric field at points A, B, C in Fig. 2.57.

**Solution**

Clearly there is spherical symmetry in the problem. Hence Gauss's law can be applied to obtain electric field.

(i) ***At point A*** $\mathbf{r} \leq \boldsymbol{R_1}$

Construct a spherical Gaussian surface with OA as radius as shown in Fig. 2.58. Applying Gauss's law

$$\oint \mathbf{E} \cdot \mathbf{ds} = \frac{q_{enc}}{\varepsilon_o}$$

Here $q_{enc} = 0$. Hence

$$\oint \mathbf{E} \cdot \mathbf{ds} = 0$$

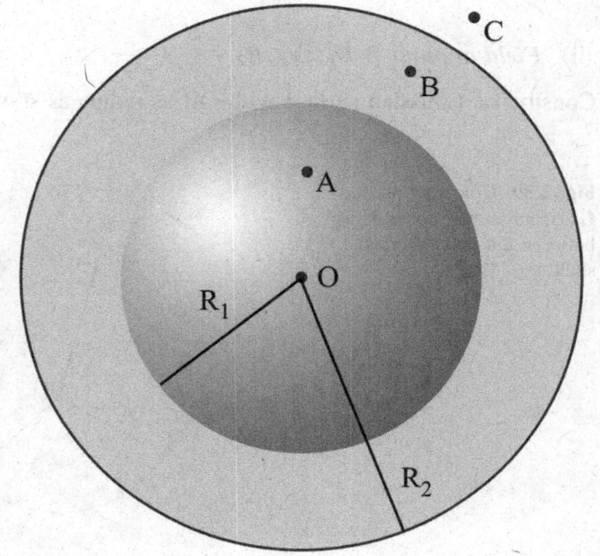


**Fig. 2.57** Two concentric spherical shells of radii $R_1$ and $R_2$ carrying charges $q_1$ and $q_2$

**Fig. 2.58** .

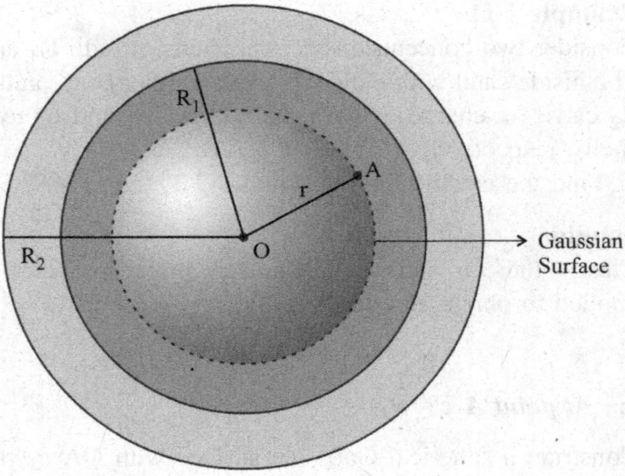


On the Gaussian surface the angle between **E** and **ds** is zero. Hence

$$\oint |\mathbf{E}||\mathbf{ds}|\cos 0 = 0 \Rightarrow \oint |\mathbf{E}||\mathbf{ds}| = 0$$

On the Gaussian surface $|\mathbf{E}|$ is constant

$$|\mathbf{E}| \oint |\mathbf{ds}| = 0 \Rightarrow |\mathbf{E}|4\pi r^2 = 0 \Rightarrow |\mathbf{E}| = 0 \text{ with } r = OA.$$

(ii) ***Field at point B*** $R_1 < r \leq R_2$

Construct a Gaussian surface with OB as radius as shown in Fig. 2.59.

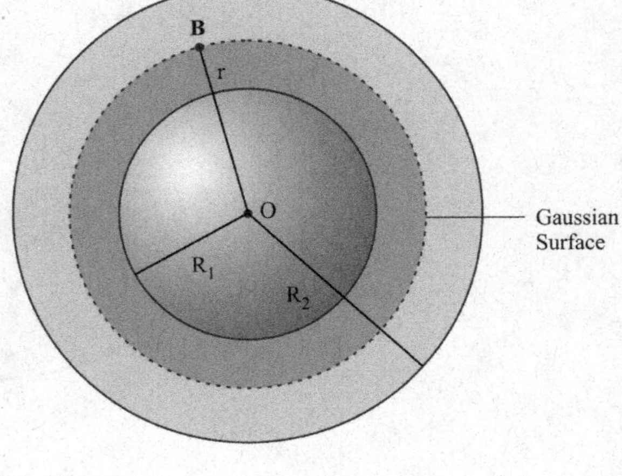


**Fig. 2.59** Construction of a Gaussian surface for a point between the two spherical shell

Applying Gauss's law

$$\oint \mathbf{E} \cdot d\mathbf{s} = \frac{q_{enc}}{\varepsilon_o}$$

Here $q_{enc} = q_1$ and r = OB.

$$\oint \mathbf{E} \cdot d\mathbf{s} = \oint |\mathbf{E}||d\mathbf{s}| \cos 0 = \oint |\mathbf{E}||d\mathbf{s}|$$

$$\Rightarrow |\mathbf{E}| \oint |d\mathbf{s}| \Rightarrow |\mathbf{E}|4\pi r^2$$

Hence Gauss's law is

$$|\mathbf{E}|4\pi r^2 = \frac{q_{enc}}{\varepsilon_o} = \frac{q_1}{\varepsilon_o} \quad \Rightarrow E = \frac{1}{4\pi\varepsilon_o}\frac{q_1}{r^2}$$

(iii) ***Field at point C r > R₂***

Consider Fig. 2.60. We construct a Gaussian surface with OC as radius.
OC = r.
Applying Gauss's law

$$\oint \mathbf{E} \cdot d\mathbf{s} = \frac{q_{enc}}{\varepsilon_o}$$

Here $q_{enc} = q_1 + q_2$ and

$$\oint \mathbf{E} \cdot d\mathbf{s} = |\mathbf{E}|4\pi r^2 \text{ with } r = OC$$

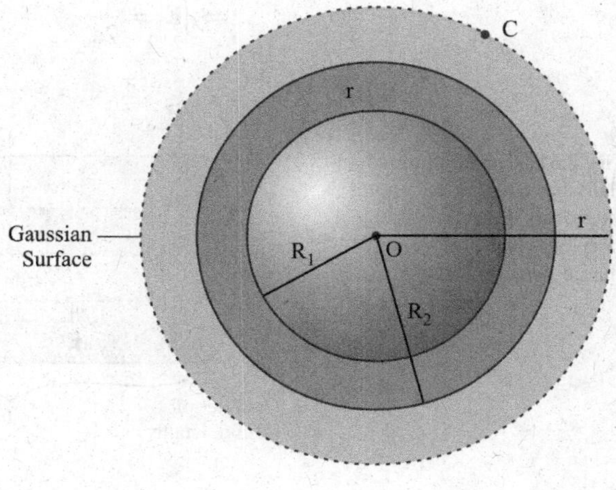


**Fig. 2.60** Construction of a Gaussian surface for a point exterior of the spherical shell

Hence Gauss's law is

$$|\mathbf{E}|4\pi r^2 = \frac{q_1 + q_2}{\varepsilon_o} \Rightarrow |\mathbf{E}| = \frac{1}{4\pi\varepsilon_o}\frac{q_1 + q_2}{r^2}$$

**Example 2.12**
Charge is uniformly distributed through out cylindrical volume of infinite length of radius R. The volume charge density is ρ. Determine the electric field intensity at a point inside the cylinder.

**Solution**
The problem has cylindrical symmetry. Hence Gauss's law can be applied to calculate electric field. Construct a Gaussian surface of radius r and length L as shown in Fig. 2.61.

$$\oint \mathbf{E}\cdot \mathbf{ds} = \frac{q_{enc}}{\varepsilon_o}$$

Here $q_{enc} = \rho[\pi r^2 L]$
where $\pi r^2 L$ is the volume of the Gaussian surface.
And

$$\oint \mathbf{E}\cdot \mathbf{ds} = \oint |\mathbf{E}||\mathbf{ds}|\cos 0 = \oint |\mathbf{E}||\mathbf{ds}|$$
$$= |\mathbf{E}|\oint \mathrm{ds} = |\mathbf{E}|\, 2\pi r L$$

Hence Gauss's law is

$$|\mathbf{E}|\cdot 2\pi r L = \rho\pi r^2 L \tag{2.287}$$

$$\Rightarrow |\mathbf{E}| = \frac{\rho r}{2\varepsilon_o} \tag{2.288}$$

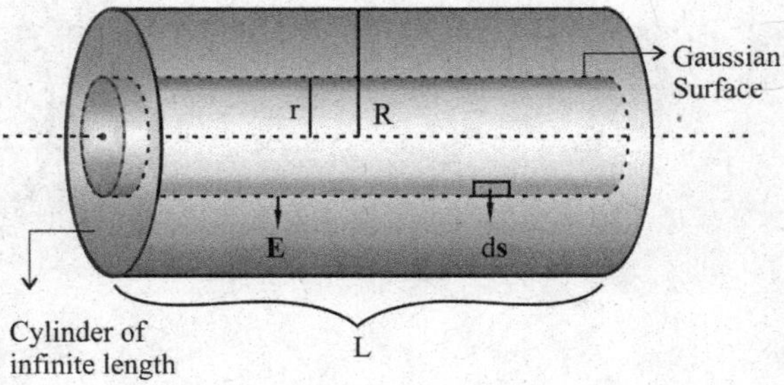


**Fig. 2.61** Construction of a Gaussian surface for a point interior of the cylinder carrying a uniform volume charge density

## 2.23 Poisson's and Laplace's Equation

We know that

$$\nabla \cdot \mathbf{E} = \frac{\rho}{\varepsilon_0} \tag{2.289}$$

$$\mathbf{E} = -\nabla V \tag{2.290}$$

From the above equations

$$\nabla \cdot (-\nabla V) = \frac{\rho}{\varepsilon_0} \tag{2.291}$$

$$\Rightarrow \nabla^2 V = \frac{-\rho}{\varepsilon_0} \tag{2.292}$$

The above equation is known as Poisson's equation. In places where there is no charge $\rho = 0$ and hence

$$\nabla^2 V = 0 \tag{2.293}$$

The above equation is known as Laplace's equation.

**Example 2.13**
Consider a uniform spherically symmetric distribution of charge in a sphere of radius R as shown in Fig. 2.62. The charge density is $\rho$. Calculate the field at a point inside and outside the sphere using Poisson's and Laplace's equation. Plot the electric field with respect to r.

**Fig. 2.62** A sphere carrying uniform volume charge density

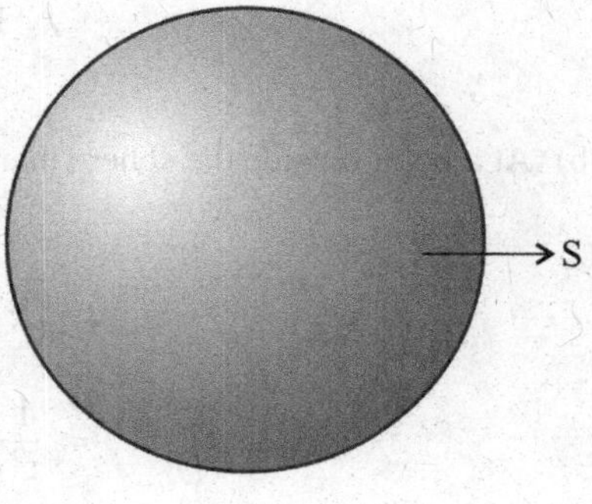

**Solution**

(a) At a point inside the sphere

$$\nabla^2 V = \frac{-\rho}{\varepsilon_o}$$

The Laplacian in spherical polar coordinates is given by

$$\frac{1}{r^2}\frac{d}{dr}\left(r^2\frac{dV}{dr}\right) = \frac{-\rho}{\varepsilon_o}$$

As the charge distribution is spherically symmetric no θ, ϕ terms appear.

$$\Rightarrow \frac{d}{dr}\left(r^2\frac{dV}{dr}\right) = \frac{-\rho}{\varepsilon_o}r^2$$

Integrating

$$r^2\frac{dV}{dr} = \frac{-\rho r^3}{3\varepsilon_o} + \kappa$$

$$\Rightarrow \frac{dV}{dr} = \frac{-\rho r}{3\varepsilon_0} + \frac{\kappa}{r^2}$$

Now $\mathbf{E}_i = -\nabla V = \frac{-\partial V}{\partial r}\hat{r} = \frac{-dV}{dr}\hat{\mathbf{r}}$

Hence $\mathbf{E}_i = \left(\frac{\rho r}{3\varepsilon_o} - \frac{\kappa}{r^2}\right)\hat{\mathbf{r}}$

At r = 0 $E_i = \infty$. An infinite electric field is impossible, and hence the constant κ = 0. Thus

$$\mathbf{E}_i = \frac{\rho r}{3\varepsilon_o}\hat{\mathbf{r}}$$

(b) At a point outside the sphere there is no charge and hence ρ = 0. Here therefore

$$\nabla^2 V = \frac{\rho}{\varepsilon_o} = 0$$

$$\Rightarrow \frac{1}{r^2}\frac{d}{dr}\left(r^2\frac{dV}{dr}\right) = 0$$

$$\Rightarrow \frac{dV}{dr} = \frac{\kappa}{r^2}$$

$$\Rightarrow \mathbf{E}_o = -\nabla V = \frac{-dV}{dr}\hat{r} = \frac{-\kappa}{r^2}\hat{r}$$

where κ is a constant.

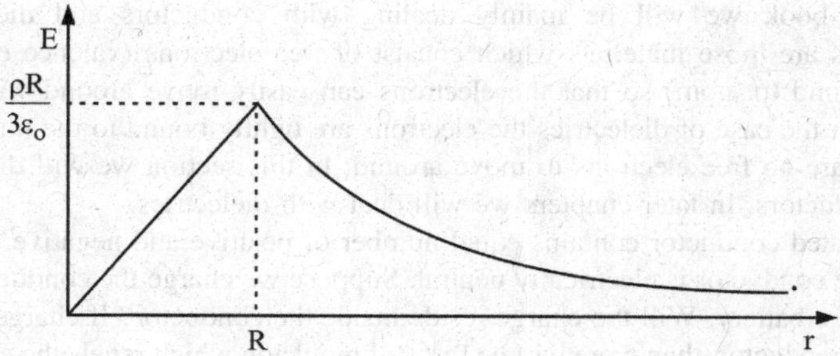


**Fig. 2.63** Electric field variation for a sphere of radius R carrying a uniform volume charge density

At $r = R \quad E_o = E_i$
Hence

$$\frac{-\kappa}{R^2} = \frac{\rho R}{3\varepsilon_0}$$
$$\Rightarrow \kappa = \frac{-\rho R^3}{3\varepsilon_0}$$

Substitute the value of $\kappa$ in $\mathbf{E}_o$.
Hence

$$\mathbf{E}_o = \frac{\rho R^3}{3\varepsilon_0} \frac{1}{r^2} \hat{r}$$
$$\Rightarrow \mathbf{E}_o = \frac{\rho R^3}{3\varepsilon_0 r^2} \hat{r}$$

The electric field of the spherical distribution of charge is plotted in Fig. 2.63.

The electric field within the spherical distribution of charge increases linearly. Outside the spherical distribution of charge the electric field falls off as inverse square law.

## 2.24 Conductors

Up to this point we have calculated the electric field of a static charge distribution in free space. However in number of practical cases we will be dealing with electric field in materials. Materials in general are of three types—conductors, semiconductors and insulators (dielectrics).

In this book we will be mainly dealing with conductors and dielectrics. Conductors are those materials which consist of free electrons (valence electrons loosely bound to atom) so that the electrons can easily move around within the material. In the case of dielectrics the electrons are tightly bound to the nucleus so that there are no free electrons to move around. In this section we will discuss in detail conductors. In later chapters we will deal with dielectrics.

An isolated conductor contains equal number of positive and negative charges; thereby the conductor is electrically neutral. Suppose we charge the conductor with the help of a battery. Will the charge reside inside the conductor? If charge resides inside the conductor, then there will be force of repulsion which repels the charge to the surface of the conductor and hence charges reside on the surface of conductor. The process talks about $10^{-14}$ s in a conductor like copper. As net charge resides on the surface of the conductor, inside the conductor there are an equal number of positive and negative charges. Hence the net charge density inside the conductor is zero.

$\rho = 0$ inside a conductor.

Suppose we place a conductor in an external electric field as shown in Fig. 2.64.

The electric field will attract the free electrons towards the right, while a net positive charge will appear on the other side. As the conductor has been placed in free space, the electrons are bounded within the conductor due to the insulating property of free space. In Fig. 2.64 we see that due to separation of charges an internal electric field $\mathbf{E}_{\text{internal}}$ is established within the conductor which is opposite to $\mathbf{E}_{\text{external}}$. Within the conductor both the fields cancel out each other and hence the field inside the conductor is zero. Charge flows inside the conductor to the surface until $\mathbf{E}_{\text{internal}} = \mathbf{E}_{\text{external}}$. So that the net field inside the conductor is zero, i.e. $\mathbf{E} = 0$. The charges residing over the surface of the conductor form a surface charge. The

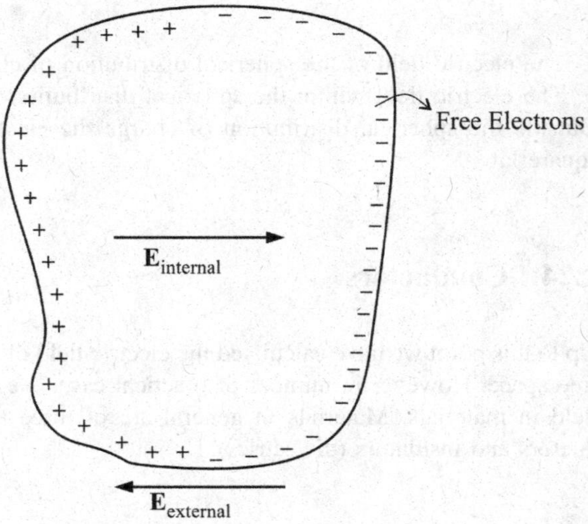


**Fig. 2.64** A conductor placed in an external electric field

surface charge produces an additional field outside the conductor which modifies the applied field.

Suppose assume that there is an empty cavity within the conductor as shown in Fig. 2.65a. Any net charge resides on the surface of the conductor as shown in the figure. However the surface of the cavity could also contain charges. Let us apply Gauss's law to surface S in Fig. 2.65a

$$\oint \mathbf{E} \cdot \mathbf{ds} = \frac{q_{enc}}{\varepsilon_o}$$

At surface S which is well within the conductor E = 0 and hence $q_{enc} = 0$ indicating that charges of single type + or – cannot be present on the cavity walls. However if the cavity surface contains an equal number of positive and negative charges as shown in Fig. 2.65b, still $q_{enc} = 0$ and such a situation cannot be ruled out using Gauss's law.

From Eq. 2.280

$$\oint \mathbf{E} \cdot \mathbf{d}\boldsymbol{l} = 0$$

Let us apply the above equation to loop $C_1$ shown in Fig. 2.65b. In loop $C_1$ we have marked three points a, b and d. The electric field points from positive to

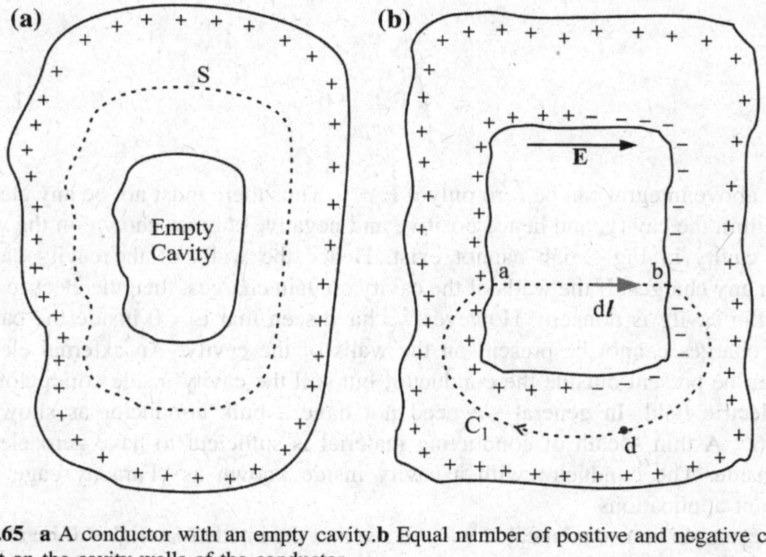


**Fig. 2.65** **a** A conductor with an empty cavity.**b** Equal number of positive and negative charges present on the cavity walls of the conductor

negative charge as shown in Fig. 2.65b. We have constructed the loop such that **E** and d$\boldsymbol{l}$ have the same direction, so that the angle between them is zero. Hence the above equation for loop $C_1$ becomes

$$\oint \mathbf{E}\cdot d\boldsymbol{l} = 0$$

$$\Rightarrow \int_a^b \mathbf{E}\cdot d\boldsymbol{l} + \int_b^d \mathbf{E}\cdot d\boldsymbol{l} + \int_d^a \mathbf{E}\cdot d\boldsymbol{l} = 0 \tag{2.294}$$

The paths bd and da are within the conductor where E = 0.
Hence

$$\int_a^b \mathbf{E}\cdot d\boldsymbol{l} + \int_b^d 0\cdot d\boldsymbol{l} + \int_d^a 0\cdot d\boldsymbol{l} = 0 \tag{2.295}$$

$$\Rightarrow \int_a^b \mathbf{E}\cdot d\boldsymbol{l} = 0 \tag{2.296}$$

$$\Rightarrow \int_a^b \mathrm{E}\, d l\cos 0 = 0 \tag{2.297}$$

$$\Rightarrow \int_a^b \mathrm{E}\, d l = 0 \tag{2.298}$$

The above integral can be zero only if E = 0. Thus there must not be any electric field within the cavity, and hence positive and negative charges shown on the walls of the cavity in Fig. 2.65b cannot exist. Hence the walls of the cavity cannot contain any charges. If the walls of the cavity contain charges, then the electric field inside the cavity is nonzero. However we have seen that E = 0 inside the cavity. Hence charges cannot be present on the walls of the cavity. An external electric field can be present outside the conductor, but still the cavity inside conductor has zero electric field. In general we need not have a bulk conductor as shown in Fig. 2.65. A thin sheath of conducting material is sufficient to have zero electric field inside. The conductor with a cavity inside known as "Faraday cage" has important applications.

For example in a coaxial cable we will see that the inside material will be always surrounded by a outside conducting sheath in order to protect the inside material from any stray electric fields present external to the conductor.

The conductor is an equipotential surface. As we have seen above the electric field inside the conductor under electrostatic condition is zero, i.e. $\mathbf{E} = 0$ inside the conductor. Hence for any two points a and b in the conductor

$$V(b) - V(a) = \int_b^a \mathbf{E} \cdot dl = 0$$

Thus conductors are equipotential surfaces.

Outside the conductor the electric field is perpendicular to the surface of the conductor. In case if there is any tangential component, then charge flows due to the action of the tangential component of the field and force of repulsion, until the tangential component of the field is zero. The surface charge is confined within the conducting object, and hence electric field is perpendicular to the surface of the conductor.

## 2.25 Boundary Conditions

Boundary conditions play an important role in calculating how much the electric field changes when we cross from one media to the other media. In this section we will see that whenever we cross a surface charge density $\sigma$, the electric field changes by an amount $\sigma/\varepsilon_0$.

Consider a surface charge density $\sigma$ as shown in Fig. 2.66. Consider a pillbox of length a. The area of top surface of the pillbox is A. Part of the pillbox lies above the surface charge, while the other part lies below the surface charge.

Applying Gauss's law to the pillbox

$$\oint \mathbf{E} \cdot \mathbf{ds} = \frac{q_{enc}}{\varepsilon_0} = \frac{\sigma A}{\varepsilon_0} \quad (2.299)$$

In the limit $a \to 0$ the contribution to the flux in Gauss's law comes only from the normal component of the electric field. Hence Eq. 2.299 can be written as

$$E_{1n}A - E_{2n}A = \frac{\sigma A}{\varepsilon_0} \quad (2.300)$$

$$\Rightarrow E_{1n} - E_{2n} = \frac{\sigma}{\varepsilon_0} \quad (2.301)$$

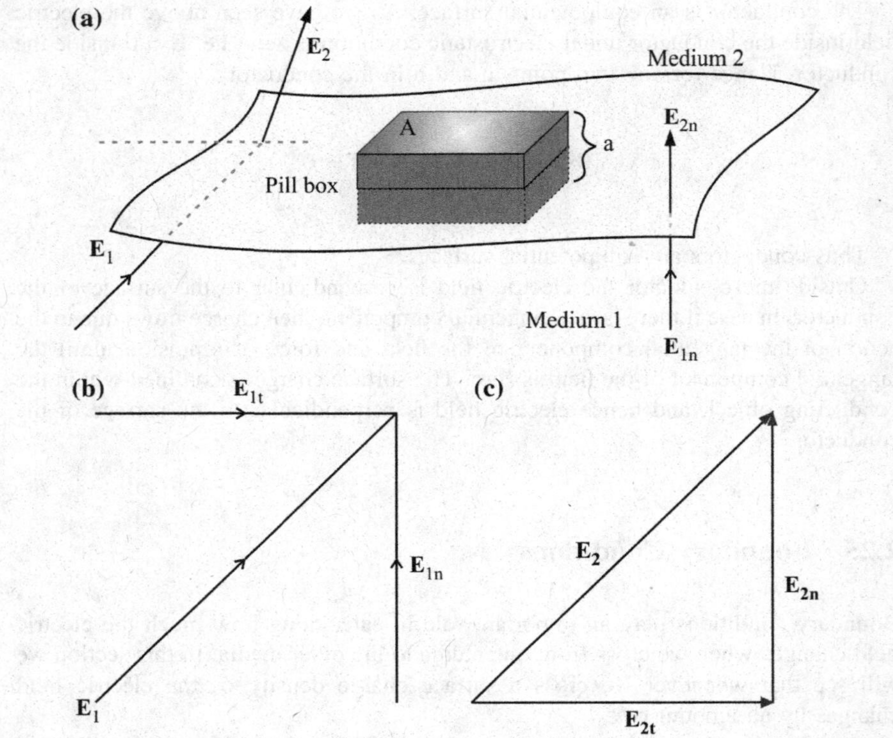


**Fig. 2.66** **a** The figure shows a surface charge density between medium 1 and medium 2 and a pill box Gaussian surface. **b, c** Resolution of electric field $E_1$ and $E_2$

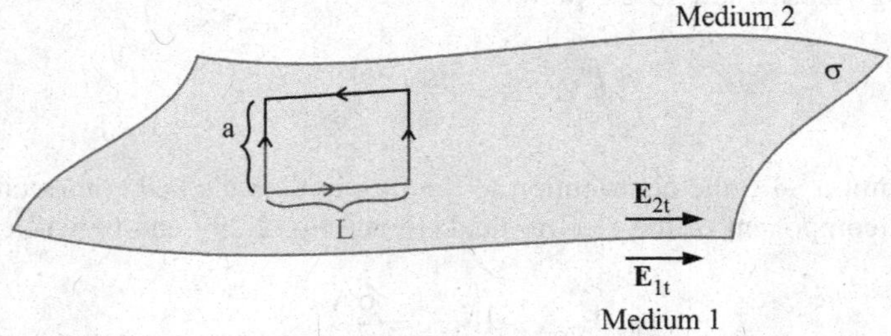


**Fig. 2.67** The figure shows a surface charge density between medium 1 and medium 2 along with a rectangular loop

Now consider a surface charge density σ as shown in Fig. 2.67. Consider a rectangle of length L and breath a as shown in Fig. 2.66. We know

$$\oint \mathbf{E} \cdot d\boldsymbol{l} = 0$$

In the limit a $\rightarrow$ 0 the contribution for the above integral is only from sides with dimension L.

Hence

$$E_{1t}L - E_{2t}L = 0 \tag{2.202}$$

$$\Rightarrow E_{1t} = E_{2t} \tag{2.303}$$

Hence the tangential components are equal at the boundary between the two media. Thus the boundary conditions are

$$E_{1n} - E_{2n} = \frac{\sigma}{\varepsilon_0}$$

$$E_{1t} = E_{2t}$$

Suppose medium 1 is free space and medium 2 is conductor as shown in Fig. 2.68. With E = 0 inside the conductor boundary conditions read

$$E_{1n} = \frac{\sigma}{\varepsilon_0} \tag{2.304}$$

$$E_{1t} = E_{2t} = 0 \tag{2.305}$$

where $\sigma$ is the "local" surface charge density.

**Fig. 2.68** At the conductor - free space boundary the electric field is normal to the surface of the conductor

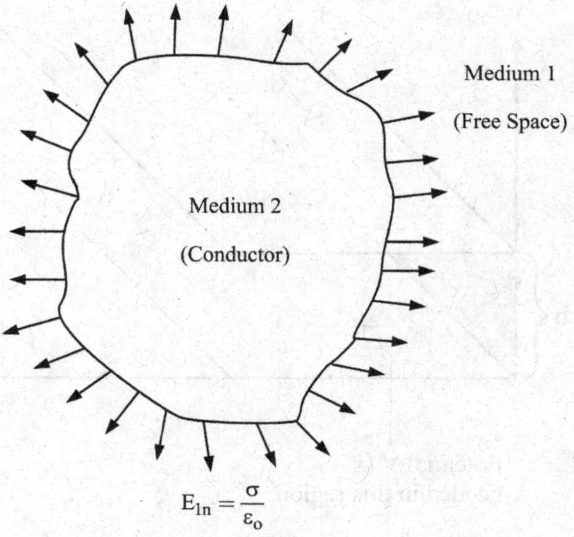

The above boundary conditions are pictorically depicted in Fig. 2.68. Thus at the conductor–free space boundary there are no tangential components and only normal component is present in the free space.

## 2.26 Uniqueness Theorem

As we have already mentioned previously that, we have considered charges distributed in free space. However in all practical cases we will be dealing with electric field in materials. We have already discussed conductors. In reality the conductors will be maintained at different potentials by connecting them to external power sources like batteries. Let us consider an example, shown in Fig. 2.69.

In this figure there are three conducting plates one at $x = 0$ maintained at potential $V_o$. The second plate is at $y = 0$ and is maintained at a potential $V = 0$. The third plate is at $y = b$ and is maintained at potential $V = 0$. Thus in Fig. 2.69 we have the following conditions.

(i) At $x = 0$, i.e. $V(0, y) = V_o$.
(ii) At $y = 0$, i.e. $V(x, 0) = 0$.
(iii) At $y = b$, i.e. $V(x, b) = 0$.

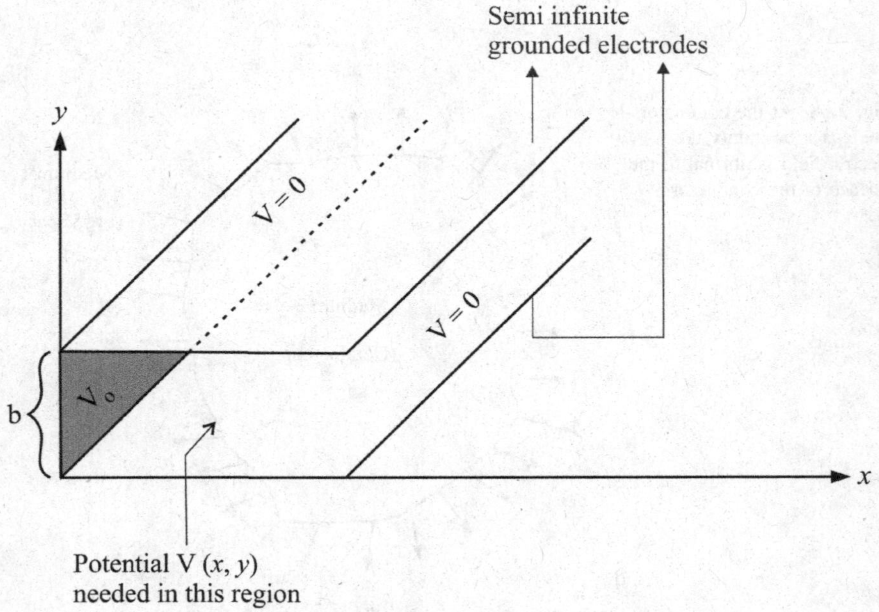


**Fig. 2.69** Three conducting plates maintained at different potentials.

These conditions are known as boundary conditions for given problem. We are interested in finding potential $V(x, y)$ in the region between three plates in Fig. 2.69. The method to find $V(x, y)$ is being found will be discussed in future chapters (see Sect. 3.9).

In Fig. 2.69 we are interested in knowing in the region between three plates whether there is only one value of potential $V(x, y)$ (or) two values of potential $V_1(x, y)$ and $V_2(x, y)$ (or) multiple values of potential. This point will be discussed in this section. The above problem is only an example. In similar problems we are interested in knowing whether the problem has single solution or two solutions or multiple solutions.

We have another important information from Fig. 2.69. Suppose assume that there are two potentials or two solutions in the region between the three plates – $V_1(x, y)$, $V_2(x, y)$.

Both the solutions $V_1$ and $V_2$ must be equal to $V_o$ at $V(0, y)$. The potential $V_o$ at $V(0, y)$ is specified by us, by connecting the conducting plate at $x = 0$ to external power supplies, say batteries. Hence at the boundary $x = 0$, $V_1 = V_o$ and $V_2 = V_o$. If we define $V_d = V_1 - V_2$, then at the boundary $x = 0$

$$V_d = V_1 - V_2 = V_o - V_o = 0.$$

Similar argument shows that

(i) $y = 0$ for the plate at $y = 0$

$$V_1 = 0, \quad V_2 = 0$$
$$V_d = V_1 - V_2 = 0$$

(ii) $y = b$ for the plate at $y = b$

$$V_1 = 0, V_2 = 0$$
$$V_d = V_1 - V_2 = 0 - 0 = 0.$$

"The same is true for any given problem. Because we are specifying the potential at the boundaries by connecting the respective boundary to an external power source, the potential, or solution obtained in the region of interest, must be equal to potential given to the boundary, at the boundary. Hence if there are multiple solutions or potentials in the region of interest, the difference between any two solutions $V_d$ must be equal to zero at the boundary".

$$V_d = 0 \text{ at the boundary} \tag{2.306}$$

"Figure 2.69 shows a specific problem in which three electrodes are maintained at specified potentials. In general i number of conductors can be maintained at different potentials as shown in Fig. 2.70a. We are interested in finding the potential in the surrounding region of the conductors. We are interested in knowing whether

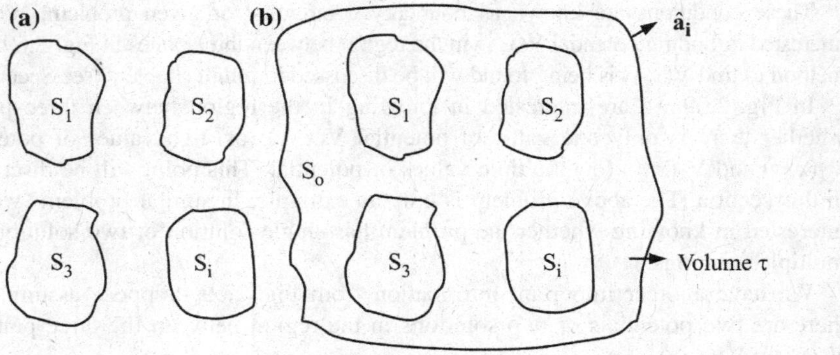


**Fig. 2.70** **a** Number of conductors maintained at specified potentials. **b** A surface $S_o$ encloses all the conductors maintained at specified potentials

the problem has no solution, one solution or more than one solution". The answer to the above question is given by uniqueness theorem.

***Uniqueness theorem states that the Poisson equation or Laplace's equation satisfying the given boundary conditions has one and only solution, i.e. unique solution.***

Let us now consider Fig. 2.70a. A number of conductors whose surfaces are $S_1$, $S_2 \ldots S_i$ are maintained at specified potentials.

Construct a surface $S_o$ which encloses all the conductors, and the total volume is $\tau$. Contrary to uniqueness let us assume there are two solutions $V_1$ and $V_2$ in the region of interest. Then

$$\nabla^2 V_1 = \frac{-\rho}{\varepsilon_o} \tag{2.307}$$

$$\nabla^2 V_2 = \frac{-\rho}{\varepsilon_o} \tag{2.308}$$

Let us denote the difference in potentials $V_1$ and $V_2$ as $V_d$. We have

$$V_d = V_1 - V_2 \tag{2.309}$$

$$\Rightarrow \nabla^2 V_d = \nabla^2 V_1 - \nabla^2 V_2$$
$$= \frac{-\rho}{\varepsilon_o} + \frac{-\rho}{\varepsilon_o} = 0$$

Hence

$$\nabla^2 V_d = 0 \tag{2.310}$$

From Sect. 1.20, using vector identity 10 we can write

$$\nabla\cdot(V_d\nabla V_d)=V_d\nabla.(\nabla V_d)+\nabla V_d.\nabla V_d \tag{2.311}$$

$$\nabla\cdot(V_d\nabla V_d)=V_d\nabla^2V_d+|\nabla V_d|^2 \tag{2.312}$$

Integrating the above equation over the volume $\tau$

$$\int_\tau \nabla.(V_d\nabla V_d)d\tau=\int_\tau V_d\nabla^2V_d d\tau+\int_\tau|\nabla V_d|^2d\tau \tag{2.313}$$

Substituting Eq. 2.310 in 2.313

$$\int_\tau \nabla\cdot(V_d\nabla V_d)d\tau=\int_\tau|\nabla V_d|^2d\tau \tag{2.314}$$

Applying Gauss's divergence theorem

$$\oint_S (V_d\nabla V_d)\cdot\hat{\mathbf{a}}_\mathbf{i}\,ds=\int_\tau|\nabla V_d|^2d\tau \tag{2.315}$$

Here the surface S encloses $S_0$, $S_1$, $S_2\ldots S_i$.
Hence

$$\oint_{S_0+(S_1+S_2+\cdots+S_i)} (V_d\nabla V_d)\cdot\hat{\mathbf{a}}_\mathbf{i}\,ds=\int_\tau|\nabla V_d|^2d\tau \tag{2.316}$$

On the conducting boundaries $S_1$, $S_2$, ... $S_i$ $V_d$ is zero from Eq. 2.306. Hence Eq. 2.316 reduces to

$$\int_{S_0} (V_d\nabla V_d)\cdot\hat{\mathbf{a}}_\mathbf{i}ds=\int_\tau|\nabla V_d|^2d\tau \tag{2.317}$$

We have constructed surface $S_0$, and we can construct the surface $S_0$ as a sphere of infinite radius. That is the surface $S_0$ lies at very large distance from the conducting boundaries. Because the surface $S_0$ is of large radii, the conducting boundaries will look like point charge to the surface $S_0$. Hence $V_1$ and $V_2$ fall off as $\frac{1}{r}$. So $V_d=V_1-V_2$ also falls off like $\frac{1}{r}$.$\nabla V_d$ then falls off like $\frac{1}{r^2}$. Thus the term $V_d\nabla V_d$ falls off like $\frac{1}{r^3}$. The surface $S_0$ for the sphere increases as $4\pi r^2$ (surface area of sphere is $4\pi r^2$). Thus the total integral $\oint_S (V_d\nabla V_d)\cdot ds$ falls off like $\frac{1}{r}$. At large distances as r goes of to infinity then $\oint_S (V_d\nabla V_d)\cdot ds$ falls off to zero.

Thus

$$\oint (V_d \nabla V_d) \cdot \mathbf{ds} = 0 \tag{2.318}$$

Hence from Eq. 2.317

$$\int_\tau |\nabla V_d|^2 d\tau = 0 \tag{2.319}$$

Because $|\nabla V_d|^2$ is positive everywhere, the above integration can be zero only when

$$|\nabla V_d| = 0 \tag{2.320}$$

for all point within $\tau$.

$$\Rightarrow V_d = \kappa \; [\text{for all points within } \tau] \tag{2.321}$$

where $\kappa$ is a constant. We see that $V_d = \kappa$ at all points within $\tau$ which includes surfaces $S_1, S_2, \ldots S_i$, but already we have seen that on the surfaces $S_1, S_2, \ldots S_i$ the value of $V_d$ is zero. Hence at the boundaries

$$V_d = \kappa = 0 \tag{2.322}$$

But from Eq. 2.321 $\kappa$ is equal to $V_d$ for all points within $\tau$. From Eq. 2.322 at the boundaries $S_1, S_2\ldots\; S_i$ which are enclosed in volume $\tau$, $V_d = \kappa = 0$. Hence $\kappa$ is zero everywhere within volume $\tau$. Hence Eq. 2.321 is

$$V_d = \kappa = 0 \text{ for all points within } \tau. \tag{2.323}$$

$$\begin{aligned} &\Rightarrow \quad V_d = 0 \\ &\Rightarrow \quad V_1 - V_2 = 0 \\ &\Rightarrow \quad V_1 = V_2 \end{aligned} \tag{2.324}$$

Hence Uniqueness theorem is proved. Thus as an example, in the region between three plates in Fig. 2.69 there is only one possible potential or solution, V $(x, y)$, for the boundary conditions stated at the beginning of this section.

## Exercises

**Problem 2.1**
Consider a infinite long cylinder of radius R. The surface of the cylinder contains uniform charge density $\sigma$. A uniform line charge of charge density $\lambda$ lies in the axis of the cylinder. Calculate the electric field inside and outside the cylinder. Which method you will use to calculate the electric field and why?

**Problem 2.2**
Consider a circular loop of radius R. Calculate the electric field at a point P as shown in Fig. 2.71. Point P lies along the axis of the loop and is at a distance of a from the centre of the ring.

**Problem 2.3**
Three charges of magnitude

$$\begin{aligned} q_1 &= -2.5 \quad \text{nC}, \\ q_2 &= 5 \quad \text{nC}, \\ q_3 &= -7 \quad \text{nC} \end{aligned}$$

are situated as shown in Fig. 2.72 with OA = 12 cm, OB = 9 cm. Calculate the force $q_1$.

**Problem 2.4**
Solve Example 2.13 using Gauss's law.

**Problem 2.5**
If three charges of magnitude q are situated at the three corners of the square, what would be the field at point P in Fig. 2.73.

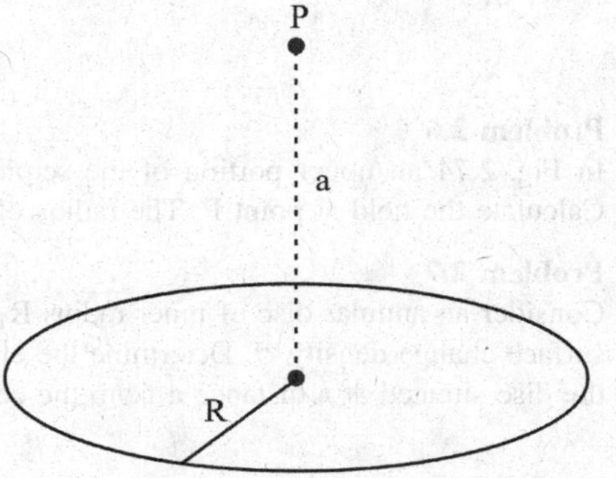


**Fig. 2.71** A circular loop. Illustration for problem 2.2

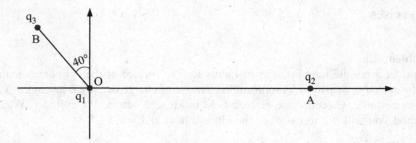

Fig. 2.72 Illustration for problem 2.3.

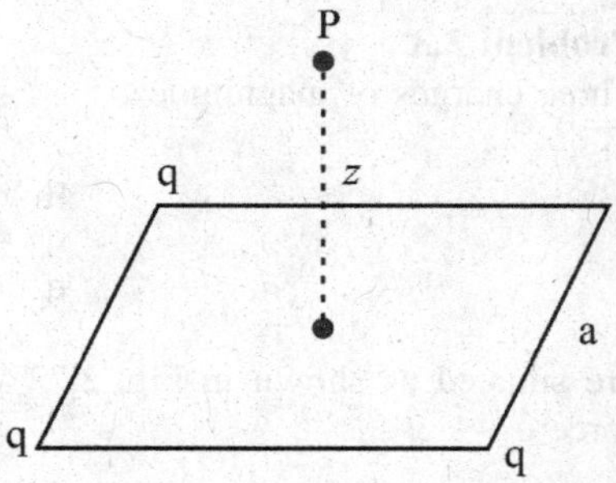


**Fig. 2.73** Three charges situated at three corners of the square. Illustration for problem 2.5

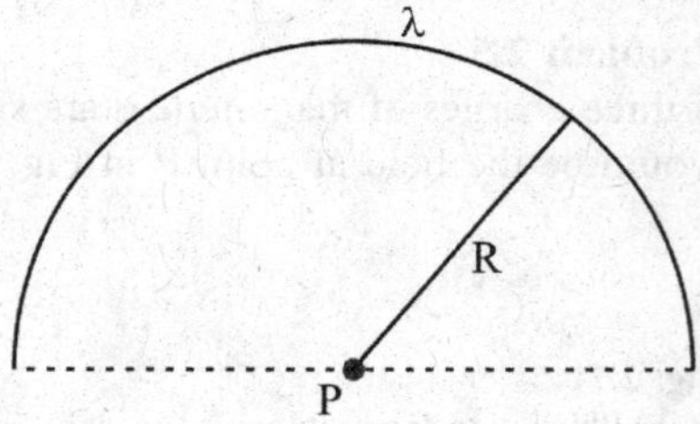


**Fig. 2.74** A semicircle carrying a uniform line charge density. Illustration for problem 2.6.

**Problem 2.6**
In Fig. 2.74 an upper portion of the semicircle carries a uniform line charge $\lambda$. Calculate the field at point P. The radius of the semicircle is R.

**Problem 2.7**
Consider an annular disc of inner radius $R_1$ and outer radius $R_2$ having a uniform surface change density $\sigma$. Determine the electric field at point P along the axis of the disc situated at a distance a from the centre of the disc as shown in Fig. 2.75.

Fig. 2.75 An annular disc carrying a uniform surface charge density. Illustration for problem 2.7

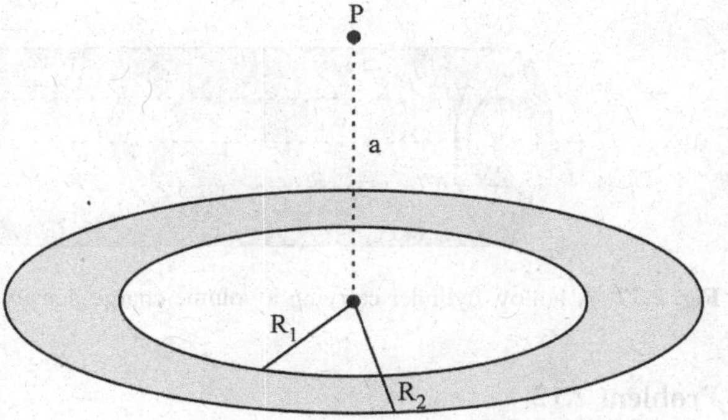


Fig. 2.76 I A hollow sphere carrying a volume charge density. Illustration for problem 2.8

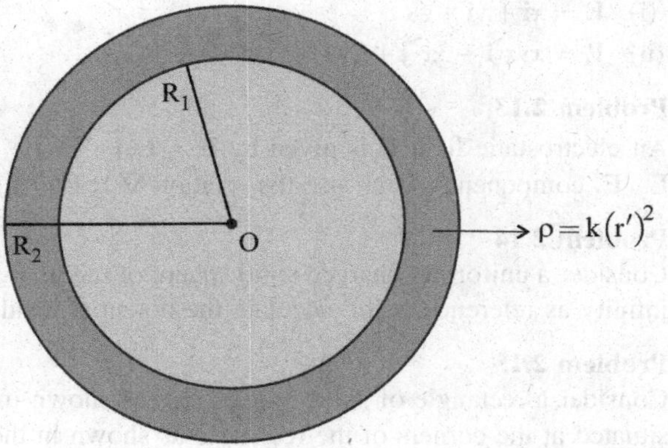


**Problem 2.8**
Consider a hollow sphere as shown in Fig. 2.76. Within the region $R_1 \le r \le R_2$ the sphere contains a volume charge density $\rho = k(r')^2$. Find the electric field at a point $r \le R_1, R_1 \le r \le R_2, r \ge R_2$ (k is a constant).

**Problem 2.9**
Calculate the field inside and outside of an infinitely long hollow cylinder whose radius is R. The cylinder carries a uniform surface charge density $\sigma$.

**Problem 2.10**
Calculate the field inside and outside of an infinitely long solid cylinder whose radius is R. The cylinder carries a uniform volume charge density $\rho$.

**Problem 2.11**
Consider infinitely long hollow cylinder as shown in Fig. 2.77. The inner and outer radii of the cylinder are $R_1$ and $R_2$, respectively. The cylinder carries a charge density of $\rho = k(r')^2$ within $R_1 \le r \le R_2$. Calculate the electric field at a point $r \le R_1, R_1 \le r \le R_2, r \ge R_2$.

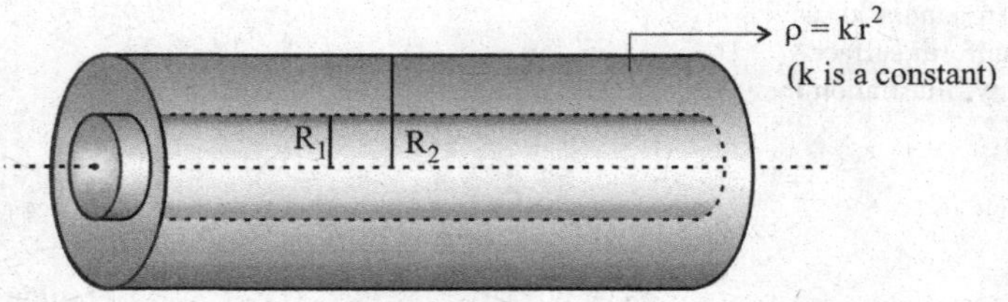


**Fig. 2.77** A hollow cylinder carrying a volume charge density. Illustration for problem 2.11

**Problem 2.12**
Check whether the following electrostatic fields are possible

(i) $\mathbf{E} = x\hat{\mathbf{i}} + y\hat{\mathbf{j}} + z\hat{\mathbf{k}}$
(ii) $\mathbf{E} = xyz^2\hat{\mathbf{i}} + 3x^2\hat{\mathbf{j}} + 2y\hat{\mathbf{k}}$

**Problem 2.13**
An electrostatic field **E** is given by $\mathbf{E} = \mathrm{E}_x\,\hat{\mathbf{i}} + \mathrm{E}_y\,\hat{\mathbf{j}} + \mathrm{E}_z\,\hat{\mathbf{k}}$. Given $\mathrm{E}_x = 3z^2y$. Find $\mathrm{E}_y$, $\mathrm{E}_z$ components (Hint: use the relation $\nabla \times \mathbf{E} = 0$).

**Problem 2.14**
Consider a uniformly charged solid sphere of radius R with charge density ρ. Using infinity as reference point calculate the potential inside and outside the sphere.

**Problem 2.15**
Consider a rectangle of sides 3 and 7 cm as shown in Fig. 2.78. Four charges are situated at the corners of the rectangle as shown in the figure. Find the magnitude and direction of the resultant force on $q_A$. Medium is free space.

**Problem 2.16**
A unit positive charge is situated at point P as shown in Fig. 2.79. Here $q_1 = 1$ nC and $q_2 = 3$ nC. Calculate the force in the unit positive charge. Medium is free space.

**Fig. 2.78** Illustration for problem 2.15

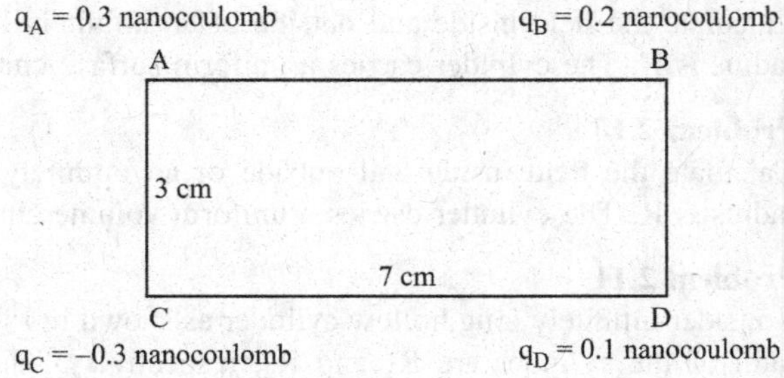

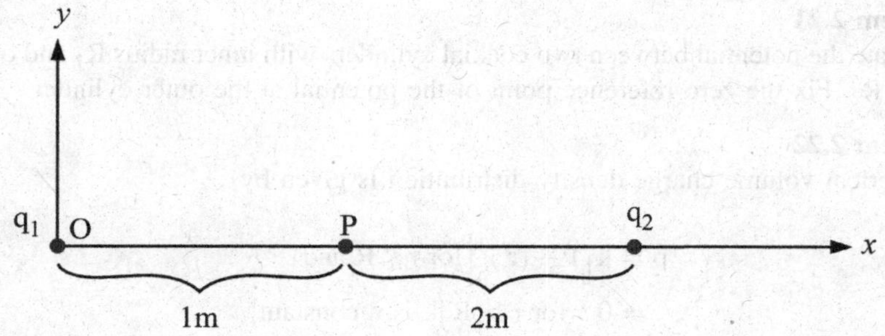


**Fig. 2.79** Illustration for problem 2.16

**Problem 2.17**
Three charges $q_A$, $q_B$ and $q_C$ each of magnitude 3, 2 and –5 nC are located at (1, 0, 0), (0, 3, 0) and (0, 0, 7). Calculate the electric field at (0, 5, 0). Medium is free space.

**Problem 2.18**
A square consists of two charges $q_A = 2$ nC and $q_B = 1$ nC as shown in Fig. 2.80. The other side of the square consists of a line charge $\lambda = 10^{-10}$coul/m as shown in the figure. Calculate the potential at point O.

**Problem 2.19**
Calculate **E** if the potential is given by (k is a constant)

$$V = \frac{k}{x^2 + y^2}$$

**Problem 2.20**
Consider a sphere of surface area 64 π m$^2$. A point charge of magnitude 1 nC is located somewhere within the sphere. Calculate the electric flux at a point located at 5 m from the centre of the sphere.

**Fig. 2.80** Illustration for problem 2.18

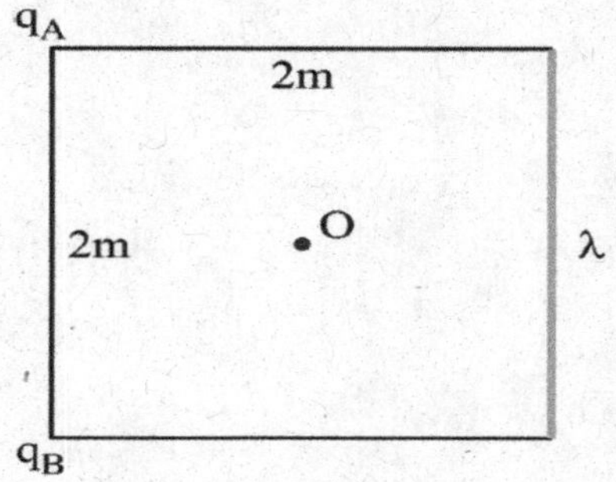

**Problem 2.21**

Calculate the potential between two coaxial cylinders with inner radius $R_1$ and outer radius $R_2$. Fix the zero reference point of the potential at the outer cylinder.

**Problem 2.22**

A spherical volume charge density distribution is given by

$$\rho = k\left[1 - (r')^2\right] \text{ for } r \leq R \text{ and}$$
$$= 0 \quad \text{for } r > R \; [k \text{ is a constant}]$$

Find the total charge q.

# Chapter 3
# Electric Charges at Rest—Part II

## 3.1 Work Done

Suppose assume that we are interested in moving a charge Q from point $a_1$ to point $a_2$ in the presence of a charge q as shown in Fig. 3.1a. The work done in the process is

$$W = \int_{a_1}^{a_2} \mathbf{F} \cdot d\boldsymbol{l} \quad (3.1)$$

Charge q will exert a force of QE to oppose the movement and hence we have exert a minimum force of –QE to make the charge Q move from $a_1$ to $a_2$. Hence Eq. 3.1 is

$$W = -Q \int_{a_1}^{a_2} \mathbf{E} \cdot d\boldsymbol{l} \quad (3.2)$$

With the aid of Eq. 2.284, Eq. 3.2 can be written as

$$W = -Q[V(a_1) - V(a_2)] \quad (3.3)$$

$$\Rightarrow W = Q[V(a_2) - V(a_1)] \quad (3.4)$$

Suppose we move charge Q from infinity to point $a_2$ as shown in Fig. 3.1b then

$$W = Q[V(a_2) - V(\infty)] \quad (3.5)$$

as $V(\infty) = 0$ we have

© Springer Nature Singapore Pte Ltd. 2020
S. Balaji, *Electromagnetics Made Easy*,
https://doi.org/10.1007/978-981-15-2658-9_3

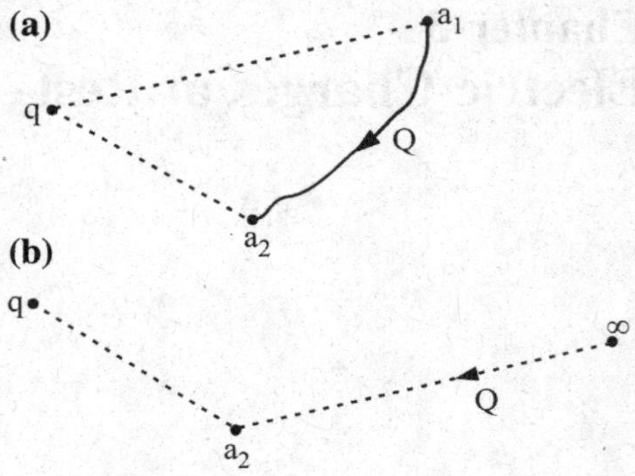


**Fig. 3.1** **a** Charge Q is moved from $a_1$ to $a_2$ in the presence of charge q. **b** Charge Q is moved from infinity to a2 in the presence of charge q

$$W = Q\,V(a_2) \tag{3.6}$$

In Sect. 2.22 we have noted that occasionally we will be changing the zero reference point for calculating potential. One interesting point to be noted is that work done doesn't change when the zero reference point is changed. Figure 2.55 is reproduced in Fig. 3.2.

From Eq. 2.288 we see that

$$V(a_1) = \kappa + V_1(a_1) \tag{3.7}$$

$$V(a_2) = \kappa + V_1(a_2) \tag{3.8}$$

where $V(a_1)$, $V(a_2)$ are the potentials at points $a_1$, $a_2$ with respect to reference point R. $V_1(a_1)$, $V_2(a_2)$ are the potentials at points $a_1$, $a_2$ with respect to reference point $R_1$ (see Fig. 3.3).

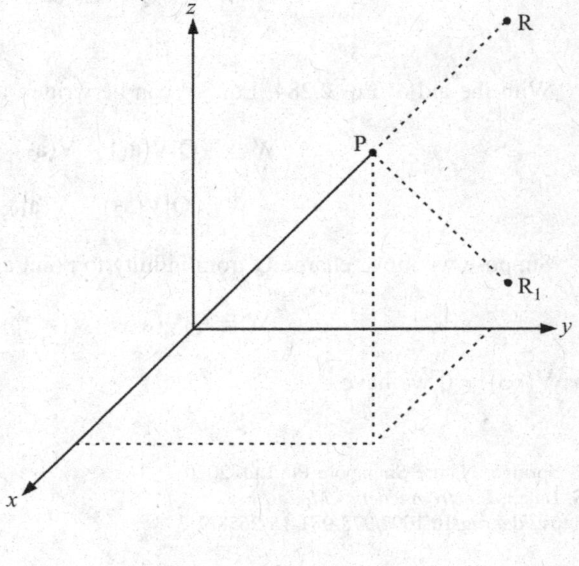


**Fig. 3.2** Figure showing change of reference point from R to R1 to measure potential at point P

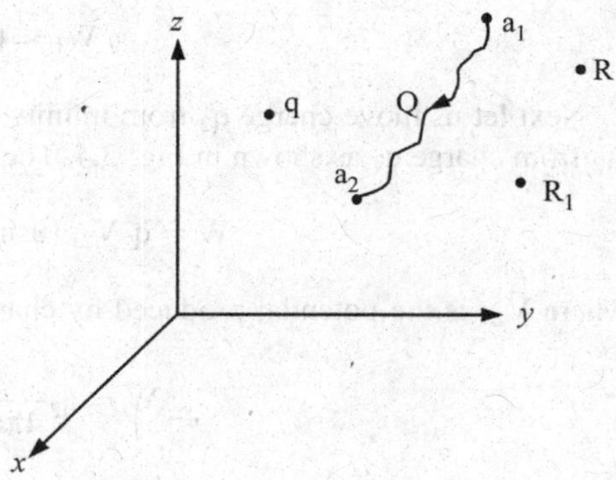


**Fig. 3.3** Charge Q is moved from $a_1$ to $a_2$ in the presence of charge q. Reference point R and $R_1$ are shown

From Eqs. 3.7 and 3.8

$$V(a_1) - V(a_2) = V_1(a_1) - V_1(a_2) \tag{3.9}$$

$$\Rightarrow Q[V(a_1) - V(a_2)] = Q[V_1(a_1) - V_2(a_2)] \tag{3.10}$$

Thus work done doesn't get altered by changing the reference point.

## 3.2 Energy in Electrostatic Fields

Suppose assume that we are interested in moving a charge $q_1$ from infinity and placing it at a point A as shown in Fig. 3.4. The amount of work done in the process is zero because there is no electric field against which we are doing work. Hence,

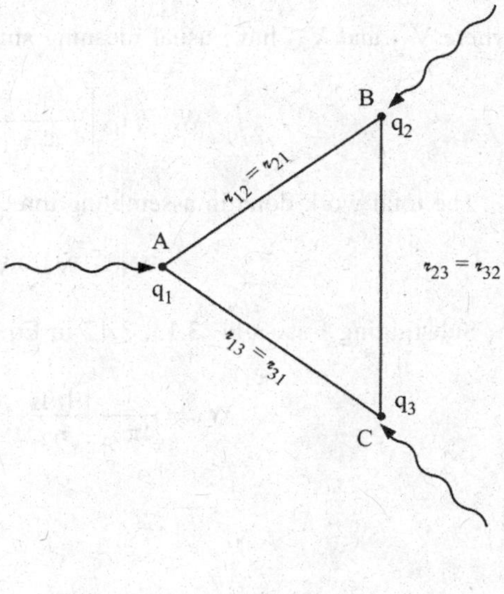


**Fig. 3.4** Three charges $q_1$, $q_2$ and $q_3$ are moved from infinity and placed at A, B and C

$$W_1 = 0 \tag{3.11}$$

Next let us move charge $q_2$ from infinity and place it at point B at a distance of $r_{12}$ from charge $q_1$ as shown in Fig. 3.4. The amount of work done in the process is

$$W = q_2 V_{21} \text{ (using Eq. 3.6)}$$

where $V_{21}$ is the potential produced by charge $q_1$ at the site of charge $q_2$

$$W = q_2 \frac{q_1}{4\pi\varepsilon_0 r_{21}} \tag{3.12}$$

Because $r_{12} = r_{21}$ in Fig. 3.4, Eq. 3.12 can be written as

$$W_2 = q_1 \frac{q_2}{4\pi\varepsilon_0 r_{12}} \tag{3.13}$$

$$W_2 = q_1 V_{12} \tag{3.14}$$

where $V_{12}$ is the potential at the site of $q_1$ produced by charge $q_2$. As $W_1 = 0$ total work done in assembling charges $q_1$ and $q_2$ is

$$\begin{aligned} W_2 &= q_2 V_{21} = q_1 V_{12} = \frac{1}{2}(q_1 V_{12} + q_2 V_{21}) \\ W_2 &= \frac{1}{2}\left\{\frac{1}{4\pi\varepsilon_0}\left(\frac{q_1 q_2}{r_{12}} + \frac{q_2 q_1}{r_{21}}\right)\right\} \end{aligned} \tag{3.15}$$

Now let us bring another charge $q_3$ from infinity and place it at point C as shown in Fig. 3.4. The work done in placing charge $q_3$ at point C is

$$W_o = q_3[V_{31} + V_{32}] \tag{3.16}$$

where $V_{31}$ and $V_{32}$ have usual meaning similar to $V_{12}$. Equation 3.16 is

$$W_o = q_3\left[\frac{q_1}{4\pi\varepsilon_0 r_{31}} + \frac{q_2}{4\pi\varepsilon_0 r_{32}}\right] \tag{3.17}$$

The total work done in assembling three charges $q_1$, $q_2$, $q_3$ is then

$$W_3 = W_1 + W_2 + W_o \tag{3.18}$$

Substituting Eqs. 3.11, 3.13, 3.17 in Eq. 3.18 we obtain

$$W_3 = \frac{1}{4\pi\varepsilon_0}\left[\frac{q_1 q_2}{r_{12}} + \frac{q_3 q_1}{r_{31}} + \frac{q_3 q_2}{r_{32}}\right]$$

The above equation can be written as

$$W_3 = \frac{1}{2}\left\{\frac{1}{4\pi\varepsilon_o}\left[\frac{q_2q_1}{r_{21}} + \frac{q_1q_2}{r_{12}} + \frac{q_3q_1}{r_{31}} + \frac{q_1q_3}{r_{13}} + \frac{q_3q_2}{r_{32}} + \frac{q_2q_3}{r_{23}}\right]\right\} \quad (3.19)$$

In general if we are assembling n number of charges we can write

$$W = \frac{1}{2}\left[\frac{1}{4\pi\varepsilon_o}\sum_{i=1}^{n}\sum_{j=1}^{n}\frac{q_iq_j}{r_{ij}}\right] \quad (3.20)$$

with $i \neq j$

Equation 3.19 can be rewritten as

$$W = \frac{1}{2}\sum_{i=1}^{n} q_i \sum_{j=1}^{n}\left(\frac{1}{4\pi\varepsilon_o}\frac{q_j}{r_{ij}}\right) \quad (3.21)$$

$$\Rightarrow W = \frac{1}{2}\sum_{i=1}^{n} q_i \sum_{j=1}^{n} V_{ij} \quad (3.22)$$

where $\sum_{j=1}^{n} V_{ij}$ is the total potential produced at the site of $q_i$ by all the remaining j charges.

For continuous charges distributions Eq. 3.22 can be written as

$$W = \frac{1}{2}\int_{\tau} \rho\, V d\tau \quad (3.23)$$

Poisson's equation (Eq. 2.292) is

$$\rho = -\varepsilon_o \nabla^2 V \quad (3.24)$$

Substituting Eq. 3.24 in Eq. 3.23

$$W = \frac{-\varepsilon_o}{2}\int_{\tau} V(\nabla^2 V) d\tau \quad (3.25)$$

From Sect. 1.20 using vector identity 10 we have

$$\nabla \cdot (u\mathbf{A}) = u\nabla \cdot \mathbf{A} + \mathbf{A} \cdot \nabla u$$

Setting $u = V$, $\mathbf{A} = \nabla V$ We get

$$\nabla \cdot (V\nabla V) = V\nabla \cdot (\nabla V) + \nabla V \cdot \nabla V \tag{3.26}$$

$$\nabla \cdot (V\nabla V) = V\nabla^2 V + \nabla V \cdot \nabla V \tag{3.27}$$

$$V\nabla^2 V = \nabla \cdot (V\nabla V) - \nabla V \cdot \nabla V \tag{3.28}$$

Noting down that $\mathbf{E} = -\nabla V$

$$V\nabla^2 V = -\nabla \cdot (V\mathbf{E}) - E^2 \tag{3.29}$$

Substituting Eq. 3.29 in Eq. 3.25

$$W = \frac{\varepsilon_0}{2}\int_\tau \nabla \cdot (V\mathbf{E})d\tau + \frac{\varepsilon_0}{2}\int_\tau E^2 d\tau \tag{3.30}$$

Applying Gauss divergence theorem

$$W = \frac{\varepsilon_0}{2}\oint_S V\mathbf{E} \cdot d\mathbf{S} + \frac{\varepsilon_0}{2}\int_\tau E^2 d\tau \tag{3.31}$$

In Eq. 3.31 we are integrating over the volume $\tau$. The volume $\tau$ is the volume which encloses all the charges. However if we increase the volume $\tau$ beyond the volume enclosing all the charges the work done given by Eq. 3.31 is not going to increase as $\tau$ increases, because $\rho = 0$ in outer volume ($\rho = 0$ in Eq. 3.23 for outer volume that doesn't contain charges). Hence let us select the sphere of very large radius. At such large distances charge density $\rho$ will look like a point charge. Thus in the first term of Eq. 3.31 V falls of as $\frac{1}{\mathscr{r}}$, **E** falls of as $\frac{1}{\mathscr{r}^2}$ while the surface S increases as $\mathscr{r}^2$ so the entire term $\oint_S V\mathbf{E} \cdot d\mathbf{S}$ falls of as $\frac{1}{\mathscr{r}}$.

Thus for $\mathscr{r}$ going to infinity $\oint_S V\mathbf{E} \cdot d\mathbf{S}$ falls of to zero.

Hence Eq. 3.31 is

$$W = \frac{1}{2}\varepsilon_0 \int_\tau E^2 d\tau \tag{3.32}$$

In the above equation $E^2$ falls of as $\frac{1}{\mathscr{r}^4}$ while the volume increases as $\mathscr{r}^3$ (volume of sphere is $\frac{4}{3}\pi\mathscr{r}^3$). So the entire integral $\int_\tau E^2\, d\tau$ must fall of as $\frac{1}{\mathscr{r}}$ and at large $\mathscr{r}$ must go to zero. But it is incorrect. Can the reader say why?

Finally for line and surface charge densities we have

$$W = \frac{1}{2}\int \lambda V dl \tag{3.33}$$

$$\text{and} \quad W = \frac{1}{2}\int \sigma V\, dS \tag{3.34}$$

The work done in assembling the charges is stored as potential energy in the system.

While Eq. 3.23 suggests electrostatic energy is stored in the charge, Eq. 3.32 suggests energy is stored in the field surrounding the charges. The exact location where the energy is stored is not answerable and only thing that can be suggested is, both equations calculate the electrostatic energy for a given charge distribution correctly.

## 3.3 Equipotential Surfaces

Equipotential surfaces are those surfaces for which all the points on the surface have same potential. From Eq. 3.4

$$W = Q[V(a_2) - V(a_1)]$$

For an equi-potential surface $V(a_2) = V(a_1)$ and hence work done on a equation potential surface is zero.

Let us move a charge Q to distance d$l$ on a equi-potential surface as shown in Fig. 3.5a.

The work done in the process is zero because the surface is an equi-potential surface.

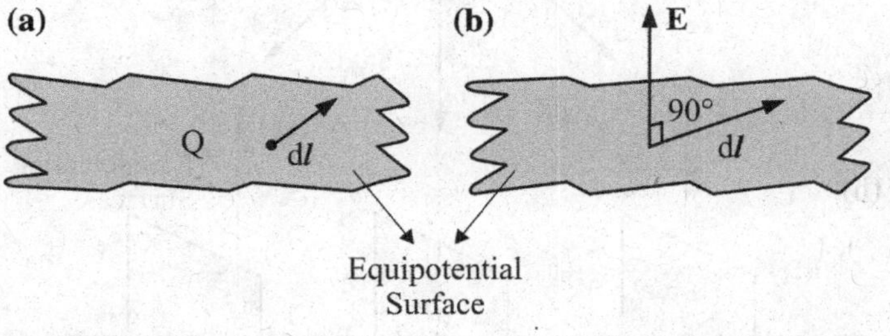


**Fig. 3.5** **a** Charge q moved on an equipotential surface through a small distance. **b** The electric field is perpendicular to the equipotential surface

Hence

$$
\begin{aligned}
& dW = \mathbf{F} \cdot d\boldsymbol{l} = 0 \\
& \Rightarrow Q\mathbf{E} \cdot d\boldsymbol{l} = 0 \\
& \Rightarrow \mathbf{E} \cdot d\boldsymbol{l} = 0 \\
& \Rightarrow E\,dl \cos\theta = 0 \\
& \text{As } E \neq 0, dl \neq 0 \\
& \cos\theta = 0. \\
& \Rightarrow \theta = 90^\circ
\end{aligned}
$$

Hence the angle between **E**, d***l*** is 90° for an equipotential surface. This is shown pictorially in Fig. 3.5b. Thus **E** is perpendicular to a equipotential surface.

Now consider a point charge q as shown in Fig. 3.6a. As the potential of the point charge is given by $\frac{q}{4\pi\varepsilon_o r}$, all points equidistant from the point charge are at same potential. Thus the equi potential surfaces for a point charge are concentric spheres as shown in Fig. 3.5a. The potential however will be different for different spheres. Because the electric field is perpendicular to the equipotential surface its direction is radial as shown in the same figure.

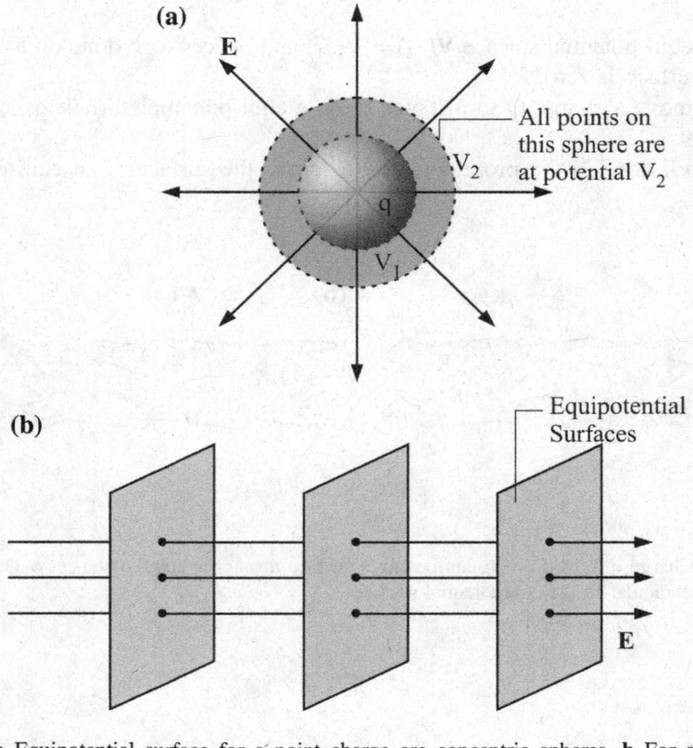


**Fig. 3.6** **a** Equipotential surface for a point charge are concentric spheres. **b** For an uniform electric field equipotential surfaces are planes

For a uniform electric field in Fig. 3.6b the equipotential are planes which lie perpendicular to the electric field.

In Sect. 2.24 we have seen that under electrostatic conditions the electric field inside a conductor is zero.

From Eq. 2.284

$$\int_{A}^{B} \mathbf{E} \cdot d\mathbf{l} = V_A - V_B$$

as $\mathbf{E} = 0$ inside a conductor

$$0 = V_A - V_B$$
$$V_A = V_B.$$

Thus any two points in a conductors are at same potential. Hence conductors are equi potential surfaces under electrostatic conditions. Thus work done to move a charge in conductor under electrostatic condition is zero.

## 3.4 A Note on Work Done

Similar to Eqs. 2.11, 2.58 and 2.77 can we write

$$W = W_1 + W_2 + \cdots + W_n \tag{3.35}$$

$$W = \sum_{i=1}^{n} W_i \tag{3.36}$$

See Problem 3.21.

**Example 3.1**
A total charge q is uniformly distributed through out a sphere of radius R and with a volume charge density of ρ. Calculate the energy stored in the sphere.

**Solution**
From Example 2.13 we see that

$$E_{out} = \frac{\rho R^3}{3\varepsilon_0 r^2} \text{ outside the sphere}$$
$$E_{in} = \frac{\rho r}{3\varepsilon_0} \text{ inside the sphere}$$

(a) ***Using the expression*** $W = \frac{\varepsilon_o}{2}\int_0^\infty E^2\, d\tau$

We have

$$W = \frac{\varepsilon_o}{2}\int E^2\, d\tau$$

$$W = \frac{\varepsilon_o}{2}\int_0^{\pi}\int_0^{2\pi}\int_0^{\infty} E^2\, d\tau$$

$$W = \frac{\varepsilon_o}{2}\int_0^{\pi}\int_0^{2\pi}\left[\int_0^{R} E_{in}^2\, d\tau + \int_R^{\infty} E_{out}^2\, d\tau\right] \tag{3.37}$$

$$W = \frac{\varepsilon_o}{2}\int_0^{\pi}\int_0^{2\pi}\int_0^{R} E_{in}^2\, d\tau + \frac{\varepsilon_o}{2}\int_0^{\pi}\int_0^{2\pi}\int_R^{\infty} E_{out}^2\, d\tau$$

Inside the sphere

$$\int_0^{\pi}\int_0^{2\pi}\int_0^{R} E_{in}^2 d\tau = \int_0^{\pi}\int_0^{2\pi}\int_0^{R} \frac{\rho^2 r^2}{9\varepsilon_0^2} r^2 \sin\theta\, dr\, d\theta\, d\phi$$

$$= \frac{\rho^2}{3\varepsilon_0^2}\int_0^{R} r^4 dr \int_0^{\pi} \sin\theta\, d\theta \int_0^{2\pi} d\phi \tag{3.38}$$

$$= \frac{4\pi\rho^2 R^5}{45\,\varepsilon_0^2}$$

Outside the sphere

$$\int_0^{\pi}\int_0^{2\pi}\int_R^{\infty} E_{out}^2\, d\tau = \int_0^{\pi}\int_0^{2\pi}\int_R^{\infty}\left(\frac{\rho R^3}{3\,\varepsilon_o\, r^2}\right)^2 r^2 \sin\theta\, dr\, d\theta\, d\phi$$

$$= \frac{\rho^2 R^6}{9\,\varepsilon_o}\int_R^{\infty}\frac{dr}{r^2}\int_0^{\pi}\sin\theta\, d\theta\int_0^{2\pi} d\phi \tag{3.39}$$

$$= \frac{4\pi\,\rho^2\, R^5}{9\,\varepsilon_0^2}$$

Substituting Eqs. 3.38, 3.39 in Eq. 3.37

$$W = \frac{\varepsilon_0}{2}\left[\frac{4\pi\,\rho^2 R^5}{45\,\varepsilon_0^2}\right] + \frac{\varepsilon_0}{2}\left[\frac{4\pi\,\rho^2 R^5}{9\,\varepsilon_0^2}\right]$$

$$\Rightarrow W = \frac{4\pi\,\rho^2 R^5}{15\,\varepsilon_0}$$

(b) ***Using equation*** $W = \frac{1}{2}\int \rho\, V\, d\tau$

Now let us use the equation $W = \frac{1}{2}\int \rho\, V\, d\tau$ to calculate the energy stored.

Let us calculate the potential V at a point inside the sphere

$$V = -\int_{\infty}^{r} \mathbf{E}\cdot d\boldsymbol{l}$$

where r is a distance of point P inside the sphere as shown in Fig. 3.7

$$\Rightarrow V = -\int_{\infty}^{R} \mathbf{E}\cdot d\boldsymbol{l} - \int_{R}^{r} \mathbf{E}\cdot d\boldsymbol{l}$$

$$\Rightarrow V = -\int_{\infty}^{R} \frac{\rho R^3}{3\varepsilon_0 r^2} d\,l - \int_{R}^{r} \frac{\rho r}{3\varepsilon_0} dl$$

$$\Rightarrow V = \frac{\rho}{6\varepsilon_0}\left[3R^2 - r^2\right]$$

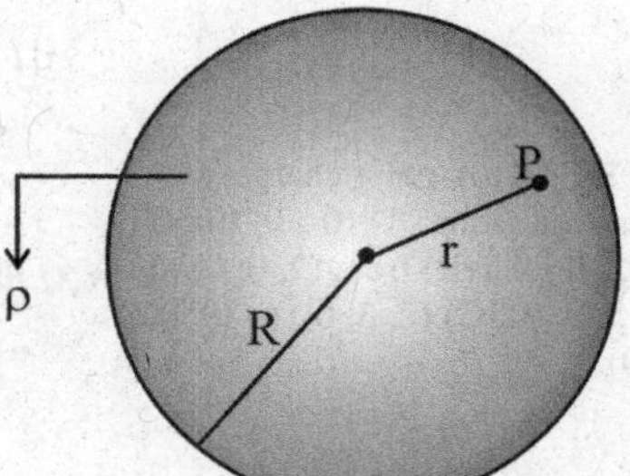


**Fig. 3.7** A sphere with uniform volume charge density

Hence the work done is given by

$$W = \frac{1}{2}\int_{\tau} \rho\, V\, d\tau$$

$$W = \frac{1}{2}\int_{\tau} \rho \frac{\rho}{3\,\varepsilon_0}\left[3R^2 - r^2\right] r^2 \sin\theta\, dr\, d\theta\, d\phi$$

$$\Rightarrow W = \frac{1}{12}\frac{\rho^2}{\varepsilon_0}\int_0^R \left[3R^2 - r^2\right] r^2\, dr \int_0^{\pi} \sin\theta\, d\theta \int_0^{2\pi} d\phi$$

$$\Rightarrow W = \frac{4\pi\,\rho^2\,R^5}{15\,\varepsilon_0}$$

**Example 3.2**

An electric field is given by $\mathbf{E} = x\hat{\mathbf{i}} + y\hat{\mathbf{j}} + z\hat{\mathbf{k}}$. Calculate the work done in moving a point charge −10 μC from (a) (0, 0, 0) to (2, 0, 0) (b) (2, 0, 0) to (2, 3, 0) (c) (2, 3, 0) to (0, 0, 0) as shown in Fig. 3.8 and hence show that electrostatic fields are conservative fields.

**Solution**

(a) We know that

$$W = -Q\int \mathbf{E}\cdot d\boldsymbol{l}$$

In Fig. 3.8 Q = −10 μC and

$$d\boldsymbol{l} = dx\,\hat{\mathbf{i}} + dy\,\hat{\mathbf{j}} + dz\,\hat{\mathbf{k}}$$

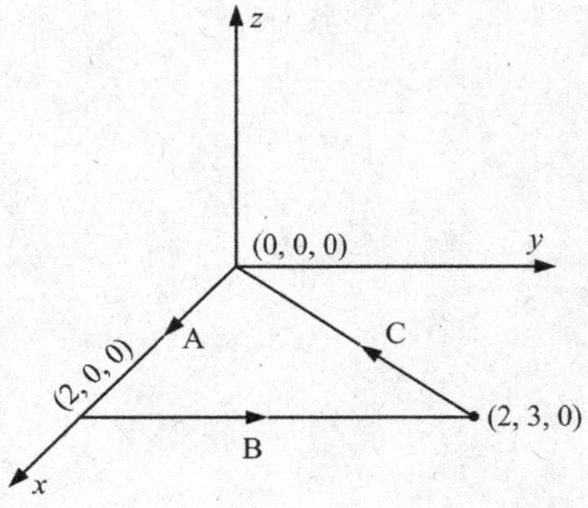


**Fig. 3.8** A point charge moved from (A) (0 , 0, 0) to (2, 0 , 0 ) (B) from (2, 0, 0) to (2, 3, 0) (C) (2, 3, 0 ) to (0, 0, 0)

For path A $dy = 0$, $dz = 0$. Hence $\mathrm{d}\boldsymbol{l} = \mathrm{d}x\,\hat{\mathbf{i}}$

$$\text{Hence W} = 10\int_0^2 (x\hat{\mathbf{i}} + y\hat{\mathbf{j}} + z\hat{\mathbf{k}}) \cdot \mathrm{d}x\,\hat{\mathbf{i}}$$

$$\mathrm{W} = 10\left[\frac{x^2}{2}\right]_0^2 = 20\,\mu\mathrm{J}$$

(b) For path B we move along *y*-axis and hence $dx = 0$, $dz = 0$. Therefore

$$\mathrm{d}\boldsymbol{l} = \mathrm{d}y\,\hat{\mathbf{j}}$$

$$\mathrm{W} = 10\int_0^3 (x\hat{\mathbf{i}} + y\hat{\mathbf{j}} + z\hat{\mathbf{k}}) \cdot \mathrm{d}y\,\hat{\mathbf{j}}$$

$$\mathrm{W} = 10\left[\frac{y^2}{2}\right]_0^3 = 45\,\mu\mathrm{J}$$

Therefore the work done in moving the charge from origin (0, 0, 0) to point (2, 3, 0) via the path A B is 20 + 45 = 65 μJ.

(c) For path C, $\mathrm{d}\boldsymbol{l} = \mathrm{d}x\,\hat{\mathbf{i}} + \mathrm{d}y\,\hat{\mathbf{j}}$

$$\text{Hence W} = -\mathrm{Q}\int_{(2,3,0)}^{(0,0,0)} (x\hat{\mathbf{i}} + y\hat{\mathbf{j}} + z\hat{\mathbf{k}}) \cdot (\mathrm{d}x\,\hat{\mathbf{i}} + \mathrm{d}y\,\hat{\mathbf{j}})$$

$$= -\mathrm{Q}\left[\frac{x^2}{2}\right]_{(2,3,0)}^{(0,0,0)} - \mathrm{Q}\left[\frac{y^2}{2}\right]_{(2,3,0)}^{(0,0,0)}$$

$$\mathrm{W} = -65\,\mu\mathrm{J}.$$

The work done in moving the charge from (0, 0, 0) to (2, 3, 0) via AB is 65 μJ and then from (2, 3, 0) to (0, 0, 0) is –65 μJ. The total work done around the closed path ABC is zero and hence the electrostatic field is conservative.

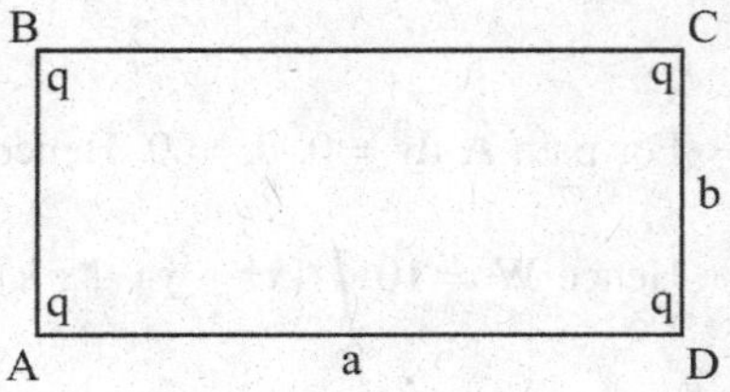


**Fig. 3.9** Four charges each of magnitude q placed in the corner of a rectangle whose sides are a, b

**Example 3.3**
Four charges each of magnitude q are placed at the corners of a rectangle of sides a, b as shown in Fig. 3.9. Calculate the energy stored in the system.

**Solution**

The total energy in the system is given by $W = \frac{1}{2}\sum_{i=1}^{4} q_i V_i$

$$W = \frac{1}{2}[qV_A + qV_B + qV_C + qV_D]$$

Setting the zero reference of the potential at point A

$$W = \frac{1}{2}[qV_B + qV_C + qV_D]$$

From Fig. 3.9

$$V_B = \frac{q}{4\pi\varepsilon_o}\left[\frac{1}{AB}\right] = \frac{q}{4\pi\varepsilon_o b}$$

$$V_C = \frac{q}{4\pi\varepsilon_o}\left[\frac{1}{AC}\right] = \frac{q}{4\pi\varepsilon_o\sqrt{a^2+b^2}}$$

$$V_D = \frac{q}{4\pi\varepsilon_o}\left[\frac{1}{AD}\right] = \frac{q}{4\pi\varepsilon_o a}$$

Hence $W = \frac{q^2}{8\pi\varepsilon_o}\left[\frac{1}{b} + \frac{1}{\sqrt{a^2+b^2}} + \frac{1}{a}\right].$

## 3.5 Method of Images

In Sects. 2.4, 2.8, 2.11 we saw how to calculate the electric field once the charge distribution is given. The charge distribution is plugged into respective Eqs. 2.17−2.19, 2.35, 2.80−2.82 and finally the electric field is calculated for the given charge distribution. However let us consider the following example and check whether the problem can be solved using previous methods namely, Coulomb's law, Gauss's law and potential formulation.

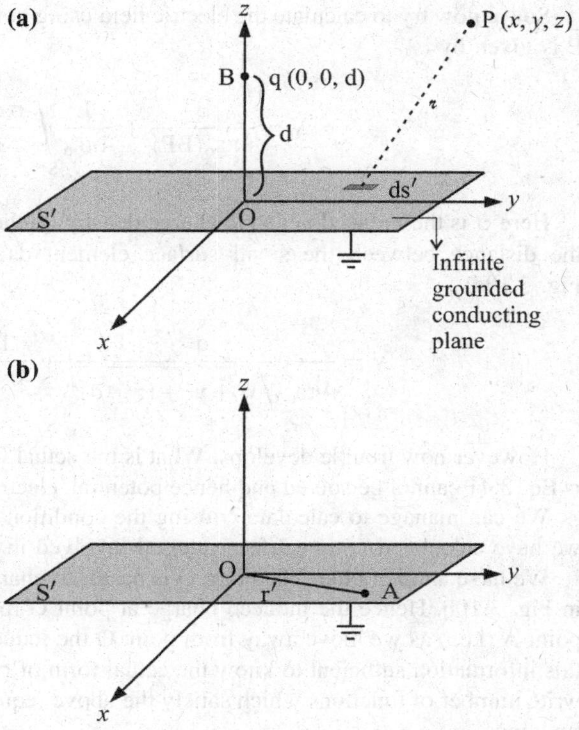


**Fig. 3.10** **a** A point charge q situated above the infinite grounded conducting plane. **b** Illustration to show that the point charge q is nearer to point O than point A

Consider a point charge q situated above the infinite grounded conducting plane at a distance d above the plane along the $z$-axis as shown in the Fig. 3.10a. The plane is situated at $z = 0$. Because the plane is grounded it must be at zero potential i.e., $V(x, y, 0) = 0$. Now our aim is to calculate the electric field above the grounded conducting plane i.e., for all points with $z > 0$. Below the grounded conducting plane (i.e., for points $z < 0$) the electric field is zero because there are no charges there.

Let us calculate the electric field at point P above the conducting plane. From Fig. 3.10a we see that $BP = \sqrt{x^2 + y^2 + (z - d)^2}$.

There are two charges responsible for electric field at point P. First one is the point charge q situated at point B in Fig. 3.10a. The point charge q at point B will induce negative charges in the conducting plane and this induced charge will produce a electric field at point P.

In order to calculate the electric field at point P our first attempt will be to check whether Gauss's law can be used to calculate the electric field at point P. However a plot of field lines for Fig. 3.10a would show you that Gauss's law is not useful here for calculating electric field.

So we now try to calculate the electric field using potential. The potential at point P is given by

$$V = \frac{q}{4\pi\varepsilon_0(BP)} + \frac{1}{4\pi\varepsilon_0}\int_{S'} \frac{\sigma\, ds'}{\mathscr{r}} \tag{3.40}$$

Here $\sigma$ is the induced negative charge density on the conducting surface and $\mathscr{r}$ is the distance between the small surface element $ds'$ and point P as shown in Fig. 3.10a.

$$V = \frac{q}{4\pi\varepsilon_0\sqrt{x^2 + y^2 + (z - d)^2}} + \frac{1}{4\pi\varepsilon_0}\int_{S'} \frac{\sigma\, ds'}{\mathscr{r}} \tag{3.41}$$

However now trouble develops. What is the actual form of $\sigma$? Without knowing $\sigma$ Eq. 3.41 cannot be solved and hence potential, electric field cannot be calculated.

We can manage to calculate $\sigma$ using the condition $V(x, y, 0) = 0$ but still once we have calculated $\sigma$ the surface integral involved in Eq. 3.40 is difficult.

We have another clue. The point O is nearer to charge q as compared to point A in Fig. 3.10b. Hence the induced charge at point O must be more as compared to point A (i.e.,) as we move away from point O the induced charge must decrease. Is this information sufficient to know the actual form of $\sigma$? The answer is no. We can write number of functions which satisfy the above requirement. For example all the functions

$$\sigma = \frac{k_1}{(r' + k_2)^2}$$

$$\sigma = \frac{k_1}{(r' + k_2)^{5/2}}$$

$$\sigma = \frac{k_1}{(r' + k_1)^3}$$

with the constant $k_1, k_2 > 0$ satisfy the condition at point O $[r' = 0]$ the induced charge $\sigma$ is maximum and as we move away from O ($r'$ increases) the induced charge decreases. So without knowing the actual form of $\sigma$ (i.e.,) how the charges are distributed over the conducting plane, Eq. 3.41 cannot be used to calculate the potential.

"***In similar problems in which*** $\lambda, \sigma, \rho$ ***are not known Coulomb's law, Gauss's law, potential formulation cannot be used to find*** **E**."

For such problems new technique's are required to calculate the electric field. Two techniques have been developed

(i) Method of images
(ii) Laplace's equation.

In this section we will see in detail about method of images and in later sections about Laplace's equation method. Now we will calculate the electric field at point P in Fig. 3.10a by method of images.

First let us write the boundary conditions for Fig. 3.10a

(i) V $(x, y, 0) = 0$ (i.e.,) the conducting plane is grounded and hence at $z = 0$, $V = 0$
(ii) At infinite point

$$V \to 0 \text{ (i.e., ) for } x^2 + y^2 + z^2 \gg d^2 \quad V \to 0.$$

As per uniqueness theorem for a given boundary condition there can be only one solution. So for the above boundary conditions for a point charge situated above the infinite grounded conducting plane there can be only one solution.

Now let us consider a completely different situation shown in Fig. 3.11. In Fig. 3.11 we have removed the grounded conducting plane at $z = 0$. Instead we have placed a –q charge at (0, 0, –d). The potential at point $P_1$ in Fig. 3.11 is given by

$$V = \frac{1}{4\pi\,\varepsilon_o}\left[\frac{q}{\sqrt{x^2 + y^2 + (z - d)^2}} - \frac{q}{\sqrt{x^2 + y^2 + (z + d)^2}}\right] \tag{3.42}$$

In Eq. 3.42 substituting $z = 0$ we see that

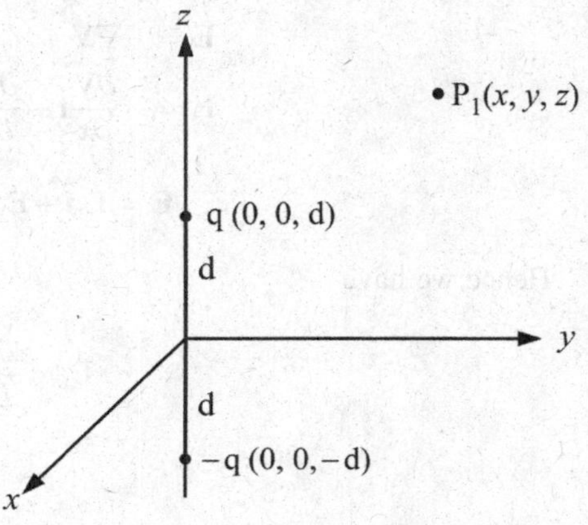


**Fig. 3.11** Two point charges q and - q situated at d and - d respectively

$$V = \frac{1}{4\pi\,\varepsilon_0}\left[\frac{q}{\sqrt{x^2+y^2+(d)^2}} - \frac{q}{\sqrt{x^2+y^2+(d)^2}}\right]$$

$$V = 0$$

Also in Eq. 3.42 we see that $V \to 0$ for $x^2+y^2+z^2 \ggg d^2$. Thus Eq. 3.42 which is the potential for the charges shown in Fig. 3.11 satisfies the boundary conditions for the charges shown in Fig. 3.10a, for $z > 0$. Hence by uniqueness theorem the only possible solution for Fig. 3.10a is Eq. 3.42.

Clearly note down the steps we have carried out. For Fig. 3.10a we constructed a image charge as shown in Fig. 3.11. We obtained the potential for the image charge in Fig. 3.11 for $z > 0$ in Eq. 3.42. Then we checked whether the potential is obtained in Eq. 3.42 for image (Fig. 3.11) satisfies the boundary conditions for the point charge situated above the infinite grounded conducting plane (Fig. 3.10a) for the region $z > 0$. It does. Then as per uniqueness theorem the potential obtained for the image charge in Eq. 3.42 is the potential for the point charge situated above the infinite grounded conducting plane. Thus we have used a indirect method for calculating potential for point P in Fig. 3.10a. We never used Coulomb's law, Gauss's law or potential formulation.

Note down the important role played by uniqueness theorem. Without uniqueness theorem we cannot guarantee that potential obtained in Eq. 3.42 for image charge in Fig. 3.11 is a solution for Fig. 3.10a for region $z > 0$.

What about $z < 0$. In Fig. 3.10a for $z < 0$ field is zero. But for Fig. 3.11 for $z < 0$ field is non zero. Our problem was to find the field for $z > 0$ in Fig. 3.10a so we need not bother for region $z < 0$.

Once we have calculated the potential let us calculate other quantities like electric field etc.

### (a) Electric field at any point above the conductor $z > 0$

Once the potential has been found it is very easy to calculate the electric field

$$\mathbf{E} = -\nabla V$$

$$\mathbf{E} = -\frac{\partial V}{\partial x}\hat{\mathbf{i}} - \frac{\partial V}{\partial x}\hat{\mathbf{j}} - \frac{\partial V}{\partial z}\hat{\mathbf{k}} \tag{3.43}$$

$$\mathbf{E} = E_x\hat{\mathbf{i}} + E_y\hat{\mathbf{j}} + E_z\,\hat{\mathbf{k}} \tag{3.44}$$

Hence we have

$$E_x = \frac{\partial V}{\partial x}$$

From Eq. 3.42 then

$$E_x = \frac{-q}{4\pi\varepsilon_o}\frac{\partial}{\partial x}\left[\frac{1}{\sqrt{x^2+y^2+(z-d)^2}} - \frac{1}{\sqrt{x^2+y^2+(z+d)^2}}\right] \tag{3.45}$$

$$E_x = \frac{q}{4\pi\varepsilon_o}\left[\frac{x}{\left[x^2+y^2+(z-d)^2\right]^{3/2}} - \frac{x}{\left[x^2+y^2+(z+d)^2\right]^{3/2}}\right] \tag{3.46}$$

Similarly

$$E_y = -\frac{\partial V}{\partial y} = \frac{q}{4\pi\varepsilon_o}\left[\frac{y}{\left[x^2+y^2+(z-d)^2\right]^{3/2}} - \frac{y}{\left[x^2+y^2+(z+d)^2\right]^{3/2}}\right] \tag{3.47}$$

$$E_z = -\frac{\partial V}{\partial z} = \frac{q}{4\pi\varepsilon_o}\left[\frac{z-d}{\left[x^2+y^2+(z-d)^2\right]^{3/2}} - \frac{z+d}{\left[x^2+y^2+(z+d)^2\right]^{3/2}}\right] \tag{3.48}$$

The electric field at point P in Fig. 3.10a is given

$$\mathbf{E} = E_x\,\hat{\mathbf{i}} + E_y\,\hat{\mathbf{j}} + E_z\,\hat{\mathbf{k}}$$

where $E_x$, $E_y$, $E_z$ are given by Eqs. 3.46, 3.47 and 3.48.

(b) **Electric field on the conductor**

On the conductor $z = 0$. From Eqs. 3.46, 3.47, 3.48 we have

$$(E_x)_{z=0} = \frac{q}{4\pi\varepsilon_o}\left[\frac{x}{\left[x^2+y^2+d^2\right]^{3/2}} - \frac{x}{\left[x^2+y^2+d^2\right]^{3/2}}\right] \tag{3.49}$$

$$\begin{aligned}(E_x)_{z=0} &= 0\\ (E_y)_{z=0} &= 0 \text{ and}\\ (E_z)_{z=0} &= \frac{-2qd}{4\pi\varepsilon_o\left[x^2+y^2+d^2\right]^{3/2}}\end{aligned} \tag{3.50}$$

Thus on the conductor

$$\begin{aligned}\mathbf{E} &= (E_x)_{z=0}\,\hat{\mathbf{i}} + (E_y)_{z=0}\,\hat{\mathbf{j}} + (E_z)_{z=0}\,\hat{\mathbf{k}} \\ \mathbf{E} &= (E_z)_{z=0}\,\hat{\mathbf{k}} \quad \text{as} \quad (E_x)_{z=0} = (E_y)_{z=0} = 0 \\ \mathbf{E} &= \frac{-2q\,d}{4\pi\varepsilon_0\left[x^2+y^2+d^2\right]^{3/2}}\,\hat{\mathbf{k}}\end{aligned} \tag{3.51}$$

(c) **Surface charge density on the conductor**

The surface charge density on the conductor which we initially were not able to find can now be calculated.

At the conductor free space interface the electric field is given by Eq. 2.304

$$E_{in} = \frac{\sigma}{\varepsilon_0}$$

where $E_{in}$ is the normal component of the electric field. The normal component of the electric field to the conductor is $E_z$ component. Hence

$$(E_z)_{z=0} = \frac{\sigma}{\varepsilon_0} \tag{3.52}$$

$$\Rightarrow \sigma = (E_z)_{z=0}\varepsilon_0 \tag{3.53}$$

From Eq. 3.50

$$\sigma = \varepsilon_0\left[\frac{-2q\,d}{4\pi\varepsilon_0\left[x^2+y^2+d^2\right]^{3/2}}\right] \tag{3.54}$$

$$\Rightarrow \sigma = \frac{-2q\,d}{4\pi\left[x^2+y^2+d^2\right]^{3/2}} \tag{3.55}$$

For problems which we have done in Chap. 2 we substituted the charge density in Eqs. 2.80–2.82 and then calculated the potential. However in this section we calculated the potential initially by indirect image method and from that we are now calculating the charge density because initially we don't know what is the actual form of charge density.

(d) **Force of attraction between the point charge q and the grounded infinite plane conductor**

Because the potential is same for the image charges in Fig. 3.11 and for the point charge above the grounded plane in Fig. 3.10a, so is the field and hence the force.

The force of attraction between the +q charge and −q charge in Fig. 3.11 is

$$\mathbf{F} = \frac{1}{4\pi\varepsilon_0}\frac{q(-q)}{(2d)^2}\hat{k} = \frac{-q^2}{16\pi\varepsilon_0 d^2}\hat{k} \tag{3.56}$$

The negative sign indicates the force is attractive.

**(e) Work done in bringing the charge from ∞ and placing it at d**

The work done in moving the charge q from ∞ and placing it at point B in Fig. 3.10a is

$$W = \int_{\infty}^{d} \mathbf{F}\cdot d\boldsymbol{l} \tag{3.57}$$

$$= \frac{q^2}{4\pi\varepsilon_0}\int_{\infty}^{d}\frac{dz}{4z^2} \tag{3.58}$$

$$W = \frac{-1}{4\pi\varepsilon_0}\frac{q^2}{4d} \tag{3.59}$$

**(f) The total induced charge on the conducting plane**

We have

$$\sigma = \frac{-q\,d}{2\pi\left[x^2+y^2+d^2\right]^{3/2}}$$

The total induced charge is

$$q_{tot} = \int_{S'} \sigma\, ds' \tag{3.60}$$

In the polar coordinates

$$(r')^2 = x^2 + y^2,\ ds' = (r')(dr')d\phi'$$

$$q_{tot} = \int_0^{2\pi}\int_0^{\infty}\frac{(-qd)(r'dr'd\phi')}{2\pi\left[(r')^2+d^2\right]^{3/2}} \tag{3.61}$$

$$q_{tot} = -q.$$

## 3.6 Point Charge Near a Grounded Conducting Sphere

Let a point charge q be placed near a grounded conducting sphere of radius R as shown in Fig. 3.12a. The distance between the center of the sphere O and the point A where the charge q is placed is d. The point charge will induce a charge on the sphere. We are interested in calculating the electric field at point P. The electric field at point P depends on point charge q and the induced charge. The actual form of induced charge on the sphere is not known and hence we go for solving the problem using method of images.

Let us solve the problem in polar coordinates. First let us identify the boundary condition for the problem.

One of the boundary conditions to the problem is that on the surface of the sphere, the potential is zero because the sphere is grounded. We will use the boundary condition to find the magnitude and location of the image charge.

In Fig. 3.12 suppose assume that the sphere is removed and a image charge $q_1$ is placed at I at a distance from the center of the sphere O. By the boundary condition the potential at point B, C must be zero. At point B we have

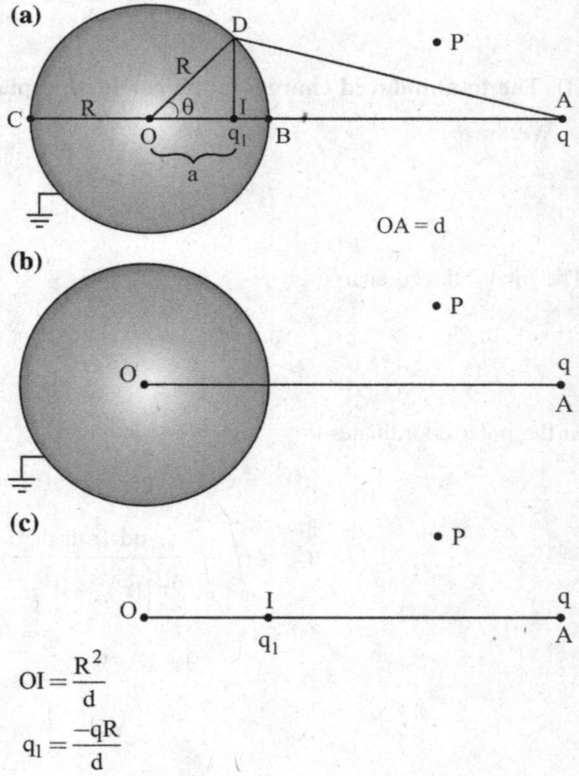


**Fig. 3.12** **a** A point charge q situated near a grounded conducting sphere. **b, c** Illustration to show point charge q and image charge $q_1$

$$\frac{q}{4\pi\varepsilon_0(AB)} + \frac{q_1}{4\pi\varepsilon_0(IB)} = 0 \quad (3.62)$$

$$\frac{q}{AB} + \frac{q_1}{IB} = 0 \quad (3.63)$$

$$\frac{q}{d-R} + \frac{q_1}{R-a} = 0 \quad (3.64)$$

At point C we have

$$\frac{q}{4\pi\varepsilon_0(AC)} + \frac{q_1}{4\pi\varepsilon_0(CI)} = 0 \quad (3.65)$$

$$\Rightarrow \frac{q}{AC} + \frac{q_1}{CI} = 0 \quad (3.66)$$

$$\Rightarrow \frac{q}{R+d} + \frac{q_1}{R+a} = 0 \quad (3.67)$$

From Eq. 3.64

$$q_1 = -q\left(\frac{R-a}{d-R}\right) \quad (3.68)$$

Substituting Eq. 3.68 in Eq. 3.67

$$\frac{q}{R+d} - q\frac{(R-a)}{(R+a)(d-R)} = 0 \quad (3.69)$$

$$\Rightarrow a = \frac{R^2}{d} \quad (3.70)$$

Substituting Eq. 3.70 in Eq. 3.68

$$q_1 = \frac{-q\left(R - \frac{R^2}{d}\right)}{(d-R)} \quad (3.71)$$

$$\Rightarrow q_1 = \frac{-qR}{d} \quad (3.72)$$

Equations 3.70, 3.72 gives the location and magnitude of the image charge respectively.

Thus a image charge of magnitude $\frac{-qR}{d}$ and located at a distance of $\frac{R^2}{d}$ from the center of the sphere satisfies the boundary condition of the grounded conducting sphere near a point charge.

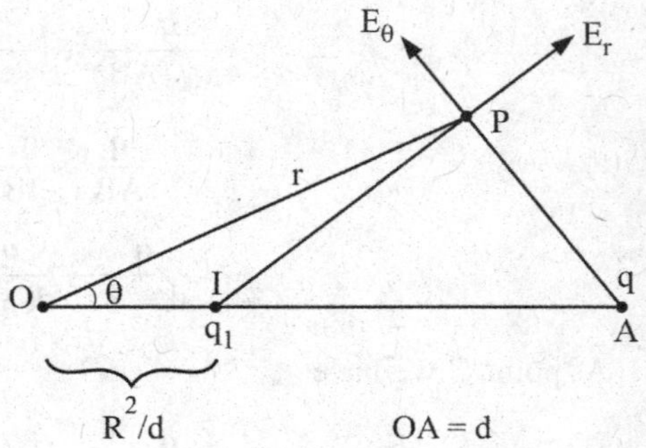


**Fig. 3.13** Illustration to calculate the potential at point P due to charges q and $q_1$

The grounded sphere along with the point charge q is shown in Fig. 3.12b. The image charge $q_1$ along with charge q with the sphere removed is shown in Fig. 3.12c.

(a) **Potential at point P**

The potential at point P in Figs. 3.12c and 3.13 is

$$V = \frac{1}{4\pi\varepsilon_o}\left[\frac{q}{AP} + \frac{q_1}{IP}\right] \tag{3.73}$$

$$V = \frac{1}{4\pi\varepsilon_o}\left[\frac{q}{\sqrt{r^2 + d^2 - 2rd\cos\theta}} - \frac{qR}{d\sqrt{r^2 + \frac{R^4}{d^2} - \frac{2R^2}{d}r\cos\theta}}\right] \tag{3.74}$$

The boundary conditions for Fig. 3.12a is

(i) At the surface of the sphere the potential must be zero because the sphere is grounded i.e., at r = R, V = 0

(ii) As $r \to \infty$ $V \to 0$.

Both these boundary conditions are satisfied by Eq. 3.74. The boundary conditions are for Fig. 3.12b—a grounded sphere near a point charge. But the potential due to the two charges q and $q_1$ in Fig. 3.13 at point P in Eq. 3.74 satisfies both the boundary conditions due to grounded sphere near the point charge in Fig. 3.12b. Hence by uniqueness theorem the potential at point P in Fig. 3.12b is given by Eq. 3.74.

(b) **Electric field intensity at point P**

The electric field at point P in Fig. 3.12a has two components $E_r$, $E_\theta$ as shown in Fig. 3.13.

Now

$$E_r = -\frac{\partial V}{\partial r} \tag{3.75}$$

$$= -\frac{1}{4\pi\varepsilon_0}\frac{\partial}{\partial r}\left[\frac{q}{\sqrt{r^2+d^2-2rd\cos\theta}} - \frac{qR}{d\sqrt{r^2+\frac{R^2}{d^2}-\frac{2R^2}{d}r\cos\theta}}\right] \tag{3.76}$$

$$E_r = \frac{1}{4\pi\varepsilon_0}\left[\frac{q(r-d\cos\theta)}{(r^2+d^2-2rd\cos\theta)^{3/2}} - \frac{qR\left(r-\frac{R^2}{d}\cos\theta\right)}{d\left(r^2+\frac{R^4}{d^2}-\frac{2R^2}{d}r\cos\theta\right)^{3/2}}\right] \tag{3.77}$$

The θ component $E_\theta$ is given by

$$E_\theta = -\frac{1}{r}\frac{\partial V}{\partial \theta} \tag{3.78}$$

$$= \frac{1}{4\pi\varepsilon_0 r}\left[\frac{\partial}{\partial\theta}\left\{\frac{q}{\sqrt{r^2+d^2-2rd\cos\theta}} - \frac{qR}{d\sqrt{r^2+\frac{R^2}{d^2}-\frac{2R^2}{d}r\cos\theta}}\right\}\right] \tag{3.79}$$

$$E_\theta = \frac{q}{4\pi\varepsilon_0}\left[\frac{d\sin\theta}{[r^2+d^2-2rd\cos\theta]^{3/2}} - \frac{d\sin\theta}{\left[\frac{r^2d^2}{R^2}+R^2-2rd\cos\theta\right]^{3/2}}\right] \tag{3.80}$$

### (c) Electric field at the surface of the sphere

At the surface of the sphere r = R. From Eqs. 3.77, 3.80

$$(E_r)_{r=R} = \frac{q}{4\pi\varepsilon_0}\left[\frac{R-d^2/R}{(d^2+R^2-2Rd\cos\theta)^{3/2}}\right] \tag{3.81}$$

$$(E_\theta)_{r=R} = 0 \tag{3.82}$$

### (d) Surface density of the induced charge

At the conductor, free space interface the electric field is given by Eq. 2.304

$$E_{in} = \frac{\sigma}{\varepsilon_0}$$

where $E_{in}$ is the normal component of electric field. The normal component of the electric field the sphere is $E_r$ component.

$$(E_r)_{r=R} = \frac{\sigma}{\varepsilon_o} \tag{3.83}$$

$$\Rightarrow \sigma = (E_r)_{r=R}\, \varepsilon_o \tag{3.84}$$

From Eq. 3.81

$$\sigma = \frac{q}{4\pi R} \frac{R^2 - d^2}{\left(d^2 + R^2 - 2Rd\cos\theta\right)^{3/2}} \tag{3.85}$$

(e) **Force between the sphere and the point charge**

The force between the sphere of radius R and point charge q is same as that between image charge $q_1$ and point charge q

$$F = \frac{1}{4\pi\varepsilon_o} \frac{q\, q_1}{(IA)^2} \tag{3.86}$$

But $IA = d - \frac{R^2}{d}$ in Fig. 3.13. Substituting the values of $q_1$ and IA in Eq. 3.86

$$F = \frac{1}{4\pi\varepsilon_o} \frac{q\left(\frac{-qR}{d}\right)}{\left(d - \frac{R^2}{d}\right)^2} \tag{3.87}$$

$$F = \frac{1}{4\pi\varepsilon_o} \frac{q^2 Rd}{\left(d^2 - R^2\right)^2} \tag{3.88}$$

## 3.7 Laplace's Equation—Separation of Variables

In the preceding sections we used method of images to calculate the electric field for the given problem. However method of images requires isolated free charges near conducting boundaries to fix a image charge. If the problem is such that there are no isolated free charges (for example see Fig. 3.14) method of images cannot be used. The potential in such situations can be obtained using Laplace's equation. Laplace's equation is a partial differential equation and can be solved by the method of separation of variables. The potential in Laplace's equation is a function of three

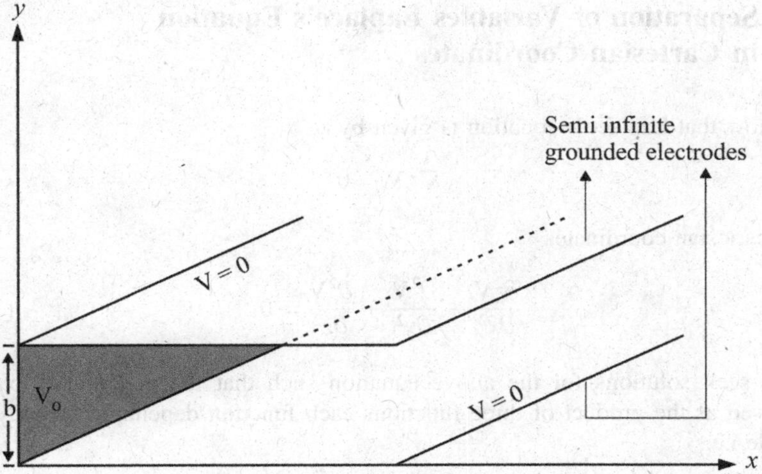


**Fig. 3.14** Three conducting plates maintained at specified potentials as shown in the figure

variables. To solve Laplace's equations by separation of variables we express the potential in terms of products of functions, each function depending on only one variable. We will see the method of separation of variables in three different coordinate systems cartesian, spherical and cylindrical coordinate system.

Another important point which we will be utilising in solving problems is that Laplace's equation is linear. Suppose assume that $V_1, V_2, V_3, V_4 \ldots$ are solutions of Laplace's equation then $V = a_1 V_1 + a_2 V_2 + a_3 V_3 + \cdots$ is also a solution of Laplace's equation ($a_1, a_2, a_3 \ldots$ are constants).

$$\nabla^2 V = a_1 \nabla^2 V_1 + a_2 \nabla^2 V_2 + a_3 \nabla^2 V_3 \ldots \tag{3.89}$$

As $V_1, V_2, V_3 \ldots$ are solutions to Laplace's equation

$$\nabla^2 V_1 = 0,\ \nabla^2 V_2 = 0,\ \nabla^2 V_3 = 0 \ldots$$

Hence

$$\begin{aligned} \nabla^2 V &= a_1(0) + a_2(0) + a_3(0) + \cdots \\ \Rightarrow \nabla^2 V &= 0 \end{aligned} \tag{3.90}$$

## 3.8 Separation of Variables Laplace's Equation in Cartesian Coordinates

We know that Laplace's equation is given by

$$\nabla^2 V = 0$$

In cartesian coordinates

$$\frac{\partial^2 V}{\partial x^2} + \frac{\partial^2 V}{\partial y^2} + \frac{\partial^2 V}{\partial z^2} = 0 \tag{3.91}$$

We seek solutions for the above equation such that the potential V can be expressed as the product of three functions each function depending on only one variable i.e.,

$$V = X(x)\,Y(y)\,Z(z) \tag{3.92}$$

Thus the functions $X$, $Y$, $Z$ each depends on only one variable $x$, $y$, $z$ respectively. Substituting Eq. 3.92 in Eq. 3.91

$$YZ\frac{d^2X}{dx^2} + XZ\frac{d^2Y}{dy^2} + XY\frac{d^2Z}{dz^2} = 0 \tag{3.93}$$

Dividing throughout by $XYZ$

$$\frac{1}{X}\frac{d^2X}{dx^2} + \frac{1}{Y}\frac{d^2Y}{dy^2} + \frac{1}{Z}\frac{d^2Z}{dz^2} = 0 \tag{3.94}$$

The first term in Eq. 3.94 $\frac{1}{X}\frac{d^2X}{dx^2}$ depends on only one variable $x$. It is independent of $y$ and $z$. The second and third terms $\frac{1}{Y}\frac{d^2Y}{dy^2}$, $\frac{1}{Z}\frac{d^2Z}{dz^2}$ are independent of $x$ and depend on $y$ and $z$ alone respectively.

Thus as we allow $x$ to vary the second and third term in Eq. 3.94 must remain unaltered because they are independent of $x$. But the first term $\frac{1}{X}\frac{\partial^2 X}{dx^2}$, depends on $x$.

Will the first term $\frac{1}{X}\frac{d^2X}{dx^2}$ vary as we vary $x$. Suppose assume that $\frac{1}{X}\frac{d^2X}{dx^2}$ varies as we vary $x$. If $\frac{1}{X}\frac{d^2X}{dx^2}$ alone varies as we vary $x$ then Eq. 3.94 cannot be equal to zero. Thus even if $X$ depends on $x$, as we vary $x$, $\frac{1}{X}\frac{d^2X}{dx^2}$ must not vary. The term $\frac{1}{X}$, $\frac{d^2X}{dx^2}$ must adjust themselves such that $\frac{1}{X}\frac{d^2X}{dx^2}$ is constant as we vary $x$. Hence $\frac{1}{X}\frac{d^2X}{dx^2}$ must be equal to a constant, say $k_x$

$$\frac{1}{X}\frac{d^2X}{dx^2} = k_x \tag{3.95}$$

Similar arguments holds good for second and third terms in Eq. 3.94

$$\frac{1}{Y}\frac{d^2Y}{dy^2} = k_y \tag{3.96}$$

$$\frac{1}{Z}\frac{d^2Z}{dz^2} = k_z \tag{3.97}$$

Substituting Eqs. 3.95, 3.96, 3.97 in Eq. 3.94

$$k_x + k_y + k_z = 0 \tag{3.98}$$

Thus instead of solving a partial differential equation in Eq. 3.91 to obtain V we now solve three ordinary differential equations 3.95, 3.96, 3.97 to obtain *X*, *Y*, *Z*. Once we have obtained *X*, *Y*, *Z* these values can be substituted in Eq. 3.92 to obtain V and hence finally electric field by

$$\mathbf{E} = -\nabla V.$$

## 3.9 Potential Between Two Grounded Semi Infinite Parallel Electrodes Separated by a Plane Electrode Held by a Potential $V_o$

Consider two semi infinite grounded parallel electrodes as shown in Fig. 3.14. The distance between the two electrodes is b. Another plane electrode lying at $x = 0$ is insulated from the semi infinite parallel electrodes and is maintained at potential $V_o$.

Because the plate at $x = 0$ is given a potential of $V_o$ (by using a external power supply) it will have a surface charge densing $\sigma$. We are now interested in finding the fields in the region between the three plates.

Coulomb's law, Gauss's law and potential formulation cannot be used to find the field in the region between three electrodes because we don't know the actual for of $\sigma$.

Method of images cannot be used here because there are no isolated free charges and hence fixing a image is not possible. The solutions is then to solve Laplace's equation

$$\nabla^2 V = 0$$

From Eq. 3.94

$$\frac{1}{X}\frac{d^2X}{dx^2} + \frac{1}{Y}\frac{d^2Y}{dy^2} = 0 \tag{3.99}$$

and the potential is independent of $z$. The potential V between the electrodes is given by

$$V = X(x)\,Y(y) \tag{3.100}$$

The boundary conditions for the potential $V(x, y)$ is

$$\left.\begin{array}{ll} \text{(i)} & V(x, 0) = 0 \\ \text{(ii)} & V(x, b) = 0 \\ \text{(iii)} & V(0, y) = V_o \text{ with } 0 < y < b \\ \text{(iv)} & V(\infty, y) = 0 \text{ with } 0 < y < b \end{array}\right\} \tag{3.101}$$

Equation 3.99 can be de coupled into ordinary differential equations as we have done in Sect. 3.8. Then similar to Eqs. 3.95, 3.96

$$\frac{1}{X}\frac{d^2X}{dx^2} = k_1 \tag{3.102}$$

$$\frac{1}{Y}\frac{d^2Y}{dy^2} = k_2 \tag{3.103}$$

Substituting Eqs. 3.102, 3.103 in Eq. 3.99

$$k_1 + k_2 = 0 \tag{3.104}$$

$$k_1 = -k_2 = k^2\,(k^2 \text{ is a constant}) \tag{3.105}$$

Substituting Eq. 3.105 in Eqs. 3.102 and 3.103 we get

$$\frac{1}{X}\frac{d^2X}{dx^2} = -k^2$$

$$\Rightarrow \frac{d^2X}{dx^2} = k^2\,X \tag{3.106}$$

$$\frac{1}{Y}\frac{d^2Y}{dy^2} = -k^2$$

$$\Rightarrow \frac{d^2Y}{dy^2} = -k^2\,Y \tag{3.107}$$

The solution for Eq. 3.106 is

$$X = A_1 e^{kx} + B_1 e^{-kx} \tag{3.108}$$

Using the boundary condition Eq. 3.101(iv) $V(\infty, y) = 0$

$$0 = A_1 e^{\infty} + B_1 e^{-\infty}$$
$$0 = A_1 e^{\infty}$$

Thus

$$A_1 = 0 \tag{3.109}$$

Substituting $A_1 = 0$ in Eq. 3.108

$$X = B_1\ e^{-kx} \tag{3.110}$$

The solution for Eq. 3.107 is

$$Y(y) = A_2 \sin ky + B_2 \cos ky \tag{3.111}$$

Using boundary condition Eq. 3.101(i) in Eq. 3.111

$$\begin{aligned} 0 &= A_2 \sin 0 + B_2 \cos(0) \\ 0 &= A_2(0) + B_2(1) \\ \Rightarrow B_2 &= 0 \end{aligned} \tag{3.112}$$

Substituting $B_2 = 0$ in Eq. 3.111 we get

$$Y(y) = A_2 \sin ky \tag{3.113}$$

Using boundary condition Eq. 3.101(ii) $V(x, b) = 0$

$$0 = A_2 \sin kb$$

Now assume $A_2 = 0$. If $A_2 = 0$ then by Eq. 3.113 $Y = 0$ and hence $V = X(x)\,Y(y) = 0$.

Thus the potential between three electrodes is zero if $A_2 = 0$ which is physically impossible. Hence the only option is

$$\begin{aligned} \sin kb &= 0 \\ kb &= n\pi \\ \Rightarrow k &= \frac{n\pi}{b} \quad n = 1, 2, 3\ldots \end{aligned} \tag{3.114}$$

We have excluded n = 0 because when n = 0 $Y = 0$ and

$V(x, y) = X(x)\ Y(y) = 0$ everywhere in between the three electrodes which is physically impossible.

Hence Eq. 3.113 is

$$Y(y) = A_2 \sin\frac{n\pi y}{b} \tag{3.115}$$

Similarly Eq. 3.110 is

$$X = B_1 e^{-n\pi x/b} \tag{3.116}$$

Substituting Eqs. 3.116, 3.115 in Eq. 3.100 we get

$$V(x, y) = B_1\, A_2\, e^{-n\pi x/b}\, \sin\frac{n\pi y}{b} \tag{3.117}$$

$$V(x, y) = D\left(e^{-n\pi x/b}\right)\left(\sin\frac{n\pi y}{b}\right) \tag{3.118}$$

where n = 1, 2, 3, 4 ... and $D = B_1\, A_2$ is a constant.

Now the final boundary condition to be satisfied is $V(0, y) = V_o$ (boundary condition Eq. 3.101(iii)). Using this boundary condition in Eq. 3.118

$$V_o = D \sin\frac{n\pi y}{b} \tag{3.119}$$

Thus boundary condition Eq. 3.101(iii) demands that $V_o$ to be of the form $\sin\frac{n\pi y}{b}$. But $V_o$ is a constant.

In order to satisfy boundary condition Eq. 3.101(iii) we note that from Eq. 3.118 there are infinite solutions possible because n = 1, 2, 3... and we have already seen in Eqs. 3.89 and 3.90 Laplace's equation is linear. Thus each solution for each n can be summed up together so that boundary condition Eq. 3.101(iii) is satisfied.

Thus similar to equation

$$V = a_1 V_1 + a_2 V_2 + a_3 V_3 \ldots = \sum_{n=1}^{\infty} a_n V_n$$

in Sect. 3.8 we can write Eq. 3.118 as

$$V(x, y) = \sum_{n=1}^{\infty} D_n\, e{-}n\pi x/b \sin\frac{n\pi y}{b} \tag{3.120}$$

The above equation satisfies the boundary condition Eq. 3.101(i), (ii), (iv). Now if we carefully select the values of $D_n$ then the boundary condition Eq. 3.101(iii) also can be satisfied.

Substituting the boundary condition Eq. 3.101(iii) in Eq. 3.120

$$V_0 = \sum_{n=1}^{\infty} D_n \sin\frac{n\pi y}{b} \tag{3.121}$$

Multiplying Eq. 3.121 by $\sin\frac{m\pi y}{b}$ and integrating between 0 and b

$$\begin{aligned}&\int_0^b V_0 \sin\frac{m\pi y}{b}\,dy \\ &= \sum_{n=1}^{\infty} D_n \int_0^b \sin\frac{n\pi y}{b}\sin\frac{m\pi y}{b}\,dy\end{aligned} \tag{3.122}$$

$$\text{But} \int_0^b \sin\frac{n\pi y}{b}\sin\frac{m\pi y}{b}\,dy = \begin{cases} 0 & \text{if } n \neq m \\ b/2 & \text{if } n = m \end{cases} \tag{3.123}$$

Thus substituting Eq. 3.123 in Eq. 3.122

$$\int_0^b V_0 \sin\frac{n\pi y}{b}\,dy = \frac{b}{2} D_n \tag{3.124}$$

$$\text{But } V_0 \int_0^b \sin\frac{n\pi y}{b}\,dy = \begin{cases} \frac{2bV_0}{n\pi} & \text{if n is odd} \\ 0 & \text{if n is even} \end{cases} \tag{3.125}$$

Substituting Eq. 3.125 in Eq. 3.124

$$\frac{2bV_0}{n\pi} = \frac{b}{2} D_n \text{ where } n = 1, 3, 5, 7\ldots \tag{3.126}$$

$$\Rightarrow D_n = \begin{cases} \frac{4V_0}{n}\pi & \text{if n is odd} \\ 0 & \text{if n is even} \end{cases} \tag{3.127}$$

Substituting Eq. 3.127 in Eq. 3.120

$$V(x, y) = \frac{4V_0}{\pi} \sum_{n=1}^{\infty} \frac{1}{n} e^{-n\pi x/b} \sin\frac{n\pi y}{b} \text{ with n odd} \tag{3.128}$$

Equation 3.128 is the desired result.

## 3.10 Potential Between Two Grounded Conducting Electrodes Separated by Two Conducting Side Plates Maintained at $V_o$ Potentials $V_o$ and $V_o$

Two parallel plane conducting electrodes of width a are maintained at zero potential and are separated by two electrodes of width b at their sides as shown in Fig. 3.15. Both the side electrodes are maintained at potentials respectively. The boundary conditions to the problem are

$$\left.\begin{array}{l}\text{(i)}\quad V(x, 0) = 0 \\ \text{(ii)}\ V(x, \mathrm{b}) = 0 \\ \text{(iii)}\ V(0, y) = V_o \text{ with } 0 < y < \mathrm{b} \\ \text{(iv)}\ V(\mathrm{a}, y) = V_o \text{ with } 0 < y < \mathrm{b}\end{array}\right\} \tag{3.129}$$

As in Sect. 3.9

$$\frac{1}{X}\frac{d^2X}{dx^2} + \frac{1}{Y}\frac{d^2Y}{dy^2} = 0$$

$$\frac{1}{X}\frac{d^2X}{dx^2} = k_1 = k^2 \tag{3.130}$$

$$\frac{1}{Y}\frac{d^2Y}{dy^2} = k_1 = -k^2 \tag{3.131}$$

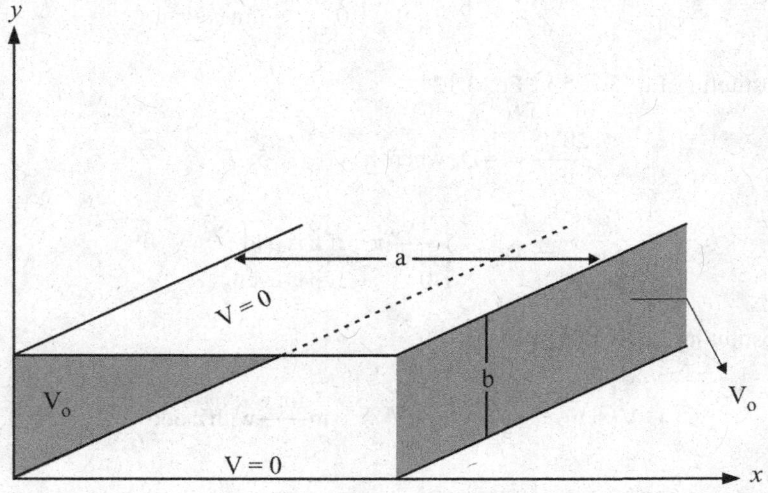


**Fig. 3.15** Four conducting plates maintained at specified potentials as shown in the figure

$$V(x, y) = X(x)\, Y(y) \tag{3.132}$$

$$X(x) = A_1\, e^{kx} + B_1\, e^{-kx} \tag{3.133}$$

$$Y(y) = A_2 \sin ky + B_2 \cos ky \tag{3.134}$$

As in Sect. 3.9 boundary condition Eq. 3.129(i) requires $B_2$ to be zero. Hence

$$Y(y) = A_2 \sin ky \tag{3.135}$$

Similar to Sect. 3.9 boundary condition ii requires

$$k = \frac{n\pi}{b} \quad n = 1, 2, 3 \ldots \tag{3.136}$$

Hence substituting Eq. 3.136 in Eqs. 3.135, 3.133

$$X(x) = A_1\, e^{n\pi x/b} + B_1\, e^{-n\pi x/b} \tag{3.137}$$

$$Y(y) = A_2 \sin \frac{n\pi y}{b} \tag{3.138}$$

Substituting Eqs. 3.137, 3.138 in Eq. 3.132 we get

$$V(x, y) = \left(A_1 e^{n\pi x/b} + B_1\, e^{-n\pi x/b}\right) \quad A_2 \sin \frac{n\pi y}{b} \tag{3.139}$$

$$\Rightarrow V(x, y) = \left(D\, e^{n\pi x/b} + F\, e^{-n\pi x/b}\right) \sin \frac{n\pi y}{b} \tag{3.140}$$

where $A_1A_2 = D$, $B_1A_2 = F$ are constants.

Because Laplace's equation is linear as in Sect. 3.9 we sum up Eq. 3.140 for all n

$$V(x, y) = \sum_{n=1}^{\infty} \left(D_n\, e^{n\pi x/b} + F_n\, e^{-n\pi x/b}\right) \sin \frac{n\pi y}{b} \tag{3.141}$$

Boundary condition Eq. 3.129(iii) requires that $V(0, y) = V_o$ substituting this boundary condition in Eq. 3.141

$$V_o = \sum_{n=1}^{\infty} (D_n + F_n) \sin \frac{n\pi y}{b} \tag{3.142}$$

Multiplying both sides by $\sin\frac{m\pi y}{b}$ and integrating between 0 and b we get

$$V_o \int_0^b \sin\frac{m\pi y}{b}\,dy = \sum_{n=1}^{\infty} \int_0^b (D_n + F_n) \sin\frac{n\pi y}{b} \sin\frac{m\pi y}{b}\,dy \tag{3.143}$$

$$\Rightarrow V_o \int_0^b \sin\frac{n\pi y}{b}\,dy = (D_n + F_n)\frac{b}{2} \tag{3.144}$$

$$\text{As} \int_0^b V_o \sin\frac{n\pi y}{b}\,dy = \begin{cases} \dfrac{2bV_o}{n\pi} & \text{if n is odd} \\ 0 & \text{if n in even} \end{cases} \tag{3.145}$$

Substituting Eq. 3.145 in Eq. 3.144 we get

$$\begin{aligned} &\Rightarrow \frac{2bV_o}{n\pi} = (D_n + F_n)\frac{b}{2} \\ &\Rightarrow D_n + F_n = \frac{4V_o}{n\pi} \text{ with } n = 1, 3, 5\ldots \end{aligned} \tag{3.146}$$

Boundary condition Eq. 3.129(iv) requires $V(a, y) = V_o$. Substituting this boundary condition in Eq. 3.141 we get

$$V_o = \sum_{n=1}^{\infty} \left(D_n\, e^{n\pi a/b} + F_n\, e^{-n\pi a/b}\right) \sin\frac{n\pi y}{b} \tag{3.147}$$

Multiplying the above equation by a $\sin\frac{m\pi y}{b}$ and integrating between 0 to b we get.

$$D_n e^{n\pi a/b} + F_n e^{-n\pi a/b} = \frac{4V_o}{n\pi} \text{ with } n = 1, 3, 5\ldots \tag{3.148}$$

Solving Eqs. 3.146, 3.148 we get

$$D_n = \frac{4V_o}{n\pi} e^{n\pi a/b} \left[\frac{1 - e^{n\pi a/b}}{1 - e^{2n\pi a/b}}\right] \tag{3.149}$$

$$F_n = \frac{4V_o}{n\pi} e^{-n\pi a/b} \left[\frac{1 - e^{-n\pi a/b}}{1 - e^{-2n\pi a/b}}\right] \tag{3.150}$$

Equation 3.141 is the desired solution with $D_n$, $F_n$ given by Eqs. 3.149, 3.150.

Selection of solution for the differential equations in Sects. 3.9, 3.10 is made by examining the given problem. In Sect. 3.9, in Fig. 3.14 we see that as we move along *y*-axis at $y = 0$, $y = b$ the potential needs to be zero. This condition is satisfied by sine function. Similarly as we move along *x*-axis from $x = 0$ we see that potential decreases gradually which is satisfied by exponential function, hence we selected $k_1$ to be positive and $k_2$ to be negative so that $X(x)$ assumes exponential function, $Y(y)$ assumes sine function.

## 3.11 Separation of Variables—Laplace's Equation in Spherical Polar Coordinates

If the given problem involves spherical conductors then spherical polar coordinates is more appropriate for the calculation of electric field. The Laplace's equation for spherical polar coordinates is given by

$$\frac{1}{r^2}\frac{\partial}{\partial r}\left[r^2\frac{\partial V}{\partial r}\right] + \frac{1}{r^2 \sin\theta}\frac{\partial}{\partial\theta}\left(\sin\theta\frac{\partial V}{\partial\theta}\right) + \frac{1}{r^2\sin^2\theta}\left(\frac{\partial^2 V}{\partial\phi^2}\right) = 0$$

Let us assume that V is independent of $\phi$ then

$$\frac{1}{r^2}\frac{\partial}{\partial r}\left(r^2\frac{\partial V}{\partial r}\right) + \frac{1}{r^2 \sin\theta}\frac{\partial}{\partial\theta}\left(\sin\theta\frac{\partial V}{\partial\theta}\right) = 0 \quad (3.151)$$

As in the previous problems we go for separation of variables

$$V = R(r)\Theta(\theta) \quad (3.152)$$

Substituting Eq. 3.152 in Eq. 3.151

$$\Theta\frac{1}{r^2}\frac{d}{dr}\left[r^2\frac{dR}{dr}\right] + \frac{R}{r^2\sin\theta}\frac{d}{d\theta}\left[\sin\theta\frac{d\Theta}{d\theta}\right] = 0 \quad (3.153)$$

$$\Rightarrow \Theta\frac{d}{dr}\left[r^2\frac{dR}{dr}\right] + \frac{R}{\sin\theta}\frac{d}{d\theta}\left[\sin\theta\frac{d\Theta}{d\theta}\right] = 0 \quad (3.154)$$

Dividing Eq. 3.154 by $V = R(r)\Theta(\theta)$

$$\Rightarrow \frac{1}{R}\frac{d}{dr}\left[r^2\frac{dR}{dr}\right] + \frac{1}{\Theta\sin\theta}\frac{d}{d\theta}\left[\sin\theta\frac{d\Theta}{d\theta}\right] = 0 \quad (3.155)$$

The first term depends on R alone and the second term on $\Theta$ alone. So each must be separately equal to a constant.

$$\frac{1}{R}\frac{d}{dr}\left[r^2\frac{dR}{dr}\right] = k_1 \tag{3.156}$$

$$\frac{1}{\Theta\ \sin\theta}\frac{d}{d\theta}\left[\sin\theta\frac{d\,\Theta}{d\theta}\right] = k_2 \tag{3.157}$$

From Eqs. 3.155, 3.156, 3.157 we have

$$k_1 + k_2 = 0 \tag{3.158}$$

$$\Rightarrow k_1 = -k_2 = n(n+1)\ (\text{say}) \tag{3.159}$$

Substituting Eq. 3.159 in Eqs. 3.156, 3.157

$$\frac{1}{R}\frac{d}{dr}\left[r^2\frac{dR}{dr}\right] = n(n+1)$$

$$\frac{1}{\Theta \sin\theta}\frac{d}{d\theta}\left[\sin\theta\frac{d\Theta}{d\theta}\right] = -n(n+1)$$

The solutions for the above equations are

$$R(r) = A_n\, r^n + \frac{B_n}{r^{n+1}} \tag{3.160}$$

$$\Rightarrow \Theta(\theta) = P_n(\cos\theta) \tag{3.161}$$

$$\text{where}\, P_n(\cos\theta) = \frac{1}{2^n n!}\left[\frac{d}{d(\cos\theta)}\right]^n (\cos^2\theta - 1)^n \tag{3.162}$$

Here $P_n(\cos\theta)$ are known as Legendre polynamials.
The potential V is given by

$$\begin{aligned} V &= R(r)\Theta\,(\theta) \\ V &= \left[A_n r^n + B_n r^{-(n+1)}\right] P_n(\cos\theta) \end{aligned} \tag{3.163}$$

The most general solution is the linear combination of individual solutions

$$V = \sum_{n=0}^{\infty}\left[A_n r^n + B_n\, r^{-(n+1)}\right] P_n(\cos\theta) \tag{3.164}$$

**Example 3.4**
Consider a uniformly charge disc with charge density σ and radius R as shown in Fig. 3.16. Calculate the potential at any point P due to the disc.

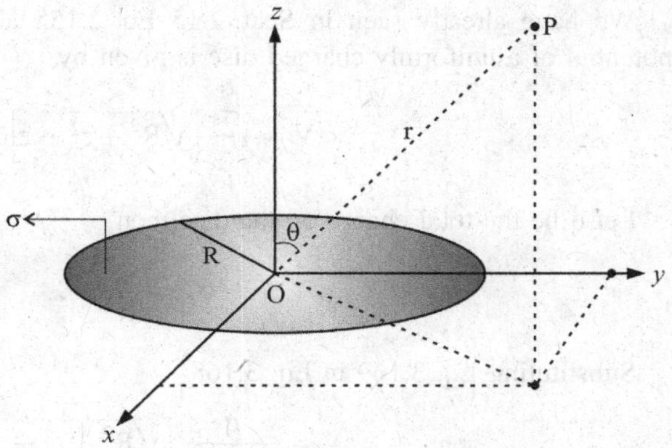


**Fig. 3.16** A circular disc of radius R carrying a uniform surface charge density

**Solution**

Let us calculate the potential at point P at a distance r from the center O of the disc. The potential at point P is

$$V = \sum_{n=0}^{\infty} \left( A_n r^n + \frac{B_n}{r^{n+1}} \right) P_n(\cos\theta) \tag{3.165}$$

There are two possible cases

(i) **r > R**

The boundary condition for the problem is

$$V \to 0 \text{ as } r \to \infty$$

As $r \to \infty$ the potential given by Eq. 3.165 tends to infinity which is physically impossible. Hence the coefficient $A_n = 0$. Thus the potential is given by

$$V(r, \theta) = \sum_n \frac{B_n}{r^{n+1}} P_n(\cos\theta) \tag{3.166}$$

Now let us evaluate $B_n$. We note that at $\theta = 0$ $P_n(\cos 0) = 1$. Hence Eq. 3.166 is

$$V(r, 0) = \sum_n \frac{B_n}{r^{n+1}} \tag{3.167}$$

We have already seen in Sect. 2.15 Eq. 2.155 along $z$-axis (i.e.,) $\theta = 0$ the potential of a uniformly charged disc is given by

$$V = \frac{\sigma}{2\varepsilon_0}\left[\sqrt{R^2 + z^2} - z\right] \tag{3.168}$$

Let q be the total charge on the disc then

$$\sigma = \frac{q}{\pi R^2} \tag{3.169}$$

Substituting Eq. 3.169 in Eq. 3.168

$$V = \frac{q}{2\pi\varepsilon_0 R^2}\left[\sqrt{R^2 + z^2} - z\right] \tag{3.170}$$

As we are measuring the potential along $z$-axis in Eq. 3.170 the point P in Fig. 3.16 lies along $z$-axis and hence r = $z$. Therefore Eq. 3.170 can be written as

$$V = \frac{q}{2\pi\varepsilon_0 R^2}\left[\sqrt{R^2 + r^2} - r\right] \tag{3.171}$$

Comparing Eqs. 3.167, 3.171

$$\sum_n \frac{B_n}{r^{n+1}} = \frac{q}{2\pi\varepsilon_0 R^2}\left[\sqrt{R^2 + r^2} - r\right] \tag{3.172}$$

$$\Rightarrow \sum_n \frac{B_n}{r^{n+1}} = \frac{q}{2\pi\varepsilon_0 R^2}\left[-r + r\left[1 + \frac{R^2}{r^2}\right]^{\frac{1}{2}}\right] \tag{3.173}$$

Applying binomial theorem

$$\begin{aligned} &\Rightarrow \frac{B_0}{r} + \frac{B_1}{r^2} + \frac{B_2}{r^3} + \cdots \\ &= \frac{q}{2\pi\varepsilon_0 R^2}\left[-r + r\left(1 + \frac{1}{2}\frac{R^2}{r^2} + \frac{\left(\frac{1}{2}\right)\left(\frac{-1}{2}\right)}{2!}\left(\frac{R^2}{r^2}\right)^2 + \cdots\right)\right] \end{aligned} \tag{3.174}$$

$$\Rightarrow \frac{B_0}{r} + \frac{B_1}{r^2} + \frac{B_2}{r^3} + \cdots = \frac{q}{2\pi\varepsilon_0 R^2}\left[\frac{1}{2}\frac{R^2}{r} - \frac{1}{6}\frac{R^4}{r^3} + \frac{1}{16}\frac{R^6}{r^5}\cdots\right] \tag{3.175}$$

Comparing the coefficients of like powers of r on both sides

$$\begin{aligned} &B_0 = \frac{q}{4\pi\varepsilon_0},\ B_2 = \frac{-qR^2}{16\,\pi\varepsilon_0},\ B_4 = \frac{qR^4}{32\pi\varepsilon_0}\ldots \text{ and} \\ &B_1 = B_3 = B_5 = \cdots = 0. \end{aligned}$$

Substituting the coefficients B in Eq. 3.165

$$V(r,\theta) = \frac{q}{4\pi\varepsilon_o}\left[\frac{1}{r}P_0(\cos\theta) - \frac{R^2}{4r^3}P_2(\cos\theta) + \frac{R^4}{8r^5}P_4(\cos\theta)\ldots\right] \tag{3.176}$$

The above equation gives the value of potential at any point P in Fig. 3.16 for $r > R$.

(ii) **r < R**

In Eq. 3.165 as $r \to 0$, $V \to \infty$ which is physically impossible. Hence the coefficients $B_n = 0$. Thus the potential at $r < R$ is given by

$$V(r,\theta) = \sum_n A_n r^n P_n(\cos\theta) \tag{3.177}$$

Consider that point P in Fig. 3.16 lies along *z*-axis. Then $\theta = 0$ and hence $P_n(\cos 0) = 1$. Hence Eq. 3.177 is

$$V(r,0) = \sum_n A_n r^n \tag{3.178}$$

From Eq. 3.171 we have

$$V(r,0) = \frac{q}{2\pi\varepsilon_o R^2}\left[-r + R\left(1 + \frac{r^2}{R^2}\right)^{\frac{1}{2}}\right] \tag{3.179}$$

Comparing Eqs. 3.178, 3.179 we have

$$\Rightarrow A_0 + A_1 r + A_2 r^2 + \cdots = \frac{q}{2\pi\varepsilon_o R^2}\left[-r + R + \frac{R}{2}\left(\frac{r}{R}\right)^2 - \frac{R}{8}\left(\frac{r}{R}\right)^4 + \cdots\right] \tag{3.180}$$

Comparing the coefficients of like powers of r on both sides

$$A_0 = \frac{q}{2\pi\varepsilon_o R},\; A_1 = \frac{-q}{2\pi\varepsilon_o R^2}$$

$$A_2 = \frac{q}{4\pi\varepsilon_o R^3},\; A_4 = \frac{-q}{10\pi\varepsilon_o R^3}\cdots$$

$$\text{and } A_3 = A_5 = A_7 = \cdots = 0.$$

Substituting the coefficients A in Eq. 3.177

$$V(r,\theta) = A_0 P_0(\cos\theta) + A_1 r P_1(\cos\theta) + A_2 r^2 P_2(\cos\theta) + \cdots \tag{3.181}$$

$$V(r,\theta) = \frac{q}{2\pi\varepsilon_o}\left[\frac{1}{R} - \frac{r}{R^2}P_1(\cos\theta) + \frac{r^2}{2R^3}P_2(\cos\theta) + \cdots\right] \tag{3.182}$$

The above equation gives the value of potential at any point P in Fig. 3.16 for $r < R$.

**Example 3.5**

Consider a conducting uncharged sphere placed in a uniform electric field $E_o\,\hat{\mathbf{k}}$. The induced positive and negative charges are shown in Fig. 3.17. Calculate the potential, dipole moment and electric field outside the sphere. The radius of the sphere is R.

**Solution**

Once the conducting sphere is placed in the uniform electric field $E_o\,\hat{\mathbf{k}}$ the positive charges are pushed in the direction of the field and the negative charges opposite to the direction of the field. Because of induced charge the uniform field lines gets distorted as shown in the Fig. 3.17.

Under electrostatic condition the sphere is an equipotential and we can take the value of equipotential to zero. Then the boundary conditions for the problem is

$$\left.\begin{array}{ll}\text{(i)} & V = 0 \text{ for } r = R\\ \text{(ii)} & V = -E_o z = -E_o\, r\cos\theta \text{ for } r \gg R\end{array}\right\} \tag{3.183}$$

The second boundary condition is needed on the grounds that the distances far away from the sphere the field is $E_o\,\hat{\mathbf{k}}$ and θ has the usual meaning as in spherical polar coordinates.

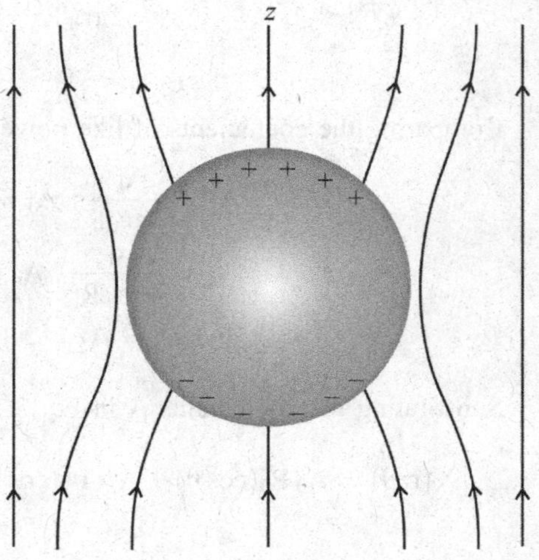


**Fig. 3.17** A conducting sphere placed in a uniform electric field. The induced positive and negative charges are shown along with the electric field lines in the figure

We know that the potential is given by

$$V(r,\theta) = \sum_n \left( A_n r^n + \frac{B_n}{r^{n+1}} \right) P_n(\cos\theta) \tag{3.184}$$

Applying the boundary conditions Eq. 3.183(i)

$$0 = A_n R^n + \frac{B_n}{R^{n+1}} \tag{3.185}$$

$$\Rightarrow B_n = -A_n R^{2n+1} \tag{3.186}$$

Substituting Eq. 3.186 in Eq. 3.184

$$V(r,\theta) = \sum_n \left[ A_n r^n - \frac{A_n R^{2n+1}}{r^{n+1}} \right] P_n(\cos\theta) \tag{3.187}$$

$$V(r,\theta) = \sum_n A_n \left[ r^n - \frac{R^{2n+1}}{r^{n+1}} \right] P_n(\cos\theta) \tag{3.188}$$

Applying the boundary condition Eq. 3.183(ii) and noting down that when $r \gg R$ the second term in Eq. 3.184 is negligible

$$\sum_{n=0}^{\infty} A_n r^n P_n(\cos\theta) = -E_o r \cos\theta \tag{3.189}$$

$$\begin{aligned} &\Rightarrow A_0 P_0(\cos\theta) + A_1 r^1 P_1(\cos\theta) + A_2 r^2 P_2(\cos\theta) + \cdots \\ &= -E_o r \cos\theta \end{aligned} \tag{3.190}$$

Comparing the coefficients of r

$$\begin{aligned} A_1 P_1(\cos\theta) &= -E_o \cos\theta \\ \Rightarrow A_1 \cos\theta &= -E_o \cos\theta \\ \Rightarrow A_1 &= -E_o \end{aligned} \tag{3.191}$$

and $A_0 = A_2 = A_3 = \cdots = 0$

Substituting the coefficients A in Eq. 3.188

$$V(r,\theta) = -E_o r\cos\theta + \frac{E_o R^3}{r^2}\cos\theta \tag{3.192}$$

The first term corresponds to the external field and hence the potential due to the induced charge is

$$V(r,\theta) = \frac{E_0 R^3 \cos\theta}{r^2} \tag{3.193}$$

In spherical polar coordinates the electric field is given by

$$\mathbf{E} = -\nabla V = -\frac{dV}{dr}\hat{\mathbf{r}} - \frac{1}{r}\frac{\partial V}{\partial \theta}\hat{\boldsymbol{\theta}} \tag{3.194}$$

as there is no $\phi$ dependence.

Now the r component is given by

$$E_r = -\frac{\partial V}{\partial r} = E_o\left[1 + 2\frac{R^3}{r^3}\right]\cos\theta \tag{3.195}$$

and the $\theta$ component is given by

$$E_\theta = -\frac{1}{r}\frac{\partial V}{\partial \theta} = -E_o\left[1 - \frac{R^3}{r^3}\right]\sin\theta \tag{3.196}$$

The electric field outside the sphere due to the induced charge on the sphere is given by

$$\mathbf{E} = E_r\,\hat{\mathbf{r}} + E_\theta\,\hat{\boldsymbol{\theta}} \tag{3.197}$$

where $E_r$ and $E_\theta$ is given by Eqs. 3.195, 3.196

The induced surface charge is given by Eq. 2.304

$$E_{in} = \frac{\sigma}{\varepsilon_o}$$

Here $E_{in} = (E_r)_{r=R}$.

Hence $(E_r)_{r=R} = \frac{\sigma}{\varepsilon_o}$

$$\Rightarrow \sigma(\theta) = \varepsilon_o (E_r)_{r=R} \tag{3.198}$$

From Eq. 3.195 we have

$$\sigma(\theta) = \varepsilon_o (E_r)_{r=R} = 3\varepsilon_o E_0 \cos\theta \tag{3.199}$$

We know that the potential due to a dipole is given by Eq. 2.256

$$V = \frac{1}{4\pi\varepsilon_o}\frac{p\cos\theta}{r^2} \tag{3.200}$$

Comparing Eqs. 3.200, 3.193

$$\frac{1}{4\pi\varepsilon_o}\frac{p\cos\theta}{r^2} = \frac{E_o R^3}{r^3}\cos\theta \tag{3.201}$$

$$\Rightarrow p = 4\pi\varepsilon_o E_o R^3 \tag{3.202}$$

is the dipole moment of the sphere.

We have found the field outside the sphere due to the induced charges on the sphere which is kept in the external electric field, in Eq. 3.197. What about the field inside the sphere. As explained in Sect. 2.24 the field inside the sphere is zero.

**Example 3.6**
Consider two conducting hemisphere's placed one on the other and separated by a small insulating ring. The radius of the sphere is R. The top hemisphere is kept at potential $V_o$ and the bottom hemisphere is kept at potential $-V_o$. Calculate the electrostatic potential inside and outside the sphere (Fig. 3.18).

**Solution**
We know that

$$V = \sum_n \left(A_n r^n + \frac{B_n}{r^{n+1}}\right) P_n(\cos\theta) \tag{3.203}$$

There are two possible cases

**Case (i) r < R**
Inside the sphere from Eq. 3.203 as $r \to 0$ $V \to \infty$ which is physically impossible. Therefore the coefficients $B_n = 0$. Hence

$$V_{in} = \sum_n A_n r^n P_n(\cos\theta) \tag{3.204}$$

Let $V_S$ be the potential at the surface of the sphere

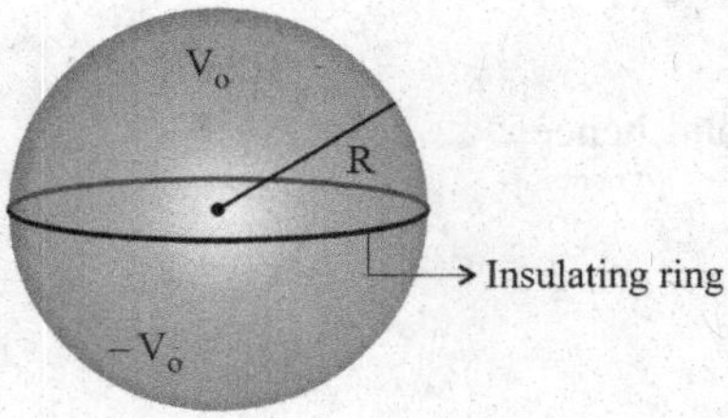


**Fig. 3.18** Two conducting hemispheres placed one on the other and separated by a small insulating ring. The top and bottom hemispheres are maintained at potentials $V_o$ and - $V_o$

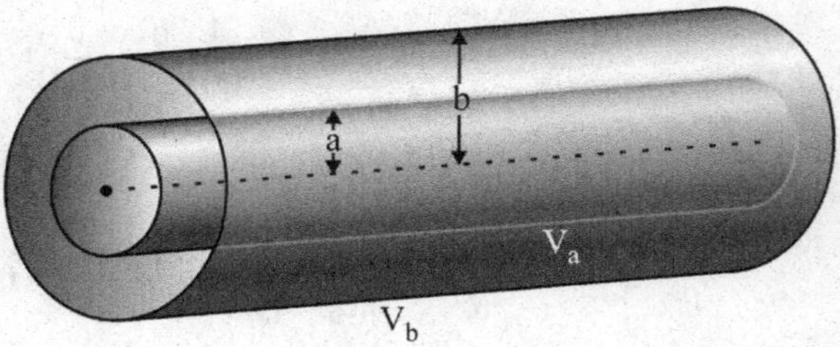


**Fig. 3.19** Two concentric cylindrical shells maintained at potentials $V_a$ and $V_b$. The radius of the inner cylinder is a and outer cylinder is b

$$V_S = \sum_n A_n R^n P_n(\cos\theta) \tag{3.205}$$

Now utilising the fact for Legendre polynamials

$$\int_0^\pi P_n(\cos\theta)\,P_m(\cos\theta)\sin\theta\,d\theta = \begin{cases} 0 & \text{if } n \neq m \\ \frac{2}{2n+1} & \text{if } n = m \end{cases} \tag{3.206}$$

Multiplying Eq. 3.205 by $P_m(\cos\theta)\sin\theta$ and integrating between 0 and $\pi$ we get

$$\int_0^\pi V_S\,P_m(\cos\theta)\sin\theta\,d\theta = \sum_n A_n R^n \int_0^\pi P_n(\cos\theta)P_m(\cos\theta)\sin\theta\,d\theta \tag{3.207}$$

using Eq. 3.206

$$\int_0^\pi V_S\;P_n(\cos\theta)\sin\theta\,d\theta = A_n R^n \frac{2}{2n+1} \tag{3.208}$$

$$\Rightarrow A_n = \frac{2n+1}{2R^n}\left[\int_0^\pi V_S\,P_n(\cos\theta)\sin\theta\,d\theta\right] \tag{3.209}$$

$$\Rightarrow A_n = \frac{2n+1}{2R^n}\left[\int_0^{\pi/2} V_o\,P_n(\cos\theta)\sin\theta\,d\theta + \int_{\pi/2}^{\pi} (-V_o)P_n(\cos\theta)\sin\theta\,d\theta\right] \tag{3.210}$$

and hence

$$A_0 = \frac{V_o}{2}\left[(-\cos\theta)_0^{\pi/2} - (-\cos\theta)_{\pi/2}^{\pi}\right] = 0$$

$$A_1 = \frac{V_o}{R}\left[\left(\frac{-\cos^2\theta}{2}\right)_0^{\pi/2} + \left(\frac{\cos^2\theta}{2}\right)_{\pi/2}^{\pi}\right] = \frac{3V_o}{2R} \quad \ldots \text{and so on.}$$

Substituting the coefficients $A_n$ in Eq. 3.204

$$V_{in} = V_o\left[\frac{3}{2}\frac{r}{R}P_1(\cos\theta) - \frac{7}{8}\frac{r^3}{R^3}P_3(\cos\theta) + \cdots\right] \tag{3.211}$$

**Case (ii) r > R**

Outside the sphere by Eq. 3.203 as $r \to \infty$ $V \to \infty$ which is physically impossible. Hence $A_n = 0$. Thus

$$V_{out} = \sum_n \frac{B_n}{r^{n+1}} P_n(\cos\theta) \tag{3.212}$$

On the surface of the sphere

$$V_S = \sum_n \frac{B_n}{R^{n+1}} P_n(\cos\theta) \tag{3.213}$$

Multiplying the above equation by $P_m(\cos\theta)\sin\theta$ and integrating between 0 to $\pi$

$$\int_0^{\pi} V_S P_m(\cos\theta)\sin\theta\, d\theta = \sum_n \int_0^{\pi} \frac{B_n}{R^{n+1}} P_n(\cos\theta) P_m(\cos\theta)\sin\theta\, d\theta \tag{3.214}$$

using Eq. 3.206

$$\int_0^{\pi} V_S P_n(\cos\theta)\sin\theta\, d\theta = \frac{B_n}{R^{n+1}}\frac{2}{2n+1} \tag{3.215}$$

$$\Rightarrow B_n = \frac{2n+1}{2}R^{n+1}\left[\int_0^{\pi} V_S P_n(\cos\theta)\sin\theta\, d\theta\right] \tag{3.216}$$

$$\Rightarrow B_n = \frac{2n+1}{2}R^{n+1}\left[\int_0^{\pi/2} V_o\, P_n(\cos\theta)\sin\theta\, d\theta + \int_{\pi/2}^{\pi} (-V_o) P_n(\cos\theta)\sin\theta\, d\theta\right] \tag{3.217}$$

so that

$$B_0 = \frac{V_o R}{2}\left[(-\cos\theta)_0^{\pi/2} + (\cos\theta)_{\pi/2}^{\pi}\right] = 0$$

$$B_1 = \tfrac{3}{2} V_o R^2 \left[\left(\frac{-\cos^2\theta}{2}\right)_0^{\pi/2} + \left(\frac{\cos^2\theta}{2}\right)_{\pi/2}^{\pi}\right] = \tfrac{3}{2} V_o R^2 \ldots \text{ and so on}$$

Substituting the values of coefficients B in Eq. 3.212

$$V_{out} = V_o\left[\frac{3}{2}\left(\frac{R}{r}\right)^2 P_1(\cos\theta) - \frac{7}{8}\left(\frac{R}{r}\right)^4 P_3(\cos\theta) + \cdots\right] \tag{3.218}$$

## 3.12 Separation of Variables—Laplace's Equation in Cylindrical Coordinates

Let us assume that the problem is independent of *z*. In cylindrical coordinates then Laplace's equation is given by (we have replaced $r_c$ by r for convenience)

$$\frac{1}{r}\frac{\partial}{\partial r}\left[r\frac{\partial V}{\partial r}\right] + \frac{1}{r^2}\frac{\partial^2 V}{\partial \phi^2} = 0 \tag{3.219}$$

Now let us apply method of separation of variables

$$V(r,\theta) = R(r)\Phi(\phi) \tag{3.220}$$

Substituting Eq. 3.220 in Eq. 3.219 and dividing throughout by $R(r)\Phi(\phi)$ we get

$$\frac{r}{R(r)}\frac{d}{dr}\left[r\frac{dR(r)}{dr}\right] + \frac{1}{\Phi(\phi)}\frac{d^2\Phi(\phi)}{d\phi^2} = 0 \tag{3.221}$$

The first term depends only on R(r) and the second term only depends only on $\Phi(\phi)$. Hence each term must be separately equal to a constant

$$\frac{r}{R(r)}\frac{d}{dr}\left[r\frac{dR(r)}{dr}\right] = k^2 \tag{3.222}$$

$$\frac{1}{\Phi(\phi)}\frac{d^2\Phi(\phi)}{d\phi^2} = -k^2 \tag{3.223}$$

where k is a constant.

The solutions for Eq. 3.223 is

$$\Phi(\phi) = A_k \cos k\phi + B_k \sin k\phi \tag{3.224}$$

The potential should be a single valued function at a given point. This condition is fulfilled when $\Phi(\phi + 2\pi) = \Phi(\phi)$. In Eq. 3.224 $\Phi(\phi + 2\pi) = \Phi(\phi)$ is satisfied only when k = 0, 1, 2, 3 …

Now k = 0 in Eq. 3.224 gives

$$\Phi(\phi) = A_0 \tag{3.225}$$

which is a constant. Substituting k = 0 in Eq. 3.223

$$\frac{d^2\Phi(\phi)}{d\phi^2} = 0 \tag{3.226}$$

Solving the above equation the solution is

$$\Phi(\phi) = a + b\phi \tag{3.227}$$

However the condition $\Phi(\phi + 2\pi) = \Phi(\phi)$ is not satisfied for the above solution and hence the term $b\phi$ is not acceptable.

The solution for Eq. 3.222 is

$$R(r) = C_k r^k + D_k r^{-k} \tag{3.228}$$

where we have already identified the values of k as 0, 1, 2, 3 …

For the case of k = 0, R(r) = 0 from Eq. 3.225; substituting k = 0 in Eq. 3.222

$$r\frac{d}{dr}\left[r\frac{dR(r)}{dr}\right] = 0 \tag{3.229}$$

Solving

$$R = a_0 + b_0 \ln r \tag{3.230}$$

From Eqs. 3.230, 3.228 and 3.224 the solution for potential can be written as

[for k = 0, and k = 1, 2, 3…]

$$V(r,\phi) = a_0 + b_0 \ln r + \sum_{k=1}^{\infty} \left[C_k r^k + D_k r^{-k}\right]\left[A_k \cos k\phi + B_k \sin k\phi\right] \tag{3.231}$$

**Example 3.7**
Consider an infinitely long conducting cylinder of radius R placed in a uniform electric field $\mathbf{E_o}$. The *z*-axis of the cylinder is perpendicular to the field $\mathbf{E_o}$. The center of the cylinder coincides with the origin. Find the potential outside the cylinder.

**Solution**
Consider a long conducting cylinder with its centre line coinciding with *z*-axis. As per the given problem the field $\mathbf{E_o}$ is perpendicular to the *z*-axis. Therefore let us assume the field is along *x*-axis. The potential outside the cylinder is given by Eq. 3.231

$$V(r,\phi) = a_0 + b_0 \ln r + \sum_{k=1}^{\infty}\left[C_k r^k + D_k r^{-k}\right]\left[A_k \cos k\phi + B_k \sin k\phi\right] \tag{3.232}$$

The cylinder is an equipotential surface and we can set its potential to zero. The boundary conditions to the problem are

$$\left.\begin{array}{ll}\text{(i)} & V = 0 \text{ at } r = R \\ \text{(ii)} & V = -E_o x = -E_o r \cos\phi \text{ at } r \gg R\end{array}\right\} \tag{3.233}$$

The second boundary condition is needed on the grounds that at distances far away from the cylinder the field is $E_o\hat{\mathbf{i}}$.

As $r \to \infty$ the potential should be finite. Hence $b_0 = 0$ in Eq. 3.232. Also at $r \gg R$ the potential should contain only terms of $\cos\phi$ and no $\sin\phi$ components as per boundary conditions (ii) in Eq. 3.233. Hence $B_k = 0$ in Eq. 3.232.

Therefore the potential in Eq. 3.232 is

$$\text{(i) } V(r,\phi) = a_0 + \sum_{k=1}^{\infty}\left[C_k r^k + D_k r^{-k}\right] A_k \cos k\phi$$

Let $A_k C_k = F_k, A_k D_k = G_k$ then

$$V(r,\phi) = a_0 + \sum_{k=1}^{\infty}\left[F_k r^k + G_k r^{-k}\right] \cos k\phi \tag{3.234}$$

Now applying the first boundary condition (i) V = 0 at r = R from Eq. 3.233

$$\begin{aligned} 0 &= F_k r^k + G_k r^{-k} \\ \Rightarrow G_k &= -F_k R^{2k} \end{aligned} \tag{3.235}$$

Substituting for $G_k$ in Eq. 3.234

$$V(r,\phi) = a_0 + \sum_{k=1}^{\infty} F_k \left[ r^k - \frac{R^{2k}}{r^k} \right] \cos k\phi \tag{3.236}$$

Substituting boundary condition (ii) from Eq. 3.233 in Eq. 3.236 and noting down that the term $\frac{R^{2k}}{r^k}$ can be neglected for $r \gg R$

$$\begin{aligned} -E_0 r \cos\phi &= a_0 + \sum_{k=1}^{\infty} F_k r^k \cos k\phi \\ &= a_0 + F_1 r \cos\phi + F_2 r^2 \cos 2\phi + \cdots \end{aligned} \tag{3.237}$$

Comparing the like coefficients of R on both sides

$$-E_0 = F_1, a_0 = F_2 = F_3 = \cdots = 0 \tag{3.238}$$

Substituting the above values in Eq. 3.236

$$V(r,\phi) = \left[ -E_0 r + \frac{E_0 R^2}{r} \right] \cos\phi \tag{3.239}$$

is the required result.

**Example 3.8**
Consider two concentric infinite cylindrical shells maintained at potentials $V_a$ and $V_b$. The radii of inner shell is a and outer shell is b. Find the potential in between the cylindrical shells.

**Solution**
The potential depends on r alone. Hence

$$\frac{1}{r}\frac{\partial}{\partial r}\left[ r \frac{\partial V}{\partial r} \right] = 0 \tag{3.240}$$

$$\Rightarrow \frac{\partial V}{\partial r} = \frac{A}{r} \tag{3.241}$$

$$\Rightarrow V = A \log_e r + B \tag{3.242}$$

where A and B are constants. The boundary conditions for the problem is

$$\left.\begin{array}{llll} \text{at} & \text{(i)} & r = a & V = V_a \\ & \text{(ii)} & r = b & V = V_b \end{array}\right\} \tag{3.243}$$

Using boundary conditions Eq. 3.243(i), (ii) in Eq. 3.242

$$V_a = A \log_e a + B \tag{3.244}$$

$$V_b = A \log_e b + B \tag{3.245}$$

Solving Eqs. 3.244, 3.245

$$A = \frac{V_a - V_b}{\log_e\left(\frac{a}{b}\right)} \tag{3.246}$$

$$B = \frac{(V_a \log_e b - V_b \log_e a)}{\log_e\left(\frac{a}{b}\right)} \tag{3.247}$$

Substituting Eqs. 3.246, 3.247 in Eq. 3.242

$$V = \frac{V_a - V_b}{\log_e\left(\frac{a}{b}\right)} \log_e r - \frac{(V_a \log_e b - V_b \log_e a)}{\log_e\left(\frac{a}{b}\right)} \tag{3.248}$$

## 3.13 Summary

In Chap. 1 we discussed in detail about vectors. In Sect. 1.1 we saw that vectors are difficult to deal with as compared to scalars. In Chap. 2 our primary aim was to calculate the electric field, for a given charge distribution. We discussed in detail about Coulomb's law, Gauss's law and potential formulation. Section 2.18 compares about this three methods and we hinted that the above three methods cannot be used to calculate the electric field if the charge distribution is not known. In such situations we calculate the electric field using method of images (or) Laplace's equations.

Section 3.5 discussed in detail about method of images. In Sect. 3.7 we saw that method of images cannot be used if there are no isolated free charges. In such cases potential can be obtained using Laplace's equation by separation of variables. However the usefulness of Laplace's equation in obtaining the potential depends on whether boundaries of the given problem coincides with the coordinate surfaces.

In three dimensions if one coordinate is held constant and other coordinates are varied then the surface obtained is called coordinate surface.

As an example consider Fig. 3.14. Consider the coordinate surface at $y = b$. Clearly it is a plane. One of the semi infinite electrode kept at $V = 0$ potential, coincides with the coordinate surface at $y = b$. Similarly another semi infinite electrode kept at $V = 0$ potential, coincides with the coordinate surface at $y = 0$. The electrode kept at potential $V = V_o$ coincides with $x = 0$ coordinate surface.

Obtaining solution in the above case was easy as the boundaries coincide with coordinate surfaces. However if the boundaries are complex obtaining solution using Laplace's equation will not be easy. In such situations we can obtain solutions using analytical methods which will not be discussed in this book.

## 3.14 Dielectrics

In Sect. 2.24 we saw that materials in general can be classified into conductors, semiconductors and insulators. We discussed in detail about conductors. In this following sections we will discuss in detail about insulators also known as dielectrics. In the case of conductors the valence electrons are free to move, however in the case of dielectrics the valence electrons are tightly bound to the atoms. When a conductor is placed in an external electric field the electrons move in a direction opposite to the field having a net positive charge on the other side of the conductor, as we have seen in Fig. 2.64. However when a dielectric is subjected to an electric field the electrons are slightly displaced from their original position as they are tightly bound to the atoms. The above state of the dielectric is called polarization of dielectric under the action of the applied electric field. If applied electric field is very strong the dielectric will breakdown—it will lose its insulating property.

In the following sections we will discuss about polarization and dielectric breakdown.

## 3.15 Dielectric in an Electric Field

A dielectric atom is shown in Fig. 3.20a. N is the nucleus surrounded by an electron cloud. The center of the electron cloud O coincides with the center of the nucleus. Suppose the dielectric atom is placed in the electric field $\mathbf{E_o}$. The electrons are attracted in the direction opposite to the field as shown in Fig. 3.20b. The center of

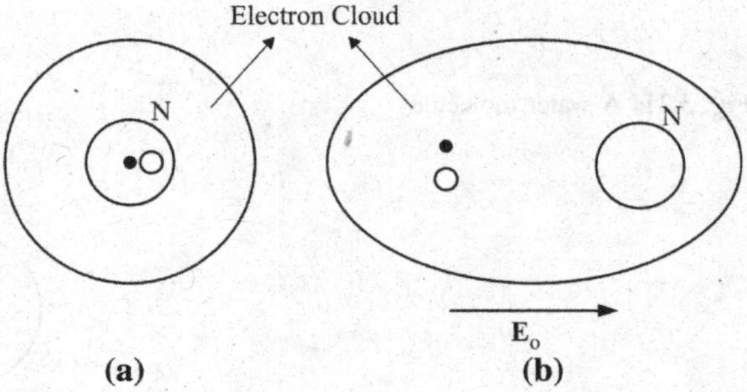


**Fig. 3.20** **a** In a given atom the center of the electron cloud O coincides with the center of the nucleus N. **b** Under the action of the external electric field the electron cloud gets displaced. The center of the electron cloud O doesn't coincide with the center of the nucleus N

the electron cloud O is thus displaced from the center of the nucleus. The displacement however is restricted by strong restoring forces due to the change in the charge configuration of the atom. Thus as we see in Fig. 3.20b the center O of the electron cloud and center of the nucleus are separated by a small distance and we know that two equal and opposite charges separated by a small distance form a electric dipole. We call this created dipole as "induced dipole" because the dipole is induced by the electric field in the dielectric atom. The electric field thus "polarizes" the dielectric atom leaving a net dipole moment.

Thus if a dielectric is placed in the external electric field all the atoms in the dielectric gets polarized. Applying an external field creates an set of "induced electric dipoles" which in turn produce an own electric field. Thus there are two distinct electric fields—one is the external electric field which we applied and the another electric field produced by the set of induced electric dipoles. In Sect. 3.17 we will calculate the electric field due to this induced electric dipoles.

## 3.16 Polar and Non-Polar Molecules

In the previous sections we considered atoms. In Fig. 3.20a the center of positive and negative charges coincide with each other leaving no dipole moment. However if such an atom is placed in a electric field as shown in Fig. 3.20b an induced dipole is produced.

Now instead of atoms let us consider molecules. Molecules in general can be classified as polar and non-polar molecules.

Consider the example, $H_2$, $N_2$ molecules. In this molecule the center of positive and negative charges coincide with each other leaving no net dipole moment. If such molecules are placed in external electric field the centers of positive and negative charges get separated by a distance giving rise to induced dipole moment. These molecules are known as non-polar molecules. Other examples of non-polar molecules are $Cl_2$, $CCl_4$, $CH_4$ etc.

On the other hand in molecules like water shown in Fig. 3.21 the centers of positive and negative charges do not coincide with each other even in the absence of electric field. That is they have permanent dipole moment. The dipole is built in the material. These molecules are called polar molecules. In a dielectric consisting

**Fig. 3.21** A water molecule

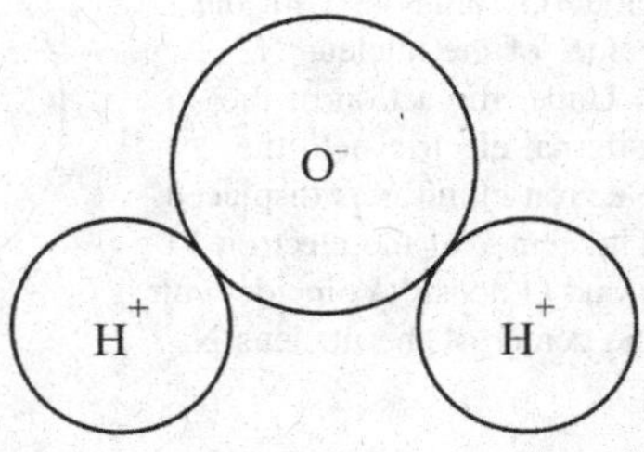

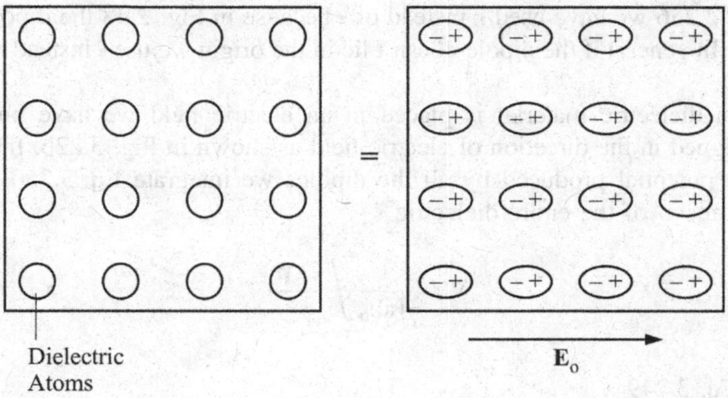


**Fig. 3.22** A dielectric material is placed in an external electric field. The net result will be a production of set of dipoles aligned in the direction of the field

of polar molecules the dipoles are randomly oriented thereby the net dipole moment is zero. If a dielectric consisting of polar molecules is placed in a electric field then the electric field tries to rotate the "permanent dipoles" in the direction of the field and align the dipoles in its direction.

Thus either it be a non-polar dielectric or polar dielectric when placed in an external electric field the net result a set of dipoles aligned in the direction of the electric field, as shown in Fig. 3.22 and the entire dielectric gets "polarized". We define polarization **P** as dipole moment per unit volume. Considering a small volume element $d\tau'$

$$\mathbf{P} = \frac{\mathbf{p}}{d\tau'} \tag{3.249}$$

where **p** is the dipole moment.

## 3.17 Potential Produced by the Polarized Dielectric

As we already noted in the case of dielectric placed in an electric field there are two distinct electric fields. The "external" electric field which caused the polarization and the electric field produced by the induced or permanent dipoles. We will calculate the field produced by the dipoles (i.e.,) due to polarization itself. The potential due to a single dipole is (Eq. 2.256)

$$V_{dip} = \frac{1}{4\pi\varepsilon_o}\frac{\hat{\mathscr{r}}\cdot\mathbf{p}}{\mathscr{r}^2} \tag{3.250}$$

(In Eq. 2.256 we have used $\hat{r}$ instead of $\hat{\mathscr{r}}$ because in Fig. 2.48 the dipole lies in the origin. In general if the dipole doesn't lie in the origin we use $\hat{\mathscr{r}}$ instead of $\hat{r}$ as in Eq. 3.250).

When a dielectric material is placed in an electric field we have number of dipoles aligned in the direction of electric field as shown in Fig. 3.22b. In order to obtain the potential produced by all the dipoles we integrate Eq. 3.250 over the entire volume $\tau'$ of the entire dielectric

$$V = \frac{1}{4\pi\varepsilon_o} \int_{\tau'} \frac{\hat{\mathscr{r}} \cdot \mathbf{p}}{\mathscr{r}^2} \tag{3.251}$$

From Eq. 3.249

$$\mathbf{p} = \mathbf{P}\,d\tau' \tag{3.252}$$

Substituting Eq. 3.252 in Eq. 3.251

$$V = \frac{1}{4\pi\varepsilon_o} \int_{\tau'} \frac{\hat{\mathscr{r}} \cdot \mathbf{P}}{\mathscr{r}^2} d\tau' \tag{3.253}$$

The above equation gives the potential produced by the dipoles in the entire volume $\tau'$ of the dielectric material at a point whose coordinates are $(x, y, z)$ as shown in Fig. 3.23.

The coordinate $(x, y, z)$ and coordinate of volume element $d\tau'$, $(x', y', z')$ is with respect to some arbitrary origin.

From Fig. 3.23

$$\mathscr{r}^2 = (x - x')^2 + (y - y')^2 + (z - z')^2 \tag{3.254}$$

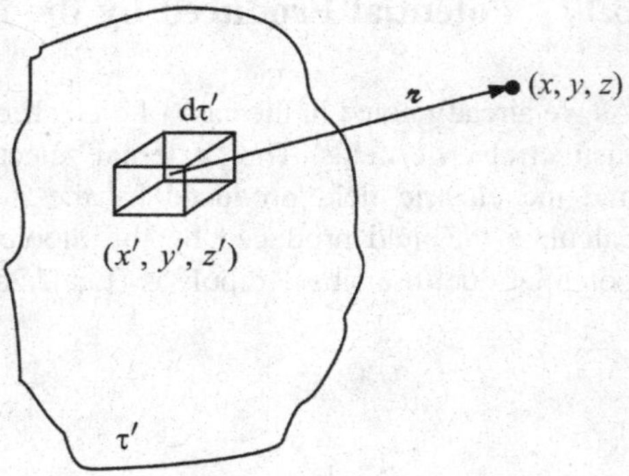


**Fig. 3.23** Illustration to show the potential produced by polarization in the dielectric material at a point **(x, y, z/Emphasis>)**

From the above equation

$$\nabla'\left(\frac{1}{\mathscr{r}}\right)=\frac{\hat{\mathscr{r}}}{\mathscr{r}^2} \tag{3.255}$$

Substituting Eq. 3.255 in Eq. 3.253

$$V=\frac{1}{4\pi\varepsilon_0}\int_{\tau'}\nabla'\left(\frac{1}{\mathscr{r}}\right)\cdot\mathbf{P}\,d\tau' \tag{3.256}$$

Now using vector identity 10 from Sect. 1.20

$$\begin{aligned}\nabla'\cdot\left(\frac{\mathbf{P}}{\mathscr{r}}\right)&=\frac{1}{\mathscr{r}}\nabla'\cdot\mathbf{P}+\mathbf{P}\cdot\nabla'\left(\frac{1}{\mathscr{r}}\right)\\ \Rightarrow\mathbf{P}\cdot\nabla'\left(\frac{1}{\mathscr{r}}\right)&=\nabla'\left(\frac{\mathbf{P}}{\mathscr{r}}\right)-\frac{1}{\mathscr{r}}\nabla'\cdot\mathbf{P}\end{aligned} \tag{3.257}$$

Substituting Eq. 3.257 in Eq. 3.256

$$V=\frac{1}{4\pi\varepsilon_0}\int_{\tau'}\nabla'\cdot\left(\frac{\mathbf{P}}{\mathscr{r}}\right)d\tau'-\frac{1}{4\pi\varepsilon_0}\int_{\tau'}\frac{1}{\mathscr{r}}\nabla'\cdot\mathbf{P}\,d\tau' \tag{3.258}$$

Using divergence theorem

$$V=\frac{1}{4\pi\varepsilon_0}\oint_{S'}\frac{1}{\mathscr{r}}\mathbf{P}\cdot d\mathbf{s}'-\frac{1}{4\pi\varepsilon_0}\oint_{\tau'}\frac{1}{\mathscr{r}}(\nabla'\cdot\mathbf{P})d\tau' \tag{3.259}$$

$$\Rightarrow V=\frac{1}{4\pi\varepsilon_0}\oint_{S'}\frac{1}{\mathscr{r}}(\mathbf{P}\cdot\hat{\mathbf{n}})ds'-\frac{1}{4\pi\varepsilon_0}\oint_{\tau'}\frac{1}{\mathscr{r}}(\nabla'\cdot\mathbf{P})\,d\tau' \tag{3.260}$$

Here $\hat{\mathbf{n}}$ is the unit vector in the direction of $ds'$.
Suppose if we replace

$$\sigma_p=\mathbf{P}\cdot\hat{\mathbf{n}} \tag{3.261}$$

$$\rho_p=-\nabla'\cdot\mathbf{P} \tag{3.262}$$

then

$$V=\frac{1}{4\pi\varepsilon_0}\oint_{S'}\frac{\sigma_p\,ds'}{\mathscr{r}}+\frac{1}{4\pi\varepsilon_0}\int_{\tau'}\frac{\rho_p\,d\tau'}{\mathscr{r}} \tag{3.263}$$

The above equation shows that the potential due to the polarized dielectric in same as that of sum of the potential of 'bound charges' $\sigma_p$ and $\rho_p$. Instead of integrating over the entire volume $\tau'$ to find the potential and electric field produced by the dipoles as in Eq. 3.253 we can calculate the potential and hence electric field of the entire dielectric by summing up the potential of $\sigma_p$ and $\rho_p$. The potential produced by $\sigma_p$ and $\rho_p$ can be calculated by the methods we discussed in Chap. 2.

## 3.18 Bound Charges $\sigma_p$ and $\rho_p$

In Sect. 3.17 we have seen that the potential due to the polarized object can be calculated by calculating potential of the bound charges $\sigma_p$ and $\rho_p$. In this section we will see the physical interpretation of those charges.

Let us redraw Fig. 3.22 in Fig. 3.24. We see that there is a set of aligned dipoles produced inside the dielectric because of the application of electric field. At points A, B in Fig. 3.24a the positive and negative charges cancel out each other. Similarly at points like C, F positive and negative charges cancel out each other. Along the given line of dipoles the positive and negtive charges cancel out each other and the result is net negative charges on one side and net positive charge on the other side of the dielectric as shown in Fig. 3.24b. Thus in the case of "uniform" polarization the application of electric field produces a surface charge density $\sigma_p$. If the polarization is "non uniform" we get volume bound charge density $\rho_p$ in addition of $\sigma_p$. With the above physical interpretation of the bound charges we will not derive expressions for these charges.

Consider a imaginary surface S inside a polarized dielectric. Let an electric field $\mathbf{E_o}$ be applied normal to the surface S. In the process of polarization for each charge q crossing the surface S a equal negative charge –q crosses the surface S in opposite

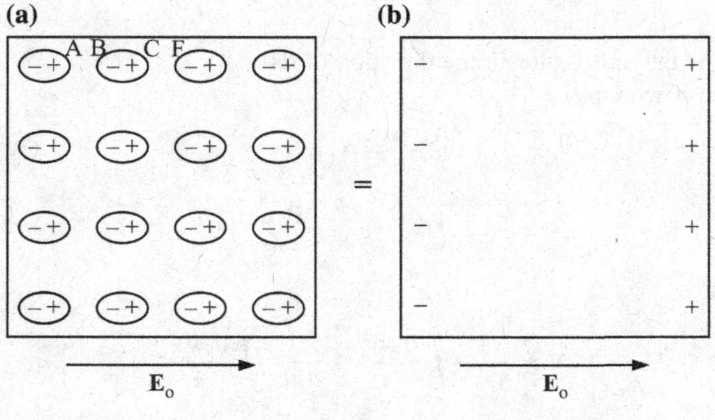


**Fig. 3.24** **a** A dielectric material is placed in an external electric field. The net result will be a production of set of dipoles aligned in the direction of the field. Adjacent positive and negative charges cancel out. **b** In an uniformly polarized dielectric cancellation of adjacent positive and negative charges results in a surface charge density

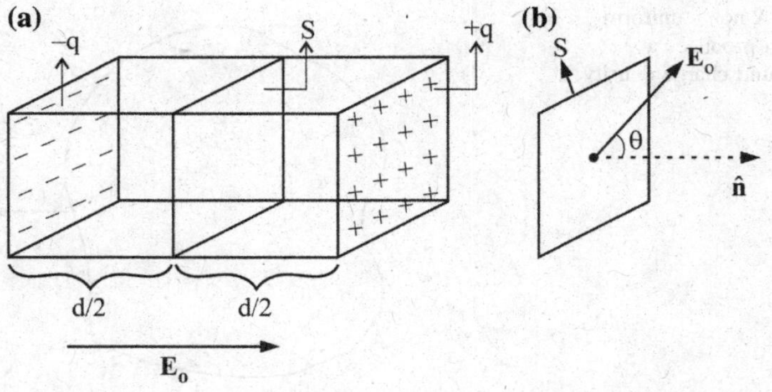


**Fig. 3.25** **a** In a polarized dielectric placed in a external electric field equal number of positive and negative charges cross a imaginary surface S. **b** In general the applied electric field can be inclined at an angle to the surface S

direction as shown in Fig. 3.25a. The positive charge +q and the negative charge −q move a distance of d/2 in opposite direction. Clearly the volume of the parallelopiped shown in Fig. 3.25a is $\mathbf{S} \cdot \mathbf{d}$.

Suppose the electric field $\mathbf{E_o}$ is inclined at an angle θ as shown in Fig. 3.25b then the effective area shown by S to the $\mathbf{E_o}$ is S cos θ (see Sect. 1.12).

$$\text{Hence the total volume is then } (S \cos\theta)d = \mathbf{S} \cdot \mathbf{d} \tag{3.264}$$

If the dielectric contains N molecules per unit volume then the total number of molecules within the volume $\mathbf{S} \cdot \mathbf{d}$ is $N(\mathbf{S} \cdot \mathbf{d})$. If $q_{net}$ is the net total charge that crosses the surface S then

$$q_{net} = N(\mathbf{S} \cdot \mathbf{d})q \tag{3.265}$$

But q**d** is the dipole moment **p**. Hence

$$q_{net} = N(\mathbf{S} \cdot \mathbf{p}) \tag{3.266}$$

As **p** is the dipole moment of single dipole, N**p** is the number of dipoles per unit volume and by the definition of polarization **P** = N**p**. Hence

$$q_{net} = \mathbf{S} \cdot \mathbf{P} = \mathbf{P} \cdot \mathbf{S} \tag{3.267}$$

$$\Rightarrow q_{net} = \mathbf{P} \cdot \hat{\mathbf{n}} S \tag{3.268}$$

where $\hat{\mathbf{n}}$ is the unit vector normal to S as shown in Fig. 3.25b. The bound surface charge density is

$$\sigma_p = \frac{q_{net}}{S} = \mathbf{P} \cdot \hat{\mathbf{n}} \tag{3.269}$$

**Fig. 3.26** A non - uniform polarization produces a volume bound charge density

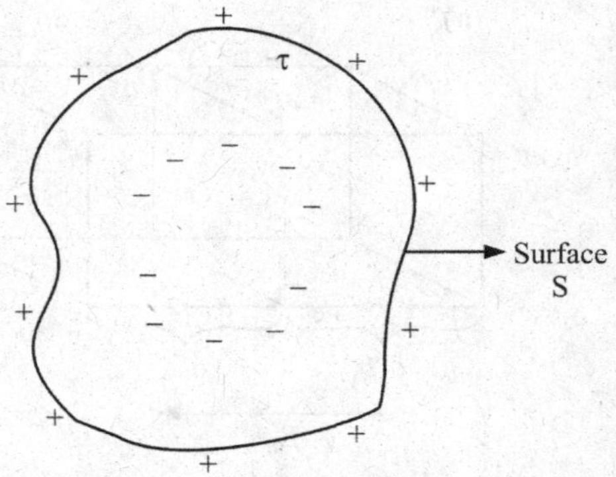


which is what we guessed in Eqs. 3.261, 3.263 and proved in Eq. 3.269.

Thus a surface bound charge $\sigma_p$ appears when the polarization is uniform. However when the polarization is non uniform in addition to bound charges $\sigma_p$ appearing at the surface a bound charge $\rho_p$ appears in the volume.

Consider an imaginary volume $\tau$ inside a dielectric whose surface is S as shown in Fig. 3.26. As positive charges are pushed out of the surface S an equal number of negative charges lie within volume $\tau$.

From Eq. 3.267 the total charges that cross out of surface S in Fig. 3.26 is

$$\oint_S \mathbf{P} \cdot \mathbf{ds} \tag{3.270}$$

An equal amount of negative charge remains within volume $\tau$ and this negative charge is given by

$$q_{net} = -\oint_S \mathbf{P} \cdot \mathbf{ds} \tag{3.271}$$

(Negative value of Eq. 3.270).

Applying divergence theorem

$$q_{net} = -\int_\tau \nabla \cdot \mathbf{P}\, d\tau \tag{3.272}$$

From Eq. 2.22

$$q_{net} = \int_\tau \rho_p\, d\tau \tag{3.273a}$$

Comparing Eqs. 3.272, 3.273a

$$\rho_p = -\nabla \cdot \mathbf{P} \tag{3.273b}$$

In Sect. 3.17 we saw that potential produced by the polarized object can be calculated by mathematical fictitious charges $\sigma_p$ and $\rho_p$. However in this section we see that $\sigma_p$ and $\rho_p$ are real accumulation of charges in the dielectric.

## 3.19 Electric Displacement Vector and Gauss Law in Dielectrics

In chapter two we calculated electric field for the given charge distribution by using Gauss's law. We constructed appropriate Gaussian surface, then evaluated $\oint \mathbf{E} \cdot \mathbf{ds}$ and finally obtained **E**. When constructing Gaussian surface we made sure that required charges are enclosed in the Gaussian surface. Whenever we apply Gauss law for dielectrics we must make sure that all charges including bound charges produced by polarization is also included in the Gaussian surface. Previously we noted there are two distinct electric fields in action in a dielectric. One the external electric field which we applied and the other true electric field produced due to polarization. The electric field produced by the polarization is due to bound charges- $\sigma_p$ and $\rho_p$. Let all the charges responsible for external electric field be represented by $\rho_f$. Let us call $\rho_f$, the free charges, in opposition to bound charges. $\rho_f$ the free charge might consist of some static charges inside the dielectric… etc., which are not due to bound charges. From Gauss law

$$\nabla \cdot \mathbf{E} = \frac{\rho}{\varepsilon_o}$$

In the case of dielectrics

$$\rho = \rho_p + \rho_f \tag{3.274}$$

Thus

$$\nabla \cdot \mathbf{E} = (\rho_p + \rho_f)/\varepsilon_o \tag{3.275}$$

Substituting Eq. 3.273b in Eq. 3.275

$$\nabla \cdot (\varepsilon_o \mathbf{E}) = -\nabla \cdot \mathbf{P} + \rho_f \tag{3.276}$$

$$\Rightarrow \nabla \cdot (\varepsilon_o \mathbf{E} + \mathbf{P}) = \rho_f \tag{3.277}$$

The quantity $\mathbf{D} = \varepsilon_o \mathbf{E} + \mathbf{P}$ is known as electric displacement (or) electric flux density

$$\nabla \cdot \mathbf{D} = \rho_f \tag{3.278}$$

is the differential form of Gauss's law in dielectrics. In integral form

$$\oint_S \mathbf{D} \cdot \mathbf{ds} = q_{f\,enc} \tag{3.279}$$

Here $q_{f\,enc}$ is the free charges enclosed in the volume bounded by surface S. The above equation refers to only to free charges. Free charges are in our control and the electric field produced by these free charges is responsible for bound charges. Once we know free charges, then we calculate **D** as we did in usual method of Gauss law in Chap. 2.

## 3.20 Linear Dielectrics

If the applied electric field is not too strong then for most substances

$$\mathbf{P} = \varepsilon_0 \chi_e \mathbf{E} \tag{3.280}$$

Here $\chi_e$ is called electric susceptibility of the material and the dielectrics which obey Eq. 3.280 are called linear dielectrics.

$$\text{Now} \quad \mathbf{D} = \varepsilon_0 \mathbf{E} + \mathbf{P} \tag{3.281}$$

For linear dielectrics

$$\mathbf{D} = \varepsilon_0 \mathbf{E} + \varepsilon_0 \chi_e \mathbf{E} \tag{3.282}$$

$$\mathbf{D} = \varepsilon_0 (1 + \chi_e) \mathbf{E} \tag{3.283}$$

$$\text{Let}\, \varepsilon = \varepsilon_0 (1 + \chi_e) \tag{3.284}$$

Then

$$\mathbf{D} = \varepsilon \mathbf{E} \tag{3.285}$$

and

$$\varepsilon_r = \frac{\varepsilon}{\varepsilon_0} = (1 + \chi_e) \tag{3.286}$$

$\varepsilon_r$ is called relative permittivity or dielectric constant of the material and $\varepsilon$ is called absolute permittivity or simply permittivity of the medium.

## 3.21 Dielectric Breakdown

We saw that whenever a dielectric is subjected to an electric field the dielectric gets polarized. The electrons are displaced from their original position but do not leave the atom. But however if the applied electric field is very strong the electrons can be permanently removed from the atom and gets accelerated by the action of the applied electric field. The accelerated electrons collide with other atoms in the dielectric thereby ionizing these atoms. The result is large currents and the dielectric starts conducting.

The dielectric has lost its insulating property and the phenomena is called dielectric breakdown. The field at which the dielectric breaks down is the maximum field the dielectric can withstand and is known as dielectric strength of the material.

## 3.22 Boundary Conditions in the Presence of Dielectrics

The boundary conditions which we derived in Sect. 2.25 must be modified in the presence of dielectrics. In Figs. 2.66, 2.67 if the media 1, 2 are two dielectrics with different dielectric primitivities $\varepsilon_1$, $\varepsilon_2$ noting down that $\mathbf{D}_1$, $\mathbf{D}_2$ point in the same direction as $\mathbf{E}_1$ and $\mathbf{E}_2$ we have

(i) **For tangential component**

$$E_{1t} = E_{2t} \tag{3.287}$$

$$\Rightarrow \frac{D_{1t}}{\varepsilon_1} = \frac{D_{2t}}{\varepsilon_2} \tag{3.288}$$

as $\mathbf{D} = \varepsilon\mathbf{E}$

(ii) **For normal component**

For medium 1 and medium 2 in Fig. 2.66, the boundary conditions in terms of **D** is

$$D_{1n} - D_{2n} = \sigma_f \tag{3.289}$$

If medium 1, medium 2 are dielectrics with no free charge in between then

$$D_{1n} = D_{2n} \tag{3.290}$$

$$\Rightarrow \varepsilon_1 E_{1n} = \varepsilon_2 E_{2n} \tag{3.291}$$

**Example 3.9**

Consider boundary between two dielectrics with permittivity $\varepsilon_1$ and $\varepsilon_2$. The electric field makes an angle $\theta_1$ and $\theta_2$ with the normal as shown in Fig. 3.27. Assume there is no charge in the boundary. Show that

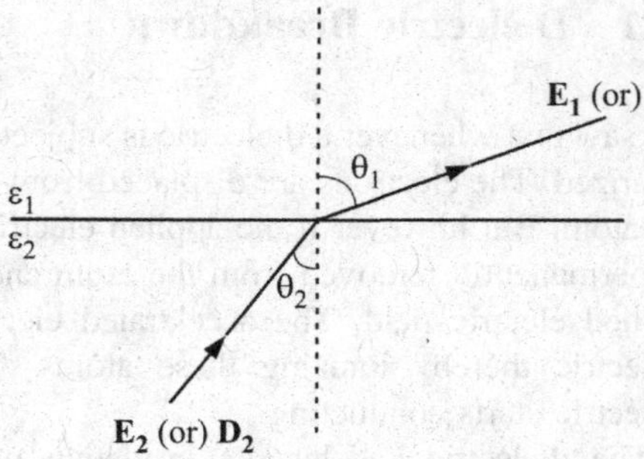


**Fig. 3.27** Boundary between two dielectrics with different permivitties

$$\frac{\tan\theta_1}{\tan\theta_2} = \frac{(\varepsilon_r)_1}{(\varepsilon_r)_2}$$

where $\varepsilon_r$ is the relative permittivity of the dielectric media.

**Solution**

Because there is no charges in the boundary from Eq. 3.290

$$D_{1n} = D_{2n}$$

From Fig. 3.27

$$D_{1n} = D_1 \cos\theta_1 \tag{3.292}$$

$$D_{2n} = D_2 \cos\theta_2 \tag{3.293}$$

As $D_{1n} = D_{2n}$

$$\Rightarrow D_1 \cos\theta_1 = D_2 \cos\theta_2 \tag{3.294}$$

$$\Rightarrow \varepsilon_1 E_1 \cos\theta_1 = \varepsilon_2 E_2 \cos\theta_2 \tag{3.295}$$

From Eq. 3.287

$$E_{1t} = E_{2t}$$

From Fig. 3.28

$$E_{1t} = E_1 \sin\theta_1 \tag{3.296}$$

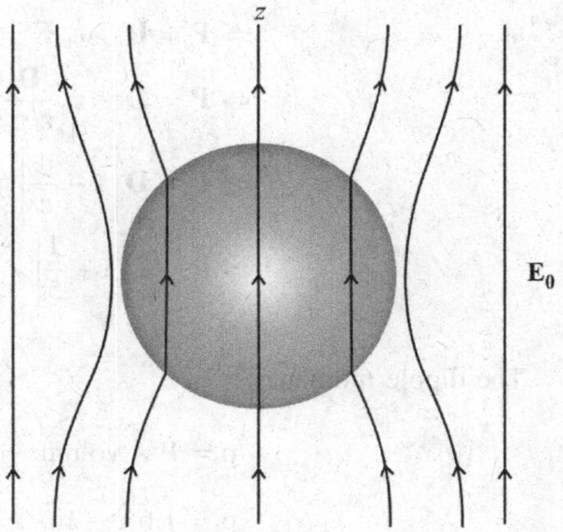


**Fig. 3.28** A linear dielectric sphere of radius R placed in an external electric field

$$E_{2t} = E_2 \sin \theta_2 \tag{3.297}$$

As $E_{1t} = E_{2t}$

$$E_1 \sin \theta_1 = E_2 \sin \theta_2 \tag{3.298}$$

From Eqs. 3.295, 3.298

$$\frac{\tan \theta_1}{\tan \theta_2} = \frac{\varepsilon_1}{\varepsilon_2} = \frac{(\varepsilon_r)_1}{(\varepsilon_r)_2} \tag{3.299}$$

**Example 3.10**

A flat linear dielectric surface whose relative permittivity is 5 is subjected to a electric field with a electric flux density of 2 C per m$^2$ and the dielectric is uniformly polarized. The electric field is perpendicular to the slab. The volume of the dielectric is 1 m$^3$. Calculate **E**, **P**, **p** for the dielectric.

**Solution**

$$\mathbf{D} = \varepsilon \mathbf{E} = \varepsilon_o \, \varepsilon_r \, \mathbf{E}$$
$$\Rightarrow \mathbf{E} = \frac{\mathbf{D}}{\varepsilon_o \, \varepsilon_r} = \frac{2}{(8.85 \times 10^{-12})\,(5)}$$
$$\Rightarrow \mathrm{E} = 4.52 \times 10^{10}\ \mathrm{V/m}$$

The polarisation $\mathbf{D} = \varepsilon_o \mathbf{E} + \mathbf{P}$

$$\Rightarrow \mathbf{P} = \mathbf{D} - \varepsilon_0 \mathbf{E}$$

$$\Rightarrow \mathbf{P} = \mathbf{D} - \varepsilon_0 \frac{\mathbf{D}}{\varepsilon_0 \varepsilon_r}$$

$$\Rightarrow \mathbf{P} = \mathbf{D}\left[1 - \frac{1}{\varepsilon_r}\right]$$

$$\Rightarrow \mathrm{P} = 2\left[1 - \frac{1}{5}\right] \quad \Rightarrow \mathrm{P} = 1.6\,\frac{\mathrm{C}}{\mathrm{m}^3}$$

The dipole moment

$$\mathrm{p} = \mathrm{P} \times \text{volume} = 2\left[1 - \frac{1}{5}\right](1)$$

$$\mathrm{p} = 1.6\ \mathrm{c-m}.$$

**Example 3.11**

Consider a linear dielectric sphere of radius R and dielectric constant $\varepsilon_r$ be placed in a electric field $\mathbf{E_o}$ as shown in Fig. 3.28. Find the electric field inside and outside the sphere. Also calculate the dipole moment and polarisation of the sphere.

**Solution**

When the dielectric sphere is placed in an electric field the sphere becomes polarised and bound charges appear on the surface.

Inside the sphere thus the electric field will be reduced although not completely zero as in the case of conducting sphere (Example 3.5).

The potential of the sphere is given by (Eq. 3.164)

$$V_{out} = \sum_n A_n r^n P_n(\cos\theta) + \sum_n \frac{B_n}{r^{n+1}} P_n(\cos\theta) \tag{3.300}$$

$$V_{in} = \sum_n C_n r^n P_n(\cos\theta) + \sum_n \frac{F_n}{r^{n+1}} P_n(\cos\theta) \tag{3.301}$$

The boundary conditions to be satisfied are

$$\left.\begin{aligned}
&\text{(i) } V_{in} = V_{out} \text{ at } r = R \\
&\text{(ii) } V = -E_o r\cos\theta \text{ at } r \gg R \\
&\text{(iii) } D_{1n} = D_{2n} \text{ as there are no free charges on the surface} \\
&\qquad \Rightarrow (E_{1n})_{r=R} = \varepsilon_r (E_{2n})_{r=R} \\
&\qquad \Rightarrow -\left(\frac{\partial V_{out}}{\partial r}\right)_{r=R} = -\varepsilon_r\left(\frac{\partial V_{in}}{\partial r}\right)_{r=R} \\
&\text{(iv) } V \text{ must be finite as } r \to 0
\end{aligned}\right\} \tag{3.302}$$

The first boundary condition is required on the physical grounds that the discontinuous potential will produce a infinite electric field.

Applying boundary condition Eq. 3.302(ii) in Eq. 3.300 and remembering that as $r \gg R$ (i.e.,) at large r the second term can be neglected

$$\begin{aligned} V_{out} &= \sum A_n r^n P_n(\cos\theta) \\ -E_o r\cos\theta &= \sum_n A_n r^n P_n(\cos\theta) \end{aligned} \tag{3.303}$$

$$\Rightarrow -E_o r P_1(\cos\theta) = A_0 P_0(\cos\theta) + A_1 r P_1(\cos\theta) + A_1 r^2 P_2(\cos\theta) + \cdots \tag{3.304}$$

Comparing the coefficients of r on both sides

$$\left.\begin{aligned} A_1 &= -E_o r \\ A_0 &= A_2 = A_3 = \cdots = 0 \end{aligned}\right\} \tag{3.305}$$

Substituting Eq. 3.305 in Eq. 3.300

$$V_{out} = -E_o r P_1(\cos\theta) + \sum_n \frac{B_n}{r^{n+1}} P_n(\cos\theta) \tag{3.306}$$

Applying boundary condition Eq. 3.302(iv) in Eq. 3.301

$$F_n = 0$$

Hence Eq. 3.301 is

$$V_{in} = \sum_n C_n r^n P_n(\cos\theta) \tag{3.307}$$

The boundary condition Eq. 3.302(i) requires

$$V_{out}(R, \theta) = V_{in}(R, \theta)$$

From Eqs. 3.306, 3.307

$$\begin{aligned} &-E_o R P_1(\cos\theta) + \frac{B_0}{R} + B_1 \frac{P_1(\cos\theta)}{R^2} + \cdots \\ &= C_o + C_1 R P_1(\cos\theta) + \cdots \end{aligned} \tag{3.308}$$

From the boundary condition Eq. 3.302(iii) and Eqs. 3.306, 3.307

$$\begin{aligned} &\Rightarrow E_o P_1(\cos\theta) + \frac{B_0}{R^2} + \frac{2B_1}{R^3} P_1(\cos\theta) + \frac{3B_2}{R^4} P_2(\cos\theta) + \cdots \\ &= \varepsilon_r[-C_1 P_1(\cos\theta) - 2C_2 R P_2(\cos\theta)\ldots] \end{aligned} \tag{3.309}$$

From Eqs. 3.308, 3.309 equating the coefficients of Legendre polynamials

$$\left.\begin{array}{l} \frac{B_0}{R} = C_0 \\ -E_o R + \frac{B_1}{R^2} = C_1 R \\ \frac{B_2}{R^2} = C_2 R^2 \\ \cdots \end{array}\right\} \tag{3.310}$$

$$\left.\begin{array}{l} \frac{B_0}{R^2} = 0 \\ E_o + \frac{2B_1}{R} = -\varepsilon_r C_1 \\ \frac{3B_2}{R^4} = -2\varepsilon_r C_2 R \\ \cdots \end{array}\right\} \tag{3.311}$$

From Eqs. 3.310, 3.311

$$\left.\begin{array}{l} B_o = C_0 = 0 \\ B_1 = \left(\frac{\varepsilon_r - 1}{\varepsilon_r + 1}\right) E_o R^3 \\ C_1 = \frac{-2E_o}{\varepsilon_r + 2} \\ B_2 = C_2 = 0 \\ B_3 = C_3 = 0 \\ \cdots \quad \cdots \quad \cdots \end{array}\right\} \tag{3.312}$$

Substituting Eqs. 3.312 in Eqs. 3.306, 3.307

$$V_{out} = -E_o r \cos\theta + \left(\frac{\varepsilon_r - 1}{\varepsilon_r + 2}\right) \frac{E_o R^3 \cos\theta}{r^2} \tag{3.313}$$

$$V_{in} = \frac{-3}{\varepsilon_r + 2} E_o r \cos\theta \tag{3.314}$$

The electric field outside the sphere is given by

$$(E_r)_{out} = \frac{-\partial V_{out}}{\partial r} = E_o \left[1 + \left(\frac{\varepsilon_r - 1}{\varepsilon_r + 1}\right) \frac{2R^3}{r^3}\right] \cos\theta \tag{3.315}$$

$$(E_\theta)_{out} = \frac{-1}{r} \frac{\partial V_{out}}{\partial \theta} = -E_o \left[1 - \left(\frac{\varepsilon_r - 1}{\varepsilon_r + 1}\right) \frac{R^3}{r^3}\right] \sin\theta \tag{3.316}$$

Field inside the sphere is given by

$$(E_r)_{in} = \frac{-\partial V_{in}}{\partial r} = \left(\frac{3}{\varepsilon_r + 2}\right) E_o \cos\theta \tag{3.317}$$

$$(E_\theta)_{in} = \frac{-1}{r}\frac{\partial V_{in}}{\partial \theta} = \left(\frac{-3}{\varepsilon_r + 2}\right) E_o \sin\theta \tag{3.318}$$

Thus

$$E_{in} = \sqrt{E_r^2 + E_\theta^2} = \left(\frac{3}{\varepsilon_r + 2}\right) E_o \tag{3.319}$$

The dipole moment can be calculated as follows. From Eq. 3.313 the potential produced by polarised charges is

$$V_{out} = \left(\frac{\varepsilon_r - 1}{\varepsilon_r + 2}\right) \frac{E_o R^3}{r^2} \cos\theta \tag{3.320}$$

The term $-E_o r \cos\theta$ in Eq. 3.313 is due to the external field.
We know that the potential due to the dipole is given by

$$V = \frac{1}{4\pi\varepsilon_o} \frac{p\cos\theta}{r^2} \tag{3.321}$$

Comparing Eqs. 3.320, 3.321

$$\begin{aligned} &\Rightarrow \frac{1}{4\pi\varepsilon_o} \frac{p\cos\theta}{r^2} = \left(\frac{\varepsilon_r - 1}{\varepsilon_r + 1}\right) \frac{E_o R^3}{r^2} \cos\theta \\ &\Rightarrow p = \left(\frac{\varepsilon_r - 1}{\varepsilon_r + 2}\right) E_o R^3 \, 4\pi\varepsilon_o \end{aligned} \tag{3.322}$$

Now the polarisation is

$$\begin{aligned} P &= \frac{p}{\text{volume}} = \frac{\left(\frac{\varepsilon_r - 1}{\varepsilon_r + 2}\right) E_o R^3 4\pi\varepsilon_o}{4/3\pi R^3} \\ P &= 3\left(\frac{\varepsilon_r - 1}{\varepsilon_r + 2}\right) \varepsilon_o E_o \end{aligned} \tag{3.323}$$

Equation 3.319 is

$$\begin{aligned} E_{in} &= \left(\frac{3}{\varepsilon_r + 2}\right) E_o \\ \Rightarrow E_{in} &= E_o - \left(\frac{\varepsilon_r - 1}{\varepsilon_r + 2}\right) E_o \\ \mathbf{E_{in}} &= \mathbf{E_o} - \frac{\mathbf{P}}{3\varepsilon_o} \end{aligned} \tag{3.324}$$

**Example 3.12**
Consider an infinite line charge which lies at the center of the cylindrical shell region as shown in the Fig. 3.29. The infinite line charge carries a uniform line charge density λ. The cylindrical shell region contains a dielectric and the inner and outer radii of the dielectric is $R_i$ and $R_o$. The cylindrical shell region extends to infinity. Calculate the electric field in the regions

(i) $r < R_i$
(ii) $R_i < r < R_o$
(iii) $r > R_o$

**Solution**
**Case (i) $r < R_i$.**
In this region Gauss's law reads

$$\oint \mathbf{E} \cdot \mathbf{ds} = \frac{q_{enc}}{\varepsilon_o}$$

Applying Gauss's law in usual manner

$$\mathbf{E} = \frac{\lambda}{2\pi\varepsilon_o r}\hat{r}$$

**Case (ii) $R_i < r < R_o$.**
In this region Gauss's law reads

$$\oint \mathbf{D} \cdot \mathbf{ds} = q_{f\,enc}$$

Constructing a cylindrical Gaussian surface within the dielectric region with radii r and length L as shown in Fig. 3.30 and noting down that the charge enclosed by the Gaussian surface is λL we get

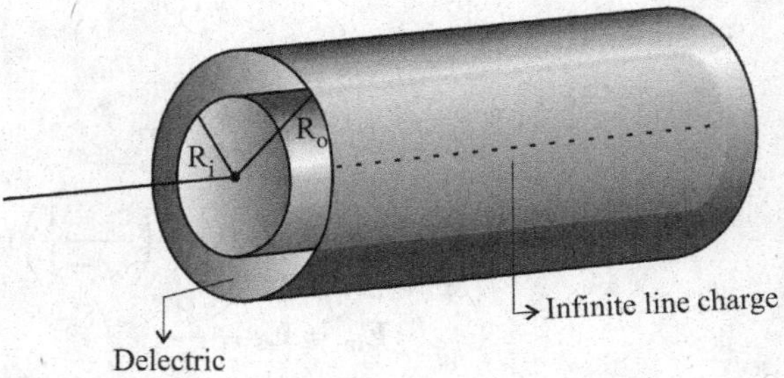


**Fig. 3.29** Illustration showing an infinite line charge lying along the center of the cylindrical shell. The infinite line charge carries an uniform line charge density. The region between the cylindrical shell contains a dielectric

$$\oint \text{Dds}\cos 0 = \lambda \text{L}$$

$$\Rightarrow \text{D}\oint \text{ds} = \lambda \text{L}$$

$$\Rightarrow \text{D}\cdot 2\pi \text{rL} = \lambda \text{L}$$

$$\Rightarrow \mathbf{D} = \frac{\lambda}{2\pi \text{r}}\hat{\mathbf{r}}$$

Inside the dielectric the electric field cannot be determined because we do not know the polarisation.

**Case (iii) For r > $R_o$.**
In this region Gauss law reads

$$\oint \mathbf{D}\cdot \mathbf{ds} = \frac{q_{f\,enc}}{\varepsilon_0}$$

Following as in case (ii) we get

$$\Rightarrow \mathbf{D} = \frac{\lambda}{2\pi \text{r}}\hat{\mathbf{r}}$$

In this region r > $R_o$ there is no polarisation and hence

$$\mathbf{D} = \varepsilon_0 \mathbf{E} + \mathbf{P} = \varepsilon_0 \mathbf{E}$$

$$\Rightarrow \mathbf{E} = \frac{\mathbf{D}}{\varepsilon_0} = \frac{\lambda}{2\pi\varepsilon_0 \text{r}}\hat{\mathbf{r}}$$

**Example 3.13**
Repeat Example 3.12 but this time the dielectric is a linear dielectric with dielectric constant $\varepsilon_r$ and electric susceptibility $\chi_e$. Also find the polarisation in the dielectric.

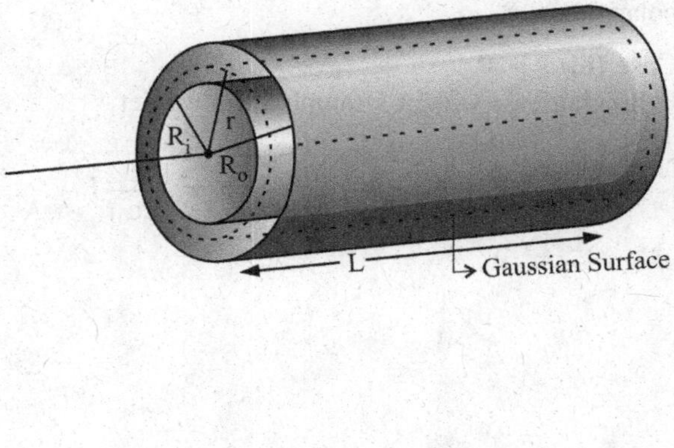


**Fig. 3.30** Construction of a Gaussian surface in the cylindrical shell region

**Solution**

**Case (i) $r < R_i$.**

Following case (i) in Example 3.12 we get

$$\mathbf{E} = \frac{\lambda}{2\pi\varepsilon_0 r}\hat{r}$$

**Case (ii) $R_i < r < R_o$.**

Following case (ii) in Example 3.12 we get

$$\mathbf{D} = \frac{\lambda}{2\pi r}\hat{r}$$

Because the dielectric is linear the electric field can be calculated using the relation

$$\mathbf{D} = \varepsilon\mathbf{E}$$
$$\Rightarrow \mathbf{E} = \frac{\lambda}{2\pi\varepsilon_0 r}\hat{r}$$

The polarisation is given by

$$\mathbf{P} = \mathbf{D} - \varepsilon_0\mathbf{E}$$
$$= \frac{\lambda}{2\pi r} - \varepsilon_0\frac{\lambda}{2\pi\varepsilon r}$$
$$\mathbf{P} = \frac{\lambda}{2\pi r}\left[1 - \frac{\varepsilon_0}{\varepsilon}\right]$$
$$\Rightarrow \mathbf{P} = \frac{\lambda}{2\pi r}\left[1 - \frac{1}{\varepsilon_r}\right]$$
$$\Rightarrow \mathbf{P} = \frac{\lambda}{2\pi r}\left[\frac{\varepsilon_r - 1}{\varepsilon_r}\right]$$

Alternatively the relation $\mathbf{P} = \varepsilon_0\chi_e\mathbf{E}$ can be directly used to calculate polarisation.

**Case (iii) $r > R_o$.**

Following case (iii) in Example 3.12 we get

$$\mathbf{E} = \frac{\lambda}{2\pi\varepsilon_0 r}\hat{r}$$

## 3.23 Capacitance and Capacitors

A capacitor is a device used for storing charges. It consists of two conductors separated by a dielectric medium. The conductors carry equal and opposite charges. The capacitance C of the capacitor is defined by the relation

$$C = \frac{q}{V} \tag{3.325}$$

Here q is the charge in coulomb in each conductor, V is the potential difference between the conductors due to equal and opposite charges on them. The unit of capacitance is farads.

## 3.24 Principle of a Capacitor

Consider two plates P, R as shown in Fig. 3.31a. The plate P is maintained at potential V and carries a charge q. Due to induction negative and positive charges get induced in plate R. However negative charges lie close to plate P and positive charges lie far away from plate P. While the negative charges tend to reduce the potential of plate P the positive charges tend to raise the potential of plate P. However as the induced positive charges on plate R are far away than the induced negative charges the effect of negative charges in lowering the potential of plate P than the effect of induced positive charges in raising the potential of plate P is more. Thus the net effect is the potential of plate P is slightly lowered. In order to bring back the plate P to potential V additional charges must be given. Suppose the plate R is earthed as shown in Fig. 3.31b. The positive charges get grounded and now only the negative induced charges remain on plate R. Thus the potential of plate P is further reduced. Additional charges must be given to retain the potential V of plate P. Thus the stored charge in P can be increased.

In other words the capacity of plate P has been increased. The arrangement in Fig. 3.31b thus forms a capacitor.

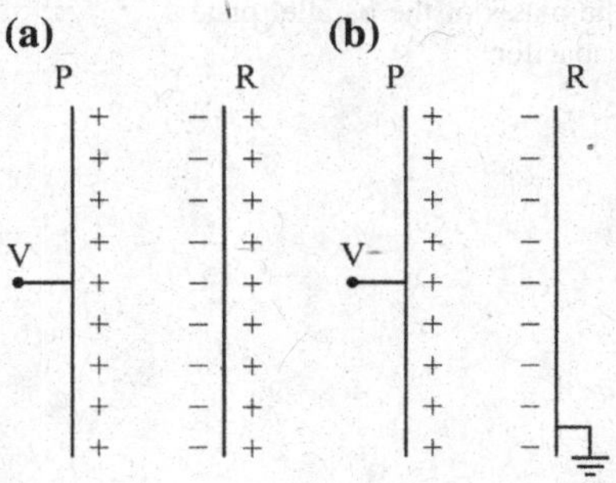


**Fig. 3.31** A parallel plate capacitor. **a** The plate P is maintained at potential V. Negative and positive charges are induced in plate R. **b** Illustration showing plate R being earthed. The positive charges are grounded

### 3.24.1 Capacity of a Parallel Plate Capacitor

Consider two parallel plates P and R separated by a distance d as shown in Fig. 3.32. Let the plate P carry a charge of q and plate R carry a charge of –q. Let the area of plates be A.

Except in the edges the electric field E is uniform between plates and is given by (Eq. 2.301)

$$E_{1n} = E = \frac{\sigma}{\varepsilon_o} \tag{3.326}$$

and tangential component is zero from Eq. 2.305. In the above equation E is the field between the plates, σ is the surface charge density on the plates.

Hence

$$E = \frac{\sigma}{\varepsilon_o} = \frac{q}{\varepsilon_o A} \tag{3.327}$$

The potential difference between the plates is equal to the work done in moving a unit positive charge form plate R to plate P

$$V = -\int_R^P \mathbf{E} \cdot d\boldsymbol{l} \tag{3.328}$$

From Fig. 3.32 $d\boldsymbol{l} = -dx$ and we have

$$\begin{aligned} \mathbf{E} \cdot d\boldsymbol{l} &= E\,dl \cos 180^\circ = -E\,dl \\ \mathbf{E} \cdot d\boldsymbol{l} &= Edx \end{aligned} \tag{3.329}$$

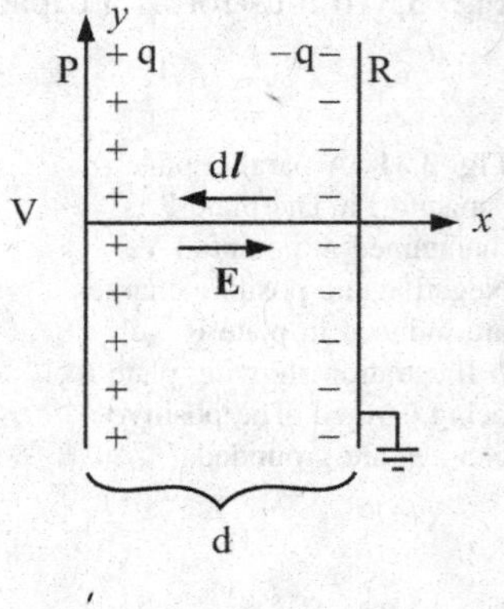


**Fig. 3.32** Illustration showing the direction of electric field vector between the plates of the parallel plate capacitor

Now $V = -\int_R^P \mathbf{E} \cdot d\boldsymbol{l}$

$$\Rightarrow V = -\int_d^0 E\,dx \qquad (3.330)$$

$$\Rightarrow V = +\frac{qd}{\varepsilon_0 A} \qquad (3.331)$$

The capacitance is given by

$$C = \frac{q}{V} = \frac{\varepsilon_0 A}{d} \qquad (3.332)$$

If the space between the capacitor is filled with a dielectric of dielectric constant $\varepsilon_r$ than the capacitance is given by

$$C = \frac{\varepsilon_0\, \varepsilon_r\, A}{d} \qquad (3.333)$$

In the above derivation we have neglected the fringing fields at the edges of the plates.

### 3.24.2 *Capacity of a Parallel Plate Capacitor with Two Dielectrics*

Consider a parallel plate capacitor consisting of two dielectrics shown in Fig. 3.33. The distance between the two plates is d and the thickness of the dielectric is t.

The potential difference between the two plates P and R is given by

$$V = \frac{q}{A\,\varepsilon_0}(d - t) + \frac{q}{A\,\varepsilon_r\,\varepsilon_0} t \text{ (From Eq. 3.331)}$$

The first term $\frac{q}{A\,\varepsilon_0}(d - t)$ corresponds to air and second term corresponds to dielectric. Hence

$$V = \frac{q}{A\,\varepsilon_0}\left[d - t + \frac{t}{\varepsilon_r}\right] \qquad (3.334)$$

Fig. 3.33 A parallel plate capacitor filled with two dielectrics

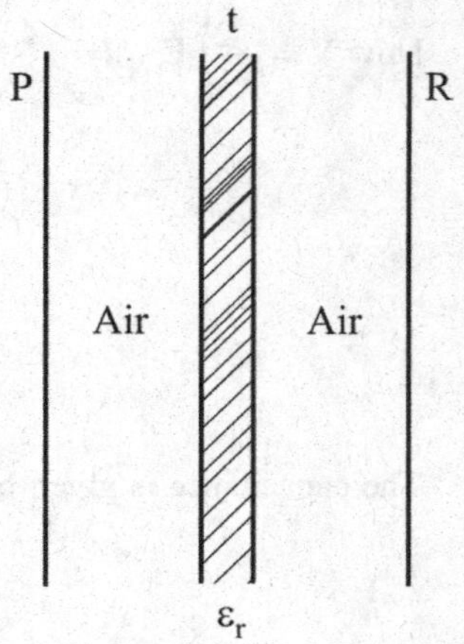


The capacitance is given by

$$C = \frac{q}{V} = \frac{\varepsilon_0 A}{d - t + \frac{t}{\varepsilon_r}} \tag{3.335}$$

## 3.25 Capacitance of a Spherical Capacitor

Consider a spherical capacitor consisting of two concentric spherical shells separated by air as shown in Fig. 3.34. Let $R_1$, $R_2$ be the radii of inner and outer shells respectively. The outer shell B is earthed. If a charge q is distributed over the outer surface of the inner shell A then an equal and opposite charge –q will be induced in the inner surface of the outer shell.

Fig. 3.34 A spherical capacitor consisting of two concentric spherical shells separated by air

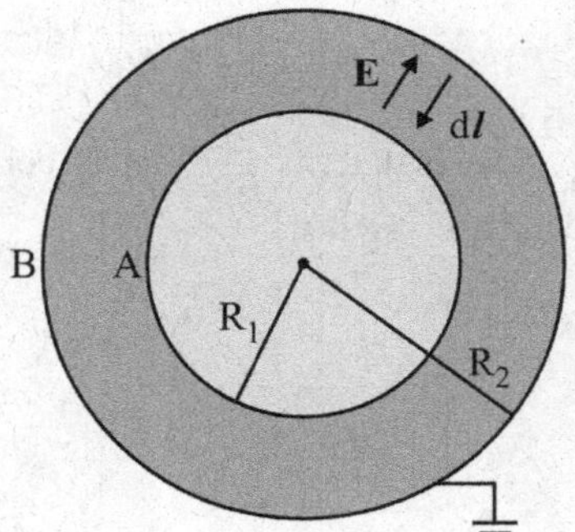

The electric field between the spheres is given by

$$E = \frac{1}{4\pi\varepsilon_0}\frac{q}{r^2} \tag{3.336}$$

from Example 2.11. The potential difference between the spheres is the work done in moving a unit positive charge from B to A

$$V_{AB} = -\int_{R_2}^{R_1} \mathbf{E}\cdot d\boldsymbol{l} \tag{3.337}$$

Now $\mathbf{E}\cdot d\boldsymbol{l} = E\,dl\cos 180° = -E\,dl$
From Fig. 3.34 $d\boldsymbol{l} = -dr$. Hence

$$\mathbf{E}\cdot d\boldsymbol{l} = -Edl = Edr \tag{3.338}$$

Now

$$V_{AB} = -\int_{R_2}^{R_1} \mathbf{E}\cdot d\boldsymbol{l} = -\int_{R_2}^{R_1} E\,dr \tag{3.339}$$

$$\Rightarrow V_{AB} = -\int_{R_2}^{R_1} \frac{q}{4\pi\varepsilon_0 r^2}dr \tag{3.340}$$

$$\Rightarrow V_{AB} = \frac{q}{4\pi\varepsilon_0}\left[\frac{R_2 - R_1}{R_1\,R_2}\right] \tag{3.341}$$

Capacitance is given by

$$C = \frac{q}{V_{AB}} \tag{3.342}$$

$$C = \frac{4\pi\varepsilon_0 R_1 R_2}{R_2 - R_1} \tag{3.343}$$

If the air between two spheres is replaced by another medium of dielectric constant $\varepsilon_r$ then

$$C = \frac{4\pi\varepsilon_r\,\varepsilon_0 R_1 R_2}{(R_2 - R_1)} \tag{3.344}$$

## 3.26 Capacitance of a Cylindrical Capacitor

The cylindrical capacitors consists of two conducting cylinders lying coaxially as shown in Fig. 3.35. The radius of inner and outer cylinders being $R_1$ and $R_2$ respectively. The outer cylinder is earthed and the length of capacitor is L. The inner cylinder A carries a charge +q and hence a negative charge –q is induced on the inner wall of the outer cylinder B (neglecting the fringing fields). Construct a cylindrical Gaussian surface of radius r and length L. Applying Gauss's law to the Gaussian surface

$$\oint \mathbf{E} \cdot \mathbf{ds} = \frac{q_{enc}}{\varepsilon_o} \tag{3.345}$$

$$\Rightarrow \oint E\,ds\cos 0 = \frac{q_{enc}}{\varepsilon_o} \tag{3.346}$$

As E is constant over the Gaussian surface

$$\Rightarrow E \oint ds = \frac{q_{enc}}{\varepsilon_o} \tag{3.347}$$

From Fig. 3.35 $\oint ds = 2\pi rL$, $q_{enc} = q$

$$E(2\pi rL) = \frac{q}{\varepsilon_o} \tag{3.348}$$

$$\Rightarrow E = \frac{q}{2\pi\varepsilon_o\, r\, L} \tag{3.349}$$

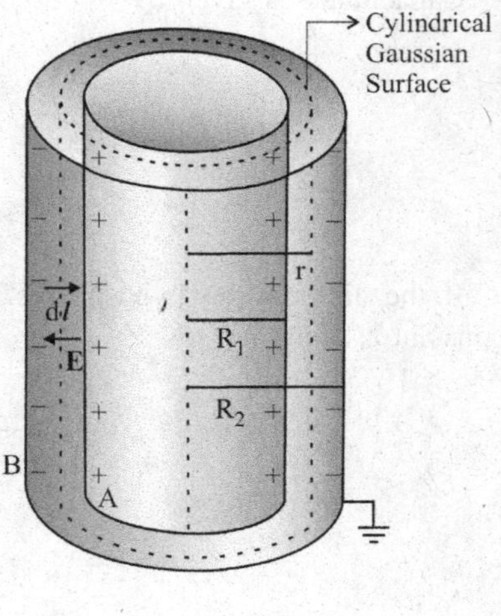


**Fig. 3.35** A cylindrical capacitor consisting of two coaxial conducting cylinders. A Gaussian surface to calculate the electric field is also shown in the figure

The potential difference between the two cylinders which is the work done in moving a unit positive charge from B to A is

$$V_{AB} = -\int_{R_2}^{R_1} \mathbf{E} \cdot d\mathbf{l} \tag{3.350}$$

From Fig. 3.35

$$\mathbf{E} \cdot d\mathbf{l} = E\,dl\cos 180^\circ = -E\,dl \text{ and } d\mathbf{l} = -dr. \text{ Hence}$$
$$\mathbf{E} \cdot d\mathbf{l} = -E\,dl = E\,dr$$

Thus

$$V_{AB} = -\int_{R_2}^{R_1} \mathbf{E} \cdot d\mathbf{l} = -\int_{R_2}^{R_1} E\,dr \tag{3.351}$$

$$V_{AB} = -\int_{R_2}^{R_1} \frac{q}{2\pi\varepsilon_0 r L} dr \tag{3.352}$$

$$V_{AB} = \frac{q}{2\pi\varepsilon_0 L} \log_e\left(\frac{R_2}{R_1}\right) \tag{3.353}$$

The capacitance is given by

$$C = \frac{q}{V_{AB}} = \frac{2\pi\varepsilon_0 L}{\log_e\left(\frac{R_2}{R_1}\right)} \tag{3.354}$$

Eq. 3.354 is based on the assumption that charge q is uniformly distributed over the inner cylinder and the space between the two cylinders is filled with air.

If the space between the two cylinders contains a medium of dielectric constant $\varepsilon_r$ then

$$C = \frac{4\pi\varepsilon_r \varepsilon_0 L}{2 \log_e\left(\frac{R_2}{R_1}\right)} \tag{3.355}$$

## 3.27 Capacitors in Parallel and Series

Let us calculate the equivalent capacity of a certain number of capacitors joined in either parallel or series. Equivalent capacitance means replacing a group of capacitors with a single capacitor such that the single capacitor has same charge for the same applied potential difference as that of the group of capacitors (Fig. 3.36).

(i) **Capacitors joined in series**

Consider N number of capacitors $C_1, C_2, C_3 \ldots C_N$ joined in series and voltage V is applied across the capacitor (Fig. 3.36). Let the left side of the plate $C_1$ is given a charge of q. As explained in Sect. 3.24 a equal negative charge and positive charge is induced on the right side plate of $C_1$. However the induced positive charge q on the right side plate of $C_1$ is given to the left side plate, of $C_2$. The entire process continues up to capacitor $C_N$ and hence each capacitor is charged to a charge q.

Because the capacitors are connected in series the sum of potential differences $V_1, V_2, V_3 \ldots V_N$ across the capacitors $C_1, C_2, C_3 \ldots C_N$ is equal to the total voltage V.

Now

$$V_1 = \frac{q}{C_1}, V_2 = \frac{q}{C_2} \ldots V_N = \frac{q}{C_N} \tag{3.356}$$

$$V = V_1 + V_2 + V_2 + \cdots + V_N \tag{3.357}$$

$$\Rightarrow V = q\left[\frac{1}{C_1} + \frac{1}{C_2} + \cdots + \frac{1}{C_N}\right] \tag{3.358}$$

C, the equivalent capacitor must carry a charge q if a voltage V is applied across it.

$$\Rightarrow V = \frac{q}{C} \tag{3.359}$$

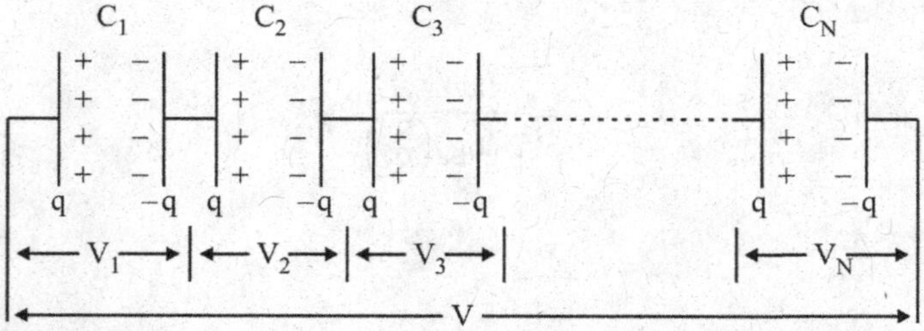


**Fig. 3.36** N number of capacitors connected in series

Comparing Eqs. 3.358 and 3.359

$$\frac{q}{C} = q\left[\frac{1}{C_1} + \frac{1}{C_2} + \cdots + \frac{1}{C_N}\right] \tag{3.360}$$

$$\Rightarrow \frac{1}{C} = \frac{1}{C_1} + \frac{1}{C_2} + \cdots + \frac{1}{C_N} \tag{3.361}$$

$$\Rightarrow \frac{1}{C} = \sum_{i=1}^{N} \frac{1}{C_i} \tag{3.362}$$

(ii) **Capacitors in parallel**

Consider N number of capacitors connected in parallel as shown in Fig. 3.37. Let a potential difference of V be applied across the capacitors. Each capacitor $C_1$, $C_2$, $C_3$ … $C_N$ carries a charge of $q_1$, $q_2$, $q_3$ … $q_N$. "The potential difference between two points is independent of the path travelled (Sect. 2.21 Eq. 2.284). Thus if we select a path between points A, B in Fig. 3.37 either via $C_1$ or $C_2$ or $C_3$ or … $C_N$ the potential difference is same. Hence all the capacitors are maintained at same potential difference V." Therefore

**Fig. 3.37** N number of capacitors joined in parallel

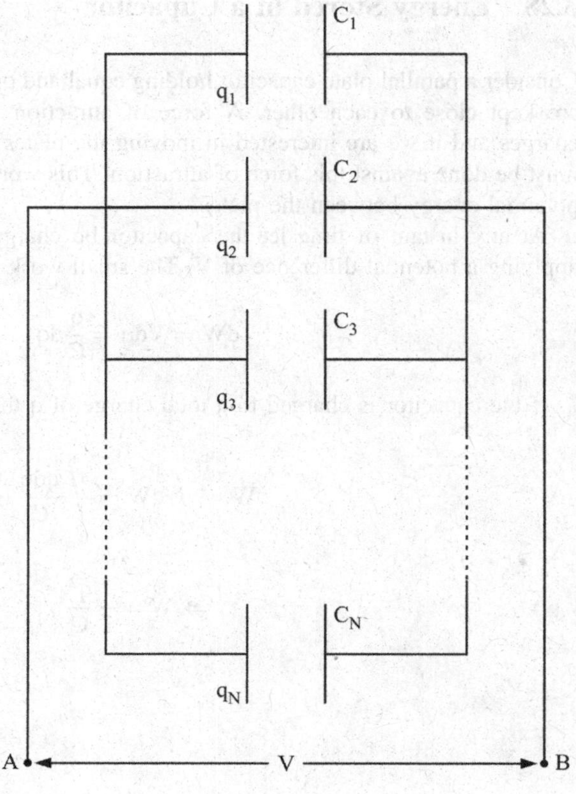

$$q_1 = VC_1, q_2 = VC_2, q_3 = VC_3 \ldots q_N = VC_N \tag{3.363}$$

The total charge on the capacitors is

$$q_{tot} = q_1 + q_2 + q_3 + \cdots + q_N \tag{3.364}$$

$$q_{tot} = VC_1 + VC_2 + VC_3 + \cdots + VC_N \tag{3.365}$$

$$q_{tot} = V[C_1 + C_2 + C_3 + \cdots + C_N] \tag{3.366}$$

C, the equivalent capacitor must carry a charge $q_{tot}$ if a voltage V is applied across it

$$q_{tot} = VC \tag{3.367}$$

Comparing Eqs. 3.366, 3.367

$$VC = V[C_1 + C_2 + C_3 + \cdots + C_N] \tag{3.368}$$

$$\Rightarrow C = C_1 + C_2 + C_3 + \cdots + C_N \tag{3.369}$$

## 3.28 Energy Stored in a Capacitor

Consider a parallel plate capacitor holding equal and opposite charges whose plates are kept close to each other. A force of attraction will exist between opposite charges and if we are interested in moving the plates a finite distance away work must be done against this force of attraction. This work done is stored as electrical potential energy between the plates.

At any instant of time let the capacitor be charged to a small charge dq by applying a potential difference of V. The small work done dW is

$$dW = Vdq = \frac{q}{C}dq \tag{3.370}$$

If the capacitor is charged to a total charge of q then

$$W = \int dW = \int_0^q \frac{qdq}{C} \tag{3.371}$$

$$\Rightarrow W = \frac{1}{2}\frac{q^2}{C} \tag{3.372}$$

as q = CV

$$\Rightarrow W = \frac{1}{2} C V^2 \tag{3.373}$$

This work done is stored as the electrical potential energy of the system.
For a parallel plate capacitor

$$C = \frac{\varepsilon_0 A}{d}, V = E d$$

Substituting the above values in Eq. 3.373

$$W = \frac{1}{2} \frac{\varepsilon_0 A}{d} E^2 d^2 = \frac{1}{2} \varepsilon_0 E^2 A d \tag{3.374}$$

Energy stored per unit volume is then

$$= \frac{1}{2} \varepsilon_0 E^2 \tag{3.375}$$

**Example 3.14**
A parallel plate capacitor filled with air has a capacitance of 0.01 nF. The distance of separation between the plates is 0.2 mm. What is the plate area?

**Solution**
For a parallel plate capacitor

$$C = \frac{\varepsilon_0 A}{d}$$

$$\Rightarrow A = \frac{Cd}{\varepsilon_0} = \frac{(0.01) \times 10^{-9} \times 0.2 \times 10^{-3}}{8.85 \times 10^{-12}}$$

$$A = 2.25 \times 10^{-4}\, m^2$$

**Example 3.15**
A parallel plate capacitor consists of air and the distance between the plates is 4 cm. A dielectric slab of thickness 2 cm and dielectric constant 5 is introduced between plates. The distance between the two plates is so charged that the capacity of the capacitor remains unchanged. Calculate the new distance between the plates.

**Solution**

In the absence of dielectric the capacity of the capacitor is given by

$$C = \frac{\varepsilon_0 A}{d} = \frac{\varepsilon_0 A}{0.04}$$

In the presence of dielectric let the new distance be d′ then

$$C = \frac{\varepsilon_0 A}{\left(d' - t + \frac{t}{\varepsilon_r}\right)} = \frac{\varepsilon_0 A}{\left[d' - 0.02 + \frac{0.02}{5}\right]}$$

Because the capacitance remains unchanged

$$\frac{\varepsilon_0 A}{0.04} = \frac{\varepsilon_0 A}{\left[d' - 0.02 + \frac{0.02}{5}\right]} \Rightarrow d' = 5.6\,\text{cm}$$

**Example 3.16**

Consider two spherical shells forming a spherical capacitor as shown in Fig. 3.34. The radius of the inner and outer sphere's are $R_1$ and $R_2$ respectively. Consider a point P between the inner and outer spheres whose distance is r from the center of the sphere. The space between the two shells is filled with two dielectrics such that from $R_1$ to r the dielectric constant is $\varepsilon_{r1}$ and from r to b the dielectric constant is $\varepsilon_{r2}$. Calculate the capacitance of the capacitor. The charge in the inner sphere is q.

**Solution**

From Eq. 3.339

$$V_{AB} = -\int_{R_2}^{R_1} E\,dr = \int_{R_1}^{R_2} E\,dr$$

$$\Rightarrow V_{AB} = \int_{R_1}^{r} E\,dr + \int_{r}^{R_2} E\,dr$$

$$\Rightarrow V_{AB} = \int_{R_1}^{r} \frac{q}{4\pi\varepsilon_0\,\varepsilon_{r1}\,r^2}\,dr + \int_{r}^{R_2} \frac{q}{4\pi\varepsilon_0\,\varepsilon_{r2}\,r^2}\,dr$$

$$\Rightarrow V_{AB} = \frac{q}{4\pi\varepsilon_0}\left[\frac{1}{\varepsilon_{r1}}\left[\frac{1}{R_1} - \frac{1}{r}\right] + \frac{1}{\varepsilon_{r2}}\left[\frac{1}{r} - \frac{1}{R_2}\right]\right]$$

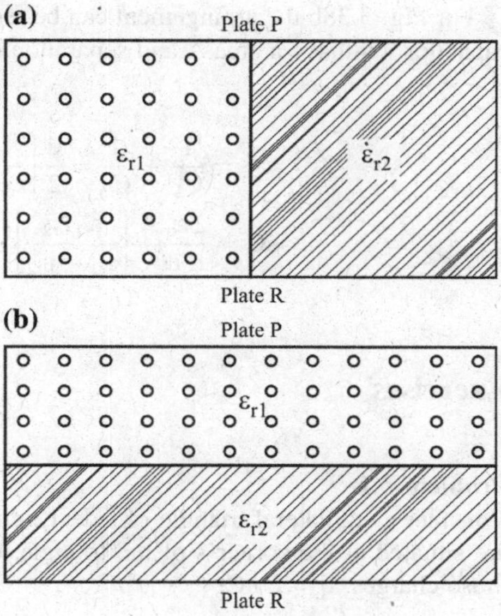


**Fig. 3.38 a, b** Two parallel plate capacitors filled with two dielectrics in two different configurations

The capacitance is

$$C = \frac{q}{V_{AB}}$$

$$C = \frac{4\pi\varepsilon_o}{\frac{1}{\varepsilon_{r1}}\left[\frac{1}{R_1} - \frac{1}{r}\right] + \frac{1}{\varepsilon_{r2}}\left[\frac{1}{r} - \frac{1}{R_2}\right]}$$

**Example 3.17**

A parallel plate capacitor is filled with two dielectrics as shown in Fig. 3.38a, b. The two dielectrics are of same dimensions with dielectric constants $\varepsilon_{r1}$, $\varepsilon_{r2}$ respectively. Find the capacitance.

**Solution**

In Fig. 3.38a the arrangement is simply two capacitors in parallel–each capacitor having area A/2. Hence total capacitance

$$C = C_1 + C_2$$

Hence

$$C = \frac{\varepsilon_o\, A/2\, \varepsilon_{r1}}{d} + \frac{\varepsilon_o\, A/2\, \varepsilon_2}{d}$$

$$\Rightarrow C = \frac{\varepsilon_o\, A}{2d}(\varepsilon_{r1} + \varepsilon_{r2})$$

For Fig. 3.38b the arrangement can be regarded as two capacitors in series. Each capacitor will have a area A and separation distance d/2. Thus the total capacitance is

$$\frac{1}{C}=\frac{1}{C_1}+\frac{1}{C_2}=\frac{d}{2}\left[\frac{1}{\varepsilon_0\, A\, \varepsilon_{r1}}+\frac{1}{\varepsilon_0\, A\, \varepsilon_{r2}}\right]$$

$$C=\frac{2\,\varepsilon_0\, A}{d}\left[\frac{\varepsilon_{r1}\,\varepsilon_{r2}}{\varepsilon_{r1}+\varepsilon_{r2}}\right]$$

## Exercises

**Problem 3.1**
Consider a equilateral triangle of side 12 cm as shown in Fig. 3.39. Three charges are situated at the corner's of the triangle. Calculate the work done is assembling those charges. q = 2 nC.

**Problem 3.2**
Consider a cube of side a. Eight charges each of magnitude q are placed at the eight corners of the cube. What is the electric potential energy stored in the system.

**Problem 3.3**
A hollow sphere consists of a uniform surface charge $\sigma$ and the total charge on the sphere is q. Find the energy in the system if the radius of the sphere is R.

**Problem 3.4**
A sphere whose radius is R carries a volume charge density $\rho = C\, r^2$ where C is a constant. Calculate the work done is assembling the charge distribution.

**Problem 3.5**
Calculate the work done in moving a point charge q = 7 nC from (0, 0, 0) to (1m, $\pi/2$, $\pi$). The electric field in spherical polar coordinates is given by

$$\mathbf{E} = 5\, r\, \hat{r} + \frac{7}{r}\, \hat{\theta}$$

Choose your own path.

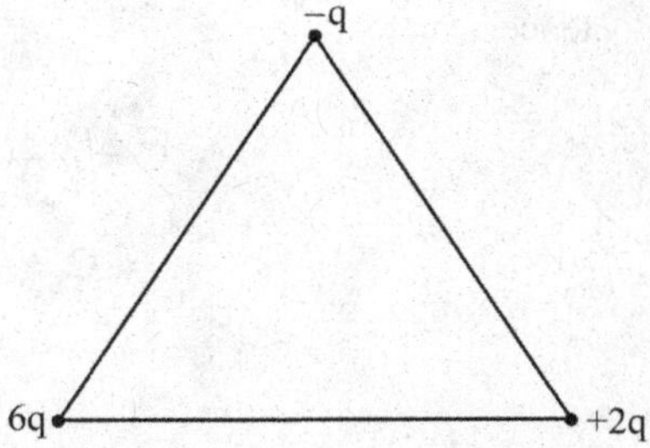


**Fig. 3.39** Three charges situated at the corners of the equilateral triangle. Illustration for problem 3.1

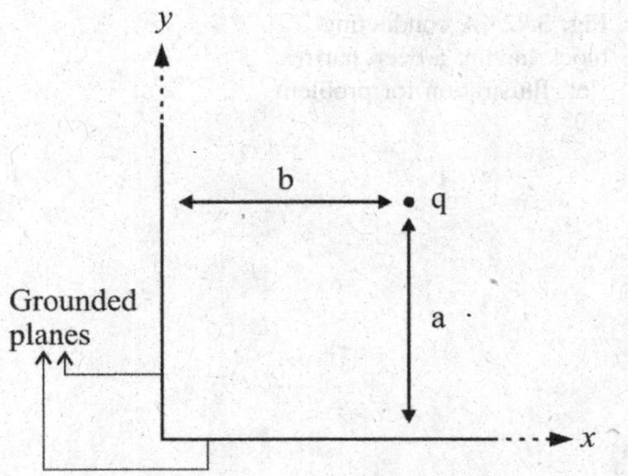


**Fig. 3.40** Illustration for problem 3.6

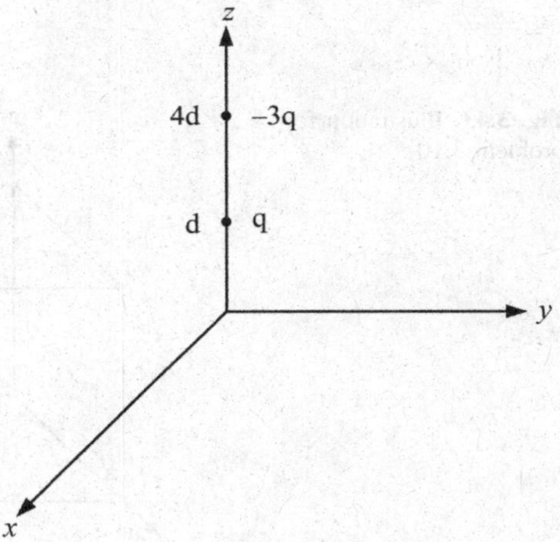


**Fig. 3.41** Illustration for problem 3.7

**Problem 3.6**
A point charge q is situated at distance a, b from two perpendicular half planes. The planes are grounded as shown in Fig. 3.40. Fix the image charge and calculate the electric field in the region $x, y > 0$.

**Problem 3.7**
In Fig. 3.41 find the potential in the region $\mathbf{z} > 0$.

**Problem 3.8**
Find the potential between the plates of a parallel plate capacitor using Laplace's equation. One of the plate is situated at $x = 0$ and the other at $x = \text{d}$ along the $x$-axis. The plates are maintained at potential $V_1$ and $V_2$.

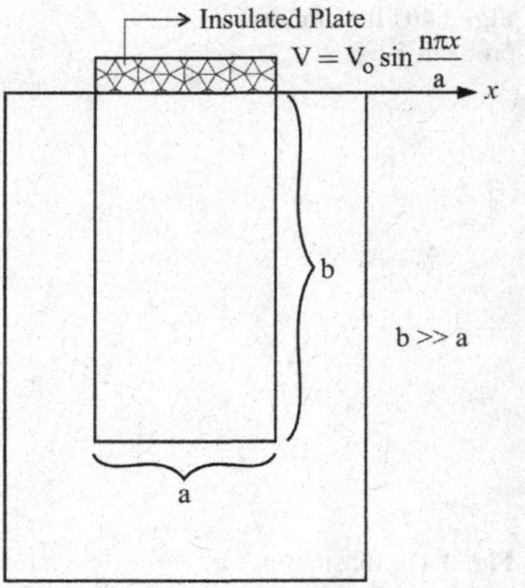


Fig. 3.42 A conducting block having a deep narrow slot. Illustration for problem 3.9

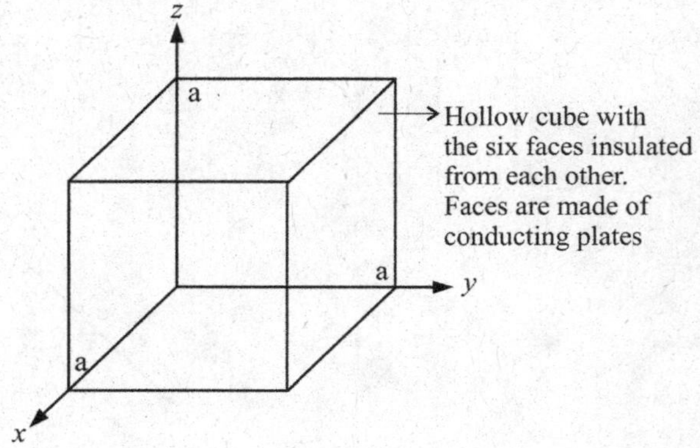


Fig. 3.43 Illustration for problem 3.10

**Problem 3.9**

A conducting block which is grounded has a very deep narrow slot as shown in Fig. 3.42. The dimensions of the slot are such that b ≫ a. The slot is covered with a plate insulated from the slot and maintained at potential

$$V = V_o \sin\left(\frac{n\pi x}{a}\right)$$

Determine the potential inside the slot.

**Problem 3.10**

In Fig. 3.43 the plate at $z = 0$ is maintained at zero potential, but the plate at $z = a$ is maintained at potential V. Calculate the potential within the hollow cube.

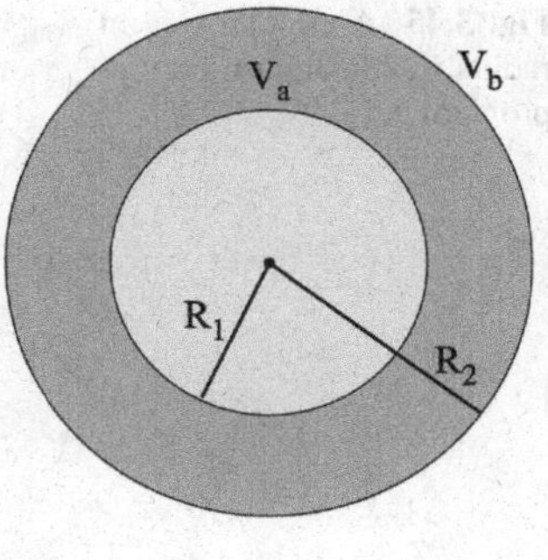


**Fig. 3.44** A spherical shell. Illustration for problem 3.12

**Problem 3.11**
In Fig. 3.43 both the plates at $z = 0$, $z = a$ are maintained at potential V. Calculate the potential within the hollow cube.

**Problem 3.12**
Using Laplace's equation calculate the potential between the plates of two spherical shells shown in Fig. 3.44.

The inner shell is maintained at a potential of $V_a$ while the outer shell is maintained at a potential of $V_b$. The radii of inner and outer shells are $R_1$ and $R_2$ respectively and $V_a > V_b$.

**Problem 3.13**
Consider a charged conducting sphere holding charge q, radius R placed in a uniform electric field $E_o\hat{k}$. By using Laplace's equation calculate the electric field inside and outside the sphere.

Calculate the dipole moment, induced charge density of the sphere.

**Problem 3.14**
Consider a circular ring of radius R carrying a total charge q. The total charge q is uniformly distributed over the ring. Calculate the potential at any point P as shown in Fig. 3.45 by using Laplace's equation.

**Problem 3.15**
Consider a hollow sphere of radius R. The potential on the sphere is given by $V(\theta) = 2\cos^2\theta/2$. Calculate the potential inside and outside the sphere using Laplace's equation.

[Hint : Use the relation $2\cos^2\theta/2 = 1 + \cos\theta = P_o(\cos\theta) + P_1(\cos\theta)$]

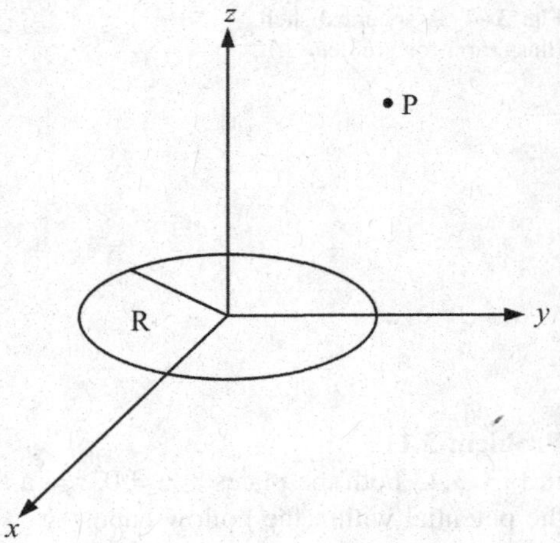


**Fig. 3.45** A circular ring of radius R. Illustration for problem 3.14

**Problem 3.16**

For Example 3.5 calculate the potential and field inside the sphere by using Laplace's equation.

**Problem 3.17**

Consider two concentric sphere's with radii $R_1$, $R_2$ as shown in Fig. 3.46. The lower hemisphere of the sphere with radius $R_2$ and upper hemisphere of the sphere with radius $R_1$ is maintained at potential $V_o$. The other hemisphere's are maintained at zero potential.

Calculate the potential within the region $R_1 \leq r \leq R_2$.

**Problem 3.18**

Consider a cylindrical shell of radius R as shown in Fig. 3.47. Find the potential at point P if the shell is maintained at potential $V_o$. The cylindrical shell is of infinite extent.

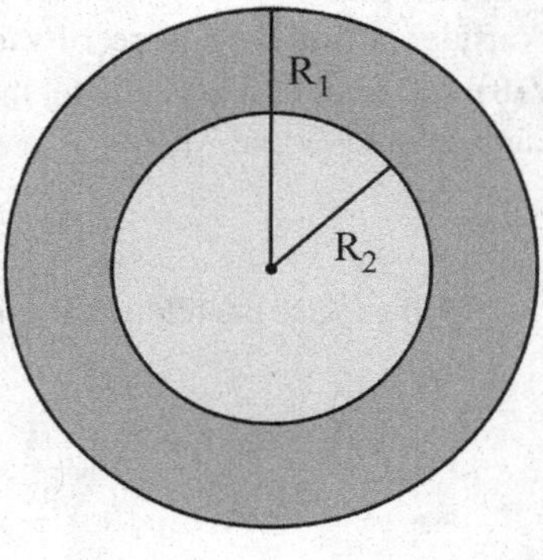


**Fig. 3.46** Two concentric spheres. Illustration for problem 3.17

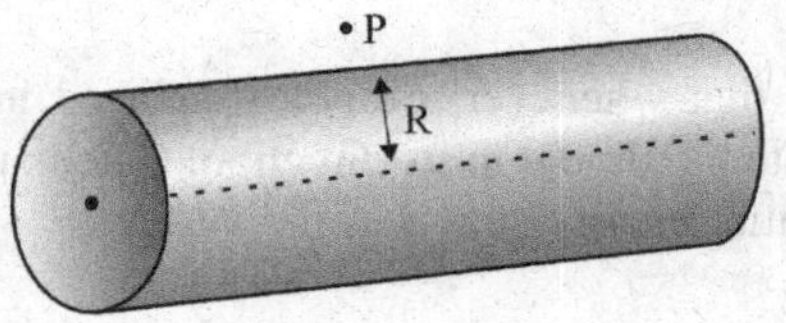


**Fig. 3.47** A cylindrical shell of radius R. Illustration for problem 3.18

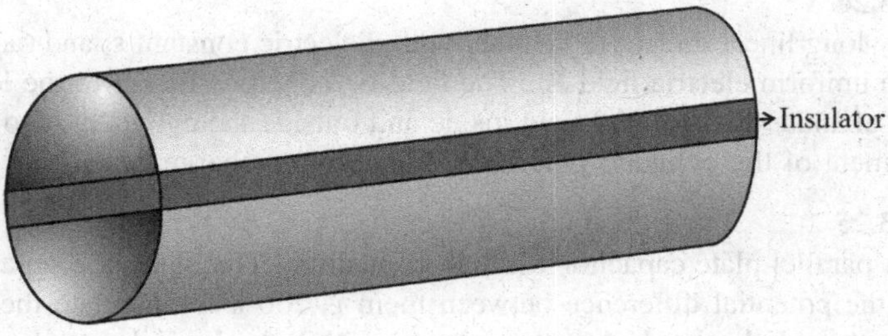


**Fig. 3.48** A cylindrical tube of radius R. Illustration for problem 3.20

**Problem 3.19**
For Example 3.7 calculate the dipole moment, electric field and surface charge density.

**Problem 3.20**
A cylindrical tube of radius R with the upper half of the tube insulated from the lower half is shown in Fig. 3.48. The upper half is maintained at potential $-V_a$ and the lower half at $V_b$. Find the potential inside and outside the tube.

**Problem 3.21**
Equations 2.11, 2.58 are

$$\mathbf{E} = \mathbf{E}_1 + \mathbf{E}_2 + \cdots + \mathbf{E}_n$$

Equation 2.77 is

$$V = V_1 + V_2 + \cdots + V_n$$

Show that for the case of work done

$$W \neq W_1 + W_2 + \cdots + W_n$$

**Problem 3.22**

In Sect. 3.16 we hinted that whenever a dipole is placed in an electric field, the electric field tries to rotate the dipole in its direction. Calculate the torque acting on the dipole and the potential energy stored in the system.

**Problem 3.23**

A dielectric sphere of radius R is uniformly polarized with polarization **P**. Find the bound charge densities, potential and field of the sphere.

**Problem 3.24**

A infinitely long linear dielectric cylinder with dielectric constant $\varepsilon_r$ and radius R is placed in a uniform electric field $\mathbf{E}_o$. The field $\mathbf{E}_o$ is perpendicular to the length of cylinder. Calculate the potential field inside and outside the cylinder. Also find the dipole moment of the cylinder, polarization and bound charge densities.

**Problem 3.25**

Consider a parallel plate capacitor with air as medium. The plates are separated by 4 cm and the potential difference between them is 200 kV. Calculate the electric field intensity and check whether the value is within the breakdown of air. What should be the minimum breakdown strength of an alternate dielectric so that the dielectric will not breakdown.

**Problem 3.26**

(a) Calculate the capacitance of a parallel plate capacitor of area 400 $cm^2$ and a separation distance of 6 mm. The dielectric medium is air. The potential difference between the plates is 400 V. Calculate the total energy stored in the capacitor and the energy density.

(b) Calculate the equivalent capacitance for the configuration shown in Fig. 3.49.

**Problem 3.27**

One of the plates of a parallel plate capacitor is maintained at zero potential. The separation distance between the plates is d. The space between the plates is filled with a dielectric. The dielectric constant of the dielectric varies uniformly from one plate to the other. Calculate the capacitance of the capacitor if the values of the dielectric constant at the two plates is $\varepsilon_{r1}$ and $\varepsilon_{r2}$ respectively. The area of the plates is A.

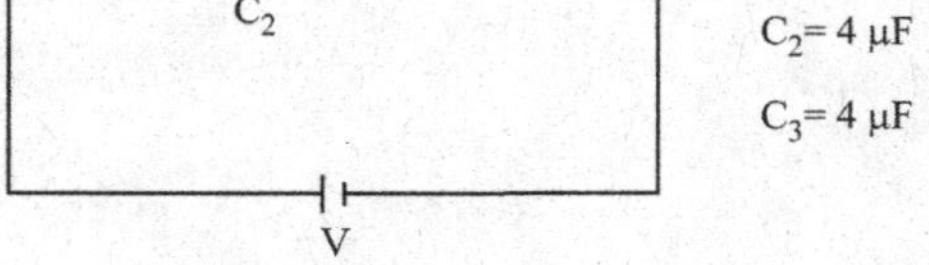

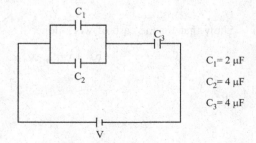

**Fig. 3.49** Illustration for problem 3.26

**Problem 3.28**
For a parallel plate capacitor (with air as dielectric) a potential difference of 200 V is applied across the plates. A dielectric slab of thickness 0.7 cm and dielectric constant 6 is introduced between the plates. Calculate

(i) Capacitance with and without dielectric.
(ii) Free Charge.
(iii) Electric field in the air gap and dielectric.

$$A = 200\,\text{cm}^2\ d = 1.5\,\text{cm}.$$

**Problem 3.29**
A cylindrical capacitor having inner and outer radii 4.5 and 6.5 cm consists of two dielectrics between them. The dielectric having contact with inner cylinder (radii 4.5 m) is having a dielectric constant of 5 and the other one is having a dielectric constant of 7. Calculate the capacitance per unit length.

**Problem 3.30**
Consider a parallel plate capacitor whose capacitance is 0.5 μF. It is connected to the voltage source so that the voltage across the capacitor is 200 V. It is then connected to a second capacitor in parallel mode. The area of the plates of the second capacitor is 5 times that of the first capacitor. Calculate the charge distribution on each capacitor. Also calculate the loss in energy.

# Chapter 4
# Magnetostatics

## 4.1 Introduction

In the previous chapters we calculated the electric field produced by static charges. In this chapter we will see that the moving charges produce not only the electric field but also the magnetic field.

In 1820 Oersted observed that whenever a compass needle is brought near a current-carrying wire the compass needle showed deflection. The direction of the deflection was reversed when the direction of the current in the wire was reversed.

"We define the space surrounding the current-carrying conductor as containing magnetic field".

As we represented electric fields by field lines we represent magnetic field by magnetic field lines.

Consider two current-carrying wires as shown in Fig. 4.1. The magnetic fields due to the current-carrying wires are shown in the figure. The magnetic lines of force are circular and are concentric. The direction of the magnetic lines of force can be obtained by right-hand rule as shown in the same figure. Let the thumb in the right hand point in the direction of the current and other fingers of the hand curl around. The direction of the fingers curling around gives the direction of the magnetic field.

We represent the cross section of a conductor (in which the current is flowing) on the paper by a circle. As shown in Fig. 4.2a the dot in the circle represents the current flowing out of the plane of the paper. The cross in the circle as shown in Fig. 4.2b represents the current flowing inside the plane of the paper.

Our approach in magnetostatics will be similar to electrostatics. In the case of electrostatics we wrote the electric force as $\mathbf{F} = Q\mathbf{E}$. Similarly we will write an expression for force in magnetostatics. Then as we went in search for easier methods like Gauss's law and potential in electrostatics, similarly we will go in search of easier methods in magnetostatics. Before discussing these methods we will see in detail about Lorentz force law.

© Springer Nature Singapore Pte Ltd. 2020

S. Balaji, *Electromagnetics Made Easy*,
https://doi.org/10.1007/978-981-15-2658-9_4

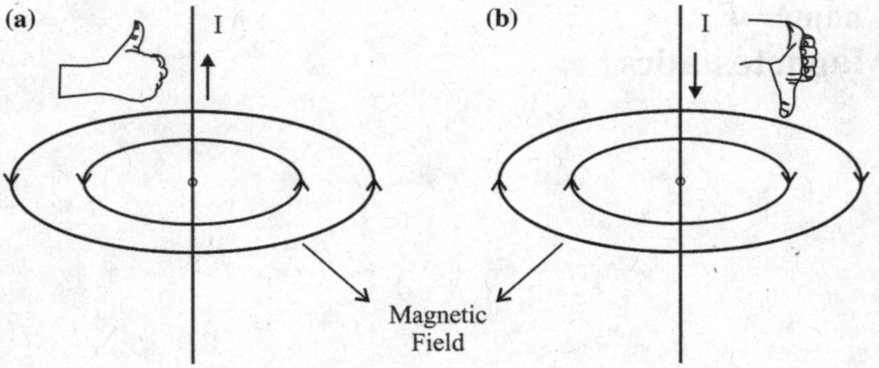


**Fig. 4.1** **a** A wire carrying current in the upward direction. **b** A wire carrying current in the downward direction (The direction of the magnetic field obtained by using right hand rule is depicted in the figure)

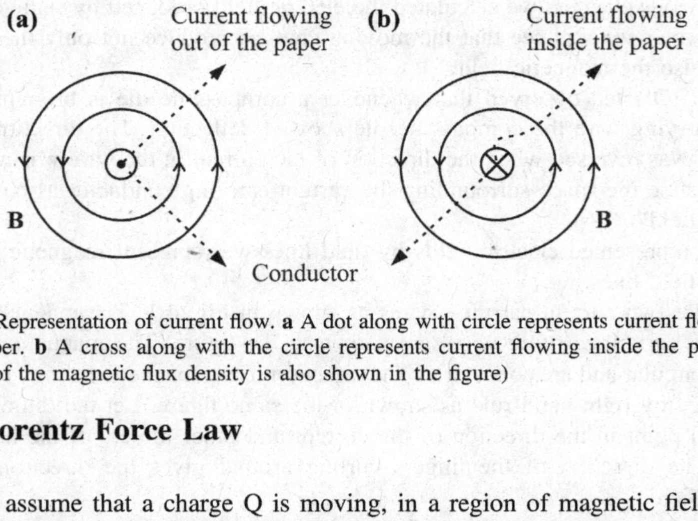


**Fig. 4.2** Representation of current flow. **a** A dot along with circle represents current flowing out of the paper. **b** A cross along with the circle represents current flowing inside the paper (The direction of the magnetic flux density is also shown in the figure)

## 4.2 Lorentz Force Law

Suppose assume that a charge Q is moving, in a region of magnetic field whose magnetic flux density is **B**. If the velocity of the charge is **v** then the magnetic force acting on the charge Q is given by

$$\mathbf{F}_{mag} = Q(\mathbf{v} \times \mathbf{B}) \tag{4.1}$$

The above equation is known as Lorentz force law. The total force acting on the charge Q in the presence of electric field **E** and magnetic flux density is given by

$$\mathbf{F}_{tot} = Q(\mathbf{E} + (\mathbf{v} \times \mathbf{B})) \tag{4.2}$$

Equation 4.1 is a postulate and cannot be derived from any fundamental law. Its validity has been established through experiments. The charged particle trajectories in cyclotron, velocity selector, etc., is determined by Lorentz force equation.

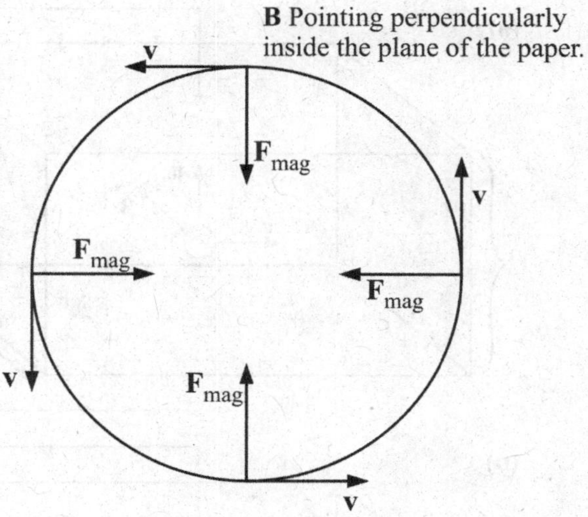


**Fig. 4.3** A charged particle moving in a circular path under the action of the magnetic field

Suppose assume that a particle with charge Q is projected in a uniform magnetic field with flux density **B** with velocity **v** as shown in Fig. 4.3. As $\mathbf{F}_{mag} = Q(\mathbf{v} \times \mathbf{B})$, the magnetic force is given by cross product of **v**, **B** the force is perpendicular to the plane containing **v**, **B**. The particle moves in a circular path.

Let us calculate the work done by magnetic force

$$\begin{aligned} dW &= \mathbf{F}_{mag} \cdot d\boldsymbol{l} = Q(\mathbf{v} \times \mathbf{B}) \cdot \mathbf{v}dt \\ dW &= 0 \end{aligned} \tag{4.3}$$

as in a scalar triple product if two vectors are equal, the scalar triple product amounts to zero. Thus magnetic forces don't do any work.

Example 6.16 discusses more about this point.

## 4.3 Applications of Lorentz Force—Hall Effect

If a conductor (or) a semiconductor material carrying a current is placed in a magnetic field that is acting perpendicular to the direction of the flow, a voltage is developed across the material in a direction perpendicular to both the direction of the current and that of the magnetic field. The voltage developed is known as Hall voltage, and the phenomenon is called Hall effect.

In Fig. 4.4 a metal or n-type semiconductor (Fig. 4.4a) and a p-type semiconductor (Fig. 4.4b) carrying current I is placed in a field of flux density **B** such that **B** is perpendicular to I.

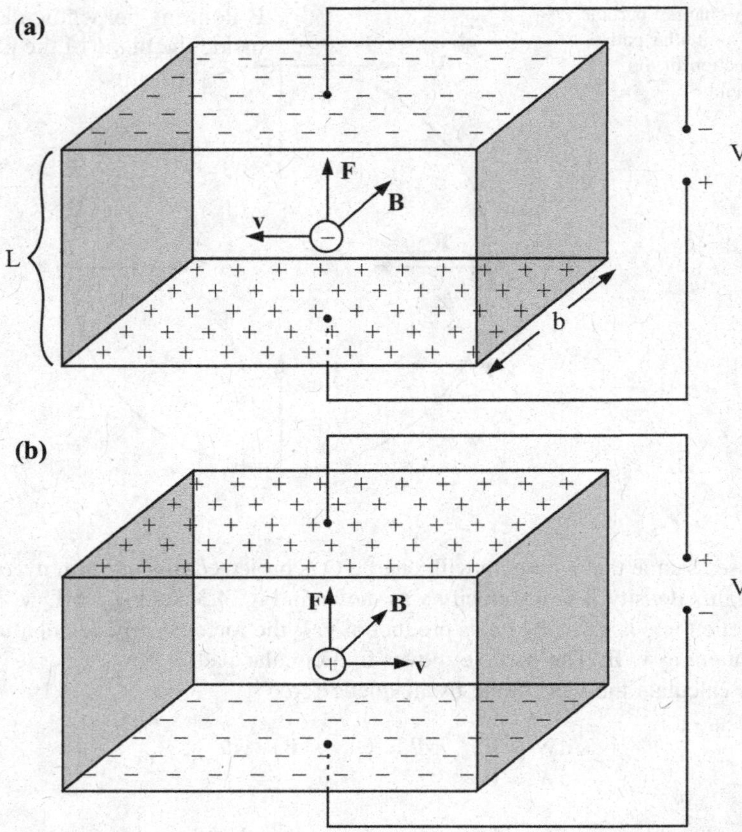


**Fig. 4.4** **a** Figure showing a) a metal or n - type semiconductor b) a p - type semiconductor carrying a current I placed in a magnetic field acting perpendicular to the direction of flow of current

In Fig. 4.4a we see that the electrons (majority carriers in n-type semiconductor) flowing with a drift velocity **v** experience a Lorentz force upward and collect upward on the top of the material leaving a net positive charge on the bottom of the material.

In Fig. 4.4b we see that the holes (majority carriers in p-type semiconductor) flowing with a velocity **v** experience a Lorentz force upward and collect upward on the top of the material leaving a net negative charge on the bottom of the material.

As shown in Fig. 4.4a, b the polarity of Hall voltage $V_H$ is different for n-type and p-type semiconductor and serves as a means to identify the type of the semiconductor.

The Lorentz force acting on the particle of charge q is given by

$$\mathbf{F} = q(\mathbf{v} \times \mathbf{B})$$

But from Fig. 4.4 **B** is perpendicular to **v** hence

$$F = qvB \sin 90° = qvB \tag{4.4}$$

In Fig. 4.4a, b application of magnetic field to a current-carrying conductor or semiconductor (**B**⊥**I**) results in a separation of charges. Hence an electric field $E_H$ is set up within the material. The separation of charges will continue until force due to $E_H$ balances the Lorentz force

$$qE_H = qvB \tag{4.5}$$

$$\Rightarrow v = \frac{E_H}{B} \tag{4.6}$$

Let n be the number of atoms per unit volume of the material and A its area of cross section. Let q be the charge moving with velocity **v**. The current flowing through the conductor is given by charge flowing per second

$$I = (nvA)q \tag{4.7}$$

But A = Lb where L, b are the dimensions of the face of the material as shown in Fig. 4.4.

Substituting the value of A and Eq. 4.6 in Eq. 4.7

$$I = n\left(\frac{E_H}{B}\right)(Lb)q \tag{4.8}$$

$$E_H = \frac{IB}{nqLb} \tag{4.9}$$

The voltage $V_H$ established between the top and bottom surfaces of the material because of $E_H$ is

$$V_H = E_H L \tag{4.10}$$

Substituting Eq. 4.9 in Eq. 4.10

$$\Rightarrow V_H = \frac{IB}{nqLb}L = \frac{IB}{nqb} \tag{4.11}$$

$$V_H = R_H \frac{IB}{b} \tag{4.12}$$

where $R_H = \frac{1}{nq}$ is called Hall coefficient and is characteristic of the material.

As mentioned previously Hall effect can be used to distinguish between n-type and p-type semiconductors and also finds applications in magnetic field measurement.

## 4.4 Sources of Magnetic Field

A magnetic field can be produced by a bar magnet or moving electrical charges (currents).

For the time being we will not discuss the magnetic field's produced by a bar magnet. We will confine our attention to the magnetic fields produced by the currents (i.e. moving electrical charges).

In "**electrostatics**" we saw that "**stationary charges**" produce "**electric fields**" that are constant in time.

Now we will deal with "**magnetostatics**" in which "**steady currents**" produce "**magnetic fields**" that are constant in time.

Suppose assume a "**steady**" current I flows in a wire (situated in free space) as shown in Fig. 4.5. The element $I d\boldsymbol{l}'$ is known as current element. We will be interested in calculating the magnetic field produced by the current element and the entire wire at a far away point.

## 4.5 Magnetic Force Between Two Current Elements

Coulomb's law gives an expression for the "**electrical force**" between two "**electrical charges**". Similarly we can write an expression for "**magnetic force**" between two "**current elements**".

Suppose assume that there are two wires carrying current $I_a$ and $I_b$ as shown in Fig. 4.6. Consider two current elements $I_a d\boldsymbol{l}'_a$ and $I_b d\boldsymbol{l}'_b$ in the two wires. The distance between the two elements is **r**.

The force between current elements $I_a d\boldsymbol{l}'_a$ and $I_b d\boldsymbol{l}'_b$

(i) varies directly as the product of magnitudes of current
(ii) depends upon the nature of the medium

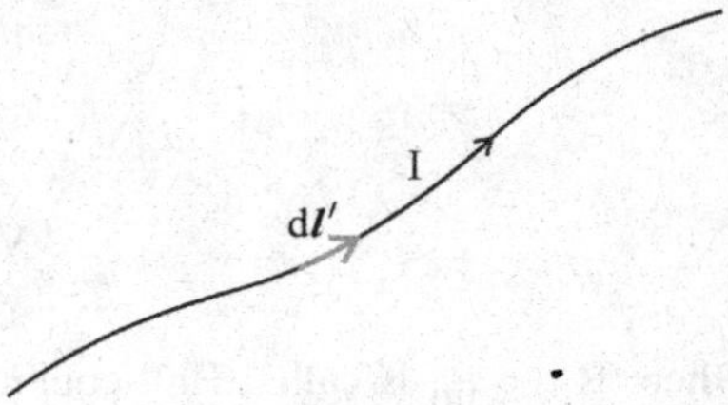


**Fig. 4.5** A wire carrying a steady current I

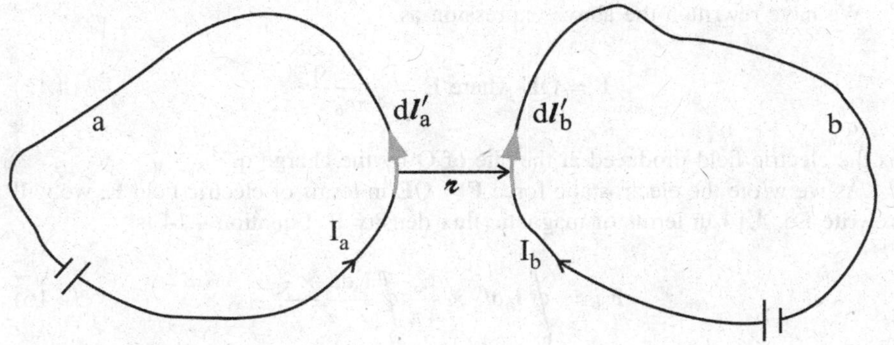


**Fig. 4.6** Two wires a, b carrying currents $I_a$, $I_b$

(iii) varies inversely as the square of distance between the two current elements
(iv) depends upon the lengths and orientations of the two current elements
(v) is attractive if the current flow is in the same direction and repulsive if they flow in opposite direction.

The force exerted by the current element $I_a dl'_a$ on $I_b dl'_b$ is

$$d\mathbf{F}_{ab} = \frac{\mu_o}{4\pi} I_a I_b \left( \frac{dl'_b \times (dl'_a \times \hat{r})}{r^2} \right) \tag{4.13}$$

$$\mathbf{F}_{ab} = \frac{\mu_o}{4\pi} I_a I_b \oint_a \oint_b \frac{dl'_b \times (dl'_a \times r)}{r^3} \tag{4.14}$$

Here $r$ is a vector pointing from current element $I_a dl'_a$ to $I_b dl'_b$.

The line integrals are evaluated over two wires. The constant $\mu_o$ is called permeability of free space and has a value of $4\pi \times 10^{-7}$ N/A$^2$.

## 4.6 Biot–Savart Law

Under electrostatic conditions the force between two charges q and Q separated by distance $r$ is given by

$$\mathbf{F} = \frac{1}{4\pi\varepsilon_o} \frac{qQ}{r^2} \hat{r}$$

We have rewritten the above expression as

$$\mathbf{F} = Q\mathbf{E} \text{ where } \mathbf{E} = \frac{1}{4\pi\varepsilon_0}\frac{q}{\boldsymbol{\imath}^2}\hat{\boldsymbol{\imath}} \tag{4.15}$$

is the electric field produced at the site of Q by the charge q.

As we wrote the electrostatic force $\mathbf{F} = Q\mathbf{E}$ in terms of electric field $\mathbf{E}$, we will rewrite Eq. 4.14 in terms of magnetic flux density $\mathbf{B}$. Equation 4.14 is

$$\mathbf{F}_{ab} = \oint_b I_b d\mathbf{l}'_b \times \frac{\mu_0}{4\pi}\oint_a \frac{I_a d\mathbf{l}'_a \times \hat{\boldsymbol{\imath}}}{\boldsymbol{\imath}^2} \tag{4.16}$$

The above equation can be written as

$$\Rightarrow \mathbf{F}_{ab} = \oint_b I_b d\mathbf{l}'_b \times \mathbf{B}_a \tag{4.17}$$

where

$$\mathbf{B}_a = \frac{\mu_0}{4\pi}\oint_a \frac{I_a d\mathbf{l}'_a \times \hat{\boldsymbol{\imath}}}{\boldsymbol{\imath}^2} \tag{4.18}$$

$\mathbf{B}_a$ is the magnetic flux density. In general for Fig. 4.7 the magnetic flux density produced at point P is

$$\mathbf{B} = \frac{\mu_0}{4\pi}\oint \frac{I d\mathbf{l}' \times \hat{\boldsymbol{\imath}}}{\boldsymbol{\imath}^2} \tag{4.19}$$

Equation 4.17 is the magnetic analogue of $\mathbf{F} = Q\,\mathbf{E}$, and Eq. 4.19 is the magnetic analogue of Eq. 2.6. $\mathbf{B}$ is a vector, and its direction can be obtained by right-hand rule as shown in Fig. 4.1. Equation 4.19 is called Biot–Savart law.

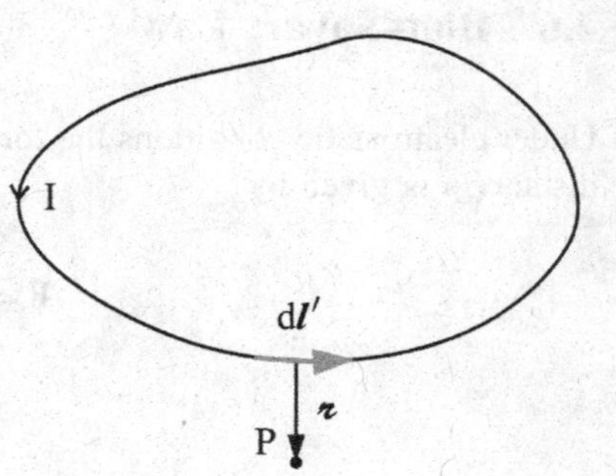


**Fig. 4.7** Illustration to show calculation of magnetic flux density at point P for a wire carrying current I

## 4.7 Current Distributions

In Figs. 4.5, 4.6, 4.7 and 4.8a we have shown currents which are distributed in a line, i.e. charge flows over a line. However charge can also flow over a surface or a volume giving rise to surface currents or volume currents.

In Fig. 4.8b we show a surface carrying a current which flows at a velocity **v**. The surface current density denoted as **K** is defined as current per unit width perpendicular to flow

$$\mathbf{K} = \frac{\mathrm{d}\mathbf{I}}{\mathrm{d}l_{\mathrm{per}}} \tag{4.20}$$

where $\mathrm{d}l_{\mathrm{per}}$ is the width (perpendicular width to the flow) a small infinitesimal element running parallel to the flow in Fig. 4.8b, and **I** is the current in the element.

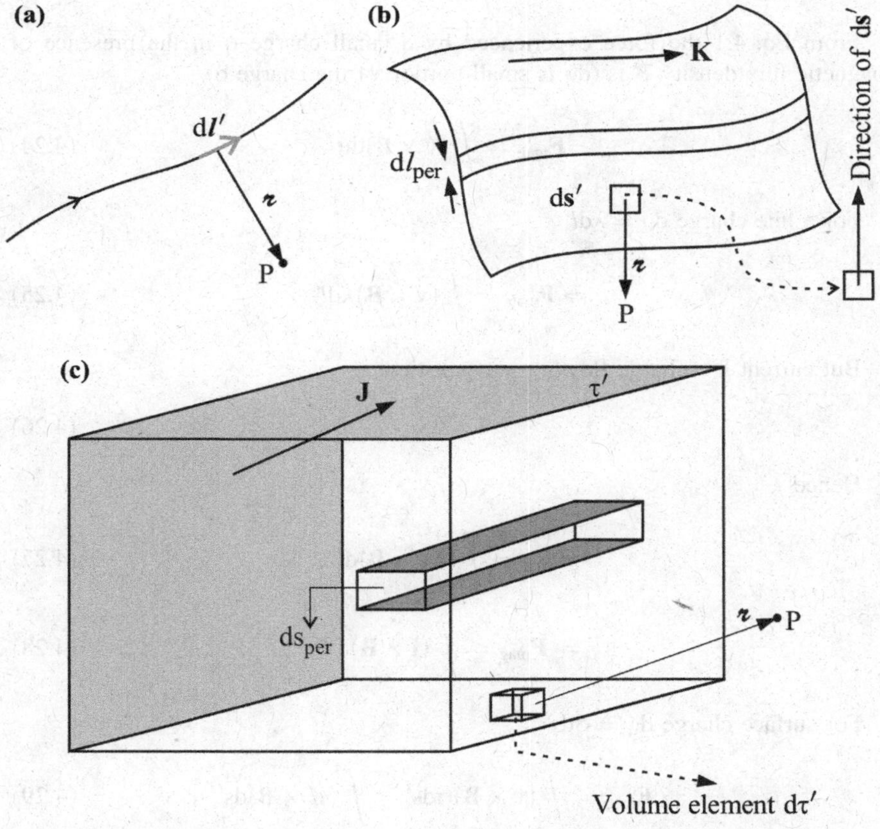


**Fig. 4.8** Figure depicting a) a current flowing along a line b) a current flowing along a surface c) a current flowing along a volume

If the surface charge density of the moving charges in Fig. 4.8b is σ then

$$\mathbf{K} = \sigma\mathbf{v} \tag{4.21}$$

Suppose assume that the charges are moving in a volume as shown Fig. 4.8c with a velocity **v**. The volume current density denoted as **J** is defined as current per unit area perpendicular to the flow

$$\mathbf{J} = \frac{\mathbf{dI}}{ds_{per}} \tag{4.22}$$

where $ds_{per}$ is a small area (perpendicular area to the flow) of a small infinitesimal element running parallel to the flow in Fig. 4.8c, and **dI** is the current in the element.

The volume charge density of the moving charges in Fig. 4.8c is ρ then

$$\mathbf{J} = \rho\mathbf{v} \tag{4.23}$$

From Eq. 4.1 the force experienced by a small charge q in the presence of magnetic flux density **B** is (dq is small portion of the charge q)

$$\mathbf{F}_{mag} = \int (\mathbf{v} \times \mathbf{B})\mathbf{dq} \tag{4.24}$$

For a line charge $dq = \lambda dl'$

$$\Rightarrow \mathbf{F}_{mag} = \int (\mathbf{v} \times \mathbf{B})\lambda dl' \tag{4.25}$$

But current I = charge flowing per unit time

$$\mathbf{I} = \lambda\mathbf{v} \tag{4.26}$$

Hence

$$\mathbf{F}_{mag} = \int (\lambda\mathbf{v} \times \mathbf{B})dl' \tag{4.27}$$

$$\Rightarrow \mathbf{F}_{mag} = \int (\mathbf{I} \times \mathbf{B})dl' \tag{4.28}$$

For surface charge $dq = \sigma ds'$

$$\Rightarrow \mathbf{F}_{mag} = \int (\mathbf{v} \times \mathbf{B})\sigma ds' = \int (\mathbf{K} \times \mathbf{B})ds' \tag{4.29}$$

For volume charge $dq = \rho d\tau'$

$$\mathbf{F}_{mag} = \int (\mathbf{v} \times \mathbf{B}) \rho d\tau' = \int (\mathbf{J} \times \mathbf{B}) d\tau' \tag{4.30}$$

The Biot–Savart law for the line, surface and volume currents as shown in Fig. 4.8a, b, c can be written as

$$\mathbf{B}(\mathbf{r}) = \frac{\mu_o}{4\pi} \int_{l'} \frac{(\mathbf{I} \times \hat{\boldsymbol{\imath}})}{\imath^2} dl' = \frac{\mu_o}{4\pi} I \int \frac{d\mathbf{l}' \times \hat{\boldsymbol{\imath}}}{\imath^2} \tag{4.31}$$

$$\mathbf{B}(\mathbf{r}) = \frac{\mu_o}{4\pi} \int_{S'} \frac{\mathbf{K}(\mathbf{r}') \times \hat{\boldsymbol{\imath}}}{\imath^2} ds' \tag{4.32}$$

$$\mathbf{B}(\mathbf{r}) = \frac{\mu_o}{4\pi} \int_{\tau'} \frac{\mathbf{J}(\mathbf{r}') \times \hat{\boldsymbol{\imath}}}{\imath^2} d\tau' \tag{4.33}$$

In Eq. 4.31 we have used the fact $\mathbf{I}dl' = Id\mathbf{l}'$ In Fig. 4.9a we see that the vectors $\mathbf{I}$ and $d\mathbf{l}'$ point in the same direction. In Sect. 1.1b we have seen that two vectors are equal to each other, only if magnitude and direction of both the vectors are equal. The vector $\mathbf{I}dl'$ points in the direction of $\mathbf{I}$ and $Id\mathbf{l}'$ points in the direction of the $d\mathbf{l}'$. Because as shown in Fig. 4.9a $\mathbf{I}$, $d\mathbf{l}'$ point in same direction the vectors $\mathbf{I}dl'$, $Id\mathbf{l}'$

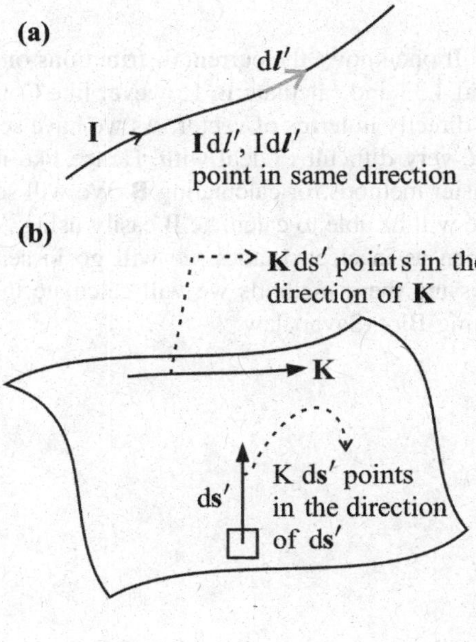


**Fig. 4.9** Figure depicting current elements and their directions for a) a line current density b) a surface current density

have same direction. Thus the magnitude and direction of $\mathbf{I}dl'$ and $\mathrm{I}d\mathbf{l}'$ are same and hence

$$\mathbf{I}dl' = \mathrm{I}d\mathbf{l}' \tag{4.34}$$

and we have used Eq. 4.34 in Eq. 4.31.

Now let us consider Fig. 4.9b. The vector $\mathbf{K}$ in Fig. 4.9b is parallel to the surface and points along the surface. Hence the vector $\mathbf{K}ds'$ having the same direction of $\mathbf{K}$ points along the surface. The vector $d\mathbf{s}'$ is perpendicular to the surface (we have selected upward normal for the surface element $d\mathbf{s}'$ for the sake of explanation). The vector $Kd\mathbf{s}'$ points in the direction of $d\mathbf{s}'$ and is perpendicular to the surface. Thus the vector $\mathbf{K}ds'$ is parallel to the surface, and the vector $\mathrm{K}d\mathbf{s}'$ is perpendicular to the surface. Hence, although vectors $\mathbf{K}ds'$, $\mathrm{K}d\mathbf{s}'$ have same magnitude but they have different directions and hence

$$\mathbf{K}ds' \neq \mathrm{K}d\mathbf{s}' \tag{4.35}$$

This issue doesn't arise with $\mathbf{J}d\tau'$ because $d\tau'$ being volume element is a scalar quantity and doesn't have direction. The vector $\mathbf{J}d\tau'$ points in the direction of $\mathbf{J}$.

If there are collection of currents $I_1$, $I_2$, $I_3$ … $I_n$ each producing its flux density $\mathbf{B}_1$, $\mathbf{B}_2$, $\mathbf{B}_3$ … $\mathbf{B}_n$ then similar to Eq. 2.11 the net magnetic flux density $\mathbf{B}$ is given by

$$\mathbf{B} = \mathbf{B}_1 + \mathbf{B}_2 + \mathbf{B}_3 + \cdots + \mathbf{B}_n \tag{4.36}$$

$$\Rightarrow \mathbf{B} = \sum_{i=1}^{n} \mathbf{B}_i \tag{4.37}$$

If one knows the current distributions one can substitute them in Eqs. 4.31, 4.32 and 4.33 and calculate $\mathbf{B}$. However like Coulomb's law, Biot–Savart law calculates $\mathbf{B}$ directly in terms of vector. As we have seen in Sect. 1.1a, Example 1.10, vectors are very difficult to deal with. Hence like in electrostatics we will go in search of easier methods for calculating $\mathbf{B}$. We will see that like Gauss's law in electrostatics we will be able to calculate $\mathbf{B}$ easily using Ampere's circuital law or Ampere's law. Then as in electrostatics we will go in search of magnetic potential. Before discussing these methods we will calculate the $\mathbf{B}$ for an infinitely long straight wire using Biot–Savart law.

## 4.8 Magnetic Flux Density Due to a Steady Current in a Infinitely Long Straight Wire

Consider an infinitely long straight wire carrying a steady current I as shown in Fig. 4.10. We are interested in calculating the **B** at point P at a distance $\mathscr{r}$ from the element $d\mathbf{l}'$. The point P lies at a distance of r from the wire. Using Biot–Savart law $d\mathbf{B}$, the flux density at point P due to the small element $d\mathbf{l}'$ is

$$d\mathbf{B} = \frac{\mu_o I}{4\,\pi}\left(\frac{d\mathbf{l}' \times \hat{\mathscr{r}}}{\mathscr{r}^2}\right) \tag{4.38}$$

$$\Rightarrow dB = \frac{\mu_o I dl' \sin\theta}{4\pi \quad \mathscr{r}^2} \tag{4.39}$$

The **B** at point P due to the entire wire is given by

$$B = \int dB = \frac{\mu_o I}{4\,\pi} \int_{l'=-\infty}^{l'=\infty} \frac{\sin\theta dl'}{\mathscr{r}^2} \tag{4.40}$$

from Fig. 4.10

$$\mathscr{r} = \left[(l')^2 + r^2\right]^{1/2} \tag{4.41}$$

and

$$\sin\theta = \sin(\pi - \theta) = \frac{r}{\mathscr{r}} = \frac{r}{\left[(l')^2 + r^2\right]^{1/2}} \tag{4.42}$$

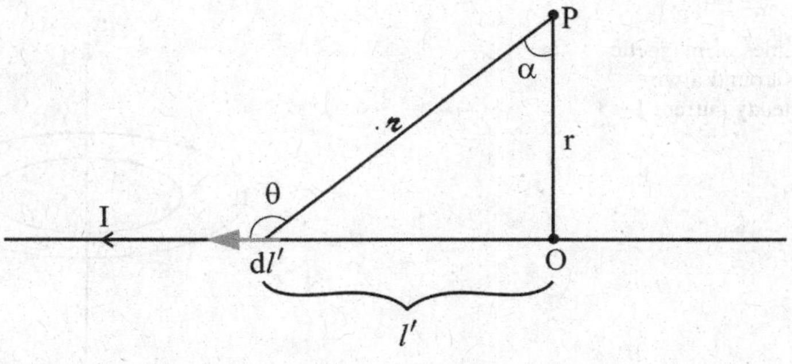


**Fig. 4.10** An infinitely long straight wire carrying a steady current I

Substituting Eqs. 4.42 and 4.41 in Eq. 4.40

$$B = \frac{\mu_o I}{4\pi} \int_{-\infty}^{\infty} \frac{r\, dl'}{\left[r^2 + (l')^2\right]^{3/2}} \tag{4.43}$$

From Fig. 4.10

$$l' = r \tan \alpha \tag{4.44}$$

$$\Rightarrow dl' = r\, \sec^2 \alpha d\alpha \tag{4.45}$$

When $l' = \infty \quad \alpha = \pi/2$
$l' = -\infty \quad \alpha = -\pi/2$

Hence

$$B = \frac{\mu_o I}{4\,\pi\, r} \int_{-\pi/2}^{\pi/2} \cos \alpha\, d\alpha \tag{4.46}$$

$$\Rightarrow B = \frac{\mu_o I}{2\,\pi\, r} \tag{4.47}$$

is the magnetic flux density at point P due to the entire wire.

The direction of **B** is given by right-hand rule as explained in Sect. 4.1. We have sketched the **B** around the wire in Fig. 4.11.

In cylindrical coordinates

$$\mathbf{B} = \frac{\mu_0 I}{2\,\pi\, r} \hat{\phi} \tag{4.48}$$

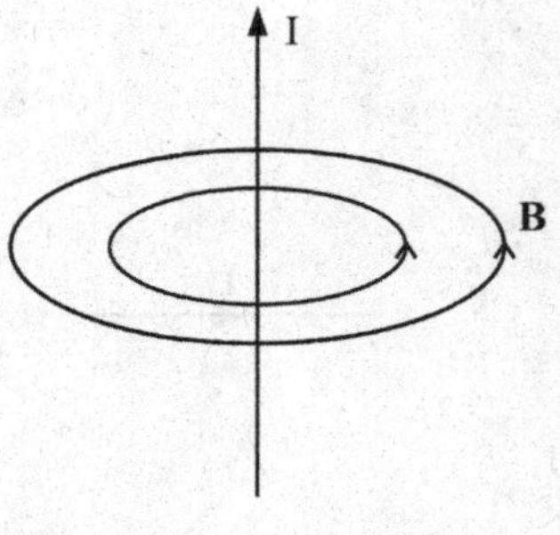


**Fig. 4.11** Lines of magnetic flux density around a wire carrying a steady current I

## 4.9 Ampere's Circuital Law

In this section we will discuss about Ampere's circuital law (or) alternatively known as Ampere's law.

Suppose assume that an infinitely long straight wire carries current as shown in Fig. 4.12a. **B** is sketched in the figure.

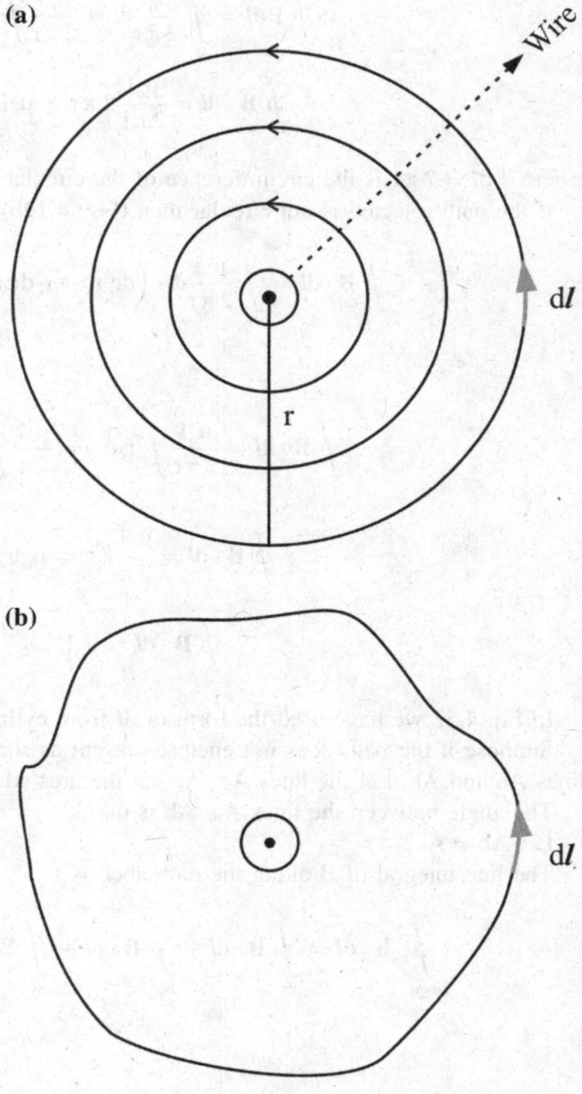


**Fig. 4.12** **a** **B** of a infinitely long straight wire. A circular path has been selected to prove Amperes law. **b** A non - circular path has been selected to prove Amperes law

Let us consider a circular path of radius r and a small element d$\boldsymbol{l}$ as shown in Fig. 4.12a. Let us calculate the line integral of **B** around the circular path. Because **B**, d$l$, point in the same direction along the circular path

$$\oint \mathbf{B}\cdot d\boldsymbol{l} = \oint B dl \tag{4.49}$$

From Eq. 4.47

$$\oint B dl = \oint \frac{\mu_o I}{2\pi r} dl = \frac{\mu_o I}{2\pi r}\oint dl \tag{4.50}$$

$$\oint \mathbf{B}\cdot d\boldsymbol{l} = \frac{\mu_o I}{2\pi r} 2\pi r = \mu_o I \tag{4.51}$$

where $\oint dl = 2\pi r$ is the circumference of the circular path.

If the path selected is not circular then (Fig. 4.12b) from Eq. 4.48

$$\oint \mathbf{B}\cdot d\boldsymbol{l} = \oint \frac{\mu_o I}{2\pi r}\hat{\boldsymbol{\phi}}\cdot\left(dr_c\mathbf{r}_c + r_c d\phi\hat{\boldsymbol{\phi}} + dz\,\hat{\mathbf{z}}\right) \tag{4.52}$$

As $r = r_c$

$$\oint \mathbf{B}\cdot d\boldsymbol{l} = \frac{\mu_o I}{2\pi r}\int_0^{2\pi} r_c d\phi = \frac{\mu_o I}{2\pi}\int_0^{2\pi} d\phi \tag{4.53}$$

$$\oint \mathbf{B}\cdot d\boldsymbol{l} = \frac{\mu_o I}{2\pi} 2\pi = \mu_o I \tag{4.54}$$

$$\oint \mathbf{B}\cdot d\boldsymbol{l} = \mu_o I \tag{4.55}$$

In Eq. 4.52 we have used the form of d$l$ from cylindrical coordinates.

Suppose if the path does not enclose current as shown in Fig. 4.13a draw two lines Aa and Ab. Let the lines Aa, Ab cut the arcs cd and ab on the path.

The angle between the lines Aa, Ab is d$\phi$.

Let Ab = $r_1$, Ad = $r_2$

The line integral of **B** along the path abcd is

$$\oint_{abcd} \mathbf{B}\cdot d\boldsymbol{l} = \int_{ab} \mathbf{B}\cdot d\boldsymbol{l} + \int_{bc} \mathbf{B}\cdot d\boldsymbol{l} + \int_{cd} \mathbf{B}\cdot d\boldsymbol{l} + \int_{da} \mathbf{B}\cdot d\boldsymbol{l} \tag{4.56}$$

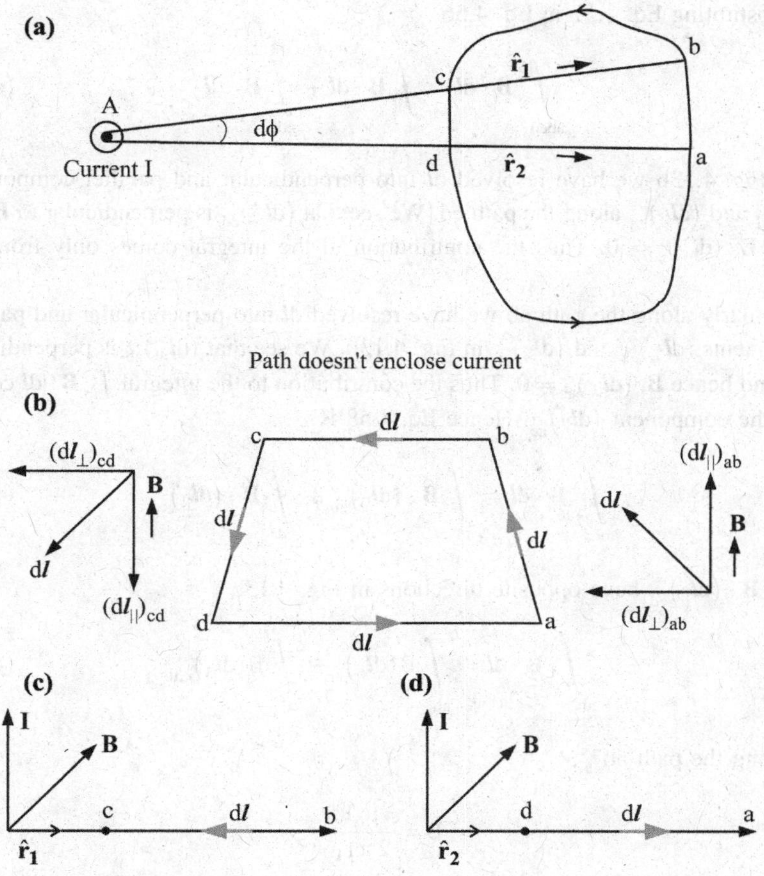


**Fig. 4.13** **a** A non - circular path that doesn't enclose current I. **b, c, d** The path abcd along with various vector components are shown in the figure

The path abcd along with d$\boldsymbol{l}$ is shown in Fig. 4.13b. Because **B** by Biot–Savart law is given by $\mathbf{I} \times \hat{\mathbf{r}}$ the direction of **B** with respect to line bc and da is shown in Fig. 4.13c, d.

Thus the angle between d$\boldsymbol{l}$ and **B** along the path bc and da is 90°. Hence

$$\int_{bc} \mathbf{B} \cdot d\boldsymbol{l} = \int_{da} \mathbf{B} \cdot d\boldsymbol{l} = 0 \tag{4.57}$$

Substituting Eq. 4.57 in Eq. 4.56

$$\int_{abcd} \mathbf{B} \cdot d\boldsymbol{l} = \int_{ab} \mathbf{B} \cdot d\boldsymbol{l} + \int_{cd} \mathbf{B} \cdot d\boldsymbol{l} \tag{4.58}$$

In Fig. 4.13b we have resolved d$\boldsymbol{l}$ into perpendicular and parallel components $(d\boldsymbol{l}_\perp)_{cd}$ and $(d\boldsymbol{l}_{||})_{cd}$ along the path cd. We see that $(d\boldsymbol{l}_\perp)_{cd}$ is perpendicular to **B** and hence **B**. $(d\boldsymbol{l}_\perp)_{cd} = 0$. Thus the contribution to the integral comes only from the $(d\boldsymbol{l}_{||})_{cd}$.

Similarly along the path ab we have resolved d$l$ into perpendicular and parallel components $(d\boldsymbol{l}_\perp)_{ab}$ and $(d\boldsymbol{l}_{||})_{ab}$ in Fig. 4.13b. We see that $(d\boldsymbol{l}_\perp)_{ab}$ is perpendicular to **B** and hence **B**. $(d\boldsymbol{l}_\perp)_{ab} = 0$. Thus the contribution to the integral $\int_{ab} \mathbf{B} \cdot d\boldsymbol{l}$ comes from the component $(d\boldsymbol{l}_{||})_{ab}$. Hence Eq. 4.58 is

$$\int_{abcd} \mathbf{B} \cdot d\boldsymbol{l} = \int_{ab} \mathbf{B} \cdot (d\boldsymbol{l}_{||})_{ab} + \int_{cd} \mathbf{B} \cdot (d\boldsymbol{l}_{||})_{cd} \tag{4.59}$$

As $\mathbf{B} \cdot (d\boldsymbol{l}_{||})_{cd}$ have opposite directions in Fig. 4.13b

$$\int_{abcd} \mathbf{B} \cdot d\boldsymbol{l} = \int_{ab} B(d\boldsymbol{l}_{||})_{ab} - \int_{cd} B(d\boldsymbol{l}_{||})_{cd} \tag{4.60}$$

Along the path ab

$$B = \frac{\mu_o I}{2\pi r_1} \tag{4.61}$$

and at path cd

$$B = \frac{\mu_o I}{2\pi r_2} \tag{4.62}$$

Substituting Eqs. 4.61 and 4.62 in Eq. 4.60

$$\int_{abcd} \mathbf{B} \cdot dl = \int_{ab} \frac{\mu_o I}{2\pi r_1}(dl_{||})_{ab} - \int_{cd} \frac{\mu_o I}{2\pi r_2}(dl_{||})_{cd} \tag{4.63}$$

$$\Rightarrow \int_{abcd} \mathbf{B} \cdot dl = \int_{ab} \frac{\mu_o I}{2\pi} \frac{(dl_{||})_{ab}}{r_1} - \int_{cd} \frac{\mu_o I}{2\pi} \frac{(dl_{||})_{cd}}{r_2} \tag{4.64}$$

From Fig. 4.13a

$$\frac{(dl_{||})_{ab}}{r_1} = \frac{(dl_{||})_{cd}}{r_2} = d\phi \tag{4.65}$$

Hence

$$\int_{abcd} \mathbf{B} \cdot dl = \int \frac{\mu_o I}{2\pi} d\phi - \int \frac{\mu_o I}{2\pi} d\phi \tag{4.66}$$

$$\int_{abcd} \mathbf{B} \cdot dl = 0 \tag{4.67}$$

Thus Ampere's law can be summarized as

$$\oint \mathbf{B} \cdot dl = \begin{cases} \mu_o I & \text{if the path encloses current} \\ 0 & \text{if the path does not enclose current} \end{cases} \tag{4.68}$$

Suppose as shown in Fig. 4.14 if the path encloses large number of currents then

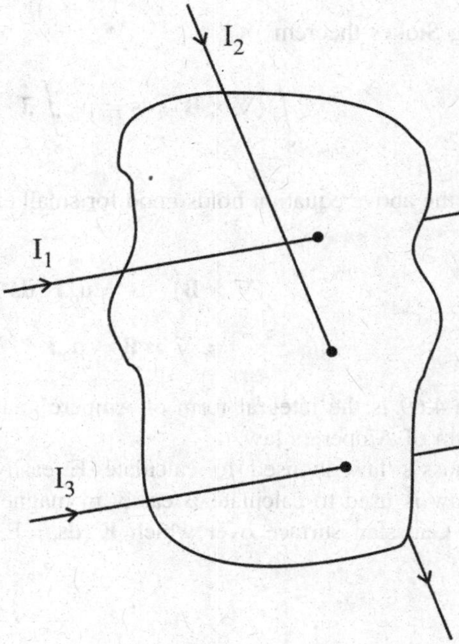


**Fig. 4.14** A non - circular path enclosing number of currents

$$\oint \mathbf{B} \cdot \mathrm{d} = \mu_o \mathbf{I}_{enc} \tag{4.69}$$

where $I_{enc}$ is the total current enclosed by the entire path.

From Eq. 4.22

$$\Rightarrow \mathrm{dI}_{enc} = \mathbf{J}\mathrm{ds}_{per} \tag{4.70}$$

$$\Rightarrow \mathrm{I}_{enc} = \int \mathbf{J}\mathrm{ds}_{per} \tag{4.71}$$

From Sect. 1.12 $\mathrm{ds}_{per} = \mathrm{ds}\cos\theta$.

Hence

$$\mathrm{I}_{enc} = \int \mathbf{J}\,\mathrm{ds}\cos\theta \tag{4.72}$$

$$\mathrm{I}_{enc} = \int \mathbf{J} \cdot \mathbf{ds} \tag{4.73}$$

Substituting Eq. 4.73 in 4.69

$$\oint \mathbf{B} \cdot \mathrm{d}\boldsymbol{l} = \mu_o \int \mathbf{J} \cdot \mathrm{ds} \tag{4.74}$$

Applying Stokes theorem

$$\int_S (\nabla \times \mathbf{B}) \cdot \mathbf{ds} = \mu_o \int_S \mathbf{J} \cdot \mathrm{ds} \tag{4.75}$$

Because the above equation holds good for small element **ds** the integration can be dropped

$$(\nabla \times \mathbf{B}) \cdot \mathrm{d}s = \mu_o \mathbf{J} \cdot \mathbf{ds} \tag{4.76}$$

$$\Rightarrow \nabla \times \mathbf{B} = \mu_o \mathbf{J} \tag{4.77}$$

Equation 4.69 is the integral form of Ampere's law while Eq. 4.77 is the differential form of Ampere's law.

Like Gauss's law is used to calculate **E** easily in electrostatics similarly Ampere's law is used to calculate **B** easily in magnetostatics. In Gauss's law we construct a Gaussian surface over which $\mathbf{E} \cdot \mathrm{ds} = \mathrm{E}\,\mathrm{ds}\cos\theta$ is evaluated easily.

Once this has been done then after ensuring $E\cos\theta$ is constant over the Gaussian surface we take $E\cos\theta$ out of the integral $\oint \mathbf{E}\cdot d\mathbf{s}$ and finally calculate E from the equation

$$\oint \mathbf{E}\cdot d\mathbf{s} = \frac{q_{enc}}{\varepsilon_o}$$

Similarly in Ampere's law we construct a Amperian loop such that over the Amperian loop $\mathbf{B}\cdot d\mathbf{l} = B\,dl\cos\theta$ is easily evaluated. Also we select the Amperian loop such that over the Amperian loop $B\cos\theta$ is constant so that $B\cos\theta$ can be taken out of the integral

$$\oint \mathbf{B}\cdot d\mathbf{l} = \oint B\,dl\cos\theta = B\cos\theta \oint dl \tag{4.78}$$

Once this has been done then B can be easily calculated using the integral

$$\oint \mathbf{B}\cdot d\mathbf{l} = \mu_o I_{enc} \tag{4.79}$$

$$\Rightarrow B\cos\theta \oint dl = \mu_o I_{enc} \tag{4.80}$$

$$\Rightarrow B = \frac{\mu_o I_{enc}}{\cos\theta \oint dl} \tag{4.81}$$

In electrostatics we found that Gauss's law is the easiest method, and successful application of Gauss's law requires $\mathbf{E}\cdot d\mathbf{s}$ to be easily evaluated over the Gaussian surface and $E\cos\theta$ to be constant over the Gaussian surface so that $E\cos\theta$ can be pulled out of the integral.

For the given problem when the above conditions are not satisfied (e.g. see Sects. 2.14 and 2.15) we went in search of new methods like potential formulation to calculate the electric field. Similarly in magnetostatics Ampere's law is the easiest method, and successful application of Ampere's law requires $\mathbf{B}\cdot d\mathbf{l}$ to be easily evaluated and $B\cos\theta$ to be constant over the Amperian loop so that $B\cos\theta$ can be pulled out of the integral. For the given problem when the above conditions are not satisfied it with be still true that $\oint \mathbf{B}\cdot d\mathbf{l} = \mu_o I_{enc}$ but the quantity which we require B is inside the integral and in such situations Ampere's law cannot be easily used to calculate **B** (see Sect. 4.21). So we will go in search of new methods like potential formulation to calculate **B**.

But before that we will calculate the divergence of **B** from Biot–Savart law directly.

## 4.10 Equation of Continuity

In a closed system it has been experimentally verified that the net amount of electric charge remains constant. Thus in a certain region, if the net charge decreases with time it implies that same quantity of charge appears in some other region.

Suppose assume that a quantity of charge located within volume τ decreases with time. The transport of charge constitutes a current

$$I = \frac{-dq}{dt} \tag{4.82}$$

Negative sign indicates that charge contained in the specified volume decreases with time. From Eq. 4.73

$$I = \oint \mathbf{J} \cdot \mathbf{ds} \tag{4.83}$$

Let S be the surface bounded by volume τ. As charge is conserved whatever charge is flowing out of surface S must come from the volume τ. Hence from Eqs. 4.82 to 4.83

$$\oint_S \mathbf{J} \cdot \mathbf{ds} = \frac{-dq}{dt} \tag{4.84}$$

Applying Gauss's divergence theorem and noting down that $q = \int_\tau \rho \, d\tau$

$$\int_\tau \nabla \cdot \mathbf{J} d\tau = \frac{-d}{dt} \int_\tau \rho \, d\tau \tag{4.85}$$

$$\Rightarrow \int_\tau \nabla \cdot \mathbf{J} \, d\tau = - \int_\tau \frac{\partial \rho}{\partial t} d\tau \tag{4.86}$$

$$\Rightarrow \int_\tau \left( \nabla \cdot \mathbf{J} + \frac{\partial \rho}{\partial t} d\tau \right) = 0 \tag{4.87}$$

The above equation must hold good for a small volume dτ and hence dropping the integral

$$\left( \nabla \cdot \mathbf{J} + \frac{\partial \rho}{\partial t} \right) d\tau = 0 \tag{4.88}$$

As $d\tau \neq 0$

$$\nabla \cdot \mathbf{J} + \frac{\partial \rho}{\partial t} = 0 \tag{4.89}$$

The above equation is known as equation of continuity and is a theoretical expression of the experimental fact that charge is conserved.

In magnetostatics we deal with steady currents. Under steady currents charge doesn't pile up anywhere and hence charge density is constant with time leading to $\frac{\partial \rho}{\partial t} = 0$.

Hence from equation of continuity for magnetostatics

$$\nabla \cdot \mathbf{J} = 0$$

That is divergence of **J** for steady current is zero.

## 4.11 The Divergence of B

In this section we will derive an expression for divergence of **B**. That is we will calculate $\nabla \cdot \mathbf{B}$. A volume current density **J** (**r**') along with a volume element is shown in Fig. 4.15.

The Biot–Savart law for volume current density **J** is given by Eq. 4.23

$$\mathbf{B}(\mathbf{r}) = \frac{\mu_0}{4\pi} \int_\tau \frac{\mathbf{J}(\mathbf{r}') \times \hat{\mathscr{r}}}{\mathscr{r}^2} d\tau' \tag{4.90}$$

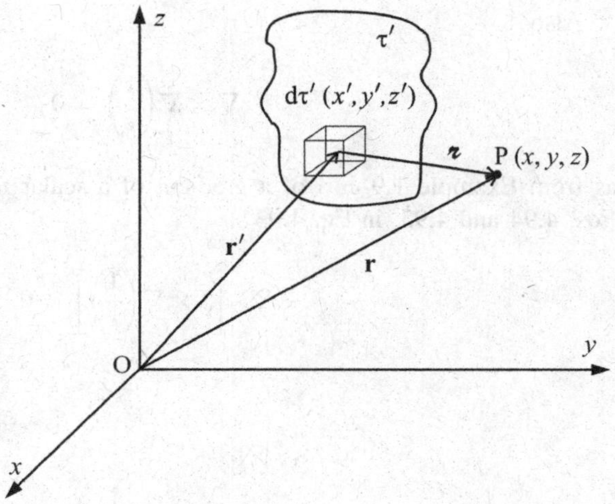


**Fig. 4.15** A volume current density J(r') along with a volume element in Cartesian coordinate system

The above expression gives the **B** at point P(*x*, *y*, *z*), i.e. **B** (**r**) is a function of (*x*, *y*, *z*). Here (*x*, *y*, *z*) are called field coordinates (see Sect. 2.4). As we are interested in calculating $\nabla \cdot \mathbf{B}$ the divergence must be carried out with respect to (*x*, *y*, *z*) the **unprimed field coordinates.**

The volume $\tau'$ contains the current density and hence is a function of $(x', y', z')$. Thus the integration in Eq. 4.90 must be carried out with respect to primed coordinates $(x', y', z')$. Here $(x', y', z')$ coordinates are the **primed source coordinates.** Keeping the above points in the mind about "unprimed field coordinates (*x*, *y*, *z*)" and "primed source coordinates $(x', y', z')$" we will proceed to calculate $\nabla \cdot \mathbf{B}$.

Using the results of Example 1.14 Eq. 4.90 can be written as

$$\mathbf{B} = \frac{\mu_o}{4\pi} \int_{\tau'} \mathbf{J} \times \left[ -\nabla \left( \frac{1}{\imath} \right) \right] d\tau' \tag{4.91}$$

Thus

$$\nabla \cdot \mathbf{B} = \frac{\mu_o}{4\pi} \int_{\tau'} \nabla \cdot \left[ -\mathbf{J} \times \nabla \left( \frac{1}{\imath} \right) \right] d\tau' \tag{4.92}$$

From Sect. 1.20 using the vector identity 8 with $\mathbf{A} = \mathbf{J}$ and $\mathbf{B} = \nabla \left(\frac{1}{\imath}\right)$

$$\nabla \cdot \left[ \mathbf{J} \times \nabla \left( \frac{1}{\imath} \right) \right] = \nabla \left( \frac{1}{\imath} \right) \cdot (\nabla \times \mathbf{J}) - \mathbf{J} \cdot \left[ \nabla \times \left[ \nabla \left( \frac{1}{\imath} \right) \right] \right] \tag{4.93}$$

But we have seen that **J** is a function of $(x', y', z')$ and in curl $\nabla \times \mathbf{J}$, $\nabla$ is carried out in unprimed coordinates *x*, *y*, *z*. As **J** is not a function of (*x*, *y*, *z*) we then have

$$\nabla \times \mathbf{J} = 0 \tag{4.94}$$

Also

$$\nabla \times \nabla \left( \frac{1}{\imath} \right) = 0 \tag{4.95}$$

as from Example 1.9 curl of a gradient of a scalar function is zero. Substituting Eqs. 4.94 and 4.95, in Eq. 4.93

$$\nabla \cdot \left[ \mathbf{J} \times \nabla \left( \frac{1}{\imath} \right) \right] = 0 \tag{4.96}$$

and hence from Eq. 4.92

$$\nabla \cdot \mathbf{B} = \frac{\mu_o}{4\pi} \int_{\tau'} \nabla \cdot \left[ -\mathbf{J} \times \nabla \left( \frac{1}{\mathscr{r}} \right) \right] d\tau' = 0 \tag{4.97}$$

$$\nabla \cdot \mathbf{B} = 0 \tag{4.98}$$

And from Eq. 4.77 we know that

$$\nabla \times \mathbf{B}(\mathbf{r}) = \mu_o \mathbf{J}(\mathbf{r}') \tag{4.99}$$

We have specifically noted that **B** is the function of unprimed field coordinate **r** and **J** is the function of primed source coordinate **r**′.

## 4.12 Magnetic Monopoles

(For better understanding reader is advised to go through Sects. 1.14 and 1.15 before continuing the section)

Thus far we have derived four important expressions from electrostatics and magnetostatics.

(i) $$\nabla \cdot \mathbf{E} = \frac{\rho}{\varepsilon_o} \tag{4.100a}$$

(ii) $$\nabla \times \mathbf{E} = 0 \tag{4.100b}$$

(iii) $$\nabla \cdot \mathbf{B} = 0 \tag{4.100c}$$

(iv) $$\nabla \times \mathbf{B} = \mu_o \mathbf{J} \tag{4.100d}$$

Recall the physical significance of divergence and curl from Sects. 1.14 and 1.15 . Divergence is the measure of how much the given vector diverges or spreads out from the point in question. Curl is the measure of how much the vector rotates about the given point in question.

From electrostatics we see that $\nabla \cdot \mathbf{E} = \frac{\rho}{\varepsilon_o}$ and $\nabla \times \mathbf{E} = 0$, i.e. in the presence of static charges $\nabla \cdot \mathbf{E} \neq 0, \nabla \times \mathbf{E} = 0$.

$\nabla \cdot \mathbf{E} \neq 0$ implies that the vector **E** diverges (or) spreads out and $\nabla \times \mathbf{E} = 0$ implies that the electrostatic field **E** doesn't curl around.

"Thus the electrostatic field **E** diverges $(\nabla \cdot \mathbf{E} \neq 0)$ and doesn't curl around $(\nabla \times \mathbf{E} = 0)$". This has been shown pictorially in Fig. 4.16a. From point charge q the electrostatic field **E** diverges or spreads out without any curl.

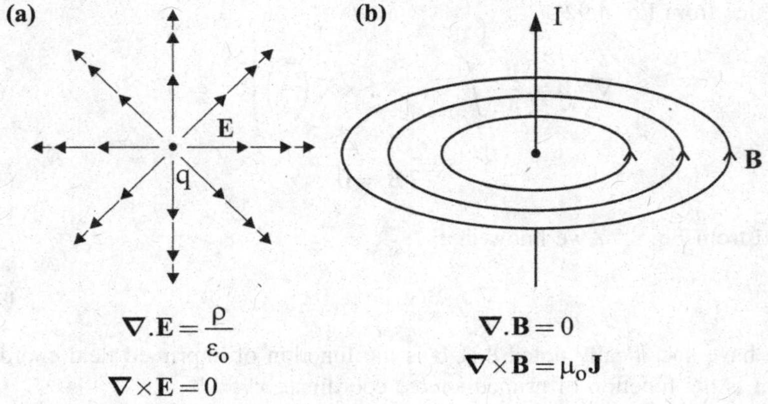


**Fig. 4.16** **a** Under electrostatic conditions the electric field diverges without any curl. **b**Under magnetostatic conditions the magnetic flux density curls around without any divergence

From magnetostatics we see that $\nabla \cdot \mathbf{B} = 0$ and $\nabla \times \mathbf{B} = \mu_0 \mathbf{J}$, i.e. in the presence of steady currents $\nabla \cdot \mathbf{B} = 0$ and $\nabla \times \mathbf{B} \neq 0 \cdot \nabla \cdot \mathbf{B} = 0$ implies that the vector **B** doesn't diverge or spread out. $\nabla \times \mathbf{B} \neq 0$ implies that the magnetic flux density **B** curls around (or) rotates around".

"Thus the **B** curls around (or) rotates around ($\nabla \times \mathbf{B} \neq 0$) and doesn't diverge $\nabla \cdot \mathbf{B} = 0$". This has been shown pictorially in 4.16b. From the steady current I the **B** curls around without any divergence.

"Thus from four Eqs. 4.100a–4.100d we get an interesting result. Nature has created **E** and **B** such that **E** diverges and doesn't curl around while **B** curls around and doesn't diverge". The above result is shown pictorially in Fig. 4.16.

Assume a vector **G** (say). In Sect. 1.14 we saw that for a given vector **G** if $\nabla \cdot \mathbf{G} > 0$ it means that there is a source from which the vector **G** emanates from, $\nabla \cdot \mathbf{G} < 0$ means that there is a sink into which the vector **G** flows in and if $\nabla \cdot \mathbf{G} = 0$ then **G** doesn't have source or sink.

Now considering electrostatic and magnetostatic cases we see that for **E**, **B**

$$\nabla \cdot \mathbf{E} = \frac{\rho}{\varepsilon_0} \Rightarrow \nabla \cdot \mathbf{E} \neq 0 \text{ and}$$

$$\nabla \cdot \mathbf{B} = 0$$

$\nabla \cdot \mathbf{E} \neq 0$ implies that there is a source or sink for **E**. For a positive charge $\nabla \cdot \mathbf{E} > 0$.

It means that a positive charge q is acting like a source from which the **E** emanates from. This is shown in Fig. 4.17a. For a negative charge $\nabla \cdot \mathbf{B} = 0$. It means that a negative charge −q is acting like a sink into which **E** flows in. This is

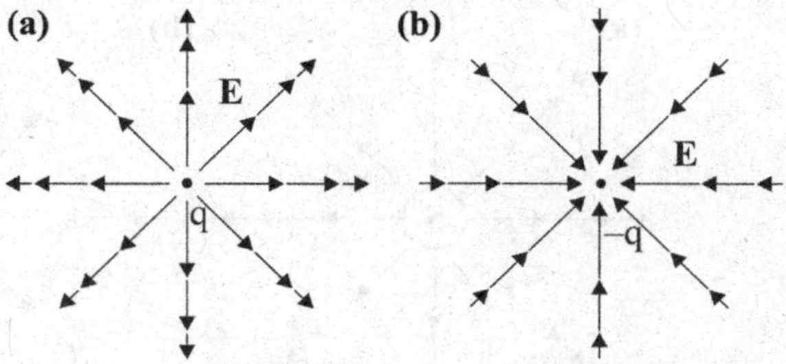


**Fig. 4.17** **a** Point charge q acting like a source for electric field vector. **b** Point charge -q acting like a sink for electric field vector

shown pictorially in Fig. 4.17b (see Figs. 1.22 and 1.23 and discussion therein). Just pipe at point p, q in Fig. 1.23a, b acts as source and sink for water, similarly charges q, −q in Fig. 4.17a, b act as source and sink for **E**. "Thus in conclusion $\nabla \cdot \mathbf{E} \neq 0$ provides us with a very interesting result that "electric monopoles q(−q)" exist in nature from which (into which) the **E** emanates from (flows in)".

Considering equation $\nabla \cdot \mathbf{B} = 0$ there is no source or sink from which or into which the vector **B** emanates from or flows into (see Figs. 1.22 and 1.23 and discussion therein).

If we see Fig. 4.16b there is no starting point (or) ending point for **B**. **B** curls around I without any beginning or end point. The vectorial fields with no starting point or ending point are called **solenoidal fields.**

Because there is no source or sink for **B** from which or into which the vector **B** emanates from or flows into, there are no magnetic monopoles.

"Thus in conclusion the equation $\nabla \cdot \mathbf{B} = 0$ provides us with a very interesting result-magnetic monopoles doesn't exist in nature".

Take a bar magnet as shown in 4.18a with north pole and south pole. In an attempt to create magnetic monopoles if the magnet is broken then we get two bar magnets as shown in Fig. 4.18b instead of isolated north and south poles. Thus magnetic monopoles shown in Fig. 4.19 do not exist in nature. It is the circulation of electric charges which produces **B** and not the magnetic monopoles.

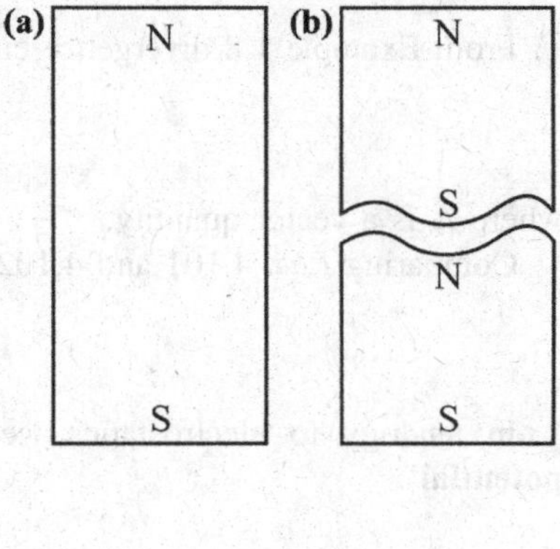


**Fig. 4.18** **a** A bar magnet with north and south poles. **b** A broken bar magnet results in two bar magnets with no isolated north and south poles

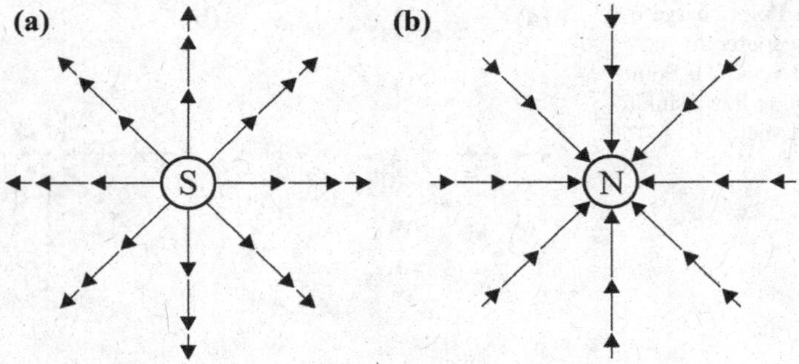


**Fig. 4.19** **a** A south pole. **b** A north pole

## 4.13 Magnetic Vector Potential

We hinted in Sect. 4.9 that Ampere's law cannot be always used to calculate **B**. When Ampere's circuital law cannot be used to calculate **B** then we need alternate methods to calculate **B**. In electrostatics we saw that when Gauss's law cannot be used to calculate **E** then we can use potential V to calculate $\mathbf{E}[\mathbf{E} = -\nabla V]$, instead of using Coulomb's law to calculate **E** because Coulomb's law calculates **E** directly in vectorial form involving direction. V being a scalar quantity doesn't involve direction.

Similarly in magnetostatics when Ampere's circuital law is not useful in calculating **B** then we are left with Biot–Savart law to calculate **B** which calculates **B** directly in vectorial form involving direction. Hence we require an easy method to calculate **B** without using Biot–Savart law when Ampere's law is not useful.

Using $\nabla \times \mathbf{E} = 0$ in electrostatics we defined an electric potential $\mathbf{E} = -\nabla V$ (see Sect. 2.11).

Similarly in magnetostatics

$$\nabla \cdot \mathbf{B} = 0 \tag{4.101}$$

From Example 1.8 divergence curl of any vector is zero

$$\nabla \cdot (\nabla \times \mathbf{A}) = 0 \tag{4.102}$$

where **A** is a vector quantity.

Comparing Eqs. 4.101 and 4.102

$$\mathbf{B} = \nabla \times \mathbf{A} \tag{4.103}$$

In analogy to electrostatics we call **A** in Eq. 4.103 as "**magnetic vector potential**".

Equation 4.91 is

$$\mathbf{B} = \frac{\mu_o}{4\pi}\int_{\tau'} \mathbf{J} \times \left[-\nabla\left(\frac{1}{\mathscr{r}}\right)\right] d\tau' \tag{4.104}$$

From Sect. 1.20 using vector identity 6 we see

$$\nabla \times \left(\frac{\mathbf{J}}{\mathscr{r}}\right) = \frac{1}{\mathscr{r}}\nabla \times \mathbf{J} + \nabla\left(\frac{1}{\mathscr{r}}\right) \times \mathbf{J} \tag{4.105}$$

However as we previously noted the curl operation is being carried out with respect to field coordinates $(x, y, z)$ while $\mathbf{J}$ depends on source coordinates $(x', y', z')$.

Hence

$$\nabla \times \mathbf{J} = 0 \tag{4.106}$$

Substituting Eq. 4.106 in 4.105

$$\nabla \times \left[\frac{\mathbf{J}}{\mathscr{r}}\right] = \nabla\left(\frac{1}{\mathscr{r}}\right) \times \mathbf{J} \tag{4.107}$$

$$\Rightarrow -\left[\mathbf{J} \times \nabla\left[\frac{1}{\mathscr{r}}\right]\right] = \nabla \times \left[\frac{\mathbf{J}}{\mathscr{r}}\right] \tag{4.108}$$

Substituting Eq. 4.108 in 4.103

$$\mathbf{B} = \frac{\mu_o}{4\pi}\int_{\tau'} \nabla \times \left[\frac{\mathbf{J}}{\mathscr{r}}\right] d\tau' \tag{4.109}$$

Because the integration is carried out with respect to source coordinates $(x', y', z')$ and curl is being carried out with respect to field coordinates $(x, y, z)$ both the operations can be interchanged.

$$\mathbf{B} = \nabla \times \left[\frac{\mu_o}{4\pi}\int \frac{\mathbf{J}d\tau'}{\mathscr{r}}\right] d\tau' \tag{4.110}$$

Comparing Eqs. 4.103 and 4.110

$$\mathbf{A} = \frac{\mu_o}{4\pi}\int_{\tau'} \frac{\mathbf{J}d\tau'}{\mathscr{r}} \tag{4.111}$$

For line and surface currents

$$\mathbf{A} = \frac{\mu_o}{4\pi}\int_{l'} \frac{\mathbf{I}dl'}{\mathscr{r}} = \frac{\mu_o}{4\pi}\int_{l'} \frac{\mathbf{I}d\mathbf{l'}}{\mathscr{r}} \tag{4.112}$$

and

$$\mathbf{A} = \frac{\mu_o}{4\pi}\int_{S'} \frac{\mathbf{K}d\mathit{hboxs}'}{\mathscr{r}} \tag{4.113}$$

**A**, the magnetic vector potential is not as much useful as electric potential V in electrostatics. Because **A** is still a vector we need to remain confused about components and directions. However the integration involved in Eqs. 4.101–4.103 is comparatively easier than the integrations in Biot–Savart law (4.31–4.33). Thus if we are interested in calculating **B**, first we use Eqs. 4.101–4.103 to calculate **A**, then substitute **A** in Eq. 4.103 to calculate **B**.

One of the interesting features of **A** is that it satisfies the Poisson's equation. Equation 4.77 is

$$\nabla \times \mathbf{B} = \mu_o\mathbf{J}$$

Substituting Eq. 4.103 in the above equation

$$\nabla \times \nabla \times \mathbf{A} = \mu_o\mathbf{J} \tag{4.114}$$

From Sect. 1.20 using vector identity 7 we see that

$$\nabla(\nabla \cdot \mathbf{A}) - \nabla^2\mathbf{A} = \mu_o\mathbf{J} \tag{4.115}$$

We have seen in electrostatics that potential V doesn't have significance, only the electric field **E** has significance. The above fact we have seen in detail in Sect. 2.22 and Eqs. 2.285 and 2.286.

For example altering the reference point in Fig. 2.55, Eqs. 2.285 and 2.286, changes the potential V, but doesn't change the electric field **E**. So the actual quantity of physical significance is **E** and not the potential V.

Similarly in magnetostatics the actual quantity of physical significance is **B** not the vector potential **A**. The value of **A** can change without altering the value of **B** as explained below. From Eq. 4.103

$$\mathbf{B} = \nabla \times \mathbf{A} \tag{4.116}$$

Because curl gradient of any scalar function is zero (Example 1.9) the above equation can be written as

$$\mathbf{B} = \nabla \times \mathbf{A} + \nabla \times \nabla u \tag{4.117}$$

where u is a scalar function. That is by adding curl gradient of scalar function u to **B**, **B** remains unchanged.

Equation 4.116 is

$$\mathbf{B} = \nabla \times [\mathbf{A} + \nabla u] \tag{4.118}$$

Let

$$\mathbf{A}_1 = \mathbf{A} + \nabla u \tag{4.119}$$

Hence

$$\mathbf{B} = \nabla \times \mathbf{A}_1 \tag{4.120}$$

Thus adding $\nabla \times \nabla u$ to **B** in Eqs. 4.116 and 4.117 doesn't alter **B** because $\nabla \times \nabla u = 0$ but alters **A** as given by Eq. 4.119. This freedom can be exploited to make divergence of **A** to be zero. That is we keep on adding any number of $\nabla$u's to **A** which will change the value of **A** such that

$$\nabla \cdot \mathbf{A} = 0 \tag{4.121}$$

Or in other words we can select the value of **A** such that its divergence is zero by adding any number of $\nabla$u's, without altering the value of **B**. Substituting Eq. 4.121 in 4.115

$$\nabla^2 \mathbf{A} = -\mu_0 \mathbf{J} \tag{4.122}$$

Comparing the above equation with its electrostatic counterpart Poisson's equation

$$\nabla^2 V = \frac{-\rho}{\varepsilon_o} \tag{4.123}$$

We see that Eq. 4.122 is the Poisson's equation in magnetostatics and **A** satisfies Poisson's equation.

Equation 4.121 is called the Coulomb gauge.

## 4.14 Magnetic Scalar Potential

As we have noted in Sect. 4.13, **A**, the magnetic vector potential is still a vector. We have already seen in Chaps. 1 and 2 that scalars are very easy to work with. So instead of magnetic vector potential **A**, it will be useful if we are able to find a magnetic scalar potential $V_m$. Because the actual quantity which we are interested in calculating is **B**, which happens to be a vector, it will be easy to calculate $V_m$ first because it is a scalar and then calculate **B**.

We know from Eq. 4.77 Ampere's law is

$$\nabla \times \mathbf{B} = \mu_o \mathbf{J}$$

In Fig. 4.8a, b, c, at points like P there is no current. In Fig. 4.8a current is restricted to a line, in Fig. 4.8b current is restricted to a plane, and in Fig. 4.8c current is restricted to a volume. Hence at point P in Fig. 4.8a, b, c, there is no current. Therefore at point P, the current-free region $\mathbf{J} = 0$. Hence in region where there is no current, Ampere's law is

$$\nabla \times \mathbf{B} = 0 \tag{4.124}$$

From Example 1.9 curl of a gradient of a scalar function is zero and hence we can define

$$\Rightarrow \nabla \times (-\mu_o \nabla V_m) = 0 \tag{4.125}$$

Comparing Eqs. 4.124 and 4.125 we get

$$\mathbf{B} = -\mu_o \nabla V_m \tag{4.126}$$

for current-free regions.

An important note must be made about $V_m$ the magnetic scalar potential. It is not a single-valued function like electric scalar potential V. Once the zero of the reference point is fixed electric scalar potential V has a unique or single value at a given point. To make the point clear let us calculate the magnetic scalar potential $V_m$ of the current-carrying infinitely long straight wire of Sect. 4.8.

Integrating Eq. 4.126

$$V_m = \frac{-1}{\mu_o} \int \mathbf{B} \cdot d\boldsymbol{l} \tag{4.127}$$

With

$$d\boldsymbol{l} = dr_c \hat{\mathbf{r}}_c + r_c d\phi \hat{\boldsymbol{\phi}} + dz \hat{\mathbf{z}}$$

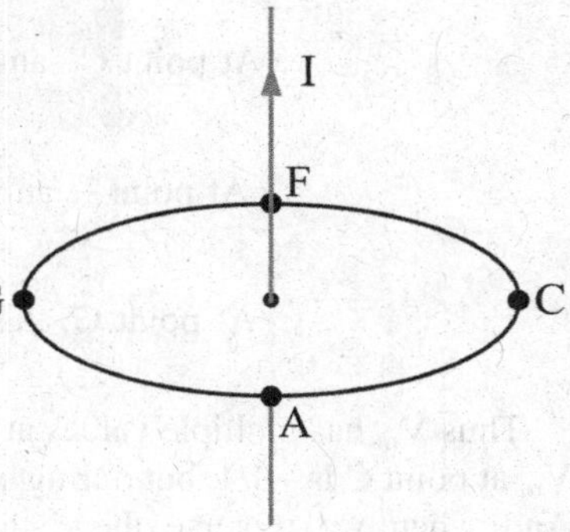


**Fig. 4.20** A circular path ACFG around a infinitely long straight wire carrying current I

in cylindrical coordinates and from Eq. 4.48 we can write

$$V_m = \frac{-1}{\mu_o}\int \frac{\mu_o I}{2\pi r}\hat{\boldsymbol{\phi}} \cdot \left(dr_c \hat{\mathbf{r}}_c + rd\phi\, \hat{\boldsymbol{\phi}} + dz\hat{\mathbf{z}}\right) \tag{4.128}$$

$$\Rightarrow V_m = \frac{-I}{2\pi}\int d\phi \tag{4.129}$$

$$\Rightarrow V_m = \frac{-I}{2\pi}\phi \tag{4.130}$$

Suppose assume that we traverse a circular path ACFG around the long straight wire as shown in Fig. 4.20.

During the first complete round or when we traverse path ACFG for the first time starting from point A

$$\text{At point A angle } \phi = 0, \text{ hence } V_{m.} = 0$$

$$\text{At point C angle } \phi = \pi/2, \text{ hence } V_m = \frac{-I}{4}$$

$$\text{At point F, angle } \phi = \pi, \text{ hence } V_m = \frac{-I}{2}$$

$$\text{At point G angle } \phi = 3\pi/2, \text{ hence } V_m = \frac{-3I}{4}$$

After completing the first round, let us traverse ACFG for the second time, i.e. path ACFG is traversed for the second time

$$\text{At point A, angle } \phi = 2\pi, \text{ hence } V_m = I$$

$$\text{At point C, angle } \phi = \frac{5\pi}{2}, \text{hence } V_m = -\frac{5I}{4}$$

$$\text{At point F, angle } \phi = 3\pi, \text{hence } V_m = -\frac{3I}{2}$$

$$\text{At point G, angle } \phi = \frac{7\pi}{2}, \text{hence } V_m = -\frac{7I}{4}$$

Thus $V_m$ has multiple values at a given point. During the first round the value of $V_m$ at point C is $-I/4$, but during second round the value of $V_m$ at point C is $-5I/4$. And when we traverse the path for third time the value of $V_m$ will change accordingly, and for each time we traverse the path the value of $V_m$ changes at a given point. The reason for multi-valuedness of $V_m$ can be explained as below.

In the case of electrostatics $\nabla \times \mathbf{E} = 0$ and $\oint \mathbf{E} \cdot d\boldsymbol{l} = 0$ (Eq. 2.280). The integral

$$V_a - V_b = -\int_a^b \mathbf{E} \cdot d\boldsymbol{l}$$

where $V_a$ and $V_b$ are potentials at points a and b are independent of the path (Sect. 2.21). But in magnetostatics $\nabla \times \mathbf{B} = 0$ (For current-free regions) and $\oint \mathbf{B} \cdot d\boldsymbol{l} = \mu_o I$ even if the path we selected is free of current. At the path ACFG in Fig. 4.20 there is no current, but still as we traverse the path for the second time the result of integration increases by I. For example at point C during the first round the potential is $-I/4$ and during the second round the potential is $-5I/4$. The difference in the potential is $-\frac{I}{4} + \frac{5I}{4} = I$. Similarly the difference between the potentials at point G is $\frac{-3I}{4} + \frac{7I}{4} = I$. Hence each time we traverse the path the result of integration increases by I. This results in multiple values of $V_m$ at a given point.

Still we can manage to get a single value of $V_m$ at a given point by restricting the path. For example in Fig. 4.21 if we traverse the circular path from A to $A_1$ along

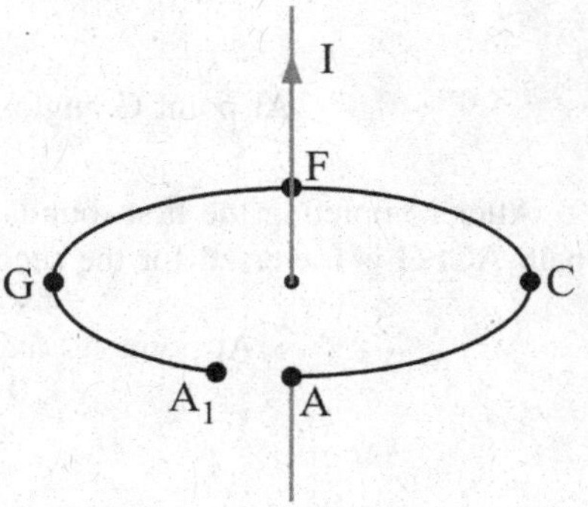


**Fig. 4.21** A circular path around a infinitely long straight wire carrying current I which doesn't enclose the current

the path ACFG $A_1$ then we do not enclose current I and hence we get a single value of $V_m$.

The magnetic potential difference between any two points a and b is given by (similar to Eq. 2.284 and from Eq. 4.127)

$$(V_m)_a - (V_m)_b = \frac{1}{\mu_o}\int_a^b \mathbf{B}\cdot d\mathbf{l} \tag{4.131}$$

## 4.15 Comments on Magnetic Vector Potential A and Magnetic Scalar Potential $V_m$

(Reader is advised to go through Sects. 2.21 and 2.22 before continuing this section).

We have seen that the magnetic vector potential happens to be a vector **A**. So if we want to calculate **B** using the relation $\mathbf{B} = \nabla \times \mathbf{A}$ then we need to remain confused about the components of **A** and the direction of **A**. Hence we went in search of magnetic scalar potential $V_m$ and found one. But as we will see below $V_m$ is not that much useful as electric potential V.

In Fig. 4.22 we have redrawn Fig. 2.56b. In Fig. 4.22 we are interested in measuring the length of the thread $A_1$ $A_2$. The length of the thread is 5 cm.

We have set the zero of the reference point as 2 cm. When we measure the length of the thread for the first time we see that initial reading is 2 cm and final reading is 7 cm. Hence the length of the thread is 7 cm − 2 cm = 5 cm. Suppose we repeat the measurement for the second time, that is we measure the length thread for the second time then once again initial reading is 2 cm and final reading is 7 cm and hence the length of the thread is 7 cm − 2 cm = 5 cm. If we repeat the measurement for the third time then once again 7 cm − 2 cm = 5 cm is the length of the thread.

"Thus the conclusion is ***once we fix the zero of the reference point*** and make repeated measurements, the ***actual physical quantity*** the length of thread remains unchanged". This is physically expected. The length of the thread cannot change

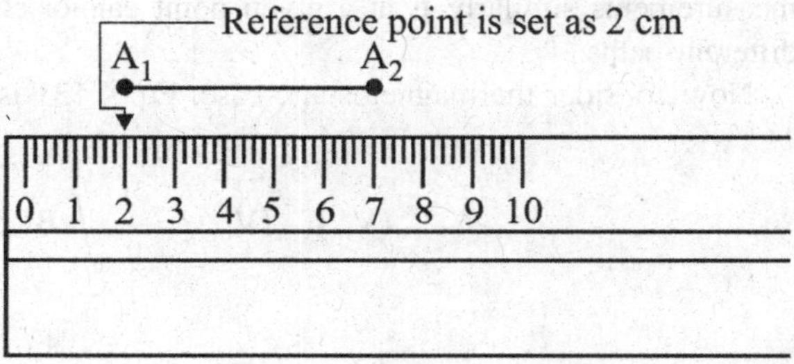


**Fig. 4.22** Measuring the length of the thread using a scale by setting the reference point at 2 cm

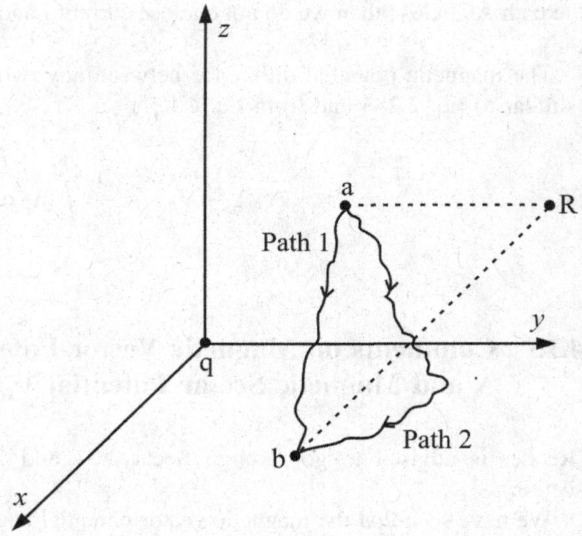


**Fig. 4.23** Moving from point a to point b along two paths - Path 1 and Path 2

when we make repeated measurements. Now consider the electrostatic case. We have redrawn Fig. 2.25 in Fig. 4.23 along with reference point R.

The potential difference between the points a and b is given by (Eq. 2.284)

$$V_a - V_b = \int_a^b \mathbf{E} \cdot d\boldsymbol{l} \tag{4.132}$$

We have seen in Sect. 2.21 Eq. 4.132 is independent of the path. That is we can travel from point a to b shown in Fig. 4.23 through path 1 or path 2 and the integral in Eq. 4.132 remains unchanged. Or in other words the potential $V_a$ and $V_b$ of points a and b have a single or unique value irrespective of the path travelled. Because $V_a$ and $V_b$ don't depend on the path we get a single unique value of **E** for the given point from Eq. 4.132. "Thus the conclusion is once we fix the zero of the reference point the single valuedness of electric potential V ensures that we get a single value of the actual physical quantity **E** at a given point". This is physically expected. Just as the length of the thread cannot change when we make repeated measurements similarly **E** at a given point cannot change, when we travel along different paths.

Now consider the magnetostatic case, Eq. 4.131 is

$$(V_m)_a - (V_m)_b = \frac{1}{\mu_o} \int_a^b \mathbf{B} \cdot d\boldsymbol{l}$$

Let us rewrite the above equation for points C, G in Fig. 4.20

$$(V_m)_C - (V_m)_G = \frac{1}{\mu_o} \int_C^G \mathbf{B} \cdot d\mathbf{l} \qquad (4.133)$$

During the first round in Fig. 4.20, after substituting $(V_m)_C$ and $(V_m)_G$ in Eq. 4.133 and evaluating the integral we can get **B**. During the second round in Fig. 4.20 as explained in Sect. 4.14 $(V_m)_C$ and $(V_m)_G$ would have changed and substituting the new value of $(V_m)_C$ and $(V_m)_G$ in Eq. 4.133 we will get another value for **B**. Thus each time the path ACFG is traversed in Fig. 4.20 the value of $V_m$ will keep changing. Because $V_m$ changes its value, for each time the path is traversed, we will get new value of **B** for each time the path is traversed from Eq. 4.133 which is physically unacceptable. The actual physical quantity **B** must not change its value even if we make measurements after travelling the path each time. But because $V_m$ is the path-dependent function each time we travel the path we get a new value of **B** which is not acceptable.

In analogy to our thread example if we measure the length of the thread repeatedly number of times and first time we get the length of the thread 5 cm, second time we get 27 cm, third time we get 101 cm, Is that acceptable? No. The length of the thread is 5 cm and any number of repeated measurements must give the same value for the length of the thread. Similarly we must get a single value of **B** (the actual physical quantity) irrespective of the path travelled. However $V_m$ is a path-dependent function and hence we get different values of **B** for each path travelled which is physically unacceptable. V, the electric potential is a scalar quantity and is independent of path. **E** is a vector quantity. Because V is a scalar quantity and is independent of path first we calculate V and then from V we calculate **E** by using the relation $\mathbf{E} = -\nabla V$.

$V_m$ the magnetic scalar potential is a scalar quantity, but is a path-dependent function. Because $V_m$ is path-dependent we get different values of **B** for each path travelled which is not acceptable. Hence $V_m$ rarely finds application in calculating **B**. As we have said in Fig. 4.21 we can restrict the path to avoid path dependency of $V_m$. A more detailed picture is shown in Fig. 4.24.

Paths $a_1ba_2$, $a_1ca_2$ and $a_1da_2$ don't enclose current I. We have restricted the path between $a_1$ and $a_2$. Hence for all these paths we get same value of **B** from Eq. 4.131. However the path $a_1ba_2\ ea_1$ encloses current I. Hence each time we travel the path $a_1ba_2\ ea_1$ the potential gets incremented by current I as explained in Sect. 4.14. Thus the potential will have different values at each given point each time the path $a_1ba_2\ ea_1$ is traversed. This will give rise to different values of **B** each time the path $a_1ba_2\ ea_1$ is traversed when Eq. 4.131 is used to calculate **B** along the path $a_1ba_2\ ea_1$.

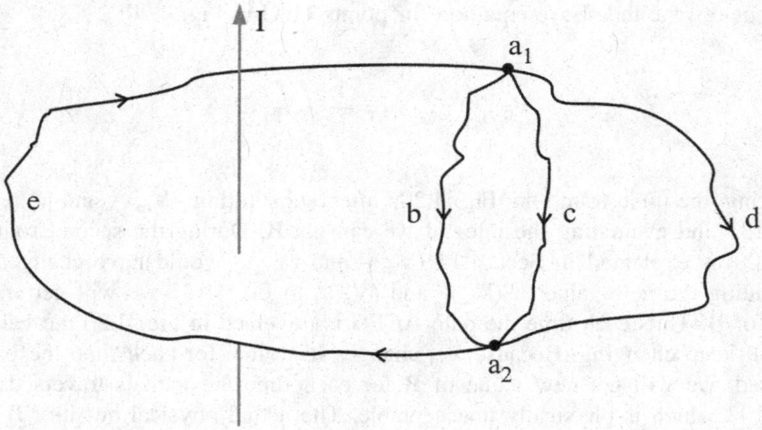


**Fig. 4.24** A infinitely long straight wire carrying current I. Figure illustrates various paths around the wire. Paths $a_1ba_2$, $a_1ca_2$, $a_1da_2$ doesn't enclose the current. While the path a1ba2ea1 encloses the current

Another point to be noted is similar to electrostatics, in magnetostatics we cannot write the following relation between $V_m$ and magnetic volume charge density $\rho_m$

$$V_m = \frac{\mu_o}{4\pi} \int_{\tau'} \frac{\rho_m}{\mathscr{r}} d\tau'$$

as magnetic monopoles do not exist.

As electrostatic forces can do work we were able to define the potential difference between two points as the work done per unit charge. However magnetic forces do not do any work and hence similar physical interpretation of **A** or $V_m$ is not possible. And **A** is still a vector and hence we need to remain confused about direction.

The above said reasons restrict the usage of **A** and $V_m$. It must be said that **A** (or) $V_m$ is not as much useful as electric scalar potential V.

However as mentioned previously, as **A** reduces the labour involved in integration to calculate **B**, it will be used to calculate **B** for the given current distribution.

Because Eq. 4.132 is independent of path and $\oint \mathbf{E} \cdot d\mathbf{l} = 0$ (Eq. 2.280) **E** is called conservative field.

However Eq. 4.131 is dependent on path and $\oint \mathbf{B} \cdot d\mathbf{l} = \mu_o I$ (Ampere's law) and hence **B** is non-conservative field.

## 4.16 B of a Current-Carrying Infinitely Long Straight Conductor

(a) **Using Biot–Savart law**. The **B** of an infinitely long straight conductor carrying current I has been calculated using Biot–Savart law in Sect. 4.8.
(b) **Using Ampere's circuital law (or) Ampere's law**. Consider Fig. 4.25. By right-hand rule we know that **B** is circumferential. We can select d*l* such that **B** and d*l* have same direction. Because of symmetry, **B** is constant over the Amperian loop in Fig. 4.25 and hence applying Ampere's law

$$\oint \mathbf{B}\cdot d\mathbf{l} = \mu_o I_{enc}$$

$$\oint B dl \cos 0 = \mu_o I \tag{4.134}$$

$$\Rightarrow \oint B dl = \mu_o I \tag{4.135}$$

Because B is constant over the Amperian loop in Fig. 4.25

$$B \oint dl = \mu_o I \tag{4.136}$$

$$\Rightarrow B 2\pi r = \mu_o I \tag{4.137}$$

$$\Rightarrow B = \frac{\mu_o I}{2\pi r} \tag{4.138}$$

By Ampere's law we calculated **B** in just few steps. Compare the above derivation with the method of calculating **B** using Biot–Savart law in Sect. 4.8.

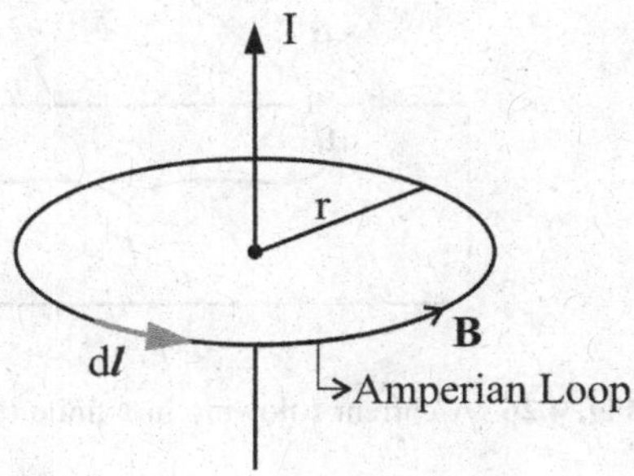


**Fig. 4.25** An Amperian loop for an infinitely long straight wire carrying current I

Biot–Savart law is laborious method because it calculates **B** directly in terms of vector. However Ampere's law calculates **B** easily by exploiting symmetry. More about this point is in Sect. 4.21.

## 4.17 B of a Current-Carrying Finite Straight Conductor

Consider a straight wire of length 2L carrying a current I. We are interested in calculating the **B** at point P at a distance $\mathscr{r}$ from the current element d$l'$. The point P lies at a distance of r from the wire as shown in Fig. 4.26. Let us calculate **B** at point P using Biot–Savart law. The dB at point P due to the small element d$l'$ is

$$dB = \frac{\mu_o I dl'}{4\pi}\frac{\sin\theta}{\mathscr{r}^2} \tag{4.139}$$

The magnetic flux density at point P due to the entire wire is given by

$$B = \int dB = \frac{\mu_o I}{4\pi}\int_{-L}^{L}\frac{\sin\theta dl'}{\mathscr{r}^2} \tag{4.140}$$

From Fig. 4.26

$$\mathscr{r} = \left[(l')^2 + r^2\right]^{1/2} \tag{4.141}$$

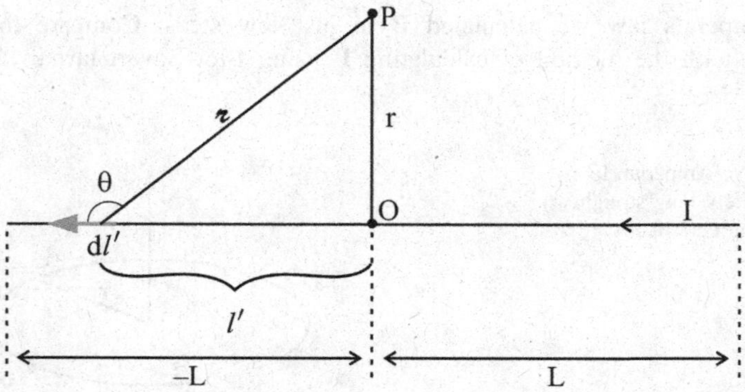


**Fig. 4.26** A current I flowing in a finite straight conductor of length 2L

and

$$\sin\theta = \sin(\pi - \theta) = \frac{r}{\mathscr{r}} = \frac{r}{\left[(l')^2 + r^2\right]^{1/2}} \tag{4.142}$$

Substituting Eqs. 4.141 and 4.142 in Eq. 4.140

$$B = \frac{\mu_o I}{4\pi} \int_{-L}^{L} \frac{r\,dl'}{\left[(l')^2 + r^2\right]^{3/2}} \tag{4.143}$$

$$B = \frac{\mu_o I r}{4\pi} \int_{-L}^{L} \frac{dl'}{\left[(l')^2 + r^2\right]^{3/2}} \tag{4.144}$$

$$B = \frac{\mu_o I r}{4\,\pi} \left[ \frac{l'}{r^2 \sqrt{r^2 + (l')^2}} \right]_{-L}^{L} \tag{4.145}$$

$$\Rightarrow B = \frac{\mu_o I}{2\,\pi r} \left[ \frac{L}{\sqrt{r^2 + L^2}} \right] \tag{4.146}$$

## 4.18 B Along the Axis of the Current-Carrying Circular Loop

Consider a circular coil of radius R carrying a current I as shown in Fig. 4.27. We are interested in calculating **B** at point P. Initially consider a small element d$\boldsymbol{l}'$ as shown in Fig. 4.27a. The d**B** at point P due to the single element d$\boldsymbol{l}'$ is

$$\mathbf{dB} = \frac{\mu_o I\, d\boldsymbol{l}' \times \hat{\mathscr{r}}}{4\,\pi\, \mathscr{r}^2} \tag{4.147}$$

In this case at point P the angle between d$\boldsymbol{l}'$ and $\mathscr{r}$ is 90° hence

$$dB = \frac{\mu_o I\, dl' \sin 90°}{4\pi\, \mathscr{r}^2} \tag{4.148}$$

$$dB = \frac{\mu_o I\, dl'}{4\pi\, \mathscr{r}^2} \tag{4.149}$$

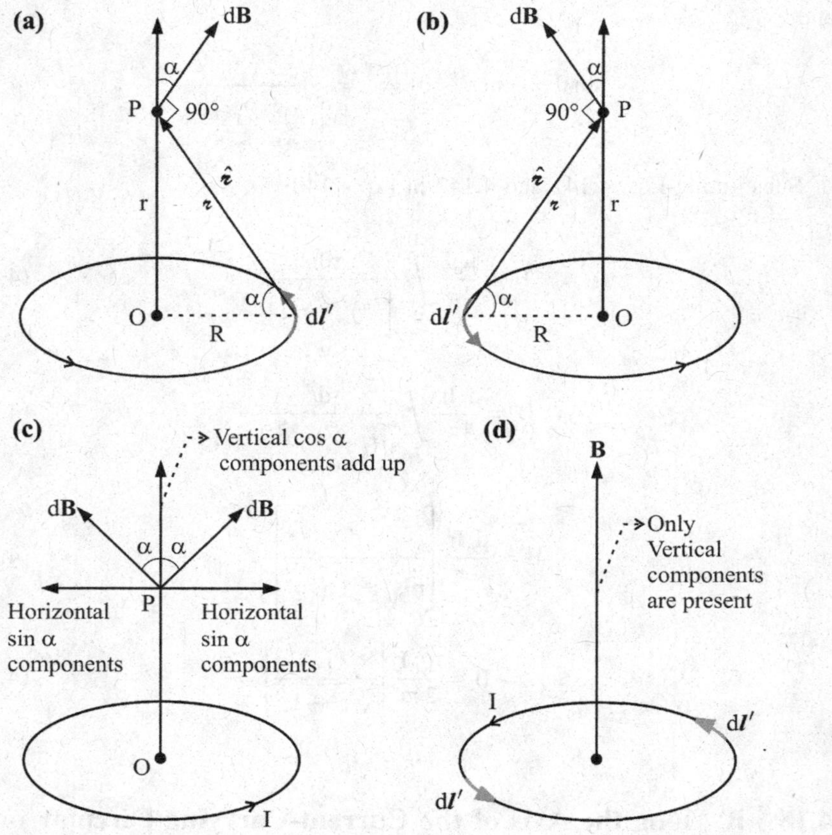


**Fig. 4.27** **a, b** A circular coil carrying current I. Two opposing current elements and their magnetic flux densities are shown in the figure. **c, d** Resolution of magnetic flux densities into its components

Let us now consider another exactly opposite element d$\boldsymbol{l}'$ as shown in Fig. 4.27b. The d$\mathbf{B}'$ produced by element's d$\boldsymbol{l}'$ in Fig. 4.27a, b is shown in Fig. 4.27c. Clearly the horizontal sin α components cancel out while the vertical cos α components add up. Similarly the entire circular loop can be divided into small elements such that for opposite elements horizontal sin α components cancel out while all the vertical cos α components add up as shown in Fig. 4.27d. Hence the **B** at point P due to the entire loop is

$$B = \int dB \cos\alpha \tag{4.150}$$

$$B = \int \frac{\mu_o I}{4\pi} \frac{dl'}{\mathscr{r}^2} \cos\alpha \tag{4.151}$$

From Fig. 4.27a

$$\cos\alpha = \frac{R}{\mathscr{r}} = \frac{R}{\left(r^2 + R^2\right)^{1/2}} \tag{4.152}$$

$$\mathscr{r}^2 = r^2 + R^2 \tag{4.153}$$

Substituting Eq. 4.152 and 4.153 in Eq. 4.151

$$B = \frac{\mu_o I}{4\pi} \int \frac{dl'}{\left(r^2 + R^2\right)} \frac{R}{\left[r^2 + R^2\right]^{1/2}} \tag{4.154}$$

$$B = \frac{\mu_o I}{4\pi} \frac{R}{\left[r^2 + R^2\right]^{3/2}} \int dl' \tag{4.155}$$

But $\int dl' = 2\pi R$ the circumference of the circular loop

$$B = \frac{\mu_o I}{4\pi} \frac{R}{\left[r^2 + R^2\right]^{3/2}} 2\pi R \tag{4.156}$$

$$B = \frac{\mu_o I}{2} \frac{R^2}{\left[r^2 + R^2\right]^{3/2}} \tag{4.157}$$

## 4.19 B Inside a Long Solenoid

**(a) Using Biot–Savart law**

A solenoid consists of a long cylindrical coil of an insulated copper wire shown in Fig. 4.28a. The cross-sectional view of the current-carrying solenoid is shown in Fig. 4.28b. From Eq. 4.157 the B along the axis of the single circular loop is

$$B = \frac{\mu_o I}{2} \frac{R^2}{\left(r^2 + R^2\right)^{3/2}}$$

If the circular loop has $N_1$ turns then

$$B = N_1 \frac{\mu_o I}{2} \frac{R^2}{\left[r^2 + R^2\right]^{3/2}} \tag{4.158}$$

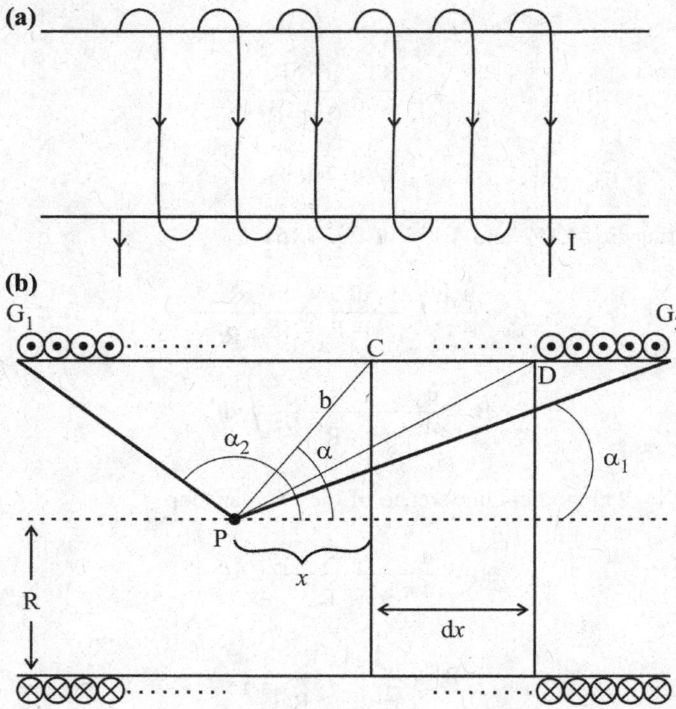


**Fig. 4.28** **a** A solenoid carrying current I. **b** A cross - sectional view of a solenoid

Consider a axial length L of the solenoid which consists of N number of turns. The number of turns per unit length of the coil is N/L. Let P be the point in Fig. 4.28b where **B** is to be determined. The edges of the solenoid $G_1$ and $G_2$ make an angle $\alpha_2$ and $\alpha_1$ with the lines $G_1P$ and $G_2P$ as shown in Fig. 4.28b. Consider the small element CD of width d$x$ situated at a distance of $x$ from point P.

The number of turns contained in the width d$x$ is

$$= \frac{N}{L^{dx}} \tag{4.159}$$

Substituting for $N_1 = \frac{N}{L}dx$ in Eq. 4.158 we get the dB at point P due to the small element CD with $x = r$

$$dB = \frac{N}{L}dx\frac{\mu_o I}{2}\frac{R^2}{\left[x^2 + R^2\right]^{3/2}} \tag{4.160}$$

where R is the radius of the solenoid as shown in Fig. 4.28. From Fig. 4.28b

$$\tan\alpha = \frac{R}{x}$$

$$\Rightarrow x = R\cot\alpha \tag{4.161}$$

$$\Rightarrow dx = -R\,\mathrm{cosec}^2\alpha\, d\alpha \tag{4.162}$$

and

$$R^2 + x^2 = b^2 \tag{4.163}$$

$$\sin\alpha = \frac{R}{b} \tag{4.164}$$

With the aid of Eqs. 4.161–4.164 Eq. 4.160 can be written as

$$dB = \frac{-N}{L}\frac{\mu_o IR^2}{2}\frac{R\,\mathrm{cosec}^2\alpha\, d\alpha}{b^3} \tag{4.165}$$

$$dB = \frac{-N}{L}\frac{\mu_o I}{2}\frac{R^3}{b^3}\mathrm{cosec}^2\alpha\, d\alpha \tag{4.166}$$

$$dB = \frac{-N}{L}\frac{\mu_o I}{2}\sin^3\alpha\,\mathrm{cosec}^2\alpha\, d\alpha \tag{4.167}$$

$$dB = \frac{-\mu_o IN}{2L}\sin\alpha\, d\alpha \tag{4.168}$$

Thus B at point P due to the entire length of the solenoid is given by

$$B = \frac{-\mu_0 IN}{2L}\int_{\alpha_2}^{\alpha_1}\sin\alpha\, d\alpha \tag{4.169}$$

$$B = \frac{\mu_o IN}{2L}[\cos\alpha_1 - \cos\alpha_2] \tag{4.170}$$

Equation 4.170 is the value of B at point P in Fig. 4.28b.

Let us consider a special case. For an infinitely long solenoid when the point P lies well within the solenoid and far away from either end

$$\alpha_1 = 0 \text{ and } \alpha_2 = \pi$$

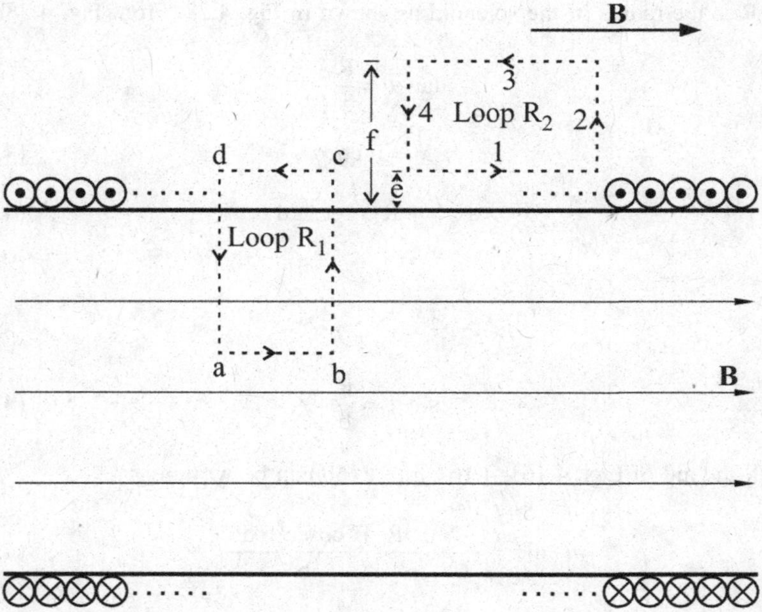


**Fig. 4.29** An infinitely long solenoid along with Amperian loops

then from Eq. 4.170

$$B = \frac{\mu_o IN}{2L}[\cos 0 - \cos \pi] \tag{4.171}$$

$$\Rightarrow B = \frac{\mu_o IN}{L} \tag{4.172}$$

(b) **Using Ampere's law**

Consider an infinite solenoid as shown in Fig. 4.29. In order to calculate the field inside and outside the solenoid we construct two Amperian loops $R_1$ and $R_2$. In the loop $R_2$ for path 2 and 4 the **B** and d$l$ are perpendicular to each other. Hence

$$\oint_{loopR_2} \mathbf{B} \cdot dl = \int_{Path1} \mathbf{B} \cdot dl + \int_{Path2} \mathbf{B} \cdot dl + \int_{Path3} \mathbf{B} \cdot dl + \int_{Path4} \mathbf{B} \cdot dl \tag{4.173}$$

$$\oint_{loopR_2} \mathbf{B} \cdot dl = \int_{Path1} \mathbf{B} \cdot dl + 0 + \int_{Path3} \mathbf{B} \cdot dl + 0 \tag{4.174}$$

$$\oint_{\text{loopR}_2} \mathbf{B} \cdot d\mathbf{l} = \int_{\text{Path1}} \mathbf{B} \cdot d\mathbf{l} + \int_{\text{Path3}} \mathbf{B} \cdot d\mathbf{l} \tag{4.175}$$

Because loop $R_2$ doesn't enclose current by Ampere's law for loop $R_2$

$$\int_{\text{loop}R_2} \mathbf{B} \cdot d\mathbf{l} = \mu_o I_{enc} = 0 \tag{4.176}$$

From Eqs. 4.175 to 4.176

$$\int_{\text{Path1}} \mathbf{B} \cdot d\mathbf{l} + \int_{\text{Path3}} \mathbf{B} \cdot d\mathbf{l} = 0 \tag{4.177}$$

In Fig. 4.29 path 1 and path 3 lie at distances e and f, respectively, from one end of the solenoid. Hence we can write Eq. 4.177 as

$$[B(e) - B(f)]L_1 = 0 \tag{4.178}$$

where $L_1$ is the path length of path 1 and 3, respectively

$$\text{As } L_1 \neq 0 \text{ we have } B(e) = B(f) \tag{4.179}$$

We know that at a point far away from the solenoid **B** is zero. If we construct loop $R_2$ such that the end f of loop $R_2$ is far away from the solenoid then $B(f) = 0$

Hence from Eq. 4.179

$$B(e) = B(f) = 0 \tag{4.180}$$

Thus for all points outside the solenoid B is zero. For the loop $R_1$ we have

$$\oint_{\text{loopR}_1} \mathbf{B} \cdot d\mathbf{l} = \mu_o I_{enc} \tag{4.181}$$

If N is the total number of turns enclosed by the loop $R_1$ in the solenoid then we can write

$$I_{enc} = NI \tag{4.182}$$

Equation 4.182 can be written as

$$\oint_{\text{loopR}_1} \mathbf{B} \cdot d\mathbf{l} = \mu_o NI \tag{4.183}$$

From Fig. 4.29

$$\int_{ab} \mathbf{B} \cdot d\boldsymbol{l} + \int_{bc} \mathbf{B} \cdot dl + \int_{cd} \mathbf{B} \cdot d\boldsymbol{l} + \int_{da} \mathbf{B} \cdot d\boldsymbol{l} = \mu_o NI \tag{4.184}$$

But for the path da and bc **B** is perpendicular to d*l*

$$\int_{ab} \mathbf{B} \cdot d\boldsymbol{l} + 0 + \int_{cd} \mathbf{B} \cdot d\boldsymbol{l} + 0 = \mu_o NI \tag{4.185}$$

$$\Rightarrow \int_{ab} \mathbf{B} \cdot d\boldsymbol{l} + \int_{cd} \mathbf{B} \cdot d\boldsymbol{l} = \mu_o NI \tag{4.186}$$

For the path cd **B** = 0 because the path cd is outside the solenoid. Hence

$$\Rightarrow \int_{ab} \mathbf{B} \cdot d\boldsymbol{l} + 0 = \mu_o IN \tag{4.187}$$

$$\Rightarrow \int_{ab} \mathbf{B} \cdot d\boldsymbol{l} = \mu_o NI \tag{4.188}$$

Along the path ab **B** is parallel to d*l*. If the length of the path ab is L then Eq. 4.188 is

$$BL = \mu_o NI \tag{4.189}$$

$$\Rightarrow B = \frac{\mu_o NI}{L} \tag{4.190}$$

The above problem demonstrates the advantage of using Ampere's law as compared to Biot–Savart law. Ampere's law uses symmetry to calculate **B** easily, while Biot–Savart law calculates **B** directly in terms of vector.

## 4.20 B of a Toroid

A toroid is solenoid in the form of a circle with the ends joined as shown in Fig. 4.30. Consider a circular Amperian loop P as shown in Fig. 4.30b. The **B** is circular within the cylindrical wire of the toroid.

Let N be the number of turns in the toroid then $I_{enc} = N\,I$ where I is the current in each turn.

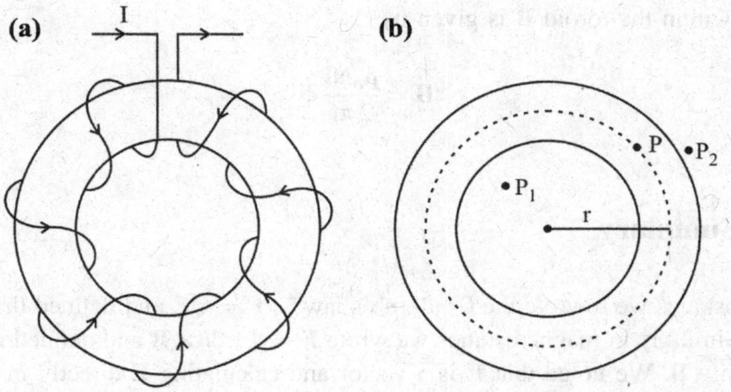


**Fig. 4.30** **a** A toroid carrying current I. **b** A toroid along with an Amperian loop is shown in the figure

Applying Ampere's law to path P

$$\oint \mathbf{B} \cdot d\boldsymbol{l} = \mu_o I_{enc}$$
$$\oint \mathbf{B} \cdot d\boldsymbol{l} = \mu_o NI \tag{4.191}$$

because **B** and d$\boldsymbol{l}$ are in same direction, $\mathbf{B} \cdot d\boldsymbol{l} = Bd\boldsymbol{l}\cos 0 = B\ d\boldsymbol{l}$ and B is constant over the Amperian loop so Eq. 4.191 is

$$B \oint d\boldsymbol{l} = \mu_o NI \tag{4.192}$$

$$\Rightarrow B\,2\pi r = \mu_o\, NI \tag{4.193}$$

$$\Rightarrow B = \frac{\mu_0 NI}{2\,\pi r} \tag{4.194}$$

For points like $P_1$ in Fig. 4.30b the Amperian loop doesn't enclose any current and hence at point $P_1$

$$\oint \mathbf{B} \cdot d\boldsymbol{l} = \mu_o I_{enc} = 0$$

$$\Rightarrow B = 0$$

In the case of solenoid in Sect. 4.19 we have seen that for external points the **B** is zero. Hence at points like $P_2$, B = 0

Thus within the toroid **B** is given by

$$\mathbf{B} = \frac{\mu_o NI}{2\pi r}\hat{\boldsymbol{\phi}} \tag{4.195}$$

## 4.21 Summary

In electrostatics we rewrote the Coulomb's law as $\mathbf{F} = Q\mathbf{E}$ and defined the electric field **E**. Similarly in magnetostatics we wrote $\mathbf{F} = \int I\, d\boldsymbol{l} \times \mathbf{B}$ and defined magnetic flux density **B**. We noted that **E** is a vector and calculating **E** directly in terms of vector using Coulomb's law is difficult. Coulomb's law calculates **E** in terms of vector and because vectors involve direction calculating **E** in terms of Coulomb's law is tedious. So we went in search of new easier methods like Gauss's law and potential to calculate **E**.

Similarly in magnetostatics calculating **B** using Biot–Savart law is a difficult process, because Biot–Savart law calculates **B** directly in terms of vector. So we went in search of new easier methods to calculate **B**.

Like Gauss's law in electrostatics we found Ampere's law in magnetostatics.

Gauss's law reads $\oint \mathbf{E}\cdot \mathbf{ds} = \frac{q_{enc}}{\varepsilon_o}$. Successful and easy application of Gauss's law requires construction of a Gaussian surface over which $\mathbf{E}\cdot\mathbf{ds} = E\, ds\cos\theta$ can be easily evaluated and to be constant over the Gaussian surface so that

$\oint \mathbf{E}\cdot \mathbf{ds} = \oint E ds\cos\theta = E\cos\theta \oint ds$ and $\oint ds$, and the calculation of area is simple mathematics. Then

$$\oint \mathbf{E}\cdot ds = \frac{q_{enc}}{\varepsilon_o}$$

$$\Rightarrow E = \frac{q_{enc}}{\cos\theta\, \varepsilon_o \oint ds} \tag{4.196}$$

Hence E can be easily evaluated. When the above conditions are not satisfied it is still true that $\oint \mathbf{E}\cdot ds = \frac{q_{enc}}{\varepsilon_o}$ for the given Gaussian surface, but the quantity **E** we are interested in calculating is inside the integral and hence E cannot be easily obtained using Gauss's law.

Similarly in magnetostatics we found Ampere's law as $\oint \mathbf{B}\cdot d\boldsymbol{l} = \mu_o I_{enc}$.

Successful and easy application of Ampere's law requires construction of an Amperian loop in which $\oint \mathbf{B}\cdot d\boldsymbol{l} = B\, dl\cos\theta$ can be easily evaluated and $B\cos\theta$ to be constant over the Amperian loop so that

$\oint \mathbf{B} \cdot \mathrm{d}\mathbf{l} = \oint \mathrm{B}\ \mathrm{d}l \cos\theta = \mathrm{B}\cos\theta \oint \mathrm{d}l$ and $\oint \mathrm{d}l$, and the calculation of line integral is simple mathematics. Then

$$\oint \mathbf{B} \cdot \mathrm{d}\mathbf{l} = \mu_o \mathrm{I}_{enc}$$

$$\Rightarrow \mathrm{B} = \frac{\mu_o \mathrm{I}_{enc}}{\cos\theta \oint \mathrm{d}l} \quad (4.197)$$

Hence B can be easily evaluated.

When the above conditions are not satisfied it is still true that $\oint \mathbf{B} \cdot \mathrm{d}\mathbf{l} = \mu_o \mathrm{I}_{enc}$ but the quantity **B** which we require is inside the integral and hence B cannot be easily obtained using Ampere's law.

Symmetry is crucial for easy application of Gauss's law for stationary charges and Ampere's law for steady currents.

In electrostatics three types of symmetry permit us to calculate $\mathbf{E} \cdot \mathrm{d}\mathbf{s}$ easily over the Gaussian surface and $\mathrm{E}\cos\theta$ to be constant over to Gaussian surface so that can be taken out of the integral $\oint \mathbf{E} \cdot \mathrm{d}\mathbf{s}$. The symmetries (which helps in calculating $\oint \mathbf{E} \cdot \mathrm{d}\mathbf{s}$ easier) are

(1) Plane symmetry (Sect. 2.16)
(2) Cylindrical symmetry (Sect. 2.13, Problems 2.1, 2.10, 2.11 and 2.9)
(3) Spherical symmetry (Sect. 2.17, Problem 2.8).

Similarly in magnetostatics four types of symmetry enable us to calculate $\mathbf{B} \cdot \mathrm{d}\mathbf{l}$ easily over the Amperian loop so that $\mathrm{B}\cos\theta$ can be taken out of the integral $\oint \mathbf{B} \cdot \mathrm{d}\mathbf{l}$

(1) Infinite straight line currents (Sect. 4.16)
(2) Infinite solenoid (Sect. 4.19)
(3) Toroids (Sect. 4.20)
(4) Infinite planes (Problem 4.1).

When Gauss's law is not useful in calculating **E** we went in search of other easy methods to calculate **E** and we found one method—potential. Potential V is a scalar quantity, and working with scalar quantities is much easier as compared to working with vector quantities. Once V has been found for the given charge distribution it is easy to calculate **E** because $\mathbf{E} = -\nabla \mathrm{V}$.

Similarly when Ampere's law is not useful we went in search of other methods. Our attempt to find easy way of calculating **B** using potential was not so much successful. The magnetic vector potential **A** still happens to be a vector. Our attempt to calculate **B** using magnetic scalar potential $\mathrm{V}_m$ was also not completely successful, because $\mathrm{V}_m$ is a path-dependent function although it is a scalar quantity. And for other reasons mentioned in Sect. 4.15 both **A** and $\mathrm{V}_m$ are not as much useful as electric potential V.

## 4.22 Magnetic Dipole

Consider a current-carrying loop as shown in Fig. 4.31. Let the current in the loop be I and we are interested in calculating the vector potential **A** at point P which is situated at a distance of **r** from the centre of the loop.

The vector potential at point P due to the entire loop is

$$\mathbf{A} = \frac{\mu_o I}{4\pi} \oint_{C'} \frac{d\boldsymbol{l}'}{\mathscr{r}} \tag{4.198}$$

From Fig. 4.31

$$\mathscr{r} = \sqrt{r^2 + (r')^2 - 2rr' \cos\theta} \tag{4.199}$$

$$\frac{1}{\mathscr{r}} = \frac{1}{\left(r^2 + (r')^2 - 2\,r\,r' \cos\theta\right)^{1/2}}$$

$$\frac{1}{\mathscr{r}} = \frac{1}{r}\left[1 + \left(\frac{r'}{r}\right)^2 - \frac{2r'}{r}\cos\theta\right]^{-1/2} \tag{4.200}$$

**Fig. 4.31** A magnetic dipole - A current carrying loop

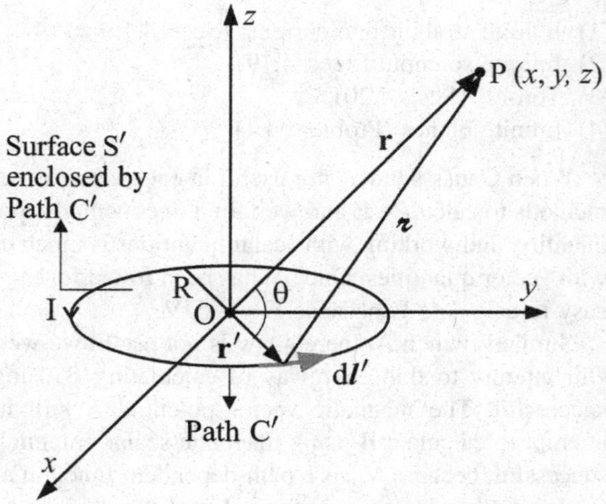

Suppose the point P is very far away from the loop then $r^2 \gg (r')^2$ and hence $\left(\frac{r'}{r}\right)^2$ term can be neglected

$$\frac{1}{\mathscr{r}} \simeq \frac{1}{r}\left[1 - 2\frac{r'}{r}\cos\theta\right]^{-1/2} \tag{4.201}$$

$$\Rightarrow \frac{1}{\mathscr{r}} \simeq \frac{1}{r}\left[1 + \frac{r'}{r}\cos\theta\right] \tag{4.202}$$

Substituting Eq. 4.202 in Eq. 4.198

$$\mathbf{A} = \frac{\mu_o I}{4\pi}\oint_{C'} \frac{1}{r}\left[1 + \frac{r'}{r}\cos\theta\right] d\boldsymbol{l}' \tag{4.203}$$

Noting down that integration is carried out with primed source coordinates we can write Eq. 4.203

$$\mathbf{A} = \frac{\mu_o I}{4\pi}\left[\frac{1}{r}\oint_{C'} d\boldsymbol{l}' + \frac{1}{r^2}\oint_{C'} r'\cos\theta\, d\boldsymbol{l}'\right] \tag{4.204}$$

$\oint_{C'} d\boldsymbol{l}'$ represents net vector displacement. Because the initial and final points are same the net vector displacement is zero.

$$\oint_{C'} d\boldsymbol{l}' = 0$$

Hence Eq. 4.204 is

$$\mathbf{A} = \frac{\mu_o I}{4\pi}\frac{1}{r^2}\oint_{C'} r'\cos\theta\, d\boldsymbol{l}' \tag{4.205}$$

$$\mathbf{A} = \frac{\mu_o I}{4\pi r^2}\oint_{C'} (\hat{\mathbf{r}}\cdot\mathbf{r}')d\boldsymbol{l}' \tag{4.206}$$

From Eq. 1.135 in Example 1.17 with $\mathbf{Q} = \hat{\mathbf{r}}$ we get

$$\oint_{C'} (\hat{\mathbf{r}}\cdot\mathbf{r}')d\boldsymbol{l}' = -\hat{\mathbf{r}} \times \int_{S'} d\mathbf{s}' \tag{4.207}$$

Substituting Eq. 4.207 in Eq. 4.206

$$\mathbf{A} = \frac{\mu_o I}{4\,\pi r^2}\left[-\hat{\mathbf{r}} \times \int\limits_{S'} ds'\right] \tag{4.208}$$

We define the magnetic dipole moment **m** as

$$\mathbf{m} = I\int\limits_{S'} ds'\,\hat{\mathbf{k}} = IS'\hat{\mathbf{k}} = I\,\pi R^2\hat{\mathbf{k}} \tag{4.209}$$

Hence Eq. 4.208 is

$$\mathbf{A} = \frac{\mu_o}{4\,\pi\, r^2}[-\hat{\mathbf{r}} \times \mathbf{m}] \tag{4.210}$$

$$\Rightarrow \mathbf{A} = \frac{\mu_o}{4\,\pi}\left[\frac{\mathbf{m} \times \hat{\mathbf{r}}}{r^2}\right] \tag{4.211}$$

The magnitude of **m** is the product of the current and the area of the loop. The direction of **m** is given by right-hand rule when the fingers of right hand curl around the direction in which the current I is flowing then the thumb of right hand gives the direction of **m**.

From Eq. 2.256 the potential of an electric dipole is given by

$$V = \frac{1}{4\,\pi\,\varepsilon_0}\frac{\mathbf{p} \cdot \hat{\mathbf{r}}}{r^2}$$

Comparing Eq. 4.211 with Eq. 2.256 we see that **A** is analogous to V. Thus we see that a small current-carrying loop is a magnetic dipole.

Now let us calculate **B** of the magnetic dipole. We know that

$$\mathbf{B} = \nabla \times \mathbf{A}$$

Substituting Eq. 4.211 in **B**

$$\mathbf{B} = \nabla \times \left[\frac{\mu_o}{4\,\pi}\left(\frac{\mathbf{m} \times \hat{\mathbf{r}}}{\hat{\mathbf{r}}^2}\right)\right] \tag{4.212}$$

$$\mathbf{B} = \frac{\mu_o}{4\,\pi}\nabla \times \left(\frac{\mathbf{m} \times \mathbf{r}}{r^3}\right) \tag{4.213}$$

Because **m** is a constant vector setting $\mathbf{Q} = \mathbf{m}$ and $\mathbf{R} = \frac{\mathbf{r}}{r^3}$ in Eq. 1.134 from Example 1.16 Eq. 4.213 can be written as

$$\mathbf{B} = \frac{\mu_o}{4\pi}\left[\mathbf{m}\left(\nabla\cdot\frac{\mathbf{r}}{r^3}\right) - (\mathbf{m}\cdot\nabla)\frac{\mathbf{r}}{r^3}\right] \tag{4.214}$$

from Fig. 4.31 $\mathbf{r} = x\hat{\mathbf{i}} + y\hat{\mathbf{j}} + z\hat{\mathbf{k}}$

We have

$$m_x\frac{\partial}{\partial x}\left(\frac{\mathbf{r}}{r^3}\right) = \frac{m_x\hat{\mathbf{i}}}{r^3} - 3m_x x\frac{\mathbf{r}}{r^5} \tag{4.215}$$

$$m_y\frac{\partial}{\partial y}\left(\frac{\mathbf{r}}{r^3}\right) = \frac{m_y\hat{\mathbf{j}}}{r^3} - 3m_y y\frac{\mathbf{r}}{r^5} \tag{4.216}$$

$$m_z\frac{\partial}{\partial z}\left(\frac{\mathbf{r}}{r^3}\right) = \frac{m_z\hat{\mathbf{k}}}{r^3} - 3m_z z\frac{\mathbf{r}}{r^5} \tag{4.217}$$

Using Eqs. 4.215, 4.216, 4.217 we can write

$$(\mathbf{m}\cdot\nabla)\frac{\mathbf{r}}{r^3} = \frac{\mathbf{m}}{r^3} - \frac{3(\mathbf{m}\cdot\mathbf{r})\mathbf{r}}{r^5} \tag{4.218}$$

and we have

$$\mathbf{m}\left(\nabla\cdot\frac{\mathbf{r}}{r^3}\right) = \mathbf{m}\left[\frac{3}{r^3} - \mathbf{r}\cdot\frac{3\mathbf{r}}{r^5}\right] = \mathbf{m}\left[\frac{3}{r^3} - \frac{3}{r^3}\right] = 0 \tag{4.219}$$

Substituting Eqs. 4.219 and 4.218 in Eq. 4.214

$$\mathbf{B} = \frac{\mu_o}{4\,\pi}\left[\frac{3(\mathbf{m}\cdot\mathbf{r})\mathbf{r}}{r^5} - \frac{\mathbf{m}}{r^3}\right] \tag{4.220}$$

In Fig. 4.32a we plot the electric field of an electric dipole and in Fig. 4.32b we plot **B** of a magnetic dipole.

## 4.23 Magnetic Boundary Conditions

In electrostatics we saw that at the interface between two media the electric field suffers a discontinuity. Similarly at the interface between two magnetic media **B** suffers a discontinuity. The two magnetic media may be two magnetic materials with different magnetic properties or a magnetic media and vacuum. Consider an interface between two magnetic media carrying a surface current **K** as shown in Fig. 4.33a.

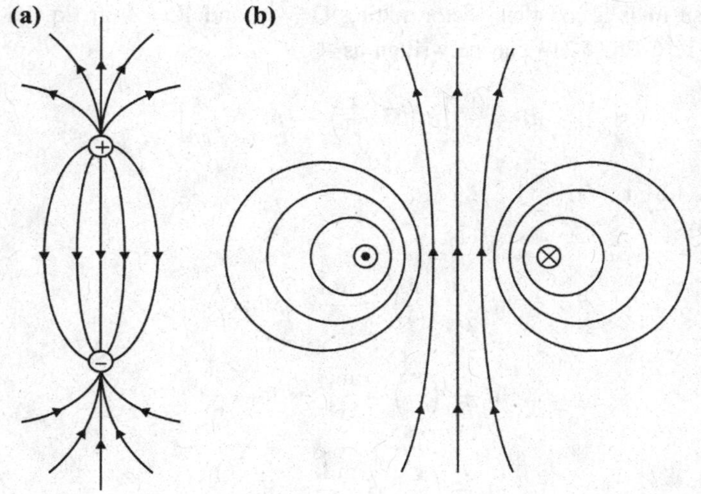


**Fig. 4.32** a Plot of electric field lines of a electric dipole. **b** Plot of magnetic flux density of a magnetic dipole

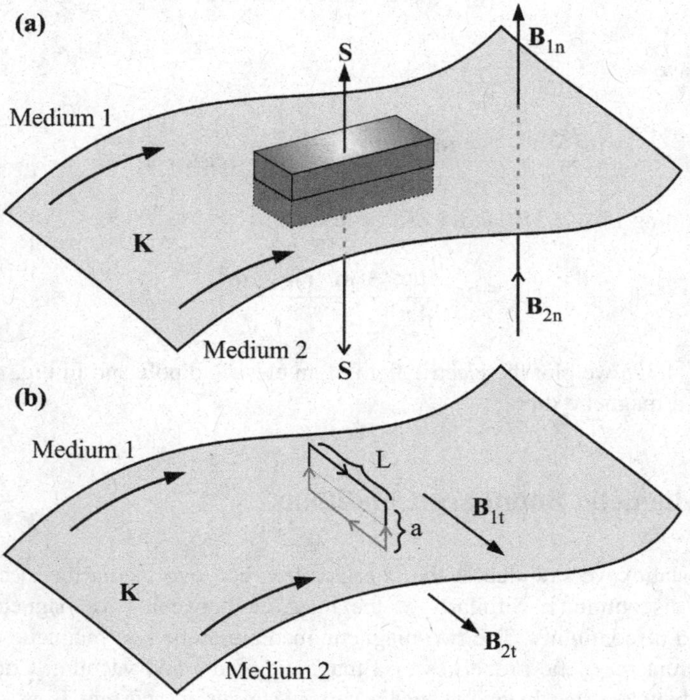


**Fig. 4.33** Interface between two magnetic media carrying a surface current density shown along. **a** with a pill box to evaluate the relationship between normal components of magnetic flux density. **b** with a Amperian loop to evaluate the relationship between tangential components of magnetic flux density

We know that $\nabla \cdot \mathbf{B} = 0$

Selecting an arbitrary volume $\tau$ and integrating the above equation over the volume $\tau$

$$\int_{\tau} \nabla \cdot \mathbf{B} d\tau = 0 \tag{4.221}$$

Applying Gauss's divergence theorem

$$\oint \mathbf{B} \cdot \mathbf{ds} = 0 \tag{4.222}$$

In Fig. 4.33a let us construct a pillbox of negligible sides, and the area of the top and bottom surface of the pillbox is S. Because the sides of the pillbox are negligible, applying Eq. 4.222 to the pill box

$$B_{1n}S - B_{2n}S = 0 \tag{4.223}$$

$$\Rightarrow (B_{1n} - B_{2n})S = 0 \tag{4.224}$$

$$\text{As } S \neq 0$$

$$B_{1n} - B_{2n} = 0$$

$$\Rightarrow B_{1n} = B_{2n} \tag{4.225}$$

In order to find the relation between tangential components construct an Amperian loop as shown in Fig. 4.33b. The Amperian loop is running perpendicular to the current. The Amperian loop is so constructed that the side of the loop a is negligible. Then applying Ampere's law

$$\oint \mathbf{B} \cdot d\boldsymbol{l} = \mu_o I_{enc}$$

Because the side a is negligible for the Amperian loop

$$B_{1t}L - B_{2t}L = \mu_o KL \tag{4.226}$$

$$\Rightarrow B_{1t} - B_{2t} = \mu_o K \tag{4.227}$$

Thus when there is a surface current then the tangential component of **B** is discontinuous by the amount given by Eq. 4.227. If K is zero across the boundary between two magnetic media then

$$B_{1t} = B_{2t} \tag{4.228}$$

## 4.24 Force Between Two Parallel Current-Carrying Conductors

Consider two long straight conductors a and b carrying currents $I_a$ and $I_b$. Let the distance of separation between the two conductors be d. The value of B at the site of conductor b produced by conductor a is given by (Eq. 4.47)

$$B_a = \frac{\mu_0 I_a}{2\pi d} \tag{4.229}$$

By right-hand rule $B_a$ points into the page as shown in Fig. 4.34. The length L of the conductor b experiences a force given by

$$F_a = I_b L B_a \tag{4.230}$$

Substituting Eq. 4.229 in 4.230

$$F_a = I_b L \frac{\mu_0 I_a}{2\pi d} \tag{4.231}$$

$$F_a = \frac{\mu_0 L}{2\pi d} I_a I_b \tag{4.232}$$

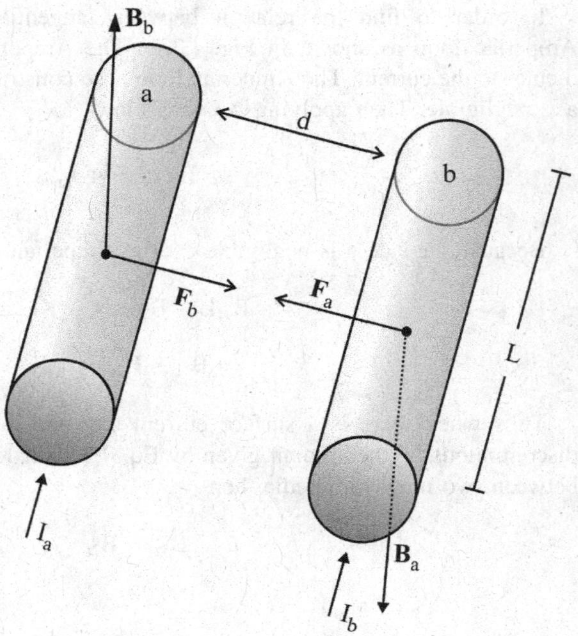


**Fig. 4.34** Two long straight conductors a, b carrying currents $I_a$ and $I_b$

Because the magnetic force is given by $\mathbf{F} = I(d\boldsymbol{l} \times \mathbf{B})$ the direction of $\mathbf{F}_a$ is towards the conductor a as shown in Fig. 4.34. The value of B at the site of conductor a produced by conductor b is given by

$$B_b = \frac{\mu_o I_b}{2\pi d} \tag{4.233}$$

By right-hand rule $B_b$ points out of the page as shown in Fig. 4.34.

The length L of the conductor a experiences a force $F_b$ given by

$$F_b = I_a L B_b$$

$$F_b = I_a L \frac{\mu_o I_b}{2\pi d} \tag{4.234}$$

$$\Rightarrow F_b = \frac{\mu_o L}{2\pi d} I_a I_b \tag{4.235}$$

The direction of $\mathbf{F}_b$ is shown in Fig. 4.34.

Thus $\mathbf{F}_a$ and $\mathbf{F}_b$ are equal in magnitude but opposite to each other. Therefore two parallel straight conductors carrying currents in same direction attract each other. Similarly it can be shown that two parallel straight conductors carrying currents in opposite direction repel each other.

## 4.25 Torque on a Current Loop in a Uniform Magnetic Field

Consider a rectangular loop PQRS placed in a uniform magnetic flux density **B** as shown in Fig. 4.35. In Fig. 4.35 PS = QR = b and PQ = SR = a. The loop is situated in the **B** such that the axis (shown as dotted line in Fig. 4.35) is perpendicular to **B**, while the plane of the loop makes an angle θ with the direction of **B**. The loop carries a current I. The force $\mathbf{F}_1$ acting on the limb Q R is given by

$$\mathbf{F}_1 = I(\mathbf{b} \times \mathbf{B}) \tag{4.236}$$

$$\Rightarrow \mathbf{F}_1 = I b B \sin\theta \tag{4.237}$$

The direction of $\mathbf{F}_1$ is downward. Similarly $\mathbf{F}_2$ given by Ib B sin θ is acting upward and hence $\mathbf{F}_1$ and $\mathbf{F}_2$ cancel out each other.

Now the force $\mathbf{F}_4$ acting on limb PQ is given by

$$\mathbf{F}_4 = I(\mathbf{a} \times \mathbf{B}) = IaB \sin 90^\circ = IaB \tag{4.238}$$

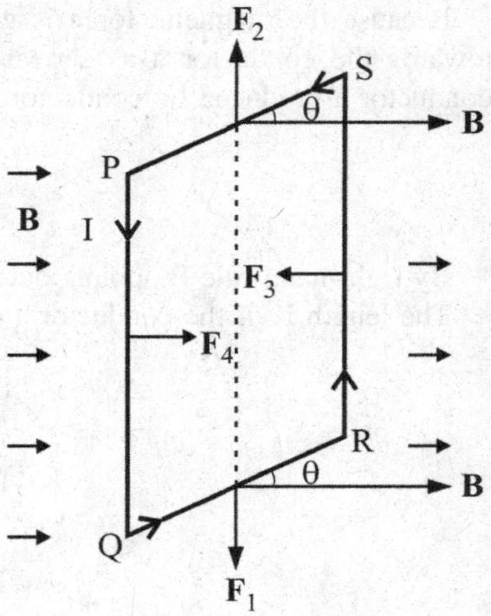


**Fig. 4.35** A rectangular loop PQRS placed in a uniform magnetic field

The direction of $\mathbf{F}_4$ is shown in Fig. 4.35. The force $\mathbf{F}_3$ acting on limb SR is IaB but has a direction opposite to $\mathbf{F}_4$ as shown in Fig. 4.35.

Thus $\mathbf{F}_3$ and $\mathbf{F}_4$ are opposite to each other and hence cancel out. However the line of action of $\mathbf{F}_3$ and $\mathbf{F}_4$ is different and constitute a torque.

The top cross-sectional view of the rectangular loop placed in $\mathbf{B}$ is shown in Fig. 4.36. The torque acting on the loop is given by

$$T = F_4(GS) = F_4 b \cos\theta \tag{4.239}$$

$$T = IaBb\cos\theta \tag{4.240}$$

$$T = IB(ab)\cos\theta \tag{4.241}$$

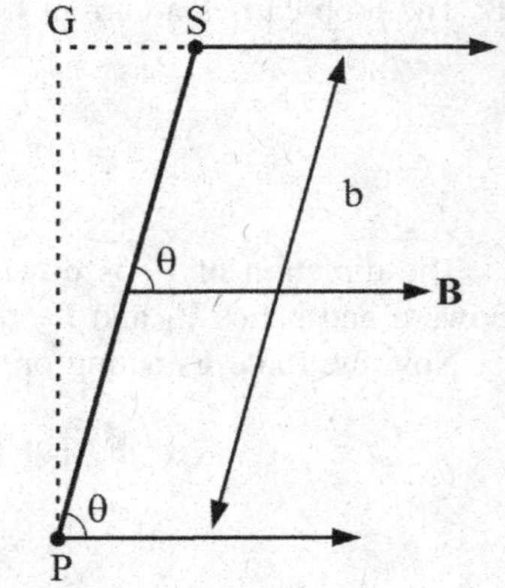


**Fig. 4.36** Top cross - sectional view of the rectangular loop placed in the uniform magnetic field

$$\Rightarrow T = IBS_1 \cos\theta \tag{4.242}$$

$$\Rightarrow \mathbf{T} = I(\mathbf{S_1} \times \mathbf{B}) \tag{4.243}$$

where $S_1 = ab$ is the area of the coil. If there are N turns in the loop then the torque is

$$\mathbf{T} = NI\mathbf{S_1} \times \mathbf{B} \tag{4.244}$$

From Eq. 4.209 $\mathbf{m} = I\,\mathbf{S_1}$ is the magnetic moment and hence

$$\mathbf{T} = N\mathbf{m} \times \mathbf{B} \tag{4.245}$$

## 4.26 Magnetic Flux

Magnetic flux is defined as the total number of lines of magnetic flux density **B** passing through the given surface and is denoted by $\Phi$. Let **B** be the magnetic flux density.

In Fig. 4.37 we see that **B** makes an angle $\theta$ with the normal to the given surface whose area is S. From Sect. 1.12 the flux passing through the surface S in Fig. 4.37 is then

$$\Phi = \oint \mathbf{B} \cdot \mathbf{ds} \tag{4.246}$$

The unit of $\Phi$ is weber.

**Example 4.1**
Find the current distribution that produces a magnetic flux density of $\mathbf{B} = \kappa \cos x\hat{\mathbf{j}}$ where $\kappa$ is a constant.

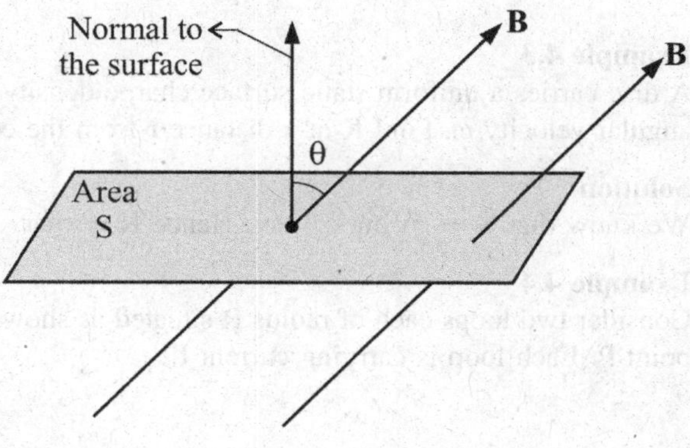


**Fig. 4.37** Lines of magnetic flux density passing through a surface area S - Magnetic flux

**Solution**
We know that

$$\nabla \times \mathbf{B} = \mu_0 \mathbf{J}$$

$$\mathbf{J} = \frac{1}{\mu_0}\nabla \times \mathbf{B}$$

$$= \frac{1}{\mu_0}\begin{vmatrix} \hat{\mathbf{i}} & \hat{\mathbf{j}} & \hat{\mathbf{k}} \\ \frac{\partial}{\partial x} & \frac{\partial}{\partial y} & \frac{\partial}{\partial z} \\ 0 & \kappa \cos x & 0 \end{vmatrix}$$

$$\mathbf{J} = -\frac{\kappa \sin x}{\mu_0}\hat{\mathbf{k}}$$

**Example 4.2**

(a) Consider a cylinder of circular cross section of radius R which carries a uniform current I perpendicular to the circular cross section. Find **J**.

(b) If the current density is such that $J = \kappa r_c^2$ then find the total current in the wire.

**Solution**

(a) We know that $J = \frac{dI}{ds_\perp}$ where $ds_\perp$ is the area perpendicular to the flow. Here $ds_\perp = \pi R^2$ Hence $J = \frac{I}{\pi R^2}$ as the current I is uniformly distributed.

(b) Here $I = \int J\, ds_\perp$ because J varies with $r_c$. Then
$I = \int \kappa\, r_c^2 r_c dr_c d\phi$ as $ds_\perp = r_c dr_c d\phi$ is the area perpendicular to the flow.

$$I = \int_0^R \kappa\, r_c^3 dr_c \int_0^{2\pi} d\phi = \frac{\pi \kappa R^4}{2}$$

**Example 4.3**
A disc carries a uniform static surface charge density. The disc is rotating with an angular velocity ω. Find K at a distance r from the centre.

**Solution**
We know that $K = \sigma v$ but $v = \omega r$. Hence $K = \sigma \omega r$.

**Example 4.4**
Consider two loops each of radius R situated as shown in Fig. 4.38. Calculate **B** at point P. Each loop is carrying current I.

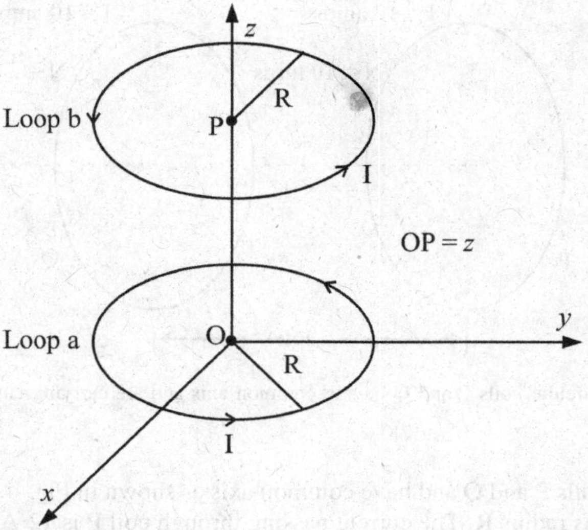


**Fig. 4.38** Two loops - loop a and loop b each of radius R carrying a current I

**Solution**

We have seen in Eq. 4.157 **B** due to a current-carrying circular loop along its axis is given by

$$\mathbf{B}_a = \frac{\mu_o I}{2}\frac{R^2}{\left[z^2 + R^2\right]^{3/2}}\hat{\mathbf{k}}$$

**B** at point P due to the loop b is

$$\mathbf{B}_b = \frac{\mu_o I}{2}\frac{R^2}{R^3}\hat{\mathbf{k}} \text{ as } z = 0 \text{ for loop b.}$$

$$\mathbf{B}_b = \frac{\mu_o I}{2}\frac{1}{R}\hat{\mathbf{k}}$$

Hence the net **B** at point P is

$$\mathbf{B} = \mathbf{B}_a + \mathbf{B}_b$$

$$\mathbf{B} = \frac{\mu_o I}{2}\left[\frac{1}{R} + \frac{R^2}{\left[z^2 + R^2\right]^{3/2}}\right]\hat{\mathbf{k}}$$

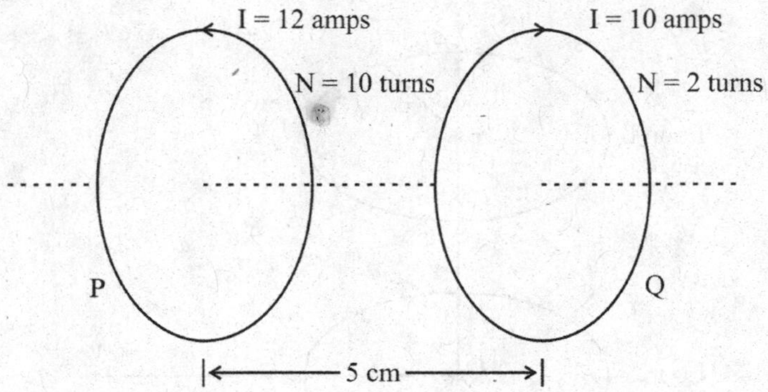


**Fig. 4.39** Two circular coils P and Q having common axis and are carrying current

**Example 4.5**
Two circular coils P and Q and have common axis as shown in Fig. 4.39. Coil P has 10 turns and has radius R. The current passing through coil P is 12 A. Coil Q has 2 turns with a radius of 7 cm, and the current passing through it is 10 A. What should be the value of R so that **B** be zero at the centre of coil P. The distance between the coils is 5 cm.

**Solution**
From Eq. 4.157

$$B = \frac{\mu_0 IN}{2} \frac{R^2}{\left[z^2 + R^2\right]^{3/2}}$$

For coil P, the value of **B** at the centre of coil P is (I = 12 A, R = R, $z$ = 0, N = 10)

$$B_P = \frac{4\pi \times 10^{-7}}{2}(12)(10)\frac{R^2}{R^3}$$

$$B_P = \left(\frac{4\,\pi \times 10^{-7}}{2}\right)(10)(12)\frac{1}{R}$$

The value of **B** due to coil Q at the centre of coil P is (I = 10 A, N = 2, R = 7 cm, $z$ = 5 cm).

$$B_Q = \frac{4\,\pi \times 10^{-7}(10)(2)}{2} \frac{7^2}{[5^2 + 7^2]}$$

For the value of **B** at the centre of coil P to be zero

$B_P = B_Q$ (Noting down that $B_P$ and $B_Q$ are in opposite directions)

$$\frac{4\pi \times 10^{-7}}{2}(10)(12)\frac{1}{R} = \frac{4\pi \times 10^{-7}}{2}(10)(2)\frac{7^2}{(5^2+7^2)}$$

$$\Rightarrow R = 9.06 \text{ cm.}$$

**Example 4.6**
Consider a long straight wire carrying current I. Calculate the magnetic vector potential **A** and hence calculate **B**.

**Solution**
Let L be the length of the wire. Let it carry current I as shown in Fig. 4.40. Let us solve the problem in Cartesian coordinates. At point P the vector potential d**A** due to element d$x$ is given by

$$d\mathbf{A} = \frac{\mu_o I\, dx}{4\pi\, r}\hat{\mathbf{i}}$$

The magnetic vector potential at P due to the whole wire is then

$$\mathbf{A} = \frac{\mu_o I}{4\pi}\int_{-L/2}^{L/2}\frac{dx}{r}\hat{\mathbf{i}}$$

$$\mathbf{A} = \frac{\mu_o I}{4\pi}\int_{-L/2}^{L/2}\frac{dx}{\sqrt{x^2+y^2}}\hat{\mathbf{i}}$$

$$\mathbf{A} = \frac{\mu_o I}{4\pi}\left[\log\left[x+\sqrt{x^2+y^2}\right]\right]_{-L/2}^{L/2}\hat{\mathbf{i}}$$

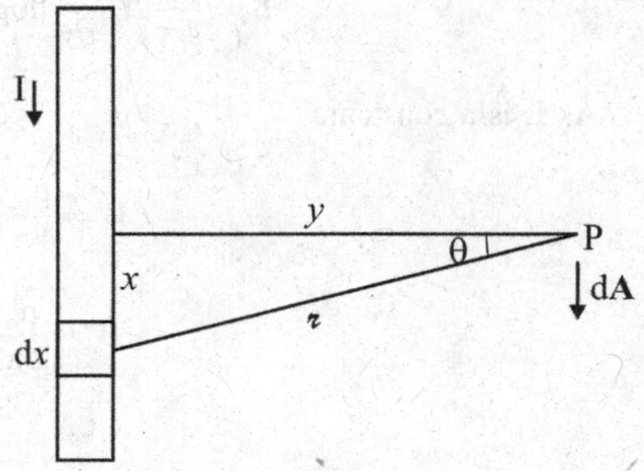


**Fig. 4.40** A long straight wire carrying a current I

$$\mathbf{A} = \frac{\mu_o \mathrm{I}}{4\pi} \log\left[\frac{\mathrm{L}/2 + \sqrt{\frac{\mathrm{L}^2}{4} + y^2}}{-\mathrm{L}/2 + \sqrt{\frac{\mathrm{L}^2}{4} + y^2}}\right]\hat{\mathbf{i}}$$

If the wire is very long so that $\mathrm{L}^2 \gg y^2$ then

$$\mathbf{A} = \frac{\mu_o \mathrm{I}}{4\pi} \log\left|\frac{1 + \left(1 + \frac{4y^2}{\mathrm{L}^2}\right)^{1/2}}{-1 + \left(1 + \frac{4y^2}{\mathrm{L}^2}\right)^{1/2}}\right|\hat{\mathbf{i}}$$

$$\mathbf{A} \simeq \frac{\mu_o \mathrm{I}}{4\pi} \log\left[\frac{1 + \left(1 + \frac{4y^2}{2\mathrm{L}^2}\right)}{-1 + \left(1 + \frac{4y^2}{2\mathrm{L}^2}\right)}\right]\hat{\mathbf{i}}$$

$$\mathbf{A} = \frac{\mu_o \mathrm{I}}{4\pi} \log\left[1 + \frac{\mathrm{L}^2}{y^2}\right]\hat{\mathbf{i}}$$

As $\mathrm{L}^2 \gg y^2$

$$\mathbf{A} = \frac{\mu_o \mathrm{I}}{4\pi} 2\log\frac{\mathrm{L}}{y}\hat{\mathbf{i}}$$

As $\mathbf{B} = \nabla \times \mathbf{A}$

$$\mathbf{B} = \frac{\mu_o}{4\pi} 2\mathrm{I}\begin{vmatrix} \hat{\mathbf{i}} & \hat{\mathbf{j}} & \hat{\mathbf{k}} \\ \frac{\partial}{\partial x} & \frac{\partial}{\partial y} & \frac{\partial}{\partial z} \\ \log\frac{\mathrm{L}}{y} & 0 & 0 \end{vmatrix}$$

$$\mathbf{B} = \frac{\mu_o}{4\pi} 2\mathrm{I}\left[-\hat{\mathbf{k}}\frac{\partial}{\partial y}\log\frac{\mathrm{L}}{y}\right]$$

$$\mathbf{B} = \frac{\mu_0}{4\pi} 2\mathrm{I}\left[\frac{-\partial}{\partial y}[\log \mathrm{L} - \log y]\right]\hat{\mathbf{k}}$$

As L is a constant.

$$\mathbf{B} = \frac{\mu_o}{4\pi}\frac{2\mathrm{I}}{y}\hat{\mathbf{k}}$$

$$\mathbf{B} = \frac{\mu_o \mathrm{I}}{2\pi y}\hat{\mathbf{k}}$$

**Example 4.7**
A vector potential is given by

(a) $\mathbf{A} = y^2 \sin x\hat{\mathbf{i}} + e^{-2x}\hat{\mathbf{k}}$ in Cartesian coordinates
(b) $\mathbf{A} = \kappa \cos\theta\,\hat{\boldsymbol{\theta}}$ in spherical polar coordinates.

Calculate **B**.

**Solution**
$\mathbf{B} = \nabla \times \mathbf{A}$

(a)
$$\mathbf{A} = y^2 \sin x\hat{\mathbf{i}} + e^{-2x}\hat{\mathbf{k}}$$

$$\mathbf{B} = \begin{vmatrix} \hat{\mathbf{i}} & \hat{\mathbf{j}} & \hat{\mathbf{k}} \\ \dfrac{\partial}{\partial x} & \dfrac{\partial}{\partial y} & \dfrac{\partial}{\partial z} \\ y^2 \sin x & 0 & e^{-2x} \end{vmatrix}$$

$$\mathbf{B} = ze^{-2x}\hat{\mathbf{j}} - 2y \sin x\hat{\mathbf{k}}$$

(b)
$$\mathbf{B} = \nabla \times \mathbf{A} = \frac{1}{r}\frac{\partial}{\partial r}(\kappa r \cos\theta)\hat{\boldsymbol{\phi}}$$

$$\mathbf{B} = \frac{\kappa \cos\theta}{r}\hat{\boldsymbol{\phi}}$$

**Example 4.8**
A wire of configuration shown in Fig. 4.41 (wire PQRS) carries a current I. Calculate the **B** at point $P_1$.

**Solution**
The point $P_1$ lies along the line PQ and along the line SR. Hence the **B** produced by PQ and SR at the site of $P_1$ is zero [$d\boldsymbol{l}, \hat{\boldsymbol{\imath}}$ in Biot–Savart law have either same or opposite direction].

The flux density at the centre of a circular loop is given by (Eq. 4.157)

$$B = \frac{\mu_o I}{2\imath}$$

where $\imath$ is the radius of the loop.

The flux density at point $P_1$ due to the segment PS is then

$$B_{PS} = \left(\frac{\theta}{2\pi}\right)\frac{\mu_o I}{2\imath_1}$$

**Fig. 4.41** A wire PQRS carrying a current I

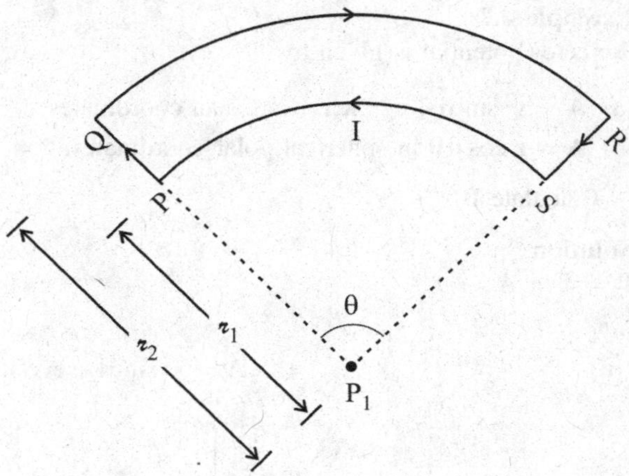


The flux density at P due to the segment BC is then

$$B_{QR} = \left(\frac{\theta}{2\pi}\right)\frac{\mu_o I}{2r_2}$$

Applying right-hand rule we see that the direction of $B_{PS}$ is out of the plane and the direction of $B_{QR}$ is into the plane of the paper. From Eq. 4.36 the value of net **B** at point $P_1$ is

$$B = B_{PS} - B_{QR}$$

$$B = \frac{\mu_o I}{4\pi}\frac{(r_2 - r_1)}{r_2 r_1}$$

As $r_1$ is shorter than $r_2$ $B_{PS} > B_{QR}$.
Hence **B** at point $P_1$ will be directed out of the plane of paper.

**Example 4.9**
Consider a long wire carrying current I. The small portion of the wire is bent into a semicircular loop as shown in Fig. 4.42. The radius of the circular loop is R. Calculate the magnetic flux density at point P due to the entire wire.

**Solution**
The **B** due to the straight portion of the wire is zero because

$$\mathbf{B} = \frac{\mu_o I}{4\pi}\int \frac{d\mathbf{l} \times \hat{r}}{r^2}$$

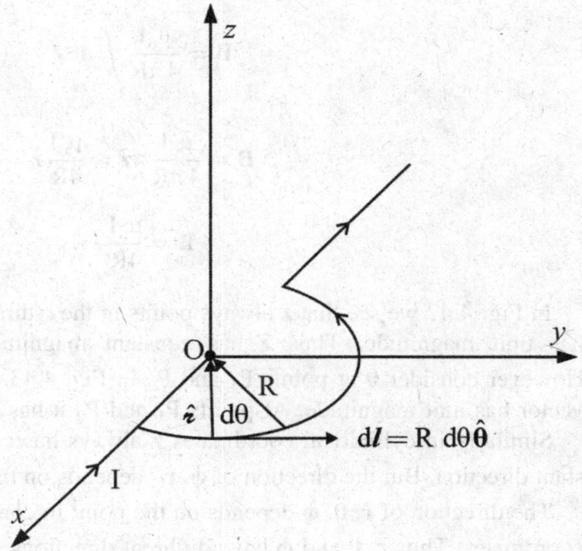


**Fig. 4.42** A long wire carrying current I. A small portion of the wire is bent into a semicircle of radius R

For the straight portion of the wire angle between $d\boldsymbol{l}$ and $\hat{\boldsymbol{\imath}}$ is zero. Hence for the straight portion

$$\mathbf{B} = \frac{\mu_o I}{4\pi}\int \frac{d\boldsymbol{l} \times \hat{\boldsymbol{\imath}}}{\imath^2}$$

$$\mathbf{B} = \frac{\mu_o I}{4\pi}\int \frac{dl' \sin 0}{\imath^2}$$

$$\mathbf{B} = 0.$$

For the curved portion

$$\mathbf{B} = \frac{\mu_o I}{4\pi}\int \frac{d\boldsymbol{l} \times \hat{\boldsymbol{\imath}}}{\imath^2}$$

From Fig. 4.42 $\imath = R$ and $d\boldsymbol{l} = R d\theta\,\hat{\boldsymbol{\theta}}$

$$\mathbf{B} = \frac{\mu_o I}{4\pi}\int_0^{\pi} \frac{R d\theta(\hat{\boldsymbol{\theta}} \times \hat{\boldsymbol{\imath}})}{\imath^2}$$

$$\mathbf{B} = \frac{\mu_o I}{4\pi R}\int_0^{\pi} d\theta\,\widehat{\mathbf{z}}$$

$$\mathbf{B} = \frac{\mu_o I}{4\pi R}\pi\widehat{\mathbf{z}} = \frac{\mu_o I}{4R}\widehat{\mathbf{z}}$$

$$\mathbf{B} = \frac{\mu_o I}{4R}\widehat{\mathbf{z}}$$

In Fig. 4.42 we see that $\hat{\mathbf{z}}$ always points in the $z$-direction. Being a unit vector it has unit magnitude. Thus $\hat{\mathbf{z}}$ has constant magnitude and constant direction. However consider $\hat{\boldsymbol{\theta}}$ at points $P_1$ and $P_2$ in Fig. 4.43. We see that $\hat{\boldsymbol{\theta}}$ being a unit vector has unit magnitude. At points $P_1$ and $P_2$ it has different directions.

Similarly in cylindrical coordinates $\hat{\mathbf{z}}$ always has constant magnitude and constant direction. But the direction of $\hat{\boldsymbol{\phi}}$, $\hat{\mathbf{r}}_C$ depends on the point in the consideration.

The direction of $\hat{\mathbf{r}}$, $\hat{\boldsymbol{\theta}}$, $\hat{\boldsymbol{\phi}}$ depends on the point in consideration in spherical polar coordinates. Thus $\hat{\mathbf{r}}$, $\hat{\boldsymbol{\theta}}$ and $\hat{\boldsymbol{\phi}}$ have different directions depending upon the point in consideration in spherical polar coordinates.

Hence whenever working with $\left(\hat{\mathbf{r}}_c, \hat{\boldsymbol{\phi}}, \hat{\mathbf{z}}\right)$ in cylindrical coordinates, $(\hat{\mathbf{r}}_c, \hat{\boldsymbol{\theta}}, \hat{\boldsymbol{\phi}})$ in spherical polar coordinates better to express these unit vectors in terms of cartesian coordinates $\hat{\mathbf{i}}, \hat{\mathbf{j}}, \hat{\mathbf{k}}$ using the following equations.

In cylindrical coordinates

$$\left.\begin{aligned} \hat{\mathbf{r}}_c &= \cos\phi\hat{\mathbf{i}} + \sin\phi\hat{\mathbf{j}} \\ \hat{\boldsymbol{\phi}} &= -\sin\phi\hat{\mathbf{i}} + \cos\phi\hat{\mathbf{j}} \\ \hat{\mathbf{z}} &= \hat{\mathbf{k}} \end{aligned}\right\} \tag{4.247}$$

In spherical polar coordinates

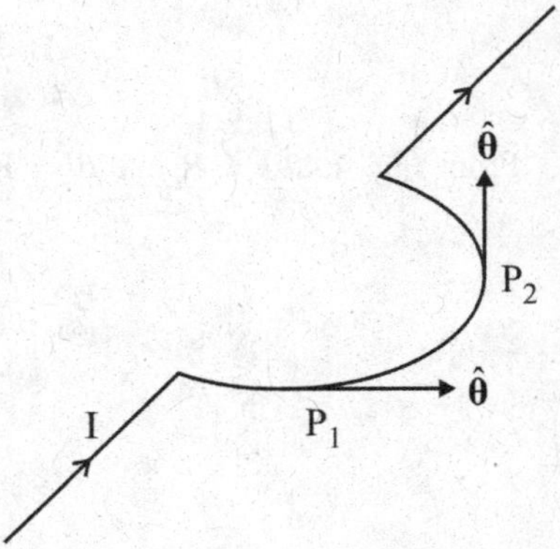


**Fig. 4.43** Figure depicts unit vectors at points $P_1$ and $P_2$ have different directions

$$\left.\begin{aligned}\hat{\mathbf{r}} &= \sin\theta\cos\phi\hat{\mathbf{i}} + \sin\theta\sin\phi\hat{\mathbf{j}} + \cos\theta\hat{\mathbf{k}}\\ \hat{\boldsymbol{\theta}} &= \cos\theta\cos\phi\hat{\mathbf{i}} + \cos\theta\sin\phi\hat{\mathbf{j}} - \sin\theta\hat{\mathbf{k}}\\ \hat{\boldsymbol{\phi}} &= -\sin\theta\hat{\mathbf{i}} + \cos\phi\hat{\mathbf{j}}\end{aligned}\right\} \tag{4.248}$$

The unit vectors $\hat{\mathbf{i}}, \hat{\mathbf{j}}, \hat{\mathbf{k}}$ have constant magnitude and direction. The unit vectors always point in *x*-, *y*- and *z*-direction, respectively, and hence are easy to work with.

**Example 4.10**
If **B** is given by $\mathbf{B} = \frac{\kappa}{r}\cos\phi\hat{\mathbf{r}}_c$

Calculate the magnetic flux in cylindrical coordinates for the surface $0 \le z \le 2\text{m}$ and $-\pi/2 \le \phi \le \pi/2$ (κ is a constant).

**Solution**
We know that

$$\Phi = \int \mathbf{B} \cdot d\mathbf{s}$$

$$\Rightarrow \Phi = \int_0^2 \int_{-\pi/2}^{\pi/2} \frac{\kappa}{r}\cos\phi\hat{\mathbf{r}}_c \cdot r d\phi dz \hat{\mathbf{r}}_c$$

$$\Rightarrow \Phi = 2\int_{-\pi/2}^{\pi/2} \cos\phi\, d\phi$$

$$\Rightarrow \Phi = 2[\sin\phi]_{-\pi/2}^{\pi/2}$$

$$\Rightarrow \Phi = 4 \text{ webers.}$$

**Example 4.11**
Consider a regular hexagon of side a carrying current I. Find the expression for **B** at the centre of the hexagon.

**Solution**
Consider a hexagon of side a as shown in Fig. 4.44. The **B** produced by side $A_1 A_2$ at point is given by (Eq. 4.146 with L = a/2)

$$B_{A_1A_2} = \frac{\mu_0 I}{2\pi r}\left[\frac{a/2}{\sqrt{r^2 + a^2/4}}\right]$$

From Fig. 4.44

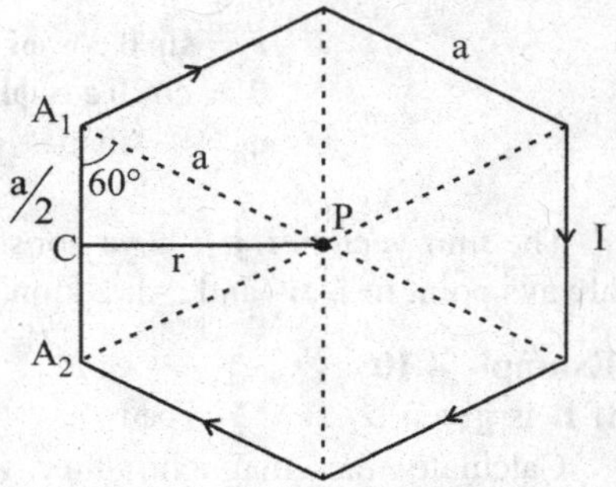


**Fig. 4.44** A regular hexagon of side a carrying current I

$$a^2 = \frac{a^2}{4} + r^2$$

$$\Rightarrow r^2 = \frac{3a^2}{4}, \quad r = \frac{\sqrt{3}a}{2}.$$

Substituting for $r^2$, r in $B_{A_1A_2}$

$$B_{A_1A_2} = \frac{\mu_o I}{2\,\pi r}\left[\frac{\frac{a}{2}}{\sqrt{\frac{3a^2}{4} + \frac{a^2}{4}}}\right]$$

$$B_{A_1A_2} = \frac{\mu_o I}{4\,\pi r}$$

$$B_{A_1A_2} = \frac{\mu_o I}{4\,\pi \frac{\sqrt{3}a}{2}} = \frac{\mu_o I}{2\,\pi\sqrt{3}a}$$

The hexagon is having six sides and hence the total **B** at point P is

$$B_{tot} = \frac{6\mu_o I}{2\,\pi\sqrt{3}a} = \frac{\sqrt{3}\mu_o I}{\pi a}$$

At point P the **B** points inward perpendicular to the plane of paper.

**Example 4.12**
Consider a long solenoid of length 2 m having 2000 turns of wire.

The mean radius is 20 cm. A current of 30 A flows through the wire. Calculate **B** at the centre of the solenoid along its axis.

**Solution**
Consider Fig. 4.45. From Eq. 4.170

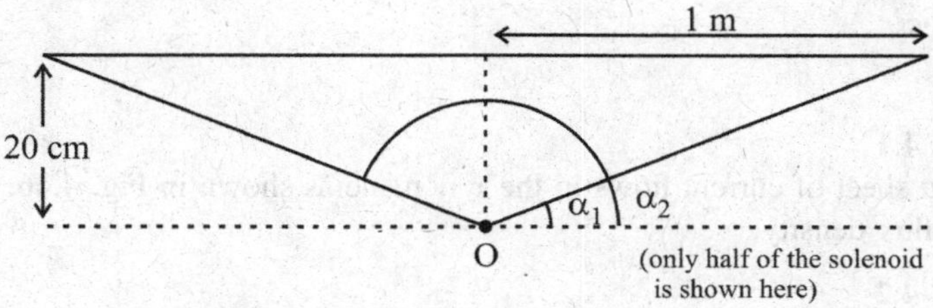


**Fig. 4.45** A long solenoid carrying current I

$$B = \frac{\mu_o IN}{2L}[\cos\alpha_1 - \cos\alpha_2]$$

$$B = \frac{\mu_o In}{2}[\cos\alpha_1 - \cos\alpha_2] \text{ with } N = \frac{n}{L}$$

Given I = 30 A, N = 1000 turns/m.

$$\mu_o = 4\pi \times 10^{-7}\frac{\text{weber}}{\text{A} \cdot \text{m}}$$

$$\cos\alpha_1 = \frac{1}{\sqrt{1+(0.2)^2}}$$

$$\cos\alpha_1 = \frac{1}{\sqrt{1.4}}$$

$$\cos\alpha_2 = -\cos(\pi - \alpha_1) = -\frac{1}{\sqrt{1.4}}$$

$$B = \frac{(4\pi \times 10^{-7})(30)(1000)}{2}\left[\frac{2}{\sqrt{1.4}}\right]$$

$$B = 3.18 \times 10^{-2}\ \text{weber/m}^2.$$

## Exercises

### Problems 4.1
An infinite sheet of current flows in the $x$–$y$ plane as shown in Fig. 4.46. Find the magnetic flux density.

### Problems 4.2
Show that for Fig. 4.47a resolving into parallel and perpendicular components $B_{1n} = B_{2n}$ and $B_{1t} - B_{2t} = \mu_o K$. Similarly for Fig. 4.47b show that $B_{1n} = B_{2n}, B_{1t} = B_{2t}$.

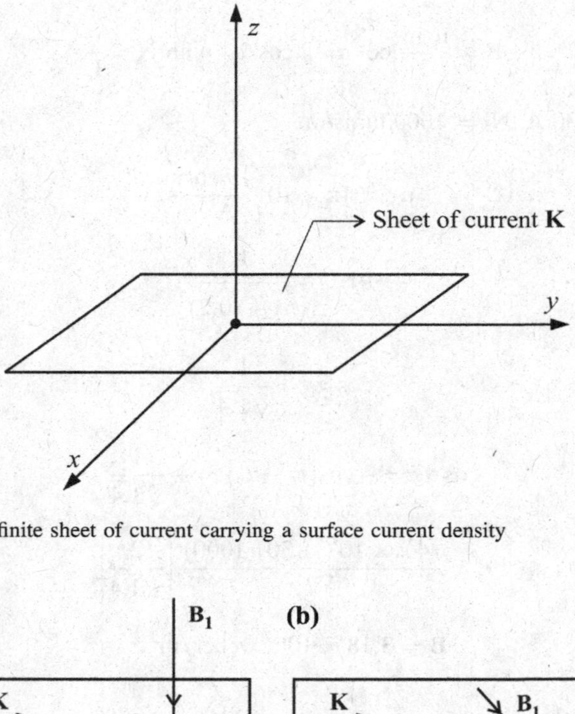


**Fig. 4.46** An infinite sheet of current carrying a surface current density

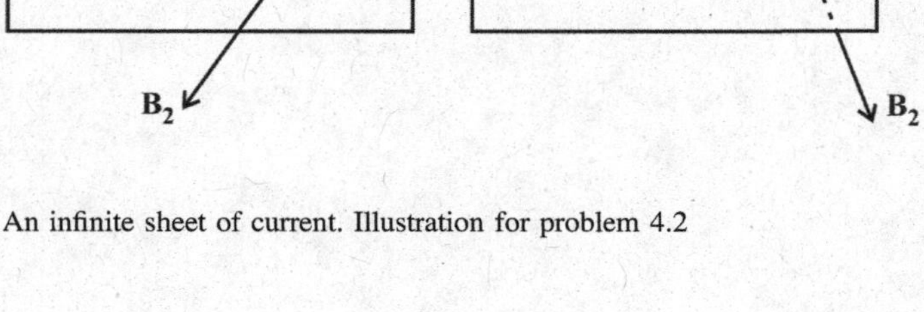

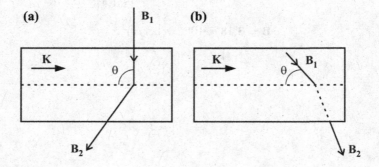

**Fig. 4.47** An infinite sheet of current. Illustration for problem 4.2

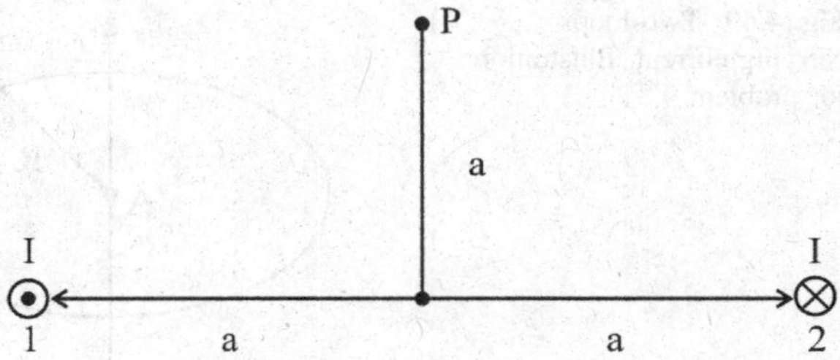


**Fig. 4.48** Two infinite wires each carrying current I. Illustration for problem 4.4

**Problem 4.3**
Consider a cylindrical conductor of radius R. The current density within the cylindrical region is given by $\mathbf{J} = Ce^{-\kappa r}\hat{\mathbf{z}}$ where C and κ are positive constants. Determine magnetic flux density everywhere.

**Problem 4.4**
Consider two infinite wires each carrying current I as shown in Fig. (4.48). Calculate **B** at point P in Fig. (4.48).

**Problem 4.5**
A solid infinite cylindrical conductor with radius R is carrying a current I. Find the **B** both inside and outside the cylinder. Sketch the variation of **B** as a function of distance from the conductor axis.

**Problem 4.6**
Consider a wire of radius R. A current I flows through the wire. Find

(a) The volume current density if the current distribution is such that it is directly proportional to distance from the axis
(b) If the charges are moving such that its volume charge density is ρ then calculate their velocity.

**Problem 4.7**
Consider circular coils carrying current $I_1$ and $I_2$ and placed apart by a distance $\boldsymbol{r}$. Calculate the force caused by current element $I_1 dl_1$ on $I_2 dl_2$ and by $I_2 dl_2$ on $I_1 dl_1$. Then calculate the force exerted by coil 1 on coil 2 and coil 2 on coil 1.

**Problem 4.8**
Consider two loops of radius 15 cm situated as shown in Fig. 4.49. Each loop is carrying a current of 10A. The distance between the centre of loop is 20 cm. Calculate the **B** at (a) $z = 10$ cm (b) $z = 20$ cm. (c) $z = 30$ cm.

**Problem 4.9**
If the current in Fig. 4.27 is reversed how will the answer to Sect. 4.18 change. Plot the magnetic flux density for Fig. 4.50.

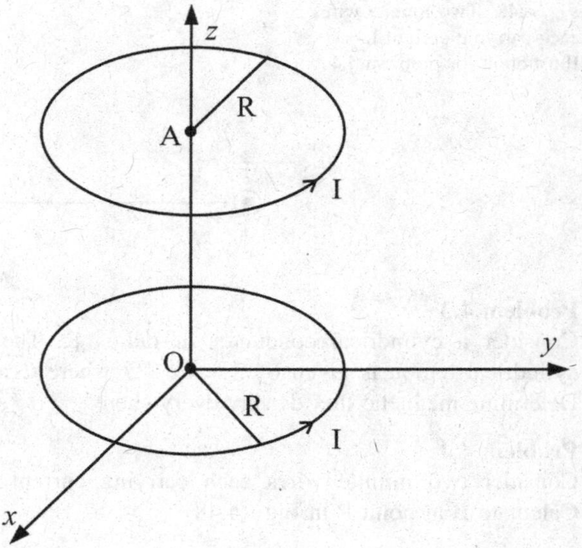


**Fig. 4.49** Two loops carrying current. Illustration for problem 4.8

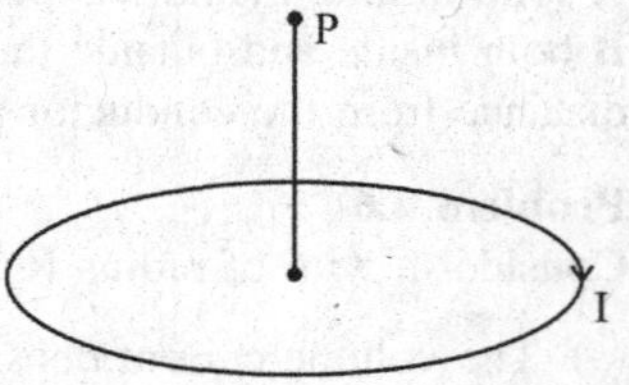


**Fig. 4.50** A circular loop carrying current I

**Problem 4.10**

Find the value of **A** and **B** for two long parallel wires separated by a distance h. Both the wires carry current I in opposite direction as shown in Fig. 4.51.

**Problem 4.11**

Consider a spherical shell of radius R. It carries a uniform surface charge. If it is rotating about its own axis find the vector potential at a point outside the spherical shell.

**Problem 4.12**

Two infinitely long wires each carrying current of 15 A are situated as shown in Fig. 4.52. Find the value of **B** at the points $P_1$, $P_2$ and $P_3$. Clearly mention the direction of **B** at each of these points. Both the wires are carrying currents in the same direction.

**Problem 4.13**

In Problem 4.12 if the wires are carrying current in opposite direction, how the answer to Problem 4.12 will get modified.

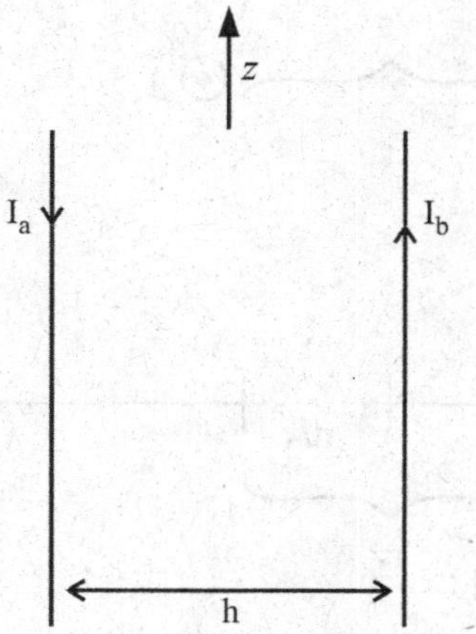


**Fig. 4.51** Two long parallel wires carrying current. Illustration for problem 4.10

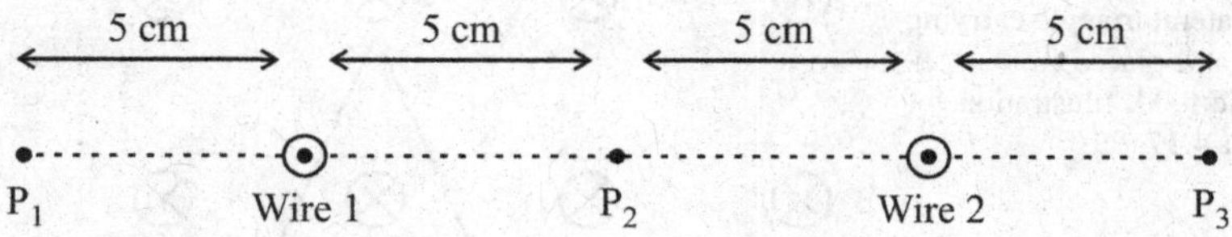


**Fig. 4.52** Illustration for problem 4.12

**Problem 4.14**

Find the **J** which will produce a vector potential

(a) $\mathbf{A} = C \sin\theta\, \hat{\boldsymbol{\theta}}$ in spherical polar coordinates.
(b) $\mathbf{A} = y^2 \cos x\, \hat{\mathbf{i}} + e^{-x}\hat{\mathbf{k}}$ in Cartesian coordinates.
(c) $\mathbf{A} = Cr^2 \hat{\boldsymbol{\phi}}$ in cylindrical coordinates.

where C is a constant.

**Problem 4.15**

Three infinitely long wires a, b and c each carrying current $I_1$, $I_2$ and $I_3$ are isolated from each other as shown in Fig. 4.53. Calculate the magnetic force on $d\boldsymbol{l}_2$ and $d\boldsymbol{l}_3$ due to $I_1$.

**Problem 4.16**

Find the vector potential above and below the plane in Problem 4.1

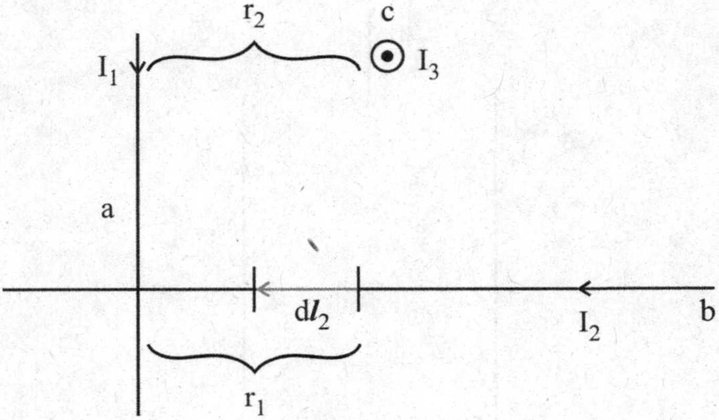


**Fig. 4.53** Three infinitely long wires carrying current. Illustration for problem 4.15

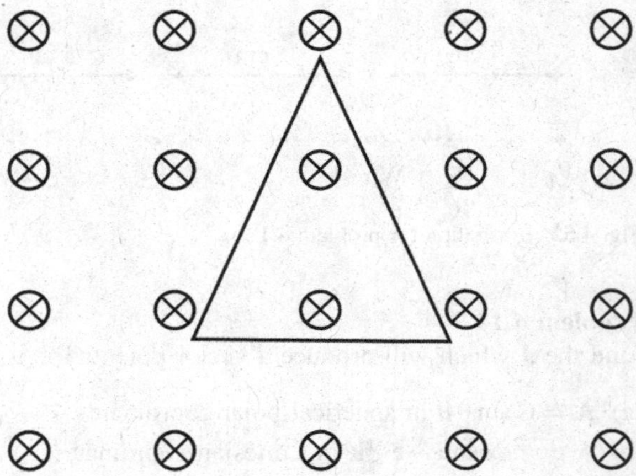

**Fig. 4.54** A wire in the form of equilateral triangle carrying current and placed in a magnetic field. Illustration for problem 4.17.

### Problem 4.17

A wire is bent in the form of an equilateral triangle. The side of the triangle is 5 cm. A current of 10 A flows in the triangle. The entire triangle is placed in a magnetic field as shown in Fig. 4.54. Find the forces on the three sides of the triangle.

### Problem 4.18

Calculate the magnetic dipole moment of spinning spherical shell in Problem 4.11.

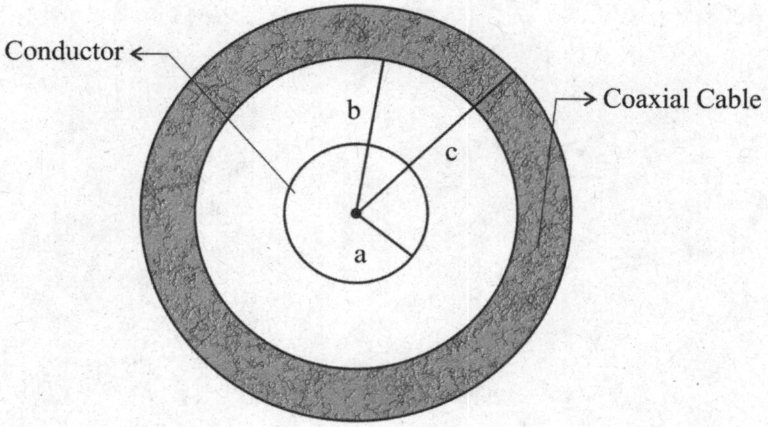


**Fig. 4.55** A cylindrical conductor placed in a cylindrical coaxial cable. Illustration for problem 4.19

**Problem 4.19**

A cylindrical conductor is placed in a cylindrical coaxial cable as shown in Fig. 4.55. The inner conductor carries a current I uniformly distributed over the conductor. The coaxial cable carries an equal current in opposite direction with uniform current density. Find the value of **B** at a distance r from the centre where (a) $r < a$ (b) $a < r < b$ (c) $b < r < c$ (d) $r > c$.

# Chapter 5
# Magnetic Fields in Materials

## 5.1 Introduction

In the previous chapter we calculated the magnetic flux density **B** for a given current distribution and the current distribution was assumed to be present in vacuum. For example we calculated the magnetic flux density **B** of a solenoid. Inside the solenoid it was assumed that there is a free space or vacuum. Now instead of free space or vacuum inside the solenoid if the free space is filled with a material, will the magnetic flux density inside the solenoid get altered? The answer is yes.

To be precise, suppose assume that we place a material in a magnetic field. How then the magnetic field gets changed? Our purpose is this chapter is to answer the above questions.

## 5.2 Diamagnetism, Paramagnetism and Ferromagnetism

In the previous chapter we saw that magnetic monopoles do not exist in nature. It is the circulation of electric charges which produce magnetic field. However if we take a bar magnet we will see it has nothing to do with currents but still it produces a magnetic field. If you examine the bar magnet at a microscopic level you will still observe very small currents, i.e. electrons orbiting about the nuclei.[1]

As we will see this electron motion gives rise to macroscopic magnetic properties of materials. The electronic current loops are small enough so that they can be treated as magnetic dipoles and in Sect. 4.22 we have seen that the magnetic dipole posses a magnetic dipole moment.

[1]Strictly speaking the motion of electrons must be studied using the theories of quantum mechanics.

© Springer Nature Singapore Pte Ltd. 2020

S. Balaji, *Electromagnetics Made Easy*,
https://doi.org/10.1007/978-981-15-2658-9_5

In the case of diamagnetic materials say for example hydrogen molecule, a number of electrons present in each atom tend to pair up. The result is magnetic effects of electrons both due to spin as well as orbital motion exactly cancel out each other and the atom as a whole is devoid of any permanent magnetic moment. "When such materials are placed in a magnetic field there will be a change in the magnetic moment and the change will be such that it is opposite to the direction of the applied magnetic field" (see Problem 5.1). These substances are repelled away from the applied magnetic field.

In the case of paramagnetic materials say for example copper, there will be at least are electron which doesn't pair up with other electrons. In such cases the atom has a permanent magnetic moment. If we take a bulk of such material due to random alignments of such atomic magnetic dipoles net magnetic moment is zero as shown in Fig. 5.1a. However when such materials are placed in an external magnetic field a "torque" (see Problem 5.2) acts on the atomic dipoles tending to align it in the direction of the magnetic field as shown in Fig. 5.1b. These substances known as paramagnetic substances are feebly attracted by external magnetic field.

Ferromagnetic materials in comparison to paramagnetic materials are distinguished by high degree of alignment of atomic dipoles. Ferromagnetic substances exhibit almost all the properties of paramagnetic materials but to a much larger extent. A strong coupling force exists between atoms in a small region of the ferromagnetic material which makes the magnetic dipole moments of all the atoms contained in the small region to orient in the same direction. This small region is called domain. A ferromagnetic material contains large number of such domains as shown in Fig. 5.2. In the absence of an external applied magnetic field the net magnetic moments of domain's point in different directions as shown in Fig. 5.2 and hence as a whole the magnetic moment of the entire material is zero. However when a ferromagnetic material is placed in a external magnetic field and the magnetic field is increased domains whose magnetization are parallel to the applied field grow and the others gets shrinked. At very high fields one domain grows over completely and then the ferromagnetic material is said to be saturated.

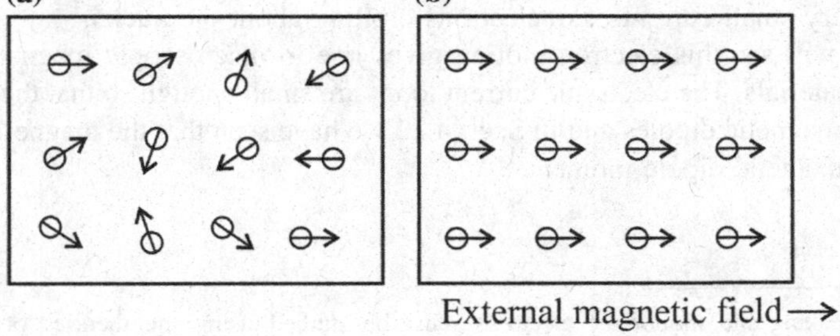


**Fig. 5.1** **a** Randomly oriented atomic magnetic dipoles in a paramagnetic material. **b** Atomic magnetic dipoles aligned in the direction of the externally applied magnetic field

**Fig. 5.2** Domains in a ferromagnetic material

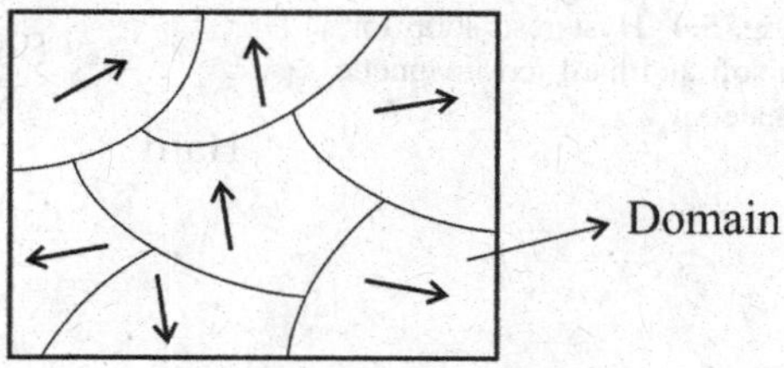


Let us now switch off the external magnetic field. It is observed that the domain growth process is irreversible. There is some return of the domains to the randomly oriented state, but the process is incomplete. Hence now the ferromagnetic material turns out to be a permanent magnet.

Now suppose you put a ferromagnetic material is a magnetic field and slowly increase the field. The magnetization and the flux density **B** increases slowly as represented by the part OA of the curve is Fig. 5.3 and reaches a maximum value at A. The ferromagnetic material is said to have reached the saturation.

If we reduce the magnetic field to zero the flux density is the magnetic material decreases at a lessor rate and does not become zero at **H** = 0. The flux density OC left is the specimen is called residual magnetism (or) remanence (or) retentivity.

If we now reverse the direction of the field **H**, the value of **B** decreases and finally becomes zero for a particular value of −**H**. This decrease is represented as CD in Fig. 5.3. The value of **H** corresponding to OD is Fig. 5.3 which renders **B** to zero is called coercive force or coercivity of the material.

With further increase in **H** in the reverse direction the part DE of the curve is obtained. Here magnetic saturation occurs at E, symmetrical to that of A. If we now reduce the magnetizing field **H** to zero and then increase **H** in positive direction we reach back A along the path EFGA.

**Fig. 5.3** Hysteresis loop for a ferromagnetic material

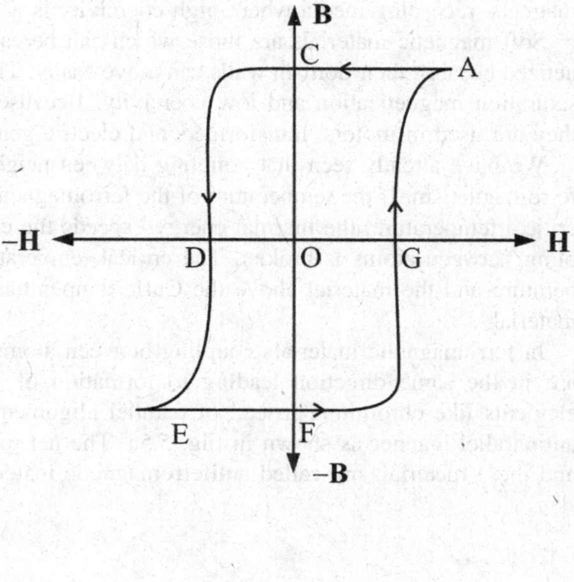

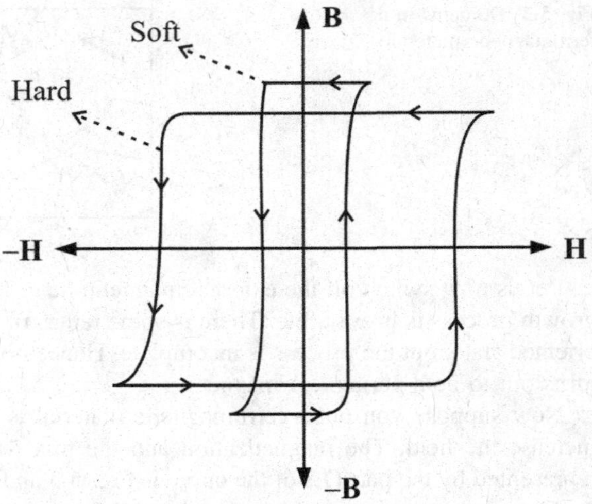


**Fig. 5.4** Hysteresis loop for a soft and hard ferromagnetic material

From Fig. 5.3 it is clear that the flux density **B** always lag behind the magnetizing field **H**, when the ferromagnetic material is taken through a complete cycle of magnetization. This loss of magnetization behind the magnetizing field is called hysteresis and the curve shown in Fig. 5.3 is called hysteresis loop or curve. The area under the curve is the energy lost per unit volume per cycle. This loss is also known as hysterestis loss and is the energy used to overcome friction encountered during domain wall motion and domain rotation.

Ferromagnetic materials in general can be classified into two types: hard magnetic materials and soft magnetic materials. The typical hysteresis loop for hard and soft ferromagnetic material is shown in Fig. 5.4.

Hard magnetic materials have high coercivity and show high resistance to magnetization and demagnetization. Hence these materials are used in permanent magnets, recording media where high coercivity is needed.

Soft magnetic materials are those which can be easily magnetized and demagnetized because their domain walls can move easily. They are characterized by high saturation magnetization and low coercivity. Because their magnetization is high they are used in motors, transformers and electric generators.

We have already seen that coupling between neighbouring atoms gives rise to ferromagnetism. If the temperature of the ferromagnetic materials is raised above a critical temperature, the thermal energy exceeds the coupling energy and the coupling between atoms is broken. The critical temperature is known as Curie temperature and the material above the Curie temperature behaves like paramagnetic material.

In ferromagnetic materials coupling between atoms is such that the alignments are in the same direction leading to formation of domains. However in some elements like chromium instead of parallel alignment atomic dipoles align in an antiparallel manner as shown in Fig. 5.5a. The net magnetic moment is thus zero and these materials are called antiferromagnetic material.

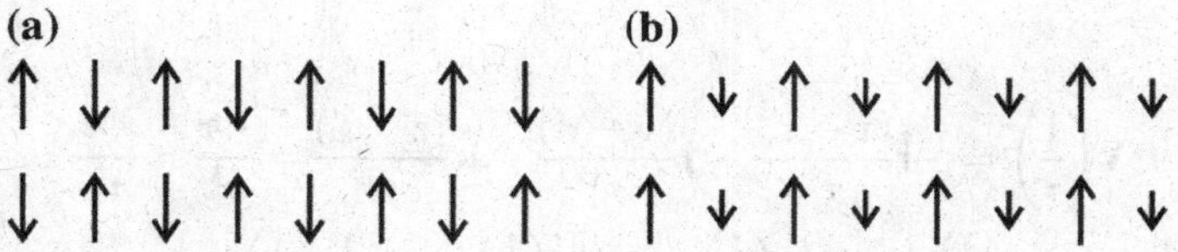


**Fig. 5.5** **a** Atomic magnetic dipoles of equal magnitude having anti-parallel arrangement in a antiferromagnetic material. **b** Atomic magnetic dipoles of unequal magnitude having anti-parallel arrangement in a ferrimagnetic material

There are another class of materials known as ferrimagnetic materials. In these materials as shown in Fig. 5.5b the magnetic moments are antiparallel but are of unequal magnitude. Hence the net magnetic moment will not be zero and a spontaneous magnetization remains. For example if the adjacent moments correspond to different ions [say $Fe^{2+}$ and $Fe^{3+}$] then they will have unequal magnetic moments leading to incomplete cancellation of dipole moments.

**Example 5.1**
Prove that

$$\nabla\left(\frac{1}{\mathscr{r}}\right) = -\nabla'\left(\frac{1}{\mathscr{r}}\right)$$

**Solution**
From Fig. 4.15

$$\boldsymbol{\mathscr{r}} = (x-x')\hat{\mathbf{i}} + (y-y')\hat{\mathbf{j}} + (z-z')\hat{\mathbf{k}}$$

$$\mathscr{r}^2 = (x-x')^2 + (y-y')^2 + (z-z')^2$$

$$\mathscr{r} = \sqrt{(x-x')^2 + (y-y')^2 + (z-z')^2}$$

$$\nabla\left(\frac{1}{\mathscr{r}}\right) = \hat{\mathbf{i}}\frac{\partial}{\partial x}\left(\frac{1}{\mathscr{r}}\right) + \hat{\mathbf{j}}\frac{\partial}{\partial y}\left(\frac{1}{\mathscr{r}}\right) + \hat{\mathbf{k}}\frac{\partial}{\partial z}\left(\frac{1}{\mathscr{r}}\right)$$

$$\frac{\partial}{\partial x}\left(\frac{1}{\mathscr{r}}\right) = \frac{-(x-x')}{\mathscr{r}^3}$$

$$\frac{\partial}{\partial y}\left(\frac{1}{\mathscr{r}}\right) = \frac{-(y-y')}{\mathscr{r}^3}$$

$$\frac{\partial}{\partial z}\left(\frac{1}{\mathscr{r}}\right) = \frac{-(z-z')}{\mathscr{r}^3}$$

Hence

$$\nabla\left(\frac{1}{\mathscr{r}}\right) = -\hat{\mathbf{i}}\frac{(x-x')}{\mathscr{r}^3} - \hat{\mathbf{j}}\frac{(y-y')}{\mathscr{r}^3} - \hat{\mathbf{k}}\frac{(z-z')}{\mathscr{r}^3} = \frac{-\boldsymbol{\mathscr{r}}}{\mathscr{r}^3} = \frac{-\hat{\boldsymbol{\mathscr{r}}}}{\mathscr{r}^2}$$

$$\text{and } \nabla'\left(\frac{1}{\mathscr{r}}\right) = \hat{\mathbf{i}}\frac{\partial}{\partial x'}\left(\frac{1}{\mathscr{r}}\right) + \hat{\mathbf{j}}\frac{\partial}{\partial y'}\left(\frac{1}{\mathscr{r}}\right) + \hat{\mathbf{k}}\frac{\partial}{\partial z'}\left(\frac{1}{\mathscr{r}}\right)$$

$$\frac{\partial}{\partial x'}\left(\frac{1}{\mathscr{r}}\right) = \frac{(x-x')}{\mathscr{r}}$$

$$\frac{\partial}{\partial y'}\left(\frac{1}{\mathscr{r}}\right) = \frac{y-y'}{\mathscr{r}}$$

$$\frac{\partial}{\partial z'}\left(\frac{1}{\mathscr{r}}\right) = \frac{z-z'}{\mathscr{r}}$$

So that

$$\nabla'\left(\frac{1}{\mathscr{r}}\right) = \frac{\hat{\boldsymbol{\mathscr{r}}}}{\mathscr{r}^2}$$

Thus $\nabla\left(\frac{1}{\mathscr{r}}\right) = -\nabla'\left(\frac{1}{\mathscr{r}}\right)$.

## 5.3 Magnetization: Bound Currents

Whenever we place a material in a magnetic field the material becomes magnetized. Two mechanisms are important for the magnetization of material:

(i) The change in orbital dipole moment due to applied magnetic field is such that it has opposite direction of the applied field—Diamagnetism.
(ii) The dipoles (or) domains in the presence of applied magnetic field experience a torque tending to align them in the direction of applied field—Paramagnetism and Ferromagnetism.

We define the magnetization **M** as "Magnetic moment per unit volume". We will not bother about the external magnetic field which produces the magnetization, but we will now calculate the magnetic field produced by the magnetization **M**.

Consider a magnetized material whose magnetization **M** is given. In order to calculate the magnetic field produced by **M** we will first write the expression for vector potential **A** for a single dipole [from Eq. 4.211].

$$\mathbf{A} = \frac{\mu_0}{4\pi}\left(\frac{\mathbf{m}\times\hat{\boldsymbol{\mathscr{r}}}}{\mathscr{r}^2}\right) \tag{5.1}$$

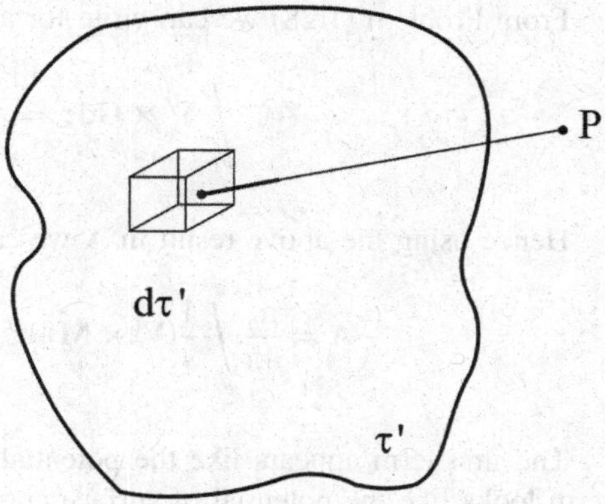


**Fig. 5.6** A magnetized material along with a small volume element is shown in the figure

A small volume element $d\tau'$ in the magnetized material carries a dipole moment $\mathbf{M}\ d\tau'$, so that the vector potential due to the entire volume is $\tau'$ (Fig. 5.6).

$$\mathbf{A} = \frac{\mu_0}{4\pi}\int \frac{\mathbf{M} \times \hat{\boldsymbol{\imath}}}{\boldsymbol{\imath}^2} d\tau' \tag{5.2}$$

Note that **M** here is a function of source coordinates while **A** is a function of field coordinaties. We know that

$$\nabla'\left(\frac{1}{\boldsymbol{\imath}}\right) = \frac{\hat{\boldsymbol{\imath}}}{\boldsymbol{\imath}^2}$$

From Example 5.1.

Hence

$$\mathbf{A} = \frac{\mu_0}{4\pi}\int \left[\mathbf{M} \times \nabla'\left(\frac{1}{\boldsymbol{\imath}}\right)\right] d\tau' \tag{5.3}$$

From Sect. 1.20 using vector identity (5.6) we get

$$\mathbf{M} \times \nabla'\left(\frac{1}{\boldsymbol{\imath}}\right) = \frac{1}{\boldsymbol{\imath}}\nabla' \times \mathbf{M} - \nabla' \times \left(\frac{\mathbf{M}}{\boldsymbol{\imath}}\right) \tag{5.4}$$

Hence substituting Eq. 5.4 in 5.3

$$\mathbf{A} = \frac{\mu_0}{4\pi}\int_{\tau'} \frac{\nabla' \times \mathbf{M}}{\boldsymbol{\imath}} d\tau' - \frac{\mu_0}{4\pi}\int_{\tau'} \nabla' \times \frac{\mathbf{M}}{\boldsymbol{\imath}} d\tau' \tag{5.5}$$

From Problem (1.28) we can write for any vector **G**

$$\int_{\tau'} \nabla' \times \mathbf{G} d\tau' = -\oint_{S'} \mathbf{G} \times d\mathbf{s}'$$

Hence using the above result in **A** we can write

$$\mathbf{A} = \frac{\mu_o}{4\pi}\int_{\tau'} \frac{1}{r}(\nabla' \times \mathbf{M})d\tau' + \frac{\mu_o}{4\pi}\oint_{S'} \frac{1}{r}(\mathbf{M} \times d\mathbf{s}') \tag{5.6}$$

The first term appears like the potential of volume current $\mathbf{J}_p$, and the second term looks like the potential of surface current $\mathbf{K}_p$. Hence with

$$\mathbf{J}_p = \nabla \times \mathbf{M}, \quad \mathbf{K}_p = \mathbf{M} \times \hat{\mathbf{n}} \tag{5.7}$$

$$A = \frac{\mu_o}{4\pi}\int_{\tau'} \frac{\mathbf{J}_p}{r} d\tau' + \frac{\mu_o}{4\pi}\oint_{S'} \frac{\mathbf{K}_p}{r} ds' \tag{5.8}$$

where $\hat{\mathbf{n}}$ is the normal unit vector to the surface.

The above expression shows that in order to calculate **A** and in turn magnetic flux density produced by magnetization **M**, instead of integrating the contributions from all dipoles we find the fields produced by $\mathbf{J}_p$ and $\mathbf{K}_p$, the bound currents and sum them up.

## 5.4 Physical Interpretation of Bound Currents

We have seen that whenever a material is placed in an external magnetic field the material gets magnetized. In Sect. 5.3 we calculated the magnetic field produced by these magnetized material. Finally we found the magnetic field produced by the magnetized material is same as that of produced by bound currents $\mathbf{J}_p$ and $\mathbf{K}_p$.

Now let us see how these currents arise physically inside the magnetized material. Consider a thin slab of uniformly magnetized material as shown in Fig. 5.7a. The dipoles are shown by tiny current loops. In the interior of the material for two neighbouring current loops the currents in the adjacent side are in opposite direction and cancel out each other. However at the edges no such cancellation is possible. Hence the result is a surface current as shown in Fig. 5.7b. When the magnetization is non-uniform the internal currents do not completely cancel out each other. In that case we get bound currents inside the volume.

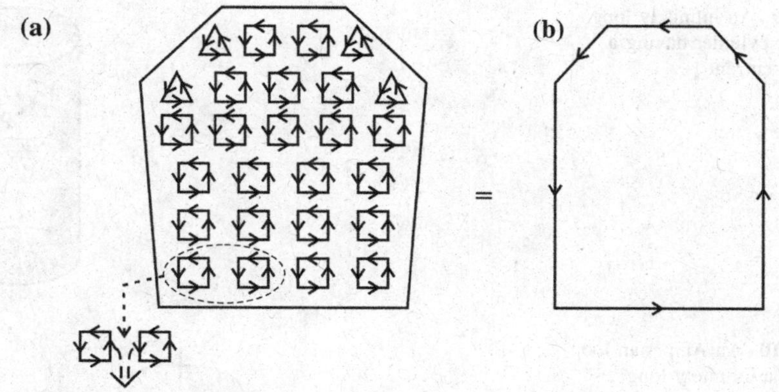


**Fig. 5.7** A thin slab of uniformly magnetized material shown in Fig. a. Currents in adjacent dipoles depicted as tiny current loops cancel out each other resulting in a surface current as shown in Fig. b

**Example 5.2**

Consider an infinitely long circular cylinder which is carrying a uniform magnetization **M** parallel to its axis. Calculate the magnetic flux density inside and outside the cylinder due to the magnetization **M**.

**Solution**

Consider the cylinder as shown in the Fig. 5.8. The volume bound current is given by $\mathbf{J}_p = \nabla \times \mathbf{M} = 0$ because the magnetization is uniform.

The surface bound current is $\mathbf{K}_p = \mathbf{M} \times \hat{\mathbf{n}}$

As shown in Fig. 5.8 if we rewrite the above expression in terms of cylindrical coordinates.

$$\mathbf{K}_p = \mathbf{M} \times \hat{\mathbf{r}}_c = (M)\hat{\mathbf{z}} \times \hat{\mathbf{r}}_c = M\hat{\boldsymbol{\phi}}$$

So finally we have the surface current over the infinite circular cylinder as shown in Fig. 5.9.

**Fig. 5.8** An infinitely long circular cylinder carrying a uniform magnetization

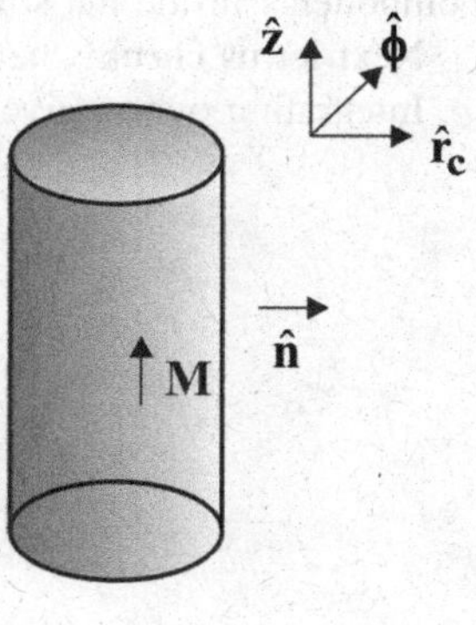

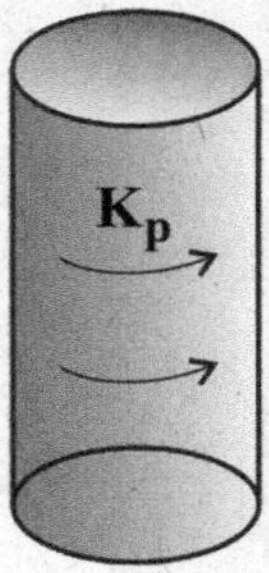


**Fig. 5.9** An infinitely long circular cylinder having a surface current

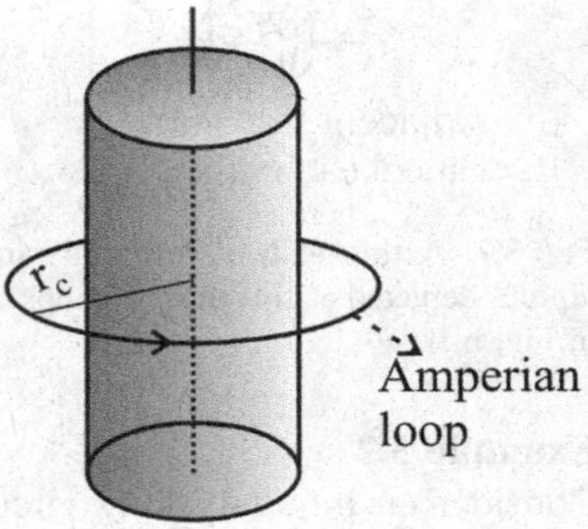


**Fig. 5.10** An Amperian loop around a infinitely long circular cylinder

Now let us apply Ampere's circuital law to calculate **B** inside and outside the cylinder in Fig. 5.9.

First let us see whether there can be any circumferential components.

In Fig. 5.10 we see that $B_\phi$ is constant around the Amperian loop concentric with the cylinder and the net current enclosed by the loop is zero. Hence for the Amperian loop shown in Fig. 5.10.

$$\oint \mathbf{B} \cdot d\boldsymbol{l} = B_\phi(2\pi r_c) = \mu_o I_{enc} = 0$$
$$\Rightarrow B_\phi = 0$$

So there are no circumferential components. For similar reasons circumferential components inside the solenoid is zero.

Next let us check whether there are $B_r$ components. We know that $\nabla \cdot \mathbf{B} = 0$

Integrating over a given volume $\tau$

$$\int \nabla \cdot \quad d = 0$$

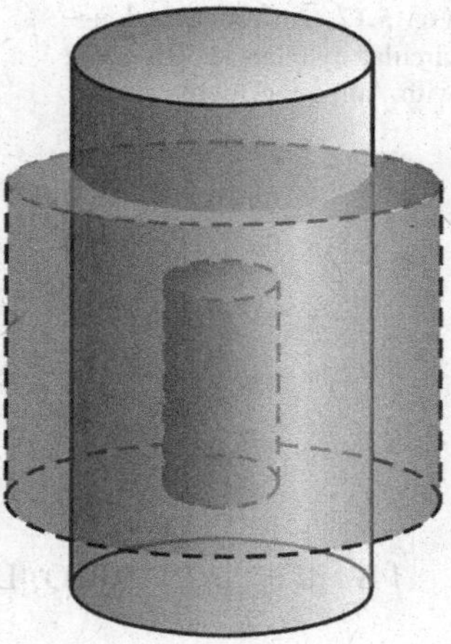

**Fig. 5.11** Two cylindrical surfaces shown in dotted lines either larger or smaller than the cylinder considered in the problem

Applying Gauss's divergence theorem

$$\Rightarrow \oint \mathbf{B} \cdot \mathbf{ds} = 0.$$

Let us apply the above equation to our cylinder shown in Fig. 5.11 where we show two cylindrical surfaces (dotted lines) either larger (or) smaller than our original cylinder. Let the length of the dotted cylindrical surfaces be L and radius be r then the top and bottom faces do not contribute to the integral $\Rightarrow \oint \mathbf{B} \cdot \mathbf{ds} = 0$ because the radial component $B_r$ is perpendicular to ds on the top and bottom faces.

Hence only the curved cylindrical surface contributes to the integral.

$$\Rightarrow \oint \mathbf{B} \cdot \mathbf{ds} = 0$$

$$\Rightarrow \oint |B||ds| = 0$$

$$\Rightarrow 2\pi r_c L B_r = 0$$

$$\Rightarrow B_r = 0 \text{ as } r_c \neq 0, L \neq 0$$

Hence both $B_\phi$ and $B_r$ are both zero inside and outside the cylinder.

The only surviving component is $B_z$.

We know that the flux density **B** must go to zero when we go far away from the cylinder. Now consider Fig. 5.12.

Applying Ampere's circuital law to loop 2 lying outside the cylinder

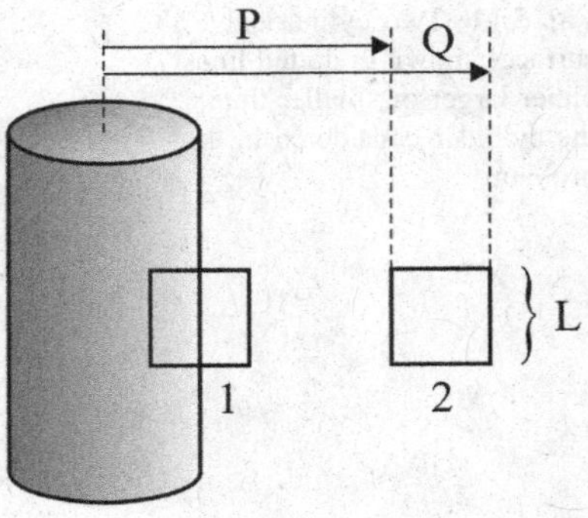


**Fig. 5.12** An infinitely long circular cylinder shown along with Amperian loops

$\oint \mathbf{B} \cdot d\mathbf{l} = [B(P) - B(Q)]L = \mu_o I_{enc} = 0$ as the loop doesn't enclose any current.

$$\Rightarrow B(P) = B(Q) = 0 \text{ as } L \neq 0$$
$$\Rightarrow B(P) = B(Q)$$

Now we can select the loop 2 such that it lies very far away from the cylinder in which case B(Q) = 0

$$\Rightarrow B(P) = B(Q) = 0.$$

Hence for all points outside the cylinder B is zero. Now let us apply Ampere's circuital law to loop 1 then

$$\oint \mathbf{B} \cdot d\mathbf{l} = B\,L = \mu_o I_{enc}$$
$$\Rightarrow B\,L = \mu_o I_{enc} = \mu_o K_p\,L$$
$$\Rightarrow B = \mu_o K_p = \mu_o M \text{ is the value of B inside the cylinder } \left[\text{As } \mathbf{K}_p = \mathbf{M} \times \hat{\mathbf{n}}\right].$$

**Example 5.3**
Consider a long circular cylinder of radius R. It carries a magnetization of where $\mathbf{M} = \kappa r_c \hat{\boldsymbol{\phi}}$ is a constant. Calculate the magnetic flux density produced by **M** both inside and outside the cylinder.

**Solution**
First let us calculate the value of $\mathbf{J}_p$ and $\mathbf{K}_p$. We have in cylindrical coordinates

$$\mathbf{M} = M_r\hat{\mathbf{r}}_c + M_z\hat{\mathbf{z}} + M_\phi\hat{\boldsymbol{\phi}}$$

As $\mathbf{M} = \kappa r_c \hat{\boldsymbol{\phi}}$ then

$$M_r = 0, M_z = 0, M_\phi = kr_c$$

The curl is cylindrical coordinates then is given by

$$\mathbf{J_p} = \boldsymbol{\nabla} \times \mathbf{M} = \frac{1}{r_c}\frac{\partial}{\partial r_c}\left[r_c M_\phi\right]\hat{\mathbf{z}}$$

$$\mathbf{J_p} = \frac{1}{r_c}\frac{\partial}{\partial r_c}\left[r_c \kappa r_c\right]\hat{\mathbf{z}}$$

$$\mathbf{J_p} = \frac{1}{r_c}\kappa\, 2 r_c \hat{\mathbf{z}} = 2\kappa\hat{\mathbf{z}}$$

$$\mathbf{J_p} = 2\kappa\hat{\mathbf{z}}$$

and $\mathbf{K_p} = \mathbf{M} \times \hat{\mathbf{n}} = \kappa r_c \hat{\boldsymbol{\phi}} \times \hat{\mathbf{r}}_c$

$\mathbf{K_p} = -\kappa R\hat{\mathbf{z}}$ as the bound current exists on the surface.

Having calculated $\mathbf{J_p}$ and $\mathbf{K_p}$ now let as calculate **B** due to $\mathbf{J_p}$ and $\mathbf{K_p}$ or equivalenty that of **M**. Consider Fig. 5.13 to find the **B** inside the cylinder for the Amperian loop 1. Applying Ampere's law for this loop.

$$B\, 2\pi r_c = \mu_o I_{enc}$$

From Eq. 4.22 we see that

$$I_{enc} = \int \mathbf{J_p} ds_{per}$$

Because $\mathbf{J_p} = 2\kappa\hat{\mathbf{z}}$, meaning $\mathbf{J_p}$ is in *z*-direction.

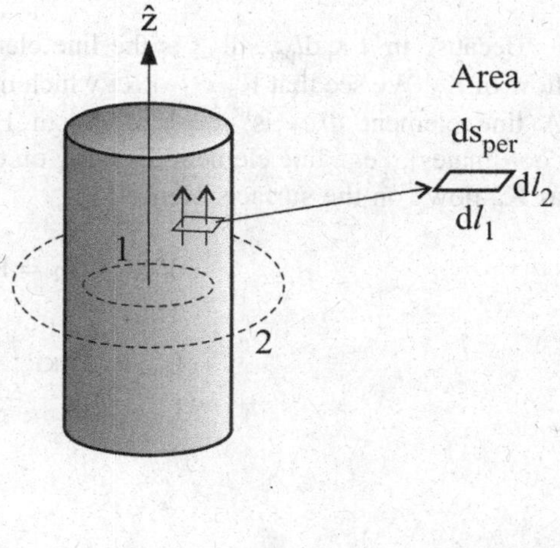


**Fig. 5.13** A long circular cylinder of radius R shown along with Amperian loops

The perpendicular area $ds_{per}$ from Fig. 5.13 is

$$ds_{per} = dl_1 dl_2$$

From Sect. 1.19 in cylindrical coordinates

$$dl_1 = dr_c, dl_2 = r_c d\phi.$$

Hence

$$ds_{per} = r_c dr_c d\phi \text{ and}$$

$$I_{enc} = \int 2\kappa\, r_c dr_c d\phi$$

$$= 2\kappa \int_o^{r_c} r_c dr_c \int_o^{2\pi} d\phi$$

$$= 2\kappa \frac{r_c^2}{2} 2\pi = 2\pi\, \kappa r_c^2$$

Substituting is Ampere's law

$$B\, 2\pi r_c = \mu_o I_{enc} = \mu_o 2\pi\, \kappa r_c^2$$

$$B = \mu_o\, \kappa\, r_c$$

To find B outside the cylinder consider Amperian loop 2 in Fig. 5.13 which encloses both $\mathbf{J}_p$ and $\mathbf{K}_p$. Now let us calculate $I_{enc}$ in Ampere's law.

$$I_{enc} = \int J_p ds_{per} + \int K_p dl_{per}$$

$$I_{enc} = \int_o^R \int_o^{2\pi} 2\kappa\, r_c dr_c d\phi + \int_0^{2\pi} -\kappa\, R(R\, d\phi)$$

Because in $\int \kappa_p dl_{per}$, $dl_{per}$ is the line element perpendicular to the direction of flow of $K_p$. We see that $K_p = -\kappa R\hat{\mathbf{z}}$ which means that $K_p$ is in negative $z$-direction. A line element $dl_{per}$ is equal to $dl_2$ in Fig. 5.13, [see Sect. 1.19—cylindrical coordinates], i.e. a line element $dl_2$ lying on the curved outer surface of the cylinder as $K_p$ flows on the surface. Hence

$$dl_{per} = dl_2 = Rd\phi$$

Thus

$$I_{enc} = 2\pi\, \kappa R^2 - \kappa R^2 2\pi.$$

$$I_{enc} = 0$$

Hence the current enclosed by the Amperian loop 2 in Fig. 5.13 is zero. Applying Ampere's law to Amperian loop 2 in Fig. 5.13

$$\oint \mathbf{B} \cdot d\boldsymbol{l} = \mu_o I_{enc}$$

$$\Rightarrow B\, 2\pi r_c = 0 \text{ as } I_{enc} = 0$$

$$\Rightarrow B = 0$$

Hence B is zero outside.

Another interesting fact to be noted is $\mathbf{J}_p = 2\kappa\hat{\mathbf{z}}$, $\mathbf{K}_p = -\kappa R\hat{\mathbf{z}}$ meaning that $J_P$ flows in $z$-direction, i.e. current flows up the cylinder and $K_p$ flows in negative $z$-direction, i.e. current returns back down the surface.

**Example 5.4**
A cylinder of radius R and length L carries a uniform magnetization as shown in Fig. 5.14. Calculate the magnetic flux density **B** at point P in Fig. 5.14. The magnetization is M $\hat{\mathbf{z}}$.

**Solution**
As in Problem 5.2

$$\mathbf{J}_p = \nabla \times \mathbf{M} = 0$$

$$\mathbf{K}_p = \mathbf{M} \times \hat{\mathbf{r}}_c = M\hat{\mathbf{z}} \times \hat{\mathbf{r}}_c = M\hat{\boldsymbol{\phi}}$$

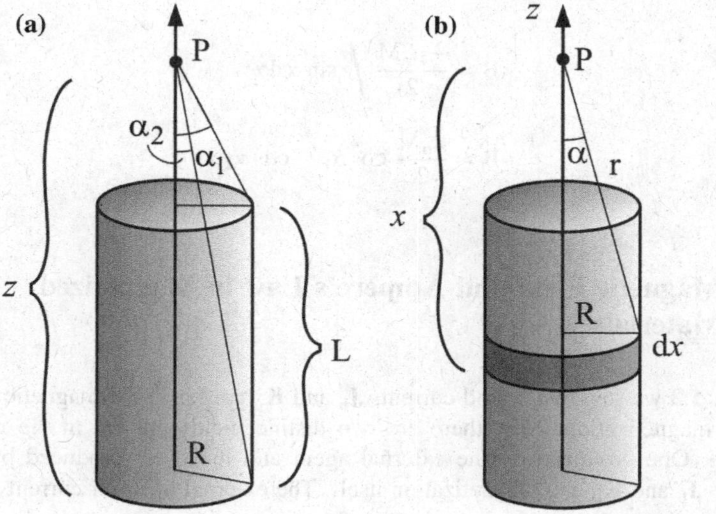


**Fig. 5.14 a, b, c** A cylinder of radius R and length L carries a uniform magnetization pointing in $z$ directionx

From Eq. 4.157 the magnetic flux density along the axis of the single circular loop is

$$B = \frac{\mu_o I}{2} \frac{R^2}{(x^2 + R^2)^{3/2}}$$

From Fig. 5.14b.

$$\tan \alpha = \frac{R}{x} \Rightarrow x = R \cot \alpha$$
$$\Rightarrow dx = -R\text{cosec}^2 \alpha d\alpha$$

and $R^2 + x^2 = r^2, \sin \alpha = \frac{R}{r}$
From Fig. 5.14b

$$I = K_p dx$$

Md and hence

$$dB = \frac{\mu_o M}{2} \frac{R^2}{(x^2 + R^2)\sqrt{x^2 + R^2}} dx$$

$$dB = \frac{-\mu_o M}{2} \sin \alpha \, d\alpha$$

Thus B due to the entire cylinder is

$$B = \frac{-\mu_o M}{2} \int_{\alpha_2}^{\alpha_1} \sin \alpha \, d\alpha$$

$$\mathbf{B} = \frac{\mu_o M}{2} [\cos \alpha_1 - \cos \alpha_2] \hat{\mathbf{z}}$$

## 5.5 Magnetic Field and Ampere's Law in Magnetized Materials

In Sect. 5.3 we saw that bound currents $\mathbf{J_p}$ and $\mathbf{K_p}$ appear in the magnetic material due to magnetization. Now there are two distinct fields present in the magentic material. One produced by the external agent and the other produced by bound currents $\mathbf{J_p}$ and $\mathbf{K_p}$, i.e. magnetization itself. The external agent or current which is responsible for the field in the magnetic material let us call it as "free current" and denote it as $\mathbf{J_f}$. Thus the total current can be written as

$$\mathbf{J} = \mathbf{J_p} + \mathbf{J_f} \tag{5.9}$$

Hence Ampere's law can be written as

$$\nabla \times \mathbf{B} = \mu_o \mathbf{J}$$

$$\Rightarrow \frac{1}{\mu_o}(\nabla \times \mathbf{B}) = \mathbf{J} = \mathbf{J_p} + \mathbf{J_f} \tag{5.10}$$

$$\Rightarrow \frac{1}{\mu_o}(\nabla \times \mathbf{B}) = \mathbf{J_f} + (\nabla \times \mathbf{M}) \tag{5.11}$$

$$\Rightarrow \nabla \times \left(\frac{1}{\mu_o}\mathbf{B} - \mathbf{M}\right) = \mathbf{J_f} \tag{5.12}$$

Let us denote

$$\mathbf{H} = \frac{1}{\mu_o}\mathbf{B} - \mathbf{M} \tag{5.13}$$

Hence

$$\nabla \times \mathbf{H} = \mathbf{J_f} \tag{5.14}$$

Here $\mathbf{H}$ is called the magnetic field. In free space $\mathbf{M} = 0$, hence.

$$\mathbf{H} = \frac{1}{\mu_o}\mathbf{B} \tag{5.15}$$

Integrating Eq. 5.14 and applying Stokes' theorem we finally get

$$\oint \mathbf{H} \cdot dl = I_{f(enclosed)} \tag{5.16}$$

where $I_{f(enclosed)}$ is the total free current passing there the Amperian loop.

In Sect. 4.9 we derived a similar expression for $\mathbf{B}$, $\oint \mathbf{B} \cdot dl = \mu_o I_{enc}$ using which we calculated $\mathbf{B}$ in Sects. 4.16–4.21. Similarly in the presence of magnetic materials $\mathbf{H}$ can be calculated using the equation:

$$\oint \mathbf{H} \cdot dl = I_{f(enclosed)}.$$

which is Ampere's law in magnetic materials. Here **H** simply is determined by $I_{f(enclosed)}$, i.e. free current. Free current is stuff which is under our control, while bound currents appear is the material on which we don't have control. Equation 5.16 enables us to calculate **H**, by considering free current alone which we control. Finally **H** vector in magnetostatics play analogous role as **D** in electrostatics.

## 5.6 Linear and Nonlinear Media

Whenever a paramagnetic or diamagnetic material is subjected to magnetic field the material gets magnetized. As the field increases the magnetization in the material also increases. Hence we write

$$\mathbf{M} \propto \mathbf{H} \tag{5.17}$$

$$\mathbf{M} = \chi_m \mathbf{H} \tag{5.18}$$

where the constant of proportionality $\chi_m$ is called magnetic susceptibility.

Materials that obey Eq. 5.17 are called linear media. Then from Eq. 5.13

$$\mathbf{B} = \mu_o(\mathbf{H} + \mathbf{M}) = \mu_o[1 + \chi_m]\mathbf{H} \tag{5.19}$$

$$\mathbf{B} = \mu \mathbf{H} \tag{5.20}$$

where $\mu = \mu_0[1 + \chi_m]$ is called the permeability of the material.

$$\Rightarrow \mu_r = \frac{\mu}{\mu_o} = 1 + \chi_m \tag{5.21}$$

where $\mu_r$ is called the relative permeability of the medium.

For vacuum $\chi_m = 0$ because there is no material to magnetize.

Hence for vacuum

$$\mu = \mu_o \tag{5.22}$$

Hence $\mu_o$ is called permeability of free space.

**Example 5.5**
Consider a wire of radius R carrying a current I and the current is uniformly distributed. The wire is made of a linear material whose susceptibility is $\chi_m$. Calculate **B** at a distance r from axis and the bound currents.

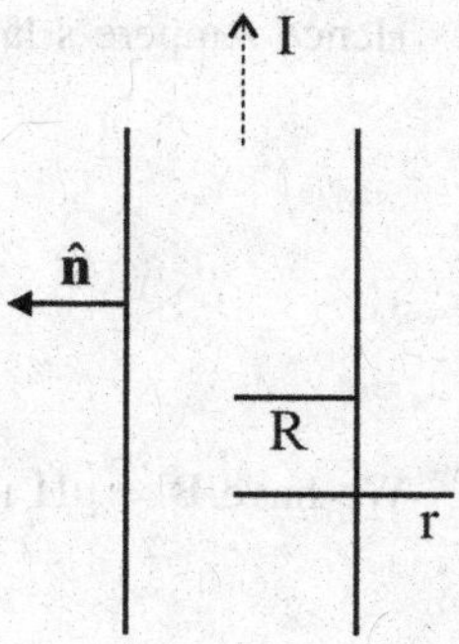


**Fig. 5.15** A wire of radius R carrying a uniform current I

**Solution**

We don't know what the bound currents are and hence directly calculating **B** is not possible. But however **B** can be calculated as follows. Applying Ampere's law (Fig. 5.15).

$$\oint \mathbf{H}\cdot d\boldsymbol{l} = I_{f(\text{enclosed})}$$

$$\oint \mathbf{H}\cdot d\boldsymbol{l} = H\,2\pi r$$

For a point outside the wire $I_{f(\text{enclosed})} = I$.
Hence Ampere's law gives

$$H\,2\pi r = I$$

$$\Rightarrow H = \frac{I}{2\pi r}$$

For a linear material outside the wire.

$$B = \mu_o H = \frac{\mu_o I}{2\pi r}$$

For a point inside the wire
Free current density in the wire is

$$J_f = \frac{I}{\pi R^2}$$

Hence $I_{f(\text{enclosed})}$ for a point $r < R$ is

$$I_{f(\text{enclosed})} = I\frac{\pi r^2}{\pi R^2} = I\frac{r^2}{R^2}$$

Hence Ampere's law gives

$$H\,2\pi r = I\frac{r^2}{R^2}$$

$$H = \frac{Ir}{2\pi R^2}$$

We have $B = \mu H$ inside the material

$$\mu = \mu_0(1+\chi_m)$$

Hence

$$B = \mu_o(1+\chi_m)H = \mu_o(1+\chi_m)\frac{Ir}{2\pi R^2}$$

Now let us estimate the bound currents. We have

$$\mathbf{J}_p = \nabla \times \mathbf{M} = \nabla \times (\chi_m \mathbf{H})$$
$$\mathbf{J}_p = \chi_m(\nabla \times \mathbf{H})$$

Applying Eq. 5.14

$$\mathbf{J}_p = \chi_m \mathbf{J}_f$$

$$J_p = \chi_m \frac{I}{\pi R^2}$$

with $\mathbf{J}_p$ having the same direction of current.

Now

$$\mathbf{K}_p = \mathbf{M} \times \hat{\mathbf{n}} = \chi_m \mathbf{H} \times \hat{\mathbf{n}}$$

$$K_p = \frac{\chi_m \mu_o I}{2\pi r}$$

with the direction opposite to that of current.

## 5.7 Boundary Conditions

The boundary conditions which we derived in Sect. 4.23 can be rewritten for the case of boundary between two magnetic materials. Noting down that $\mathbf{H}_1$ and $\mathbf{H}_2$ point in same direction as $\mathbf{B}_1$ and $\mathbf{B}_2$

Equation 4.225 is

$$\mathbf{B_{1n}} = \mathbf{B_{2n}}$$
$$\Rightarrow \mu_1 H_{1n} = \mu_2 H_{2n} \tag{5.23}$$

where $\mu_1$, $\mu_2$ are the permeability of the two magnetic mediums. In Fig. 4.33b if medium 1, 2 are magnetic mediums then applying Eq. (5.16) to the Amperian loop in Fig. 4.33b we obtain.

$$\oint \mathbf{H} \cdot d\boldsymbol{l} = I_{f(\text{enclosed})}$$

Because the side a is negligible for the Amperian loop (Fig. 4.33b)

$$H_{1t}L - H_{2t}L = K_f L \tag{5.24}$$

where $H_{1t}$, $H_{2t}$ are the tangential components of magnetic fields is medium 1,2. $K_f$ is the free surface current in Fig. 4.33b

$$\Rightarrow H_{1t} - H_{2t} = K_f \tag{5.25}$$

**Example 5.6**

We have a magnetic flux density

$$\mathbf{B} = 5e^{-2y}\hat{\mathbf{k}}\frac{\text{mwb}}{\text{m}^2}$$

Where $\mu = 4.7\ \mu_0$. Find (a) **H** (b) M

**Solution**

(a) $H = \dfrac{B}{\mu} = \dfrac{5 \times 10^{-3} e^{-2y}}{4\pi \times 10^{-7} \times 4.5}$

$H = 884e^{-2y}\hat{\mathbf{k}}\ \text{A/m}$

(b) $\mu_r = \frac{\mu}{\mu_o} = 4.7$

$\chi_m = \mu_r - 1 = 4.7 - 1 = 3.7$

$M = \chi_m H = (3.7)(884)e^{-2y}\ \text{A/m}$

$M = 3270.8\ \text{A/m}$

**Example 5.7**

At a point of a boundary between two magnetic materials (conductors) the magnetic field makes an angle $\theta_1$ and $\theta_2$ with the normal of the media of permeability $\mu_1$ and $\mu_2$, respectively, as shown in Fig. 5.16. Show that

**Fig. 5.16** **a** Boundary between two magnetic material. **b** Resolution of magnetic flux density vector into component

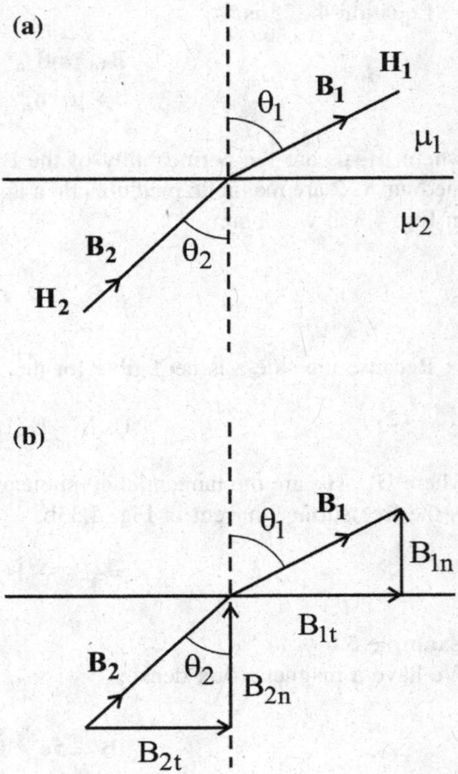


$$\frac{\tan \theta_1}{\tan \theta_2} = \frac{\mu_1}{\mu_2}$$

**Solution**
Resolving into components

$$\begin{aligned}
B_{1n} &= B_1 \cos \theta_1 \\
B_{1t} &= B_1 \sin \theta_1 \\
B_{2n} &= B_2 \cos \theta_2 \\
B_{2t} &= B_2 \sin \theta_2
\end{aligned}$$

and

$$\begin{aligned} H_{1n} &= H_1 \cos\theta_1 \\ H_{1t} &= H_1 \sin\theta_1 \\ H_{2n} &= H_2 \cos\theta_2 \\ H_{2t} &= H_2 \sin\theta_2 \end{aligned}$$

Applying the boundary conditions.

$$\begin{aligned} B_{1n} &= B_{2n} \\ \Rightarrow B_1 \cos\theta_1 &= B_2 \cos\theta_2 \end{aligned} \tag{5.26}$$

As the materials are conductors the boundary doesn't contain any surface current hence

$$\begin{aligned} H_{1t} &= H_{2t} \\ H_1 \sin\theta_1 &= H_2 \sin\theta_2 \end{aligned} \tag{5.27}$$

From Eqs. 5.26 and 5.27

$$\begin{aligned} \frac{H_1 \sin\theta_1}{B_1 \cos\theta_1} &= \frac{H_2 \sin\theta_2}{B_2 \cos\theta_2} \\ \Rightarrow \frac{1}{\mu_1}\tan\theta_1 &= \frac{1}{\mu_2}\tan\theta_2 \\ \Rightarrow \frac{\tan\theta_1}{\tan\theta_2} &= \frac{\mu_1}{\mu_2} \end{aligned} \tag{5.28}$$

where $B_1 = \mu_1 H_1$, $B_2 = \mu_2 H_2$

From the above equation we can derive the following conclusions.

(a) If $\theta_1 = 0$, then $\theta_2$ is also zero. That is the magnetic field lines are normal to the boundary in both the regions.

(b) If region 2 has high permeability as compared to region 1 and let $\theta_2$ is less than 90°, then the above equation shows that $\theta_1$ is small. That is the magnetic field lines are normal.

## Exercises

**Problem 5.1**

Consider an electron orbiting around the nucleus with radius R. The orbiting electron is placed in a magnetic field **B**. Using the fact that the electron speeds up or slows down depending on the orientation of **B** show that the change in dipole moment is opposite to the direction of **B**.

**Problem 5.2**

A circular loop carrying a current I is placed is a magnetic field **B** as shown in Fig. 5.17.

Show that the torque acting on the loop is $\mathbf{N} = \mathbf{m} \times \mathbf{B}$.

**Problem 5.3**

Show that the hysteresis loss in the magnetic material is equal to the area of the hysteresis loop.

**Problem 5.4**

In a region specified by $0 \leq z \leq 3\text{m}$ lies an infinite slab of magnetic material. The value of $\mu_r$ is 2.4. The value of $\mathbf{B} = 2y\hat{\mathbf{i}} - 7x\hat{\mathbf{j}}\,\frac{\text{mwb}}{\text{m}^2}$ inside the slab. Find (a) $\mathbf{J_P}$ (b) $\mathbf{K_P}$ at $z = 0$.

**Problem 5.5**

The average circumference of a metal ring is 140 cm. Its cross-sectional area is 9 cm$^2$. It is carrying a current of 2.5 A and the corresponding flux is found to be 2 milli-webers. It is wound with 500 turns. Calculate the relative permeability of the material.

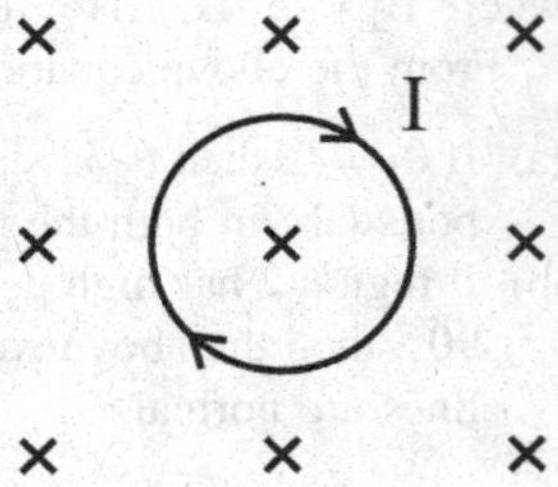


**Fig. 5.17** A circular loop carrying a current is placed in a magnetic field. Illustration for problem 5.2

**Fig. 5.18** Boundary between two magnetic materials. Illustration for problem 5.6

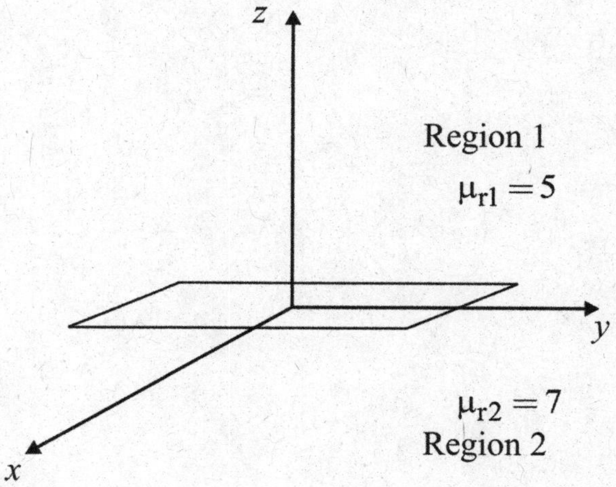


**Problem 5.6**

Two magnetic materials which are conductors are present as shown in Fig. 5.18.

In region 1 lying at $z > 0$, $\mathbf{H}_1 = 5\hat{\mathbf{i}} + 4\hat{\mathbf{j}} - 2\hat{\mathbf{k}}$ and $\mu_{r1} = 5$. In region 2 lying at $z < 0$, $\mu_{r2} = 7$

Find $\theta_2$ and $\mathbf{H}_2$.

# Chapter 6
# Time-Varying Fields and Maxwell's Equation

## 6.1 Introduction—Ohm's law

In metals like copper the outermost valence electrons are not attached to a particular atom and are called as free electrons. These free electrons are capable of moving through the entire crystal lattice. The free electrons acquire thermal energy and wander inside the metal in all possible directions. The thermal velocities of the free electrons are in random direction and average out to zero. As a result there is no net current in an isolated conductor.

However when a negative terminal of the battery is connected to one end of the conductor and the positive terminal is connected to the other end of the conductor, free electrons start flowing from the negative terminal of the battery to the positive terminal through the conductor and hence constitutes a current in the conductor.

When the battery is connected to the conductor as said above, an electric field **E** is established inside the conductor and the free electrons in the conductor accelerate towards the positive terminal under the action of the electric field. However the electrons undergo collision with the ions in the lattice. The collision slows down the electron and changes its direction. The electric field once again accelerates the electrons towards the positive terminal. The electron suffers numerous collisions with the ions of the lattice as the electric field accelerates them towards the positive terminal of the battery (see Problem 6.2 and Fig. 6.29). The electrons slowly drift towards the positive terminal with an average velocity known as drift velocity.

### 6.1.1 *Ohm's law*

When an electric field is maintained across the material, electrons in the material flow in the direction opposite to the electric field **E** because the electrons with charge –e experience a force given by

© Springer Nature Singapore Pte Ltd. 2020
S. Balaji, *Electromagnetics Made Easy*,
https://doi.org/10.1007/978-981-15-2658-9_6

$$\mathbf{F} = -e\mathbf{E} \tag{6.1}$$

Motion of these charges produces current. In a given material how effectively charges move to produce current depends upon the nature of the material. For most materials, the current density **J** is proportional to **E**

$$\mathbf{J} \propto \mathbf{E} \tag{6.2}$$

$$\Rightarrow \mathbf{J} = \sigma\mathbf{E} \tag{6.3}$$

Here σ the proportionality constant varies from one material to the other material and is known as conductivity of the material. Equation 6.3 is called Ohm's law.

The reciprocal of σ,ρ is known as resistivity of the material

$$\sigma = \frac{1}{\rho} \tag{6.4}$$

Conductors are materials in which σ is very high. Resistors are materials in which the resisitivity ρ is very high.

Consider a cylindrical resistor of length L whose ends are maintained at a potential difference of V (Fig. 6.1). The area of cross section of the cylindrical resistor is A. The current that flows in the cylindrical resistor is

I = J A (see Eq. 4.22)

Substituting Eq. 6.3 in the above equation

$$I = \sigma E A \tag{6.5}$$

$$\Rightarrow I = \sigma \frac{V}{L} A \tag{6.6}$$

where σ is the conductivity of the resistor. Equation 6.6 can be written as

$$V = \frac{L}{\sigma A} I = \frac{\rho L}{A} I \tag{6.7}$$

$$\Rightarrow V = RI \tag{6.8}$$

where $R = \frac{\rho L}{A}$ is called the resistance of the material and Eq. 6.8 is the familiar form of Ohm's law.

$$R = \frac{\rho L}{A} \tag{6.9}$$

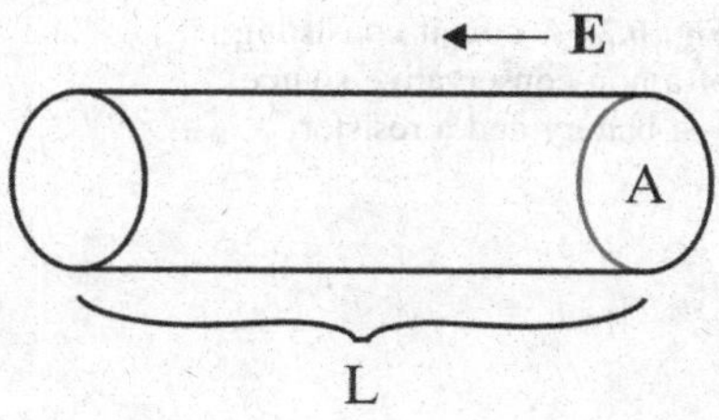


**Fig. 6.1** A cylindrical resistor with cross sectional area A and length L

## 6.2 Electromotive Force

We know that work done by a conservative electrostatic force around a closed path is zero (from Eq. 2.280)

$$\oint \mathbf{E} \cdot d\mathbf{l} = 0 \tag{6.10}$$

Thus if there is circuit carrying current, conservative electrostatic forces alone will not be enough to maintain current in the circuit. That is, as the charges flow down the wire they collide with atoms and dissipate energy. The dissipated energy must be supplied to the charges if current is to be maintained in the circuit. The electrostatic forces cannot do any work and hence cannot supply energy to the charges. Thus in order to make up for the dissipated energy and maintain current in the circuit, non-conservative sources like batteries, thermocouples, electric generators, etc., must be present in the circuit.

Consider the circuit in Fig. 6.2. The circuit contains a non-conservative source—the battery. The non-conservative source—the battery pushes the charge inside the wire when the key K is closed. As the charges are pushed inside the wire at some point (say point B in Fig. 6.2), charges are accumulating. Hence the electrostatic force of repulsion builds up at point B. Charges flowing towards B are repelled and charges flowing away from B are promoted. This electrostatic force of repulsion evens out the flow of charges. The process takes place at each point of the circuit and hence results in even flow of current I over the entire circuit. Hence as the battery pushes the charge inside the wire, in the remaining part of the circuit the electrostatic force of repulsion maintains the current. Thus there are two fields responsible for maintaining current in the circuit–one non-conservative field and the other conservative field, the electrostatic field.

Let the electric field due to the non-conservative source be denoted by $\mathbf{E}_s$ and the electric field due to the conservative electrostatic field be denoted by $\mathbf{E}$. The total electric field is then

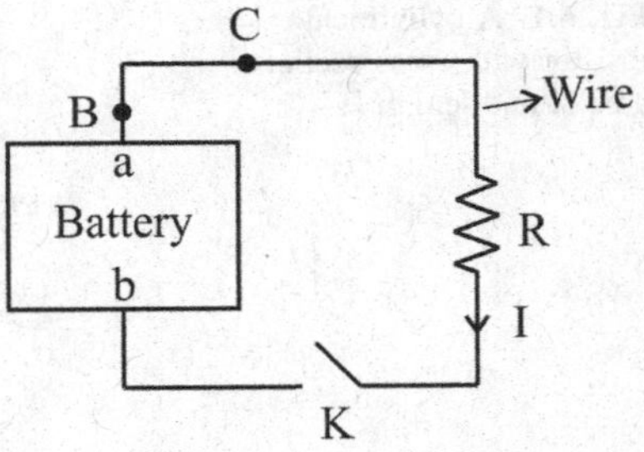


**Fig. 6.2** A circuit consisting of a non-conservative source —a battery and a resistor

$$\mathbf{E_T} = \mathbf{E} + \mathbf{E_s} \tag{6.11}$$

Let us define the line integral of $\mathbf{E_T}$ around the closed path as

$$\oint \mathbf{E_T} \cdot d\boldsymbol{l} = \oint \mathbf{E} \cdot d\boldsymbol{l} + \oint \mathbf{E_s} \cdot d\boldsymbol{l} \tag{6.12}$$

As $\mathcal{E} = \oint \mathbf{E} \cdot d\boldsymbol{l} = 0$

$$\mathcal{E} = \oint \mathbf{E_s} \cdot d\boldsymbol{l} \tag{6.13}$$

The term $\mathcal{E}$ is called electromotive force or emf and represents the measure of strength of the non-conservative source.

In an open-circuited battery, $\mathbf{E_T}$ is zero inside the battery, as no current flows inside the open-circuited battery. Hence from Eq. 6.11 for a open-circuited battery

$$\mathbf{E_T} = 0 = \mathbf{E} + \mathbf{E_s} \tag{6.14}$$

$$\Rightarrow \mathbf{E} = -\mathbf{E_s} \tag{6.15}$$

In Fig. 6.2 the potential difference between the two terminals a and b of the battery is

$$\mathrm{V} = -\int_a^b \mathbf{E} \cdot d\boldsymbol{l} = \int_a^b \mathbf{E_s} \cdot d\boldsymbol{l} \tag{6.16}$$

$\mathbf{E_s}$ the electric field of the source is present within the source alone and $\mathbf{E_s}$ in the wire in Fig. 6.2 is zero. Hence we can write

$$\mathrm{V} = \int_a^b \mathbf{E_s} \cdot d\boldsymbol{l} = \oint \mathbf{E_s} \cdot d\boldsymbol{l} = \mathcal{E} \tag{6.17}$$

where the closed path in the integral $\oint \mathbf{E_s} \cdot d\boldsymbol{l}$ represents path aBCRKba in Fig. 6.2. In Eq. 6.17 we have used Eq. 6.13. Thus the potential difference across the battery in open circuit is equal to the emf of the battery.

We know that $\mathbf{E}_s$ is the electric field of the non-conservative source. By definition, the electric field is the force per unit charge. Then the integral $\oint \mathbf{E}_s \cdot dl$ with $\mathbf{E}_s$ as the force per unit charge represents work done per unit charge. Hence from Eq. 6.13 we can write

$$\mathcal{E} = \frac{dW}{dq} = \oint \mathbf{E}_s \cdot dl \tag{6.18}$$

## 6.3 Motional emf

In this section we will discuss about motional emf's. Motional emf's arise when we move a wire in a magnetic field. As shown in Fig. 6.3 a magnetic field of magnetic induction **B** is pointing into the page. A wire is connected to the resistor and part of the wire abcd lies in the magnetic field with bc = $l$. Suppose assume that the wire is pulled in the $x$-direction. The charges in the segment bc experience a magnetic force, thereby a current I flows in the circuit. The charges in the segment ab and bc experience a magnetic force perpendicular to the wire and hence do not contribute to the current. Thus when the charges are pulled along the $x$-direction by an external agency a magnetic force acting along $y$-direction which drives the charges along $y$-direction responsible for a current I in the circuit develops. Therefore the charges acquire two velocity components one in $x$-direction (due to external agency) and the other along $y$-direction (due to magnetic force).

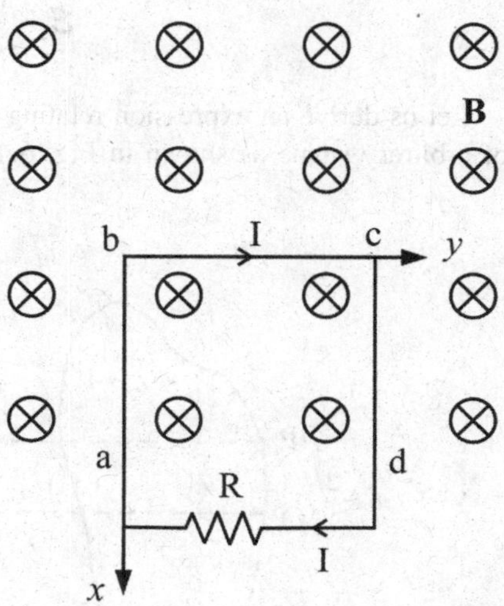


**Fig. 6.3** A wire abcd lying in a magnetic field, connected to a resistor and pulled in the $x$-direction by an external agency

Because the charges acquire new velocity component along *y*-direction responsible for current I another magnetic force component acting along *x*-direction develops and is given by

$$\mathbf{F}_m = \mathbf{I}\boldsymbol{l} \times \mathbf{B}$$

$\mathbf{F}_m$ acting along negative *x*-direction opposes the motion of the wire along *x*-direction. In order to make the wire to move in the *x*-direction, the external agency must exert the force

$$\mathbf{F}_{ext} = -\mathbf{F}_m = -\mathbf{I}\boldsymbol{l} \times \mathbf{B} \tag{6.19}$$

$$\Rightarrow \mathbf{F}_{ext} = -\mathrm{I}(l\mathrm{B}(-\hat{\mathbf{i}})) \tag{6.20}$$

$$\Rightarrow \mathbf{F}_{ext} = \mathrm{IB}l\hat{\mathbf{i}} \tag{6.21}$$

The work done by the external force is

$$\mathrm{dW} = \mathrm{IB}l\mathrm{d}x = \mathrm{IB}l\mathrm{udt} \tag{6.22}$$

where d*x* is the distance moved by the wire ab in a time dt when it is pulled with a velocity **u** along *x*-direction.

But I dt = dq the charge transferred in time dt. Hence

$$\mathrm{dW} = \mathrm{Budq}l \tag{6.23}$$

The emf $\mathcal{E}$ is given by

$$\mathcal{E} = \frac{\mathrm{dW}}{\mathrm{dq}} = \mathrm{Bu}l \tag{6.24}$$

Let us derive an expression relating $\mathcal{E}$, the emf and the flux $\phi$. Consider a wire of arbitrary shape as shown in Fig. 6.4 being pulled by an external agency in the

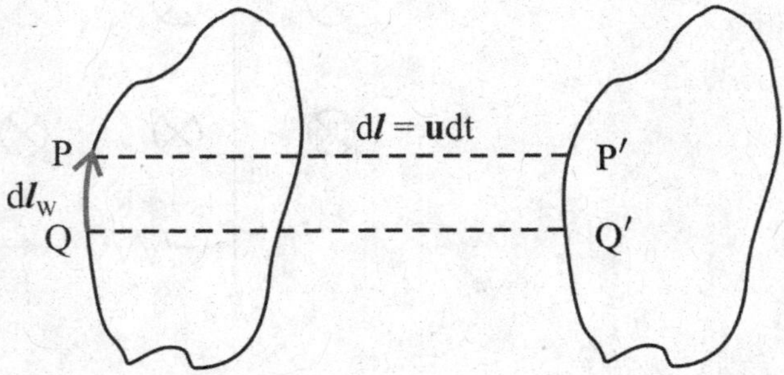


**Fig. 6.4** A wire of arbitrary shape being pulled in a non-uniform magnetic field by an external agency

presence of an non-uniform magnetic field with flux density **B**. Divide the wire into small elements $d\boldsymbol{l}_W$ as shown in the figure. The area PP′QQ′ is

$$\mathbf{ds} = (d\boldsymbol{l} \times d\boldsymbol{l}_w) \tag{6.25}$$

$$\Rightarrow \mathbf{ds} = (\mathbf{u} \times d\boldsymbol{l}_w)dt \tag{6.26}$$

where **u** is the velocity with which wire is pulled. Here we need to take $(d\boldsymbol{l} \times d\boldsymbol{l}_w)$ and not $(d\boldsymbol{l}_w \times d\boldsymbol{l})$ because by convention the direction of area vector is outside normal.

From Eq. 6.19 the force exerted by the external agency is

$$\mathbf{F}_{ext} = -\oint_C \mathbf{I}(d\boldsymbol{l}_w \times \mathbf{B}) \tag{6.27}$$

Here **C** is the path of integration in the direction of current in the wire. With PP′ $= d\boldsymbol{l}$, the work done is

$$dW = \mathbf{F}_{ext} \cdot d\boldsymbol{l} \tag{6.28}$$

Substituting Eq. 6.27 in 6.28

$$dW = -Id\boldsymbol{l} \cdot \oint d\boldsymbol{l}_w \times \mathbf{B} \tag{6.29}$$

With $I = \frac{dq}{dt}$ and $\mathbf{u} = \frac{d\boldsymbol{l}}{dt}$ Eq. 6.29 can be written as

$$dW = -\mathbf{u} \cdot \oint_C d\boldsymbol{l}_w \times \mathbf{B}(dq) \tag{6.30}$$

The motional emf is then

$$\mathcal{E} = \frac{dW}{dq} = -\mathbf{u} \cdot \oint_C (d\boldsymbol{l}_w \times \mathbf{B}) \tag{6.31}$$

(or)

$$\mathcal{E} = \frac{dW}{dq} = -\oint_C \mathbf{u} \cdot (d\boldsymbol{l}_w \times \mathbf{B}) \tag{6.32}$$

(or)

$$\mathcal{E} = \frac{dW}{dq} = -\oint_C \mathbf{B} \cdot (\mathbf{u} \times d\boldsymbol{l}_w) \tag{6.33}$$

The change in flux associated with the area **ds**, is given by

$$= \mathbf{B} \cdot \mathbf{ds} \,(\text{from Eq. 4.246}) \tag{6.34}$$

Substituting Eq. 6.26 in 6.34

$$= \mathbf{B} \cdot (\mathbf{u} \times d\boldsymbol{l}_w)dt \tag{6.35}$$

The total change in flux over the entire wire is

$$d\phi = \oint_C \mathbf{B} \cdot (\mathbf{u} \times d\boldsymbol{l}_w)dt \tag{6.36}$$

$$\Rightarrow \frac{d\phi}{dt} = \oint_C \mathbf{B} \cdot (\mathbf{u} \times d\boldsymbol{l}_w) \tag{6.37}$$

Comparing Eqs. 6.33 and 6.37

$$\mathcal{E} = \frac{-d\phi}{dt} \tag{6.38}$$

## 6.4 Faraday's Law

Michael Faraday in 1831 demonstrated that apart from current flowing in the circuit, when the wire is pulled in the magnetic field in Fig. 6.3, a current also flows in a stationary circuit, if the magnetic flux associated with the wire varies with respect to time.

In Fig. 6.3 we have seen that when a wire is pulled in the presence of the magnetic field in the *x*-direction a current flows in the circuit. Instead let the wire be held stationary and let the magnetic field be time varying so that a time-varying magnetic flux is associated with the wire loop shown in Fig. 6.3. In that case once again current flows in the circuit and clearly it is not the magnetic force responsible for the current flow. The wire is held stationary and hence magnetic force cannot be responsible for the flow of current [$\mathbf{v} = 0$ in $\mathbf{F} = q(\mathbf{v} \times \mathbf{B})$]. Faraday stated that "An induced electric field is responsible for the flow of current in the circuit whenever the wire is held stationary". The induced emf due to the changing magnetic field is once again given by

$$\mathcal{E} = \oint \mathbf{E} \cdot d\boldsymbol{l} = \frac{-d\phi}{dt} \tag{6.39}$$

We know that $\phi = \int_S \mathbf{B} \cdot d\mathbf{s}$ (From Eq. 4.246).
Hence

$$\oint \mathbf{E} \cdot d\boldsymbol{l} = \frac{-d}{dt} \int_S \mathbf{B} \cdot d\mathbf{s} \tag{6.40}$$

$$\Rightarrow \oint \mathbf{E} \cdot d\boldsymbol{l} = -\int_S \frac{\partial \mathbf{B}}{\partial t} \cdot d\mathbf{s} \tag{6.41}$$

Equation 6.41 is Faraday's law in integral form.
Applying Stoke's theorem

$$\int_S (\nabla \times \mathbf{E}) \cdot d\mathbf{s} = -\int_S \frac{\partial \mathbf{B}}{\partial t} \cdot d\mathbf{s} \tag{6.42}$$

Because the above equation is valid for an small area ds, for the area ds, we need not integrate

$$(\nabla \times \mathbf{E}) \cdot d\mathbf{s} = \frac{-\partial \mathbf{B}}{\partial t} \cdot d\mathbf{s} \tag{6.43}$$

$$\Rightarrow \nabla \times \mathbf{E} = \frac{-\partial \mathbf{B}}{\partial t} \tag{6.44}$$

Equation 6.44 is Faraday's law in differential form.
Let us compare Eq. 6.44 with the electrostatic counterpart

$$\nabla \times \mathbf{E} = 0$$

That is the induced electric field given by Eq. 6.44 curl's around (or) $\mathbf{E}$ associated with changing magnetic flux forms closed loops. However the lines of $\mathbf{E}$ associated with static charges do not form closed loops, but instead they originate on a positive charge and end on a negative charge or go of the infinity.

## 6.5 Lenz's Law

Faraday's law relates the induced electric field to the rate of change of magnetic flux density. Now we will discuss about the direction of the induced emf and current which is given by Lenz's law. It state's that the direction of the induced current or emf in a closed circuit is such that it opposes the very cause that produces it. In other words the law states that current induced is in the direction, such that it produces a magnetic flux tending to oppose the original change of flux.

Suppose assume that the applied flux density **B** is decreasing in magnitude, then the current induced in the loop will be in a direction such that it will produce a field to increase **B**. If on the other hand when **B** is increasing the current induced in the loop will be in a direction such that the field produced by the induced current will tend to decrease **B**.

Suppose assume as shown in Fig. 6.5 a magnet approaches a loop. When the magnet approaches the loop the induced current $I_i$ set up in the loop has to produce a field so that the induced field produced by the incoming magnet in the loop opposes the magnetic field of the incoming magnet. Hence the induced current flows as shown in Fig. 6.5 so that field produced by the current acts leftward to oppose the increase in field produced by the incoming magnet. On the other hand, if the magnet is withdrawn then the field in the loop decreases and the induced current flows in the opposite direction as shown in Fig. 6.6 so that the field produced by the induced current is rightward in order to oppose the decrease in magnetic field due to withdrawal of magnet.

**Example 6.1**

Consider a coaxial cable of length L with the intermediate material whose conductivity is σ. The inner radius of the cylinder is $r_{c1}$ and the outer radius is $r_{c2}$ as shown in Fig. 6.7. A current I flows from the inner conductor to the outer conductor. Find the resistance R between the inner and outer conductors?

**Solution**

The current density at an arbitrary point $r_c$ between the inner and outer conductors is $J = \frac{I}{A} = \frac{I}{2\pi r_c L}$.

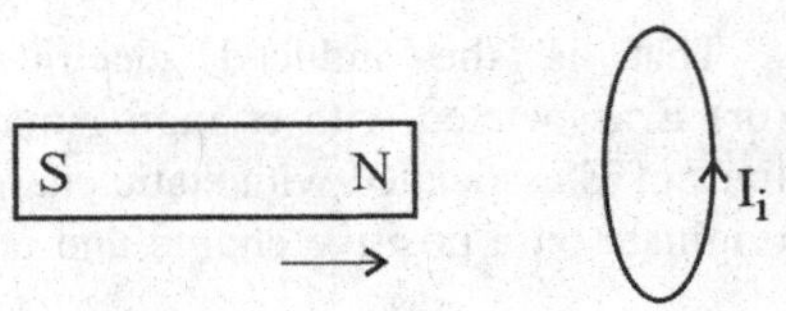


**Fig. 6.5** Magnet moving towards the loop. The direction of the induced current in the loop is shown

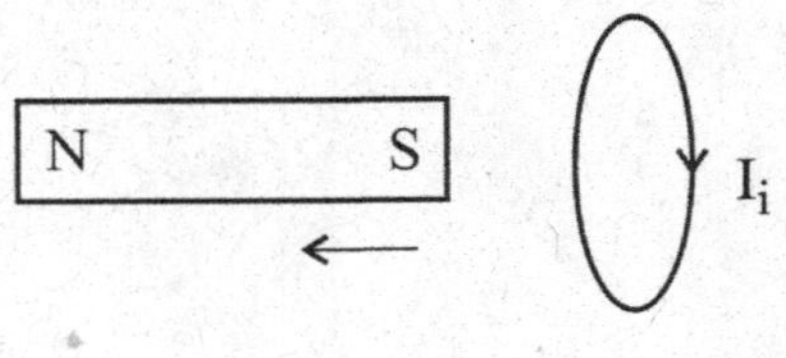


**Fig. 6.6** Magnet withdrawn from the loop. The direction of the induced current in the loop is shown

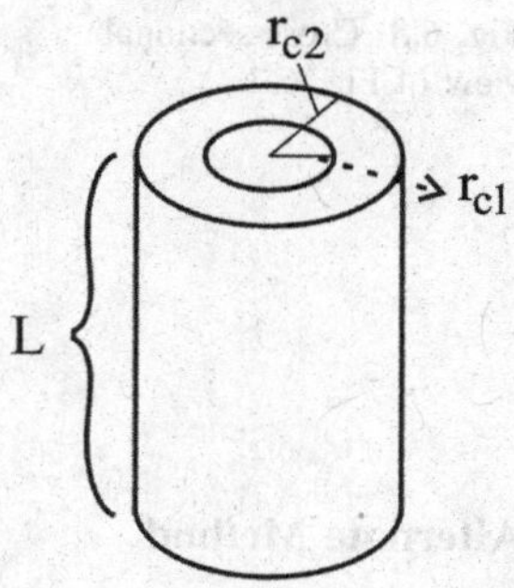


**Fig. 6.7** A coaxial cable of length L, filled with an intermediate material and carrying a current I

The electric field **E** is then

$$\mathbf{J} = \sigma \mathbf{E}$$

$$\Rightarrow \mathbf{E} = \frac{\mathbf{J}}{\sigma}$$

$$\mathbf{E} = \frac{I}{2\pi\sigma r_c L}\hat{\mathbf{r}}_c$$

The potential difference between the conductors is

$$V = -\int_{r_{c2}}^{r_{cl}} \mathbf{E}\cdot d\boldsymbol{l}$$

$$= -\int_{r_{c2}}^{r_{cl}} E d\boldsymbol{l}$$

$$= -\int_{r_{c2}}^{r_{cl}} \frac{I}{2\pi\sigma r_c L} = \frac{I}{2\pi r_c L} ln\left(\frac{r_{c2}}{r_{cl}}\right)$$

Then the resistance is

$$R = \frac{V}{I} = \frac{1}{2\pi\sigma L} ln\left(\frac{r_{c2}}{r_{c1}}\right)$$

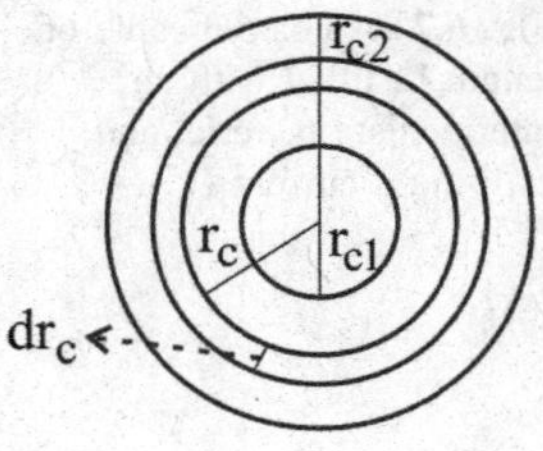


**Fig. 6.8** Cross-sectional view of Fig. 6.7

**Alternate Method**

Consider Fig. 6.8.

Let $r_c$ and $r_c + dr_c$ be the radii of the concentric layer of the intermediate material. The surface area of the concentric layer is $2\pi r_c L$. From Eq. 6.9

$$R = \frac{\rho L}{A}$$

where $\rho = \frac{1}{\sigma}$ is the resistivity or specific resistance of the material.

Equation 6.9 can be written for the concentric layer of intermediate material as

$$dR = \frac{\rho dr_c}{2\pi r_c L}$$

For the entire material

$$R = \rho \int_{r_{c1}}^{r_{c2}} \frac{dr_c}{2\pi r_c L}$$

$$R = \frac{\rho}{2\pi L} \int_{r_{c1}}^{r_{c2}} \frac{dr_c}{r_c}$$

$$R = \frac{1}{2\pi\sigma L} ln\left(\frac{r_{c2}}{r_{c1}}\right)$$

**Example 6.2** In Example 6.1 $r_{c1}$ = 0.9 cm and $r_{c2}$ = 1.4 cm. The intermediate material is having a resistance per unit length of 700 MΩ/km. What is the resistivity of the intermediate material. If the resistance is supposed to be increased to 1600 MΩ/km by increasing the thickness of the same intermediate material, how much should be the increase in thickness?

**Solution**

We know that

$$R = \frac{\rho}{2\pi L} ln\left(\frac{r_{c2}}{r_{c1}}\right)$$

Now $\frac{r_{c2}}{r_{c1}} = \frac{1.4}{0.9}$

L = 1 km = $10^5$ cm.

R = 700 MΩ.

Hence

$$700 = \frac{\rho}{2\pi 10^5} ln\left(\frac{1.4}{0.9}\right)$$

$$4396 \times 10^5 = \rho(0.44183)$$

$$\Rightarrow \rho = 9.9 \times 10^8 \text{ M}\Omega\text{-cm.}$$

Total insulation has to be 1600 MΩ/km. Hence the remaining insulation that needs to be increased is 1600–700 = 900 MΩ/km. Now

$$900 = \frac{9.9 \times 10^8}{2 \times 3.14 \times 10^5} ln\left(\frac{r_{c2} + x}{r_{c2}}\right)$$

Here $r_{c2}$ is the outer radius of the existing intermediate material (or) inner radius of the additional intermediate material added. $r_{c2} + x$ is the outer radii of the additional intermediate material added.

$r_{c2}$ = 1.4 cm

Hence

$$900 = \frac{9.9 \times 10^8}{2 \times 3.14 \times 10^5} ln\left(\frac{1.4 + x}{1.4}\right)$$

$$570 \times 10^{-3} = ln\left(\frac{1.4 + x}{1.4}\right)$$

$$1.769 = \frac{1.4 + x}{1.4}$$

$$x = 1.078 \text{ cm}$$

**Example 6.3**

A metal disc revolves about 15 revolutions per second about the horizontal axis passing through its centre. The disc placed vertically has a diameter of 30 cm. A uniform magnetic flux density of strength 150 Gauss acts perpendicular to the plane of the disc? Calculate the induced emf.

**Solution**
We know that

$$\mathcal{E} = \frac{-d\phi}{dt} = \frac{-d}{dt}(BA)$$
$$\mathcal{E} = B\frac{dA}{dt} \text{ (Numerically)}$$

Given B = 150 Gauss = $150 \times 10^{-4}$ Wb/m$^2$
r = 15 cm = 0.15 m

$$\begin{aligned}\frac{dA}{dt} &= \text{Area swept by the disc in unit time.}\\ &= \pi r^2 \times \text{No. of revolutions per second}\\ &= (3.14)(0.15)^2 \times 15\\ &= 1.05975\end{aligned}$$

Hence $\mathcal{E} = B\frac{dA}{dt}$

$$\mathcal{E} = 150 \times 10^{-4} \times 1.05975$$
$$\mathcal{E} = 15.896 \text{ mV.}$$

**Example 6.4**
A circular path of initial radius 50 cms is centreed at the origin of the *x–y* plane. The circular radius gradually increases at a rate of 70 m/s. A uniform magnetic flux density of 100 Gauss is directed along the *z*-axis. Determine the induced emf is the circular path as a function of time.

**Solution**
The induced emf is

$$\mathcal{E} = \frac{-d\phi}{dt} = \frac{d\phi}{dt} \text{(Numerically)}$$
$$\phi = B(\pi r^2)$$
$$\mathcal{E} = B\pi r\frac{dr}{dt}$$
$$B = 100 \text{ Gauss} = 10^{-2}\,\text{Wb/m}^2$$

$$r = 0.5\,\text{m} \quad \frac{dr}{dt} = 70\,\text{m/s}$$
$$\mathcal{E} = 10^{-2}(3.14)(0.5)(70)$$
$$\mathcal{E} = 1.099\,\text{V}$$

**Example 6.5**

A metal wire of length 1 m lies perpendicular to the *x*–*y* plane. It is moving with a velocity $\mathbf{u} = (4\hat{\mathbf{i}} + 3\hat{\mathbf{j}} + 2\hat{\mathbf{k}})$ m/s. The magnetic flux density in the region is $\mathbf{B} = 3\hat{\mathbf{i}} + 4\hat{\mathbf{j}}$ Wb/m$^2$. Determine the potential difference the ends of the wire.

**Solution**

From Eq. 6.33 for the entire length of the wire L

$$\mathcal{E} = (\mathbf{u} \times \mathbf{B}) \cdot \mathbf{L}$$

$$\mathbf{u} \times \mathbf{B} = \begin{vmatrix} \hat{\mathbf{i}} & \hat{\mathbf{j}} & \hat{\mathbf{k}} \\ 4 & 3 & 2 \\ 3 & 4 & 0 \end{vmatrix}$$

$$= \hat{\mathbf{i}}(0 - 8) - \hat{\mathbf{j}}(0 - 6) + \hat{\mathbf{k}}(16 - 9)$$

$$\mathbf{u} \times \mathbf{B} = -8\hat{\mathbf{i}} + 6\hat{\mathbf{j}} + 7\hat{\mathbf{k}}$$

$$\mathcal{E} = (\mathbf{u} \times \mathbf{B}) \cdot \mathbf{L}$$

$$= (-8\hat{\mathbf{i}} + 6\hat{\mathbf{j}} + 7\hat{\mathbf{k}})\hat{\mathbf{k}}$$

$$\Rightarrow \mathcal{E} = 7\,\text{V}$$

**Example 6.6**

Consider Fig. 6.9 A metal rod AB moves on fixed conductors CA and DB with a velocity of 5 m/s along the *x*-direction. A uniform magnetic flux density B of strength 0.7 Wb/m$^2$ acts along the negative *z*-direction.

(a) Calculate the magnitude and direction of the induced emf.
(b) Calculate the current in the circuit if the resistance of ABCD is 0.25 Ω.
(c) Calculate the force required so that the motion of metal rod AB is maintained.

**Fig. 6.9** A metal rod Ab moves on a fixed conductors CA and DB in a uniform magnetic field

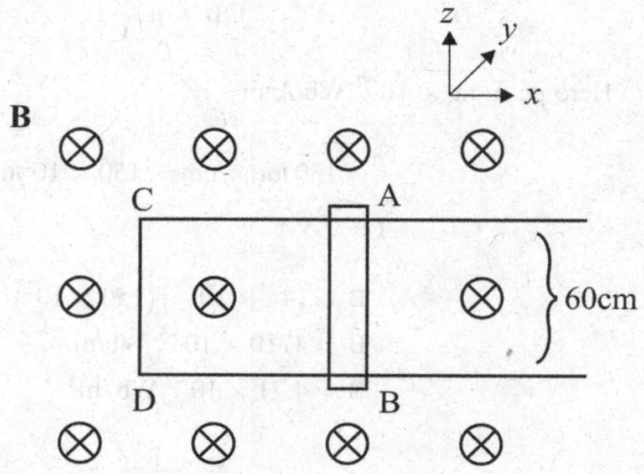

**Solution**

$\mathcal{E} = (\mathbf{u} \times \mathbf{B}) \cdot \mathbf{L}$ [Numerically]

$\mathbf{u} = 5\hat{\mathbf{i}}\ \mathbf{B} = -0.7\hat{\mathbf{k}}\ \mathbf{L} = 0.6\hat{\mathbf{j}}$

(a)
$$\mathbf{u} \times \mathbf{B} = \begin{vmatrix} \hat{\mathbf{i}} & \hat{\mathbf{j}} & \hat{\mathbf{k}} \\ 5 & 0 & 0 \\ 0 & 0 & -0.7 \end{vmatrix}$$

$$\mathbf{u} \times \mathbf{B} = \hat{\mathbf{i}}(0-0) - \hat{\mathbf{j}}(-3.5-0) + \hat{\mathbf{k}}(0-0)$$

$$\mathbf{u} \times \mathbf{B} = 3.5\hat{\mathbf{j}}$$

$$\Rightarrow \mathcal{E} = (\mathbf{u} \times \mathbf{B}) \cdot \mathbf{L} \Rightarrow \mathcal{E} = 2.1\,\text{V}$$

The direction of induced emf is from B to A in Fig. 6.9 as per Lenz's law.

(b) $I = \frac{V}{R} = \frac{2.1}{0.25} = 8.4\,\text{A}$

(c) The force required is

$F = ILB = (8.4)(0.6)(0.7)$

$F = 3.528\,\text{N}.$

**Example 6.7**

A very lengthy solenoid has 150 closely and evenly wound turns per cm of its length. The diameter of the solenoid is 7 cm. Exactly at the centre of the solenoid a small coil of 60 turns and 5 cm diameter is placed such that the magnetic field produced by the solenoid is parallel to the axis of the coil. The current in the solenoid is changed from 2.5 A to zero and then raised to 2.5 A in the opposite direction at a steady rate over a time period of 0.1 s. Calculate the induced emf in the coil during this change.

**Solution**

For a solenoid we know that

$$B = \mu_o \frac{N}{L} I$$

Here $\mu_o = 4\pi \times 10^{-7}$ Wb/A m

$$\frac{N}{L} = 150\,\text{turns/cm} = 150 \times 10^2\,\text{turns/m}$$

$$I = 2.5\,\text{A}$$

$$B = \left(4\pi \times 10^{-7}\right)\left(150 \times 10^2\right)(2.5)$$

$$B = 4710 \times 10^{-5}\ \text{Wb/m}^2$$

$$B = 4.71 \times 10^{-2}\,\text{Wb/m}^2$$

$$\text{Area of central coil} = (\text{No. of turns}) \times \text{area of each turn}$$
$$= (60)(\pi)\left(2.5 \times 10^{-2}\right)^2$$
$$= 0.11775.$$

The initial flux is $\phi = BA$

$$\phi = 4.71 \times 10^{-2} \times 0.11775$$
$$\phi = 0.5546 \times 10^{-2}\,\text{Wb}.$$

When the current is changed from 2.5 to –2.5 A the flux will change from $0.5546 \times 10^{-2}$ to $-0.5546 \times 10^{-2}$ Wb. The net change in flux is then

$$d\phi = [0.5546 - (-0.5546)] \times 10^{-2}$$
$$d\phi = 1.1092 \times 10^{-2}\ \text{Wb}.$$

The magnitude of induced emf is then

$$\mathcal{E} = \frac{d\phi}{dt}\,[\text{Numerically}]$$
$$\mathcal{E} = \frac{1.1092 \times 10^{-2}}{0.1}$$
$$\mathcal{E} = 11.092 \times 10^{-2}\,\text{V}$$
$$\mathcal{E} = 1.1092\,\text{m V}.$$

## 6.6 Magnetic Circuits

We come across number of instruments in which magnetism finds application. Few instruments to mention are transformers, magnetic recording devices, motors, generators, etc. Similar to electric circuit problems where we find voltages and currents in various branches of the circuit, we are required to find magnetic fluxes, magnetic field intensities is various parts of the so-called magnetic circuits. Magnetic circuit problems appear is above-mentioned devices.

As an example of a magnetic circuit consider the arrangement shown in Fig. 6.10. A toroidal coil of N turns is wound on a ring of iron of uniform permeability $\mu$. The toroidal coil carries a current of I amperes and the cross-sectional area of iron ring is A. The magnetic flux $\phi$ is confined to the iron ring and has the same value all around the ring, which is similar to current I in the electric circuit. That is the current I is confined to the electric circuit and has the same value all along the circuit. Hence it can be considered that magnetic flux $\phi$ is the magnetic circuit is analogous to current I is the electric circuit.

The magnetic counterpart to the electromotive force (e.m.f) is the magnetomotive force (m.m.f).

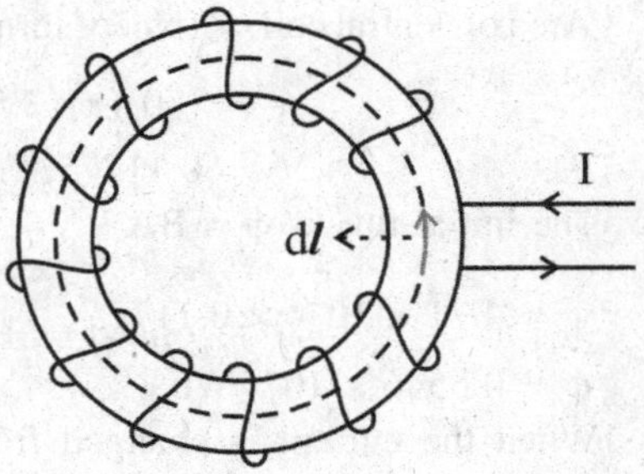


**Fig. 6.10** A Toroid carrying a current I

From Eq. 6.39 we have

$$\mathcal{E} = \oint \mathbf{E} \cdot d\boldsymbol{l} \tag{6.45}$$

where $\mathcal{E}$ is the e.m.f

By analogy we define m.m.f as

$$\text{m.m.f} = \oint \mathbf{H} \cdot d\boldsymbol{l} \tag{6.46}$$

Applying Ampere's circuital law to the dotted path shown in Fig. 6.10 we get

$$\oint \mathbf{H} \cdot d\boldsymbol{l} = \text{NI} \tag{6.47}$$

Comparing Eqs. 6.46 and 6.47

$$\text{m.m.f} = \oint \mathbf{H} \cdot d\boldsymbol{l} = \text{NI} \tag{6.48}$$

Along the dotted path **H** and d*l* are parallel and also we have $\phi = \text{BA}$, hence

$$\text{B} = \mu\text{H} \tag{6.49}$$

$$\Rightarrow \text{H} = \frac{\text{B}}{\mu} = \frac{\phi}{\mu\text{A}} \tag{6.50}$$

From Eqs. 6.46 and 6.50

$$\text{m.m.f} = \oint \mathbf{H} \cdot d\boldsymbol{l} = \oint \text{H}d\boldsymbol{l} \tag{6.51}$$

$$\text{m.m.f} = \oint \frac{\phi}{\mu A} dl \tag{6.52}$$

As we expect $\phi$ to be constant all over circuit

$$\text{m.m.f} = \phi \oint \frac{dl}{\mu A} \tag{6.53}$$

$$\Rightarrow \phi = \frac{\text{m.m.f}}{\oint \frac{dl}{\mu A}} \tag{6.54}$$

From Eq. 6.48 we can write

$$\Rightarrow \phi = \frac{\text{m.m.f}}{\oint \frac{dl}{\mu A}} = \frac{NI}{\oint \frac{dl}{\mu A}} \tag{6.55}$$

In a pure electric circuit we can write Ohm's law as

$$\text{Current} = \text{emf/Resistance} \tag{6.56}$$

From 6.55 and 6.56 we see that the quantity $\oint \frac{dl}{\mu A}$ in magnetic circuit plays equivalent role of resistance is electric circuit. Then

$$R = \oint \frac{dl}{\mu A} \tag{6.57}$$

is called reluctance of magnetic circuit. Suppose assume that the area of cross section is constant then

$$R = \frac{\oint dl}{\mu A} = \frac{L}{\mu A} \tag{6.58}$$

where L is the total length.

Reluctance behaves similar to resistance. The total reluctance in series combination is

$$R = \sum_i \frac{L_i}{\mu_i A_i} = \sum_i R_i \tag{6.59}$$

The total reluctance is parallel is

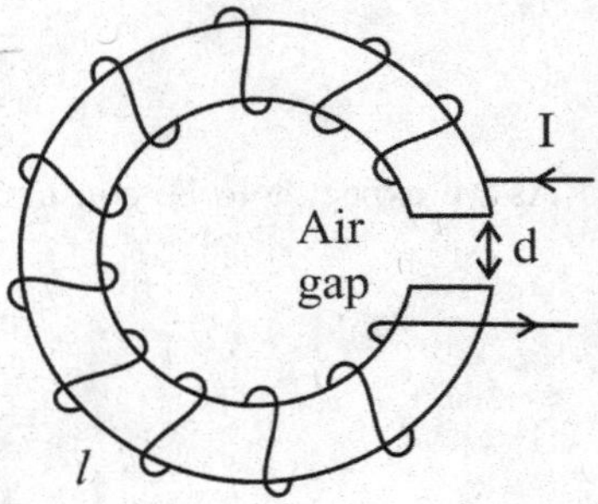


**Fig. 6.11** A toroidal coil wound on a iron ring with a small air gap d carries a current I

$$\frac{1}{R} = \sum_i \frac{1}{R_i} \tag{6.60}$$

As mentioned previously the magnetic circuit like transformers, relays finds importance in practical applications. The magnetic flux stored in the particular part of the machine can be calculated using the relation

$$\phi = \frac{NI}{\frac{L}{\mu A}} \tag{6.61}$$

Now as an example let us investigate the effect of air gap left in the magnetic circuit. Consider Fig. 6.11. A iron ring of cross-sectional area a with a small air gap d wound with a toroidal coil with N turns carries current I as shown in the figure.

$$\text{The reluctance of air gap} = \frac{d}{\mu_o a} \tag{6.62}$$

$$\text{The reluctance of iron} = \frac{l-d}{\mu a} \tag{6.63}$$

where $l$ is the total length of air gap and iron.

The total reluctance is

$$\text{Total reluctance} = \frac{d}{\mu_o a} + \frac{l-d}{\mu a}\ (\text{Eq. 6.59}) \tag{6.64}$$

The magnetic flux is then given by

$$\phi = \frac{NI}{\frac{d}{\mu_o a} + \frac{l-d}{\mu a}} \tag{6.65}$$

The above expression applies to a very narrow gap. In the case of wide gap, the lines of force spread out and the area of cross section varies considerably. Suppose assume that a magnetic circuit consists of two or more sections of magnetic materials as shown in Fig. 6.12a.

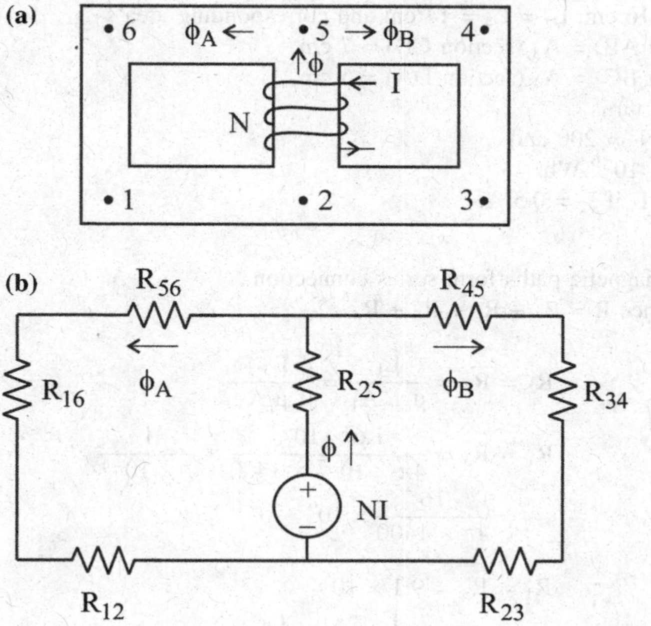


**Fig. 6.12** **a** A magnetic circuit consisting of many sections of magnetic material. **b** Equivalent circuit for Fig. 6.12a.

The circuit shown in Fig. 6.12a can be represented in terms of reluctance is equivalent form as shown in Fig. 6.12b. The total reluctance of the circuit in Fig. 6.12b can be obtained by series and parallel combination of reluctances.

**Example 6.8**

Consider a magnetic circuit as shown in Fig. 6.13. The entire magnetic circuit is made of cast iron with $\mu_r = 1400$. The length and area of the different sections is (see Fig. 6.13)

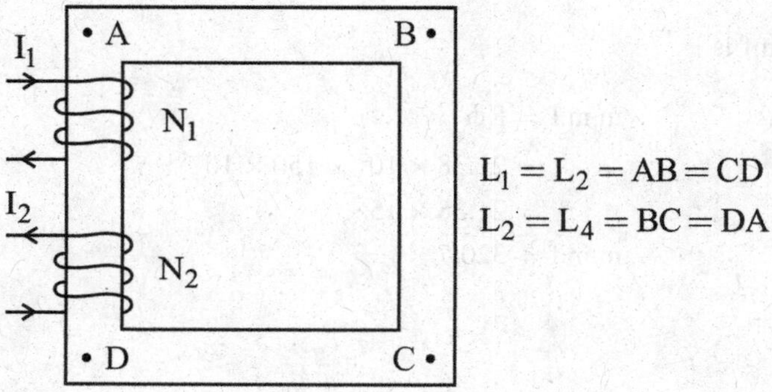


**Fig. 6.13** A magnetic circuit made of cast iron. Illustration for Example 6.8

$L_1 = L_3 = 16$ cm, $L_2 = L_4 = 14$ cm and corresponding area's
$A_1$ (section AB) = $A_3$ (section CD) = 7 cm$^2$
$A_2$ (section BC) = $A_4$ (section DA) = 5 cm$^2$
Number of turns
$N_1 = 300$ $N_2 = 200$ and
$\phi = 150 \times 10^{-6}$ Wb.
Determine $I_1$ if $I_2 = 0.5$ A.

**Solution**

The four magnetic paths form series connection

Reluctance $R = R_1 + R_2 + R_3 + R_4$

$$R_1 = R_3 = \frac{L_1}{\mu_o \mu_r A_1} = \frac{L_3}{\mu_o \mu_r A_3}$$

$$R_1 = R_3 = \frac{1.6 \times 10^{-2}}{4\pi \times 10^{-7} \times 1400} \times \frac{1}{7 \times 10^{-4}}$$

$$= \frac{16}{4\pi \times 1400} \times 10^9$$

$$R_1 = R_3 = 9.1 \times 10^5$$

$$R_2 = R_4 = \frac{L_2}{\mu_o \mu_r A_2} = \frac{L_4}{\mu_o \mu_r A_4}$$

$$R_2 = R_4 = \frac{14 \times 10^{-2}}{(4\pi \times 10^{-7})(1400)(5 \times 10^{-4})}$$

$$R_2 = R_4 = \frac{14}{(4\pi)(1400)(5)} \times 10^9$$

$$R_2 = R_4 = 1.59 \times 10^5$$

Total reluctance

$$R = 9.1 \times 10^5 + 1.59 \times 10^5 + 9.1 \times 10^5 + 1.59 \times 10^5$$

$$R = 21.38 \times 10^5$$

Total mmf is

$$\text{m.m.f} = R\phi$$

$$= 21.38 \times 10^5 \times 150 \times 10^{-6}$$

$$= 21.38 \times 15$$

$$\text{m.m.f} = 320.7$$

The currents $I_1$ and $I_2$ produce m.m.f of opposite signs, hence

$$\text{m.m.f} = N_1 I_1 - N_2 I_2$$
$$320.7 = 300 I_1 - 200(0.5)$$
$$\Rightarrow I_1 = 1.4\,\text{A}.$$

**Example 6.9**
Calculate the magnetic flux in an electromagnet.

**Solution**
An electromagnet consists of four parts.

(1) A yoke
(2) Two limbs
(3) Two pole pieces
(4) An air gap as shown in Fig. 6.14.

Let $L_1$ be the effective length, $\mu_1$ permeability and $A_1$ is the cross-sectional area of the yoke. The reluctance of the yoke is

$$R_1 = \frac{L_1}{\mu_1 A_1}$$

If $L_2$, $\mu_2$, $A_2$ are the length, permeability and area of cross section of the limbs then reluctance of the limbs is

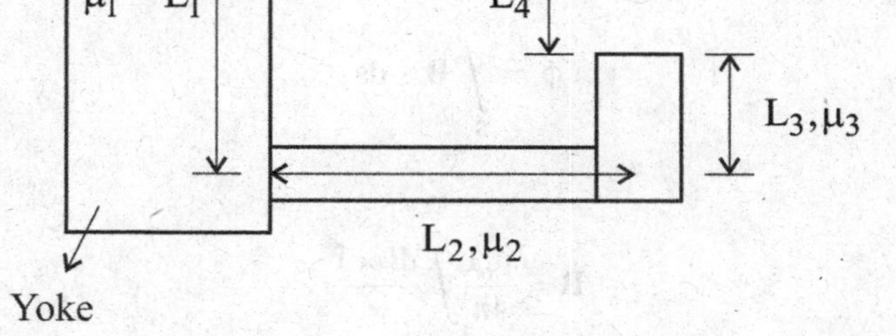

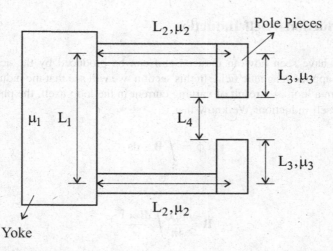

**Fig. 6.14** An electromagnet

$$R_2 = \frac{2L_2}{\mu_2 A_2}$$

If $L_3$, $\mu_3$, $A_3$ are the length, permeability and area of cross section of the pole pieces

$$R_3 = \frac{2L_3}{\mu_3 A_3}$$

Similarly for air gap

$$R_4 = \frac{L_4}{\mu_o A_4}$$

Total reluctance

$$R = R_1 + R_2 + R_3 + R_4$$
$$R = \frac{L_1}{\mu_1 A_1} + \frac{2L_2}{\mu_2 A_2} + \frac{2L_3}{\mu_3 A_3} + \frac{L_4}{\mu_o A_4}$$

Hence magnetic flux

$$\phi = \frac{NI}{R}$$
$$\phi = \frac{NI}{\frac{L_1}{\mu_1 A_1} + \frac{2L_2}{\mu_2 A_2} + \frac{2L_3}{\mu_3 A_3} + \frac{L_4}{\mu_o A_4}}$$

## 6.7 Induction—Self Induction

So far we have seen how an induced emf can be produced by the action of an externally applied magnetic field. In this section, we will see that the induced emf is produced in a loop as a result of varying current in the loop itself, the phenomenon known as self-induction. We know that

$$\phi = \int_S \mathbf{B} \cdot d\mathbf{s} \tag{6.66}$$

and

$$\mathbf{B} = \frac{\mu_o I}{4\pi} \oint \frac{d\boldsymbol{l} \times \hat{\mathbf{r}}}{\boldsymbol{r}^2} \tag{6.67}$$

by Biot–Savart law. Suppose assume a loop carries current I producing a magnetic flux density **B**. If I changes then **B** changes, and hence $\phi$ linked to the same loop changes. As $\phi$ linked to the loop changes, due to the change in the current I carried by the same loop, then according to the equation

$$\mathcal{E} = \frac{-d\phi}{dt} \tag{6.68}$$

an induced emf is produced in the loop which is responsible for establishing **B** or $\phi$. The induction is called self-induction because the emf is in the same loop which is responsible for the establishing flux $\phi$.

From Eq. 6.67 B is proportional to I and hence from 6.66, because $\phi$ is proportional to B we have

$$\phi \propto I \tag{6.69}$$

(or)

$$\phi = LI \tag{6.70}$$

From Eq. 6.68 then

$$\mathcal{E} = -L\frac{dI}{dt} \tag{6.71}$$

Here the constant of proportionality L is known as coefficient of self-inductance. It depends upon the geometry of the circuit and the permeability of the medium is which the loop is immersed.

## 6.8 Induction—Mutual Induction

In Fig. 6.15 we show two loops in which loop 1 is carrying current $I_1$, and hence a magnetic flux density $\mathbf{B}_1$ is produced at the site of loop 2. The total flux through the loop 2 is then

$$\phi_{21} = \int_{S_2} \mathbf{B}_1 \cdot d\mathbf{s}_2 \tag{6.72}$$

where $S_2$ is the area of loop 2. We have

$$\mathbf{B}_1 = \frac{\mu_o I_1}{4\pi} \oint \frac{d\boldsymbol{l}_1 \times \hat{\mathbf{r}}}{\boldsymbol{\imath}^2} \tag{6.73}$$

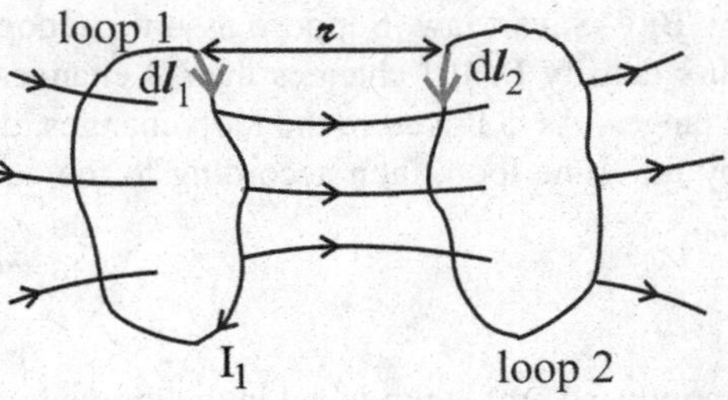


**Fig. 6.15** Mutual Induction: Two loops loop 1, loop 2 situated as shown in the figure and loop 1 carrying current $I_1$

by Biot–Savart law. Thus as $I_1$ changes $\mathbf{B}_1$ changes and $\phi_{21}$ the flux linked to the loop 2 due to flux density $\mathbf{B}_1$ changes. Thus $\phi_{21}$ produced by the loop 1 is proportional to $I_1$

$$\phi_{21} \propto I_1 \tag{6.74}$$

(or)

$$\phi_{21} = M_{21} I_1 \tag{6.75}$$

The constant of proportionality $M_{21}$ is known as coefficient of mutual inductance.

An emf is induced in the loop 2 given by

$$\mathcal{E}_{21} = \frac{-d\phi_{21}}{dt} \tag{6.76}$$

From Eq. 6.75

$$\mathcal{E}_{21} = -M_{21}\frac{dI_1}{dt} \tag{6.77}$$

We have

$$\phi_{21} = \int_{S_2} \mathbf{B}_1 \cdot d\mathbf{s}_2 = \int_{S_2} (\nabla \times \mathbf{A}_1) \cdot d\mathbf{s}_2 \tag{6.78}$$

Applying Stokes theorem

$$\phi_{21} = \oint \mathbf{A}_1 \cdot d\mathbf{l}_2 \tag{6.79}$$

From Eq, 4.112

$$\mathbf{A}_1 = \frac{\mu_o I_1}{4\pi} \oint \frac{dl_1}{\mathscr{r}} \tag{6.80}$$

Substituting 6.80 in 6.79

$$\phi_{21} = \frac{\mu_o I_1}{4\pi} \oint \oint \frac{dl_1}{\mathscr{r}} \cdot dl_2 \tag{6.81}$$

Comparing 6.81 to 6.75

$$M_{21} = \frac{\mu_o}{4\pi} \oint \oint \frac{dl_1 \cdot dl_2}{\mathscr{r}} \tag{6.82}$$

If we pass the same current in loop 2 instead of loop 1 we can show that

$$M_{12} = \frac{\mu_o}{2\pi} \oint \oint \frac{dl_1 \cdot dl_2}{\mathscr{r}} \tag{6.83}$$

or in other words

$$M_{12} = M_{21} = M \tag{6.84}$$

That is irrespective of whatever loop carries the same current, same flux is linked with the other loop.

**Example 6.10**
Consider a solenoid of length L with air core. The total number of turns is N and area of each turn is A. Calculate the self-inductance of the solenoid if the current I flows in the solenoid.

**Solution**
We know that the magnetic flux density of a solenoid is given by

$$B = \mu_o \frac{N}{L} I$$

Then the flux through each turn is BA.
$= \frac{\mu_o NI}{L} A$ [By substituting the value of B]
Then the total flux through the entire solenoid is

$$\phi = \frac{\mu_o NIA}{L} N$$

$$\phi = \frac{\mu_o N^2 IA}{L}$$

Now

$$\mathcal{E} = \frac{-d\phi}{dt} = \frac{-d}{dt}\left(\frac{\mu_o N^2 IA}{l}\right)$$

$$\mathcal{E} = \frac{-\mu_o N^2 A}{l}\frac{dI}{dt}$$

Comparing the above equation with equation $\mathcal{E} = -L\frac{dI}{dt}$ we get

$$L = \frac{\mu_o N^2 A}{l}$$

**Example 6.11**
Consider a circuit consisting of two parallel wires P and Q separated by a distance d as shown in Fig. 6.16. The current I goes through P and returns via Q. The radius of each wire is r. Calculate the self-inductance.

**Solution**
Consider a small element d$y$ at a distance $y$ from P and d – $y$ from Q. Noting down the B due to a thin wire at a distance R carrying current I is $\frac{\mu_o I}{2\pi R}$.

B due to the wires P and Q at d$y$ is

$$\mathbf{B} = \frac{\mu_0 I}{2\pi y} + \frac{\mu_0 I}{2\pi(d-y)}$$

d$\phi$ through d$y$ is

$$d\phi = Bdy$$

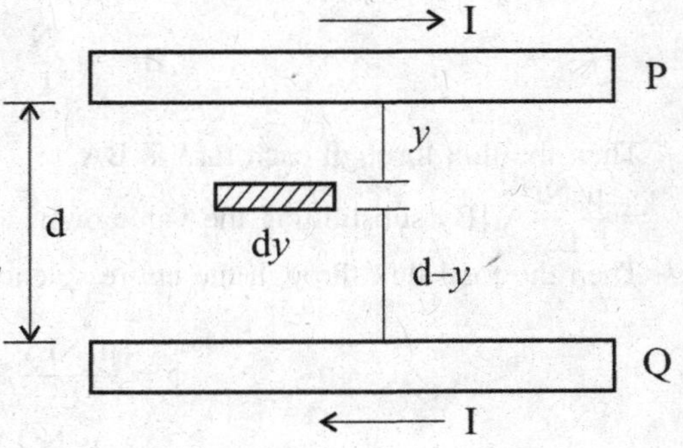


**Fig. 6.16** Two parallel wires P and Q carrying current I and separated by distance d

and then the total flux is

$$\phi = \int_0^d d\phi = \int_r^{d-r} \left(\frac{1}{y} + \frac{1}{d-y}\right) \frac{\mu_o I}{2\pi} dy$$

$$\phi = \frac{\mu_o I}{2\pi} \int_r^{d-r} \left(\frac{1}{y} + \frac{1}{d-y}\right) dy$$

$$\phi = \frac{\mu_o I}{\pi} ln\left(\frac{d-r}{r}\right)$$

Now $\mathcal{E} = \frac{-d\phi}{dt} = \frac{\mu_o}{\pi} ln\left(\frac{d-r}{r}\right) \frac{dI}{dt}$

Comparing with Eq. 6.71 $\mathcal{E} = \frac{-LdI}{dt}$

$$L = \frac{\mu_o}{4\pi} \ln\left(\frac{d-r}{r}\right)$$

**Example 6.12**

Consider a toroid of rectangular cross section as shown in Fig. 6.17. A coil of N number of turns is wound on it and a current I flows in it. Let a and b are the internal and external radius, respectively. Calculate the self-inductance of the toroid.

**Solution**

Applying Ampere's law

$$\oint \mathbf{B} \cdot d\mathbf{\mathit{l}} = \mu_o NI$$

**Fig. 6.17** A toroid of rectangular cross-section

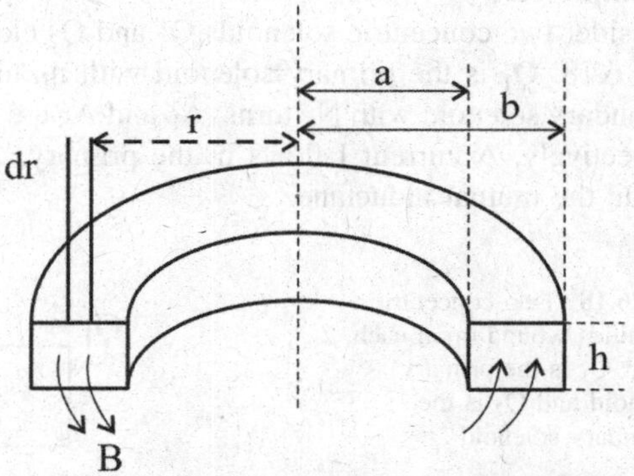


Coil would on the toroid and currents flows in it.

For the path of radius r

$$B(2\pi r) = \mu_o NI$$
$$\Rightarrow B = \frac{\mu_o NI}{2\pi r}$$

Considering a rectangular strip of width dr and height h at a distance r from the centre the magnetic flux linked with the strip is

$$d\phi = BdA$$
$$= \frac{\mu_o NI}{2\pi r} hdr$$

as dA = hdr

Thus flux linked with one turn of the toroid is

$$\phi = \int_a^b d\phi = \frac{\mu_o NIh}{2\pi} \int_a^b \frac{dr}{r}$$
$$\phi = \frac{\mu_o NIh}{2\pi} \log_e b/a$$

Total flux with N turns

$$= N\phi$$
$$= \frac{\mu_o N^2 Ih}{2\pi} \log_e b/a$$

Also the total flux with N turns is = L I

Hence $L = \frac{\mu_o N^2 h}{2\pi} \log_e b/a$

**Example 6.13**

Consider two concentric solenoids $Q_1$ and $Q_2$ closely wound upon each other as in Fig. 6.18. $Q_1$ is the primary solenoid with $n_P$ turns per unit length and $Q_2$ is the secondary solenoid with $N_S$ turns. $A_P$ and $A_S$ are the area of cross of the $Q_1$ and $Q_2$, respectively. A current I flows in the primary. Assuming no leakage in flux calculate the mutual inductance.

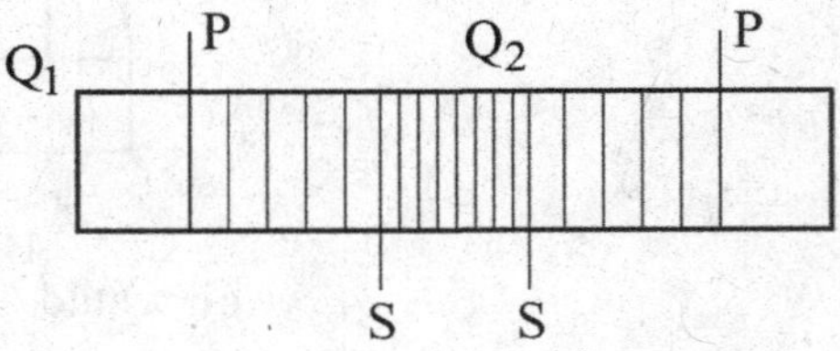


**Fig. 6.18** Two concentric solenoids wound upon each other. $Q_1$ is the primary solenoid and $Q_2$ is the secondary solenoid

**Solution**

The magnetic flux density due to the current I in the primary is

$$B = \mu_0 n_p I$$

The magnetic flux is the product of B and area of cross section. Hence $\mu_0 n_P A_P I$ is the flux linked with each turn of the secondary. As there is no leakage of flux the total flux linked with $N_S$ turns of the secondary is $\phi = \mu_0 n_P A_P I N_S$

The induced emf

$$\mathcal{E} = \frac{-d\phi}{dt}$$
$$= -\mu_0 n_P N_S A_P \frac{dI}{dt}$$

We know that

$$\mathcal{E} = -M\frac{dI}{dt}$$

Hence comparing $M = \mu_0 n_P N_S A_P$ is the mutual inductance.

**Example 6.14**

Suppose assume that there are two windings in a toroid with one being primary having $N_1$ turns and the other being secondary with $N_2$ turns. Then show that mutual inductance between both the windings is

$$M = \sqrt{L_1 L_2}$$

where $L_1$ and $L_2$ are the self-inductance of primary and secondary, respectively. The radius of the toroid is R. Assume no flux leakage.

**Fig. 6.19** A toroid with two windings—A primary winding and a secondary winding

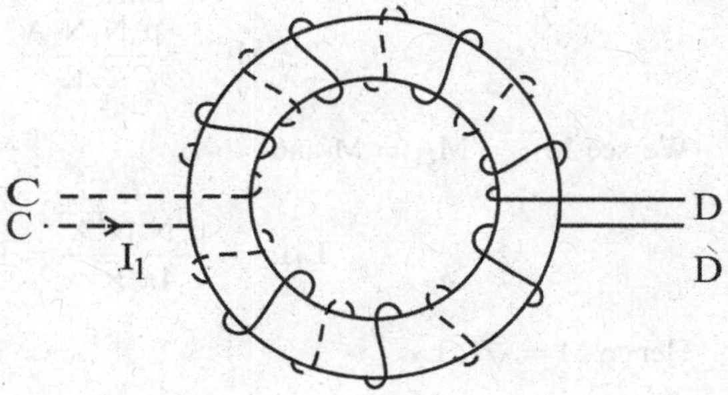

**Solution**
Suppose assume that in Fig. 6.19 CC is the primary and DD is the secondary and a current $I_1$ flows in primary. The magnetic flux density due to the first winding is

$$B_1 = \frac{\mu_o N_1 I_1}{2\pi R}$$

Let $\phi_{11}$ be the flux through the primary winding then

$$\phi_{11} = B_1 N_1 A = \frac{\mu_o N_1^2 A I_1}{2\pi R},$$

Also $\phi_{11} = L_1 I_1$
Comparing the above two equations

$$L_1 = \frac{\mu_o N_1^2 A}{2\pi R}$$

The flux through the secondary winding is

$$\phi_{21} = B_1 N_2 A = \frac{\mu_o N_1 N_2 A I_1}{2\pi R}$$

Also $\phi_{21} = M_{21} I_1$
Comparing the above two equations

$$M_{21} = \frac{\mu_o N_1 N_2 A}{2\pi R}$$

Suppose a current $I_2$ flows is the secondary instead of $I_1$ in primary then following the same procedure

$$L_2 = \frac{\mu_o N_2^2 A}{2\pi R}$$

$$M_{12} = \frac{\mu_o N_1 N_2 A}{2\pi R}$$

We see $M_{12} = M_{21} = M$ and

$$L_1 L_2 = \frac{\mu_o^2 N_1^2 N_2^2 A^2}{4\pi^2 R^2} = M^2$$

Hence $M = \sqrt{L_1 L_2}$.

## 6.9 Energy Stored in the Circuit in Terms of Self-Inductance

We have seen in Sect. 6.7 that a loop possesses self-inductance. According to Lenz's law this self-inductance will oppose any change in current in the loop. Hence this self-inductance is called back emf. Suppose assume a circuit connected to a battery and a key and the key is just closed. The current builds up in the circuit.

The current is charging in the circuit and as per Lenz's law a back emf is induced which oppose the growth in current. In order to build up the current up to a steady value I, work has to be done against this back emf. This work done remains hidden in the circuit. When the key in the circuit is opened current is decreasing and once again a back emf is produced which tries to maintain the current in the circuit. Thus the work done by the current during the growth is recovered.

The work done against the back emf on a unit charge is $-\mathcal{E}$ on a single round around the circuit. The negative sign is included to note that work is done by the external agency against the back emf. Work done per unit time is

$$\frac{dW}{dt} = -\mathcal{E}I \tag{6.85}$$

From Eq. 6.71 $\mathcal{E} = -L\frac{dI}{dt}$

$$\frac{dW}{dt} = LI\frac{dI}{dt} \tag{6.86}$$

The total work done in increasing the current from zero to maximum value of current I is

$$W = \int_0^I LI\frac{dI}{dt}dt \tag{6.87}$$

$$W = \frac{1}{2}LI^2 \tag{6.88}$$

## 6.10 Energy Stored in Magnetic Fields

The work done expressed in terms of self-inductance in Eq. 6.88 can be expressed in terms of magnetic field. We know that

$$\phi = \int_S \mathbf{B} \cdot \mathbf{ds} = \int_S (\nabla \times \mathbf{A}) \cdot \mathbf{ds} \tag{6.89}$$

Applying Stokes theorem

$$\phi = \oint \mathbf{A} \cdot \mathrm{d}\boldsymbol{l} \tag{6.90}$$

As $\phi = \mathrm{LI}$

$$\mathrm{LI} = \oint \mathbf{A} \cdot \mathrm{d}\boldsymbol{l} \tag{6.91}$$

Substituting 6.91 in 6.88

$$\mathrm{W} = \frac{1}{2}\mathrm{I}\oint \mathbf{A} \cdot \mathrm{d}\boldsymbol{l} \tag{6.92}$$

As **I**, d*l* always point in same direction

$$\mathrm{W} = \frac{1}{2}\oint (\mathbf{A} \cdot \mathbf{I})\mathrm{d}l \tag{6.93}$$

The above equation is for line current. For volume current

$$\mathrm{W} = \frac{1}{2}\int_\tau (\mathbf{A} \cdot \mathbf{J})\mathrm{d}\tau \tag{6.94}$$

With Ampere's law $\nabla \times \mathbf{B} = \mu_o \mathbf{J}$

$$\mathrm{W} = \frac{1}{2\mu_o}\int_\tau \mathbf{A} \cdot (\nabla \times \mathbf{B})\mathrm{d}\tau \tag{6.95}$$

From Sect. 1.20 identity 8 is

$$\begin{aligned}
&\nabla \cdot (\mathbf{A} \times \mathbf{B}) = \mathbf{B} \cdot (\nabla \times \mathbf{A}) - \mathbf{A} \cdot (\nabla \times \mathbf{B}) \\
&\Rightarrow \mathbf{A} \cdot (\nabla \times \mathbf{B}) = \mathbf{B} \cdot (\nabla \times \mathbf{A}) - \nabla \cdot (\mathbf{A} \times \mathbf{B}) \\
&\text{As } \mathbf{B} = \nabla \times \mathbf{A} \\
&\Rightarrow \mathbf{A} \cdot (\nabla \times \mathbf{B}) = \mathbf{B} \cdot \mathbf{B} - \nabla \cdot (\mathbf{A} \times \mathbf{B}) \\
&\Rightarrow \mathbf{A} \cdot (\nabla \times \mathbf{B}) = \mathrm{B}^2 - \nabla \cdot (\mathbf{A} \times \mathbf{B})
\end{aligned} \tag{6.96}$$

Substituting Eq. 6.96 in Eq. 6.95

$$W = \frac{1}{2\mu_o}\left[\int_\tau B^2 d\tau - \int \nabla \cdot (\mathbf{A} \times \mathbf{B}) d\tau\right] \tag{6.97}$$

Applying Gauss's divergence theorem

$$W = \frac{1}{2\mu_o}\left[\int_\tau B^2 d\tau - \oint_S (\mathbf{A} \times \mathbf{B}) \cdot \mathbf{ds}\right] \tag{6.98}$$

In Eq. 6.94 the integration is to be taken over volume $\tau$ which encloses current $\mathbf{J}$. But larger volumes are also acceptable because $\mathbf{J}$ is zero over there. Hence a larger volume can be selected. When we select a large volume, however, surface integral in Eq. 6.98 vanishes. Because $|\mathbf{A}|$ falls of as $\frac{1}{r}$, $|\mathbf{B}|$ falls off as $\frac{1}{r^2}$, and hence magnitudes of $\mathbf{A} \times \mathbf{B}$ decreases as $\frac{1}{r^3}$ while the surface S increases as $r^2$. The entire term $\oint_S (\mathbf{A} \times \mathbf{B}) \cdot \mathbf{ds}$ falls off as $\frac{1}{r}$ hence when we select entire volume then the surface integral goes to zero leaving

$$W = \frac{1}{2\mu_o}\int_\tau B^2 d\tau \tag{6.99}$$

**Example 6.15**
Consider two coils of self-inductance $L_1$ and $L_2$, respectively, as shown in Fig. 6.20. A current $I_1$ and $I_2$ flow in the each coil. Let M be the mutual inductance between the coils. Calculate the energy stored in the system. Consider both cases when the fluxes due to $I_1$ and $I_2$ are in either in same or opposite direction.

**Solution**
Let $\mathcal{E}_1$ and $\mathcal{E}_2$ be the induced emf is coil 1 and coil 2, respectively. Then

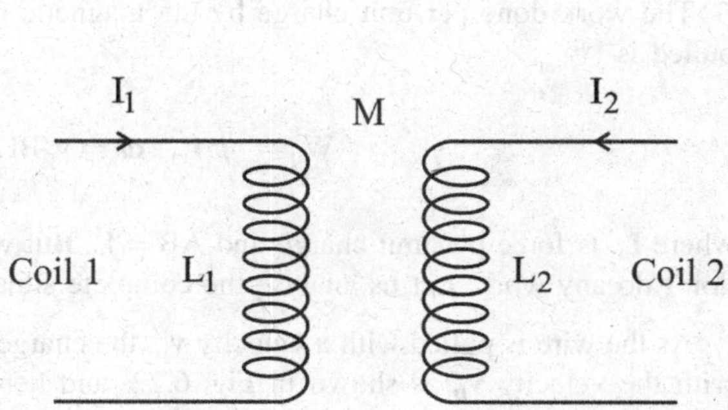


**Fig. 6.20** Two coils of self inductance $L_1$ and $L_2$ respectively placed close to each other and carrying currents $I_1$ and $I_2$ respectively

$$\mathcal{E}_1 = L_1\frac{dI_1}{dt} \pm M\frac{dI_2}{dt} \text{ (Numerically)}$$
$$\mathcal{E}_2 = L_2\frac{dI_2}{dt} \pm M\frac{dI_1}{dt} \text{ (Numerically)}$$

where +M stands for when the fluxes due to $I_1$ and $I_2$ are in same direction in each coil and –M when the fluxes due to $I_1$ and $I_2$ are in opposite direction.

We know that from Eq. 6.85

$$\frac{dW}{dt} = \mathcal{E}_1 I_1 + \mathcal{E}_2 I_2 \text{ (Numerically)}$$

Substituting for $\mathcal{E}_1, \mathcal{E}_2$

$$\frac{dW}{dt} = \left(L_1\frac{dI_1}{dt} \pm M\frac{dI_2}{dt}\right)I_1 + \left(L_2\frac{dI_2}{dt} \pm M\frac{dI_1}{dt}\right)I_2$$
$$\Rightarrow dW = (L_1 dI_1 \pm M dI_2)I_1 + (L_2 dI_2 \pm M dI_1)I_2$$
$$dW = L_1 I_1 dI_1 + L_2 I_2 dI_2 \pm d(MI_1I_2)$$

Integrating we get the energy stored in the system

$$= \frac{1}{2}L_1I_1^2 + \frac{1}{2}L_2I_2^2 \pm MI_1I_2$$

The first two terms corresponds to the self-inductance of each coil and the third term corresponds to mutual inductance between two coils.

**Example 6.16**
In Sect. 4.2 we saw that magnetic forces don't do any work. We will discuss this point in detail here. Suppose assume that a wire ABCD is pulled in the presence of magnetic field, the field pointing into the plane of the paper. The wire is pulled with a velocity $\mathbf{v_p}$ as shown in Fig. 6.21. The segments BC and AD experience magnetic forces in opposite direction which cancel out.

The charges in the segment AB experience a magnetic force from A to B and hence a current I flows in the circuit. The entire picture is shown in Fig. 6.22.

The work done per unit charge by the magnetic force $\mathbf{f}_m$ at the instant wire is pulled is

$$W_q = \oint \mathbf{f}_m \cdot d\boldsymbol{l} = v_p BL \tag{6.100}$$

where $\mathbf{f}_m$ is force per unit charge and AB = L. But we know that magnetic forces don't do any work. Let us analyse the complete situation.

As the wire is pulled with a velocity $\mathbf{v_p}$, the charges in the wire are pulled along with the velocity $\mathbf{v_p}$ as shown in Fig. 6.22, and hence a magnetic force per unit charge $\mathbf{f}_m$ acts on the charges.

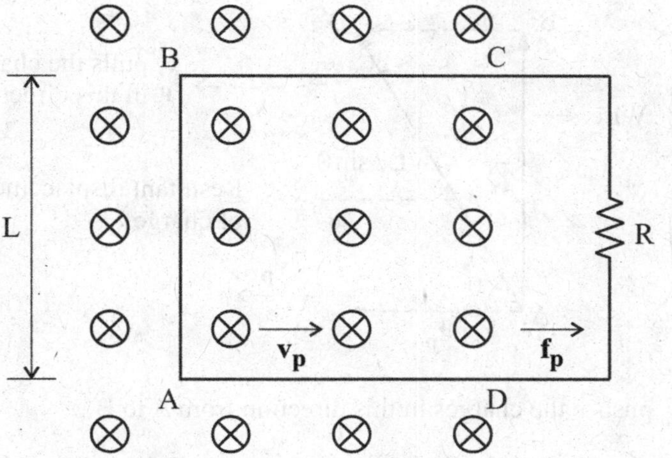


**Fig. 6.21** A wire ABCD is pulled in the presence of the magnetic field

**Fig. 6.22** The magnetic force acting on the charges in the segment AB of the wire

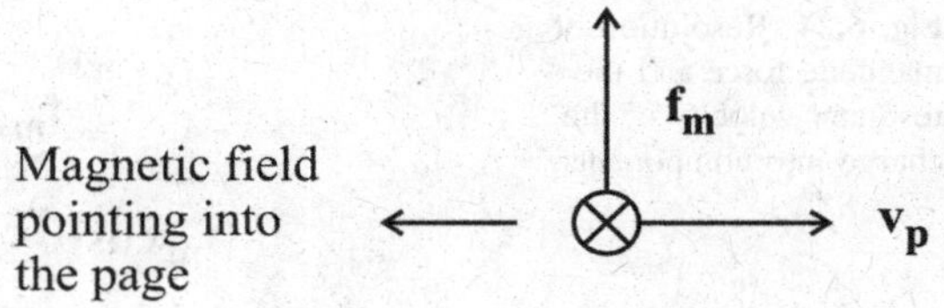


The charges in the wire are pulled horizontally by the force $f_p$ where $f_p$ is the pulling force per unit charge. But, however, the magnetic force pushes the charges from A to B vertically. Hence the resultant displacement $\frac{L}{\sin\theta}$ of the charges will be as shown in Fig. 6.23 as the wire moves.

The resultant velocity of the charges $\mathbf{v_q}$ will be along resultant displacement $\frac{L}{\sin\theta}$ of the charges as shown in Fig. 6.24 and makes an angle $\theta$ with $\mathbf{f_p}$. Because the magnetic force $\mathbf{f_m}$ is perpendicular to the velocity it no longer points in the vertical direction as shown in Fig. 6.22 but tilts by the same angle $\theta$ as shown in Fig. 6.24.

From Fig. 6.24 we see that $\mathbf{f_m}$ has a component $f_m \sin\theta$ which opposes $f_p$. Hence in order to make the wire to move in its direction $f_p$ has to exert a minimum force of

$$f_p = f_m \sin\theta \tag{6.101}$$

From Fig. 6.24 we see that

$$v_p = v_q \cos\theta \tag{6.102}$$

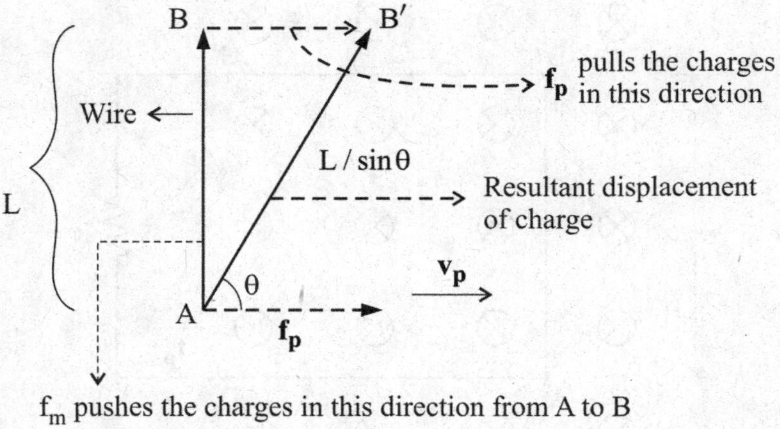


**Fig. 6.23** Resultant displacement of the charge in the segment AB due to different forces acting on the charge

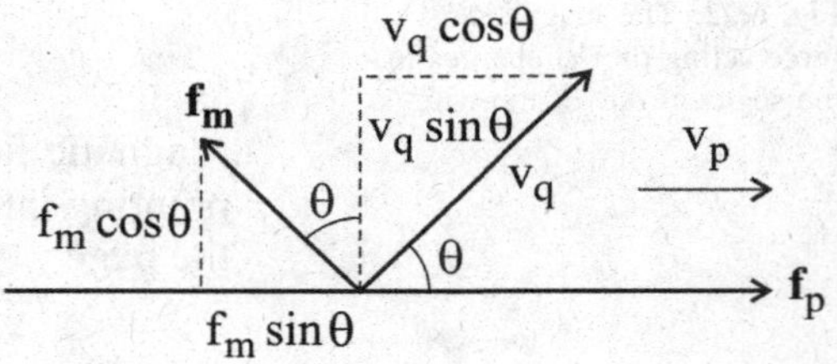


**Fig. 6.24** Resolution of magnetic force and the resultant velocity of the charge into components

The work done per unit charge by the pulling force $f_p$ is

$$W_p = \oint \mathbf{f_p} \cdot d\boldsymbol{l} \tag{6.103}$$

As $f_p = f_m \sin\theta$

$$\Rightarrow W_p = f_m \sin\theta \left(\frac{L}{\sin\theta}\right) \cos\theta \tag{6.104}$$

$$\Rightarrow W_p = f_m L \cos\theta \tag{6.105}$$

Substituting for $f_m$

$$W_p = (v_q B)(L \cos\theta) \tag{6.106}$$

$$W_p = (v_q \cos\theta)(BL) \tag{6.107}$$

$$W_p = v_p BL \tag{6.108}$$

(In equation for $W_q$ we used $\mathbf{v_p}$ for $\mathbf{f_m}$ as in that instant the charges are moving with velocity $\mathbf{v_p}$. But in the equation for $W_p$ we are using velocity $\mathbf{v_q}$ in $\mathbf{f_m}$ because now charges are moving with velocity $\mathbf{v_q}$).

Comparing Eqs. 6.100 and 6.108

$$W_p = W_q$$

Thus the actual work is done by the pulling agent and magnetic force takes credit for it. Then what is the role of magnetic force. When the wire is pulled by $f_p$ the charges move with a horizontal velocity $\mathbf{v_p}$. Magnetic force simply adds a vertical component $v_q \sin\theta$ to the horizontal component $v_q \cos\theta = v_p$, thereby producing a current I in the wire.

## 6.11 Maxwell's Equation

We have four important laws

(a)
$$\nabla \cdot \mathbf{E} = \frac{\rho}{\varepsilon_0} \tag{6.109}$$

(b)
$$\nabla \cdot \mathbf{B} = 0 \tag{6.110}$$

(c)
$$\nabla \times \mathbf{E} = \frac{-\partial \mathbf{B}}{\partial t} \tag{6.111}$$

(d)
$$\nabla \times \mathbf{B} = \mu_o \mathbf{J} \tag{6.112}$$

The equation a, the Gauss's law and equation b state the divergence of electric and magnetic field's, respectively, while equation c the Faraday's law and equation d the Ampere's law state their curls.

Taking divergence of equation c

$$\nabla \cdot (\nabla \times \mathbf{E}) = \nabla \cdot \left(\frac{-\partial \mathbf{B}}{\partial t}\right) = \frac{-\partial}{\partial t}(\nabla \cdot \mathbf{B}) \tag{6.113}$$

The left land side of the above equation is zero because the divergence curl of any vector is zero (Example 1.8). The right-hand side of the equation is zero by virtue of equation b

Taking divergence of equation d

$$\nabla \cdot (\nabla \times \mathbf{B}) = \mu_o \nabla \cdot \mathbf{J} \tag{6.114}$$

The left-hand side of the above equation is zero because divergence of curl of the vector is zero. However the right-hand side of the equation is zero only for steady currents. Because from equation of continuity (Eq. 4.89)

$$\nabla \cdot \mathbf{J} + \frac{\partial \rho}{\partial t} = 0 \tag{6.115}$$

$$\Rightarrow \nabla \cdot \mathbf{J} = \frac{-\partial \rho}{\partial t} \tag{6.116}$$

The above equation is zero if $\frac{\partial \rho}{\partial t} = 0$, the steady currents, i.e. only in a magnetostatic condition. In order to generalize for non-steady currents Maxwell modified the equation as follows

We have

$$\nabla \cdot \mathbf{J} = \frac{-\partial \rho}{\partial \mathbf{t}} \tag{6.117}$$

Using Gauss's law

$$\nabla \cdot \mathbf{J} = \frac{-\partial}{\partial t}(\varepsilon_o \nabla \cdot \mathbf{E}) \tag{6.118}$$

$$\Rightarrow \nabla \cdot \mathbf{J} = -\nabla \cdot \left(\varepsilon_o \frac{\partial \mathbf{E}}{\partial t}\right) \tag{6.119}$$

Thus if we add the above term in Eq. 6.112 to **J** then Eq. 6.112 will be valid for non-steady currents. Hence Eq. 6.112 now becomes

$$\nabla \times \mathbf{B} = \mu_o \mathbf{J} + \mu_o \varepsilon_o \frac{\partial \mathbf{E}}{\partial t} \tag{6.120}$$

Taking divergence of above equation

$$\nabla \cdot (\nabla \times \mathbf{B}) = \mu_o \nabla \cdot \mathbf{J} + \mu_o \varepsilon_o \frac{\partial}{\partial t}(\nabla \cdot \mathbf{E}) \tag{6.121}$$

Using Gauss's law

$$\nabla \cdot (\nabla \times \mathbf{B}) = \mu_o \left[ \nabla \cdot \mathbf{J} + \frac{\partial \rho}{\partial t} \right] \tag{6.122}$$

The left-hand side is zero as usual and the right-hand side is zero by equation of continuity. Apart from satisfying for non-steady currents Maxwell equation gives an interesting information. "Faraday's law says that changing magnetic field induces an electric field, while Maxwell's equation states that changing electric field induces an magnetic field".

Maxwell named the extra term in Eq. 6.121 as displacement current

$$\mathbf{J}_d = \varepsilon_o \frac{\partial \mathbf{E}}{\partial t} \tag{6.123}$$

## 6.12 The Displacement Current

The concept of displacement current was a great theoretical discovery. It helped to retain the notion that flow of current in a circuit is continuous.

As an example consider the plates of a parallel-plate capacitor connected to a voltage source as shown in Fig. 6.25. Conduction current flows in the leads of the capacitor charging the plates of the capacitor as shown in Fig. 6.25. But the conduction current I is not continuous across the gap between the plates as there is no transfer of charge between the plates. But from Maxwell's equation we see that the changing electric field between the plates serves the purpose of the conduction current inside the gap. Within the gap the changing electric field produces a displacement current which is exactly equal to conduction current as shown below.

If q is the charge given to one of the plates of the capacitor at any given instant of time and if A is the area of the plates then the electric field between the plates of the capacitor from Eq. 3.331 is given by

$$\mathbf{E} = \frac{V}{d} = \frac{q}{\varepsilon_o A} \tag{6.124}$$

$$\Rightarrow \frac{\partial \mathbf{E}}{\partial t} = \frac{1}{\varepsilon_o A} \frac{\partial q}{\partial t} = \frac{I}{\varepsilon_o A} \tag{6.125}$$

**Fig. 6.25** A capacitor connected to a voltage source

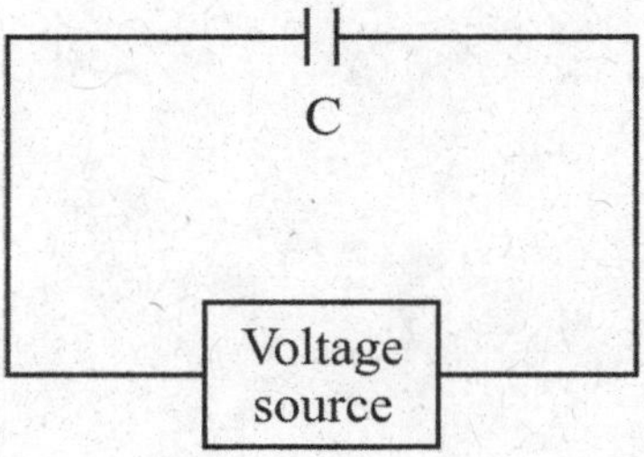

$$\Rightarrow \varepsilon_o \frac{\partial \mathbf{E}}{\partial t} = \frac{I}{A} \tag{6.126}$$

$$\Rightarrow \frac{\partial}{\partial t}(\varepsilon_o \mathbf{E}) = \frac{I}{A} \tag{6.127}$$

As $\mathbf{D} = \varepsilon_o \mathbf{E}$ in free space ($\mathbf{P} = 0$ in $\mathbf{D} = \varepsilon_o \mathbf{E} + \mathbf{P}$)

$$\Rightarrow \frac{\partial}{\partial t}(\mathbf{D}) = \frac{I}{A} \tag{6.128}$$

As $\mathbf{J}_d = \frac{\partial \mathbf{D}}{\partial t}$

$$\Rightarrow \mathbf{J}_d = \frac{I}{A} \tag{6.129}$$

$$\Rightarrow \mathbf{J}_d A = I \tag{6.130}$$

As $\mathbf{J}_d A = I_d$

$$\Rightarrow I_d = I \tag{6.131}$$

Thus the displacement current in the gap between the capacitor plates is found to be equal to the conduction current flowing in the leads of the capacitor.

The following are the four Maxwell's equation

(a) $$\nabla \cdot \mathbf{E} = \frac{\rho}{\varepsilon_o} \tag{6.132}$$

(b) $$\nabla \cdot \mathbf{B} = 0 \tag{6.133}$$

(c) $$\nabla \times \mathbf{E} = \frac{-\partial \mathbf{B}}{\partial t} \tag{6.134}$$

(d) $$\nabla \times \mathbf{B} = \mu_o \mathbf{J} + \mu_o \varepsilon_o \frac{\partial \mathbf{E}}{\partial t} \tag{6.135}$$

In free space with $\mathbf{D} = \varepsilon_o \mathbf{E}$ and $\mathbf{B} = \mu_o \mathbf{H}$ we can write the above equations as

$$\nabla \cdot \mathbf{D} = \rho \tag{6.136}$$

$$\nabla \cdot \mathbf{H} = 0 \tag{6.137}$$

$$\nabla \times \mathbf{E} = \frac{-\partial \mathbf{B}}{\partial t} \tag{6.138}$$

$$\nabla \times \mathbf{H} = \mathbf{J} + \frac{\partial \mathbf{D}}{\partial t} \tag{6.139}$$

## 6.13 Maxwell's Equation in Matter

In the presence of materials Maxwell's equation can be derived as follows. In the materials in the presence of electric and magnetic fields the material gets polarized and magnetized. With polarization we have seen that in Sects. 3.15–3.18, in addition to free charge density $\rho_f$ a bound charge density appears. Hence

$$\rho = \rho_f + \rho_p \tag{6.140}$$

In the case of magnetization we have seen that in Sect. 5.3 in addition to free current $\mathbf{J}_f$ a bound current $\mathbf{J}_p$ appears. Hence

$$\mathbf{J} = \mathbf{J}_f + \mathbf{J}_p \tag{6.141}$$

Equations 6.140 and 6.141 correspond to electrostatic and magnetostatic conditions. However in a non-static case the bound charge density noted down in Eq. 6.140 may change with time giving rise to currents. Hence under non-static conditions there will be additional currents those arise from the change of polarization with time given by

$$\mathbf{J}_1 = \frac{\partial \mathbf{P}}{\partial t} \tag{6.142}$$

In order to arrive at Eq. 6.142 consider a dielectric of arbitrary volume $\tau$ bounded by surface subjected to a time-varying electric field. The dielectric is polarized giving rise to bound charges. Under the action of time-varying electric field the bound charges flow out of volume $\tau$ giving rise to polarization current density $\mathbf{J}_1$. The rate at which bound charge flows out through S must be equal to the rate of decrease of bound charge within $\tau$.

Hence

$$\int_S \mathbf{J}_1 \cdot \mathbf{ds} = \frac{-\partial}{\partial t} \int_\tau \rho_p d\tau \tag{6.143}$$

The right-hand side of the above equation signifies that the bound charge density $\int_\tau \rho_p d\tau$ changes with time as per $\frac{\partial}{\partial t} \int_\tau \rho_p d\tau$ and the negative sign indicates charge is decreasing within volume τ. The flow of bound charges outside τ gives rise to current $\int_S \mathbf{J}_1 \cdot \mathbf{ds}$.

Applying Gauss's divergence theorem and from Eq. 3.262 $\rho_P = -\nabla \cdot \mathbf{P}$

$$\int_\tau \nabla \cdot \mathbf{J}_1 d\tau = \frac{\partial}{\partial t}\int_\tau \nabla \cdot \mathbf{P} d\tau \tag{6.144}$$

$$\Rightarrow \int_\tau \nabla \cdot \mathbf{J}_1 d\tau = \int_\tau \nabla \cdot \frac{\partial \mathbf{P}}{\partial t} d\tau \tag{6.145}$$

The above equation must be valid even for small volume dτ and in that case we need not integrate

$$\nabla \cdot \mathbf{J}_1 d\tau = \nabla \cdot \frac{\partial \mathbf{P}}{\partial t} d\tau \tag{6.146}$$

$$\Rightarrow \nabla \cdot \mathbf{J}_1 = \nabla \cdot \frac{\partial \mathbf{P}}{\partial t} \tag{6.147}$$

$$\Rightarrow \mathbf{J}_1 = \frac{\partial \mathbf{P}}{\partial t} \tag{6.148}$$

which is Eq. 6.142.

Hence under non-static conditions Eq. 6.141 can be written as

$$\mathbf{J} = \mathbf{J}_f + \mathbf{J}_p + \mathbf{J}_1 \tag{6.149}$$

Thus variation of bound charge density with time gives rise to current $\mathbf{J}_1$. However variation of magnetization with time doesn't give rise to similar currents. As M varies with time, bound current given by Eq. 6.151 varies with time accordingly.

From Eqs. 3.262 and 5.7, we see that

$$\rho_p = -\nabla \cdot \mathbf{P} \tag{6.150}$$

$$\mathbf{J}_p = \nabla \times \mathbf{M} \tag{6.151}$$

Hence Eqs. 6.140 and 6.149 can be written as

$$\rho = \rho_f - \nabla \cdot \mathbf{P} \tag{6.152}$$

$$\mathbf{J} = \mathbf{J}_f + \nabla \times \mathbf{M} + \frac{\partial \mathbf{P}}{\partial t} \tag{6.153}$$

Maxwell's first equation is

$$\nabla \cdot \mathbf{E} = \frac{\rho}{\varepsilon_o} \tag{6.154}$$

Using Eq. 6.152

$$\nabla \cdot \mathbf{E} = \frac{1}{\varepsilon_o}(\rho_f - \nabla \cdot \mathbf{P}) \tag{6.155}$$

with $\mathbf{D} = \varepsilon_o \mathbf{E} + \mathbf{P}$ in the presence of materials Eq. 6.155 is

$$\nabla \cdot \mathbf{D} = \rho_f \tag{6.156}$$

Maxwell's second equation is as usual

$$\nabla \cdot \mathbf{B} = 0 \tag{6.157}$$

Maxwell's third equation is

$$\nabla \times \mathbf{E} = \frac{-\partial \mathbf{B}}{\partial t} \tag{6.158}$$

Maxwell's fourth equation is

$$\nabla \times \mathbf{B} = \mu_o \mathbf{J} + \varepsilon_o \frac{\partial \mathbf{E}}{\partial t} \tag{6.159}$$

From Eq. 6.153

$$\nabla \times \mathbf{B} = \mu_o \left[ \mathbf{J}_f + \nabla \times \mathbf{M} + \frac{\partial \mathbf{P}}{\partial t} \right] + \varepsilon_o \frac{\partial \mathbf{E}}{\partial t} \tag{6.160}$$

We know

$$\mathbf{H} = \frac{\mathbf{B}}{\mu_o} - \mathbf{M} \tag{6.161}$$

and

$$\mathbf{D} = \varepsilon_o \mathbf{E} + \mathbf{P} \tag{6.162}$$

Using the above equations in Eq. 6.160

$$\nabla \times \mathbf{H} = \mathbf{J}_f + \frac{\partial \mathbf{D}}{\partial t} \tag{6.163}$$

Equations (6.132–6.135), (6.136–6.139) are the Maxwell's equation in free space.

Equations 6.156, 6.157, 6.158 and 6.161 written below are Maxwell's equation in the presence of materials

$$\nabla \cdot \mathbf{D} = \rho_f \tag{6.164}$$

$$\nabla \cdot \mathbf{B} = 0 \tag{6.165}$$

$$\nabla \times \mathbf{E} = \frac{-\partial \mathbf{B}}{\partial t} \tag{6.166}$$

$$\nabla \times \mathbf{H} = \mathbf{J}_f + \frac{\partial \mathbf{D}}{\partial t} \tag{6.167}$$

## 6.14 Maxwell's Equation in Integral Form and Boundary Conditions

Maxwell's Eqs. 6.164–6.167 can be written in integral form as

(i)
$$\oint_S \mathbf{D} \cdot d\mathbf{s} = q_{f(\text{enclosed})} \tag{6.168}$$

(ii)
$$\oint_S \mathbf{B} \cdot d\mathbf{s} = 0 \tag{6.169}$$

(iii)
$$\oint_l \mathbf{E} \cdot d\mathbf{l} = \frac{-d}{dt} \int_S \mathbf{B} \cdot d\mathbf{s} \tag{6.170}$$

(iv)
$$\oint_l \mathbf{H} \cdot d\mathbf{l} = I_{f(\text{enclosed})} + \frac{d}{dt} \int_S \mathbf{D} \cdot d\mathbf{s} \tag{6.171}$$

Consider Fig. 6.26. Using Eq. 6.168 and following procedures mentioned in Sects. 2.25 and 3.22

$$D_{1n} - D_{2n} = \sigma_f \tag{6.172}$$

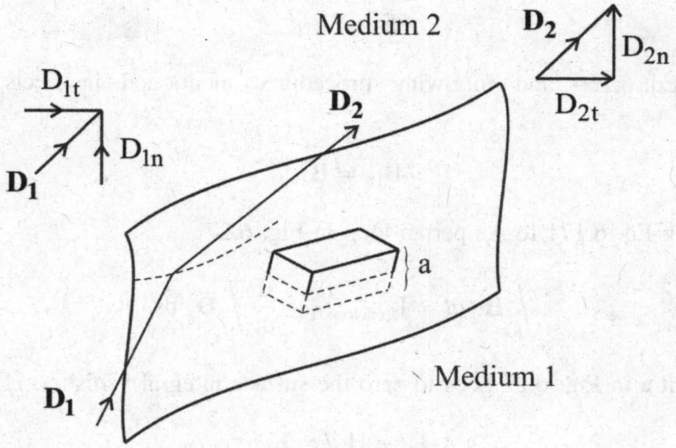


**Fig. 6.26** Boundary between two mediums shown along with a pill box

**Fig. 6.27** Boundary between two mediums shown along with a Amperian loop

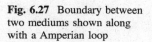

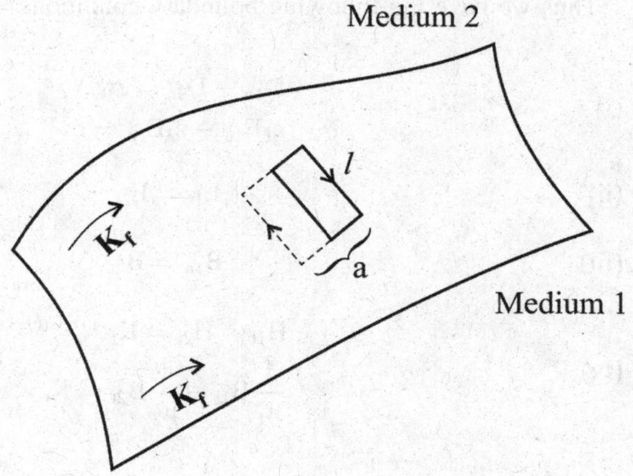


Consider Fig. 6.27. Using Eq. 6.170

$$\oint_{l} \mathbf{E} \cdot \mathrm{d}l = \frac{-\mathrm{d}}{\mathrm{dt}} \int_{S} \mathbf{B} \cdot \mathbf{ds} \tag{6.173}$$

In the limit a in Fig. 6.27 goes to zero the surface integral vanishes. Hence

$$E_{1t}l - E_{2t}l = 0 \tag{6.174}$$

$$E_{1t} = E_{2t} \tag{6.175}$$

Using Eq. 6.169 and following procedures mentioned in Sects. 5.6 and Sect. 4.23

$$B_{1n} = B_{2n} \tag{6.176}$$

Applying Eq. 6.171 to Amperian loop in Fig. 6.27

$$\oint_l \mathbf{H} \cdot dl = I_{f(enclosed)} + \frac{d}{dt}\int_S \mathbf{D} \cdot ds \tag{6.177}$$

In a limit a in Fig. 6.27 goes to zero the surface integral vanishes. Hence

$$H_{1t}l - H_{2t}l = K_f l \tag{6.178}$$

$$H_{1t} - H_{2t} = K_f \tag{6.179}$$

Thus we have the following boundary conditions

(i)
$$\begin{aligned} D_{1n} - D_{2n} &= \sigma_f \\ \varepsilon_1 E_{1n} - \varepsilon_2 E_{2n} &= \sigma_f \end{aligned} \tag{6.180}$$

(ii)
$$E_{1t} = E_{2t} \tag{6.181}$$

(iii)
$$B_{1n} = B_{2n} \tag{6.182}$$

(iv)
$$\begin{aligned} H_{1t} - H_{2t} &= K_f \\ \frac{1}{\mu_1}B_{1t} - \frac{1}{\mu_2}B_{2t} &= K_f \end{aligned} \tag{6.183}$$

## 6.15 Potential Functions

For electrostatic field we know that

$$V = \frac{1}{4\pi\varepsilon_o}\int_\tau \frac{\rho d\tau}{r} \tag{6.184}$$

and for magnetostatic field

$$\mathbf{A} = \frac{\mu_0}{4\pi}\int_\tau \frac{\mathbf{J}d\tau}{\mathscr{r}} \tag{6.185}$$

Now what will happen to these potentials when the fields are varying with time.

Our fundamental aim is to find the value of electric and magnetic fields. In electrostatic case electric field $\mathbf{E}(\mathbf{r})$ is given by Coulomb's law. In magnetostatic case the magnetic flux density $\mathbf{B}(\mathbf{r})$ is given by Biot–Savart law. But for time-varying fields $\mathbf{E}$ ($\mathbf{r}$, t), $\mathbf{B}$ ($\mathbf{r}$, t) are buried in Eqs. (6.132)–(6.135).

As we have seen in Sect. 1.1 it is easy to work with scalars instead of vectors. As $\mathbf{E}(\mathbf{r}, t)$ and $\mathbf{B}(\mathbf{r}, t)$ are vectors we will now try to simplify the problem by working with potentials.

In magnetostatics $\nabla \cdot \mathbf{B} = 0$ invited us to write B as $\nabla \times \mathbf{A}$ similarly from Eq. 6.133 we can write

$$\mathbf{B} = \nabla \times \mathbf{A} \tag{6.186}$$

Substituting the above Eq. (6.186) in Eq. (6.134)

$$\nabla \times \mathbf{E} = \frac{-\partial}{\partial t}(\nabla \times \mathbf{A}) \tag{6.187}$$

$$\Rightarrow \nabla \times \left[\mathbf{E} + \frac{\partial \mathbf{A}}{\partial t}\right] = 0 \tag{6.188}$$

Because curl gradient of any scalar function is zero, we can write from the above equation

$$\mathbf{E} + \frac{\partial \mathbf{A}}{\partial t} = -\nabla V \tag{6.189}$$

$$\Rightarrow \mathbf{E} = -\nabla V - \frac{\partial \mathbf{A}}{\partial t} \tag{6.190}$$

Now substituting the above Eq. 6.190 in Eq. 6.132

$$\nabla \cdot \left[-\nabla V - \frac{\partial \mathbf{A}}{\partial t}\right] = \frac{\rho}{\varepsilon_0} \tag{6.191}$$

$$\Rightarrow \nabla^2 V + \frac{\partial}{\partial t}(\nabla \cdot \mathbf{A}) = \frac{-\rho}{\varepsilon_0} \tag{6.192}$$

With the help of Eqs. 6.186, 6.191 and 6.135 can be written as

$$\nabla \times (\nabla \times \mathbf{A}) = \mu_o \mathbf{J} - \mu_o \varepsilon_o \nabla \left( \frac{\partial V}{\partial t} \right) - \mu_o \varepsilon_o \frac{\partial^2 \mathbf{A}}{\partial t^2} \tag{6.193}$$

From Sect. 1.20 identity 7 gives

$$\nabla \times (\nabla \times \mathbf{A}) = \nabla(\nabla \cdot \mathbf{A}) - \nabla^2 \mathbf{A} \tag{6.194}$$

Hence

$$\nabla \cdot (\nabla \cdot \mathbf{A}) - \nabla^2 \mathbf{A} = \mu_o \mathbf{J} - \mu_o \varepsilon_o \nabla \left( \frac{\partial V}{\partial t} \right) - \mu_o \varepsilon_o \frac{\partial^2 \mathbf{A}}{\partial t^2} \tag{6.195}$$

$$\Rightarrow \left( \nabla^2 \mathbf{A} - \mu_o \varepsilon_o \frac{\partial^2 \mathbf{A}}{\partial t^2} \right) - \nabla \left( \nabla \cdot \mathbf{A} + \mu_o \varepsilon_o \frac{\partial V}{\partial t} \right) = -\mu_o \mathbf{J} \tag{6.196}$$

We have to extract **A**, V from Eqs. 6.192, 6.196 and then substituting them in Eqs. 6.186, 6.190 we can obtain **E**(**r**, t), **B**(**r**, t). It seems that solving Eqs. 6.192, 6.196 to get **A**, V is a tedious task. However as we will see in next section gauge transformations simplify this problem.

## 6.16 Gauge Transformations

Equation 6.186 is

$$\mathbf{B} = \nabla \times \mathbf{A}$$

We know curl gradient of any scalar function is zero. Suppose $\alpha$ is any scalar function then

$$\mathbf{B} = \nabla \times \mathbf{A} + \nabla \times \nabla \alpha \tag{6.197}$$

$$\mathbf{B} = \nabla \times [\mathbf{A} + \nabla \alpha] \tag{6.198}$$

Let the new vector function be

$$\mathbf{A}_1 = \mathbf{A} + \nabla \alpha \tag{6.199}$$

Hence

$$\mathbf{B} = \nabla \times \mathbf{A}_1 \tag{6.200}$$

"Thus changing the vector potential by adding gradient of any scalar function (Eq. 6.197) doesn't alter the value of **B**".

If we change vector potential then what happens to **E**. Equation 6.190 is

$$\mathbf{E} = -\nabla V - \frac{\partial \mathbf{A}}{\partial t}$$

Substituting for **A** from Eq. 6.199

$$\mathbf{E} = -\nabla V - \frac{\partial}{\partial t}(\mathbf{A}_1 - \nabla\alpha) \tag{6.201}$$

$$\Rightarrow \mathbf{E} = -\nabla V - \frac{\partial \mathbf{A}_1}{\partial t} + \nabla\frac{\partial \alpha}{\partial t} \tag{6.202}$$

$$\Rightarrow \mathbf{E} = -\nabla\left(V - \frac{\partial \alpha}{\partial t}\right) - \frac{\partial \mathbf{A}_1}{\partial t} \tag{6.203}$$

With

$$V_1 = V - \frac{\partial \alpha}{\partial t} \tag{6.204}$$

$$\Rightarrow \mathbf{E} = -\nabla V_1 - \frac{\partial \mathbf{A}_1}{\partial t} \tag{6.205}$$

From Eqs. 6.205, 6.197 and 6.200 we observe that changing vector potential and scalar doesn't alter the electric and magnetic fields. We have

$$\mathbf{A}_1 = \mathbf{A} + \nabla\alpha \tag{6.206}$$

$$V_1 = V - \frac{\partial \alpha}{\partial t} \tag{6.207}$$

That is if we add $\nabla\alpha$ to **A** and simultaneously substract $\frac{\partial \alpha}{\partial t}$ from V, then the values of **E**, **B** doesn't get altered.

We know that the actual physical quantities of interest are **E**, **B** and not the potentials **A**, V. We can change **A**, V for our advantage without changing **E**, **B**.

That is we are free to select any form of vector potential and scalar potential without affecting **E**, **B**.

Because we can change vector potential and scalar potential without affecting **E**, **B** we select the **A** such that

$$\nabla \cdot \mathbf{A} = -\mu_0 \varepsilon_0 \frac{\partial V}{\partial t} \tag{6.208}$$

What is the advantage of Eq. 6.208. To check, substitute 6.208 in 6.196 then

$$\nabla^2 \mathbf{A} - \mu_o \varepsilon_o \frac{\partial^2 \mathbf{A}}{\partial t^2} = -\mu_o \mathbf{J} \tag{6.209}$$

and in 6.192

$$\nabla^2 V - \mu_o \varepsilon_o \frac{\partial^2 V}{\partial t^2} = \frac{-\rho}{\varepsilon_o} \tag{6.210}$$

Thus from Eqs. 6.192, 6.196 we have uncoupled **A** and V in Eqs. 6.209, 6.210 using the Eq. 6.208.

Equations 6.206 and 6.207 are called gauge transformations.

Equations 6.209 and 6.210 are inhomogeneous wave equations for **A** and V, and Eq. 6.208 is called Lorentz condition or Lorentz gauge for potentials.

## 6.17 Retarded Potentials

Solving Eqs. 6.209 and 6.210 we can get **A** and V for time-varying conditions. However there is one important point to be considered.

First we shall write the potentials for static case, i.e. from electrostatics and magnetostatics

$$V(\mathbf{r}) = \frac{1}{4\pi\varepsilon_o} \int_{\tau'} \frac{\rho(\mathbf{r}')}{\mathscr{r}} d\tau' \tag{6.211}$$

$$\mathbf{A}(\mathbf{r}) = \frac{\mu_o}{4\pi} \int_{\tau'} \frac{\mathbf{J}(\mathbf{r})}{\mathscr{r}} d\tau' \tag{6.212}$$

Consider Fig. 6.28 where we show either $\rho(\mathbf{r}')$ or $\mathbf{J}(\mathbf{r}')$

In the static case, $\rho(\mathbf{r}')$ or $\mathbf{J}(\mathbf{r}')$ doesn't change with time. But when $\rho(\mathbf{r}')$ or $\mathbf{J}(\mathbf{r}')$ changes with time then the situation is slightly different. As we will see the electromagnetic waves travel with velocity c the speed of light. At time $t_r$ a change produced in $\rho(\mathbf{r}', t_r)$ or $\mathbf{J}(\mathbf{r}', t_r)$ travels with the velocity c and takes time t to reach point P in Fig. 6.28.

Suppose assume that we are interested in knowing the time-varying potentials and hence time-varying fields at point P at time t. But the potentials and fields at point P at time t depends on $\rho(\mathbf{r}')$ or $\mathbf{J}(\mathbf{r}')$ at time $t_r$.

The situation is similar to astronomy because stars are few hundreds of light years away from us. Whatever condition of stars we see today is the information which happened in stars few hundred light years before.

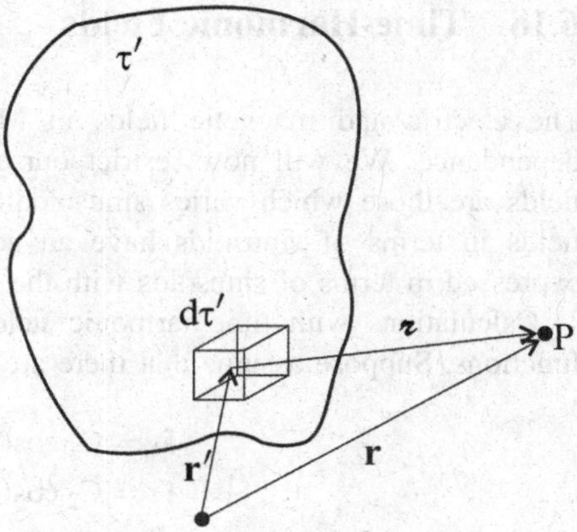


**Fig. 6.28** Volume charge density or volume current density is shown in the figure. Changes produced in volume charge density or volume current density at a given time $t_r$ travels a long distance and reaches point P at a later time t

Thus at point P whatever we measure at time t is that corresponds to that of $\rho(\mathbf{r}')$ or $\mathbf{J}(\mathbf{r}')$ at time $t_r$. Thus

$$t_r = t - \frac{\mathscr{r}}{c} \tag{6.213}$$

where $\frac{\mathscr{r}}{c}$ is the time taken for the change produced in either $\rho(\mathbf{r}', t_r)$ or $\mathbf{J}(\mathbf{r}', t_r)$ to reach point P from the source. $t_r$ is called retarded time.

Equations 6.211 and 6.212 can then be generalized to non-static case as

$$V(\mathbf{r}, t) = \frac{1}{4\pi\varepsilon_o} \int_{\tau'} \frac{\rho(\mathbf{r}', t_r)}{\mathscr{r}} d\tau' \tag{6.214}$$

$$\mathbf{A}(\mathbf{r}, t) = \frac{\mu_o}{4\pi} \int_{\tau'} \frac{\mathbf{J}(\mathbf{r}', t_r)}{\mathscr{r}} d\tau' \tag{6.215}$$

Once $V(\mathbf{r}, t)$ and $\mathbf{A}(\mathbf{r}, t)$ has been calculated from Eqs. 6.214 and 6.215 $\mathbf{E}(\mathbf{r}, t)$, $\mathbf{B}(\mathbf{r}, t)$ can be calculated using equations

$$\mathbf{E}(\mathbf{r}, t) = -\nabla V - \frac{\partial \mathbf{A}}{\partial t} \tag{6.216}$$

$$\mathbf{B}(\mathbf{r}, t) = \nabla \times \mathbf{A} \tag{6.217}$$

## 6.18 Time-Harmonic Fields

The electric and magnetic fields in Maxwell's equations have arbitrary time dependence. We will now restrict ourselves to time-harmonic fields. Harmonic fields are those which varies sinusoidally or periodically with time. Expressing fields in terms of sinusoids have an advantage that most wave forms can be expressed in terms of sinusoids with the Fourier's techniques.

Calculations with time-harmonic fields can be easily done with exponential functions. Suppose assume that there are two time-harmonic functions given by

$$\mathrm{f}_1 = \mathrm{C}_1 \cos(\omega \mathrm{t} + \delta_1) \text{ and}$$
$$\mathrm{f}_2 = \mathrm{C}_2 \cos(\omega \mathrm{t} + \delta_2)$$

We are interested in adding them

$$\mathrm{f}_3 = \mathrm{f}_1 + \mathrm{f}_2$$
$$\mathrm{f}_3 = \mathrm{C}_1 \cos(\omega \mathrm{t} + \delta_1) + \mathrm{C}_2 \cos(\omega \mathrm{t} + \delta_2)$$

However instead of adding $\mathrm{f}_1$ and $\mathrm{f}_2$ directly, if we carry out the addition of $\mathrm{f}_1$ and $\mathrm{f}_2$ as stated below calculations become easier.

$$\begin{aligned} \mathrm{f}_3 &= \mathrm{f}_1 + \mathrm{f}_2 \\ &= \mathrm{C}_1 \cos(\omega \mathrm{t} + \delta_1) + \mathrm{C}_2 \cos(\omega \mathrm{t} + \delta_2) \\ \mathrm{f}_3 &= \mathrm{Re}\left[\mathrm{C}_1 \mathrm{e}^{\mathrm{i}(\omega \mathrm{t} + \delta_1)}\right] + \mathrm{Re}\left[\mathrm{C}_2 \mathrm{e}^{\mathrm{i}(\omega \mathrm{t} + \delta_2)}\right] \end{aligned}$$

Similarly if we want to add sine functions $\mathrm{C}_1 \sin(\omega \mathrm{t} + \delta_1), \mathrm{C}_2 \sin(\omega \mathrm{t} + \delta_2)$ then calculations become easier if we carry out the addition as follows

$$\begin{aligned} &= \mathrm{C}_1 \sin(\omega \mathrm{t} + \delta_1) + \mathrm{C}_2 \sin(\omega \mathrm{t} + \delta_2) \\ &= \mathrm{Im}\left[\mathrm{C}_1 \mathrm{e}^{\mathrm{i}(\omega \mathrm{t} + \delta_1)}\right] + \mathrm{Im}\left[\mathrm{C}_2 \mathrm{e}^{\mathrm{i}(\omega \mathrm{t} + \delta_2)}\right] \end{aligned}$$

The point that, working with exponential functions is much easier than working with time-harmonic functions, will be more clear to the reader when the reader completes Sect. 7.6 in Chap. 7.

Having seen that it is much easier to work with exponential functions than with harmonic functions, we will now discuss phasors.

Let $\mathbf{A}(x, y, z, t)$ be a time-harmonic field. The phasor form of $\mathbf{A}$ is given by $\mathbf{A}_\mathrm{s}(x, y, z)$ which depends on space coordinates $x$, $y$, $z$ alone. Then we can write

$$\mathbf{A}(x, y, z, \mathrm{t}) = \mathrm{Re}\left[\mathbf{A}_\mathbf{s}(x, y, z)\mathrm{e}^{\mathrm{i}\omega \mathrm{t}}\right] \tag{6.218}$$

or

$$\mathbf{A} = \mathrm{Re}\left[\mathbf{A}_s e^{i\omega t}\right] \tag{6.219}$$

Here

$$\mathbf{A}_s = \mathbf{A}_s(x, y, z) \tag{6.220}$$

is the phasor form of $\mathbf{A}(x, y, z, t) \cdot \mathbf{A}_s$ the phasor, is independent of time and depends on $x$, $y$, $z$ only.

As an example consider the electric field given by

$$\mathbf{E}(z, t) = \mathbf{E}_o \cos(\omega t - \beta z) \tag{6.221}$$

where the meaning of β will be clear to the reader in Chap. 7.

We can write Eq. 6.221 as

$$\mathbf{E}(z, t) = \mathrm{Re}\left[\mathbf{E}_o e^{i(\omega t - \beta z)}\right] \tag{6.222}$$

$$\Rightarrow \mathbf{E}(z, t) = \mathrm{Re}\left[\mathbf{E}_o e^{-i\beta z} e^{i\omega t}\right] \tag{6.223}$$

We define

$$\mathbf{E}_s = \mathbf{E}_o e^{-i\beta z} \tag{6.224}$$

$\mathbf{E}_s$ in Eq. 6.224 is the phasor form of $\mathbf{E}$ ($z$, t) in Eq. 6.223. Thus Eq. 6.223 can be written as

$$\mathbf{E}(z, \mathbf{t}) = \mathrm{Re}\left[\mathbf{E}_s e^{i\omega t}\right] \tag{6.225}$$

In Eq. 6.224 we have thus separated the spatial part in the form of phasors. Phasors of the time-harmonic fields are a complex quantities which are function of space only. As seen in Eq. 6.225 the time dependency is present in $e^{i\omega t}$. Thus we have separated spatial part in the form of phasors and time part in $e^{i\omega t}$.

In summary in this section we have seen that we use time-harmonic fields because all other wave forms can be expressed in the form of time-harmonic fields. Then we have expressed time-harmonic fields in exponential form as it is easy to work with exponentials. Finally we separated the exponential form into space part (phasors) and time part $e^{i\omega t}$.

Reader must take care to differentiate between i in $e^{i\omega t}$ which denotes the complex number and $\hat{\mathbf{i}}$ where $\hat{\mathbf{i}}$ is the unit vector along $x$-direction.

**Example 6.17**

If the electric field is given by $\tilde{\mathbf{E}} = \tilde{\mathbf{E}}_o e^{i\omega t}$ calculate the ratio between the conduction current $J_c$ and displacement current $J_d$.

**Solution**
We know that the conduction current is given by

$$\mathbf{J}_c = \sigma \mathbf{E}$$

The displacement current is given by

$$\mathbf{J}_d = \varepsilon \frac{\partial \mathbf{E}}{dt} = \varepsilon \frac{\partial}{dt}\left[\tilde{\mathbf{E}}_o e^{i\omega t}\right]$$

$$\Rightarrow \mathbf{J}_d = i\omega\varepsilon \mathbf{E}$$

$$\Rightarrow \frac{|\mathbf{J}_c|}{|\mathbf{J}_d|} = \frac{\sigma}{\omega\varepsilon}$$

**Example 6.18**
Using the results of Example 6.17 obtain the condition under which the material medium can acts as an conductor or an insulator when an alternating field is present.

**Solution**
In a conductor, the conduction current density is very high as compared to displacement current density.

$$J_c \gg J_d$$

Then

$$\sigma \gg \omega\varepsilon$$

$$\Rightarrow \omega \ll \frac{\sigma}{\varepsilon}$$

In a insulator the conduction current density is lessor than displacement current density

$$J_c \ll J_d$$

$$\Rightarrow \sigma \ll \omega\varepsilon$$

$$\Rightarrow \omega \gg \frac{\sigma}{\varepsilon}$$

**Example 6.19**
Check whether at 1 MHz aluminium is a good conductor and Teflon is a good insulator. Given

(a) For aluminium
$\sigma = 35.3$ Mega-Mho/m
$\varepsilon = \varepsilon_o$

(b) For Teflon
$\sigma$ = 30 nano-Mho/m
$\varepsilon = 2.1\ \varepsilon_o$

*Note* Mho/m is the unit of $\sigma$.

**Solution**

(a) For aluminium

$$\frac{\sigma}{\omega\varepsilon} = \frac{35.3 \times 10^6}{2(3.14)10^6(8.854 \times 10^{-2})}$$
$$\frac{\sigma}{\omega\varepsilon} = 6.4 \times 10^{11}$$

which is very high than unity or in other words

$$\omega \ll \frac{\sigma}{\varepsilon}$$

Hence aluminium is a conductor. Even at very high frequencies say 100 GHz the following condition is satisfied in the conductors

$$\omega \ll \frac{\sigma}{\varepsilon} \text{ or } J_c \gg J_d$$

Thus conduction current dominates displacement current by a very large value. Hence in general, in conductors, displacement current can be neglected.
For Teflon

$$\frac{\sigma}{\omega\varepsilon} = \frac{30 \times 10^{-9}}{2\pi \times 10^6 \times 2.1 \times 8.854 \times 10^{-12}}$$
$$= 2.57 \times 10^{-4}$$

which is lessor than unity. (or)

$$J_c \ll J_d$$

Hence at 1 MHz Teflon is an insulator. Conduction current density is negligible as compared to displacement current density. However at lower frequencies the ratio $\frac{\sigma}{\omega\varepsilon}$ may approach unity and the conduction current may become comparable to displacement current.

**Example 6.20**
Calculate the frequency at which $J_c = J_d$ in distilled water for which $\varepsilon = 81\varepsilon_o$ and $\sigma = 2 \times 10^{-4}$ Mho/m.

**Solution**
For $J_c = J_d$

$$\sigma = \omega\varepsilon$$

Frequency

$$f = \frac{\omega}{2\pi} = \frac{\sigma}{2\pi\varepsilon}$$
$$f = \frac{2\times 10^{-4}}{2\pi(8.854\times 10^{-12})(81)}$$
$$\Rightarrow f = 44.4\,\text{kHz}$$

**Example 6.21**
The electric field **E** in free space is given by

$$\mathbf{E} = E_o\cos(\omega t - kz)\hat{\mathbf{i}}$$

Find **B**.

**Solution**
We know

$$\nabla\times\mathbf{E} = \frac{-\partial\mathbf{B}}{\partial t} = \begin{vmatrix} \hat{\mathbf{i}} & \hat{\mathbf{j}} & \hat{\mathbf{k}} \\ \frac{\partial}{\partial x} & \frac{\partial}{\partial y} & \frac{\partial}{\partial z} \\ E_o\cos(\omega t - kz) & 0 & 0 \end{vmatrix}$$
$$\Rightarrow \frac{-\partial\mathbf{B}}{\partial t} = kE_o\sin(\omega t - kz)\hat{\mathbf{j}}$$
$$\Rightarrow \mathbf{B} = \frac{kE_o}{\omega}\cos(\omega t - kz)\hat{\mathbf{j}}$$

**Example 6.22**
In a source free region if

$$\mathbf{A} = x^4\hat{\mathbf{i}} + z^2t^2\hat{\mathbf{k}}$$

Find $\mathbf{E}(\mathbf{r}, t), \mathbf{B}(\mathbf{r}, t)$

**Solution**
In source free region as $\rho(\mathbf{r}', t_r) = 0$ and hence.

$$V(\mathbf{r},t) = \frac{1}{4\pi\varepsilon_0}\int_{\tau'} \frac{\rho(\mathbf{r}', t_r)}{\mathscr{r}} d\tau'$$

$$V(\mathbf{r},t) = 0$$

$$\mathbf{E} = -\nabla V - \frac{\partial \mathbf{A}}{\partial t} = \frac{-\partial \mathbf{A}}{\partial t}$$

$$\Rightarrow \mathbf{E} = -2z^2 t\hat{\mathbf{k}}$$

and

$$\mathbf{B} = \nabla \times \mathbf{A} = 0$$

## 6.19 Maxwell's Equation in Phasor Form

From Eq. 6.218 an electric field $\mathbf{E}$ $(x, y, z, t)$ can be written as

$$\mathbf{E}(x,y,z,t) = \mathrm{Re}\left[\mathbf{E}_s(x,y,z)e^{i\omega t}\right] \tag{6.226}$$

From Eq. 6.226

$$\frac{\partial \mathbf{E}}{\partial t} = \mathrm{Re}\left[i\omega \mathbf{E}_s e^{i\omega t}\right] \tag{6.227}$$

$$\frac{\partial \mathbf{E}}{\partial t} = \mathrm{Re}[i\omega \mathbf{E}] \tag{6.228}$$

Hence from Eq. 6.228 we can write

$$\frac{\partial}{\partial t} \rightarrow i\omega \tag{6.229}$$

Similar to Eq. 6.226 writing expressions for $\mathbf{D}$ $(x, y, z, t)$, $\mathbf{H}$ $(x, y, z, t)$, $\mathbf{B}$ $(x, y, z, t)$, $\mathbf{J}$ $(x, y, z, t)$ and $\rho$ $(x, y, z, t)$ we can express Maxwell's equations 6.136–6.139, 6.164–6.167 in phasor form. $e^{i\omega t}$ being present in both sides of the equations cancel out.

$$\nabla \cdot \mathbf{D}_s = \rho_s \tag{6.230}$$

$$\nabla \cdot \mathbf{B}_s = 0 \tag{6.231}$$

$$\nabla \times \mathbf{E}_s = -i\omega \mathbf{B}_s \tag{6.232}$$

$$\nabla \times \mathbf{H}_s = \mathbf{J}_s + i\omega \mathbf{D}_s \tag{6.233}$$

The above four Maxwell's equation in phasor form depends on space part ($x$, $y$, $z$) only and is independent of time.

The integral form of Maxwell's equations 6.168–6.171 can also be derived in phasor form. See Problem 6.20.

## Exercises

**Problem 6.1**

Suppose assume that in Problem 3.8 the space between parallel-plate capacitors is filled with a medium which is homogeneous with finite conductivity $\sigma$. Find the resistance of the region between the plates.

**Problem 6.2**

In a material when electric field is applied the electrons get accelerated and flow towards the positive terminal. As they move towards the positive terminal they collide with the atoms. Because of these collisions the electron undergoes random motion in all directions as shown in Fig. 6.29. The net result is that electrons acquire a slow drift velocity $\mathbf{v_d}$ towards the positive terminal.

Show that the drift velocity is related to current density as

$$\mathbf{J} = \mathrm{ne}\mathbf{v_d}$$

where n is the number of electrons per unit volume.

**Problem 6.3**

Repeat Example 6.1 with $\sigma = \frac{C_1}{\rho} + C_2$ where $C_1$ and $C_2$ are constants.

**Problem 6.4**

A silver wire of 0.9 mm diameter carries a charge of 75 C in 1 h. Calculate the drift velocity of the electrons in the wire using the results of Problem 6.2 if n for silver is $5.8 \times 10^{22}$ electrons/cm$^3$.

**Problem 6.5**

A Nickel wire of diameter 0.5 cm is welded to a copper wire of diameter 0.4 cm. Both the wires as such carry a current of 9 A. What is the current density in each wire. Assuming one free electron per atom in copper calculate drift speed of the

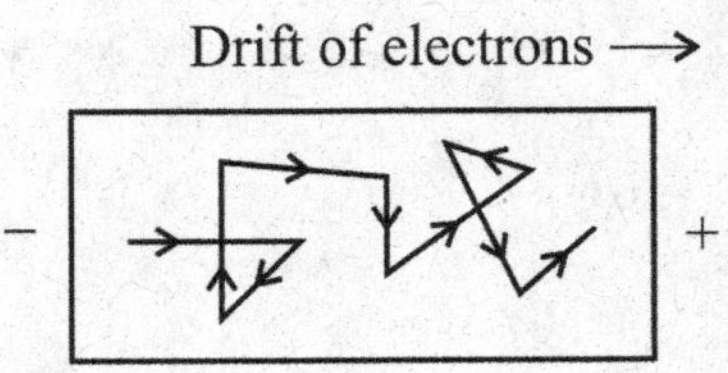


**Fig. 6.29** Drift of electrons towards the positive terminal in material subjected to an electric field. Illustration for Problem 6.2

electrons in copper wire. Given atomic weight of copper 64 and density of copper is 9 gm/cc. Hint: Use the results of Problem 6.2.

**Problem 6.6**

Calculate the induced emf of a metal aeroplane vertically diving at a speed of 500 km/s in an earth's magnetic field where the horizontal component is $0.5 \times 10^{-4}$ Wb/m$^2$. The wingspan of the aeroplane is 40 m.

**Problem 6.7**

Repeat example 6.4 but this time the magnetic field is pointing at an angle of 45° to the *z*-axis.

**Problem 6.8**

A metal rod is rotating with a constant angular velocity $\omega$ in a uniform magnetic flux density **B**. The length of the rod is L. The magnetic flux density **B** is perpendicular to the length of the rod. What is the emf induced between the ends of the rod?

**Problem 6.9**

A train passes on a railway track whose rails are assumed to be insulated from each other. The velocity of the train is 150 km/h. The distance between the rails in the railway track is 1.6 m. If the vertical component of the earth's magnetic field at the points of the rail is 0.373 Wb/m$^2$ calculate the emf that will exist between the rails?

**Problem 6.10**

A steel core of rectangular cross section 15 mm × 10 mm and a mean length of 160 mm is shown in Fig. 6.30. The air gap length is 1 mm.

The flux in the air gap is 156 μ Wb.

(a) Determine the mmf if the value of H is 200 A/m for B = 1.04 T. Using the relation mmf $= \oint \mathbf{H} \cdot d\mathbf{l}$.

(b) Draw the equivalent circuit for Fig. 6.30, calculate reluctance and hence compute mmf.

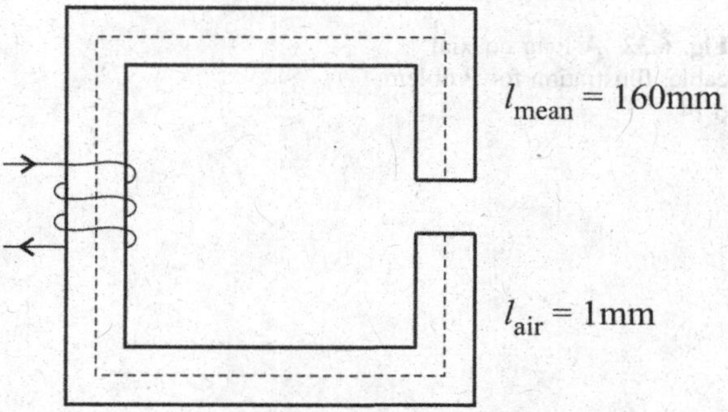


**Fig. 6.30** A steel core of rectangular cross-section. Illustration for Problem 6.10

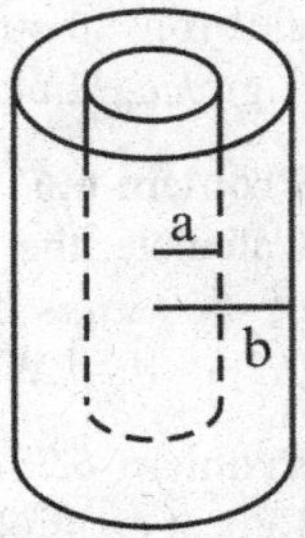


**Fig. 6.31** Two coaxial cylinders. Illustration for Problem 6.11

**Problem 6.11**
Consider two coaxial cylinders as shown in Fig. 6.31. The inner cylinder is a solid cylinder carrying a current I distributed uniformly over its cross section. Calculate the self-inductance per unit length.

**Problem 6.12**
In Example 6.10 calculate the energy stored in the solenoid.

**Problem 6.13**
In Example 6.12 calculate the energy stored in the toroid.

**Problem 6.14**
Consider a long coaxial cable as shown in Fig. 6.32. The current I flows down the surface of the inner radius and comes back along the outer cylinder. Find the self-inductance and magnetic energy stored in the section of length L.

**Problems 6.15**
Consider a material with $\varepsilon_r = 1$ and conductivity $\sigma = 4.3$ Mho/m. The value of **E** is $\mathbf{E} = 200 \sin 10^9 t\,\hat{\mathbf{i}}\,\mathrm{V/m}$. Calculate the conduction and displacement current densities. Also find the frequency at which they have equal magnitude.

**Problem 6.16**
Consider a coaxial capacitor. The inner and outer radius of the coaxial capacitor is 5 mm and 7 mm, respectively. The length of the coaxial capacitor is 700 mm. The

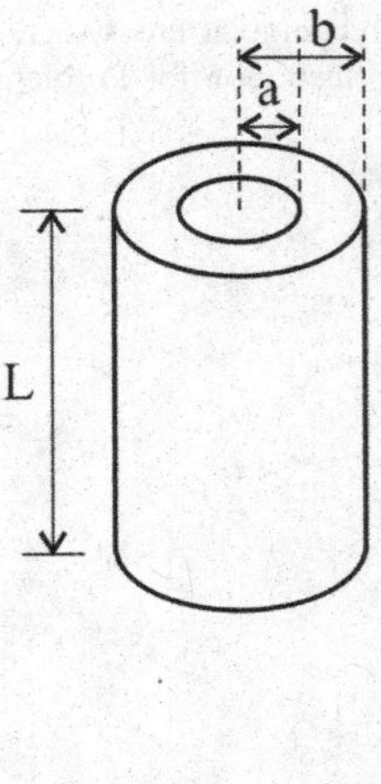


**Fig. 6.32** A long coaxial cable. Illustration for Problem 6.14

space between capacitor is filled with dielectric with $\varepsilon_r = 5$. The capacitor is subjected to a field.

$$\mathbf{E} = \frac{-2 \times 10^2}{r} \sin 400t\, \hat{\mathbf{r}} \frac{V}{m}$$

Calculate the displacement current and conduction current.

**Problem 6.17**
Consider a parallel-plate capacitor filled with a dielectric with $\varepsilon_r = 6$. The area of the plates separated by a distance of 4 mm is 20 $cm^2$. The capacitor is connected to a voltage source of 400 V whose frequency is 2 MHz. Calculate the displacement current.

**Problem 6.18** In free space show that the following field vectors satisfy Maxwell's equation

$$\mathbf{E} = E_o \cos(\omega t - kz)\hat{\mathbf{i}}$$

$$\mathbf{H} = \frac{E_o}{\eta} \cos(\omega t - kz)\hat{\mathbf{j}}$$

Calculate $\eta$.

**Problem 6.19**
Consider a spherical capacitor consisting of two concentric spheres of inner and outer radius a, b, respectively. The region between two spheres is filled with air. Find the displacement current.

**Problem 6.20**
For time-harmonic fields show that

$$\int \mathbf{E}\partial \mathbf{t} \rightarrow \frac{\mathbf{E}}{i\omega}$$

and hence obtain integral form of Maxwell's equation 6.168–6.171 in phasor form

$$\oint \mathbf{D}_s \cdot \mathbf{ds} = \int \rho_s d\tau$$

$$\oint \mathbf{B}_s \cdot \mathbf{ds} = 0$$

$$\oint \mathbf{E}_s \cdot \mathbf{d}\boldsymbol{l} = -i\omega \int \mathbf{B}_s \cdot \mathbf{ds}$$

$$\oint \mathbf{H}_s \cdot \mathbf{d}\boldsymbol{l} = \int (\mathbf{J}_s + i\omega \mathbf{D}_s) \cdot \mathbf{ds}$$

The reader should note that subscript s denotes phasor form while s in ds denotes surface.

# Chapter 7
# Plane Electromagnetic Waves

## 7.1 Introduction

In this chapter we will discuss about plane electromagnetic waves. What is a wave? A wave is a disturbance passing through a medium. The wave carries energy and momentum with it. As an example of a wave assume there are some bricks arranged in a line as shown in Fig. 7.1a. Suppose we push one of the bricks at one of the end, then the bricks fall in sequence, one by one, as shown in Fig. 7.1b. The hand shown in Fig. 7.1b produces a disturbance by pushing one of the brick at the end. The disturbance then travels through the medium—the bricks.

There are quite a few number of classifications of a wave. One of the important classifications of wave worth mentioning here is, longitudinal wave and transverse wave.

In a longitudinal wave the displacement of the medium is parallel to the direction of the propagation of the wave. An example of the longitudinal wave is shown in Fig. 7.1. The medium, that is the bricks, is displaced parallel to the direction of propagation of the wave. In Fig. 7.1b the bricks are displaced in $x$-direction and wave also propagates in the $x$-direction.

In a transverse wave the displacement of the medium is perpendicular to the direction of propagation of the wave. An example of a transverse wave is shown in Fig. 7.2. A person holds one end of the rope. He moves the rope up and down thereby producing a disturbance. The disturbance travels through the medium—the rope. In this example the medium, the rope, is displaced perpendicular to the direction of wave propagation. In Fig. 7.2 the medium, the rope, is displaced in $y$-direction, while the wave propagates in the $x$-direction.

Let us see the mathematical description for the wave. As an example consider a wave as shown in Fig. 7.3. The wave travels with a velocity $\mathbf{v}$ in the $z$-direction. u ($z$, t) is the function representing the wave. For example in Fig. 7.3 u($z$, t) represents the displacement of the rope at point $z$ at time t. Then the wave equation is given by

© Springer Nature Singapore Pte Ltd. 2020

S. Balaji, *Electromagnetics Made Easy*,
https://doi.org/10.1007/978-981-15-2658-9_7

**Fig. 7.1** (**a**) Bricks arranged in a line. (**b**) On of the bricks is pushed and one by one the bricks fall thereby the disturbance travels through the bricks

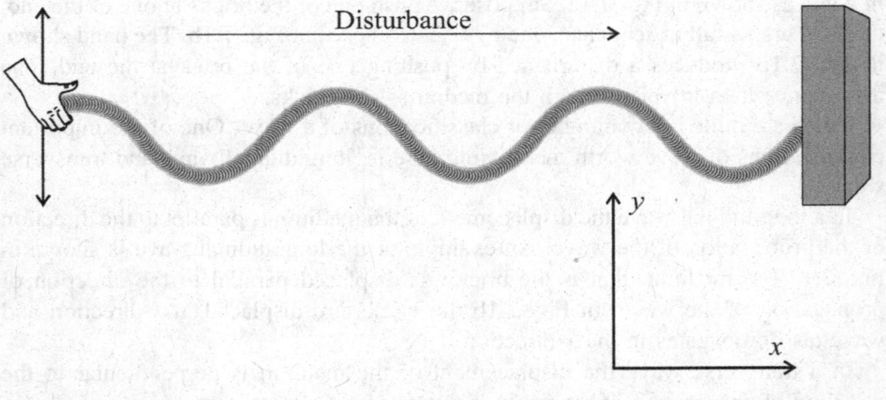


**Fig. 7.2** A rope is moved up and down thereby a disturbance passes through the rope

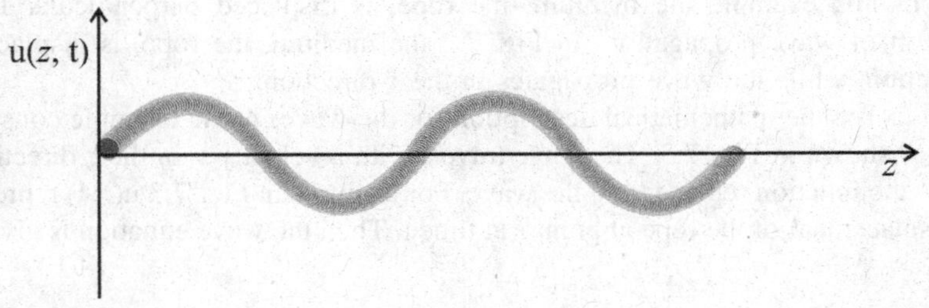


**Fig. 7.3** A wave is propagating along the *z* direction and u (*z*,t) is the function representing the wave.

$$\frac{\partial^2 u}{\partial z^2} = \frac{1}{v^2}\frac{\partial^2 u}{\partial t^2} \tag{7.1}$$

The solution of the above equation is

$$u(z, t) = u_o e^{i(\omega t - \beta z)} \tag{7.2}$$

where $u_o$ is the amplitude of the wave and $\omega$ is the angular frequency and $\beta$ is given by

$$\beta = \frac{2\pi}{\lambda}$$

The reader is referred to any standard book on waves to learn complete theory of waves.

We will now discuss about propagation of electromagnetic waves in different media and use phasors to our advantage. That is we will discuss initially about the propagation of electromagnetic waves in spatial coordinates-($x$, $y$, $z$), with the help of phasors. Then we will insert the time part $e^{i\omega t}$ to explain the propagation of electromagnetic waves with respect to time.

## 7.2 The Wave Equation

Let us now derive the wave equation from Maxwell's equation.

Consider a uniform linear medium with permittivity $\varepsilon$, permeability $\mu$ and conductivity $\sigma$. Let the medium be source free. That is the medium doesn't contain any charge and currents other than that determined by Ohm's law. Then we have

$$\mathbf{D} = \varepsilon\mathbf{E},\ \mathbf{J}_f = \sigma\mathbf{E},\ \mathbf{B} = \mu\mathbf{H} \text{ and } \rho_f = 0 \tag{7.3}$$

Hence from Eqs. 6.164–6.167

$$\left.\begin{aligned} \nabla\cdot\mathbf{D} &= 0 \\ \nabla\cdot\mathbf{B} &= 0 \\ \nabla\times\mathbf{E} &= \frac{-\partial\mathbf{B}}{\partial t} \\ \nabla\times\mathbf{H} &= \mathbf{J}_f + \frac{\partial\mathbf{D}}{\partial t} \end{aligned}\right\} \tag{7.4}$$

Substituting Eq. 7.3 in 7.4.

$$\left.\begin{aligned} &\text{(a)} \quad \nabla\cdot\mathbf{E} = 0 \\ &\text{(b)} \quad \nabla\cdot\mathbf{H} = 0 \\ &\text{(c)} \quad \nabla\times\mathbf{E} = -\mu\frac{\partial\mathbf{H}}{\partial t} \\ &\text{(d)} \quad \nabla\times\mathbf{H} = \sigma\mathbf{E} + \varepsilon\frac{\partial\mathbf{E}}{\partial t} \end{aligned}\right\} \tag{7.5}$$

Taking curl of Eq. 7.5c we have

$$\nabla \times \nabla \times \mathbf{E} = -\mu \frac{\partial}{\partial t}(\nabla \times \mathbf{H}) \tag{7.6}$$

Substituting Eq. 7.5d in Eq. 7.6

$$\nabla \times \nabla \times \mathbf{E} = -\mu \frac{\partial}{\partial t}\left(\sigma \mathbf{E} + \varepsilon \frac{\partial \mathbf{E}}{\partial t}\right) \tag{7.7}$$

Now using vector identify 7 from Sect. 1.20

$$\nabla(\nabla \cdot \mathbf{E}) - \nabla^2 \mathbf{E} = -\mu \frac{\partial}{\partial t}\left(\sigma \mathbf{E} + \varepsilon \frac{\partial \mathbf{E}}{\partial t}\right) \tag{7.8}$$

Substituting Eq. 7.5a in Eq. 7.8

$$\nabla^2 \mathbf{E} = \mu \frac{\partial}{\partial t}\left(\sigma \mathbf{E} + \varepsilon \frac{\partial \mathbf{E}}{\partial t}\right) \tag{7.9}$$

$$\Rightarrow \nabla^2 \mathbf{E} - \sigma\mu \frac{\partial \mathbf{E}}{\partial t} - \varepsilon\mu \frac{\partial^2 \mathbf{E}}{\partial t^2} = 0 \tag{7.10}$$

Taking curl of Eq. 7.5d

$$\nabla \times (\nabla \times \mathbf{H}) = \sigma(\nabla \times \mathbf{E}) + \varepsilon\left(\frac{\partial}{\partial t}(\nabla \times \mathbf{E})\right) \tag{7.11}$$

Substituting Eq. 7.5c in Eq. 7.11 and using vector identity 7 from Sect. 1.2 and using Eq. 7.5b we have

$$\nabla^2 \mathbf{H} - \sigma\mu \frac{\partial \mathbf{H}}{\partial t} - \varepsilon\mu \frac{\partial^2 \mathbf{H}}{\partial t^2} = 0 \tag{7.12}$$

Equations 7.10, 7.12 are called general wave equations.

## 7.3 Plane Electromagnetic Wave in Free Space

Let us assume a wave travelling in *z*-direction without having *x* or *y* dependance. The fields are uniform over the entire plane perpendicular to the *z*-direction, i.e. the direction of wave propagation. Such a wave is called as uniform plane wave. In order to create a uniform field over the entire plane a source of infinite extent will be required. However, such sources are practically impossible because the sources are of finite extent. On the other hand if we are very far away from the source then the wave front which is surface of constant phase is almost spherical and a small portion of a sphere is a plane.

For free space

$$\sigma = 0, \varepsilon = \varepsilon_o, \mu = \mu_o \tag{7.13}$$

Substituting Eq. 7.13 in Eqs. 7.10, 7.12

$$\nabla^2 \mathbf{E} - \varepsilon_o \mu_o \frac{\partial^2 \mathbf{E}}{\partial t^2} = 0 \tag{7.14}$$

$$\nabla^2 \mathbf{H} - \varepsilon_o \mu_o \frac{\partial^2 \mathbf{H}}{\partial t^2} = 0 \tag{7.15}$$

The above two equations can be written as

$$\left(\frac{\partial^2}{\partial x^2} + \frac{\partial^2}{\partial y^2} + \frac{\partial^2}{\partial z^2}\right)\mathbf{E} - \varepsilon_o \mu_o \frac{\partial^2 \mathbf{E}}{\partial t^2} = 0 \tag{7.16a}$$

$$\left(\frac{\partial^2}{\partial x^2} + \frac{\partial^2}{\partial y^2} + \frac{\partial^2}{\partial z^2}\right)\mathbf{H} - \varepsilon_o \mu_o \frac{\partial^2 \mathbf{H}}{\partial t^2} = 0 \tag{7.16b}$$

For uniform plane waves travelling in the $z$-direction Eqs. 7.16a, 7.16b can be written as

$$\frac{\partial^2 \mathbf{E}}{\partial z^2} - \varepsilon_o \mu_o \frac{\partial^2 \mathbf{E}}{\partial t^2} = 0 \tag{7.16c}$$

$$\frac{\partial^2 \mathbf{H}}{\partial z^2} - \varepsilon_o \mu_o \frac{\partial^2 \mathbf{H}}{\partial t^2} = 0 \tag{7.16d}$$

Comparing Eqs. 7.16c, 7.16d to Eq. 7.1

$$\mathrm{v} = \frac{1}{\sqrt{\varepsilon_o \mu_o}} \tag{7.17}$$

Substituting the values of $\varepsilon_o, \mu_o$ we get

$$\mathrm{v} = \mathrm{c} = 3 \times 10^8 \mathrm{m/sec}$$

Hence electromagnetic waves travel with the velocity of light.

Let us now assume that **E** and **H** have harmonic time dependency. Hence we write the electric field as

$$\mathbf{E} = \mathbf{E}_s e^{i\omega t} \tag{7.18a}$$

where $\mathbf{E}_s$ is the usual phasor which depends on space coordinates. Substituting Eqs. 7.17, 7.18a in Eq. 7.16c

$$\frac{\partial^2 \mathbf{E}}{\partial t^2} = v^2 \frac{\partial^2 \mathbf{E}}{\partial z^2} \tag{7.18b}$$

$$\Rightarrow \frac{\partial^2}{\partial t^2}\left[\mathbf{E}_s e^{i\omega t}\right] = v^2 \frac{\partial^2}{\partial z^2}\left(\mathbf{E}_s e^{i\omega t}\right)$$

$$\Rightarrow \mathbf{E}_s \frac{\partial^2}{\partial t^2}\left(e^{i\omega t}\right) = v^2 e^{i\omega t}\left(\frac{\partial^2 \mathbf{E}_s}{\partial z^2}\right)$$

$$\Rightarrow \mathbf{E}_s = -\frac{v^2}{\omega^2}\frac{\partial^2 \mathbf{E}_s}{\partial z^2}$$

$$\Rightarrow \frac{\partial^2 \mathbf{E}_s}{\partial z^2} = -\frac{\omega^2}{v^2}\mathbf{E}_s$$

We define

$$\beta = \frac{\omega}{v}$$

Hence

$$\begin{aligned} \frac{\partial^2 \mathbf{E}_s}{\partial z^2} &= -\beta^2 \mathbf{E}_s \\ \Rightarrow \frac{\partial^2 \mathbf{E}_s}{\partial z^2} &+ \beta^2 \mathbf{E}_s = 0 \end{aligned} \tag{7.19}$$

Similarly with

$$\mathbf{H} = \mathbf{H}_s e^{i\omega t} \tag{7.20a}$$

Equation 7.16d can be written as

$$\frac{\partial^2 \mathbf{H}_s}{\partial z^2} + \beta^2 \mathbf{H}_s = 0 \tag{7.20b}$$

The solution of the differential Eq. 7.19 is

$$\mathbf{E}_s = \mathbf{E}_o e^{-i\beta z} + \mathbf{E}'_o e^{i\beta z} \tag{7.21a}$$

where the term $\mathbf{E_o}e^{-i\beta z}$ represents a wave travelling in +$z$-direction. The term $E_o'e^{i\beta z}$ represents a wave travelling in–$z$-direction. Because we consider a plane wave travelling in the +$z$-direction $\mathbf{E_o'} = 0$. Hence

$$\mathbf{E}_s = \mathbf{E_o}e^{-i\beta z} \tag{7.21b}$$

Substituting Eq. 7.21b in Eq. 7.18a

$$\mathbf{E}(z, t) = \mathbf{E_o}e^{i(\omega t-\beta z)} \tag{7.22}$$

Similarly we get $\left[\text{with } \mathbf{H}_s = \mathbf{H}_o e^{-i\beta z}\right]$

$$\mathbf{H}(z, t) = \mathbf{H}_o e^{i(\omega t-\beta z)} \tag{7.23}$$

The actual physical fields are given by real parts of **E** and **H**. $\mathbf{E}_o$, $\mathbf{H}_o$ are the amplitudes of the fields and are in general complex numbers. The propagation vector β is given by

$$\boldsymbol{\beta} = \beta\hat{\mathbf{k}} = \frac{2\pi}{\lambda}\hat{\mathbf{k}} = \frac{2\pi\nu}{v}\hat{\mathbf{k}} = \frac{\omega}{v}\hat{\mathbf{k}} \tag{7.24}$$

Here $\hat{\mathbf{k}}$ is the unit vector in the direction of wave propagation, the $z$-direction. Let $\mathbf{E}_o$ in Eq. 7.21b be written as

$$\mathbf{E}_o = (E_o)_{xs}\hat{\mathbf{i}} + (E_o)_{ys}\hat{\mathbf{j}} + (E_o)_{zs}\hat{\mathbf{k}} \tag{7.25}$$

In Eq. 7.16c we have assumed that the wave propagates in the $z$-direction. Let the wave be uniform plane wave characterized by uniform electric field over plane surfaces perpendicular to $z$ with no electric field component in $z$-direction (see Fig. 7.4a).

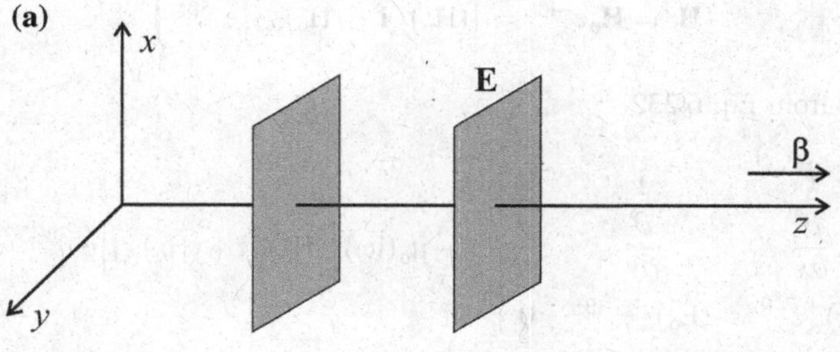


**Fig. 7.4** Uniform plane wave with uniform electric field over a plane surface perpendicular $z$

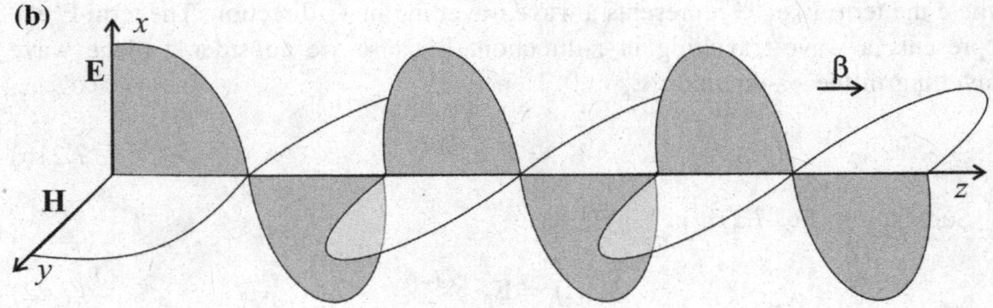


**Fig. 7.4** An electromagnetic wave propagating in free space in the $z$-direction

Hence

$$(\mathrm{E_o})_{zs=0} \tag{7.26}$$

Thus Eq. 7.25 is

$$\mathbf{E}_\mathrm{o} = (\mathrm{E_o})_{xs}\hat{\mathbf{i}} + (\mathrm{E_o})_{ys}\hat{\mathbf{j}} \tag{7.27}$$

Similarly we can write

$$\mathbf{H_o} = (\mathrm{H_o})_{xs}\hat{\mathbf{i}} + (\mathrm{H_o})_{ys}\hat{\mathbf{j}} \tag{7.28}$$

From Eq. 6.232

$$\nabla \times \mathbf{E_s} = -\mathrm{i}\omega\,\mu_\mathrm{o}\mathbf{H}_\mathrm{s} \tag{6.232}$$

We have from Eq. 7.21b

$$\left.\begin{array}{l} \mathbf{E_s} = \mathbf{E_o}\mathrm{e}^{-\mathrm{i}\beta z} = \left[(\mathrm{E_o})_{xs}\hat{\mathbf{i}} + (\mathrm{E_o})_{ys}\hat{\mathbf{j}}\right]\mathrm{e}^{-\mathrm{i}\beta z} \\ \text{and similarly} \\ \mathbf{H_s} = \mathbf{H_o}\mathrm{e}^{-\mathrm{i}\beta z} = \left[(\mathrm{H_o})_{xs}\hat{\mathbf{i}} + (\mathrm{H_o})_{ys}\hat{\mathbf{j}}\right]\mathrm{e}^{-\mathrm{i}\beta z} \end{array}\right\} \tag{7.29}$$

Hence from Eq. 6.232

$$\begin{vmatrix} \hat{\mathbf{i}} & \hat{\mathbf{j}} & \hat{\mathbf{k}} \\ \dfrac{\partial}{\partial x} & \dfrac{\partial}{\partial y} & \dfrac{\partial}{\partial z} \\ (\mathrm{E_o})_{xs}\mathrm{e}^{-\mathrm{i}\beta z} & (\mathrm{E_o})_{ys}\mathrm{e}^{-\mathrm{i}\beta z} & 0 \end{vmatrix} = -\mu_\mathrm{o}(\mathrm{i}\omega)\left[(\mathrm{H_o})_{xs}\hat{\mathbf{i}} + (\mathrm{H_o})_{ys}\hat{\mathbf{j}}\right]\mathrm{e}^{-\mathrm{i}\beta z} \tag{7.30}$$

$$\left.\begin{array}{rl} \Rightarrow -\beta(\mathrm{E_o})_{ys} & = \mu_\mathrm{o}\omega(\mathrm{H_o})_{xs} \\ \beta(\mathrm{E_o})_{xs} & = \mu_\mathrm{o}\omega(\mathrm{H_o})_{ys} \end{array}\right\} \tag{7.31}$$

Equation 7.31 can be written as

$$\mu_o \mathbf{H_o} = \frac{\beta}{\omega}\left[\hat{\mathbf{k}} \times \mathbf{E_o}\right] \tag{7.32}$$

Multiplying Eq. 7.32 by $e^{-i\beta z}$ on both sides and inserting the time factor $e^{i\omega t}$ then using Eqs. 7.22, 7.23 we can write

$$\mu_o \mathbf{H} = \frac{\beta}{\omega}(\hat{\mathbf{k}} \times \mathbf{E}) \tag{7.33}$$

So **E**, **H** are mutually perpendicular. As the wave is travelling in *z*-direction, if **E** is pointing in *x*-direction then **H** must point in *y*-direction. (Hence electromagnetic waves are transverse waves) (Further see Example 7.1). In such situation the magnitude of the real part in Eq. 7.32

$$(H_o)_y = \frac{\beta}{\mu_o \omega}(E_o)_x \tag{7.34}$$

From the above equation the ratio of **E**, **H** is

$$\eta = \frac{(E_o)_x}{(H_o)_y} = \frac{\omega \mu_o}{\beta} = \sqrt{\frac{\mu_o}{\varepsilon_o}} \tag{7.35}$$

where we have used Eqs. 7.17, 7.24.

The unit of $\eta$ is ohms, as **E**, is in volt/m and **H** is in amp-turn/m. Thus

$$\frac{|\mathbf{E}|}{|\mathbf{H}|} = \frac{(E_o)_x}{(H_o)_y} \text{ is in volt/amp or ohms.}$$

Hence $\eta$ is called intrinsic impedance of the medium. Substituting the values of $\mu_o$ and $\varepsilon_o$ in Eq. 7.35

$$\eta = 376.6 \text{ ohms} \tag{7.36}$$

for free space. From Eq. 7.35 we see that the ratio

$\dfrac{|\mathbf{E}|}{|\mathbf{H}|} = \dfrac{(E_o)_x}{(H_o)_y}$ is a constant. This means that the field vectors **E** and **H** are in phase.

Because **E**, **H** are pointing in x-direction and y-direction Eqs. 7.22, 7.23 can be written as

$$\mathbf{E}(z,t) = (E_o)_x e^{i(\omega t - \beta z)}\hat{\mathbf{i}} \tag{7.37}$$

$$\mathbf{H}(z,t) = (H_o)_y e^{i(\omega t - \beta z)}\hat{\mathbf{j}} \tag{7.38}$$

The actual electric and magnetic fields are given by real parts of equations

$$\mathbf{E}(z,t) = (E_o)_x \cos(\omega t - \beta z)\hat{\mathbf{i}} \quad (7.39)$$

$$\mathbf{H}(z,t) = (H_o)_y \cos(\omega t - \beta z)\hat{\mathbf{j}} \quad (7.40)$$

The results we have obtained in this section are pictorially depicted in Fig. 7.4b. In the figure **E**, **H** are in phase. **E**, **H**, **β** are perpendicular to each other.

In Fig. 7.1 the medium is brick through which the wave or disturbance propagates. In Fig. 7.2 the medium is a rope through which the wave or disturbance propagates.

In Fig. 7.4b there is no **physical** medium. It is a free space. The **variations** of electric field in Fig. 7.4 are in $x$-direction. The **variations** of magnetic field in Fig. 7.5 are in $y$-direction.

*The propagation of the variations of electric and magnetic fields is in z-direction*. It is this propagation we call as a wave. Hence the electromagnetic waves travel without a need for an external and physical medium.

***Note***:

(1) From Eqs. 7.21b and 7.27 we can write

$$\mathbf{E_s} = \mathbf{E_o} e^{-i\beta z}$$
$$= \left[(E_o)_{xs}\hat{\mathbf{i}} + (E_o)_{ys}\hat{\mathbf{j}}\right] e^{-i\beta z}$$
$$\mathbf{E}_s = E_{xs}\hat{\mathbf{i}} + E_{ys}\hat{\mathbf{j}}$$

where

$$E_{xs} = (E_o)_{xs} e^{-i\beta z}$$
$$E_{ys} = (E_o)_{ys} e^{-i\beta z}$$

Suppose the electric field is along $x$-axis then $(E_o)_{ys} = E_{ys} = 0$. Hence

$$\mathbf{E}_s = E_{xs}\hat{\mathbf{i}} = (E_o)_{xs} e^{-i\beta z}\hat{\mathbf{i}}$$

(2) In the presence of a medium with $\sigma = 0$ and permeability $\mu$, permittivity $\varepsilon$ Eqs. 7.10, 7.12, 7.17, 7.35 can be written as

$$\nabla^2\mathbf{E} - \varepsilon\mu\frac{\partial^2\mathbf{E}}{\partial t^2} = 0,$$

$$\nabla^2\mathbf{H} - \varepsilon\mu\frac{\partial^2\mathbf{H}}{\partial t^2} = 0,$$

$$v = \frac{1}{\sqrt{\varepsilon\mu}}, \eta = \sqrt{\frac{\mu}{\varepsilon}}$$

**Example 7.1**
If an electromagnetic wave is travelling in $z$-direction in free space with no charge density with an electric and magnetic fields given by

$$\mathbf{E} = \mathbf{E_o} e^{i(\omega t - \beta z)}$$
$$\mathbf{H} = \mathbf{H_o} e^{i(\omega t - \beta z)}$$

Then prove that electromagnetic waves are transverse waves.

**Solution**
Let $\mathbf{E} = (E_o)_x \hat{\mathbf{i}} + (E_o)_y \hat{\mathbf{j}} + (E_o)_z \hat{\mathbf{k}}$
Then

$$\nabla \cdot \mathbf{E} = \left[\hat{\mathbf{i}}\frac{\partial}{\partial x} + \hat{\mathbf{j}}\frac{\partial}{\partial y} + \hat{\mathbf{k}}\frac{\partial}{\partial z}\right] \cdot \left[(E_o)_x \hat{\mathbf{i}} + (E_o)_y \hat{\mathbf{j}} + (E_o)_z \hat{\mathbf{k}}\right]$$
$$\Rightarrow \nabla \cdot \mathbf{E} = (i\beta) e^{i(\omega t - \beta z)} (E_o)_z = 0$$

by using Maxwell's equation 6.132 for charge free space.

$$\Rightarrow (E_o)_z = 0$$

Similarly using Eq. 6.133 we can show that

$$(H_o)_z = 0.$$

Thus as there are no components of electric and magnetic fields in the direction of propagation, the z-direction, electromagnetic waves are transverse waves.

## 7.4 Poynting Theorem

Energy is transported through the electromagnetic wave whenever the wave travels from the transmitter to the receiver. In this section we will derive a relation regarding the energy transport in the electromagnetic wave.

From Eq. 6.163

$$\nabla \times \mathbf{H} = \mathbf{J}_f + \frac{\partial \mathbf{D}}{\partial t}$$
$$\Rightarrow \mathbf{J}_f = \nabla \times \mathbf{H} - \frac{\partial \mathbf{D}}{\partial t} \tag{7.41}$$

Taking the dot product of **E** on both sides

$$\mathbf{E}\cdot\mathbf{J}_f = \mathbf{E}\cdot(\nabla\times\mathbf{H}) - \mathbf{E}\cdot\frac{\partial\mathbf{D}}{\partial t} \tag{7.42}$$

Using vector identity 8 from Sect. 1.20

$$\mathbf{E}\cdot(\nabla\times\mathbf{H}) = \mathbf{H}\cdot(\nabla\times\mathbf{E}) - \nabla\cdot(\mathbf{E}\times\mathbf{H}) \tag{7.43}$$

Substituting Eq. 7.43 in 7.42

$$\mathbf{E}\cdot\mathbf{J}_f = \mathbf{H}\cdot(\nabla\times\mathbf{E}) - \nabla\cdot(\mathbf{E}\times\mathbf{H}) - \mathbf{E}\cdot\frac{\partial\mathbf{D}}{\partial t} \tag{7.44}$$

Substituting Eq. 6.166 in Eq. 7.44

$$\mathbf{E}\cdot\mathbf{J}_f = -\mathbf{H}\cdot\frac{\partial\mathbf{B}}{\partial t} - \nabla\cdot(\mathbf{E}\times\mathbf{H}) - \mathbf{E}\cdot\frac{\partial\mathbf{D}}{\partial t} \tag{7.45}$$

For a linear, homogeneous and isotropic medium

$$\mathbf{B} = \mu\mathbf{H} \text{ and } \mathbf{D} = \varepsilon\mathbf{E}.$$

Hence

$$\begin{aligned}\mathbf{H}\cdot\frac{\partial\mathbf{B}}{\partial t} &= \mu\mathbf{H}\cdot\frac{\partial\mathbf{H}}{\partial t} = \frac{\mu}{2}\frac{\partial(\mathbf{H}\cdot\mathbf{H})}{\partial t}\\ \Rightarrow \mathbf{H}\cdot\frac{\partial\mathbf{B}}{\partial t} &= \frac{\mu}{2}\frac{\partial \mathrm{H}^2}{\partial t}\end{aligned} \tag{7.46}$$

Similarly

$$\mathbf{E}\cdot\frac{\partial\mathbf{D}}{\partial t} = \varepsilon\mathbf{E}\cdot\frac{\partial\mathbf{E}}{\partial t} = \frac{\varepsilon}{2}\frac{\partial \mathrm{E}^2}{\partial t} \tag{7.47}$$

Substituting Eqs. 7.46, 7.47 in 7.45

$$\mathbf{E}\cdot\mathbf{J}_f = \frac{-1}{2}\frac{\partial}{\partial t}\left(\frac{\mu}{2}\mathrm{H}^2 + \frac{\varepsilon}{2}\mathrm{E}^2\right) - \nabla\cdot(\mathbf{E}\times\mathbf{H}) \tag{7.48}$$

Integrating over a volume τ

$$\int_\tau \mathbf{E}\cdot\mathbf{J}_f d\tau = -\frac{\partial}{\partial t}\int_\tau \left(\frac{\mu}{2}\mathrm{H}^2 + \frac{\varepsilon}{2}\mathrm{E}^2\right)d\tau - \int_\tau \nabla\cdot(\mathbf{E}\times\mathbf{H})d\tau \tag{7.49}$$

Applying Gauss's divergence theorem to the second term in the right-hand side of Eq. 7.49 we get

$$\int_\tau \mathbf{E}\cdot\mathbf{J}_f d\tau = \frac{-\partial}{\partial t}\int_\tau \left(\frac{\mu}{2}H^2 + \frac{\varepsilon}{2}E^2\right)d\tau - \oint_S (\mathbf{E}\times\mathbf{H})\cdot ds \qquad (7.50)$$

where S is the surface bounded by volume τ. Now, we will interpret Eq. 7.50 term by term.

(i) The term $\frac{-\partial}{\partial t}\int_\tau \left(\frac{\mu}{2}H^2 + \frac{\varepsilon}{2}E^2\right)d\tau$

For the electrostatic case the energy stored is given by (Eq. 3.32)

$$\int_\tau \frac{1}{2}\varepsilon E^2 d\tau$$

For the magnetostatic case the energy stored is given by (Eq. 6.99)

$$\int_\tau \frac{1}{2}\mu H^2 d\tau$$

Assuming that the above equations are valid for time-varying case the total electromagnetic energy stored per unit volume is

$$\int_\tau \left(\frac{\mu}{2}H^2 + \frac{\varepsilon}{2}E^2\right)d\tau$$

Then the equation

$$\frac{-\partial}{\partial t}\int_\tau \left(\frac{\mu}{2}H^2 + \frac{\varepsilon}{2}E^2\right)d\tau$$

along with the negative sign represents the rate at which the stored energy in volume τ is decreasing.

(ii) The term $\int_\tau \mathbf{E}\cdot\mathbf{J}_f d\tau$

Let **E** and **B** be the time-varying fields. The force **F** experienced by a charged particle q due to this time-varying field is

$$\mathbf{F} = q[\mathbf{E} + (\mathbf{v}\times\mathbf{B})]$$

Due to the force the charge q moves a distance $d\boldsymbol{l}$ in time dt.

Then the work done is

$$\begin{aligned} dW &= \mathbf{F} \cdot d\boldsymbol{l} \\ &= q[\mathbf{E} + (\mathbf{v} \times \mathbf{B})] \cdot d\boldsymbol{l} \end{aligned} \tag{7.51}$$

The power supplied by the fields is then

$$P = \frac{dW}{dt} = q[\mathbf{E} + (\mathbf{v} \times \mathbf{B})] \cdot \frac{d\boldsymbol{l}}{dt} \tag{7.52}$$

$$\Rightarrow P = q[\mathbf{E} + (\mathbf{v} \times \mathbf{B})] \cdot \mathbf{v} \tag{7.53}$$

$$\Rightarrow P = q\mathbf{E} \cdot \mathbf{v} \tag{7.54}$$

But $q = \rho\, d\tau$ and $\mathbf{J_f} = \rho\mathbf{v}$ where we have assumed charge q is distributed over a volume $\tau$ with volume charge density $\rho$. The entire charge moves with an average velocity $\mathbf{v}$ producing current $\mathbf{J}_f$.

$$\Rightarrow P = \rho d\tau \mathbf{E} \cdot \mathbf{v} \tag{7.55}$$

$$\Rightarrow P = \mathbf{E} \cdot \rho \mathbf{v} d\tau \tag{7.56}$$

$$\Rightarrow P = \mathbf{E} \cdot \mathbf{J}_f d\tau \tag{7.57}$$

Hence the power dissipated per unit volume is

$$= \mathbf{E} \cdot \mathbf{J}_f \tag{7.58}$$

Thus the term

$$\int_\tau \mathbf{E} \cdot \mathbf{J}_f d\tau$$

represents power dissipated over the total volume $\tau$.

(iii) The term $-\oint_S (\mathbf{E} \times \mathbf{H}).\mathbf{ds}$

The term $\int_\tau \mathbf{E} \cdot \mathbf{J}_f d\tau$ represents the rate of energy dissipated in volume $\tau$. The term $\frac{-\partial}{\partial t}\int_\tau \left(\frac{\mu}{2}H^2 + \frac{\varepsilon}{2}E^2\right)d\tau$ represents the rate at which stored energy in the volume $\tau$ is decreasing. Thus as per law of conservation of energy considering a volume $\tau$ bounded by surface S.

"The rate of electromagnetic energy dissipation in a volume $\tau$ must equal to the sum of the rate at which the electromagnetic stored energy is decreasing in the volume $\tau$ and the rate at which energy is entering volume $\tau$ through the surface S".

Hence $-\oint_S (\mathbf{E} \times \mathbf{H}) \cdot \mathbf{ds}$ must represent the rate of flow of energy inward of volume τ. Thus the expression $-\oint_S (\mathbf{E} \times \mathbf{H}) \cdot \mathbf{ds}$ represents the rate of outward flow of energy through the surface S which bounds a volume τ.

Thus $(\mathbf{E} \times \mathbf{H})$ is energy flow per unit time per unit area and is known as Poynting vector, denoted as $\mathcal{P}$.

$$\mathcal{P} = \mathbf{E} \times \mathbf{H} \tag{7.59}$$

Equation 7.50 is known as Poynting theorem.

## 7.5 Average Poynting Vector

We know the electric and magnetic fields are given by

$$\begin{aligned} \mathbf{E} &= \mathbf{E}_o e^{i(\omega t - \beta z)} \\ &= (\mathbf{E}_1 + i\mathbf{E}_2) e^{i(\omega t - \beta z)} \end{aligned} \tag{7.60}$$

as $\mathbf{E}_o$ is a complex number. Similarly

$$\begin{aligned} \mathbf{H} &= \mathbf{H}_o e^{i(\omega t - \beta z)} \\ &= (\mathbf{H}_1 + i\mathbf{H}_2) e^{i(\omega t - \beta z)} \end{aligned} \tag{7.61}$$

where $\mathbf{E}_1$, $\mathbf{E}_2$, $\mathbf{H}_1$ and $\mathbf{H}_2$ are real.

Then

$$\mathbf{E} = (\mathbf{E}_1 + i\mathbf{E}_2)[\cos(\omega t - \beta z) - i \sin(\omega t - \beta z)] \tag{7.62}$$

$$\begin{aligned} \mathbf{E} = {} & [\mathbf{E}_1 \cos(\omega t - \beta z) + \mathbf{E}_2 \sin(\omega t - \beta z)] \\ & - i[\mathbf{E}_1 \sin(\omega t - \beta z) - \mathbf{E}_2 \cos(\omega t - \beta z)] \end{aligned} \tag{7.63}$$

From Eq. 7.63

$$\text{Real}(\mathbf{E}) = \mathbf{E}_1 \cos(\omega t - \beta z) + \mathbf{E}_2 \sin(\omega t - \beta z) \tag{7.64}$$

Similarly

$$\text{Real}(\mathbf{H}) = \mathbf{H}_1 \cos(\omega t - \beta z) + \mathbf{H}_2 \sin(\omega t - \beta z) \tag{7.65}$$

Now

$$\langle \boldsymbol{\mathcal{P}} \rangle = \langle \mathrm{Re}\,\mathbf{E} \times \mathrm{Re}\,\mathbf{H} \rangle \tag{7.66}$$

$$\begin{aligned} \mathrm{Re}\,\mathbf{E} \times \mathrm{Re}\,\mathbf{H} &= [\mathbf{E}_1 \cos(\omega t - \beta z) + \mathbf{E}_2 \sin(\omega t - \beta z)] \\ &\quad \times [\mathbf{H}_1 \cos(\omega t - \beta z) + \mathbf{H}_2 \cos(\omega t - \beta z)] \\ &= (\mathbf{E}_1 \times \mathbf{H}_1) \cos^2(\omega t - \beta z) \\ &\quad + (\mathbf{E}_2 \times \mathbf{H}_2) \sin^2(\omega t - \beta z) \\ &\quad + [(\mathbf{E}_1 \times \mathbf{H}_2) + (\mathbf{E}_2 \times \mathbf{H}_1)] \sin(\omega t - \beta z) \cos(\omega t - \beta z) \end{aligned} \tag{7.67}$$

$$\begin{aligned} &= (\mathbf{E}_1 \times \mathbf{H}_1) \cos^2(\omega t - \beta z) \\ &\quad + (\mathbf{E}_2 \times \mathbf{H}_2) \sin^2(\omega t - \beta z) \\ &\quad + [(\mathbf{E}_1 \times \mathbf{H}_2) + (\mathbf{E}_2 \times \mathbf{H}_1)] \sin(\omega t - \beta z) \cos(\omega t - \beta z) \end{aligned} \tag{7.68}$$

As with $\omega = 2\pi/\mathrm{T}$ (see Eq. 7.79a)

$$\frac{1}{\mathrm{T}} \int_0^{\mathrm{T}} \sin^2(\omega t - \beta z) dt = \frac{1}{\mathrm{T}} \int_0^{\mathrm{T}} \cos^2(\omega t - \beta z) dt = \frac{1}{2}$$

and

$$\frac{1}{\mathrm{T}} \int_0^{\mathrm{T}} \sin(\omega t - \beta z) \cos(\omega t - \beta z) dt = 0$$

We have

$$\langle \mathrm{Re}\mathbf{E} \times \mathrm{Re}\mathbf{H} \rangle = \frac{1}{2} [(\mathbf{E}_1 \times \mathbf{H}_1) + (\mathbf{E}_2 \times \mathbf{H}_2)] \tag{7.69}$$

The complex conjugate of **H** is

$$\mathbf{H}^* = (\mathbf{H}_1 - i\mathbf{H}_2) e^{-i(\omega t - \beta z)} \tag{7.70}$$

$$\mathbf{E} \times \mathbf{H}^* = \left[ (\mathbf{E}_1 + i\mathbf{E}_2) e^{i(\omega t - \beta z)} \times (\mathbf{H}_1 - i\mathbf{H}_2) e^{-i(\omega t - \beta z)} \right] \tag{7.71}$$

$$\mathbf{E} \times \mathbf{H}^* = [(\mathbf{E}_1 \times \mathbf{H}_1) + (\mathbf{E}_2 \times \mathbf{H}_2)] + i[(\mathbf{E}_2 \times \mathbf{H}_1) - (\mathbf{E}_1 \times \mathbf{H}_2)] \tag{7.72}$$

$$\mathrm{Re}(\mathbf{E} \times \mathbf{H}^*) = (\mathbf{E}_1 \times \mathbf{H}_1) + (\mathbf{E}_2 \times \mathbf{H}_2) \tag{7.73}$$

Substituting Eq. 7.73 in Eq. 7.69,

$$\langle \mathrm{Re}\,\mathbf{E} \times \mathrm{Re}\,\mathbf{H} \rangle = \frac{1}{2}\mathrm{Re}\,(\mathbf{E} \times \mathbf{H}^*) \tag{7.74}$$

Substituting Eq. 7.74 in Eq. 7.66

$$\langle \boldsymbol{\mathcal{P}} \rangle = \frac{1}{2}\boldsymbol{\mathcal{P}}(\mathbf{E} \times \mathbf{H}^*) \tag{7.75}$$

The above equation can be used to calculate the average value of Poynting vector directly from **E**, **H**.

## 7.6 Poynting Vector for Wave Propagation in Free Space

We know that the Poynting vector is given by

$$\boldsymbol{\mathcal{P}} = \mathbf{E} \times \mathbf{H} \tag{7.76}$$

For plane waves travelling in $z$-direction with electric and magnetic fields in x-direction and y-direction, respectively, from Eqs. 7.39, 7.40

We have

$$\mathbf{E}(z, t) = \mathrm{E_o}\cos(\omega t - \beta z)\hat{\mathbf{i}} \tag{7.77}$$

$$\mathbf{H}(z, t) = \frac{\mathrm{E_o}}{\eta}\cos(\omega t - \beta z)\hat{\mathbf{j}} \tag{7.78}$$

Here we have used $(\mathrm{E_o})_x = \mathrm{E_o}$, $(\mathrm{H_o})_y = \mathrm{H_o}$ and Eq. 7.35.

Substituting Eqs. 7.75, 7.76 in 7.59

$$\boldsymbol{\mathcal{P}} = \frac{(\mathrm{E_o})^2}{\eta}\cos^2(\omega t - \beta z)(\hat{\mathbf{i}} \times \hat{\mathbf{j}}) \tag{7.79}$$

$$\boldsymbol{\mathcal{P}} = \frac{\mathrm{E_o^2}}{\eta}\cos^2(\omega t - \beta z)\hat{\mathbf{k}} \tag{7.80}$$

For the case of time-varying fields $\boldsymbol{\mathcal{P}} = \mathbf{E} \times \mathbf{H}$ gives the instantaneous value of Poynting vector. For time-varying electric and magnetic fields whenever we make a measurement it will be over large number of cycles if the frequency is very high. In such instances average value will be of interest. We denote the average value by < > brackets,

Thus

$$\langle \boldsymbol{\mathcal{P}} \rangle = \frac{E_o^2}{\eta} \langle \cos^2(\omega t - \beta z) \rangle \hat{\mathbf{k}} \tag{7.80a}$$

$$\langle \cos^2(\omega t - \beta z) \rangle = \frac{1}{T} \int_0^T \cos^2(\omega t - \beta z) dt$$

where T is the time period of the wave.

Now,

$$\begin{aligned}
\langle \cos^2(\omega t - \beta z) \rangle &= \frac{1}{T} \int_0^T \left( \frac{\cos 2(\omega t - \beta z) + 1}{2} \right) dt \\
&= \frac{1}{2T} \int_0^T \cos 2(\omega t - \beta z) dt + \frac{1}{2T} \int_0^T dt \\
&= \frac{1}{2T} \int_0^{2\pi/\omega} \cos 2(\omega t - \beta z) dt + \frac{1}{2T} [T] \\
&= \frac{1}{2T} \left[ \frac{\sin 2(\omega t - \beta z)}{\omega} \right]_0^{2\pi/\omega} + \frac{1}{2} \\
&= \frac{1}{2T} \left[ \frac{\sin(4\pi - 2\beta z)}{\omega} - \frac{\sin(-2\beta z)}{\omega} \right] + \frac{1}{2} \\
&= \frac{1}{2T} \left[ \frac{\sin 4\pi \cos 2\beta z - \cos 4\pi \sin 2\beta z}{\omega} + \frac{\sin 2\beta z}{\omega} \right] + \frac{1}{2} \\
&= \frac{1}{2T} \left[ \frac{-\sin 2\beta z}{\omega} + \frac{\sin 2\beta z}{\omega} \right] + \frac{1}{2}
\end{aligned}$$

Hence

$$\langle \cos^2(\omega t - \beta z) \rangle = \frac{1}{2} \tag{7.80b}$$

Hence Eq. 7.80a is then

$$\langle \boldsymbol{\mathcal{P}} \rangle = \frac{1}{2} \frac{|E_o^2|}{\eta} \hat{\mathbf{k}} \tag{7.81a}$$

Let us illustrate the advantage of working with exponentials than with time-harmonic fields as discussed in Sect. 6.18.

From Eqs. 7.37, 7.38 with

$(E_o)_x = E_o$, $(H_o)_y = H_o$ and Eq. 7.35 we can write

$$\mathbf{E}(z,t) = E_o e^{i(\omega t - \beta z)}\hat{\mathbf{i}}$$
$$\mathbf{H}^*(z,t) = \frac{E_o^*}{\eta} e^{-i(\omega t - \beta z)}\hat{\mathbf{j}}.$$

Substituting the above equations in (7.74)

$$\langle \boldsymbol{\mathcal{P}} \rangle = \frac{1}{2}\left[E_o \frac{E_o^*}{\eta} e^{i(\omega t - \beta z)} e^{-i(\omega t - \beta z)}\right](\hat{\mathbf{i}} \times \hat{\mathbf{j}})$$

$$\Rightarrow \langle \boldsymbol{\mathcal{P}} \rangle = \frac{1}{2}\frac{|E_o|^2}{\eta}\hat{\mathbf{k}} \tag{7.81b}$$

Compare the above equation with Eq. 7.80a. Directly working with time-harmonic fields in Eqs. 7.75, 7.76 so many lengthy and laborious steps were involved in arriving at Eqs. 7.80a. Using the exponential form the same result was arrived at in just few steps in Eq. 7.80b.

It is encouraged that the reader can solve Problem 7.4 to verify further the advantage of working with exponential functions than with time-harmonic fields.

## 7.7 Plane Electromagnetic Waves in Lossy Dielectrics

In the previous sections we discussed the plane wave propagation in free space. In this section we will discuss the plane wave propagation in lossy dielectrics. For a lossy dielectric

$$\sigma \neq 0, \quad \varepsilon = \varepsilon_r \varepsilon_o \quad \mu = \mu_r \mu_o \tag{7.82}$$

A lossy dielectric is a partially conducting dielectric in which the electromagnetic wave losses energy as it propagates in the medium. A lossy dielectric can be an imperfect dielectric or imperfect conductor. Let the medium be free of charge $\rho_f = 0$.

Using Eq. 7.82 we can write Eqs. 7.10, 7.12 as

$$\nabla^2 \mathbf{E} = \sigma\mu \frac{\partial \mathbf{E}}{\partial t} + \varepsilon\mu \frac{\partial^2 \mathbf{E}}{\partial t^2} \tag{7.83a}$$

$$\nabla^2 \mathbf{H} = \sigma\mu \frac{\partial \mathbf{H}}{\partial t} + \varepsilon\mu \frac{\partial^2 \mathbf{H}}{\partial t^2} \tag{7.83b}$$

Now using the phasor form of the electric field vector (Eq. 6.219)

$\mathbf{E} = \mathbf{E}_s e^{i\omega t}$ Eq. 7.83a becomes

$$\nabla^2\left(\mathbf{E}_s e^{i\omega t}\right) = \sigma\mu\frac{\partial}{\partial t}\left(\mathbf{E}_s e^{i\omega t}\right) + \varepsilon\mu\frac{\partial^2}{\partial t^2}\left(\mathbf{E}_s e^{i\omega t}\right) \tag{7.84a}$$

$$\Rightarrow \nabla^2\mathbf{E}_s = i\omega\mu(\sigma + i\omega\varepsilon)\mathbf{E}_s$$

Similarly with $\mathbf{H} = \mathbf{H}_s e^{i\omega t}$ Eq. 7.82b becomes

$$\nabla^2\mathbf{H}_s = i\omega\mu(\sigma + i\omega\varepsilon)\mathbf{H}_s \tag{7.84b}$$

Let

$$\gamma^2 = (i\omega\mu)(\sigma + i\omega\varepsilon) \tag{7.85}$$

$\gamma$ is known as propagation constant.
Then Eqs. 7.83a, 7.83b can be written as

$$\nabla^2\mathbf{E_s} - \gamma^2\mathbf{E_s} = 0 \tag{7.86a}$$

$$\nabla^2\mathbf{H_s} - \gamma^2 H_s = 0 \tag{7.86b}$$

Let us assume the wave propagates in the $z$-direction with the electric field along $x$-axis and the magnetic field along $y$-axis. Then Eqs. 7.86a, 7.86b can be written as

$$\frac{\partial^2\mathbf{E_s}}{\partial z^2} - \gamma^2\mathbf{E_s} = 0 \tag{7.87a}$$

$$\frac{\partial^2\mathbf{H_s}}{\partial z^2} - \gamma^2\mathbf{H_s} = 0 \tag{7.87b}$$

Because $\gamma$ is complex we can write

$$\gamma = \alpha + i\beta \tag{7.88}$$

$\alpha$ is called attenuation constant, $\beta$ the phase constant.
From Eq. 7.88

$$\gamma^2 = (\alpha + i\beta)^2 = \alpha^2 - \beta^2 + 2\alpha i\beta \tag{7.89}$$

Equating the real and imaginary parts of Eqs. 7.88, 7.85

$$\beta^2 - \alpha^2 = \omega^2\mu\varepsilon \tag{7.90a}$$

$$2\alpha\beta = \omega\mu\sigma \tag{7.90b}$$

From Eq. 7.88

$$\begin{aligned} |\gamma^2| &= \sqrt{(\alpha^2 - \beta^2)^2 + (2\alpha\beta)^2} \\ \Rightarrow |\gamma^2| &= \alpha^2 + \beta^2 \end{aligned} \tag{7.91}$$

From Eq. 7.85

$$|\gamma^2| = \omega\mu\sqrt{\sigma^2 + \omega^2\varepsilon^2} \tag{7.92}$$

Comparing Eqs. 7.91, 7.92

$$\alpha^2 + \beta^2 = \omega\mu\sqrt{\sigma^2 + \omega^2\varepsilon^2} \tag{7.93}$$

Solving Eqs. 7.90a, 7.93, 7.90b

$$\alpha = \omega\sqrt{\frac{\mu\varepsilon}{2}\left[\sqrt{1 + \left(\frac{\sigma}{\omega\varepsilon}\right)^2} - 1\right]} \tag{7.94a}$$

$$\beta = \omega\sqrt{\frac{\mu\varepsilon}{2}\left[\sqrt{1 + \left(\frac{\sigma}{\omega\varepsilon}\right)^2} + 1\right]} \tag{7.94b}$$

Because we assume the wave propagates in $z$-direction and the electric field is pointing in $x$-direction

$$\mathbf{E}_s = \mathrm{E}_{xs}(z)\hat{\mathbf{i}} \tag{7.95}$$

Substituting Eq. 7.95 in Eq. 7.87a

$$\left[\frac{d^2}{dz^2} - \gamma^2\right]\mathrm{E}_{xs}(z) = 0 \tag{7.96}$$

The solution of the above equation is

$$E_{xs}(z) = E_o e^{-\gamma z} + E_0' e^{\gamma z} \tag{7.97}$$

The term $e^{\gamma z}$ denotes a wave travelling in the $-z$-direction while we assume a wave travelling in the $+z$-diection. Hence $E_o' = 0$. Thus

$$E_{xs}(z) = E_o e^{-\gamma z} \tag{7.98a}$$

Substituting for $E_{xs}(z)$ in Eq. 7.95

$$E_s = E_o e^{-\gamma z}\hat{\mathbf{i}} \tag{7.98b}$$

Now inserting the time factor $e^{i\omega t}$ and using Eq. 7.87 and then taking the real part

$$\begin{aligned} \mathbf{E}(z,t) &= \mathrm{Re}\left[\mathbf{E}_s e^{i\omega t}\right] = \mathrm{Re}\left[E_o e^{-\gamma z} e^{i\omega t}\hat{\mathbf{i}}\right] \\ &= \mathrm{Re}\left[E_o e^{-\alpha z} e^{i(\omega t - \beta z)}\hat{\mathbf{i}}\right] \\ \mathbf{E}(z,t) &= E_o e^{-\alpha z}\cos(\omega t - \beta z)\hat{\mathbf{i}} \end{aligned} \tag{7.99}$$

Now using the Maxwell's Eq. 6.232 (phasor form)

$$\mathbf{H}_s = \frac{1}{-i\omega\mu}\nabla \times \mathbf{E}_s$$

Because the electric field has only $x$-component

$$\begin{aligned} \nabla \times \mathbf{E}_s &= \begin{vmatrix} \hat{\mathbf{i}} & \hat{\mathbf{j}} & \hat{\mathbf{k}} \\ \frac{\partial}{\partial x} & \frac{\partial}{\partial y} & \frac{\partial}{\partial z} \\ E_{xs} & 0 & 0 \end{vmatrix} \\ &= \frac{\partial E_{xS}}{\partial z}\hat{\mathbf{j}} \end{aligned} \tag{7.100}$$

Hence $\mathbf{H}_s = \dfrac{1}{-i\omega\mu}\left(\dfrac{\partial E_{xs}}{\partial z}\right)\hat{\mathbf{j}}$
with $E_{sx} = E_o e^{-\gamma z}$

$$\mathbf{H}_s = \frac{\gamma}{i\omega\mu}E_o e^{-\gamma z}\hat{\mathbf{j}}$$

with $\mathbf{H} = \mathbf{H}_S e^{i\omega t}$

$$\mathbf{H} = \frac{\gamma}{i\omega\mu} E_o e^{-\gamma z} e^{i\omega t} \hat{\mathbf{j}}$$

Substituting for $\gamma$ from Eq. 7.85 and simplifying

$$\mathbf{H}(z,t) = \sqrt{\frac{\sigma + i\omega\varepsilon}{i\omega\mu}} E_o e^{-\alpha z} e^{i(\omega t - \beta z)} \hat{\mathbf{j}}$$

Let

$$\eta = \sqrt{\frac{i\omega\mu}{\sigma + i\omega\varepsilon}} = \frac{i\omega\mu}{\gamma} \text{ and } H_o = \frac{E_o}{\eta} \tag{7.101a}$$

then

$$\mathbf{H}(z,t) = H_o e^{-\alpha z} e^{i(\omega t - \beta z)} \hat{\mathbf{j}} \tag{7.101b}$$

Here $\eta$, the intrinsic impedance of the medium, is complex and hence we can write

$$\eta = |\eta| e^{i\theta_\eta} \tag{7.102}$$

With Eqs. 7.101a, 7.102 we can write Eq. 7.101b as

$$\mathbf{H} = \frac{E_o}{|\eta|} e^{-\alpha z} e^{i\left(\omega t - \beta z - \theta_\eta\right)} \hat{\mathbf{j}} \tag{7.103}$$

Now taking the real part

$$\mathbf{H} = \frac{E_o}{|\eta|} e^{-\alpha z} \cos\left(\omega t - \beta z - \theta_\eta\right) \hat{\mathbf{j}} \tag{7.104}$$

From Eqs. 7.99, 7.103 we see that **E**, **H** are out of phase by angle $\theta_\eta$, because intrinsic impedance is complex.

The factor $e^{-\alpha z}$ in Eqs. 7.99, 7.104 is called attenuation factor. It shows that the amplitude of **E**, **H** decreases as $z$ increases. The SI unit of the attenuation constant is Neper per metre (Np/m). Thus **E**, **H** attenuate as the electromagnetic wave propagates inside the medium as shown in Fig. 7.5a. From Eqs. 7.102, 7.103 we observe that **E** leads **H** (or) **H** lags **E** by an angle of $\theta_\eta$ which is shown pictorially in Fig. 7.5a.

In Problem 7.4 we calculate the average Poynting vector for the case electromagnetic wave propagation in lossy dielectrics.

From Eq. 7.84

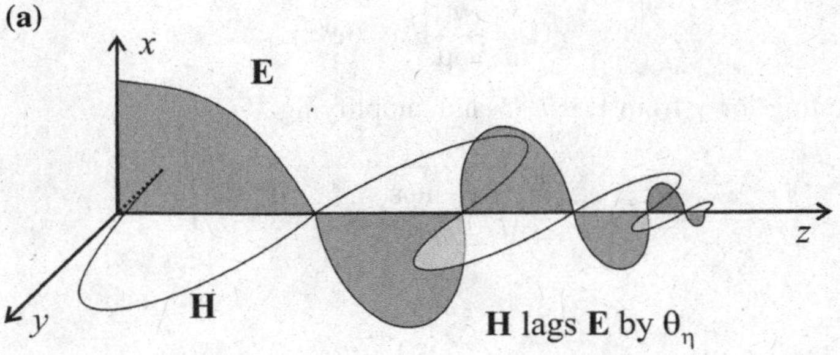


**Fig. 7.5** (**a**) Propagation of an electromagneic wave in a lossy dielectric. Electric and magnetic fields attenuate as they propagate inside the material

$$i\omega\mu(\sigma + i\sigma\varepsilon) = -\omega^2\mu\varepsilon\left[1 - i\frac{\sigma}{\omega\varepsilon}\right] = -\omega^2\mu\varepsilon_c \tag{7.105}$$

where

$$\varepsilon_c = \varepsilon\left[1 - i\frac{\sigma}{\sigma\varepsilon}\right] \tag{7.106}$$

$$\Rightarrow \varepsilon_c = \varepsilon' - i\varepsilon'' \tag{7.107}$$

where

$$\varepsilon' = \varepsilon, \varepsilon'' = \frac{\sigma}{\omega} \tag{7.108}$$

$\varepsilon_c$ is called the complex permittivity of the medium. β in Eq. 7.99, 7.104 is the usual wave number given by Eq. 7.24.

We know that the conduction current density and displacement current density are given by

$$\mathbf{J}_{cs} = \sigma\mathbf{E}_s \tag{7.109}$$

$$\mathbf{J}_d = \varepsilon\frac{\partial\mathbf{E}}{\partial t} \tag{7.110a}$$

$$\Rightarrow \mathbf{J}_{ds} = i\omega\varepsilon\mathbf{E}_s \tag{7.110b}$$

where we have used equations $\mathbf{E} = \mathbf{E_s}e^{i\omega t}$ and $\mathbf{J_d} = \mathbf{J_{ds}}e^{i\omega t}$ in Eqs. 7.110a, 7.110b. Now taking the ratio

$$\Rightarrow \frac{J_{cs}}{J_{ds}} = \frac{\sigma}{i\omega\varepsilon} \tag{7.111}$$

Hence the two vectors point in the same direction in space, but they are 90° out of phase in time. This is shown pictorially in Fig. 7.5b.

In the figure we have used $\mathbf{E}_s$ as reference.

The total current density is (adding vectorically).

$$\mathbf{J}_{ts} = \mathbf{J}_{cs} + \mathbf{J}_{ds} \tag{7.112}$$

$\mathbf{J}_{ts}, \mathbf{J}_{cs}, \mathbf{J}_{ds}$ is plotted in Fig. 7.5c. $\mathbf{J}_{ds}$ makes on angle δ with $\mathbf{J}_{ts}$. From the figure it is clear that

$$\tan\delta = \frac{|\mathbf{J}_{cs}|}{|\mathbf{J}_{ds}|} = \frac{|\sigma \mathbf{E}_s|}{|j\omega\varepsilon \mathbf{E}_s|} = \frac{\sigma}{\omega\varepsilon} \tag{7.113}$$

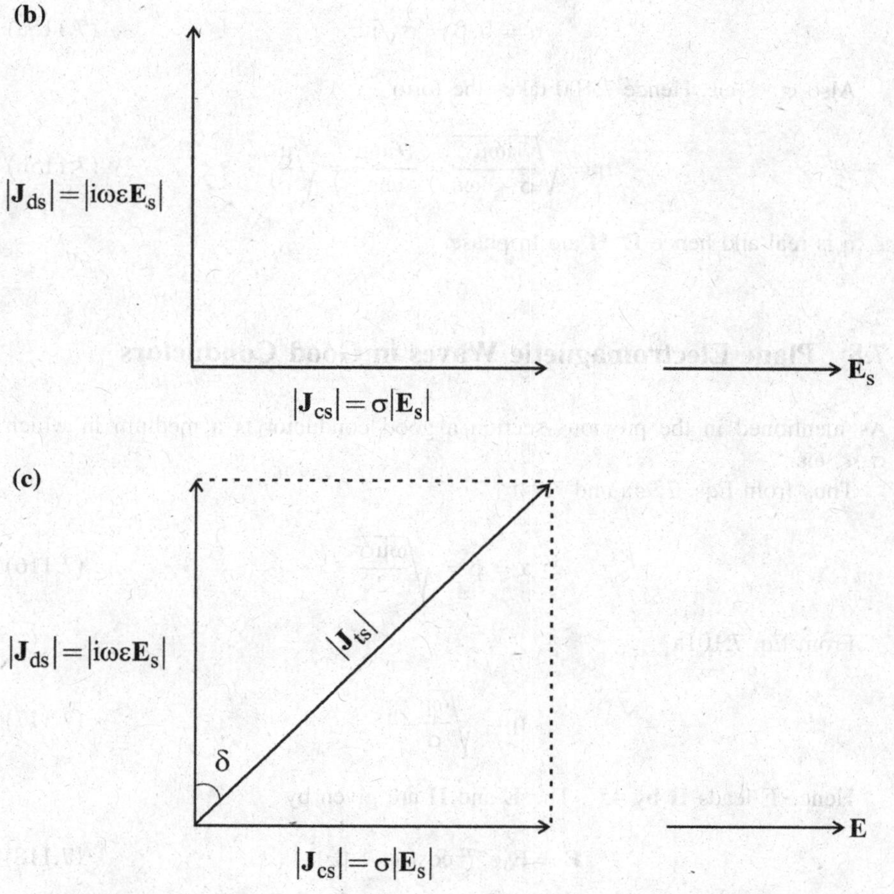


**Fig. 7.5** (**b**) Relationship between conduction current density and displacement current density in a lossy dielectric. (**c**) Relationship between displacement current density, conduction current density and total current density in a lossy dielectric

Here tan δ is known as loss tangent. δ is the loss angle of the medium. A perfect dielectric is one in which tan δ is very small, i.e. $\sigma \ll \omega\varepsilon$. For a good conductor tan δ is very large, i.e. $\sigma \ll \omega\varepsilon$. It should be noted that tan δ depends on frequency of the wave. Hence if a medium is a good conductor at low frequencies it shall be a good dielectric at high frequencies.

The calculation of Poynting vector for the case of lossy dielectric is given as an exercise (see Problem 7.4).

**Special Case**:

For a perfect lossless dielectric

$$\sigma \simeq 0, \varepsilon = \varepsilon_0\varepsilon_r, \mu = \mu_0\mu_r \tag{7.114}$$

Hence

$$\alpha = 0, \beta = \sigma\sqrt{\mu\varepsilon} \tag{7.115a}$$

Also $\sigma \ll \omega\varepsilon$. Hence 7.100 takes the form.

$$\eta = \sqrt{\frac{i\omega\mu}{\sigma + i\omega\varepsilon}} = \frac{\sqrt{i\omega\mu}}{i\omega\varepsilon} = \sqrt{\frac{\mu}{\varepsilon}} \tag{7.115b}$$

η is real and hence **E**, **H** are in phase.

## 7.8 Plane Electromagnetic Waves in Good Conductors

As mentioned in the previous section a good conductor is a medium in which $\sigma \ll \omega\varepsilon$.

Thus from Eqs. 7.94a and 7.94b

$$\alpha = \beta = \sqrt{\frac{\omega\mu\sigma}{2}} \tag{7.116}$$

From Eq. 7.101a

$$\eta = \sqrt{\frac{\omega\mu}{\sigma}}\angle 45^\circ \tag{7.117}$$

Hence **E** leads **H** by 45°. The **E** and **H** are given by

$$\mathbf{E} = E_0 e^{-\alpha z}\cos(\omega t - \beta z)\hat{\mathbf{i}} \tag{7.118}$$

$$\mathbf{H} = \frac{H_o}{|\eta|} e^{-\alpha z} \cos(\omega t - \beta z - 45°)\hat{\mathbf{j}} \tag{7.119}$$

Thus as the electromagnetic wave propagates through the conducting medium the amplitude of **E** and **H** gets attenuated by a factor of $e^{-\alpha z}$.

We define the penetration depth or skin depth of the electromagnetic wave in the conducting medium as, the distance travelled by the wave in the conducting medium such that the wave amplitude decreases by a factor of $e^{-1}$ (about 37%). We denote the skin depth as $\delta$.

Hence

$$E_o e^{-\alpha\delta} = E_o e^{-1} \tag{7.120}$$

$$\Rightarrow \delta = \frac{1}{\alpha} \tag{7.121}$$

Substituting Eq. 7.116 in 7.121

$$\delta = \frac{1}{\sqrt{\frac{\omega\mu\sigma}{2}}} = \sqrt{\frac{2}{\omega\mu\sigma}} \tag{7.122}$$

With $\omega = 2\pi\nu$ where $\nu$ is the frequency of the em wave we have

$$\delta = \sqrt{\frac{1}{\pi\nu\mu\sigma}} \tag{7.123}$$

Let us calculate the skin depth for silver in microwave frequencies. For silver $\sigma \simeq 10^7$ mho/m. At a typical microwave frequency of $10^8$ Hz the skin depth is $10^{-4}$ cm. Thus the depth of penetration of the em wave in silver at microwave frequencies is very small. Depending upon the frequency and conducting material the em wave attenuates within a small distance of the conductor. Thus electromagnetic energy doesn't penetrate deeper inside the conductor. It travels in the region surrounding the conductor. Hence the conductor merely guides the waves.

Suppose we wish to make a waveguide made of silver at microwave frequencies. As we know that the skin depth of em wave at microwaves frequency in silver is $\approx 10^{-4}$ cm we can coat silver on brass or glass with the silver coating thickness of $\approx 10^{-4}$ cm. Thus the silver coating on brass or glass will serve the purpose. No need to make the entire waveguide of silver. Hence lot of silver is saved; thereby reducing the cost of waveguide.

From 7.116, 7.121, 7.123 and with

$$\beta = \frac{2\pi}{\lambda}$$

The wavelength can be written as

$$\lambda = 2\pi\delta \tag{7.124}$$

We know the velocity of the wave is

$$v = \frac{\omega}{\beta} \tag{7.125}$$
$$\Rightarrow v = \omega\delta$$

**Example 7.2**
A plane wave with frequency of 9 GHz is travelling in free space with an electric field of amplitude 1 V/m.

(a) Find the velocity and the propagation constant of the em wave.
(b) Calculate the characteristic impedance of the medium.
(c) Find the amplitude of the magnetic field intensity.

**Solution**

(a) For free space

$$v = \frac{1}{\sqrt{\mu_o \varepsilon_o}}$$
$$v = 3 \times 10^8 \text{m/s}.$$
$$\lambda = \frac{v}{\nu} = \frac{3 \times 10^8}{9 \times 10^9} = \frac{1}{3} \times 10^{-1}\text{m}.$$
$$= 3.3\text{cm}.$$

Now $\beta = \frac{2\pi}{\lambda}$

$$\beta = \frac{2\pi}{3.3} = 1.903\,\text{rad/cm}$$

(b) The characteristic impedance of the medium is

$$\eta = \sqrt{\frac{\mu_o}{\varepsilon_o}} = 377\Omega$$

(c) The amplitude of the magnetic field intensity is given by

$$\eta = \frac{(E_o)_x}{(H_o)_y}$$

$$\Rightarrow (H_o)_y = \frac{(E_o)_x}{\eta}$$

$$(H_o)_y = \frac{1}{377} = 2.65 \times 10^{-3}\,\text{A/m.}$$

## 7.9 Plane Electromagnetic Waves in Lossy Dielectrics Using Maxwell's Equation in Phasor Form

The results which we derived in Sect. 7.7 for the case of lossy dielectrics can be derived directly from Maxwell's equation in phasor form as follows:

From Eqs. 6.230–6.233 we can write

(a) $$\nabla \cdot \mathbf{D_s} = \rho_s = 0 \text{ as } \rho_s = 0 \quad (7.126)$$

$$\text{With } \mathbf{D_s} = \varepsilon \mathbf{E_s} \text{ we have then } \nabla \cdot \mathbf{E_s} = 0 \quad (7.127)$$

(b) $$\nabla \cdot \mathbf{B_s} = 0 \text{ With } \mathbf{B_s} = \mu \mathbf{H_s} \quad \nabla \cdot \mathbf{H_s} = 0 \quad (7.128)$$

(c) $$\nabla \times \mathbf{E_s} = -i\omega \mathbf{B_s} \Rightarrow \nabla \times \mathbf{E_s} = -i\omega\mu \mathbf{H_s} \quad (7.129)$$

(d) $$\nabla \times \mathbf{H_s} = \mathbf{J_s} + i\omega \mathbf{D_s} \quad (7.130)$$

$$\Rightarrow \nabla \times \mathbf{H_s} = \sigma \mathbf{E_s} + i\omega\varepsilon \mathbf{E_s} \quad (7.131)$$

$$\Rightarrow \nabla \times \mathbf{H_s} = (\sigma + i\omega\varepsilon)\mathbf{E_s} \quad (7.132)$$

Taking curl on both sides of Eq. 7.129

$$\nabla \times \nabla \times \mathbf{E_s} = -i\omega\mu\nabla \times \mathbf{H_s} \tag{7.133}$$

Using vector identity 7 from Sect. 1.20 and using Eqs. 7.132, 7.127

$$\nabla^2\mathbf{E_s} = i\omega\mu(\sigma + i\omega\varepsilon)\mathbf{E_s} \tag{7.134}$$

Let

$$\gamma^2 = i\omega\mu(\sigma + i\omega\varepsilon) \tag{7.135}$$

$$\Rightarrow \nabla^2\mathbf{E}_s = \gamma^2\mathbf{E}_s \tag{7.136}$$

Here $\gamma$ is the propagation constant of the medium.

Similarly using Eqs. 7.132, 7.128 it can be shown that

$$\nabla^2\mathbf{H_s} = \gamma^2\mathbf{H_s} \tag{7.137}$$

Equations 7.136, 7.137 are nothing but Eqs. 7.86a, 7.86b.

Hence all the derivations done in Sect. 7.7 from Eqs. 7.87a, 7.87b onwards can be derived from Eqs. 7.136, 7.137 that is from Maxwell's equation in phasor form.

## 7.10 Wave Polarization

In Fig. 7.4 we saw that when the wave is propagating along $z$-axis, the electric field **E** is pointing in $x$-direction and the magnetic field **H** is pointing in $y$-direction.

"The wave polarization is defined as its electric field vector orientation as a function of time, at a fixed position in space"

Because the direction of **H** is related to **E** by Maxwell's equation specifying the direction of **E** alone is sufficient.

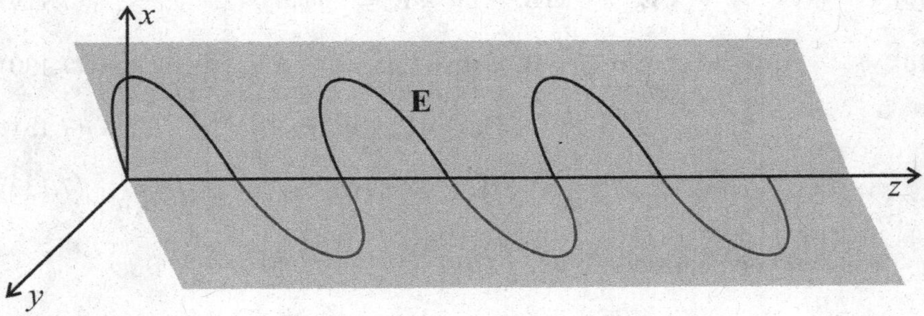


**Fig. 7.6** A linearly polarized electromagnetic wave

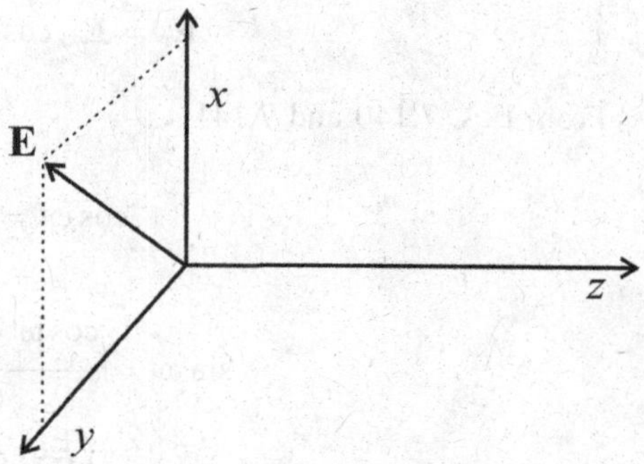


**Fig. 7.7** An electric field vector of an polarized electromagnetic wave

In Fig. 7.4 **E** is in fixed straight orientation in all times and positions. Such a wave is said to be linearly polarized. In Fig. 7.4 **E** always lies in the $x$-$z$ plane. But **E** can be oriented in any fixed direction in the $x$-$y$ plane and be linearly polarized. This is shown in Fig. 7.6. **H** is not shown in Fig. 7.6.

If the tip of the electric field traces a circle the wave is said to be circularly polarized. If the tip of the electric field traces an ellipse the wave is said to be elliptically polarized.

Consider Fig. 7.7 in which we show an electric field vector **E**. The electric field vector may correspond to a linearly or circularly or elliptically polarized wave. The conditions under which the electric field vector corresponds to linearly or circularly or elliptically polarized wave can be derived as follows:

In general the electric field vector **E** shown in Fig. 7.7 will be a superposition of two linearly polarized waves-one polarized wave in $x$-direction and the other polarized wave in the $y$-direction and lagging in phase by angle $\phi$ at any value of $z$. Then we can write

$$\mathbf{E} = \mathrm{E}_{xo}\cos(\omega t - \beta z)\hat{\mathbf{i}} + \mathrm{E}_{yo}\cos(\omega t - \beta z + \phi)\hat{\mathbf{j}} \tag{7.138}$$

Let us examine the direction change of **E** at $z = 0$ as t changes. Hence Eq. 7.138 becomes [with $z = 0$].

$$\mathbf{E} = \mathrm{E}_{xo}\cos\omega t\hat{\mathbf{i}} + \mathrm{E}_{yo}\cos(\omega t + \phi)\hat{\mathbf{j}} \tag{7.139}$$

The above equation can be written as

$$\mathbf{E} = \mathrm{E}_x\hat{\mathbf{i}} + \mathrm{E}_y\hat{\mathbf{j}} \tag{7.140}$$

with

$$\mathrm{E}_x = \mathbf{E}_{xo}\cos\omega t \tag{7.141}$$

$$\mathbf{E}_y = \mathbf{E}_{yo}\cos(\omega t + \phi) \tag{7.142}$$

From Eqs. 7.140 and 7.141

$$\cos\omega t = \frac{E_x}{E_{xo}} \tag{7.143}$$

$$\sin\omega t = \frac{\cos\omega t\cos\phi - \frac{E_y}{E_{yo}}}{\sin\phi}$$

$$\sin\omega t = \frac{\left(\frac{E_x}{E_{xo}}\right)\cos\phi - \frac{E_y}{E_{yo}}}{\sin\phi} \tag{7.144}$$

Now

$$\sin^2\omega t + \cos^2\omega t = 1$$

$$= \left(\frac{E_x}{E_{xo}}\right)^2 + \frac{(E_x/E_{xo})^2\cos^2\phi + \left(\frac{E_y}{E_{yo}}\right)^2 - 2\frac{E_xE_y}{E_{xo}E_{yo}}\cos\phi}{\sin^2\phi} \tag{7.145}$$

$$\Rightarrow \left(\frac{E_x}{E_{xo}}\right)^2 + \left(\frac{E_y}{E_{yo}}\right)^2 - \frac{2E_xE_y}{E_{xo}E_{yo}}\cos\phi = \sin^2\phi \tag{7.146}$$

(a) **Linear Polarization**

If $E_x$ and $E_y$ are in phase so that $\phi = 0$ Eq. 7.145 reduces to

$$\left(\frac{E_x}{E_{xo}} - \frac{E_y}{E_{yo}}\right)^2 = 0 \tag{7.147}$$

$$\Rightarrow \frac{E_y}{E_x} = \frac{E_{yo}}{E_{xo}} \tag{7.148}$$

Because $E_{yo}/E_{xo}$ the ratio between amplitudes is constant Eq. 7.148 represents a straight line as shown in Fig. 7.8.

$E_x$, $E_y$ are given by Eqs. 7.141, 7.142 with $\phi = 0$

$$E_x = E_{xo}\cos\omega t \tag{7.149}$$

$$E_y = E_{yo}\cos\omega t \tag{7.150}$$

The above discussion is at $z = 0$. If we allow $z$ to vary then the linearly polarized wave will be as shown in Fig. 7.6. In Fig. 7.8

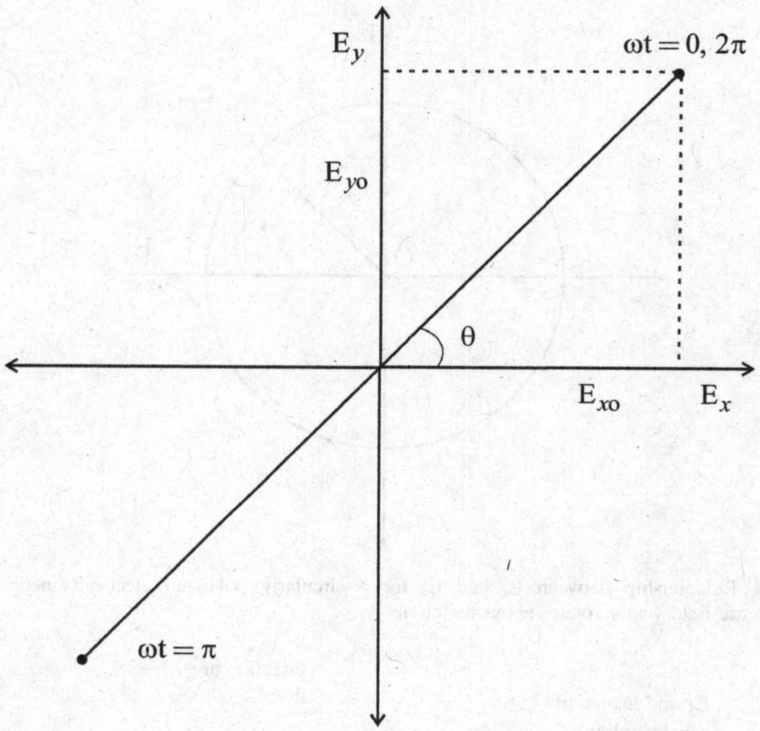


**Fig. 7.8** Relationship between $E_x$ and $E_y$ for a linearly polarized electromagnetic wave

$$\tan\theta = \frac{E_{yo}}{E_{xo}} \tag{7.151}$$

(b) **Circularly Polarized wave**

If $\phi = \pi/2$ and $E_{xo} = E_{yo} = E_o$ then Eq. 7.146 reduces to

$$E_x^2 + E_y^2 = E_o^2 \tag{7.152}$$

which is the equation of the circle. The electric field **E** rotates about as shown in Fig. 7.9.

Figure 7.9 is a circle. The tip of the **E** rotates about the circle thereby the wave is a circularly polarized wave.

The above discussion is at $z = 0$ as the time varies. On the other hand if $z$ also varies in addition to time then the circularly polarised wave will be as shown in Fig. 7.10.

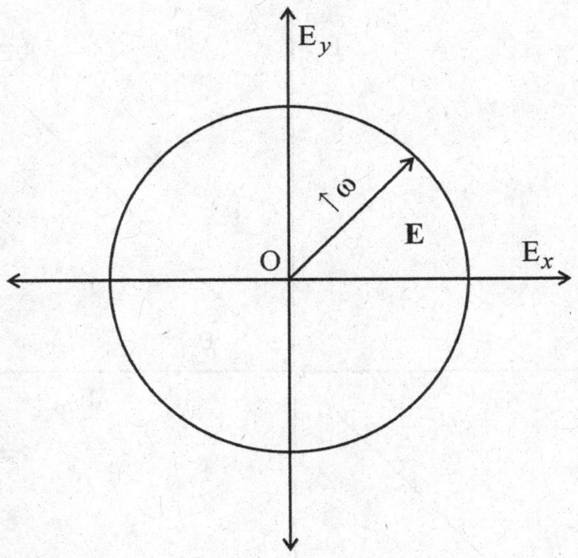


**Fig. 7.9** Relationship between $E_x$ and $E_y$ for a circularly polarized electromagnetic wave. The electric field vector rotates about the circle

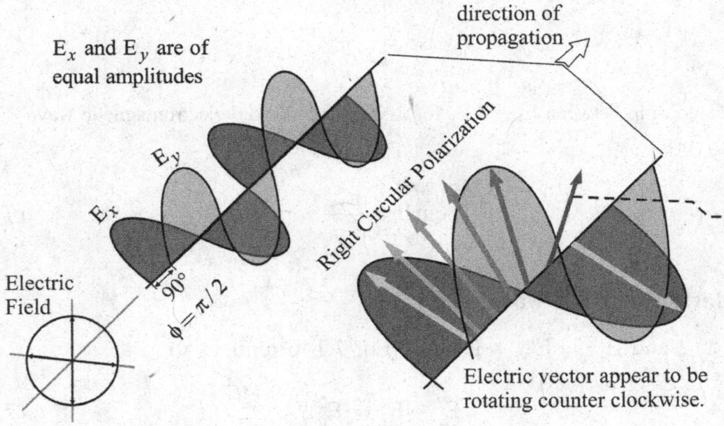


**Fig. 7.10** Circularly polarized electromagnetic wave

Figure 7.9 shows that **E** rotates at a uniform rate with an angular velocity ω in a counter clockwise direction. By right-hand rule when fingers of the right hand point in the direction of rotation of **E** then the thumb points in the direction of the wave. This is right hand circularly polarized wave. In Fig. 7.9 thus the wave propagates in the direction of outward drawn normal to the plane of the paper at point O and is a right circularly polarized wave. Hence when $E_y$ lags $E_x$ by $\pi/2$ in time phase we get right circularly polarized wave.

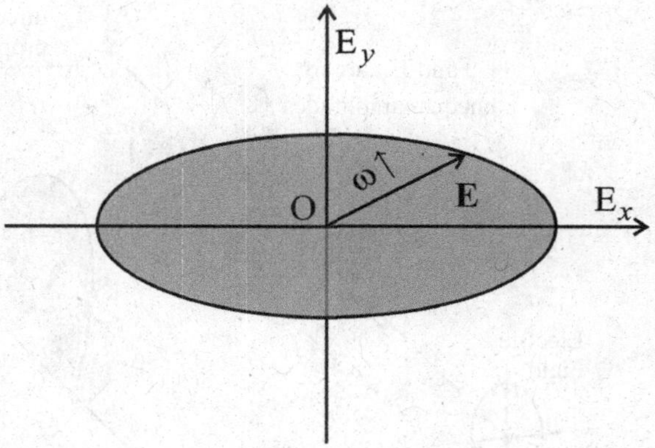


**Fig. 7.11** Relationship between $E_x$ and $E_y$ for a elliptically polarized electromagnetic wave. The electric field vector rotates about the ellipse.

(c) **Elliptically Polarized wave**

If $\phi = \pi/2$ and $E_{ox} \neq E_{oy}$ then Eq. 7.146 reduces to

$$\left(\frac{E_x}{E_{ox}}\right)^2 + \left(\frac{E_y}{E_{oy}}\right)^2 = 1 \quad (7.153)$$

which is the equation of ellipse. The tip of the **E** then rotates about in the form of the ellipse as shown in figure 7.11. Similar to circularly polarized wave we can define right and left elliptically polarized wave. An elliptically polarized wave is shown in figure 7.12

**Example 7.3**
A plane wave is travelling in a perfect dielectric medium. Its electric field intensity is given by

$$E_x(z,t) = 5\cos\left(2\pi \times 10^7 t - 0.2\pi z\right)\mathrm{V/m}$$

(a) Calculate the velocity of propagation.
(b) Find the magnetic field intensity if $\mu = \mu_o$.

**Solution**
We know that

$$E_x(z,t) = E_o\cos(\omega t - \beta z)$$

Hence

$$\omega = 2\pi\nu = 2\pi \times 10^7 \Rightarrow \nu = 10^7 \mathrm{Hz}.$$

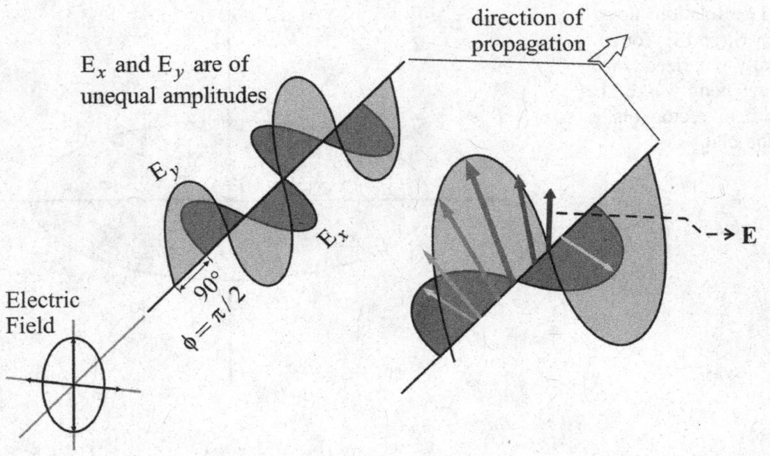


**Fig. 7.12** Elliptically polarized electromagnetic wave

Also $\beta = \frac{2\pi}{\lambda} = 0.2\pi$.
Hence $\lambda = 10$m.
The velocity of propagation is then

$$v = \nu\lambda = 10^7 \times 10 = 10^8 \text{m/sec.}$$

Velocity in free space = $3 \times 10^8$ m/sec.

Hence $\sqrt{\varepsilon_r \mu_r} = \frac{3 \times 10}{2 \times 10} = 3.$

as $\mu_r = 1, \varepsilon_r = 3 = 9.$

Now $\eta = \sqrt{\frac{\mu}{\varepsilon}} = \frac{120\pi}{3} = 40\pi$

As $\eta = \frac{E_x}{H_y}$

$$H_y = \frac{5}{40\pi} \cos\left(2\pi \times 10^7 t - 0.2\pi z\right) \text{A/m.}$$

Here $E_x$ and $H_y$ are the components of electric and magnetic fields, respectively.

## 7.11 Reflection and Transmission of the Plane Wave at Normal Incidence

In the previous section we discussed the propagation of uniform plane waves in unbounded, homogeneous medium. We now consider a monochromatic uniform plane wave travelling from one medium and entering another medium of infinite

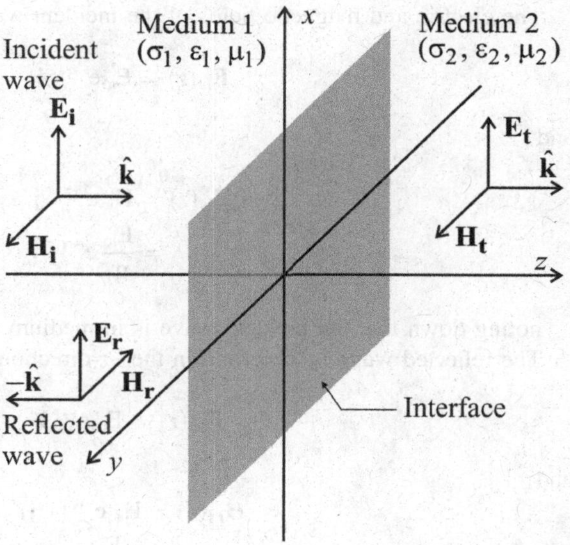


**Fig. 7.13** The figure shows medium 1 and medium 2. An electromagnetic wave is incident normal to the boundary between two mediums. Part of the incident wave is reflected and remaining part is transmitted

extent. Whenever an electromagnetic wave from one medium enters a different medium it is partly reflected and partly transmitted. The parameters ε, μ, σ determine the proportion of the wave reflected and transmitted.

In this section we will restrict our discussion for normal incidence of the plane wave. In later sections we shall discuss about oblique incidence of the plane wave.

Consider a linearly polarized uniform plane wave travelling in the *z*-direction. The wave is incident at the interface between two media, the interface being situated at $z = 0$. Medium 1 present at $z < 0$ is characterized by $\sigma_1$, $\varepsilon_1$, $\mu_1$ and medium 2 present at $z > 0$ is characterized by $\sigma_2$, $\varepsilon_2$, $\mu_2$ as shown in Fig. 7.13. As shown in the same figure incident wave ($\mathbf{E_i}$, $\mathbf{H_i}$) is travelling in the *z*-direction in medium 1. The transmitted wave ($\mathbf{E_t}$, $\mathbf{H_t}$) is travelling in the *z*-direction in medium 2. The reflected wave ($\mathbf{E_r}$, $\mathbf{H_r}$) is travelling in the $-z$-direction.

We define the reflection coefficient Γ as the ratio between the amplitude of the reflected wave $E_{ro}$ and amplitude of the incident wave $E_{io}$ at the interface $z = 0$

$$\Gamma = \frac{E_{ro}}{E_{io}} \tag{7.154}$$

The transmission coefficient T is defined as the ratio between the amplitude of the transmitted wave $E_{to}$ and amplitude of the incident wave $E_{io}$ at the interface $z = 0$.

$$T = \frac{E_{to}}{E_{io}} \tag{7.155}$$

The electric and magnetic fields of the incident wave in phasor form are

$$\mathbf{E_{is}}(z) = \mathrm{E_{io}} e^{-\gamma_1 z}\,\hat{\mathbf{i}} \tag{7.156}$$

and

$$\begin{aligned}\mathbf{H_{is}}(z) &= \mathrm{H_{io}} e^{-\gamma_1 z}\,\hat{\mathbf{j}}\\ &= \frac{\mathrm{E_{io}}}{\eta_1} e^{-\gamma_1 z}\,\hat{\mathbf{j}}\end{aligned} \tag{7.157}$$

noting down that the incident wave is in medium 1.

The reflected wave is travelling in the $-z$-direction in medium 1. Hence

$$\mathbf{E_{rs}}(z) = \mathrm{E_{ro}} e^{\gamma_1 z}\,\hat{\mathbf{i}} \tag{7.158}$$

and

$$\begin{aligned}\mathbf{H_{rs}}(z) &= \mathrm{H_{ro}} e^{\gamma_1 z}(-\hat{\mathbf{j}})\\ &= \frac{-\mathrm{E_{ro}}}{\eta_1} e^{\gamma_1 z}\,\hat{\mathbf{j}}\end{aligned} \tag{7.159}$$

because in Fig. 7.13 $\mathbf{H_r}$ is pointing in the $-y$-direction.

The transmitted wave is travelling in the $+z$-direction in medium 2. Thus

$$\mathbf{E_{ts}}(z) = \mathrm{E_{to}} e^{-\gamma_2 z}\,\hat{\mathbf{i}} \tag{7.160}$$

and

$$\begin{aligned}\mathbf{H_{ts}}(z) &= \mathrm{H_{to}} e^{-\gamma_2 z}\,\hat{\mathbf{j}}\\ &= \frac{\mathrm{E_{to}}}{\eta_2} e^{-\gamma_2 z}\,\hat{\mathbf{j}}\end{aligned} \tag{7.161}$$

In medium 1 we have both incident and reflected fields, whereas in medium 2 only transmitted field exists. Denoting the fields in medium 1 and medium 2 as $\mathrm{E_{1s}}$, $\mathrm{H_{1s}}$ and $\mathrm{E_{2s}}$, $\mathrm{H_{2s}}$, respectively, we have

$$\mathbf{E_{1s}} = \mathbf{E_{is}} + \mathbf{E_{rs}} \tag{7.162}$$

$$\mathbf{H_{1s}} = \mathbf{H_{is}} + \mathbf{H_{rs}} \tag{7.163}$$

$$\mathbf{E_{2s}} = \mathbf{E_{ts}} \tag{7.164}$$

$$\mathbf{H_{2s}} = \mathbf{H_{ts}} \tag{7.165}$$

Now the boundary conditions require that tangential components to be continuous at the interface [assuming no charges and currents at the interface (Eqs. 6.181, 6.183)

Thus

$$\mathbf{E}_{\mathbf{1tan}} = \mathbf{E}_{\mathbf{2tan}} \tag{7.166}$$

$$\mathbf{H}_{\mathbf{1tan}} = \mathbf{H}_{\mathbf{2tan}} \tag{7.167}$$

at the interface at $z = 0$.

As the electromagnetic waves are transverse both the electric and magnetic fields are tangential at the interface without any normal components. Hence substituting Eqs. 7.162–7.165 in Eqs. 7.166, 7.167

$$\mathbf{E}_{\mathbf{1s}}(0) = \mathbf{E}_{\mathbf{2s}}(0) \tag{7.168}$$

$$\mathbf{H}_{\mathbf{1s}}(0) = \mathbf{H}_{\mathbf{2s}}(0) \tag{7.169}$$

From Eq. 7.168

$$\mathbf{E}_{\mathbf{is}}(0) + \mathbf{E}_{\mathbf{rs}}(0) = \mathbf{E}_{\mathbf{ts}}(0) \tag{7.170}$$

Using Eqs. 7.156, 7.158, 7.160 then

$$\mathrm{E_{io}} + \mathrm{E_{ro}} = \mathrm{E_{to}} \tag{7.171}$$

From Eq. 7.169

$$\mathbf{H}_{\mathbf{is}}(0) + \mathbf{H}_{\mathbf{rs}}(0) = \mathbf{H}_{\mathbf{ts}}(0) \tag{7.172}$$

Using Eqs. 7.157, 7.159, 7.161 then

$$\frac{1}{\eta_1}(\mathrm{E_{io}} - \mathrm{E_{ro}}) = \frac{\mathrm{E_{to}}}{\eta_2} \tag{7.173}$$

where we have already noted that $\mathrm{E_{io}}$, $\mathrm{E_{ro}}$ and $\mathrm{E_{to}}$ are the amplitudes of the incident, reflected and transmitted waves at the interface $z = 0$.

Dividing Eq. 7.171 throughout by $\mathrm{E_{io}}$ and using Eqs. 7.154, 7.155

$$1 + \Gamma = \mathrm{T} \tag{7.174}$$

Dividing Eq. 7.173 throughout by $\mathrm{E_{io}}$ and using Eqs. 7.154, 7.155

$$1 - \Gamma = \frac{\eta_1}{\eta_2}\mathrm{T} \tag{7.175}$$

From Eqs. 7.174, 7.175

$$\Gamma = \frac{\eta_2 - \eta_1}{\eta_2 + \eta_1} \tag{7.176}$$

and

$$T = \frac{2\eta_2}{\eta_1 + \eta_2} \tag{7.177}$$

In general Γ and T are complex.

**Special Cases**

(a) **Perfect Dielectric–Conductor Boundary**

Let us consider that medium 1 is a perfect dielectric and medium 2 is a perfect conductor. Then for medium 1 $\sigma_1 = 0$ and for medium 2 $\sigma_2 = \infty$.

From Eq. 7.117

$$\eta_2 = \sqrt{\frac{\omega\mu_2}{\sigma_2}} = 0 \tag{7.178}$$

as $\sigma_2$ is infinity. Hence from Eqs. 7.176, 7.177

$$\left.\begin{aligned}\Gamma &= -1\\ T &= 0\end{aligned}\right\} \tag{7.179}$$

Thus the wave is totally reflected back into medium 1 without any transmission into medium 2. On physical grounds the same result should be expected. We know that the electric field inside the conductor is zero. Hence whatever energy is brought by the incident wave is simply reflected back. The amplitudes of the incident and reflected wave are thus equal and together form a standing wave pattern. The standing wave consists of two travelling waves (incident and reflected) which are travelling in opposite direction.

The electric field of the standing wave can be obtained from Eq. 7.162

$$\mathbf{E_{1s}} = \mathbf{E_{is}} + \mathbf{E_{rs}} \tag{7.180}$$

Substituting Eqs. 7.156, 7.158 in Eq. 7.180

$$\mathbf{E_{1s}} = [E_{io}e^{-\gamma_1 z} + E_{ro}e^{\gamma_1 z}]\hat{\mathbf{i}} \tag{7.181}$$

Substituting Eqs. 7.154, 7.179 in Eq. 7.181

$$\mathbf{E_{1s}} = E_{io}[e^{-\gamma_1 z} + \Gamma e^{\gamma_1 z}]\hat{\mathbf{i}} \tag{7.182}$$

$$\mathbf{E_{1s}} = E_{io}[e^{-\gamma_1 z} - e^{\gamma_1 z}]\hat{\mathbf{i}} \tag{7.183}$$

From Eqs. 7.88, 7.115a [$\alpha_1 = 0$]

$$\gamma_1 = \alpha_1 + i\beta_1 = i\beta_1 \tag{7.184}$$

Substituting Eq. 7.184 in 7.183

$$\mathbf{E_{1s}} = E_{io}\left[e^{-i\beta_1 z} - e^{i\beta_1 z}\right]\hat{\mathbf{i}} \tag{7.185}$$

$$\Rightarrow \mathbf{E_{1s}} = -2iE_{io}\, \sin \beta_1\, z\,\hat{\mathbf{i}} \tag{7.186}$$

Now inserting the time part

$$\mathbf{E_1} = \mathrm{Re}\left[\mathbf{E_{1s}} e^{i\omega t}\right] \tag{7.187}$$

Hence

$$\mathbf{E}_1 = 2\,E_{io}\sin \beta_1 z \sin \omega t\hat{\mathbf{i}} \tag{7.188}$$

By following similar procedure

$$\mathbf{H}_1 = \frac{2E_{io}}{\eta_1}\cos \beta_1 \cos \omega t\hat{\mathbf{j}} \tag{7.189}$$

In Fig. 7.14 we sketch the electric and magnetic fields given by Eqs. 7.188, 7.189. In Fig. 7.14a we plot the electric field along *x*-axis. In Fig. 7.14b we plot the magnetic field along *y*-axis. *x*-axis and *y*-axis are orthogonal to each other in Fig. 7.14. For the sake of explanation they are drawn in the same direction. Figure 7.14 shows that the fields oscillate with time and they do not travel. Thus **E** and **H** are mutually orthogonal to each other and represent standing waves.

(b) **Perfect Dielectric–Perfect Dielectric Boundary**

Let the medium 1 and medium 2 be perfect lossless dielectric. In that case $\sigma_1 = \sigma_2 = 0$. Hence from Eqs. 7.87, 7.114a

$$\alpha_1 = 0, \alpha_2 = 0, \gamma_1 = i\beta_1, \gamma_2 = i\beta_2 \tag{7.190}$$

Hence from Eq. 7.115b $\eta_1$ and $\eta_2$ are real and hence from Eqs. 7.176, 7.177 $\Gamma$ and T are real.

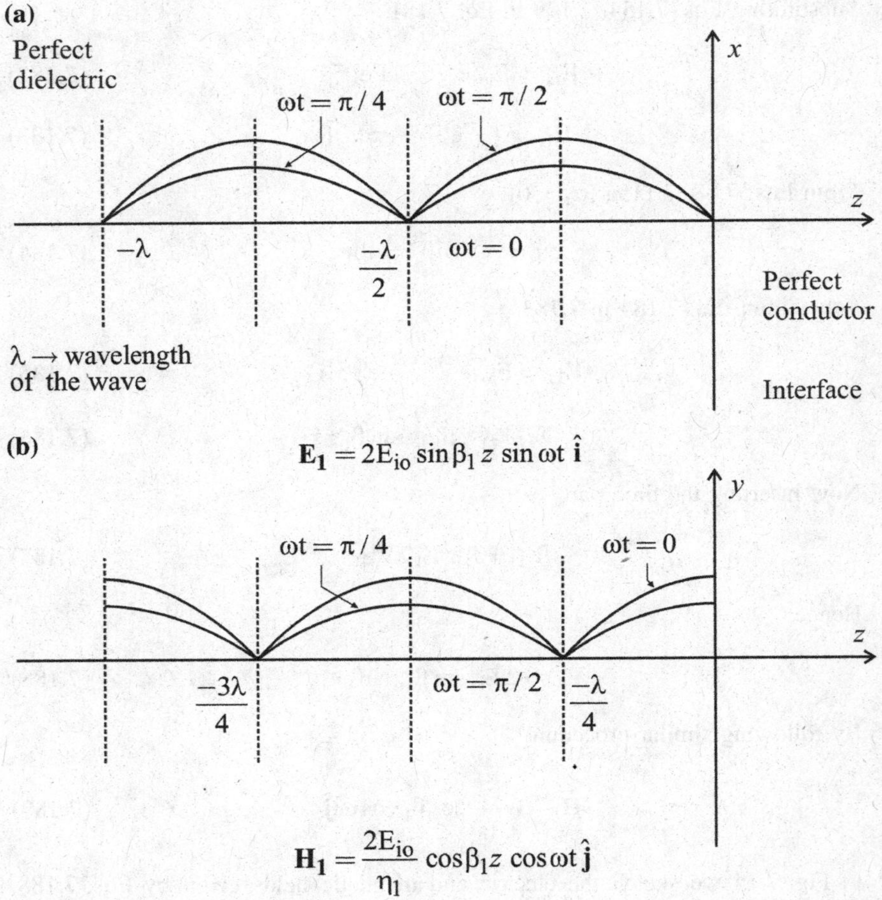


**Fig. 7.14** Standing waves pattern for (**a**) electric field and (**b**) magnetic fields for an electromagnetic wave reflected at the perfect dielectric - conductor boundary

Because both the mediums are perfect dielectrics part of the incident wave is transmitted into medium 2 and part of the wave is reflected back into medium 1. Hence whatever energy is brought by the incident wave is divided between the reflected wave and transmitted wave. Thus in this case the amplitudes of the incident and reflected waves are unequal. Energy division between the reflected and transmitted waves is determined by the nature of the dielectric mediums.

Let us calculate the electric field in the medium 1. From Eq. 7.162

$$\mathbf{E_{1s}} = \mathbf{E_{is}} + \mathbf{E_{rs}}$$

Using Eqs. 7.156, 7.158

$$\mathbf{E_{1s}} = E_{io}e^{-\gamma_1 z}\,\hat{\mathbf{i}} + E_{ro}e^{\gamma_1 z}\,\hat{\mathbf{i}} \tag{7.191}$$

Using Eq. 7.154

$$\mathbf{E_{1s}} = E_{io}\left[e^{-\gamma_1 z} + \Gamma e^{\gamma_1 z}\right]\hat{\mathbf{i}} \tag{7.192}$$

Using Eqs. 7.190

$$\mathbf{E_{1s}} = E_{io}\left[e^{-i\beta_1 z} + \Gamma e^{i\beta_1 z}\right]\hat{\mathbf{i}} \tag{7.193}$$

Using the Euler's formula $e^{i\theta} = \cos\theta + i\sin\theta$ in Eq. 7.194

$$\mathbf{E_{1s}} = E_{io}[(1+\Gamma)\cos\beta_1 z - i(1-\Gamma)\sin\beta_1 z]\hat{\mathbf{i}} \tag{7.194}$$

Considering the magnitude

$$|\mathbf{E_{1s}}| = E_{io}\sqrt{(1+\Gamma)^2\cos^2\beta_1 z + (1-\Gamma)^2\sin^2\beta_1 z} \tag{7.195}$$

Now let us consider two cases:
**Case (i)**:
Let $\eta_2 > \eta_1$, $\Gamma > 0$ then the maximum values of $|\mathbf{E_{1s}}|$ occur at

$$-\beta_1 z_{max} = n\pi \tag{7.196}$$

where n = 0, 1, 2, 3…

The negative sign indicates the wave is travelling in the *z*-direction. With the help of Eq. 7.24 we can write Eq. 7.196 as

$$z_{max} = \frac{-n\lambda_1}{2} \tag{7.197}$$

Minimum values of $|\mathbf{E_{1s}}|$ occur at

$$-\beta_1 z_{min} = (2n+1)\pi/2 \tag{7.198}$$

$$\Rightarrow z_{min} = \frac{-(2n+1)}{4}\lambda_1 \tag{7.199}$$

with n = 0, 1, 2, 3…

Hence from Eqs. 7.195

$$|\mathbf{E_{1s}}|_{max} = E_{io}(1 + \Gamma) \tag{7.200}$$

$$|\mathbf{E_{1s}}|_{min} = E_{io}(1 - \Gamma) \tag{7.201}$$

**Case (ii)**:

Let $\eta_2 < \eta_1$ then $\Gamma < 0$. Then using same procedure from Eqs. 7.196–7.201 we can show that

$$|\mathbf{E_{1s}}|_{max} = E_{io}(1 - \Gamma) \tag{7.202}$$

$$|\mathbf{E_{1s}}|_{min} = E_{io}(1 + \Gamma) \tag{7.203}$$

Hence in this case the locations of $|\mathbf{E_{1s}}|_{max}$ and $|\mathbf{E_{1s}}|_{min}$ are interchanged with respect to case (i) $-\Gamma > 0$.

The incident and reflected waves of unequal amplitudes form a standing wave in the medium 1. The ratio $|\mathbf{E_{1s}}|_{max}$ to $|\mathbf{E_{1s}}|_{min}$ is called standing wave ratio S. Hence

$$S = \frac{|\mathbf{E_{1s}}|_{max}}{|\mathbf{E_{1s}}|_{min}} = \frac{1 + |\Gamma|}{1 - |\Gamma|} \tag{7.204}$$

$$\Rightarrow |\Gamma| = \frac{S - 1}{S + 1} \tag{7.205}$$

Calculation of $\mathbf{H}_1$ and Poynting vector is given as a problem. See Problem 7.6.

## 7.12 Reflection and Transmission of the Plane Wave at Oblique Incidence

In the previous section we discussed the reflection of the plane wave at normal incidence. In this section we will discuss the reflection of the plane wave at oblique incidence. Oblique incidence refers to the case in which the electromagnetic wave is incident on a plane boundary at some arbitrary angle.

Consider two mediums separated by an interface at $z = 0$ as shown in Fig. 7.15. The plane containing the normal to the interface and direction of propagation of incident wave is called "Plane of incidence". The plane of incidence is shown in Fig. 7.15.

In general the electric field vector of the incident wave $\mathbf{E_i}$ can make an arbitrary angle with plane of incidence.

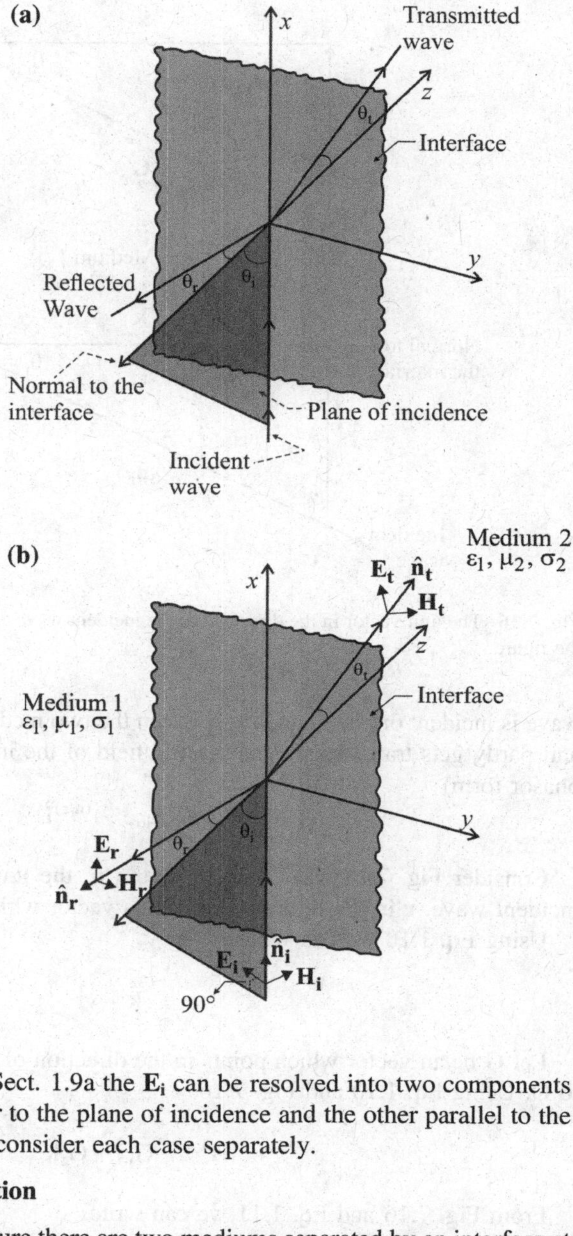


**Fig. 7.15** (**a**) Oblique incidence of an electromagnetic wave: An electromagnetic wave is on the boundary between two mediums. Part of the incident wave is reflected and remaining part is transmitted. (**b**) Perpendicular polarization: The electric and magnetic fields of the incident, reflected and transmitted wave is shown in the figure

However as discussed in Sect. 1.9a the $\mathbf{E_i}$ can be resolved into two components one component perpendicular to the plane of incidence and the other parallel to the plane of incidence. We will consider each case separately.

(a) **Perpendicular Polarization**

Consider Fig. 7.15. In this figure there are two mediums separated by an interface at $z = 0$. In Fig. 7.15b the electric field of the incident wave is perpendicular to the plane of incidence. The case in which $\mathbf{E_i}$ is perpendicular to the plane of incidence is called perpendicular polarization or s-polarized wave. Let the two media be linear, isotropic and homogeneous but finitely conducting media. The electromagnetic

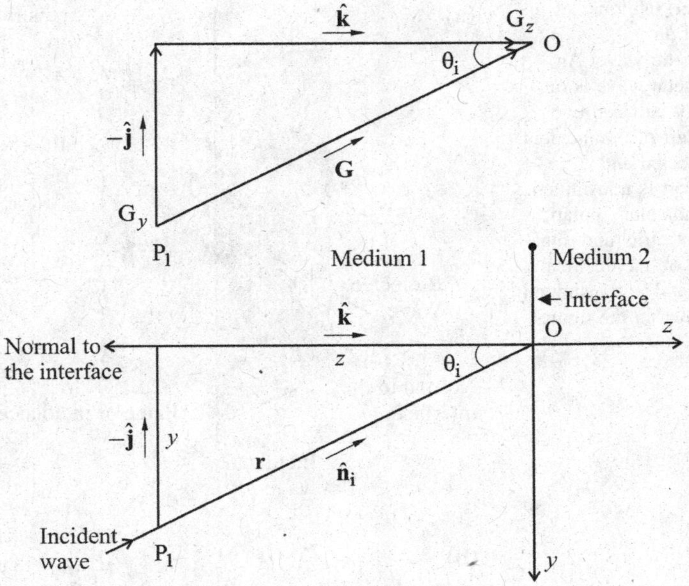


**Fig. 7.16** The unit vector in the direction of the incident wave and the radius vector **r** is shown in the figure

wave is incident on the boundary between the two medium and partly gets reflected and partly gets transmitted. The electric field of the incident wave is given by (in phasor form)

$$\mathbf{E}_{is} = E_{io}e^{-\gamma_1(\hat{\mathbf{n}}_i\mathbf{r})}\hat{\mathbf{i}} \tag{7.206}$$

Consider Fig. 7.16 where we show the $\hat{\mathbf{n}}_i$ the unit vector in the direction of incident wave. **r** in the figure is the radius vector which points from O to $P_1$.

Using Eq. 1.10 we can write

$$\mathbf{r} = z\hat{\mathbf{k}} + y\hat{\mathbf{j}} \tag{7.207}$$

Let **G** be an vector which points in the direction of $\hat{\mathbf{n}}_i$. That is **G** points from $P_1$ to O. Using Eq. 1.10 and Fig. 7.16

$$\mathbf{G} = -G_y\hat{\mathbf{j}} + G_z\hat{\mathbf{k}} \tag{7.208}$$

From Fig. 7.16 and Eq. 1.11 we can write

$$\cos\theta_i = \frac{G_z}{|\mathbf{G}|}, \sin\theta_i = \frac{G_y}{|\mathbf{G}|} \tag{7.209}$$

Substituting Eq. 7.209 in 7.208

$$\mathbf{G} = -|\mathbf{G}| \sin \theta_i \hat{\mathbf{j}} + |\mathbf{G}| \cos \theta_i \hat{\mathbf{k}} \quad (7.210)$$

$$\Rightarrow |\mathbf{G}|\hat{\mathbf{n}}_i = -|\mathbf{G}| \sin \theta_i \hat{\mathbf{j}} + |\mathbf{G}| \cos \theta_i \hat{\mathbf{k}} \quad (7.211)$$

$$\Rightarrow \hat{\mathbf{n}}_i = -\sin \theta_i \hat{\mathbf{j}} + \cos \theta_i \hat{\mathbf{k}} \quad (7.212)$$

Substituting Eqs. 7.207, 7.212 in Eq. 7.206

$$\mathbf{E}_{is} = E_{io} e^{-\gamma_1 (z \cos \theta_i - y \sin \theta_i)} \hat{\mathbf{i}} \quad (7.213)$$

From the Maxwell's equation

$$\nabla \times \mathbf{E}_s = -i\omega\mu_1 \mathbf{H}_s$$

We obtain the magnetic field as

$$\mathbf{H}_{is} = \frac{E_{io}}{\eta_1} \left[ \hat{\mathbf{k}} \sin \theta_i + \hat{\mathbf{j}} \cos \theta_i \right] e^{-\gamma_1 (z \cos \theta_i - y \sin \theta_i)} \quad (7.214)$$

where $\eta_1 = \frac{i\omega\mu_1}{\gamma_1}$ is the intrinsic impedance of the medium (See Eq. 7.100a).

Now consider the reflected wave. Let us assume the electric field is still polarized in the $x$-direction. Let $\hat{\mathbf{n}}_r$ be the unit vector in the direction of propagation of the reflected wave as shown in Fig. 7.17.

The electric field of the reflected wave is given by

$$\mathbf{E}_{rs} = E_{ro} e^{-\gamma_1 (\hat{\mathbf{n}}_r . \mathbf{r})} \hat{\mathbf{i}} \quad (7.215)$$

In Fig. 7.17 $\mathbf{r}$ is the radius vector pointing from O to $P_2$. Hence

$$\mathbf{r} = z\hat{\mathbf{k}} + y\hat{\mathbf{j}} \quad (7.216a)$$

Let $\mathbf{N}$ be a vector which points in the direction of $\hat{\mathbf{n}}_r$. That is $\mathbf{N}$ points from O to $P_2$. Hence from Fig. 7.17

$$\mathbf{N} = -N_y \hat{\mathbf{j}} + N_z \hat{\mathbf{k}} \quad (7.216b)$$

From Fig. 7.17 and Eq. 1.11 we can write

$$\cos \theta_r = \frac{N_z}{|\mathbf{N}|}, \sin \theta_r = \frac{N_y}{|\mathbf{N}|} \quad (7.217)$$

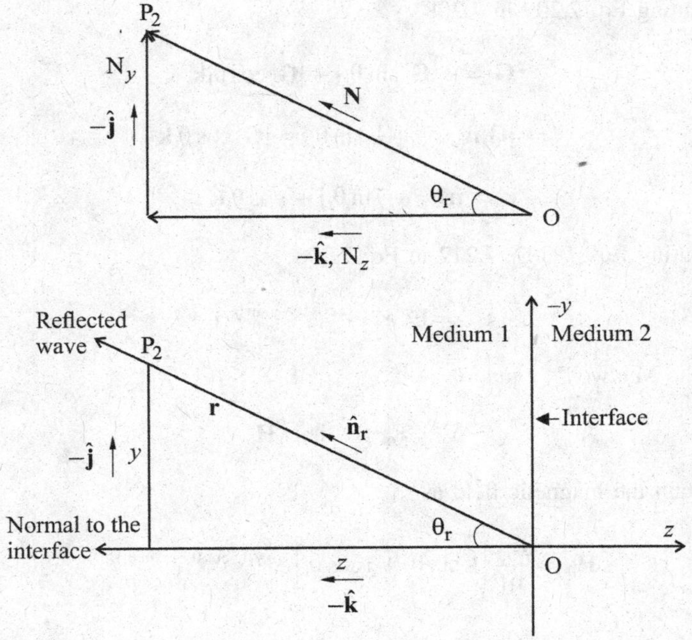


**Fig. 7.17** The unit vector in the direction of the reflected wave and the radius vector **r** is shown in the figure

Substituting Eq. 7.217 in Eq. 7.216a we obtain

$$\mathbf{N} = -|\mathbf{N}| \sin\theta_r \hat{\mathbf{j}} - |\mathbf{N}| \cos\theta_r \hat{\mathbf{k}} \tag{7.218}$$

$$\Rightarrow |\mathbf{N}|\hat{\mathbf{n}}_r = -|\mathbf{N}| \sin\theta_r \hat{\mathbf{j}} - |\mathbf{N}| \cos\theta_r \hat{\mathbf{k}} \tag{7.219}$$

$$\Rightarrow \hat{\mathbf{n}}_r = -\sin\theta_r \hat{\mathbf{j}} - \cos\theta_r \hat{\mathbf{k}} \tag{7.220}$$

Substituting Eqs. 7.220, 7.216a in Eq. 7.215

$$\mathbf{E}_{rs} = \mathrm{E}_{ro} e^{-\gamma_1(-y\sin\theta_r - z\cos\theta_r)} \hat{\mathbf{i}} \tag{7.221}$$

From the Maxwell's equation (Eq. 6.232)

$$\nabla \times \mathbf{E}_s = -i\omega\mu_1 \mathbf{H}_s$$

We obtain the magnetic field as

$$\mathbf{H}_{rs} = \frac{\mathrm{E}_{ro}}{\eta_1}\left[\sin\theta_r \hat{\mathbf{k}} - \cos\theta_r \hat{\mathbf{j}}\right] e^{-\gamma_1(-y\sin\theta_r - z\cos\theta_r)} \tag{7.222}$$

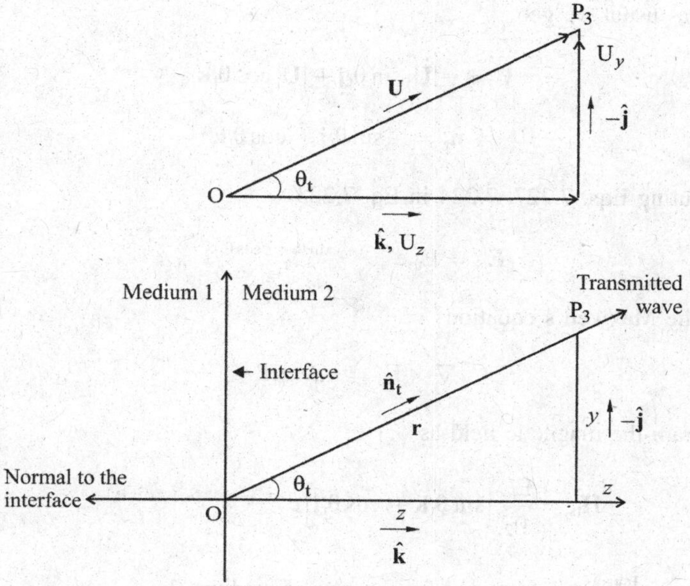


**Fig. 7.18** The unit vector in the direction of the transmitted wave and the radius vector **r** is shown in the figure

Let us now consider the transmitted wave. Let us assume that the electric field of the transmitted wave is polarized in the $x$-direction.

Let $\hat{\mathbf{n}}_t$ be the unit vector in the direction of propagation of the transmitted wave as shown in Fig. 7.18. The electric field of the transmitted wave is given by

$$\mathbf{E}_{ts} = E_{to}e^{-\gamma_2(\hat{\mathbf{n}}_t.\mathbf{r})}\hat{\mathbf{i}} \tag{7.223}$$

In Fig. 7.18 the radius vector **r** points from O to $P_3$. Hence

$$\mathbf{r} = z\hat{\mathbf{k}} + y\hat{\mathbf{j}} \tag{7.224}$$

Let **U** be an vector which points in the direction of $\hat{\mathbf{n}}_t$. That is **U** points from O to $P_3$ in Fig. 7.18 Hence

$$\mathbf{U} = -U_y\hat{\mathbf{j}} + U_z\hat{\mathbf{k}} \tag{7.225}$$

From Fig. 7.18 and Eq. 1.11 we can write

$$\cos\theta_t = \frac{U_z}{|\mathbf{U}|},\ \sin\theta_t = \frac{U_y}{|\mathbf{U}|} \tag{7.226}$$

Hence as usual we get

$$\mathbf{U} = -|\mathbf{U}| \sin\theta_t \hat{\mathbf{j}} + |\mathbf{U}| \cos\theta_t \hat{\mathbf{k}} \tag{7.227}$$

$$\Rightarrow \hat{\mathbf{n}}_t = -\sin\theta_t \hat{\mathbf{j}} + \cos\theta_t \hat{\mathbf{k}} \tag{7.228}$$

Substituting Eqs. 7.227, 7.224 in Eq. 7.223

$$\mathbf{E}_{ts} = \mathrm{E}_{to} e^{-\gamma_2(-y\sin\theta_t + z\cos\theta_t)} \hat{\mathbf{i}} \tag{7.229}$$

From the Maxwell's equation

$$\nabla \times \mathbf{E}_s = -i\omega\mu \mathbf{H}_s$$

We obtain the magnetic field as

$$\mathbf{H}_{ts} = \frac{\mathrm{E}_{to}}{\eta_2}\left[\sin\theta_t \hat{\mathbf{k}} + \cos\theta_t \hat{\mathbf{j}}\right] e^{-\gamma_2(-y\sin\theta_t + z\cos\theta_t)} \tag{7.230}$$

Here $\eta_2 = \frac{i\omega\mu_2}{\gamma_2}$ is the intrinsic impedance of medium 2.

At the boundary $z = 0$ the boundary condition is

$$\mathbf{E}_{1\,\tan} = \mathbf{E}_{2\,\tan} \tag{7.231}$$

where $\mathbf{E}_{1\,\tan}, \mathbf{E}_{2\,\tan}$ are the tangential components of the electric fields in medium 1, medium 2, respectively. Noting down that in medium 1 there are incident and reflected waves and in medium 2 only transmitted wave using Eqs. 7.213, 7.215, 7.221 in 7.230

$$\mathrm{E}_{io} e^{\gamma_1(y\sin\theta_i)} + \mathrm{E}_{ro} e^{\gamma_1(y\sin\theta_r)} = \mathrm{E}_{to} e^{\gamma_2(y\sin\theta_t)} \tag{7.232}$$

Using Eqs. 7.154, 7.155 in Eq. 7.239

$$e^{\gamma_1(y\sin\theta_i)} + \Gamma e^{\gamma_1(y\sin\theta_r)} = \mathrm{T} e^{\gamma_2(y\sin\theta_t)} \tag{7.233}$$

The above equation must be valid for all values of $y$ that is every where on the interface. At $y = 0$ Eq. 7.232 becomes

$$1 + \Gamma = \mathrm{T} \tag{7.234}$$

For Eqs. 7.233, 7.231 to hold for all values of $y$

$$\gamma_1 \sin\theta_i = \gamma_1 \sin\theta_r = \gamma_2 \sin\theta_t \tag{7.235}$$

In Eq. 7.234 from the first equality

$$\theta_i = \theta_r \tag{7.236}$$

Hence angle of incidence is equal to angle of reflection. This is the well known Snell's law of reflection.

Also from Eq. 7.234

$$\gamma_1 \sin\theta_i = \gamma_2 \sin\theta_t \tag{7.237}$$

which is well known Snell's law of refraction.

Assuming that there are no currents at the boundary at $z = 0$ the boundary condition for **H** is

$$\mathbf{H}_{1\,\tan} = \mathbf{H}_{2\,\tan} \tag{7.238}$$

Noting down that the $y$ components are the only tangential components for **H** in Eqs. 7.214, 7.222, 7.230 we can then write Eq. 7.238 as

$$\frac{[\cos\theta_i]}{\eta_1} e^{\gamma_1 y \sin\theta_i} - \frac{\Gamma[\cos\theta_r]}{\eta_1} e^{\gamma_1 y \sin\theta_r} = \frac{T[\cos\theta_t]}{\eta_2} e^{\gamma_2 y \sin\theta_t} \tag{7.239}$$

where we have used Eqs. 7.154, 7.155 in Eq. 7.239.

Equation 7.239 can be simplified with Eq. 7.235 as

$$\frac{\cos\theta_i}{\eta_1} - \frac{\Gamma\cos\theta_r}{\eta_1} = \frac{T}{\eta_2}\frac{\cos\theta_t}{\cos\theta_i} \tag{7.240}$$

With $\theta_i = \theta_r$ we can write the above equation as

$$1 - \Gamma = \frac{\eta_1}{\eta_2}\frac{\cos\theta_t}{\cos\theta_i} T \tag{7.241}$$

From Eqs. 7.234, 7.241 we get

$$\Gamma_\perp = \frac{\eta_2\cos\theta_i - \eta_1\cos\theta_t}{\eta_2\cos\theta_i + \eta_1\cos\theta_t} \tag{7.242}$$

$$T_\perp = \frac{2\eta_2\cos\theta_i}{\eta_2\cos\theta_i + \eta_1\cos\theta_t} \tag{7.243}$$

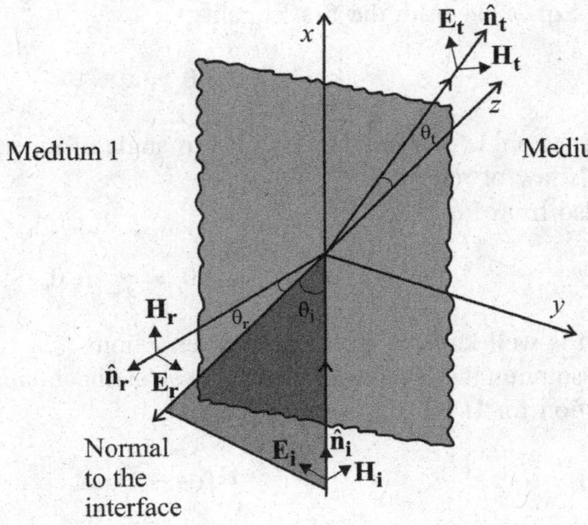


**Fig. 7.19** Parallel polarization: The electric and magnetic fields of the incident, reflected and transmitted wave is shown in the figure.

where $\perp$ in the above equations denote that the reflection and transmission coefficients corresponding to perpendicular polarizations.

(b) **Parallel Polarization**

Let us now consider the case in which the electric field is parallel to the plane of incidence as shown in Fig. 7.19. When **E** is parallel to the plane of incidence the situation is known as parallel polarization or p-polarized.

The magnetic field $\mathbf{H}_{is}$ of the incident wave is given by (in phasor form).

$$\mathbf{H}_{is} = H_{io} e^{-\gamma_1 (\hat{\mathbf{n}}_i . \mathbf{r})} \hat{\mathbf{i}} \tag{7.244}$$

Following the similar procedures as from Eqs. 7.206–7.213 we get

$$\mathbf{H}_{is} = H_{io} e^{-\gamma_1 (z \cos\theta_i - y \sin\theta_i)} \hat{\mathbf{i}} \tag{7.245}$$

Now consider the Maxwell's equation (in current-free region (Eq. 6.233)

$$\nabla \times \mathbf{H}_i = i\omega\varepsilon_c \mathbf{E}_i$$

where $\varepsilon_c$ the complex permittivity of the dielectric is given by Eq. 7.105. From the above Maxwell's equation

$$\mathbf{E}_{is} = \frac{-\gamma_1}{i\omega\varepsilon_{c1}} (H_{io}) \left[\cos\theta_i \hat{\mathbf{j}} + \sin\theta_i \hat{\mathbf{k}}\right] e^{-\gamma_1 [z\cos\theta_i - y \sin\theta_i]} \tag{7.246}$$

From Eqs. 7.85, 7.106 we get

$$\frac{\gamma_1}{i\omega\varepsilon_{c1}} = \frac{\sqrt{i\omega\mu_1(\sigma_1 + i\omega\varepsilon_1)}}{(i\omega)\left(\frac{\sigma_1 + i\omega\varepsilon_1}{i\omega}\right)}$$
$$\frac{\gamma_1}{i\omega\varepsilon_{c1}} = \sqrt{\frac{i\omega\mu_1}{\sigma_1 + i\omega\varepsilon_1}} = \eta_1 \tag{7.247}$$

from Eq. 7.101a.

Substituting Eq. 7.247 in 7.246

$$\mathbf{E}_{is} = -H_{io}\eta_1\left[\cos\theta_i\hat{\mathbf{j}} + \sin\theta_i\hat{\mathbf{k}}\right]e^{-\gamma_1[z\cos\theta_i - y\sin\theta_i]} \tag{7.248}$$

In the similar manner we obtain the fields associated with the reflected and transmitted waves as

$$\mathbf{H}_{rs} = H_{ro}e^{-\gamma_1(-y\sin\theta_r - z\cos\theta_r)}\hat{\mathbf{i}} \tag{7.249}$$

$$\mathbf{E}_{rs} = -H_{ro}\eta_1\left[-\cos\theta_r\hat{\mathbf{j}} + \sin\theta_r\hat{\mathbf{k}}\right]e^{-\gamma_1(-y\sin\theta_r - z\cos\theta_r)} \tag{7.250}$$

$$\mathbf{H}_{ts} = H_{to}e^{-\gamma_2(-y\sin\theta_t + z\cos\theta_t)}\hat{\mathbf{i}} \tag{7.251}$$

$$\mathbf{E}_{ts} = -(H_{to})\eta_2\left[\cos\theta_t\hat{\mathbf{j}} + \sin\theta_t\hat{\mathbf{k}}\right]e^{-\gamma_2(-y\sin\theta_t + z\cos\theta_t)} \tag{7.252}$$

Let

$$\left.\begin{aligned} E_{io} &= H_{io}\eta_1, E_{ro} = H_{ro}\eta_1 \\ E_{to} &= H_{to}\eta_2 \end{aligned}\right\} \tag{7.253}$$

and using Eqs. 7.154, 7.155 we can write

$$\mathbf{H}_{is} = \frac{E_{io}}{\eta_1}e^{-\gamma_1(z\cos\theta_i - y\sin\theta_i)}\hat{\mathbf{i}} \tag{7.254}$$

$$\mathbf{E}_{is} = -E_{io}\left[\cos\theta_i\hat{\mathbf{j}} + \sin\theta_i\hat{\mathbf{k}}\right]e^{-\gamma_1(z\cos\theta_i - y\sin\theta_i)} \tag{7.255}$$

$$\mathbf{H}_{rs} = \frac{E_{io}\Gamma}{\eta_1}e^{-\gamma_1(-y\sin\theta_r - z\cos\theta_r)}\hat{\mathbf{i}} \tag{7.256}$$

$$\mathbf{E}_{rs} = -\Gamma E_{io}\left[-\cos\theta_r\hat{\mathbf{j}} + \sin\theta_r\hat{\mathbf{k}}\right]e^{-\gamma_1(-y\sin\theta_r - z\cos\theta_r)} \tag{7.257}$$

$$\mathbf{H}_{ts} = \frac{E_{io}T}{\eta_2} e^{-\gamma_2(-y\sin\theta_t + z\cos\theta_t)} \tag{7.258}$$

$$\mathbf{E}_{ts} = -E_{io}T\left[\cos\theta_t\hat{\mathbf{j}} + \sin\theta_t\hat{\mathbf{k}}\right] e^{-\gamma_2(-y\sin\theta_t + z\cos\theta_t)} \tag{7.259}$$

Now using Eqs. 7.254, 7.256, 7.258 in the boundary condition for **H** at $z = 0$, i.e. Eq. 7.238

$$e^{\gamma_1(y\sin\theta_i)} + \Gamma e^{\gamma_1(y\sin\theta_r)} = \frac{\eta_1 T}{\eta_2} e^{\gamma_2(y\sin\theta_t)} \tag{7.260}$$

The above equation must be valid for all values of $y$. Hence for $y = 0$

$$1 + \Gamma = \frac{\eta_1}{\eta_2} T \tag{7.261}$$

For Eq. 7.260 to be valid for all values of $y$

$$\gamma_1 \sin\theta_i = \gamma_2 \sin\theta_r = \gamma_2 \sin\theta_t \tag{7.262}$$

which simply reproduces Snell's law of reflection and refraction.

Noting down that the $y$ components are the only tangential components for **E** in Eqs. 7.255, 7.257, 7.259 applying the boundary condition 7.231 at the boundary at $z = 0$ and using Eq. 7.262

$$1 - \Gamma = T\frac{\cos\theta_t}{\cos\theta_i} \tag{7.263}$$

From Eqs. 7.261, 7.263 we get

$$\Gamma_{\parallel} = \frac{\eta_1\cos\theta_i - \eta_2\cos\theta_t}{\eta_1\cos\theta_i + \eta_2\cos\theta_t} \tag{7.264}$$

$$T_{\parallel} = \frac{2\eta_2\cos\theta_i}{\eta_1\cos\theta_i + \eta_2\cos\theta_t} \tag{7.265}$$

where $\parallel$ in the above equations denote that the reflection and transmission coefficients corresponding to parallel polarization.

## 7.13 Total Reflection and Total Transmission

Let us consider that medium 2 is a conductor. In that case $\eta_2 = \frac{\mathbf{E}_2}{\mathbf{H}_2} = 0$ as there are no fields inside the conductor. Then from Eqs. 7.242, 7.264 $\Gamma_\perp = -1, \Gamma_\parallel = 1$. Hence all the wave is reflected without any transmission. Irrespective of the incident angle or polarization total reflection occurs.

We define the refractive index or index of refraction as

$$n = \sqrt{\varepsilon_r \mu_r} \tag{7.266}$$

Consider the case of dielectric–dielectric interface with $\sigma_1 = 0, \sigma_2 = 0$ and $\mu_1 \simeq \mu_2 \simeq \mu_o$.

In that case

$$\frac{\gamma_1}{\gamma_2} = \frac{i\beta_1}{i\beta_2} = \sqrt{\frac{\varepsilon_{r1}}{\varepsilon_{r2}}} = \frac{n_1}{n_2} \tag{7.267}$$

where $n_1$, $n_2$ are the refractive index of medium 1, medium 2, respectively.

From Eqs. 7.266, 7.237

$$\sin\theta_t = \frac{n_1}{n_2}\sin\theta_i \tag{7.268}$$

Now

$$\cos\theta_t = \left[1 - \sin^2\theta_t\right]^{1/2} = \left[1 - \frac{\sin^2\theta_i}{\left(\frac{n_2}{n_1}\right)^2}\right] \tag{7.269}$$

If $\sin\theta_i = \frac{n_2}{n_1}$ in that case we observe that $\cos\theta_i = 0$. Hence $\Gamma_\perp = 1, \Gamma_\parallel = 0$ meaning that the incident wave suffers total reflection.

Another case is when $\sin\theta_i > \frac{n_2}{n_1}$ and hence $\cos\theta_t$ is imaginary. Consider Eq. 7.242

$$\Gamma_\perp = \frac{a_2 - i|a_1|}{a_2 + i|a_1|} \tag{7.270}$$

where

$$a_2 = \eta_2 \cos\theta_i \tag{7.271}$$

$$a_1 = i\eta_1 \cos\theta_t \tag{7.272}$$

$\Rightarrow \left|\Gamma_{\perp}^{2}\right| = 1$. The same is true for $\Gamma_{\parallel}$ as the reader may observe from Eq. 7.264. This means that the wave suffers total power reflection whenever $\sin\theta_i > n_2/n_1$.

From the above discussion we can write the incident wave suffers total reflection when

$$\sin\theta_i \geq \frac{n_2}{n_1} \tag{7.273}$$

We define the critical angle for total reflection as

$$\theta_i \geq \theta_c \tag{7.274}$$

where $\theta_c$ is given by

$$\sin\theta_c = \frac{n_2}{n_1} \tag{7.275}$$

For the case of total transmission, reflection coefficient must be zero. That is $\Gamma = 0$. Hence from Eq. 7.264 for $\Gamma_{\parallel}$ to be zero

$$\eta_1\cos\theta_i = \eta_2\cos\theta_t \tag{7.276}$$

The angle $\theta_i = \theta_{B\parallel}$ when the value of $\Gamma_{\parallel}$ goes to zero is called Brewster angle. At this angle the parallel component of the field of wave suffers total transmission.

Using Eq. 7.267 and with $\mu_1 = \mu_2$ Eq. 7.276 can be written as

$$\sin^2\theta_{B\parallel} = \frac{1 - \left(\frac{\varepsilon_1}{\varepsilon_2}\right)}{1 - \left(\frac{\varepsilon_1}{\varepsilon_2}\right)^2} \tag{7.277}$$

$$\sin\theta_{B\parallel} = \frac{1}{\sqrt{1 + \left(\frac{\varepsilon_1}{\varepsilon_2}\right)}} \tag{7.278}$$

For the case of perpendicular polarization for total reflection to be zero that is $\Gamma_{\perp} = 0$ from Eq. 7.242

$$\eta_2\cos\theta_i = \eta_1\cos\theta_t \tag{7.279}$$

The corresponding angle $\theta_i = \theta_{B\perp}$ is the Brewster angle for no reflection for the case of perpendicular polarization.

From Eqs. 7.279, 7.276 we get

$$\sin^2 \theta_{B\perp} = \frac{1 - \mu_1\varepsilon_2/\mu_2\varepsilon_1}{1 - \left(\frac{\mu_1}{\mu_2}\right)^2} \tag{7.280}$$

Brewster's angle given by Eqs. 7.277, 7.280 for the case parallel and perpendicular polarizations are completely different. Hence if an unpolarized wave is incident on a media it is possible to separate parallel and perpendicular polarizations from the unpolarized wave. Suppose assume that the light incident on a medium satisfies Eq. 7.277 then the perpendicular polarization component will be reflected. The above facts have applications in reducing glare, controlling polarization of laser light, etc.

Thus still now we have discussed perpendicular and parallel polarizations, where **E** is perpendicular and parallel to the plane of incidence as shown in Fig. 7.15b and 7.19, respectively, thereby deriving their reflection and transmission coefficients.

Now consider Fig. 7.20 where **E** is neither parallel nor perpendicular to the plane of incidence. In that case we can resolve **E** into two components, $\mathbf{E}_\perp$ the perpendicular component and $\mathbf{E}_\perp$ the parallel component, the reflection and transmission coefficients are given by Eqs. 7.242, 7.243, respectively, and that of $\mathbf{E}_\parallel$ is given by Eqs. 7.264, 7.265.

The results for **E** in Fig. 7.20 can then be obtained by superposing the result of $\mathbf{E}_\perp$ and $\mathbf{E}_\parallel$.

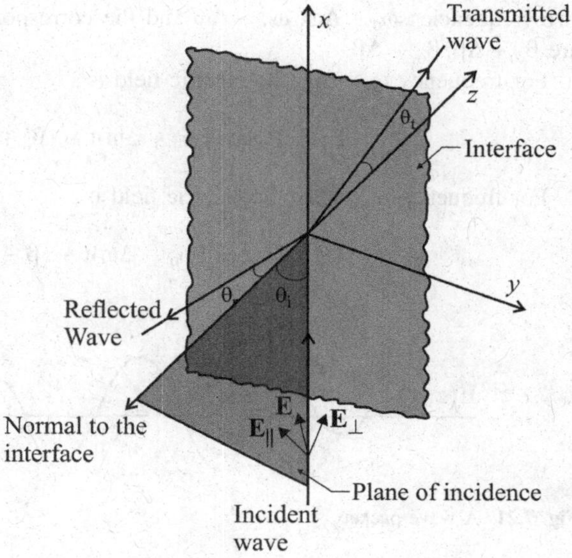


**Fig. 7.20** Oblique incidence of an electromagnetic wave: The electric field is neither parallel nor perpendicular to the plane of incidence. In that case both reflection and transmission coefficients can be obtained by resolution of electric field

## 7.14 Dispersion—Group Velocity

In the previous sections we discussed the propagation of electromagnetic waves in conductors, dielectrics, reflection and transmission of these waves at the boundaries. The reader would have observed that depending upon σ, ε, μ the electromagnetic waves get reflected, transmitted, attenuated, etc., in the medium. However to some extent the parameters σ, ε, μ depend on the frequency of the incident wave. Hence depending upon the frequency of the wave the wave behaviour inside the medium gets modified. We know that the velocity of the wave is given by

$$v = \frac{\omega}{\beta} = \frac{1}{\sqrt{\varepsilon\mu}}$$

v is called phase velocity of the wave. Because now ε, μ depend on ω, waves of different frequency travel with different phase velocities in a medium. This phenomenon is known as dispersion.

A given carrier wave will have its main frequency and will be associated with a minor spread in frequencies. Therefore the electromagnetic wave has a group of frequencies and forms a wave packet. The envelope of the wave packet travels with a single velocity known as group velocity as shown in Fig. 7.21.

Let the electric field of the electromagnetic wave be given by

$$E(z,t) = E_o \cos(\omega t - \beta z) \tag{7.281}$$

Consider the above wave to be wave packet having equal amplitudes but with two frequencies $\omega_o + \Delta\omega$, $\omega_o - \Delta\omega$ and the corresponding propagation constants are $\beta_o + \Delta\beta, \beta_o - \Delta\beta$.

For frequency $\omega_o + \Delta\omega$ the electric field is

$$E_1 = E_o \cos[(\omega_o + \Delta\omega)t - (\beta_o + \Delta\beta)z] \tag{7.282}$$

For frequency $\omega_o - \Delta\omega$ the electric field is

$$E_2 = E_o \cos[(\omega_o - \Delta\omega)t - (\beta - \Delta\beta)z] \tag{7.283}$$

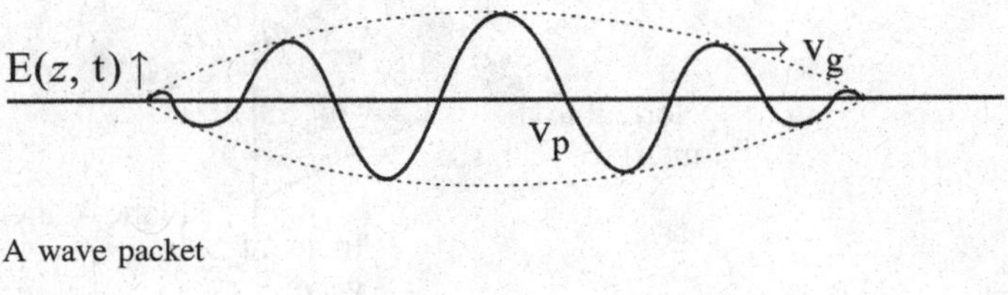


**Fig. 7.21** A wave packet

The net electric field is

$$\begin{aligned} E &= E_1 + E_2 \\ &= 2E_o \cos(\omega_o t - \beta z)\cos(\Delta\omega t - \Delta\beta z) \end{aligned} \quad (7.284)$$

The phase velocity $v_p$ shown in Fig. 7.21 is the carrier velocity. That is velocity of the carrier wave inside the envelope and is given by

$$v = \frac{dz}{dt} = \frac{\omega_o}{\beta_o} \quad (7.285)$$

The group velocity is the velocity of envelope of the wave packet as shown in Fig. 7.21 and is given by

$$v_g = \frac{dz}{dt} = \frac{\Delta\omega}{\Delta\beta} \quad (7.286)$$

**Example 7.4**

Consider a electromagnetic wave travelling in the $z$-direction with peak amplitude of $E_{io} = 700\text{V/m}$.

The wave is incident normally form medium 1 to medium 2 the interface being situated at $z = 0$.

Given

$$\begin{aligned} &\varepsilon_1 = 12 \times 10^{-12}\text{F/m}, \mu_1 = 2.5 \times 10^{-06}\text{H/m}, \sigma_1 = 0. \\ &\varepsilon_2 = 7 \times 10^{-12}\text{F/m}, \mu_2 = 5 \times 10^{-6}\text{H/m}, \sigma_2 = 0, \omega = 10^8\text{rad/s}. \end{aligned}$$

Find $\mathbf{E}_{is}, \mathbf{E}_{rs}$

**Solution**

Let us calculate the intrinsic impedance of the medium

$$\eta_1 = \sqrt{\frac{\mu_1}{\varepsilon_1}} = \sqrt{\frac{2.5 \times 10^{-6}}{12 \times 10^{-12}}} = 456.4$$

$$\eta_2 = \sqrt{\frac{\mu_2}{\varepsilon_2}} = \sqrt{\frac{5 \times 10^{-6}}{7 \times 10^{-12}}} = 845.2$$

Now

$$
\begin{aligned}
v_1 &= \frac{1}{\sqrt{\varepsilon_1 \mu_1}} \\
&= \frac{1}{\sqrt{12 \times 10^{-12} \times 2.5 \times 10^{-6}}} \\
v_1 &= 1.8 \times 10^8 \mathrm{m/s}.
\end{aligned}
$$

Hence

$$
\begin{aligned}
\beta_1 &= \frac{\omega_1}{v_1} = \frac{10^8}{1.8 \times 10^8} \\
\beta_1 &= 0.6
\end{aligned}
$$

Hence

$$
\begin{aligned}
\mathbf{E}_{is} &= E_{io} \cos(\omega t - \beta_1 z)\hat{\mathbf{i}} \\
&= 700 \cos\left(10^8 t - 0.6z\right)\hat{\mathbf{i}} \mathrm{V/m}
\end{aligned}
$$

Assuming the electric field is in the *x*-direction

Hence $\Gamma = \dfrac{E_{ro}}{E_{io}} = \dfrac{\eta_2 - \eta_1}{\eta_2 + \eta_1}$

$$
\Gamma = \frac{845.2 - 456.4}{845.2 + 456.4} = 0.3
$$

Hence

$$
E_{ro} = \Gamma E_{io} = 210 \mathrm{V/m}.
$$

$$
\mathbf{E}_{rs} = 210 \cos\left(10^8 t - 0.6z\right)\hat{\mathbf{i}} \mathrm{V/m}.
$$

**Example 7.5**

Consider a electromagnetic wave travelling in a medium. The electric field of the wave is given by

$$
\mathbf{E} = 0.7 e^{-z/2} \cos\left(10^8 t - \beta z\right)\hat{\mathbf{i}} \mathrm{V/m}.
$$

Determine

(a) $\beta$

(b) the loss tangent.

Given $\varepsilon_r = 9, \mu_r = 2.5$.

**Solution**
Because

$$\mathbf{E} = E_o e^{-\alpha z} \cos(\omega t - \beta z)\hat{\mathbf{i}}$$

we get

$$\alpha = \frac{1}{2}, \quad \omega = 10^8$$

(a) The velocity of the wave is

$$v = \frac{1}{\sqrt{\mu\varepsilon}} = \frac{1}{\sqrt{\mu_o \mu_r \varepsilon_o \varepsilon_r}} = 0.63 \times 10^8 \text{m/s}$$

$$\beta = \frac{\omega}{v} = \frac{10^8}{0.63 \times 10^8}$$
$$= 1.59 \text{rad/m}$$

(b)
$$\alpha^2 = \omega^2 \left\{ \frac{\mu\varepsilon}{2} \left[ \sqrt{\left(\frac{\sigma}{\omega\varepsilon}\right)^2 + 1} - 1 \right] \right\}$$

$$\Rightarrow \left(\frac{1}{2}\right)^2 = (10^8)^2 \left\{ \frac{1}{2 \times (0.63 \times 10^8)^2} \left[ \sqrt{\left(\frac{\sigma}{\omega\varepsilon}\right)^2 + 1} - 1 \right] \right\}$$

$$\Rightarrow \frac{\sigma}{\omega\varepsilon} = 0.66 \text{ is the loss tangent.}$$

**Example 7.6**
For a medium $\varepsilon_r = 52, \mu_r = 1, \sigma = 25\mho/\text{m}$.

Calculate the propagation constant, attenuation constant, phase constant for the medium at $\omega = 10^{11}$ rad/s. The medium is a lossy dielectric.

**Solution**

$$\frac{\sigma}{\omega\varepsilon} = \frac{25}{10^{11} \times 52 \times 8.85 \times 10^{-12}}$$
$$= 0.5432.$$

Now

$$\gamma^2 = i\omega\mu(\sigma + i\omega\varepsilon)$$
$$\Rightarrow \gamma = i\omega\sqrt{\mu\varepsilon}\sqrt{\left(1 - i\frac{\sigma}{\omega\varepsilon}\right)}$$

Now

$$\begin{aligned}\sqrt{\mu\varepsilon} &= \sqrt{\mu_r\varepsilon_r\mu_o\varepsilon_o}\\ &= \sqrt{1 \times 52 \times 4\pi \times 10^{-7} \times 8.85 \times 10^{-12}}\\ &= 24.042 \times 10^{-9}.\end{aligned}$$

Hence

$$\omega\sqrt{\mu\varepsilon} = 2404.2$$

Thus

$$\gamma = i\,2404.2\sqrt{1 - i\,0.5432}$$

Now

$$\begin{aligned}&1 - i\,0.5432 = re^{i\theta} = 1.138e^{-i\,28.51}\\ &\Rightarrow (1 - i0.5432)^{1/2} = (1.138)^{1/2}e^{-i\frac{28.51}{2}}\\ &= 1.0339 - i\,0.262\end{aligned}$$

Hence

$$\gamma = i\,2404.2(1.0339 - i\,0.262)$$
$$\Rightarrow \gamma = 629.9 + i\,2485.7 = \alpha + i\beta$$

is the propagation constant.

Thus

$$\alpha = \text{attenuation constant} = 626.9$$
$$\beta = \text{phase constant} = 2485.7$$

**Example 7.7**
In Example 7.6 calculate the wavelength of the wave and the intrinsic impedance of the medium.

**Solution**

$$\lambda = \frac{2\pi}{\beta} = \frac{2 \times 3.14}{2485.7}$$
$$\lambda = 2.52\text{mm}.$$

Now

$$\eta = \sqrt{\frac{i\omega\mu}{\sigma + i\omega\varepsilon}} = \sqrt{\frac{\mu}{\varepsilon}}\sqrt{\frac{1}{1 - i(\sigma/\omega\varepsilon)}}$$
$$\eta = \sqrt{\frac{4\pi \times 10^{-7}}{8.85 \times 10^{-12} \times 52}}\sqrt{\frac{1}{1 - i0.5432}}$$
$$\Rightarrow \eta = 47.4 - 12.02i$$

is the intrinsic impedance.

## Exercises

**Problem 7.1**
For water $\varepsilon_r = 78$ at 300 MHz. If water is assumed to be a lossless medium and 300 MHz of em wave is propagating through water calculate **H**. Given $E_o = 0.25\text{V/m}$ $\mu_1 = 1 \cdot E_o$ is the amplitude of the electric field.

**Problem 7.2**
An em wave of frequency 70 Hz is propagating in copper whose conductivity is $5.8 \times 10^7\text{s/m}$. Calculate v, $\lambda$ attenuation constant and $\beta$.

**Problem 7.3**
A plane TEM wave travelling in z-direction and **E** and **H** in x-direction and y-direction have a power density of 1.4 W/m$^2$ in a medium with $\mu_r = 1$ and $\varepsilon_r = 3$. Find the amplitude of **E** and **H**.

**Problem 7.4**
Calculate the average value of Poynting vector for the case of lossy dielectric using Eqs. 7.59 and 7.74 separately and verify the statement that exponential is easier to work with than time-harmonic functions.

**Problem 7.5**

From the results obtained in Sect. 7.7 for the case of lossy dielectric obtain the results for the case of free space that we derived in Sect. 7.3 with $\sigma = 0$, $\varepsilon = \varepsilon_o$, $\mu = \mu_o$.

**Problem 7.6**

For the case of two perfect dielectrics discussed in Eqs. 7.189–7.204 calculate $\mathbf{H}_1$ and hence Poynting vector.

**Problem 7.7**

A plane wave is propagating in a medium. The electric field of the wave is

$$\mathbf{E} = 0.7e^{-z/2}\cos\left(10^9 t - \beta z\right)\hat{\mathbf{i}}\ \text{V/m and}\ \varepsilon_r = 7, \mu_r = 2.$$

Calculate the

(a) loss tangent
(b) wave velocity
(c) wave impedance.

**Problem 7.8**

A electromagnetic wave is travelling in free space in $z$-direction with a frequency of 2 MHz. Its electric field is pointing in the $x$-direction and the peak value of the field is given by $1.4\pi(\text{mv/m})$. Obtain the expressions for $\mathbf{E}(z, t)$ and $\mathbf{H}(z, t)$.

**Problem 7.9**

A 1000 MHz standing wave pattern exists in a non-magnetic dielectric. The distance between maximum and its adjacent minimum is 1.6 cm. Compute the value of $\varepsilon_r$.

**Problem 7.10**

A perpendicularly polarised electromagnetic wave travels from medium 1 to medium 2. Medium 2 is vacuum. For the case of medium 1. $\varepsilon_{r1} = 7.0, \mu_r = 1, \sigma = 0$. The angle of incidence is 16°. If $E_{io} = 2\mu\text{V/m}$. Calculate $E_{ro}$, $H_{ro}$, $H_{to}$.

**Problem 7.11**

Calculate the reflection and transmission coefficients when an electromagnetic wave is incident normally on a dielectric from free space.

Given $\mu_2 = \mu_o, \varepsilon_{r2} = 4$.

**Problem 7.12**

For a dielectric $\varepsilon_r = 2.5$, $\mu = \mu_o$. From free space air is incident on the dielectric with $\theta_i = 29°$. Calculate the angle of transmission $\theta_t$.

# Chapter 8
# Transmission Lines

## 8.1 Introduction

In the previous chapter we saw plane electromagnetic wave propagation in free space. The wave propagating in free space is an unguided wave and propagates in all possible directions. These type of unguided wave propagation will be helpful when there are multiple receivers located at different points. Good examples of such wave propagation are radio or TV broadcasting systems.

On the other hand for a point (source) to point (load) transmission such unguided wave propagation will be inefficient. For such transmission, electromagnetic energy should be transferred in a well-defined path without any spreading.

A good example where such transmission is helpful is in a situation like telephone conversation-wave propagation in such transmission is said to be guided.

Typical examples in which the wave propagation is guided from source to load in a well-defined path are transmission line and waveguides. Transmission lines are discussed in this chapter and waveguides in the next chapter.

A transmission line consists of two or more parallel conductors used to transmit electromagnetic energy. The most common types of transmission lines are shown in Fig. 8.1.

Figure 8.1a shows a coaxial cable which consists of a inner conductor and a conducting sheath separated by a dielectric medium. Figure 8.1b shows a parallel-plate transmission line separated by a dielectric slab of uniform thickness. Two-wire transmission line shown in Fig. 8.1c consists of two wires separated by a uniform distance. Figure 8.1d shows a microstrip line.

© Springer Nature Singapore Pte Ltd. 2020

S. Balaji, *Electromagnetics Made Easy*,
https://doi.org/10.1007/978-981-15-2658-9_8

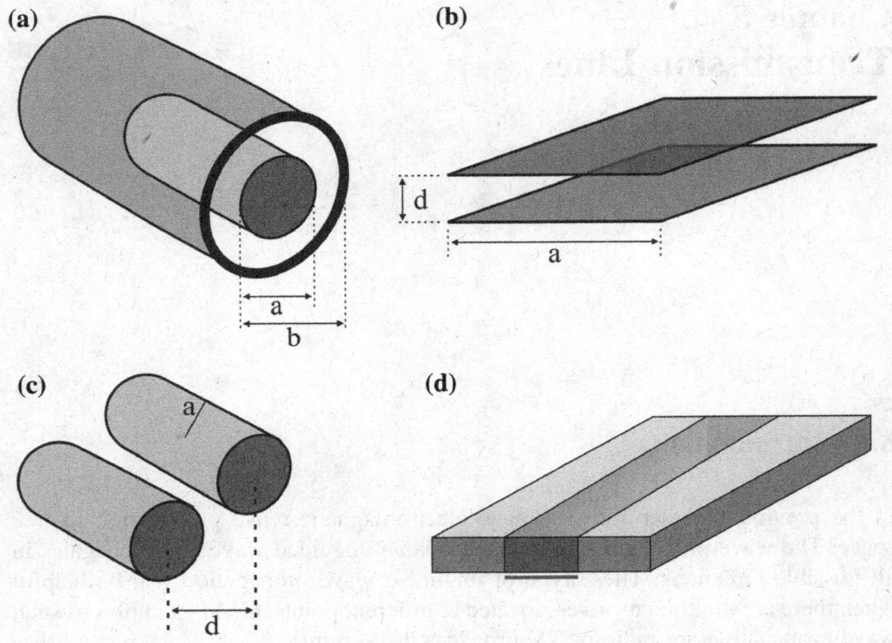


**Fig. 8.1** Different types of transmission lines **a** A coaxial cable. **b** A parallel plate transmission line. **c** A two-wire transmission line. **d** Microstrip line

## 8.2 Description of a Transmission Line

In Sect. 6.2 we saw that when the key K is closed (Fig. 6.2) a current flows in the circuit. It takes a small time duration for the current to flow in the circuit as soon as the key K is closed. However due to small dimensions of the circuit the time delay is negligible, and for all practical purposes the current flow is instantaneous.

On the other hand assume the resistance R in Fig. 6.2 lies thousands of miles away and is connected to the battery. When the key K is closed a current will start flowing in the circuit, but now time delay effects become appreciable. From the point of closing the key and for the current to flow in the resistance R a propagation delay is introduced due to large dimensions of the circuit. The time duration required for the flow of current in the circuit thus becomes appreciable.

Let us assume that in Fig. 6.2 the resistance lying at a thousands of miles away is absent. Or in other words the circuit is having large dimensions, but the other end is open. Will there be current in the circuit. To answer the above question the following points should be noted. A transmission line used in circuits of large dimensions has capacitance, inductance, resistance and a finite conductivity.

(1) Any pair of two conductors separated by a insulating medium forms a capacitor.
(2) The current flowing through the conductors produces a magnetic field around it. The flux of the magnetic field is linked to the conductor through inductance.
(3) Each conductor has a resistance.
(4) The dielectric separating the two conductors in the transmission line has finite conductivity.

Due to large dimensions of the transmission line we characterize the above parameters on per unit length basis.

(1) Let C be the capacitance per unit length of the transmission line.
(2) Let L be the inductance per unit length of the transmission line.
(3) Let R be the resistance per unit length of the transmission line. [Not to be confused with R in Fig. 6.2]
(4) Let G be the conductance per unit length of the dielectric medium separating the conductors. Note that

$$G \neq \frac{1}{R}$$

Like resistors, capacitors have specific symbols in electronic circuits, similarly transmission lines have specific symbols. Irrespective of whatever be the shape of the transmission line its symbolic representation is parallel wire configuration as shown in Fig. 8.2a.

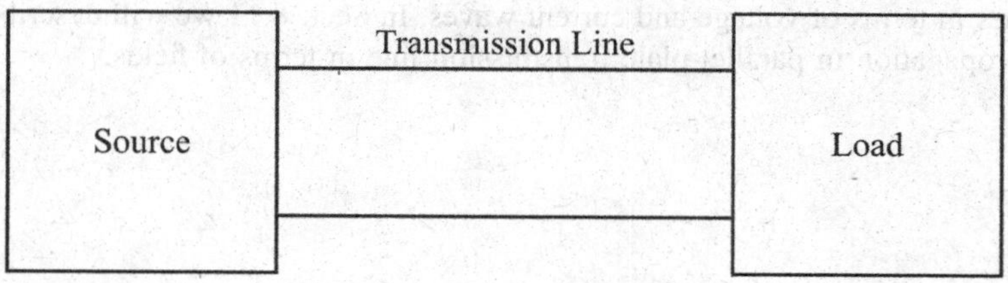


(a) Parallel wire representation

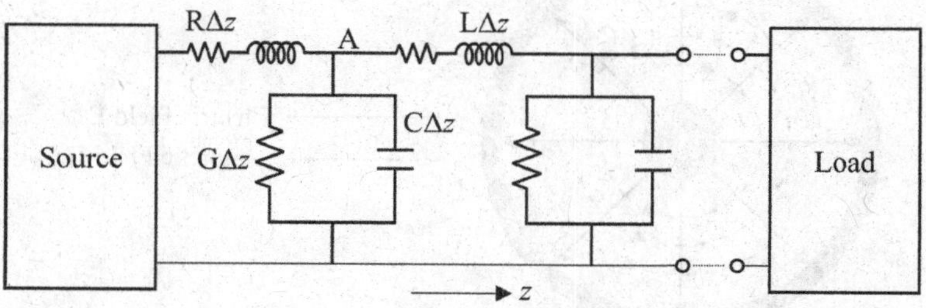


(b) Lumped Circuit Model

**Fig. 8.2** **a** Symbolic representation of a transmission line. **b** Circuit model of a transmission line

## 8.3 Wave Propagation in a Transmission Line

Transmission lines allow transverse electromagnetic wave (TEM) to propagate along their length. The electric field, magnetic field and direction of propagation in such TEM waves are mutually perpendicular to each other. Consider that in Fig. 8.2a in the parallel wire representation the transmission line is a coaxial cable. Suppose in Fig. 8.2a the source is switched on and it generates a signal.

A TEM wave propagates along the transmission line towards the load in Fig. 8.2a.

The electric field and magnetic field of coaxial line is shown in Fig. 8.3. The electric field points in the radial direction from the inner conductor to the outer conductor. The magnetic field circles the inner conductor.

Switching on the source in Fig. 8.2a establishes a voltage and current in the transmission line. The voltage and current established in the transmission line are related to the electric and magnetic fields through Eqs. 2.65 and 5.16.

$$V = -\int \mathbf{E} \cdot d\boldsymbol{l}$$

$$I = \oint \mathbf{H} \cdot d\boldsymbol{l}$$

Thus the TEM wave propagation in the transmission line can be described in terms of electric and magnetic fields or in terms of voltage and current (electric circuit theory). As we will see in future sections the lumped parameter model leading to transmission line equations describe the wave propagation in transmission lines in terms of voltage and current waves. In Sect. 8.11 we will describe the wave propagation in parallel-plate transmission line in terms of fields.

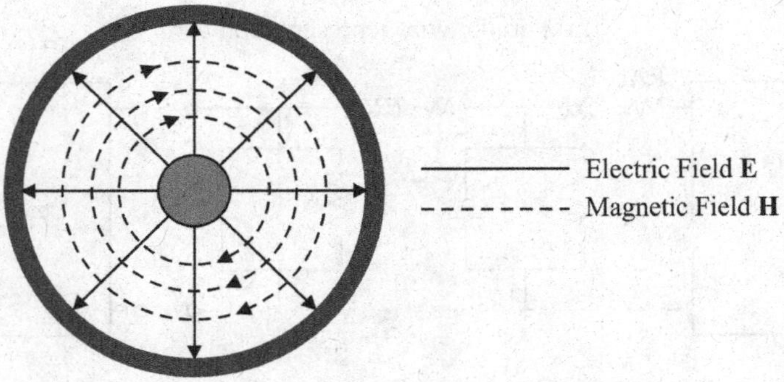


**Fig. 8.3** Electric and magnetic fields for a TEM wave propagation in a coaxial transmission line

## 8.4 Transmission Line Equations

As explained, transmission line problems can be solved using electromagnetic theory and electric circuit theory. In this section we will solve the transmission line problem using electric circuit theory. For the circuit shown in Fig. 6.2 and similar circuits the dimensions are so small compared to the wavelength $\lambda$ of the electromagnetic wave they can be considered as lumped parameter circuits. But for the circuit shown in Fig. 8.2a, b their dimensions are very large as compared to the wavelength of the electromagnetic wave and their circuit parameters are distributed over a length. In those distributed parameter circuits we divide the length of the transmission line of small elements $\Delta z$ such that $\Delta z << \lambda$. and analyse them using usual methods of circuit theory. One such division is shown in Fig. 8.4.

Noting down that Kirchhoff's law cannot be applied to distributed circuits but can be applied to lumped parameter circuits we use Kirchhoff's voltage law (KVL) and Kirchhoff's current law (KCL) to analyse Fig. 8.4.

Applying KVL to Fig. 8.4

$$V(z,t) = (R\Delta z)I(z,t) + (L\Delta z)\frac{\partial I(z,t)}{\partial t} + V(z+\Delta z,t) \tag{8.1}$$

where V(z, t), I (z, t) and V (z + Δz, t), I (z + Δz, t) are the instantaneous voltages and currents at z, z + Δz, respectively.

The above equation can be written as

$$-\left[\frac{V(z+\Delta z,t) - V(z,t)}{\Delta z}\right] = R\,I(z,t) + L\frac{\partial I(z,t)}{\partial t} \tag{8.2}$$

In the limit $\Delta z \to 0$

$$-\frac{\partial V(z,t)}{\partial z} = R\,I(z,t) + L\frac{\partial I(z,t)}{\partial t} \tag{8.3}$$

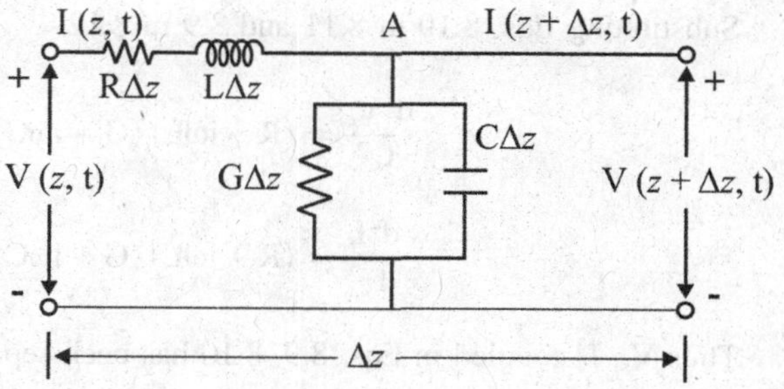


**Fig. 8.4** Equivalent circuit for small length of the two conductor transmission line

Applying KCL to point A in Fig. 8.4

$$I(z,t) = (G\Delta z)V(z+\Delta z,t) + (C\Delta z)\frac{\partial V(z+\Delta z,t)}{\partial t} + I(z+\Delta z,t) \tag{8.4}$$

$$-\left[\frac{I(z+\Delta z,t) - I(z,t)}{\Delta z}\right] = G\,V(z+\Delta z,t) + C\frac{\partial V(z+\Delta z,t)}{\partial t} \tag{8.5}$$

In the limit $\Delta z \to 0$

$$-\frac{\partial I(z,t)}{\partial z} = G\,V(z,t) + C\frac{\partial V(z,t)}{\partial t} \tag{8.6}$$

Assuming harmonic time dependence and using the phasor notation

$$V(z,t) = \mathrm{Re}\left[V_s(z)e^{i\omega t}\right] \tag{8.7}$$

$$I(z,t) = \mathrm{Re}\left[I_s(z)e^{i\omega t}\right] \tag{8.8}$$

With the aid of Eqs. 8.7 and 8.8, Eqs. 8.3 and 8.6 become

$$-\frac{dV_s}{dz} = (R + i\omega L)I_s \tag{8.9}$$

$$-\frac{dI_s}{dz} = (G + i\omega C)V_s \tag{8.10}$$

In Eqs. 8.9 and 8.10 the quantities $V_s$ and $I_s$ are coupled. From Eqs. 8.9 and 8.10

$$-\frac{d^2V_s}{dz^2} = (R + i\omega L)\frac{dI_s}{dz} \tag{8.11}$$

$$-\frac{d^2I_s}{dz^2} = (G + i\omega C)\frac{dV_s}{dz} \tag{8.12}$$

Substituting Eqs. 8.10 in 8.11 and 8.9 in 8.12

$$\frac{d^2V_s}{dz^2} = (R + i\omega L)\,(G + i\omega C)V_s \tag{8.13}$$

$$\frac{d^2I_s}{dz^2} = (R + i\omega L)\,(G + i\omega C)I_s \tag{8.14}$$

Thus $V_s$, $I_s$ coupled in Eqs. 8.9, 8.10 has been separated in Eqs. 8.13 and 8.14.

## 8.5 Wave Propagation in the Transmission Line—Circuit Model

Equations 8.13 and 8.14 can be written as

$$\frac{d^2V_s}{dz^2} - \gamma^2 V_s = 0 \tag{8.15}$$

$$\frac{d^2I_s}{dz^2} - \gamma^2 I_s = 0 \tag{8.16}$$

where

$$\gamma = \alpha + i\beta = \sqrt{(R + i\omega L)\,(G + i\omega C)} \tag{8.17}$$

Equations 8.15 and 8.16 are having similar form as that of Eqs. 7.87a, 7.87b. Hence we conclude Eqs. 8.15 and 8.16 are wave equations for voltage and current, respectively. $\gamma$ in Eq. 8.17 is the propagation constant where $\alpha$ is the attenuation constant (Np/m) and $\beta$ is the phase constant (rad/m). $\alpha$ and $\beta$ are given by

$$\begin{aligned} \alpha &= \mathrm{Re}[\gamma] \\ &= \mathrm{Re}\left[\sqrt{(R + i\omega L)\,(G + i\omega C)}\right] \end{aligned} \tag{8.18}$$

$$\begin{aligned} \beta &= \mathrm{Im}[\gamma] \\ &= \mathrm{Im}\left[\sqrt{(R + i\omega L)\,(G + i\omega C)}\right] \end{aligned} \tag{8.19}$$

The solutions for Eqs. 8.15 and 8.16 are similar to Eq. 7.97

$$V_s(z) = V_o^+\, e^{-\gamma z} + V_o^-\, e^{\gamma z} \tag{8.20}$$

and

$$I_s(z) = I_o^+\, e^{-\gamma z} + I_o^-\, e^{\gamma z} \tag{8.21}$$

The $e^{-\gamma z}$ term represents a forward travelling wave in the $z$-direction, and $e^{\gamma z}$ term represents a backward travelling wave in the $-z$-direction. $V_o^+$, $V_o^-$, $I_o^+$, $I_o^-$ are the wave amplitudes. The forward and backward travelling waves are shown in Fig. 8.5.

The instantaneous expression for the voltage is then,

$$V(z,t) = \mathrm{Re}\left[V_s(z)e^{i\omega t}\right] \tag{8.22}$$

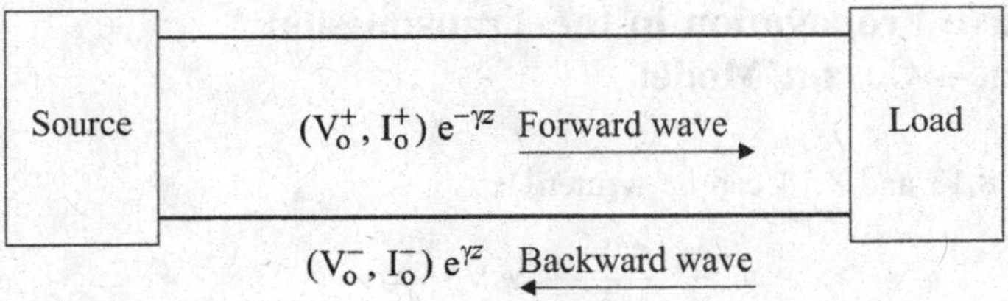


**Fig. 8.5** Forward and backward travelling wave in a transmission line

$$= \mathrm{Re}\left[\left(V_o^+ e^{-\gamma z} + V_o^- e^{\gamma z}\right) e^{i\omega t}\right] \tag{8.23}$$

Noting down that $\gamma = \alpha + i\beta$ and in general $V_o^+$ and $V_o^-$ are complex

$$V(z,t) = \mathrm{Re}\left\{\left(V_o^+ e^{-(\alpha + i\beta)z} + V_o^- e^{-(\alpha + i\beta)z}\right) e^{i\omega t}\right\} \tag{8.24}$$

**Characteristic impedance**

As the source in Fig. 8.2 is switched on, voltage and current waves start propagating in the transmission line. Let us assume that at a given instant of time the disturbance produced travels and reaches point A in Fig. 8.2. At that instant the source "feels" the impedance of the transmission line only and not the impedance of the load. The impedance of the transmission line is called the characteristic impedance.

The characteristic impedance $Z_o$ of the transmission line is defined as the ratio of voltage to current for the wave propagating in the $z$-direction.

Substituting Eq. 8.20 in Eq. 8.9

$$\frac{-d}{dz}\left(V_o^+ e^{-\gamma z} + V_o^- e^{\gamma z}\right) = (R + i\omega L) I_s \tag{8.25}$$

$$\Rightarrow I_s = \frac{\gamma V_o^+}{(R + i\omega L)} e^{-\gamma z} - \frac{\gamma V_o^-}{(R + i\omega L)} e^{\gamma z} \tag{8.26}$$

Comparing Eq. 8.26 with 8.21

$$I_o^+ = \frac{\gamma V_o^+}{R + i\omega L} \tag{8.27}$$

$$I_o^- = \frac{-\gamma V_o^-}{R + i\omega L} \tag{8.28}$$

From Eq. 8.27 and 8.28 we see that

$$Z_o = \frac{V_o^+}{I_o^+} = -\frac{V_o^-}{I_o^-} = \frac{R + i\omega L}{\gamma} \tag{8.29}$$

Substituting Eq. 8.17 in Eq. 8.29

$$Z_o = \sqrt{\frac{R + i\omega L}{G + i\omega C}} \tag{8.30}$$

The wavelength λ and wave velocity v are given by,

$$\lambda = \frac{2\pi}{\beta} \tag{8.31}$$

$$v = \frac{\omega}{\beta} = \upsilon\lambda \tag{8.32}$$

We will now consider special cases of wave propagation in transmission lines.

## 8.6 Lossless Line

In the case of lossless propagation whatever power is launched by the source into the transmission line arrives to load without any loss. There is no dissipation along the transmission line, and the wave propagation in the transmission is said to be lossless. For no dissipation to occur the resistance of the conductor R and the conductivity of the dielectric G should be zero. Thus R = 0, G = 0. The lossless transmission line is shown in figure 8.6 with R = G = 0.

Initially when the source is switched on in Fig. 8.6 no instant voltage appears at the load. Instead current increases in $L_a$ until $C_a$ is charged, next current in $L_b$ increases charging $C_b$, and this sequential charging process continues until the voltage appears at the load. Thus as the source is switched on the voltage doesn't appear instantaneously everywhere on the line but rather travels from the source to the load at a certain velocity. The velocity of this wave can be estimated as follows.

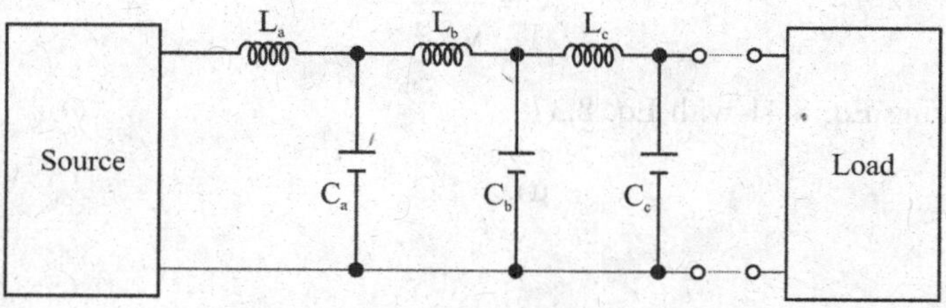


**Fig. 8.6** Lumped circuit model of a lossless transmission line

For a lossless line,

$$R = G = 0 \tag{8.33}$$

From Eq. 8.17

$$\alpha + i\beta = i\omega\sqrt{LC} \tag{8.34}$$

Comparing real and imaginary parts on both sides

$$\alpha = 0 \tag{8.35}$$

$$\gamma = i\beta = i\omega\sqrt{LC} \tag{8.36}$$

Thus the velocity of the wave is

$$v = \frac{\omega}{\beta} = \frac{1}{\sqrt{LC}} \tag{8.37}$$

The characteristic impedance of the lossless transmission line is

$$Z_o = \sqrt{\frac{L}{C}} \tag{8.38}$$

We have discussed about the propagation of plane electromagnetic waves in lossy dielectric in Sect. 7.7. From Eq. 7.83a,

$$\nabla^2 \mathbf{E} - \sigma\mu \frac{\partial \mathbf{E}}{\partial t} = \varepsilon\mu \frac{\partial^2 \mathbf{E}}{\partial t^2} \tag{8.39}$$

For a lossless dielectric the conductivity σ is zero. Hence in one dimension the above equation can be written as

$$\frac{\partial^2 \mathbf{E}}{\partial z^2} - \varepsilon\mu \frac{\partial^2 \mathbf{E}}{\partial t} = 0 \tag{8.40}$$

Comparing Eq. 8.40 with Eq. 7.1

$$v = \frac{1}{\sqrt{\varepsilon\mu}} \tag{8.41}$$

Comparing Eq. 8.41 with Eq. 8.37

$$\mu\varepsilon = LC \tag{8.42}$$

## 8.7 Low-loss Line

In Sect. 8.5 we discussed the propagation of the wave in a lossy transmission line. $\alpha$, attenuation constant given by Eq. 8.18 and the phase velocity given by Eq. 8.32 are in general dependent on frequency.

In the case of lossless propagation (Sect. 8.6) $\alpha$, and phase velocity given by Eqs. 8.35 and 8.37, respectively, are independent of frequency. There is an another special case in which

$$R \ll \omega L,\ G \ll \omega C \tag{8.43}$$

The above case is called low-loss propagation.

Under this condition Eq. 8.17 is

$$\begin{aligned}\gamma &= (R+i\omega L)^{\frac{1}{2}}(G+i\omega C)^{\frac{1}{2}}\\ &= (i\omega L)^{\frac{1}{2}}(i\omega C)^{\frac{1}{2}}\left(1+\frac{R}{i\omega L}\right)^{\frac{1}{2}}\left(1+\frac{G}{i\omega C}\right)^{\frac{1}{2}}\end{aligned} \tag{8.44}$$

$$\gamma = i\omega\sqrt{LC}\left(1+\frac{R}{i\omega L}\right)^{\frac{1}{2}}\left(1+\frac{G}{i\omega C}\right)^{\frac{1}{2}}$$

As $R \ll \omega L$, $G \ll \omega C$ applying binomial approximation

$$\gamma \simeq i\omega\sqrt{LC}\left(1+\frac{R}{2i\omega L}\right)\left(1+\frac{G}{2i\omega C}\right) \tag{8.45}$$

$$\gamma \simeq i\omega\sqrt{LC}\left(1+\frac{R}{2i\omega L}+\frac{G}{2i\omega C}+\frac{RG}{4\omega^2 LC}\right) \tag{8.46}$$

Neglecting the $\frac{1}{\omega^2}$ term

$$\gamma \approx i\omega\sqrt{LC}\left(1+\frac{R}{2i\omega L}+\frac{G}{2i\omega C}\right) \tag{8.47}$$

$$\gamma \approx i\omega\sqrt{LC}+\frac{1}{2}\left(R\sqrt{\frac{C}{L}}+G\sqrt{\frac{L}{C}}\right) \tag{8.48}$$

Noting down that $\gamma = \alpha + i\beta$

$$\alpha = \frac{1}{2}\left(R\sqrt{\frac{C}{L}}+G\sqrt{\frac{L}{C}}\right) \tag{8.49}$$

$$\beta = \omega\sqrt{LC} \tag{8.50}$$

The phase velocity is given by

$$v = \frac{\omega}{\beta} = \frac{1}{\sqrt{LC}} \tag{8.51}$$

From Eq. 8.30 the characteristic impedance is given by

$$Z_o = \frac{(R + i\omega L)^{\frac{1}{2}}}{(G + i\omega C)^{\frac{1}{2}}}$$

$$Z_o = \frac{\sqrt{i\omega L}}{\sqrt{i\omega C}} \left( \frac{1 + \frac{R}{i\omega L}}{1 + \frac{G}{i\omega C}} \right)^{\frac{1}{2}} \tag{8.52}$$

$$Z_o = \sqrt{\frac{L}{C}} \left(1 + \frac{R}{i\omega L}\right)^{\frac{1}{2}} \left(1 + \frac{G}{i\omega C}\right)^{\frac{-1}{2}} \tag{8.53}$$

Using binomial approximation

$$Z_o \approx \sqrt{\frac{L}{C}} \left(1 + \frac{R}{2i\omega L}\right) \left(1 - \frac{G}{2i\omega C}\right) \tag{8.54}$$

Neglecting the $\frac{1}{\omega^2}$ term

$$Z_o \approx \sqrt{\frac{L}{C}} \left(1 + \frac{1}{2i\omega} \left(\frac{R}{L} - \frac{G}{C}\right)\right) \tag{8.55}$$

## 8.8 Distortionless Line

Transmission lines in general are used to transmit signals from source to receiver. The signal which carries the information usually has a band of frequencies.

For the case of lossless propagation $\alpha$ the attenuation constant is zero (Eq. 8.35). So there is no attenuation. The phase velocity given by Eq. 8.37 is independent of frequency. It means that all the waves with different frequencies travel at the same velocity.

For the case of low-loss propagation the attenuation constant $\alpha$ given by Eq. 8.49 is approximately independent of frequency. Hence all waves with different frequencies are attenuated to the same extent.

The phase velocity v given by Eq. 8.51 is approximately independent of frequency. Thus waves with different frequencies travel with same velocity. However under low-loss condition when the transmission line is lengthy, loss becomes

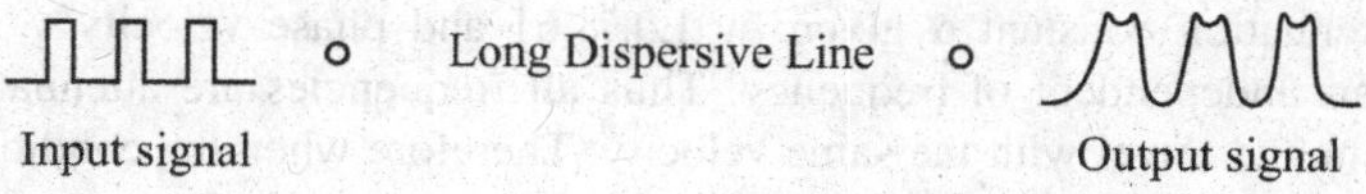


**Fig. 8.7** Distortion of the input signal in a long dispersive line

appreciable and α, v depends on frequency. The result is waves of different frequencies get attenuated to a different extent and travel with different velocities. Therefore the input signal carrying information in a band of frequencies get distorted as shown in Fig. 8.7.

Even though the line is lossy if the following condition is satisfied then the transmission line would become distortionless

$$\frac{R}{L} = \frac{G}{C} \tag{8.56}$$

The above condition is known as Heaviside condition.

The propagation constant as given by Eq. 8.17 is

$$\gamma = \sqrt{(R + i\omega L)(G + i\omega C)} \tag{8.57}$$

$$\gamma = \sqrt{LC\left(\frac{R}{L} + i\omega\right)\left(\frac{G}{C} + i\omega\right)} \tag{8.58}$$

Using Eq. 8.56

$$\gamma = \sqrt{LC}\left(\frac{R}{L} + i\omega\right) \tag{8.59}$$

$$\gamma = R\sqrt{\frac{C}{L}} + i\omega\sqrt{LC} \tag{8.60}$$

As $\gamma = \alpha + i\beta$, from Eq. 8.60 we can write

$$\alpha = R\sqrt{\frac{C}{L}} \tag{8.61}$$

$$\beta = \omega\sqrt{LC} \tag{8.62}$$

The phase velocity is then

$$v = \frac{\omega}{\beta} = \frac{1}{\sqrt{LC}} \tag{8.63}$$

The attenuation constant α given by Eq. 8.61 and phase velocity v given by Eq. 8.63 are independent of frequency. Thus all frequencies are attenuated to an equal extent and travel with the same velocity. Therefore when the condition given by 8.56 is met then the line is distortionless.

For a distortionless line the characteristic impedance given by Eq. 8.30 is

$$Z_0 = \sqrt{\frac{R + i\omega L}{\frac{RC}{L} + i\omega C}} \tag{8.64}$$

$$Z_0 = \sqrt{\frac{L}{C}} \tag{8.65}$$

## 8.9 Relationship Between G and C

For a given transmission line the inductance per unit length L and capacitance per unit length C are related by Eq. 8.42. Assuming that the conductor in the transmission line is having very high conductivity we can set R = 0 in Eq. 8.17. Then

$$\gamma = \sqrt{i\omega L(G + i\omega C)} \tag{8.66}$$

$$\gamma = i\omega\sqrt{LC}\left(1 + \frac{G}{i\omega C}\right)^{\frac{1}{2}} \tag{8.67}$$

For a lossy dielectric from Eq. 7.85

$$\gamma = \sqrt{(i\omega\,\mu)(\sigma + i\omega\varepsilon)} \tag{8.68}$$

The above equation can be written as

$$\gamma = i\omega\,\sqrt{\mu\varepsilon}\left(1 + \frac{\sigma}{i\omega\varepsilon}\right)^{\frac{1}{2}} \tag{8.69}$$

Comparing Eqs. 8.67 and 8.69

$$LC = \mu\varepsilon \tag{8.70}$$

$$\frac{G}{C} = \frac{\sigma}{\varepsilon} \tag{8.71}$$

Equation 8.71 gives the relationship between conductance per unit length G and capacitance per unit length C of the transmission line.

## 8.10 Determination of Transmission Line Parameters

**(A) Parallel-plate transmission line**

The parallel-plate transmission line is shown in Fig. 8.1b. Let d be the separation between the two plates in Fig. 8.1b. Let "a" be the width of the transmission line. Then from the Eq. 3.333 the capacitance per unit length of the transmission line can be written as

$$C = \frac{\varepsilon a}{d} \text{ F/m} \tag{8.72}$$

In the above equation the fringing fields of the transmission line have been neglected by assuming $d \ll a$.

Substituting Eq. 8.72 in 8.70.

The inductance per unit length of the transmission line is

$$L = \frac{\mu d}{a} \text{ H/m} \tag{8.73}$$

Inserting Eq. 8.72 into 8.71 the conductance per unit length of the transmission line is

$$G = \frac{\sigma a}{d} \text{ S/m} \tag{8.74}$$

**(B) Coaxial transmission line**

A coaxial transmission line with inner radius a and outer radius b is shown Fig. 8.1a. From Eq. 3.355 the capacitance per unit length of the transmission line can be written as

$$C = \frac{2\pi\varepsilon}{\ln(b/a)} \tag{8.75}$$

Substituting Eq. 8.75 in 8.70.

The inductance per unit length of the transmission line can be written as

$$L = \frac{\mu}{2\pi}\ln(b/a) \tag{8.76}$$

The conductance per unit length of the transmission line can be obtained from Eqs. 8.75 and 8.71 as

$$G = \frac{2\pi\sigma}{\ln(b/a)} \tag{8.77}$$

The transmission line parameters R, L, C and G for various types of transmission lines is shown in Table 8.1. The resistance per unit length of the transmission line has been calculated by assuming the conductors are thick compared to the skin

**Table 8.1** Transmission line parameters for common transmission lines

| Parallel-plate transmission line | Coaxial line | Two-wire line |
|---|---|---|
| $R = \frac{2}{a\delta\sigma_c}$ | $R = \frac{1}{2\pi\delta\sigma_c}\left[\frac{1}{a} + \frac{1}{b}\right]$ | $R = \frac{1}{\pi a\delta\sigma_c}$ |
| $L = \frac{\mu d}{a}$ | $L = \frac{\mu}{2\pi}\ln(b/a)$ | $L = \frac{\mu}{\pi}\cos h^{-1}\frac{d}{2a}$ |
| $G = \frac{\sigma a}{d}$ | $G = \frac{2\pi\sigma}{\ln(b/a)}$ | $G = \frac{\pi\sigma}{\cos h^{-1}\frac{d}{2a}}$ |
| $C = \frac{\varepsilon a}{d}$ | $C = \frac{2\pi\varepsilon}{\ln(b/a)}$ | $C = \frac{\pi\varepsilon}{\cos h^{-1}\frac{d}{2a}}$ |

depth $\delta$, and current in the transmission line is restricted to $\delta$. The conductors in the transmission line are characterized by $\sigma_c$, $\mu_c$, $\varepsilon_c$.

The homogeneous dielectric separating the conductors is characterized by $\sigma$, $\mu$, $\varepsilon$. For the parallel-plate transmission line shown in Fig. 8.1b the separation distance between the plates is d and the width of the plates is a. For the coaxial transmission line shown in Fig. 8.1a the inner radius is a and outer radius is b. While for the two-wire transmission line shown in Fig. 8.1c a is the radius of the conductor and d is the distance between the conductor.

## 8.11 Field Approach—TEM Waves in a Parallel-Plate Transmission Line

In Sect. 8.5 we discussed the wave propagation in transmission lines in terms of voltage and current variables. As stated in the Sect. 8.3 the same wave propagation in transmission lines can be described in terms of electric and magnetic fields.

Taking a parallel-plate transmission line as an example we will elaborate the electromagnetic wave propagation in this transmission line in terms of field quantities. Consider a parallel-plate transmission line as shown in Fig. 8.8 with a width a and separation between the plates d, such that $d \ll a$. As $d \ll a$ the fringing fields can be neglected. The plates are separated by a dielectric. Assume that the conductor is perfectly conducting and the dielectric is lossless.

a) **Boundary conditions**

The boundary conditions to be satisfied at the conductor dielectric interface at $x = 0$ and $x = d$ are as follows

From Eq. 6.181

$$E_{1t} = E_{2t} \tag{8.78}$$

We know that in a conductor the electric field is zero. If medium 2 is conductor then $E_{2t} = 0$. Therefore from Eq. 8.78

$$E_{1t} = E_{2t} = 0 \tag{8.79}$$

From Eq. 6.182

$$B_{1n} = B_{2n} \tag{8.80}$$

From Maxwell's Eq. 6.166

$$\nabla \times \mathbf{E} = -\frac{\partial \mathbf{B}}{\partial t} \tag{8.81}$$

As medium 2 is a conductor, **E** = 0. Hence from Eq. 8.81

$$\frac{\partial \mathbf{B}}{\partial t} = 0 \tag{8.82}$$

The above equation implies magnetic flux density is constant in time. Assuming the initial magnetic flux density is zero then as per Eq. 8.82

$$\mathbf{B} = 0 \tag{8.83}$$

Therefore in medium 2 inside the conductor

$$B_{2n} = 0 \tag{8.84}$$

And hence from Eq. 8.80

$$B_{1n} = B_{2n} = 0 \tag{8.85}$$

$$\begin{aligned} &\text{As } \mathbf{B} = \mu\mathbf{H} \\ &\Rightarrow H_{1n} = 0 \end{aligned} \tag{8.86}$$

From Eq. 8.79 we see that tangential components of the electric field are zero. Hence only normal components of electric field are present. Thus in Figs. 8.8 and 8.9b the electric field between the plates is perpendicular to the plane of the plates.

From Eq. 8.86 the normal components of **H** are zero. Hence only tangential components of **H** are present. Thus in Figs. 8.8 and 8.9d the **H** between the plates is parallel to the plane of the plates.

(b) **Wave equation**

Consider a TEM wave propagating in the *z*-direction along the uniform parallel-plate transmission line in Fig. 8.8. Let the electric field be along *x*-axis and magnetic field be along *y*-axis. As per the assumption the dielectric between the plates is lossless. Hence $\sigma = 0$. Then from Eq. 7.85.

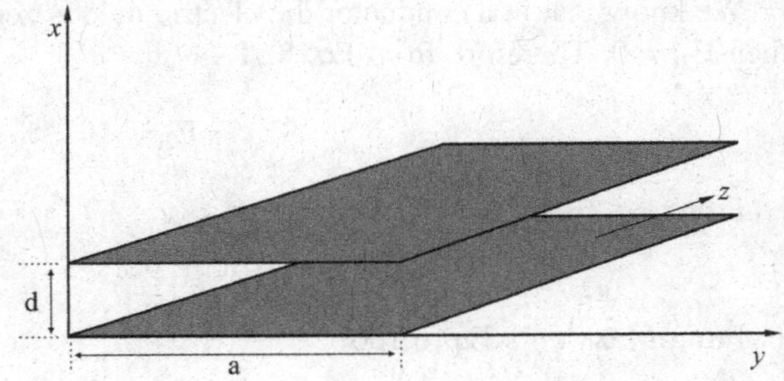


**Fig. 8.8** A parallel plate transmission line

$$\gamma^2 = -\omega^2 \mu \varepsilon \tag{8.87}$$

The dielectric between the plates is sourceless, and the wave equation to be satisfied is then given by Eqs. 7.86a, 7.86b

$$\nabla^2 \mathbf{E_s} - \gamma^2 \mathbf{E_s} = 0 \tag{8.88}$$

$$\nabla^2 \mathbf{H_s} - \gamma^2 \mathbf{H_s} = 0 \tag{8.89}$$

where $\gamma^2$ is given by Eq. 8.87.

The solutions for Eqs. 8.88 and 8.89 are

$$\mathbf{E_s} = E_{xs}\hat{\mathbf{i}} = E_o\, e^{-\gamma z}\hat{\mathbf{i}} \tag{8.90}$$

$$\begin{aligned}\mathbf{H_s} &= H_{ys}\hat{\mathbf{j}} = H_o e^{-\gamma z}\hat{\mathbf{j}} \\ &= \frac{E_o}{\eta} e^{-\gamma z}\,\hat{\mathbf{j}}\end{aligned} \tag{8.91}$$

Here η is the intrinsic impedance of the dielectric medium, and its value is given by [from Eq. 7.101a with $\sigma = 0$].

$$\eta = \sqrt{\frac{\mu}{\varepsilon}} \tag{8.92}$$

From Eq. 8.87

$$\gamma = i\beta = i\omega\sqrt{\mu\,\varepsilon} \tag{8.93}$$

Substituting Eq. 8.93 in Eqs. 8.90 and 8.91

$$\mathbf{E_s} = E_o\, e^{-i\beta z}\,\hat{\mathbf{i}} \tag{8.94}$$

$$\mathbf{H_s} = \frac{E_o}{\eta}\, e^{-i\beta z}\hat{\mathbf{j}} \tag{8.95}$$

In the time domain

$$\mathbf{E} = \mathbf{E}_s e^{i\omega t} \tag{8.96}$$

$$\mathbf{H} = \mathbf{H}_s e^{i\omega t} \tag{8.97}$$

Substituting Eqs. 8.94 and 8.95 in Eqs. 8.96 and 8.97

$$\mathbf{E} = E_o e^{i(\omega t - \beta z)} \hat{\mathbf{i}} \tag{8.98}$$

$$\mathbf{H} = \frac{E_o}{\eta} e^{i(\omega t - \beta z)} \hat{\mathbf{j}} \tag{8.99}$$

Taking the real part of Eqs. 8.98 and 8.99

$$\mathbf{E} = E_o \cos(\omega t - \beta z) \hat{\mathbf{i}} \tag{8.100}$$

$$\mathbf{H} = \frac{E_o}{\eta} \cos(\omega t - \beta z) \hat{\mathbf{j}} \tag{8.101}$$

(c) **Induced charges and currents on the plates of the transmission line**
Consider Problem 6.21 where the boundary conditions are expressed in scalar and vector form. From the results of the problem the boundary conditions for **D** and **H** are

$$D_{1n} - D_{2n} = \sigma_f \tag{8.102}$$

$$\hat{\mathbf{n}}.(\mathbf{D_1} - \mathbf{D_2}) = \sigma_f \tag{8.103}$$

$$H_{1t} - H_{2t} = K_f \tag{8.104}$$

$$\hat{\mathbf{n}} \times (\mathbf{H_1} - \mathbf{H_2}) = \mathbf{K_f} \tag{8.105}$$

The medium 2 is a conductor where the electric and magnetic fields are zero; hence, $\mathbf{D_2}$, $\mathbf{H_2}$ are zero inside the conductor. Thus the only surviving components will be $\mathbf{D_1}$, $\mathbf{H_1}$ in the dielectric. Dropping the subscripts then we can write Eqs. 8.102 and 8.104 as

$$D_n = \sigma_f \tag{8.106}$$

$$\hat{\mathbf{n}}.\mathbf{D} = \sigma_f \tag{8.107}$$

$$H_t = K_f \tag{8.108}$$

$$\hat{\mathbf{n}} \times \mathbf{H} = \mathbf{K_f} \tag{8.109}$$

**D**, **H** are the fields in the dielectric.

**Plate at $x$ = 0**

Consider the lower plate at $x = 0$. At $x = 0$ $\hat{\mathbf{n}} = \hat{\mathbf{i}}$. Therefore Eq. 8.107 becomes

$$\hat{\mathbf{i}} \cdot \mathbf{D} = \sigma_f \tag{8.110}$$

Noting down that $\hat{\mathbf{i}} \cdot \mathbf{D}$ selects the normal components of **D** we have

$$D_n = \sigma_f \Rightarrow \varepsilon E_n = \sigma_f \tag{8.111}$$

From Fig. 8.8 and Eq. 8.94 the normal component $E_n$ is equal to $E_s$ as $E_s$ is pointing in $x$-direction and is normal to the plate at $x = 0$.

Therefore from Eq. 8.94

$$\sigma_f = \varepsilon E_s = \varepsilon E_o\, e^{i\beta z} \tag{8.112}$$

Equation 8.109 is

$$\hat{\mathbf{i}} \times \mathbf{H} = \mathbf{K_f} \tag{8.113}$$

The cross product $\hat{\mathbf{i}} \times \mathbf{H}$ selects the tangential component $|\mathbf{H_t}|$.

From Fig. 8.8 and Eq. 8.95 the tangential component $H_t$ is equal to $H_s$ as $H_s$ is pointing in $y$-direction and is tangential to the plate at $x = 0$. Hence from Eq. 8.113

$$\mathbf{K_f} = H_z\, \hat{\mathbf{k}} = H_s \hat{\mathbf{k}} \tag{8.114}$$

Substituting Eq. 8.95

$$\mathbf{K_f} = \frac{E_o}{\eta} e^{-i\beta z}\, \hat{\mathbf{k}} \tag{8.115}$$

**Plate at $x$ = d**

Considering the upper plate at $x = d$. At $x = d$ $\hat{\mathbf{n}} = -\hat{\mathbf{i}}$. Therefore Eq. 8.107 becomes

$$-\hat{\mathbf{i}} \cdot \mathbf{D} = \sigma_f \tag{8.116}$$

Following the same procedure as detailed above

$$\sigma_f = -\varepsilon E_o e^{i\beta z} \tag{8.117}$$

Equation 8.109 becomes

$$\hat{\mathbf{i}} \times \mathbf{H} = K_f \tag{8.118}$$

Following the same procedure as detailed above

$$\mathbf{K_f} = -\frac{E_o}{\eta} e^{i\beta z}\,\hat{\mathbf{k}} \tag{8.119}$$

Substituting the space part from Eq. 8.94 in Eq. 8.96 we obtain Eq. 8.100. Following exactly same procedure we can obtain $\sigma_f$ and $K_f$.

For the plate at $x = 0$ from Eqs. 8.112 and 8.115

$$\sigma_f = \varepsilon E_o \cos(\omega t - \beta z) \tag{8.120}$$

$$\mathbf{K_f} = \frac{E_o}{\eta} \cos(\omega t - \beta z)\hat{\mathbf{k}} \tag{8.121}$$

For the plate at $x = d$ from Eqs. 8.117 and 8.119

$$\sigma_f = \varepsilon E_o \cos(\omega t - \beta z) \tag{8.122}$$

$$\mathbf{K_f} = \frac{E_o}{\eta} \cos(\omega t - \beta z)\hat{\mathbf{k}} \tag{8.123}$$

The charge densities, current densities and field distribution are as shown in Fig. 8.9.

(d) **Inductance per unit length of the transmission line**

From Eq. 8.95 we see that the **H** points in the $y$-direction. In Fig. 8.10 consider a area PQRS which is perpendicular to $y$-axis (i.e. the magnetic field **H**). From Fig. 8.10 the cross section per unit length along the $z$-direction and perpendicular to **H** is d. Therefore the magnetic flux per unit length passing through such a cross-sectional area is Bd or $\mu$Hd (from Eq. 4.246). As only tangential components of the magnetic field are present the flux per unit length is $\mu H_t d$.

From Eq. 4.20 the surface line current flowing in the parallel-plate transmission line is $K_f$ a. Using Eq. 8.114 the line current can be written as $H_t a$.

From Eq. 6.70 the inductance per unit length of the transmission line in the $z$-direction is the ratio between the magnetic flux per unit length at a given value of $z$ to the line current at that value of $z$.

$$L = \frac{\mu H_t d}{H_t a} = \frac{\mu d}{a} \text{H/m} \tag{8.124}$$

(e) **Capacitance per unit length of the transmission line**

From Eq. 8.94 we see that the electric field points in the $x$-direction.

In Fig. 8.11 consider a area ABCD which is perpendicular to the $x$-axis, (i.e. the electric field). From Fig. 8.11 the cross-sectional area per unit length along the $z$-direction and perpendicular to **E** is a.

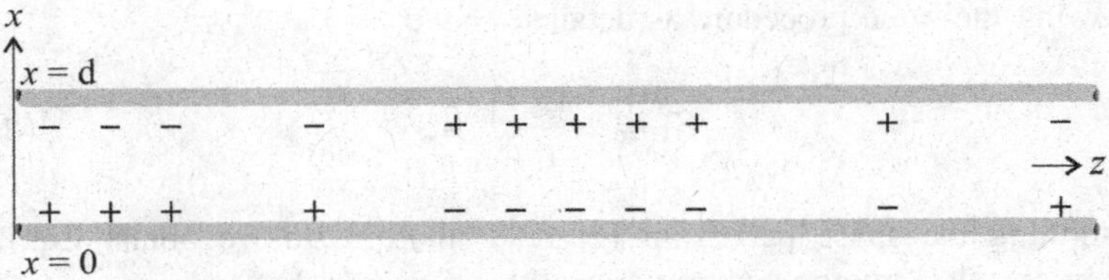


**(a)** Surface charge density distribution in the parallel plate transmission line as per equations 8.120, 8.122

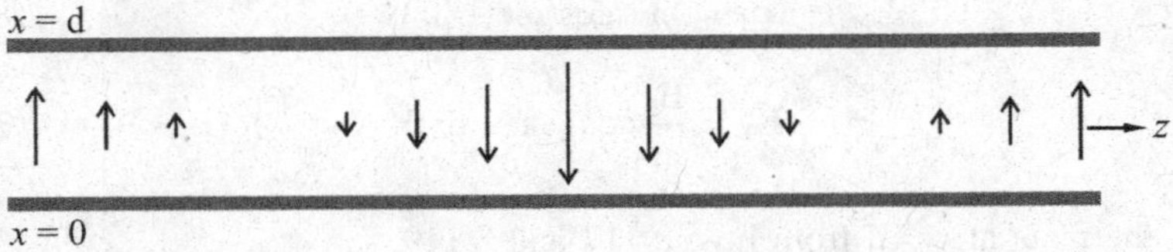


**(b)** Electric field distribution in the parallel plate transmission line as per equation 8.100

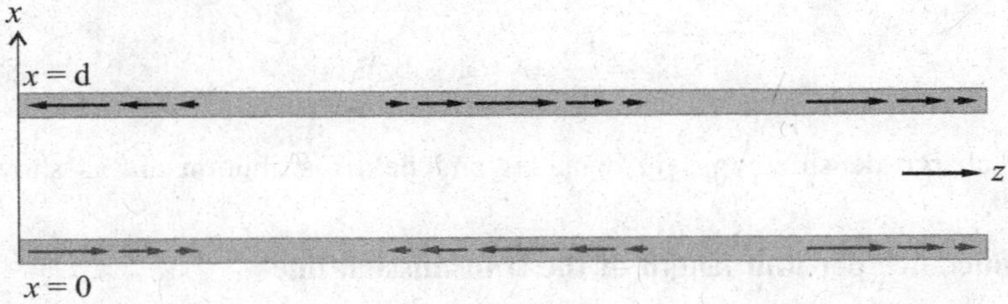


**(c)** Surface current density distribution on the parallel plate transmission line as per equations 8.122 and 8.123

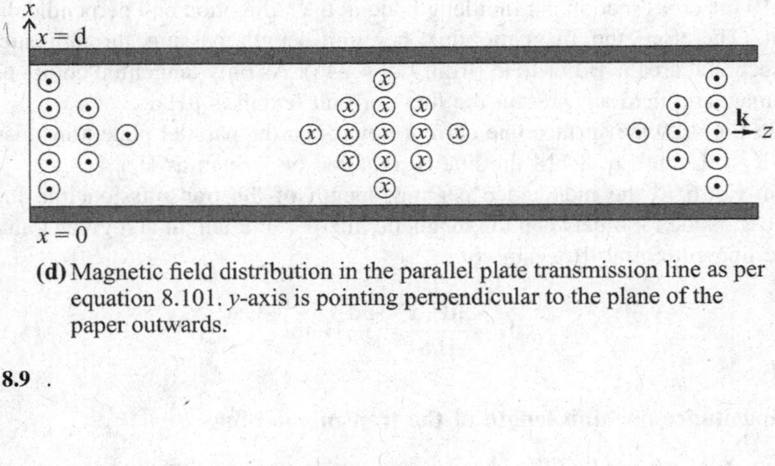


**(d)** Magnetic field distribution in the parallel plate transmission line as per equation 8.101. *y*-axis is pointing perpendicular to the plane of the paper outwards.

**Fig. 8.9** .

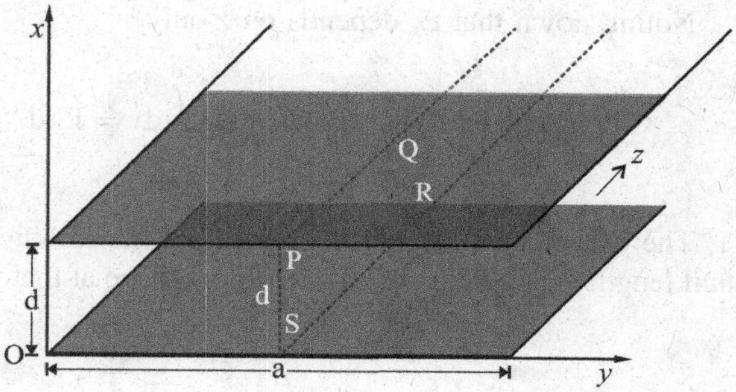


**Fig. 8.10** A parallel plate transmission line—Area PQRS perpendicular to *y*-axis

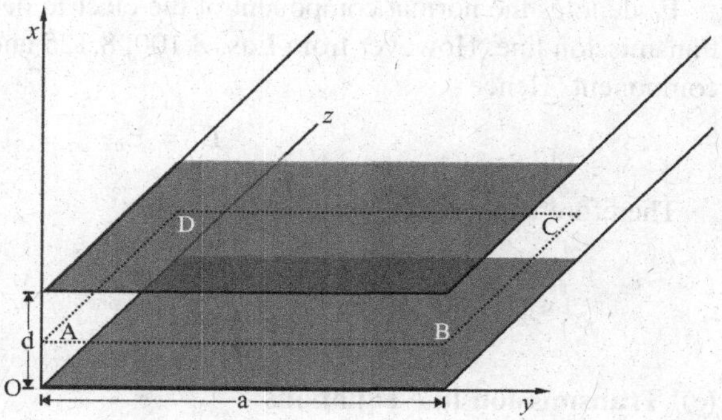


**Fig. 8.11** A parallel plate transmission line—Area ABCD perpendicular to *x*-axis

Thus the surface charge per unit length on the plates of the transmission line is $\sigma_f a$. Using Eq. 8.111 the surface charge per unit length is then $\varepsilon E_n a$.

We know that

$$\mathbf{E} = E_x\hat{\mathbf{i}} + E_y\hat{\mathbf{j}} + E_z\hat{\mathbf{k}} \tag{8.125}$$

Comparing Eqs. 8.100 and 8.125

$$E_x = E_o \cos(\omega t - \beta z),\ E_y = 0,\ E_z = 0 \tag{8.126}$$

Thus only $E_x$ components are present and the $E_x$ components depend on $z$ only and not on $x$, $y$.

The potential difference between the plates is equal to work done in moving a unit positive charge against the electric field.

Following exactly the same procedure from Eqs. 3.328 to 3.330

$$V = \int_0^d E_x \, dx \tag{8.127}$$

Noting down that $E_x$ depends on $z$ only

$$V = E_x \int_0^d dx = E_x d \tag{8.128}$$

The capacitance per unit length of the transmission line is equal to charge per unit length at any value of $z$ to the line voltage at that value of $z$ (Eq. 3.325). Hence

$$C = \frac{\varepsilon E_n a}{E_x d} \tag{8.129}$$

$E_n$ denotes the normal component of the electric field to the parallel plates of the transmission line. However from Eqs. 8.100, 8.125 and 8.126 $E_x$ is the only normal component. Hence

$$E_x = E_n \tag{8.130}$$

Therefore capacitance per unit length is

$$C = \frac{\varepsilon a}{d} \tag{8.131}$$

(e) **Transmission line equations**

The transmission line equations derived in Eq. 8.4 based on circuit model can be derived from Maxwell's equation as well. Maxwell's Eqs. 6.166 and 6.167 in the absence of free currents can be written as

$$\nabla \times \mathbf{E} = -\frac{\partial \mathbf{B}}{\partial t} = -\mu \frac{\partial \mathbf{H}}{\partial t} \tag{8.132}$$

$$\nabla \times \mathbf{H} = -\frac{\partial \mathbf{D}}{\partial t} = \varepsilon \frac{\partial \mathbf{H}}{\partial t} \tag{8.133}$$

From Eqs. 8.94, 8.95 and 8.126, Eq. 8.132 can be written as

$$\begin{vmatrix} \hat{\mathbf{i}} & \hat{\mathbf{j}} & \hat{\mathbf{k}} \\ \frac{\partial}{\partial x} & \frac{\partial}{\partial y} & \frac{\partial}{\partial z} \\ E_x & 0 & 0 \end{vmatrix} = -\mu \frac{\partial H_y}{\partial t} \tag{8.134}$$

Noting down that $E_x$ depends on $z$ only and is independent of $x$, $y$

$$\frac{\partial E_x}{\partial z} = -\mu \frac{\partial H_y}{\partial t} \tag{8.135}$$

From Eqs. 8.94, 8.95 and 8.126 Eq. 8.132 can be written as

$$\begin{vmatrix} \hat{\mathbf{i}} & \hat{\mathbf{j}} & \hat{\mathbf{k}} \\ \dfrac{\partial}{\partial x} & \dfrac{\partial}{\partial y} & \dfrac{\partial}{\partial z} \\ 0 & \mathrm{H}_y & 0 \end{vmatrix} = -\varepsilon \frac{\partial \mathrm{H}_x}{\partial \mathrm{t}} \tag{8.136}$$

Noting down that $\mathrm{H}_y$ depends on $z$ only and is independent of $x$, $y$

$$\frac{\partial \mathrm{H}_x}{\partial z} = -\varepsilon \frac{\partial \mathrm{E}_y}{\partial \mathrm{t}} \tag{8.137}$$

From the discussion above the surface line current flowing in the parallel plates is

$$\mathrm{I} = \mathrm{K_f\, a} = \mathrm{H_t\, a} = \mathrm{H}_y \mathrm{a} \tag{8.138}$$

Substituting Eqs. 8.128 and 8.138 in Eqs. 8.135 and 8.137

$$\frac{\partial \mathrm{V}}{\partial z} = -\frac{\mu \mathrm{d}}{\mathrm{a}} \frac{\partial \mathrm{I}}{\partial \mathrm{t}} \tag{8.139}$$

$$\frac{\partial \mathrm{I}}{\partial z} = -\frac{\varepsilon \mathrm{a}}{\mathrm{d}} \frac{\partial \mathrm{V}}{\partial \mathrm{t}} \tag{8.140}$$

Using Eqs. 8.124 and 8.131 in 8.139 and 8.140

$$\frac{\partial \mathrm{V}}{\partial z} = -\mathrm{L} \frac{\partial \mathrm{I}}{\partial \mathrm{t}} \tag{8.141}$$

$$\frac{\partial \mathrm{I}}{\partial z} = -\mathrm{C} \frac{\partial \mathrm{V}}{\partial \mathrm{t}} \tag{8.142}$$

For a lossless line noting down that R = 0, G = 0 Eq. 8.6 reduces to Eqs. 8.141 and 8.142

(f) **Poynting vector**

The power flow along the line in the $z$-direction can be evaluated by considering a transverse plane as shown in Fig. 8.12 and calculating the Poynting vector over the plane

$$\mathrm{P}(z, \mathrm{t}) = \int_{x=0}^{\mathrm{d}} \int_{y=0}^{\mathrm{d}} (\mathbf{E} \times \mathbf{H}).\mathrm{ds} \tag{8.143}$$

Noting down that $\mathbf{E} = \mathrm{E}_x \hat{\mathbf{i}}$, $\mathbf{H} = \mathrm{H}_y \hat{\mathbf{j}}$ then

$$\mathbf{E} \times \mathbf{H} = \mathrm{E}_x \mathrm{H}_y \hat{\mathbf{k}} \tag{8.144}$$

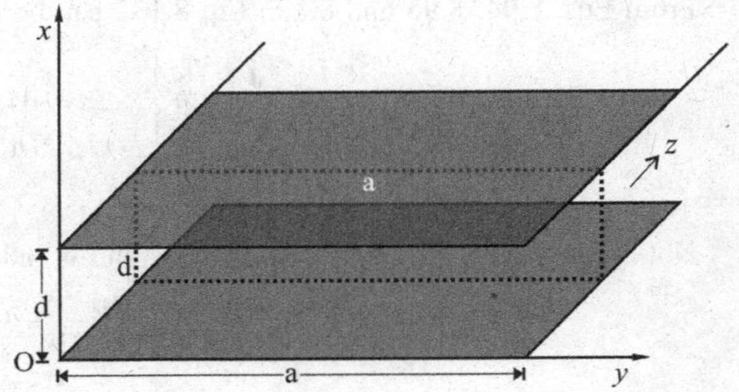


**Fig. 8.12** A parallel plate transmission line—A plane area perpendicular to $z$-axis

Also from Fig. 8.12

$$d\mathbf{s} = dx\,dy\,\hat{\mathbf{k}} \tag{8.145}$$

Substituting Eqs. 8.144 and 8.145 in Eq. 8.143

$$P(z, t) = \int_{x=0}^{d}\int_{y=0}^{a} E_x H_y \hat{\mathbf{k}} \cdot dx\,dy\,\hat{\mathbf{k}} \tag{8.146}$$

Substituting Eqs. 8.128 and 8.138

$$P(z, t) = \int_{x=0}^{d}\int_{y=0}^{a} \frac{V(z, t)}{d}\frac{I(z, t)}{a} dx\,dy \tag{8.147}$$

$$P(z, t) = V(z, t)\, I(z, t) \tag{8.148}$$

is the power flow.

**Example 8.1**
The characteristic impedance of a transmission line is 40 Ω. The capacitance per unit length of the transmission line is 0.2 nanofarad/m. Calculate the velocity of the propagation of the electromagnetic wave in the transmission line.

The transmission line is lossless.

**Solution**
The characteristic impedance of the transmission line is

$$Z = \sqrt{\frac{L}{C}} \Rightarrow \sqrt{L} = Z\sqrt{C}$$

$$\Rightarrow L = Z^2 C$$

$$\Rightarrow L = (40)^2 \times 0.2 \times 10^{-9}$$

$$\Rightarrow L = 320 \times 10^{-9}\ \text{Henry/m}$$

The phase velocity is given by

$$v = \frac{1}{\sqrt{LC}}$$

Substituting the values

$$v = \frac{1}{\sqrt{320 \times 10^{-9} \times 0.2 \times 10^{-9}}}$$
$$v = 1.25 \times 10^{8}\ \text{m/sec}$$

**Example 8.2**
For a distortionless transmission line it is observed that $R = 0.06\,\Omega$ and $G = 23\ \mu S/m$. Calculate characteristic impedance Z and attenuation constant $\alpha$.

**Solution**
Given $G = 23\ \mu S/m$, $R = 0.06\ \Omega$

The condition for distortionless transmission line is

$$\frac{R}{L} = \frac{G}{C}$$

From the above equation

$$\frac{G}{R} = \frac{C}{L}$$

Substituting the values of G and R

$$\frac{23 \times 10^{-6}}{0.06} = \frac{C}{L}$$

Thus the characteristic impedance is

$$Z = \sqrt{\frac{L}{C}} = 51\text{Ohms.}$$

The attenuation constant

$$\alpha = R\sqrt{\frac{C}{L}} = \frac{0.06}{51}$$
$$\alpha = 0.0012\,\text{dB/m}$$

## 8.12 The Infinite Transmission Line

In Eq. 8.20 the first term $V_o^+ e^{-\gamma z}$ represents a wave travelling in the $z$-direction in Fig. 8.2. While the second term $V_o^- e^{\gamma z}$ represents a wave travelling in the $-z$-direction. Thus the first term represents a forward travelling wave and the second term represents a backward travelling wave.

In Sects. 7.11 and 7.12 we saw that when an electromagnetic wave encounters a different medium part of the wave is transmitted while the other part is reflected. In the case of transmission line when there is a change in impedance along the line the wave is partly transmitted and partly reflected. For example in Fig. 8.2 if the load impedance is not equal to the characteristic impedance of the transmission line part of the wave is reflected at the load and another part is transmitted.

In the case of infinite transmission line there will be no reflected wave because the line is of infinite extent with no change in impedance along the line. Hence Eqs. 8.20 and 8.21 for infinite transmission line can be written as

$$V_s(z) = V_o^+ e^{-\gamma z} \tag{8.149}$$

$$I_s(z) = I_o^+ e^{-\gamma z} \tag{8.150}$$

## 8.13 Finite Transmission Line

In the previous section we saw about wave propagation in an infinite transmission line. There is no reflected wave as the transmission line is of infinite extent. Let us now consider the nature of wave propagation in a finite transmission line.

Consider Fig. 8.13 in which the transmission line is of finite extent with length $l$. The $z$-coordinate whose positive direction is from source to load is shown in the figure. A voltage source with internal impedance $Z_g$ is connected to the line at $z = 0$. The characteristic impedance of the transmission line is $Z_0$, and the load impedance is $Z_L$. We require a reference point from which all discussions about the transmission line can be carried out. The normal convention is that the load is taken as reference point. As shown in Fig. 8.13 the reference point $z' = 0$ is set at the load. That is we have introduced a new coordinate $z'$ and the positive direction of $z'$ from load is shown in the figure. $l$ is the length of the transmission line. The voltage and current waves as per Eqs. 8.20 and 8.21 are

$$V_s(z) = V_o^+ e^{-\gamma z} + V_o^+ e^{\gamma z} \tag{8.151}$$

$$I_s(z) = I_o^+ e^{-\gamma z} + I_o^- e^{\gamma z} \tag{8.152}$$

Here the first term in the above Eqs. 8.151 and 8.152 represents a forward travelling wave from source to load in the positive $z$-direction. As seen in Sect. 8.12 whenever there is change in impedance along the line the wave is reflected.

In case, the impedance of the load $Z_L$ is not equal to the characteristic impedance $Z_0$ of the transmission line then the incident wave is partly reflected at the load. In such situations a backward wave travels in the negative $z$-direction.

Now let us consider the coordinate $z'$. From Fig. 8.13 the forward propagating wave from source to load is travelling in the negative $z'$-direction. The backward reflected wave is travelling in the positive $z'$-direction. Therefore Eqs. 8.151 and 8.152 for the $z'$-coordinate can be written as

$$V_s(z') = V_o^+ e^{\gamma z'} + V_o^- e^{-\gamma z'} \tag{8.153}$$

$$I_s(z') = I_o^+ e^{\gamma z'} + I_o^- e^{-\gamma z'} \tag{8.154}$$

From Fig. 8.13 $z' = l - z$
Therefore Eqs. 8.153 and 8.154 can be written as

$$V_s(z') = V_o^+ e^{\gamma(l-z)} + V_o^- e^{-\gamma(l-z)} \tag{8.155}$$

$$I_s(z') = I_o^+ e^{\gamma(l-z)} + I_o^- e^{-\gamma(l-z)} \tag{8.156}$$

The load impedance is defined as

$$Z_L = \frac{V_L}{I_L} \tag{8.157}$$

The load is located at $z' = 0$
Hence

$$Z_L = \frac{V_L}{I_L} = \frac{V_s(z'=0)}{I_s(z'=0)} \tag{8.158}$$

From Eqs. 8.153 and 8.154 with $z' = 0$ we have

$$V_L = V_s(z'=0) = V_o^+ + V_o^- \tag{8.159}$$

$$I_L = I_s(z'=0) = I_o^+ + I_o^- \tag{8.160}$$

Substituting Eqs. 8.159 and 8.160 in Eq. 8.158

$$Z_L = \frac{V_o^+ + V_o^-}{I_o^+ + I_o^-} \tag{8.161}$$

Noting down that the characteristic impedance of the transmission line is the ratio between forward travelling voltage wave and forward travelling current wave, from Eqs. 8.153 and 8.154

$$Z_0 = \frac{V_o^+ e^{\gamma z'}}{I_o^+ e^{\gamma z'}} = \frac{V_o^+}{I_o^+} = -\frac{V_o^-}{I_o^-} \tag{8.162}$$

Substituting Eq. 8.162 in 8.161

$$Z_L = \frac{V_o^+ + V_o^-}{\frac{V_o^+}{Z_0} - \frac{V_o^-}{Z_0}} = Z_0 \frac{V_o^+ + V_o^-}{V_o^+ - V_o^-} \tag{8.163}$$

The above equation can be rearranged as

$$V_o^- = V_o^+ \left( \frac{Z_L - Z_0}{Z_L + Z_0} \right) \tag{8.164}$$

The generalized reflection coefficient ρ is defined as

$$\rho = \frac{V_{\text{reflected}}}{V_{\text{incident}}} \tag{8.165}$$

ρ is the generalized reflection coefficient which denotes the reflection at any point on the transmission line.

In case if there is a mismatch of impedance between the transmission line and load, reflection occurs at load. We denote the load reflection coefficient as $\rho_L$

$$\rho_L = \frac{V_{\text{reflected}}}{V_{\text{incident}}} = \frac{V_o^-}{V_o^+} \tag{8.166}$$

Substituting Eq. 8.164 in 8.166

$$\rho_L = \frac{Z_L - Z_0}{Z_L + Z_0} \tag{8.167}$$

Since $\rho_L$ is the ratio of complex amplitudes $V_o^-$ and $V_o^+$, $\rho_L$ is a complex number.

Now let us derive an expression for line impedance at any point $z'$ on the transmission line in Fig. 8.13.

From Eq. 8.153 and 8.166,

$$V_s(z') = V_o^+ \left( e^{\gamma z'} + \rho_L e^{-\gamma z'} \right) \tag{8.168}$$

From Eqs. 8.154, 8.162 and 8.166,

$$\begin{aligned} I_s(z') &= I_o^+ e^{\gamma z'} + I_o^- e^{-\gamma z'} \\ &= \frac{V_o^+}{Z_o} e^{\gamma z'} - \frac{V_o^-}{Z_o} e^{-\gamma z'} \end{aligned} \tag{8.169}$$

$$I_s(z') = \frac{V_o^+}{Z_o}\left[e^{\gamma z'} - \rho_L e^{-\gamma z'}\right] \tag{8.170}$$

The line impedance at point $z'$ is given by

$$\begin{aligned} Z(z') &= \frac{V_s(z')}{I_s(z')} \\ &= Z_o \frac{(e^{\gamma z'} + \rho_L e^{-\gamma z'})}{(e^{\gamma z'} - \rho_L e^{-\gamma z'})} \end{aligned} \tag{8.171}$$

Substituting for $Z_L$ from Eq. 8.167

$$\begin{aligned} Z(z') &= Z_o\left[\frac{(Z_L + Z_0)e^{\gamma z'} + (Z_L - Z_0)e^{-\gamma z'}}{(Z_L + Z_0)e^{\gamma z'} - (Z_L - Z_0)e^{-\gamma z'}}\right] \\ Z(z') &= Z_o\left[\frac{Z_L(e^{\gamma z'} + e^{-\gamma z'}) + Z_0(e^{\gamma z'} - e^{-\gamma z'})}{Z_0(e^{\gamma z'} + e^{-\gamma z'}) + Z_L(e^{\gamma z'} - e^{-\gamma z'})}\right] \end{aligned} \tag{8.172}$$

Noting down that $(e^{\gamma z'} + e^{-\gamma z'})/2 = \cosh \gamma z'$ and $(e^{\gamma z'} - e^{-\gamma z'})/2 = \sin h\, \gamma z'$ Eq. 8.172 can be written as

$$Z(z') = Z_o\left[\frac{Z_L \cos h\, \gamma z' + Z_0 \sin h\, \gamma z'}{Z_0 \cos h\, \gamma z' + Z_L \sin h\, \gamma z'}\right] \tag{8.173}$$

The above equation can be written as

$$Z(z') = Z_o\left[\frac{Z_L + Z_0 \tan h\, \gamma z'}{Z_0 + Z_L \tan h\, \gamma z'}\right] \tag{8.174}$$

From Eq. 8.168 $V_o^+ e^{\gamma z'}$ is the incident wave and $V_o^+ e^{-\gamma z'}$ is the reflected wave. The generalized reflection coefficient at any point $z'$ of the transmission line is given by

$$\rho(z') = \frac{V_{\text{reflected}}}{V_{\text{incident}}} = \frac{V_o^+ \rho_L e^{-\gamma z'}}{V_o^+ e^{\gamma z'}} \tag{8.175}$$

$$\rho(z') = \rho_L e^{-2\gamma z'} \tag{8.176}$$

The reflection coefficient given by Eq. 8.166 in general is complex, and hence we can write

$$\rho_L = |\rho_L| e^{i\theta_L} \tag{8.177}$$

Now consider a lossless line for which $\alpha = 0$, and hence $\gamma = \alpha + i\beta = i\beta$. Therefore Eqs. 8.168 and 8.170 can be written as

$$V_s(z') = V_o^+ \left( e^{i\beta z'} + \rho_L e^{-i\beta z'} \right) \tag{8.178}$$

$$I_s(z') = \frac{V_o^+}{Z_o} \left( e^{i\beta z'} - \rho_L e^{-i\beta z'} \right) \tag{8.179}$$

The line impedance given by Eq. 8.174 can be written as

$$Z(z') = Z_o \left[ \frac{Z_L + iZ_0 \tan \beta z'}{Z_0 + iZ_L \tan \beta z'} \right] \tag{8.180}$$

where we have used

$$\tan h\, \gamma z' = \tan i\beta z' = i \tan \beta z' \tag{8.181}$$

The quantity $\beta z'$ is called electrical length of the transmission line and can be expressed in degrees or radians.

A transmission line transfers power from source to load, and the average input power delivered the load can be calculated as follows

$$< P > = \frac{1}{2} \mathrm{Re} \left[ V_s(z') I_s^*(z') \right] \tag{8.182}$$

where $z'$ is the distance of the given point on the transmission line from the load as shown in Fig. 8.13

Substituting Eqs. 8.168 and 8.169 in Eq. 8.182

$$< P > = \frac{1}{2} \mathrm{Re} \left[ V_o^+ \left( e^{i\beta z'} + \rho_L e^{-i\beta z'} \right) \frac{V_o^{+*}}{Z_o} \left( e^{i\beta z'} - \rho_L^* e^{-i\beta z'} \right) \right] \tag{8.183}$$

$$< P > = \frac{1}{2} \mathrm{Re} \left[ \frac{|V_o^+|^2}{Z_o} \left( 1 - |\rho_L|^2 + \rho_L e^{-2i\beta z'} - \rho_L^* e^{2i\beta z'} \right) \right] \tag{8.184}$$

$$< P > = \frac{|V_o^+|^2}{2Z_o} \left( 1 - |\rho_L|^2 \right) + \frac{|V_o^+|^2}{2Z_o} \mathrm{Re} \left( \rho_L e^{-2i\beta z'} - \rho_L^* e^{2i\beta z'} \right) \tag{8.185}$$

Substituting Eq. 8.177 in 8.185

$$< P > = \frac{|V_o^+|^2}{2Z_o}\left(1 - |\rho_L|^2\right) + \frac{|V_o^+|^2}{2Z_o}\mathrm{Re}\left(|\rho_L|e^{-i(2\beta z' - \theta_L)} - |\rho_L|e^{i(2i\beta z' - \theta_L)}\right) \tag{8.186}$$

Taking the real parts

$$< P > = \frac{|V_o^+|^2}{2Z_o}\left(1 - |\rho_L|^2\right) + \frac{|V_o^+|^2}{2Z_o}\mathrm{Re}[|\rho_L|\{\cos(2\beta z' - \theta_L) - \cos(2i\beta z' - \theta_L)\}] \tag{8.187}$$

$$< P > = \frac{|V_o^+|^2}{2Z_o}\left[1 - |\rho_L|^2\right] \tag{8.188}$$

## 8.14 Standing Waves

As seen in the previous sections there is a forward travelling wave and a backward travelling wave in Fig. 8.13. The forward and backward travelling waves interfere with each other to produce standing waves. Equations 8.178 and 8.179 written below are the expressions for voltage and current standing waves along the lossless transmission line.

$$V_s(z') = V_o^+\left(e^{i\beta z'} + \rho_L e^{-i\beta z'}\right) \tag{8.189}$$

$$I_s(z') = \frac{V_o^+}{Z_o}\left(e^{i\beta z'} - \rho_L e^{-i\beta z'}\right) \tag{8.190}$$

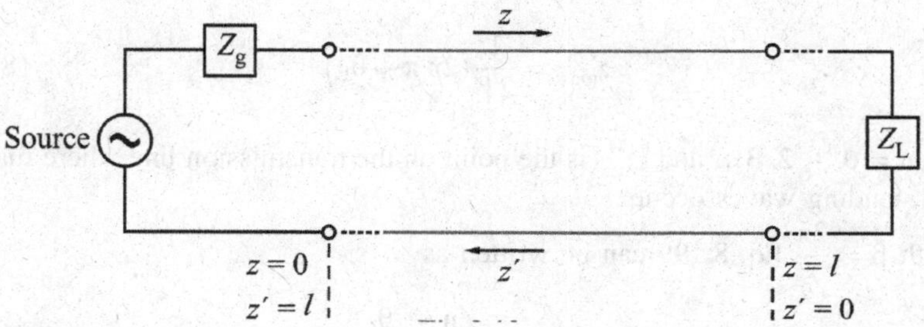


**Fig. 8.13** A finite transmission line of length *l*. The *z* coordinate which is from source to load and *z'* coordinate which is from load to source is shown in the figure

Substituting Eqs. 8.177 in Eq. 8.189 and 8.190

$$V_s(z') = V_o^+ \left(e^{i\beta z'} + |\rho_L| e^{-i(\beta z' - \theta_L)}\right) \tag{8.191}$$

$$I_s(z') = \frac{V_o^+}{Z_o} \left(e^{i\beta z'} - |\rho_L| e^{-i(\beta z' - \theta_L)}\right) \tag{8.192}$$

Noting down that $V_S(z')$ and $I_S(z')$ are complex numbers we have

$$|V_s(z')| = \left[V_s(z') V_s^*(z')\right]^{1/2} \tag{8.193}$$

where $V_S^*(z')$ is the complex conjugate of $V_S(z')$. Substituting Eqs. 8.191 in 8.193

$$|V_s(z')| = \left\{V_o^+ \left(e^{i\beta z'} + |\rho_L| e^{-i(\beta z' - \theta_L)}\right) V_o^{+*} \left(e^{-i\beta z'} + |\rho_L| e^{i(\beta z' - \theta_L)}\right)\right\}^{1/2} \tag{8.194}$$

Simplifying

$$|V_s(z')| = |V_o^+| \left[1 + |\rho_L|^2 + |\rho_L| \left\{e^{i(2\beta z' - \theta_L)} + e^{-i(2\beta z' - \theta_L)}\right\}\right]^{1/2} \tag{8.195}$$

$$\Rightarrow |V_s(z')| = |V_o^+| \left[1 + |\rho_L|^2 + 2|\rho_L| \cos(2\beta z' - \theta_L)\right]^{1/2} \tag{8.196}$$

Following same procedure for Eq. 8.192

$$|I_s(z')| = \frac{|V_o^+|}{Z_o} \left[1 + |\rho_L|^2 - 2|\rho_L| \cos(2\beta z' - \theta_L)\right]^{1/2} \tag{8.197}$$

The variations of $V_S(z')$ and $I_S(z')$ in Eq. 8.196 and 8.197 gives rise to standing wave pattern. The maximum value of $V_S(z')$, $V_{max}$ occurs when

$$2\beta z'_{max} - \theta_L = 2n\pi \tag{8.198}$$

(or)

$$z'_{max} = \frac{1}{2\beta}(2n\pi + \theta_L) \tag{8.199}$$

where n = 0, 1, 2, 3... and $z'_{max}$ is the point on the transmission line where maxima of the standing waves occurs.

With $\beta = \frac{2\pi}{\lambda}$ Eq. 8.199 can be written as

$$z'_{max} = \lambda\left(\frac{n}{2} + \frac{\theta_L}{4\pi}\right) \tag{8.200}$$

Substituting the values of n = 0, 1, 2, 3 … in Eq. 8.200, the points on the transmission line where the maxima occurs can be identified.

$$z'_{max} = \frac{\lambda\theta_L}{4\pi}, \lambda\left(\frac{1}{2} + \frac{\theta_L}{4\pi}\right), \lambda\left(1 + \frac{\theta_L}{4\pi}\right), \ldots \quad (8.201)$$

Thus the maxima of the standing wave are separated by $\lambda/2$.

Using Eq. 8.198 in 8.196

$$V_{max} = |V_o^+|[1 + |\rho_L|] \quad (8.202)$$

The minimum value of $V_S(z')$ and $V_{min}$ in Eq. 8.196 occurs when

$$2\,\beta z'_{min} - \theta_L = (2n+1)\pi \quad (8.203)$$

(or)

$$z'_{min} = \frac{1}{2\,\beta}[(2n+1)\pi + \theta_L] \quad (8.204)$$

where n = 0, 1, 2, 3 … and $z'_{min}$ is the point on the transmission line where minima of the standing waves occurs.

With $\beta = \frac{2\pi}{\lambda}$ Eq. 8.204 can be written as

$$z'_{min} = \lambda\left[\frac{2n+1}{4} + \frac{\theta_L}{4\pi}\right] \quad (8.205)$$

Substituting the values of n = 0, 1, 2, 3 … in Eq. 8.205 the points on the transmission line where the minima occurs can be identified

$$z'_{min} = \lambda\left[\frac{1}{4} + \frac{\theta_L}{4\pi}\right], \lambda\left[\frac{3}{4} + \frac{\theta_L}{4\pi}\right], \lambda\left[\frac{5}{4} + \frac{\theta_L}{4\pi}\right] \ldots \quad (8.206)$$

Thus the minima of the standing waves are separated by $\lambda/2$.

Substituting Eq. 8.203 in 8.196

$$V_{min} = |V_o^+|[1 - |\rho_L|] \quad (8.207)$$

From Eq. 8.197 the maximum value of $I_S(z')$ and $I_{max}$ occurs when

$$2\,\beta z'_{max} - \theta_L = (2n+1)\pi \quad (8.208)$$

(or)

$$z'_{max} = \lambda\left[\frac{2n+1}{4} + \frac{\theta_L}{4\pi}\right] \quad (8.209)$$

And the minimum value of $I_S(z')$ and $I_{min}$ occurs when

$$2\beta z'_{min} - \theta_L = 2n\pi$$

(or)

$$z'_{min} = \lambda\left[\frac{n}{4} + \frac{\theta_L}{4\pi}\right] \tag{8.210}$$

Comparison of Eqs. 8.201, 8.204, 8.209 and 8.210 shows that the maximum of voltage wave corresponds to minimum of current wave and minimum of voltage wave corresponds to maximum of current wave. This is shown pictorially in Fig. 8.14. In the figure the voltage and current waves are in phase opposition. We observe that when the voltage wave is maximum current wave is minimum and vice versa. Using Eq. 8.197 $I_{max}$ and $I_{min}$ can be written as

$$I_{max} = \frac{V_{max}}{Z_o} = \frac{|V_o^+|}{Z_o}[1 + |\rho_L|] \tag{8.211}$$

$$I_{min} = \frac{V_{min}}{Z_o} = \frac{|V_o^+|}{Z_o}[1 - |\rho_L|] \tag{8.212}$$

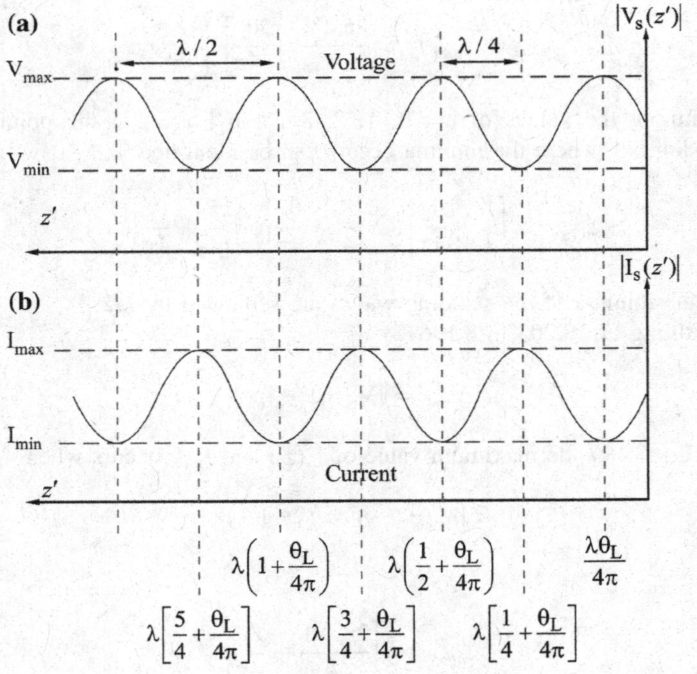


**Fig. 8.14** Standing waves of voltage and current in a finite transmission line. The maxima's and minima's of voltage and current are shown in the figure

Standing wave pattern of $|V_S(z')|$ and $|I_S(z')|$ is shown in Fig. 8.14, Two adjacent maxima and two adjacent minima are separated by $\lambda/2$. The adjacent maxima and minima are separated by $\lambda/4$. $V_{max}$, $V_{min}$ and $I_{max}$, $I_{min}$ given by Eqs. 8.202, 8.207, 8.204 and 8.205 are shown in the figure.

We define standing wave ratio (SWR) to estimate the degree of mismatch of the impedance of the load to the impedance of the lossless transmission line.

$$\text{SWR} = \frac{V_{max}}{V_{min}} = \frac{I_{max}}{I_{min}} \tag{8.213}$$

Substituting Eqs. 8.202, 8.207 and 8.204, 8.205 in 8.206

$$\text{SWR} = \frac{1 + |\rho_L|}{1 - |\rho_L|} \tag{8.214}$$

From Eq. 8.166 when there is no reflection $|\rho_L| = 0$. In that case SWR is 1. When the incident wave is completely reflected then $|\rho_L| = 1$. In that case SWR is $\infty$. Thus SWR varies between 1 to $\infty$.

We will now consider special cases.

(a) **Lossless, shorted transmission line**

For a shorted line $Z_L = 0$, then from Eq. 8.180

$$Z(z') = iZ_0 \tan \beta z' \tag{8.215}$$

And from Eqs. 8.167 and 8.214

$$\rho_L = -1, \quad \text{SWR} = \infty \tag{8.216}$$

$\tan \beta z'$ in Eq. 8.215 can alternate between $-\infty$ to $\infty$. Thus the variation of $Z(z')$ in Eq. 8.215 is shown in Fig. 8.15. $Z(z')$ is pure reactance, and from the figure we notice that $Z(z')$ could be capacitive or inductive depending upon $z'$.

(b) **Lossless open-circuited transmission line**

For a open-circuited transmission line $Z_L \to \infty$. As $Z_L >> Z_0$ from Eq. 8.180

$$Z(z') = z_0 \left[\frac{Z_L}{iZ_L \tan \beta z'}\right] \tag{8.217}$$

$$Z(z') = -iZ_o \cot \beta z' \tag{8.218}$$

And from Eqs. 8.167 and 8.214

$$\rho_L = 1, \quad \text{SWR} = \infty \tag{8.219}$$

The variation of $Z(z')$ with $\beta z'$ is shown in Fig. 8.16, and as in shorted transmission line the impedance is capacitive or inductive.

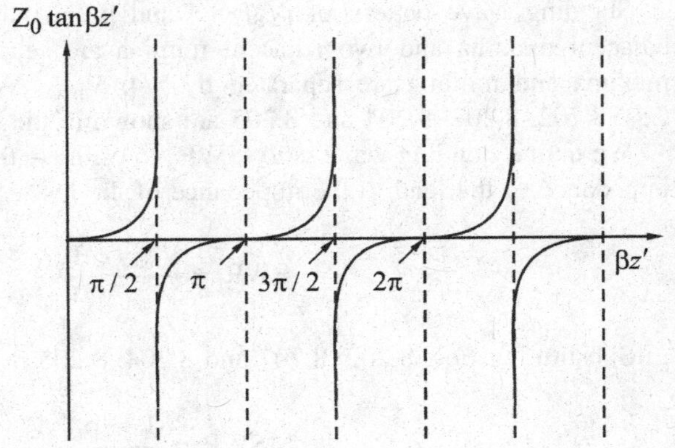


**Fig. 8.15** Impedance of a shorted transmission line

(c) **Lossless matched transmission line**

For a matched transmission line the load impedance is equal to the characteristic impedance of the transmission line.

$$Z_L = Z_0 \tag{8.220}$$

The line impedance on any given point $z'$ of the line is

$$Z(z') = Z_0 \left[ \frac{Z_0 + iZ_0 \tan \beta z'}{Z_0 + iZ_0 \tan \beta z'} \right] \tag{8.221}$$

From Eqs. 8.167 and 8.214

$$\rho_L = 0, \; \text{SWR} = 1 \tag{8.222}$$

signifying that there is no reflected wave and hence no standing wave in the transmission line. All the power in the transmission line is transmitted to the load.

**Example 8.3**
Consider a parallel-plate transmission line with a width 20 mm, and separation between the plates is 1 mm. The applied voltage of the transmission line is V (t) = $120\cos 10^5 t$ V. The plates are separated by free space. Determine the variation of surface charge and surface current density on the plate.

**Solution**

$$\beta = \frac{\omega}{v} = \frac{10^5}{3 \times 10^8} = 3.3 \times 10^{-4} \text{ rad/m}$$

and for free space

$$\eta = 377\,\Omega$$

Neglecting fringing fields and assuming the charge distribution on the lower plate is positive with respect to the upper plate at a given instant the electric field intensity between the two plates at $z = 0$ is

$$E_o = \frac{V}{d} = \frac{120\cos 10^5 t}{1 \times 10^{-3}}$$

$$E_o = 120 \times 10^3 \cos 10^5 t$$

The surface charge density on the lower plate is

$$\sigma_f = \varepsilon_0 E_0 \cos(\omega t - \beta z)$$

$$\sigma_f = 8.85 \times 10^{-12} \times 120 \times 10^3 \cos\left[10^5 t - 3.3 \times 10^{-4} z\right]$$

$$\sigma_f = 1.06\cos\left[10^5 t - 3.3 \times 10^{-4} z\right]\mu C/m^2$$

The surface current density of the lower plate is

$$K_f = \frac{E_0}{\eta}\cos[\omega t - \beta z]$$

$$K_f = \frac{120 \times 10^3}{377}\cos[\omega t - \beta z]$$

$$K_f = 318\cos[10^5 t - 3.3 \times 10^{-4} z]$$

**Example 8.4**

The characteristic impedance of a lossless transmission line is 200 Ω. It is connected to a load whose impedance is 200 + i314 Ω. At 5 MHz calculate

(a) the reflection coefficient
(b) standing wave ratio
(c) location of the voltage maximum nearest to the load
(d) location of the current maximum nearest to the load.

**Solution**

Given

$$Z_L = 200 + i314\,\Omega$$

$$Z_0 = 200\,\Omega$$

(a) the reflection coefficient

$$\rho_L = \frac{Z_L - Z_0}{Z_L + Z_0} = \frac{i314\,\Omega}{400 + i314\,\Omega}$$
$$= 0.381 + i0.486$$
$$= 0.618e^{i51.9}$$

(b)

$$SWR = \frac{1 + |\rho_L|}{1 - |\rho_L|}$$
$$= \frac{1 + 0.618}{1 - 0.618}$$
$$SWR = 4.24$$

(c) The first voltage maximum nearest to the load

$$z'_{max} = \frac{\lambda\theta_L}{4\pi}$$
$$= \left[\frac{75}{4\pi}\right]\left[\frac{51.9^\circ\pi}{180^\circ}\right]$$
$$z'_{max} = 5.4\text{m}$$

(d) The first current maximum nearest to the load occurs at

$$z'_{max} = \frac{\lambda}{4} + \frac{\lambda\theta_L}{4\pi}$$
$$z'_{max} = \frac{75}{4} + \left\{\left[\frac{75}{4\pi}\right]\left[\frac{51.9^\circ\pi}{180^\circ}\right]\right\}$$
$$z'_{max} = 24.15\,\text{m}$$

**Example 8.5**
Consider a lossless transmission line whose length is equal to λ/4. Calculate the input impedance of the transmission line at the sending end and comment on the results.

**Solution**
The input impedance [Eq. 8.180] at the sending end of the lossless transmission line is

$$Z_{in} = Z_0\left[\frac{Z_L + iZ_0\tan\left(\frac{2\pi}{\lambda}\frac{\lambda}{4}\right)}{Z_0 + iZ_L\tan\left(\frac{2\pi}{\lambda}\frac{\lambda}{4}\right)}\right] \tag{8.223}$$

$$\text{where } z' = \frac{\lambda}{4} \text{ and } \beta = \frac{2\pi}{\lambda}$$

$$Z_{in} = Z_0 \left[\frac{Z_L + iZ_0 \tan \pi/2}{Z_0 + iZ_L \tan \pi/2}\right] \tag{8.224}$$

$$Z_{in} = \frac{Z_0^2}{Z_L} \tag{8.225}$$

A line with length $\lambda/4$ is called quarter-wave transmission line. Suppose assume that at the receiving end the line is shortened then as per Eq. 8.225 the input impedance becomes infinite. Thus when a quarter wavelength transmission line is shortened at the receiving end it behaves as open-circuit at the input terminal. Also from Eq. 8.225 we observe that the quarter wavelength transmission line terminated into a open-circuit acts like a short-circuit as viewed from the input terminals. The above conclusions hold good for a transmission line of given length, for the length of the transmission line can be equal to quarter wavelength at one frequency only.

**Example 8.6**

Consider a lossless transmission line whose length is equal to $\lambda/2$. Calculate the input impedance of the transmission line at the sending end.

**Solution**

For lossless transmission line with length $\lambda/2$ the input impedance as per Eq. 8.180 is given by

$$Z(0) = Z_0 \left[\frac{Z_L + iZ_o \tan \frac{2\pi}{\lambda}\frac{\lambda}{2}}{Z_o + iZ_L \tan \frac{2\pi}{\lambda}\frac{\lambda}{2}}\right] \tag{8.226}$$

$$Z(0) = Z_L \tag{8.227}$$

A line with length $\lambda/2$ is called half wavelength line. As per Eq. 8.227 when the length of the transmission line is equal to half the wavelength the load impedance is transferred to the input terminals without change.

## 8.15 Lossless Transmission Lines with Resistive Transmission

Suppose in Fig. 8.13 the load is purely resistive load $R_L$ and the transmission line is lossless. Then the load $Z_L = R_L$ is the real quantity. For a lossless transmission the characteristic impedance given by Eq. 8.38 is also real. The reflection coefficient from Eq. 8.127 can then be written as

$$\rho_L = \frac{R_L - Z_o}{R_L + Z_o} \tag{8.228}$$

Thus the reflection coefficient is a real quantity. There are two possible special cases to consider

**Case (a)**

$R_L > Zo$. In this case $\rho_L$ is real and positive so that we can write

$$\rho_L = \frac{R_L - Z_o}{R_L + Z_o} \Rightarrow \rho_L = |\rho_L| e^{i0} \tag{8.229}$$

From Eqs. 8.156 and 8.157 the voltage and current waves can be written as

$$|V_s(z')| = |V_o^+| \left[1 + |\rho_L|^2 + 2|\rho_L| \cos(2\beta z' - \theta_L)\right]^{1/2} \tag{8.230}$$

$$|I_s(z')| = \frac{|V_o^+|}{Z_o} \left[1 + |\rho_L|^2 - 2|\rho_L| \cos(2\beta z' - \theta_L)\right]^{1/2} \tag{8.231}$$

Noting down that at the load $z' = 0$ and from Eq. 8.229 $\theta_L = 0$ the voltage and current at the load can be written from Eqs. 8.230 and 8.231as

$$|V_L| = |V_o^+| [1 + |\rho_L|] \tag{8.232}$$

$$|I_L| = \frac{|V_o^+|}{Z_o} [1 - |\rho_L|] \tag{8.233}$$

From Eqs. 8.230 and 8.231 the maximum and minimum values of voltage and current occur when

$$2\beta z' - \theta_L = 2n\pi \tag{8.234}$$

Therefore using Eqs. 8.230, 8.231, 8.232 and 8.233 we can write

$$|V_{max}| = |V_o^+| [1 + |\rho_L|] = V_L \tag{8.235}$$

$$|I_{min}| = \frac{|V_o^+|}{Z_o} [1 - |\rho_L|] = I_L \tag{8.236}$$

Similarly from Eqs. 8.230 and 8.231 the minimum and maximum values of voltage and current occur at

$$2\beta z' - \theta_L = (2n+1)\pi \tag{8.237}$$

Therefore from Eqs. 8.229, 8.230, 8.231, 8.232 and 8.233 we can write

$$|V_{min}| = |V_o^+|[1 - |\rho_L|] = V_L \frac{Z_o}{R_L} \tag{8.238}$$

$$|I_{max}| = \frac{|V_o^+|}{Z_o}[1 + |\rho_L|] = I_L \frac{R_L}{Z_o} \tag{8.239}$$

Thus the SWR is given by

$$\left|\frac{V_{max}}{V_{min}}\right| = \left|\frac{I_{max}}{I_{min}}\right| = SWR = \frac{R_L}{Z_o} \tag{8.240}$$

From Eqs. 8.229 and 8.234 with $\theta_L = 0$, $\beta = \frac{2\pi}{\lambda}$ the voltage maximum occurs at

$$z'_{max} = \frac{\lambda}{4\pi}(2n\pi) \tag{8.241}$$

This means the first voltage maximum occurs at terminating resistance n = 0, $z'_{max} = 0$. Other maximum of the voltage standing wave (Minimum of current standing wave) occurs at intervals of $\frac{\lambda}{2}$.

**Case (b)**

$R_L < Z_o$. From Eq. 8.228 $\rho_L$ is real and negative. Hence

$$\rho_L = \frac{R_L - Z_o}{R_L + Z_o} \Rightarrow \rho_L = -|\rho_L| = |\rho_L|e^{-i\pi} \tag{8.242}$$

Using Eqs. 8.242, 8.156 and 8.157 the voltage and current waves can be written as

$$|V_s(z')| = |V_o^+|[1 + |\rho_L|^z + 2|\rho_L|\cos(2\beta z' - \pi)] \tag{8.243}$$

$$|I_s(z')| = \frac{V_o^+}{Z_o}[1 + |\rho_L|^z - 2|\rho_L|\cos(2\beta z' - \pi)] \tag{8.244}$$

The above equations can be simplified to

$$|V_s(z')| = |V_o^+|[1 + |\rho_L|^z - 2|\rho_L| \cos 2\beta z'] \quad (8.245)$$

$$|I_s(z')| = \frac{|V_o^+|}{Z_o}[1 + |\rho_L|^z + 2|\rho_L| \cos 2\beta z'] \quad (8.246)$$

The voltage and current at the load can be obtained from Eqs. 8.245 and 8.246 by setting $z' = 0$.

$$V_L = |V_o^+|[1 - |\rho_L|] \quad (8.247)$$

$$I_L = \frac{|V_o^+|}{Z_o}[1 + |\rho_L|] \quad (8.248)$$

The minimum value of voltage and maximum value of current along the transmission line occur in Eqs. 8.245 and 8.246 when

$$2\beta z' = 2n\,\pi \quad (8.249)$$

Therefore using Eqs. 8.245, 8.246, 8.247, 8.248 and 8.249

$$|V_{min}| = |V_o^+|[1 - |\rho_L|] \quad (8.250)$$

$$I_{max} = \frac{|V_o^+|}{Z_o}[1 + |\rho_L|] = I_L \quad (8.251)$$

Similarly from Eqs. 8.245 and 8.246 the maximum value of voltage and minimum value of current occur when

$$2\beta z' = (2n + 1)\pi \quad (8.252)$$

Therefore using Eqs. 8.247, 8.248, 8.245, 8.246 and 8.252 and noting $\rho_L = -|\rho_L|$ in Eq. 8.242 we can write

$$|V_{max}| = |V_o^+|[1 + |\rho_L|] = V_L \frac{Z_o}{R_L} \quad (8.253)$$

$$|I_{min}| = \frac{|V_o^+|}{Z_o}[1 - |\rho_L|] = I_L \frac{R_L}{Z_o} \quad (8.254)$$

Thus the standing wave ratio is then

$$\left|\frac{V_{max}}{V_{min}}\right| = \left|\frac{I_{max}}{I_{min}}\right| = SWR = \frac{Z_o}{R_L} \quad (8.255)$$

As per Eq. 8.245 the voltage maximum occurs at

$$z'_{\min} = \frac{\lambda}{\pi}(2n\,\pi) \qquad n = 0, 1, 2\ldots \tag{8.256}$$

Thus the first voltage minimum (current maximum) occurs at terminating resistance n = 0, $z'_{\min} = 0$. Other minimum of voltage standing wave (maximum of current standing wave) occurs at intervals of $\lambda/2$.

From Eq. 8.245 the voltage maximum occurs at

$$2\beta z' = (2n+1)\pi \quad n = 0, 1, 2, 3\ldots \tag{8.257}$$

The line impedance for both the cases is given by

$$Z(z') = Z_o\left[\frac{R_L + iZ_o \tan \beta z'}{Z_o + iR_L \tan \beta z'}\right] \tag{8.258}$$

Thus the impedance varies with respect to $z'$.

## 8.16 Smith Chart

In the discussion of transmission lines in the previous sections the reader might have noticed that the entire study depends on calculations involving complex numbers. The determination of impedance, reflection coefficient, etc., of the transmission line involves complex calculations of complex numbers. To ease out the calculation graphical methods were invented out of which Smith chart is widely used.

We will see how to construct a Smith chart. From Eq. 8.167

$$\rho_L = \frac{Z_L - Z_o}{Z_L + Z_o} \tag{8.259}$$

The above equation can be written as

$$Z_L = Z_o\left[\frac{1+\rho_L}{1-\rho_L}\right] \tag{8.260}$$

$\rho_L$ being a complex number we can write

$$\rho_L = |\rho_L|e^{i\theta_L} = u + iv \tag{8.261}$$

From the above equation

$$|\rho_L|^2 = u^2 + v^2 \tag{8.262}$$

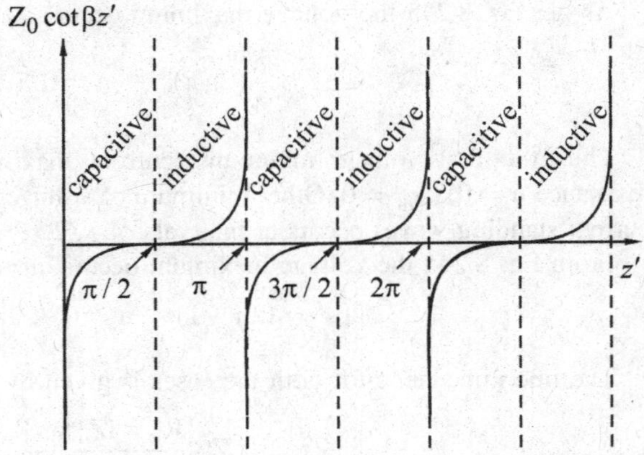


**Fig. 8.16** Impedance of a open—circuited transmission line

which is the equation of the circle with radius $|\rho_L|$ and centred at origin. From Eq. 8.166 when there is no reflection $|\rho_L| = 0$. On the other hand when the incident wave is completely reflected then $|\rho_L| = 1$. Thus $|\rho_L|$ varies between 0 to 1.

Therefore we construct a Smith chart within a circle of unit radius $|\rho_L| = 1$ as shown in Fig. 8.17, and u and v are the projections on the real and imaginary axis as shown in the same figure.

We will use normalized impedance, as normalizing the impedance allows the Smith chart to be used for problems involving any characteristic impedance. All impedances are normalized with respect to characteristic impedance as follows. From Eq. 8.260

$$\mathfrak{z}_L = \frac{Z_L}{Z_o} = \frac{1 + \rho_L}{1 - \rho_L} \tag{8.263}$$

$\mathfrak{z}_L$ being the complex number we can write

$$\mathfrak{z}_L = r + ix \tag{8.264}$$

From Eqs. 8.261, 8.263 and 8.264, we can write

$$r + ix = \frac{1 + u + iv}{1 - u - iv} \tag{8.265}$$

Simplifying the above equation

$$r = \frac{1 - u^2 - v^2}{(1 - u)^2 + v^2} \tag{8.266}$$

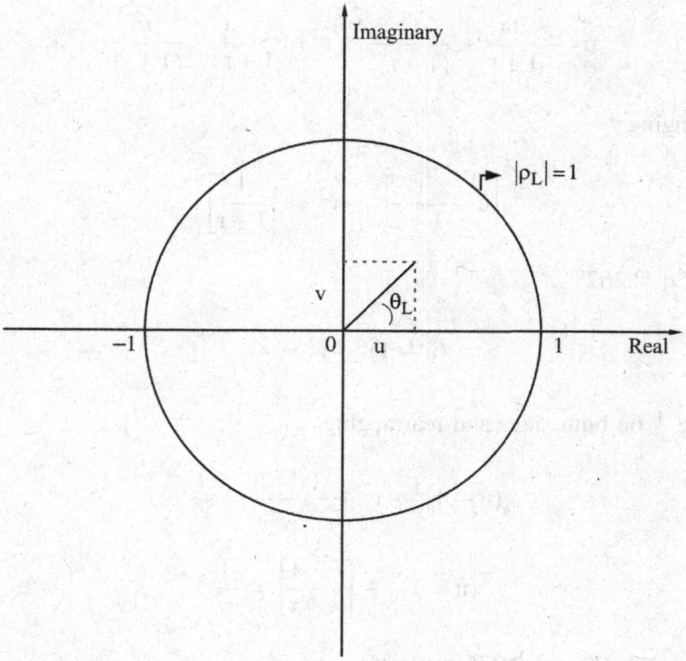


**Fig. 8.17** Smith chart—a circle of unit radius

$$x = \frac{2v}{(1-u)^2 + v^2} \tag{8.267}$$

From Eq. 8.266

$$\left[1 + u^2 - 2u + v^2\right] r = 1 - u^2 - v^2 \tag{8.268}$$

Rearranging

$$u^2(r+1) - 2ur + v^2[1+r] = 1 - r \tag{8.269}$$

Dividing by 1 + r

$$u^2 - \frac{2ur}{1+r} + v^2 = \frac{1-r}{1+r} \tag{8.270}$$

Adding $\frac{r^2}{(1+r)^2}$ on both sides

$$u^2 - \frac{2ur}{1+r} + \frac{r^2}{(1+r)^2} + v^2 = \frac{1-r}{1+r} + \frac{r^2}{(1+r)^2} \quad (8.271)$$

Rearranging

$$\left[u - \frac{r}{1+r}\right]^2 + v^2 = \left[\frac{1}{1+r}\right]^2 \quad (8.272)$$

From Eq. 8.267

$$(u-1)^2 + v^2 = \frac{2v}{x} \quad (8.273)$$

Adding $\frac{1}{x^2}$ on both sides and rearranging

$$(u-1)^2 + v^2 + \frac{1}{x^2} - \frac{2v}{x} = \frac{1}{x^2} \quad (8.274)$$

$$(u-1)^2 + \left[v - \frac{1}{x}\right]^2 = \frac{1}{x^2} \quad (8.275)$$

Equations 8.272 and 8.275 are thus

$$\left(u - \frac{r}{1+r}\right)^2 + v^2 = \frac{1}{(1+r)^2} \quad (8.276)$$

$$(u-1)^2 + \left[v - \frac{1}{x}\right]^2 = \frac{1}{x^2} \quad (8.277)$$

Equations 8.276 and 8.277 are similar to the equation

$$(x-a)^2 + (y-b)^2 = R^2 \quad (8.278)$$

The above equation is the general equation of two circles with radius R centred at (a, b).

Equation 8.276 is thus r-circle—the resistance circle with

$$\text{radius} = \frac{1}{1+r} \quad (8.279)$$

centre at

$$(u, v) = \left(\frac{1}{1+r}, 0\right) \tag{8.280}$$

The r-circle for a given r is shown in Fig. 8.18.

In Fig. 8.19 the resistance circles for various values of r are shown.

As an example for r = 5 the radius is $\frac{1}{6}$ and the centre is $\left(\frac{5}{6}, 0\right)$.

Similarly Eq. 8.277 is a $x$-circle the resistance circle with

$$\text{radius} = \frac{1}{x} \tag{8.281}$$

$$\text{center at } (u, v) = \left(1, \frac{1}{x}\right) \tag{8.282}$$

The $x$-circle for a given $x$ is shown in Fig. 8.20.

In Fig. 8.21 the $x$-circles for various values of $x$ are shown.

The reader might notice that while r is always positive, $x$ can be positive (for inductive impedance) or negative (for capacitive impedance). Accordingly there are two possible circles for a given $x$ one centred at u = 1, $v = -\frac{1}{x}$ for positive $x$ and other centred at u = 1, $v = -\frac{1}{x}$ for negative $x$ as shown in Fig. 8.21.

As $|\rho| \leq 1$, only that part of r-circle and $x$-circle within $|\rho_L| \leq 1$ is shown in Figs. 8.19 and 8.21. Remaining part of circles lying outside $|\rho_L| \leq 1$ is meaningless.

The following points about the r-circle and $x$-circle can be noted.

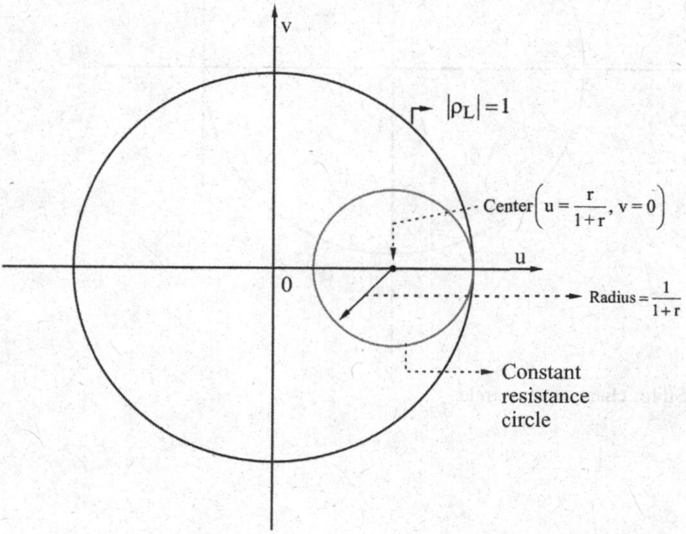


**Fig. 8.18** Smith chart—the r-circle

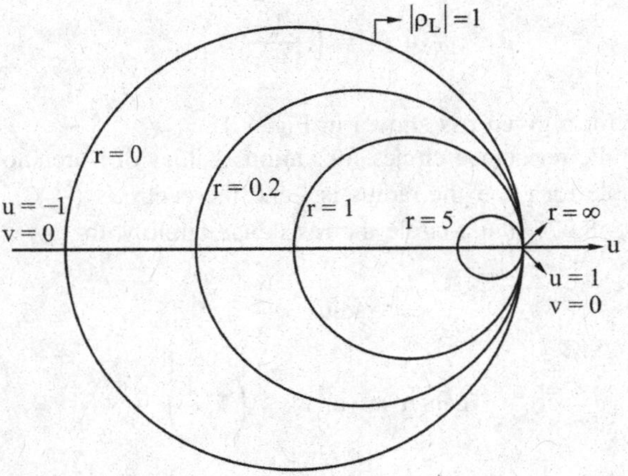


**Fig. 8.19** Smith chart—resistance circles for various values of r

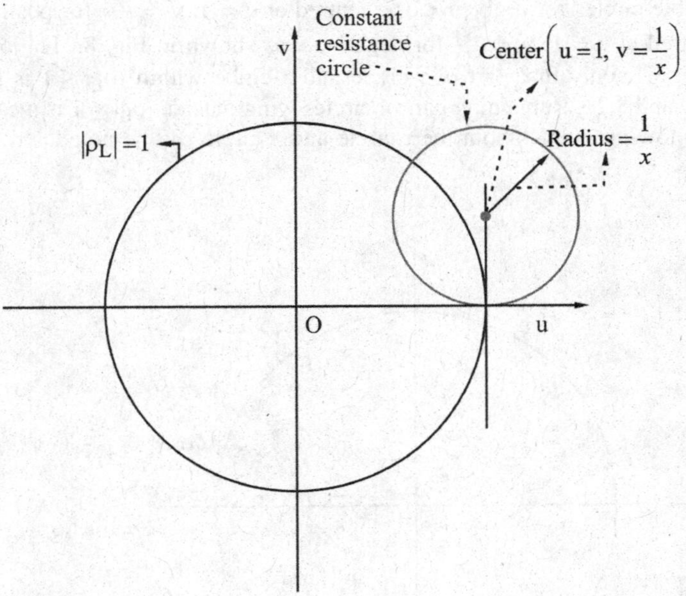


**Fig. 8.20** Smith chart—the *x*-circle

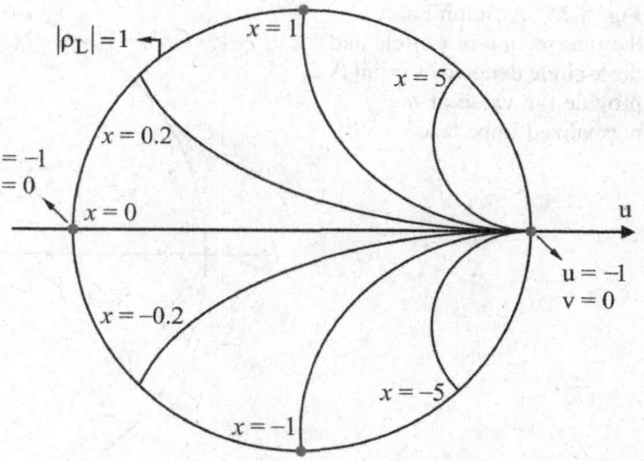


**Fig. 8.21** Smith chart-$x$-circle for various values of $x$

(1) It can be observed from Fig. 8.19 that r-circle becomes progressively smaller as r increases from 0 to ∞ ending at u = 1, $v$ = 0 point.
(2) From Fig. 8.21 it is observed that $x$ = 0 circle becomes the u-axis. As $|x|$ increases from 0 towards ∞ the $x$-circles becomes progressively smaller ending at u = 1, $v$ = 0 point.
(3) Both r-circles and $x$-circles pass through the u = 1, $v$ = 0 point.
(4) The circles for $x$ and $-x$ are images of each other reflected about the u-axis.

A Smith chart is obtained by superposing the r-circles and the $x$-circles as shown in Fig. 8.22.

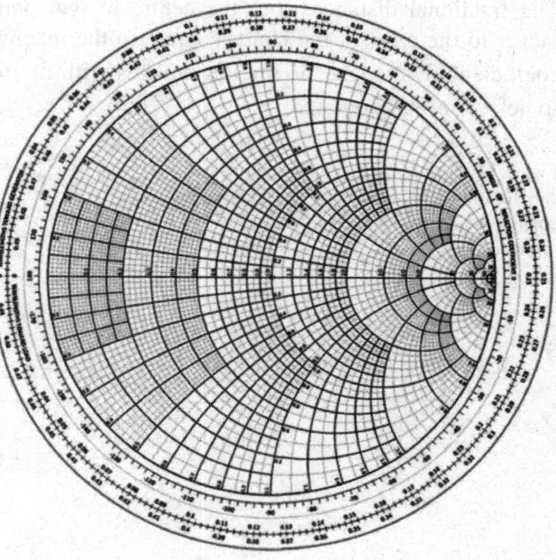

Smith Chart

**Fig. 8.22** A Smith chart

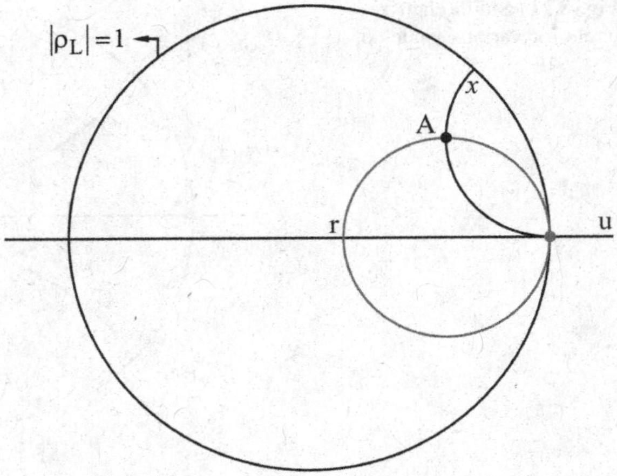


**Fig. 8.23** A Smith chart—the intersection of r-circle and the $x$-circle denoted as point A provide the value of the normalized impedance

The intersection of the r-circle and the $x$-circle defines a point where the normalized impedance is given by $\mathfrak{z} = \mathrm{r} + \mathrm{i}x$. For the point u = 1, $v = 0$ we have r = ∞ giving $\mathfrak{z} = \infty$ which is the impedance of the open transmission line. For the point u = –1, $v = 0$ we have r = 0, $x = 0$ giving $\mathfrak{z} = 0$ which is the impedance of the short circuit transmission line.

As noted above the point if intersection of the r-circle and the $x$-circle noted as point A in Fig. 8.23 provides the value of normalized impedance $\mathfrak{z}_L = \mathrm{r} + \mathrm{i}x$

The intersection point marked as A is shown in Fig. 8.24.

Draw a circle centred at origin [dotted line in Fig. 8.24] passing through point A. The fractional distance from the centre to two point A as compared to the unit radius to the edge of the chart is equal to the magnitude $|\rho_L|$ of the load reflection coefficient. The angle the line OA makes with the real axis is $\theta_L$. Therefore $\rho_L = |\rho_L|e^{i\theta_L}$ can be computed.

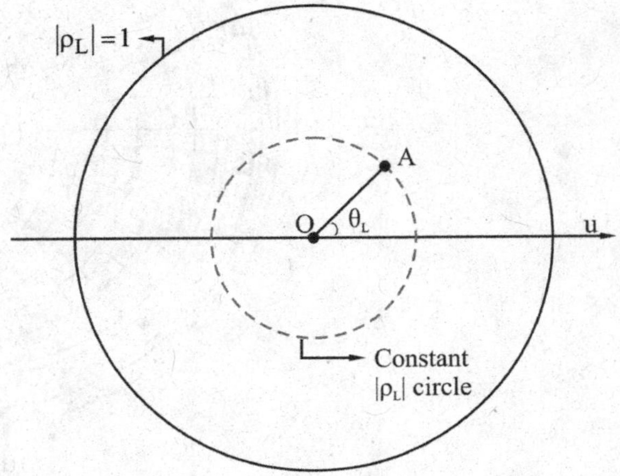


**Fig. 8.24** Smith chart—calculation of load reflection coefficient

## (a) Calculation of input impedance

Smith chart can be used to calculate the input impedance at a distance $z'$ from the load. As $\alpha = 0$ for a lossless line we have $r = \alpha + i\beta = i\beta$. Therefore from Eq. 8.171 we can write the input impedance as

$$Z_i(z') = Z_o \frac{[e^{i\beta z'} + \rho_L e^{-i\beta z'}]}{[e^{i\beta z'} - \rho_L e^{-i\beta z'}]} \tag{8.283}$$

$$Z_i(z') = Z_o \left[\frac{1 + \rho_L e^{-2i\beta z'}}{1 - \rho_L e^{-2i\beta z'}}\right] \tag{8.284}$$

The normalized input impedance is then given by

$$\mathfrak{z}i = \frac{Z_i}{Z_o} = \frac{1 + \rho_L e^{-2i\beta z'}}{1 - \rho_L e^{-2i\beta z'}} \tag{8.285}$$

Substituting for $\rho_L$ from Eq. 8.177

$$\mathfrak{z}i = \frac{1 + |\rho_L|\, e^{i(\theta_L - 2\beta z')}}{1 - |\rho_L|\, e^{i(\theta_L - 2\beta z')}} \tag{8.286}$$

The above equation is similar to Eq. 8.263

$$\mathfrak{z}L = \frac{1 + |\rho_L|}{1 - |\rho_L|} = \frac{1 + |\rho_L| e^{i\theta_L}}{1 - |\rho_L| e^{i\theta_L}} \tag{8.287}$$

where Eq. 8.177 is used.

Comparing Eqs. 8.286, 8.287 we observe the $\mathfrak{z}_L$ is replaced by $\mathfrak{z}_i$ and $\theta_L$ is replaced by $\theta_L - 2\beta z'$. This means that the term $|\rho_L| e^{i(\theta_L - 2\beta z')}$ (Eqs. 8.286) behaves exactly as $|\rho_L| e^{i\theta}_L$(Eq. 8.287) in the Smith chart.

The magnitude of the reflection coefficient $|\rho_L|$ and hence the standing wave ratio SWR are not changed by the additional length $z'$. Hence as we find $|\rho_L|$ and $\theta_L$for a given $\mathfrak{z}_L$ at the load (Figs. 8.23 and 8.24) Similarly we can utilize the Smith chart to find $\mathfrak{z}_{in}$ as follows. Keep $|\rho_L|$ constant and subtract from $\theta_L$ an angle equal to $2\beta z' = \frac{4\pi z'}{\lambda}$.

This will locate the point for $|\rho_L| e^{i(\theta_L - 2\beta z')}$ which determines $\mathfrak{z}_i$ the normalized input impedance at the point. As an example let us calculate the input impedance for the case shown in Figs. 8.23 and 8.24

**Step 1**

As stated in Figs. 8.23 and 8.24 and discussion therein calculate $\mathfrak{z}_L$ and $\theta_L$.

**Step 2**

From $\mathfrak{z}_L$ subtract an angle equal to $2\beta z' = \frac{4\pi z'}{\lambda}$ as shown in Fig. 8.25. Thus the point A moves to A′ in the clockwise direction thereby reducing the angle $\theta_L$by $2\beta z'$.

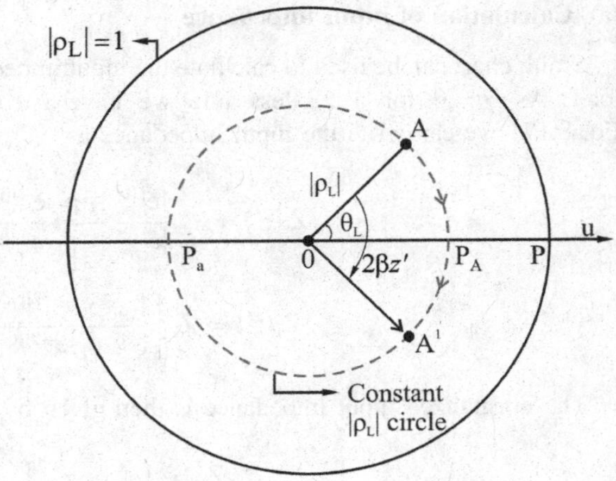


**Fig. 8.25** Smith chart—calculation of input impedance

**Step 3**

Corresponding to the point A′ find r′ and $x'$.

**Step 4**

Thus the normalized input impedance for the point A′ is given by

$$\mathfrak{z}_i = \mathfrak{z}_o[r' + ix'] \tag{8.288}$$

**Step 5**

Finally the input impedance is given by

$$Z_i = Z_o[r' + ix'] \tag{8.289}$$

the constant $|\rho_L|$ circle shown in Fig. 8.25 is also a constant SWR circle because of the relation

$$\text{SWR} = \frac{1 + |\rho_L|}{1 - |\rho_L|} \tag{8.290}$$

Additional three scales are provided around the periphery of the Smith chart as shown in Figs. 8.22 and 8.26. The outer most scale marked "wavelength towards generator" (WTG) is noted in the circle 1.

In some transmission line problems it becomes essential to move from a given point on the transmission line towards generator. This is accomplished on the Smith chart by moving of WTG circle. As the reader might observe circle I denoted as wavelength towards generator corresponds to the movement on the transmission line towards generator. The movement is marked in units is λ.

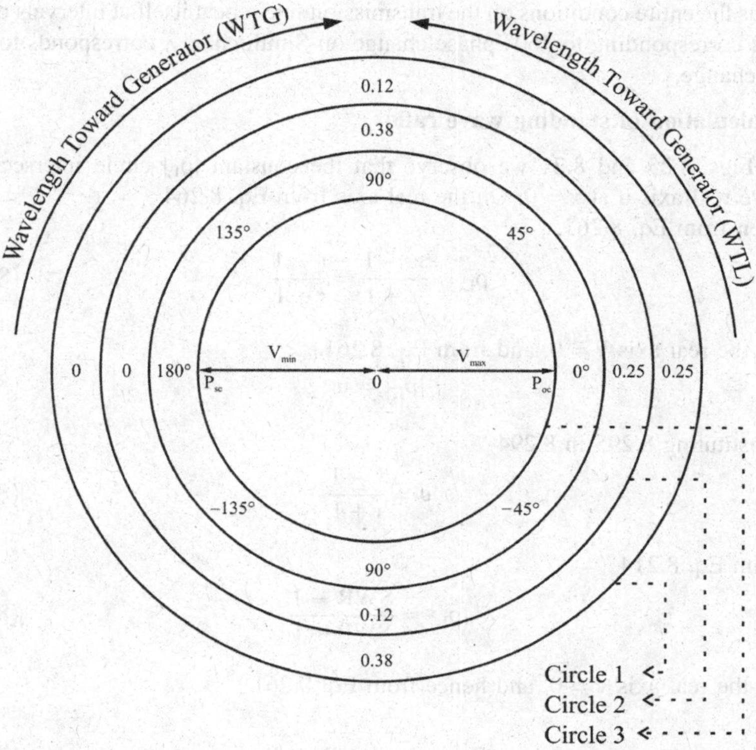


**Fig. 8.26** Smith chart—illustrating the scales around the periphery

On the other hand in some circumstances it becomes essential to move from the given point on the transmission line towards load. This is accomplished by circle 2 which is noted as "wavelength towards load" (WTL) marked in units of $\lambda$.

The innermost circle denoted as circle 3 and marked in angles is used to determine $\theta_L$.

A complete rotation around the Smith chart shown in Fig. 8.22 corresponds to a phase change of 360° or $2\pi$ in $\rho$. From Eq. 8.286 any change in phase of $\rho$ is given by $2\beta\, z'$. Hence

$$2\beta z' = 2\,\pi \tag{8.291}$$

$$\Rightarrow 2.\frac{2\pi}{\lambda}.z' = 2\pi \tag{8.292}$$

$$\Rightarrow z' = \frac{\lambda}{2} = 0.5\,\lambda \tag{8.293}$$

Thus the entire conditions on the transmission line repeat itself at intervals of $\lambda/2$. As $\lambda/2$ corresponding to 360° phase change on Smith chart $\lambda$ corresponds to 720° phase change.

(b) **Calculation of standing wave ratio**

From Figs. 8.25 and 8.21 we observe that the constant $|\rho_L|$ circle intersects the positive real axis u at $x = 0$. On the real axis from Eq. 8.264 $\mathfrak{z}_L = r$.

Then from Eq. 8.263,

$$\rho_L = \frac{\mathfrak{z}_L - 1}{\mathfrak{z}_L + 1} = \frac{r - 1}{r + 1} \quad (8.294)$$

On the real axis v = 0, and from Eq. 8.261

$$|\rho_L| = u \quad (8.295)$$

Substituting 8.295 in 8.294

$$u = \frac{r - 1}{r + 1} \quad (8.296)$$

From Eq. 8.214

$$|\rho_L| = \frac{\text{SWR} - 1}{\text{SWR} + 1} \quad (8.297)$$

On the real axis v = 0, and hence from Eq. 8.261

$$|\rho_L| = \sqrt{u^2 + v^2} = u \quad (8.298)$$

Substituting 8.298 in 8.297

$$u = \frac{\text{SWR} - 1}{\text{SWR} + 1} \quad (8.299)$$

Comparing Eqs. 8.296 and 8.299

$$\frac{r - 1}{r + 1} = \frac{\text{SWR} - 1}{\text{SWR} + 1} \quad (8.300)$$

Solving

$$\text{SWR} = r \quad (8.301)$$

Thus the standing wave ratio equals the value of r at the location of intersection of the constant reflection coefficient circle and the positive real axis point $P_A$ in Fig. 8.25. One of the important properties of the r-circles shown in Fig. 8.19 is—Consider a value of $r = r''$. The intersection of the r-circle with the real axis for $r = r''$ and $r = \frac{1}{r''}$ occurs at points symmetric about the centre the chart (u = 0. v = 0).

For example in Fig. 8.19 the intersection of r = 5 circle and $r = \frac{1}{5} = 0.2$ circle with the real -u-axis is equidistant from the centre of the chart (u = 0. v = 0).

From the above property, point $P_a$ in Fig. 8.25 corresponds to the intersection of reflection coefficient circle with negative real axis at $\frac{1}{r''}$, and thus the value of the r-circle passing through the point $P_a$ gives $\frac{1}{r} = \frac{1}{SWR}$ [from Eq. 8.301].

**(c) Calculation of $V_{max}$ and $V_{min}$**

The condition for appearance of $V_{max}$ from Eq. 8.198

$$2\beta z'_{max} - \theta_L = 2n\pi \tag{8.302}$$

where n = 0, 1, 2, 3 ...

From Eq. 8.286 let us denote $|\rho_L|e^{i(\theta_L - 2\beta d)}$ as $\rho_d$.

$$\rho_d = |\rho_L|e^{i(\theta_L - 2\beta d)} \tag{8.303}$$

From Fig. 8.25 at point $P_A$ the phase of $\rho_d$, $\theta_L - 2\beta z'$ is zero or $-2n\pi$ where n being a positive integer. Therefore the point $P_A$ in Fig. 8.25 corresponds to occurrence of $V_{max}$.

From Eq. 8.203 the condition for occurence of $V_{min}$ is

$$2\beta z'_{min} - \theta_L = (2n+1)\pi \tag{8.304}$$

which corresponds to point $P_a$ in Fig. 8.25 where the total phase of $\rho_d$ corresponds to $-(2n + 1)\pi$.

As we have already seen $V_{max}$ corresponds to $I_{min}$ and $V_{min}$ corresponds to $I_{max}$, Smith chart can be used to locate all maxima and minima.

**(d) Calculation of admittance from Smith Chart**

The point A in Fig. 8.25 is drawn in Fig. 8.27. The point A corresponds to the normalized impedance $\mathfrak{z}_L = r + ix$. The admittance point corresponds to diametrically opposite point on the reflection coefficient circle. In Fig. 8.27 the admittance point is found by drawing a line which passes through the point A centre O and intersects the reflection coefficient circle point B. Point B represents the normalized admittance of the load, $y = \frac{1}{\mathfrak{z}}$.

$$y = \frac{1}{\mathfrak{z}} = \frac{1}{r + ix} = \frac{1}{r + ix} \cdot \frac{r - ix}{r - ix} \tag{8.305}$$

$$y = \frac{r}{\sqrt{r^2 + x^2}} - \frac{ix}{\sqrt{r^2 + x^2}} = g - ib \tag{8.306}$$

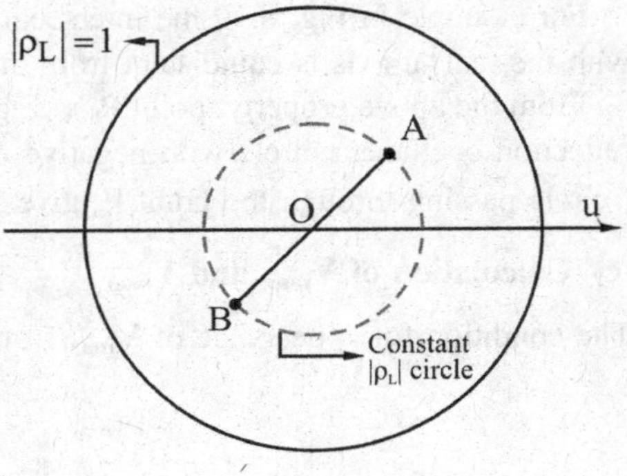


**Fig. 8.27** Smith chart—calculation of admittance

**Example 8.7**
A transmission line with characterize impedance of 50Ω is operating at 4 MHz and is terminated with a load of $Z_L = 75 + 25i\,\Omega$.

Find the

(a) Reflection coefficient
(b) Standing wave ratio
(c) The input impedance

The length of the line is 50 m.

(a) **Method 1**
(i) $\rho = \frac{Z_L - Z_o}{Z_L + Z_o}$

$$\rho = \frac{75 + 25i - 50}{75 + 25i + 50} = \frac{25 + 25i}{125 + 25i}$$

$$\rho = \frac{1 + i}{5 + i} = \frac{(1 + i)\,(5 - i)}{(5 + i)\,(5 - i)}$$

$$= \frac{5 + 5i - i + 1}{5^2 + 1^2} = \frac{6 + 4i}{26}$$

$$\rho = 0.231 + 0.154i$$

$$|\rho| = \sqrt{(0.231)^2 + (0.154)^2}$$

$$|\rho| = \sqrt{0.0534 + 0.0237}$$

$$|\rho| = \sqrt{0.07712} = 0.28$$

$$\theta = \tan^{-1}\left(\frac{0.154}{0.231}\right) = \tan^{-1}(0.667)$$

$$\theta = 37^\circ$$

**Standing wave ratio**

$$S = \frac{1+|\rho|}{1-|\rho|} = \frac{1+0.28}{1-0.28}$$

$$S = \frac{1.2777}{0.7223} = 1.8$$

**Input impedance**

$$\beta z' = \frac{\omega}{v} z' = \frac{2\pi \times 4 \times 10^6 \times 50}{3 \times 10^8}$$

$$\beta z' = 240.$$

$$Z_{in} = Z_o \left[\frac{Z_L + iZ_o \tan \beta z'}{Z_0 + iZ_L \tan \beta z'}\right]$$

$$\tan \beta z' = \tan 240^\circ = 1.73$$

$$Z_o \tan \beta z' = 50 \tan 240 = (50)(1.73) = 86.6$$

$$iZ_o \tan \beta z' = i86.6$$

$$\begin{aligned} Z_L \tan \beta z' &= (75+25i)(1.73) \\ &= (75)(1.73) + (25)(1.73)i \\ &= 129.75 + 43.25i \end{aligned}$$

$$\begin{aligned} \frac{Z_L + iZ_o \tan \beta z'}{Z_o + iZ_L \tan \beta z'} &= \frac{75 + 25i + 86.6i}{50 + 129.75i - 43.25} \\ &= \frac{75 + 111.6i}{6.75 + 129.75i} \end{aligned}$$

$$\begin{aligned} \frac{Z_L + iZ_o \tan \beta z'}{Z_o + iZ_L \tan \beta z'} &= \frac{75 + 111.6i}{6.75 + 129.75i} \\ &= \left(\frac{75 + 111.6i}{6.75 + 129.75i}\right)\left(\frac{6.75 - 129.75i}{6.75 - 129.75i}\right) \\ &= \frac{(75)(6.75) + (111.6)i(6.75) - (75)(129.75)i + (111.6)(129.75)}{(6.75)^2 + (129.75)^2} \\ &= \frac{506.25 + 753.3i - 9731.25i + 14480.1}{45.46 + 16835} \end{aligned}$$

$$\frac{Z_L + iZ_o \tan \beta z'}{Z_o + iZ_L \tan \beta z'} = \frac{14986.35 - 8977.95i}{16881}$$
$$= 0.887 - 0.532i$$

$$Z_{in} = Z_o \left[ \frac{Z_L + iZ_o \tan \beta z'}{Z_o + iZ_L \tan \beta z'} \right]$$
$$= 50[0.887 - 0.532i]$$
$$Z_{in} = 44.4 - 26.6i$$

(b) **By Smith chart method**

(a) The normalized impedance is

$$\mathfrak{z}L = \frac{Z_L}{Z_o} = \frac{75}{50} + \frac{25i}{50}$$
$$\mathfrak{z}L = 1.5 + 0.5i$$

Now as shown in Fig. 8.28 locate the point where r = 1.5 circle and $x = 0.5$ circle meet. This corresponds to point P in Fig. 8.28. In order to find ρ at $\mathfrak{z}_L$ extend the line OP to meet the r = 0 circle at S. Measure the length of the lines OP and OS. In the enlarged Fig. 8.28 OP = 1.9 cm and OS = 6.7 cm As OS corresponds to | ρ | = 1 then at P

$$|\rho| = \frac{OP}{OS} = \frac{1.9}{6.7} = 0.283$$

The lengths OP = 1.9 cm and OS = 6.7 cm were measured by the author on the Smith chart.
Irrespective of the size of Smith chart in Fig. 8.29 the ratio OP/OS remains the same.
The angle $\theta_L$ is measured on the Smith chart. It is the angle between OP and OQ

$$\theta_L = 37°$$

Therefore $\rho = |\rho| \underline{|\theta_L}. = 0.283 \underline{|37°}.$

(b) Now let us calculate the SWR. With radius OP and centre O draw a circle. This circle corresponds to constant SWR or constant |ρ| circle. The circle intersects the real axis and the u-axis at point Q. At this point r = 1.83. This is the value of the standing wave ratio
SWR = 1.83

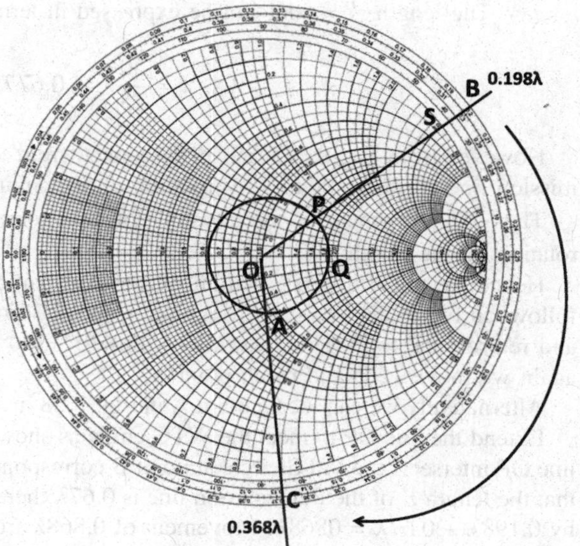


**Fig. 8.28** Illustration for example 8.7

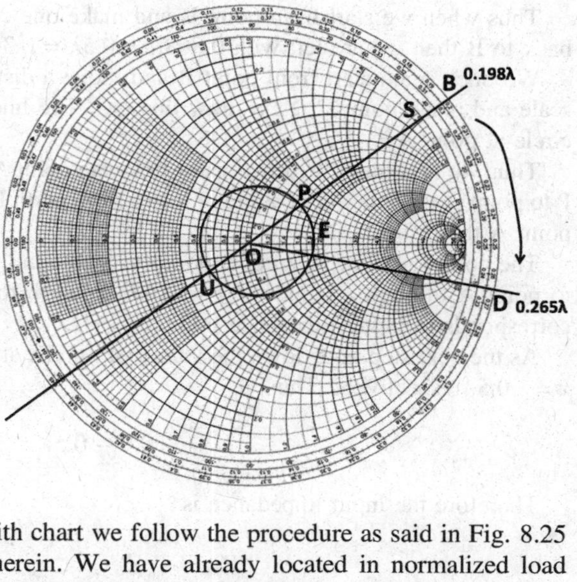


**Fig. 8.29** Illustration for example 8.8

(c) To find $Z_{in}$ using Smith chart we follow the procedure as said in Fig. 8.25 and the discussion therein. We have already located in normalized load impedance at point P. The length of the transmission line is $z' = 50$ m. The wavelength of the electromagnetic wave is

$$\lambda = \frac{3 \times 10^8}{4 \times 10^6} = 75 \text{ m}$$

The length $z' = 50$m can be expressed in terms of $\lambda$ as

$$l = \frac{50}{75}\lambda = 0.67\,\lambda$$

However from Eq. 8.293 and discussion therein a movement $\lambda$ on the transmission line corresponds to 720° rotation on the Smith chart.

Thus $\frac{50}{75}\lambda$ movement on the transmission line corresponds to $\left(\frac{50}{75}\right)(720) = 480°$ rotation on the Smith chart.

Hence we make a 480° rotation on the constant SWR circle from point P as follows in Fig. 8.28. We start from point P move 360° on the constant SWR circle and return back to P. This leaves us with 480°−360° = 120°. From point P once again we move 120° and reach point A.

Alternatively we can locate point A in Fig. 8.28 in a different method as follows.

Extend the line OP to meet the WTG circle as shown in Fig. 8.28. The extended line OP intersects the WTG circle at point B corresponding to 0.198$\lambda$. Noting down that the length $z'$ of the transmission line is 0.67$\lambda$ then we move on the WTG circle by 0.198$\lambda$ + 0.67$\lambda$ = 0868$\lambda$. Movement of 0.868$\lambda$ from point B(= 0.198$\lambda$) on the WTG scale is accomplished as follows. From Eq. 8.293 and discussion therein we note that the entire circumference of the Smith chart corresponds to 0.5$\lambda$.

Thus when we start from point B and make one completed notation and return back to B then we are left with 0.868$\lambda$ − 0.5$\lambda$ = 0.368$\lambda$.

We once again start from point B and move a distance of 0.368$\lambda$ on the WTG scale and reach point C. Next draw line OC. The line OC cuts the constant SWR circle at point A.

Thus we have described two ways to locate point A. Either we move from point P to point A via rotation of angle or move from point B to point C and finally locate point A the result is same.

The point P in Fig. 8.28 corresponds to load point and as we move from point P to point A through a length of $z'$ = 0.67$\lambda$ = 50 m towards the generator point A corresponds to input impedance point.

As the next step note down the values of r and $x$ at point A. We find r = 0.9 and $x$ = −0.5, respectively. Thus

$$\mathfrak{z}_{in} = 0.9 - 0.5i$$

Therefore the input impedance is

$$\begin{aligned} Z_{in} &= \mathfrak{z}_{in} Z_o \\ &= 50[0.9 - 0.5i] \\ Z_{in} &= 45 - 25i \end{aligned}$$

As the reader might observe the results obtained by Smith chart are approximate and are close to the exact values.

**Example 8.8**
For the Example 8.7 calculate

(a) the load admittance
(b) $Z_{in}$ at 5 m from the load.
(c) The location of $V_{max}$ and $V_{min}$ from the load.

**Solution**

(a) In order to find the load admittance extend the line PO to meet the constant SWR circle at U as shown in Fig. 8.29. Note down the values of r and $x$ corresponding to point U. We find r = 0.65 and $x = -0.21$

$$y_L = 0.65 - 0.21i$$

Therefore the load admittance

$$Y_L = Y_O y_L = \frac{1}{50}[0.65 - 0.21i]$$
$$Y_L = 0.013 - 0.0042i\ S$$

We can verify the results as follows

$$Y_L = \frac{1}{Z_L} = \frac{1}{75+25i}\left(\frac{75-25i}{75-25i}\right)$$
$$Y_L = 0.012 - 0.004i\ S$$

(b) Given $z' = 5$m
Writing $z'$ in terms of λ

$$z' = 5\,\text{m} = \frac{5}{75}\lambda = 0.067\lambda$$

Therefore in Fig. 8.29 we move from point B to point D [0.198λ + 0.067λ = 0.265λ] on the WTG scale which corresponds to 5 m movement on transmission line towards the generator from load. Next we draw the line OD which intersects the constant SWR circle at point E. At point E the r and $x$ values are r = 1.8 and $x = -0.22$. Hence

$$\mathfrak{z}_{in} = 1.8 - 0.22i$$

Then

$$Z_{in} = Z_0\,\mathfrak{z}_{in} = 50\ [1.8 - 0.22i]$$
$$Z_{in} = 90 - 11i$$

We can check the above result as follows:

$$\beta z' = \frac{2\pi}{\lambda} \cdot \frac{5}{75}\lambda = 24°$$

Now

$$Z_{in} = Z_o \left[\frac{Z_L + iZ_o \tan \beta z'}{Z_o + iZ_L \tan \beta z'}\right]$$
$$\tan \beta z' = \tan 24° = 0.45$$

$$\frac{Z_L + iZ_o \tan 24°}{Z_L + iZ_o \tan 24°} = \frac{75 + 25i + i50(0.45)}{50 + i(75 + 25i)0.45}$$
$$\text{With } Z_L = 75 + 25i, \ Z_o = 50$$

Solving

$$\frac{Z_L + iZ_o \tan 24°}{Z_L + iZ_o \tan 24°} = 1.7 - 0.26i$$
$$Z_{in} = Z_o(1.7 - 0.26i)$$
$$= 50(1.7 - 0.26i) = 85 - 13i$$

(c) As we have already seen in Eqs. 8.302, 8.303 and 8.304 and discussion therein and Fig. 8.25 the maxima and minima of voltage are located at $P_A$ and $P_a$ respectively. The points $P_A$ and $P_a$ for the present problem is shown in Fig. 8.30. Therefore the nearest voltage maxima from the load is found from moving point P to point $P_A$ as described below

From Eq. 8.293 and the WTG scale on Smith chart we observe that a complete rotation on Smith chart corresponds to 0.5λ movement on the transmission line. Therefore half the rotation corresponds to 0.25λ as shown in Fig. 8.30.

Thus to move from P [load] to point $P_A$ [nearest voltage maximum] we subtract 0.198λ [Point B on the WTG scale corresponding to point P in Fig. 8.30] from 0.25λ.

$$0.25\lambda - 0.198\lambda = 0.052\lambda = 0.052(75) = 3.9\,\text{m}$$

Thus the first voltage maximum is located at 0.052λ (or) 3.9 m from the load. The first voltage minimum is located at $P_a$. To locate the first $V_{min}$ then we move from point P to point $P_a$. Thus

$$0.25\lambda + 0.198\lambda = 0.448\lambda = (0.448)\ (75) = 33.6\,\text{m}$$

Thus the first voltage minimum is located at 0.448λ (or) 33.6 m from the load.

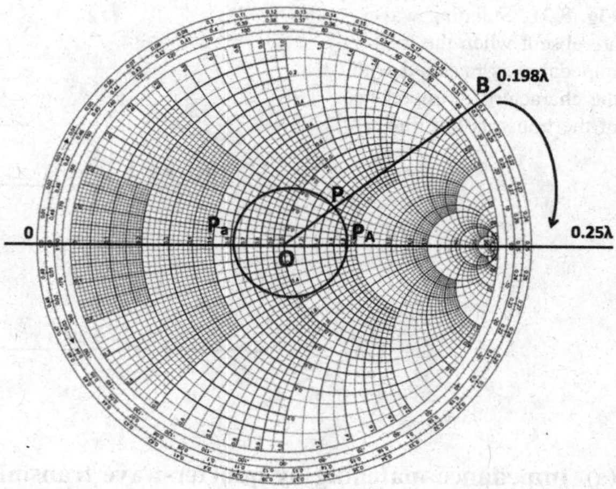


**Fig. 8.30** Illustration for example 8.8

## 8.17 Impedance Matching

In general the characteristic impedance $Z_o$ of the transmission line is not equal to the impedance of the load. As we have already seen in such situations the incident wave is reflected and gives rise to standing wave on the line.

However, a perfect match between $Z_o$ and $Z_L$ [$Z_L = Z_o$] is desirable because in such situations all the incident power can be delivered to the load. When $Z_L = Z_o$ from Eqs. 8.167 and 8.214 we have $\rho_L = 0$ (no reflection) and SWR = 1. Thus the standing wave pattern shown in Fig. 8.14 doesn't exist and for a matched condition the voltage wave pattern is shown in Fig. 8.31

From Eq. 8.196 then

$$|V_s(z')| = |V_o^+| \tag{8.307}$$

as $|\rho_L| = 0$, and hence when $Z_L = Z_o$ the amplitude of the wave is same all along the line.

Figure 8.31 shows the voltage wave pattern has a constant amplitude of $V_o^+$ for all values $z'$.

Any mismatch [$Z_L \neq Z_o$] between the impedances of load and transmission line might be detrimental to the circuitry as significant power which is not delivered to the load exists on the line. Therefore whenever there is a mismatch $Z_L \neq Z_o$ the situation is corrected by adding a matching network as shown in Fig. 8.32.

Although standing waves might exist in the matching network in Fig. 8.32 no standing waves will be present in the main transmission line due to matching condition. Various methods of impedance matching are described below.

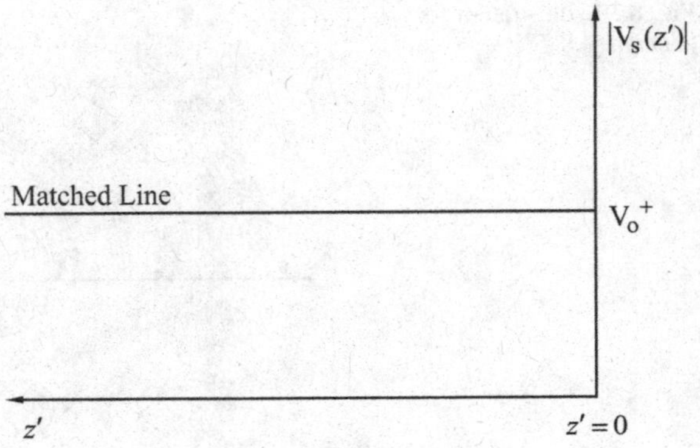


**Fig. 8.31** Standing waves are absent when the load impedance is matched with the characteristic impedance of the transmission line

(a) **Impedance matching by quarter-wave transmission line**
A quarter-wave transmission line discussed in Example 8.5 can be used as a matching network as shown in Fig. 8.33. We know that the input impedance of a quarter-wave transformer from Eq. 8.225 is

$$Z_{in} = \frac{(Z_T)^2}{Z_L} \tag{8.308}$$

where $Z_T$ is the impedance of the quarter-wave transformer shown in Fig. 8.33.

If we want $Z_{in}$ to be numerically equal to $Z_o$ then

$$Z_{in} = \frac{(Z_T)^2}{Z_L} = Z_o \tag{8.309}$$

$$\Rightarrow Z_T = \sqrt{Z_o Z_L} \tag{8.310}$$

Therefore if $Z_T$ is equal to $\sqrt{Z_o Z_L}$ then matching condition can be achieved. Thus the quarter-wave transmission line with impedance $Z_T = \sqrt{Z_o Z_L}$ will match a main transmission line with characteristic impedance $Z_o$ to the load impedance $Z_L$. However it should be noted that perfect matching condition can be achieved only at one frequency.

(b) **Single stub matching**
Stubs can be used to match a load impedance to the characteristic impedance of the transmission line. A open- or short-circuit stub can be used for matching the transmission line with the load. The stub has same characteristic impedance as that of the main transmission line. Assume a transmission line connected to the load as shown Fig. 8.34.

Suppose we are interested in introducing a stub at point D to match the transmission line with the load. Assume at point D on the line the impedance is $Z_{in}$.

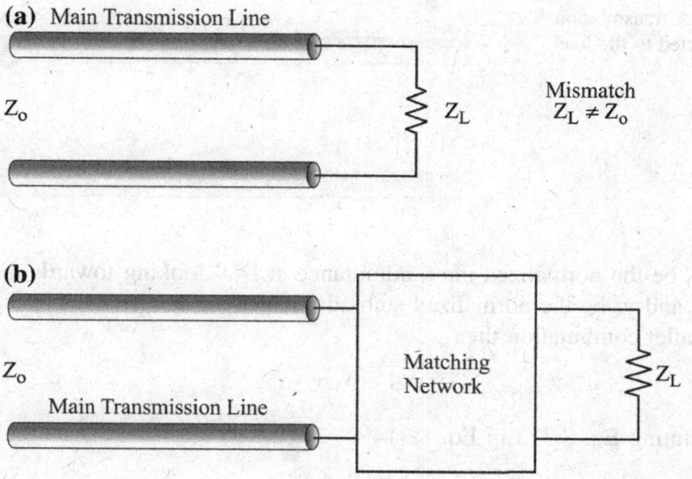


**Fig. 8.32** **a** Mismatch between load impedance and characteristic impedance of the transmission line. **b** Any mismatch between load impedance and the characteristic impedance is corrected by adding a matching network

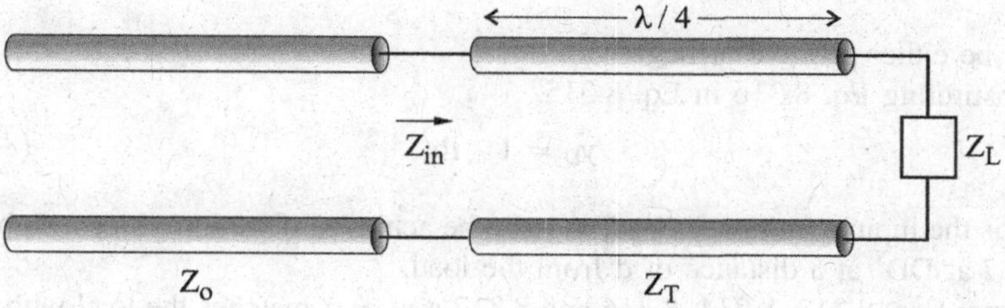


**Fig. 8.33** Matching network using quarter wave transmission line

For matching condition $Z_{in}$ should be equal to $Z_o$ the characteristic impedance of the main transmission line.

$$Z_{in} = Z_o \tag{8.311}$$

$$z_{in} = 1 \tag{8.312}$$

$$\text{or } y_{in} = 1 \tag{8.313}$$

Therefore if $y_{in}$ at point DD' is made 1 then matching condition is achieved.

In Fig. 8.35 we have connected a stub parallel to the main transmission line. The stub is distance d from the load and is of length $l$.

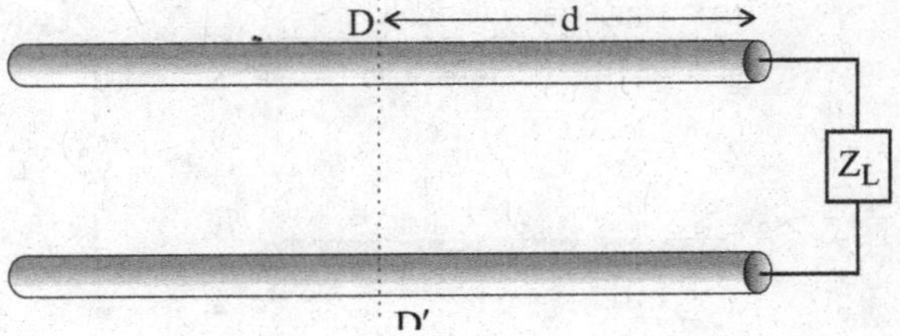


**Fig. 8.34** A transmission line connected to the load

Let $y_D$ be the normalized input admittance at DD' looking towards the load in Fig. 8.35 and $y_s$ be the normalized stub admittance

In parallel combination then

$$y_{in} = y_s + y_D \tag{8.314}$$

Substituting Eq. 8.313 in Eq. 8.314

$$1 = y_s + y_D \tag{8.315}$$

Noting down that the input admittance of the stub is purely susceptive and hence imaginary we can write

$$y_s = -ib_s \tag{8.316}$$

$b_s$ can be either positive or negative.

Substituting Eq. 8.316 in Eq. 8.315,

$$y_D = 1 + ib_s \tag{8.317}$$

Thus the input admittance of $y_{in} = 1$ can be achieved if we introduce a stub with length $l$ at DD' at a distance of d from the load.

As per Eqs. 8.313, 8.314, 8.316 and 8.317 a $y_{in} = 1$ matches the load with main transmission line.

In Fig. 8.35 we have used a short-circuited stub. However a open-circuited stub is also possible. But open-circuited stubs radiate energy at their open ends and hence are not preferable.

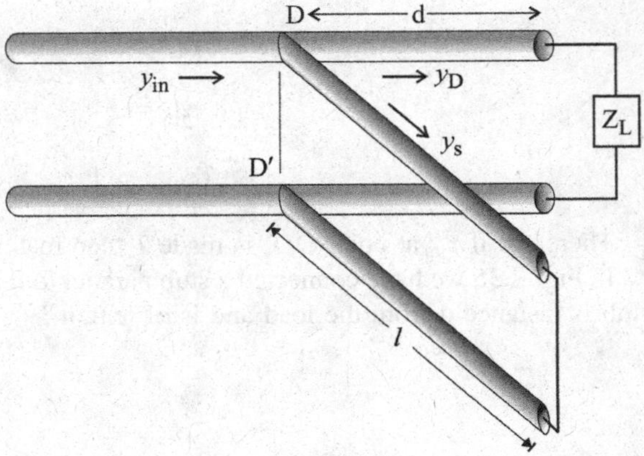


**Fig. 8.35** Single stub matching

A Smith chart can be used to find $l$ and d. Example 8.10 illustrates this part.

**Example 8.9**

A quarter-wave transformer is used to match a transmission line whose characteristic impedance is 152 Ω. The load impedance is resistive with 38 Ω resistance.

(a) What is the value of impedance of quarter-wave transformer for perfect matching condition.

(b) Find the SWR on the quarter-wave transmission line.

**Solution**

(a) $Z_T = \sqrt{R_L R_o} = \sqrt{38 \times 152}$

$Z_T = 76\ \Omega$

(b) $\rho = \dfrac{Z_L - Z_T}{Z_L + Z_T} = \dfrac{38 - 76}{38 + 76} = \dfrac{-38}{114}$

$\rho = -0.333$

$$\text{SWR} = \frac{1 + |\rho|}{1 - |\rho|} = \frac{1 + 0.333}{1 - 0.333}$$

$$\text{SWR} = 1.998$$

**Example 8.10**

A lossless transmission line $Z_o = 50\ \Omega$ is terminated with a load impedance of $Z_L = 19 - 60i\ \Omega$. A parallel stub is to be connected to the main transmission line for impedance matching. Calculate the location and length of the short-circuited stub.

**Solution**

We have $Z_L = 19 - 160\ \Omega$, $Z_o = 50\ \Omega$. The normalized impedance is

$$\mathfrak{z}_L = \frac{19 - i60}{50} = 0.38 - 1.2i\ \Omega$$

(a) Enter $\mathfrak{z}_L$ on Smith chart as shown in Fig. 8.36. It corresponds to point P on Smith chart.

(b) Next with OP as radius draw $|\rho|$ circle.

(c) Extend the line OP to cut the $|\rho|$ circle at Q. The part Q corresponds to the admittance of

$$y_L = 0.24 + 0.76i$$

(d) Extend the line POQ which cuts the WTG scale at $0.16\lambda$ at point P′. This corresponds to the load point on WTG scale.

(e) The $|\rho|$ circle cuts the g = 1 circle in Smith chart on two points A and B. At these points the admittance is

$$y_A = 1 + ib_A,\ y_B = 1 + ib_B.$$

Both are the points which corresponds to the two possible locations of the stub to provide perfect matching.
In our present problem at points A and B the admittance y is

$$\text{at B}\ y_B = 1 + 2.8i$$
$$\text{at A}\ y_A = 1 - 2.8i$$

(f) Solution for the position of the stub d is
**At B** The load is located at P′ in the Smith chart while the point B is located at B′ on the Smith chart on WTG scale. Thus the location of point B is [from P′ to B′ on WTG scale]

$$d_B = 0.21\lambda - 0.61\lambda = 0.05\lambda$$

**At A** The load is located at P′ on the Smith chart while the point A is located at A′ on the Smith chart on WTG scale. Thus the location of point A is [from P′ to A′ on the WTG scale]

$$d_A = 0.3\lambda - 0.16\lambda = 0.14\lambda$$

(g) Solution for length of the short-circuited stubs.
We require the length $l$ of the stub which will provide a admittance of $-ib_B$ and $-ib_A$.

**Fig. 8.36** Illustration for example 8.10

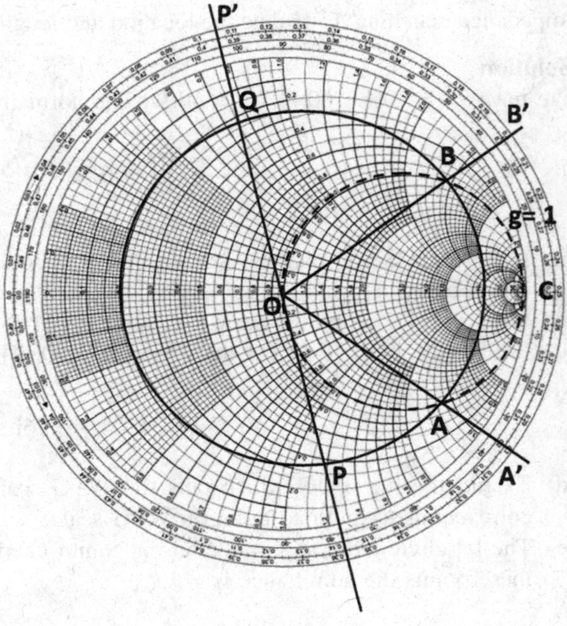

The point C on the Smith chart in Fig. 8.36 where the WTG scale is $0.25\lambda$ corresponds to short-circuit. At this point $r = 0$ and $x = 0$. Therefore the normalized impedance is $\mathfrak{z} = 0 + i0$ from Eq. 8.264. We move from the short-circuit point C to the points representing $-ib_B$ and $-ib_A$ to get $l_B$ and $l_A$.

In Fig. 8.36 we will be moving from short-circuited stub to the load. Hence on the Smith chart we will be moving from wavelength towards load on WTL scale.

To final $l_B$ we move from point C to point B′ on WTL scale.

$$l_B = 0.299\lambda - 0.25\lambda = 0.049\lambda$$

To find $l_A$ we move from point C to point A′ on WTL scale

$$l_A = 0.2\lambda + 0.25\lambda = 0.045\lambda.$$

## Exercises

**Problem 8.1**

Derive Eq. 8.30 from Eqs. 8.10 and 8.21.

**Problem 8.2**

Consider a airline with a phase constant of 25 rad/m at 650 MHz. The characteristic impedance of the transmission line is 70 Ω. Calculate the inductance per metre and capacitance per metre of the line.

**Problem 8.3**

Show that the resistance per unit length of the transmission line is given by

$$R = \frac{2}{a\delta\sigma_c}$$

where $\delta$ is the skin depth and $\sigma_c$ is the conductivity of the conductor. Assume that the conductor is thick than $\delta$ and current in the transmission line is restricted to $\delta$.

**Problem 8.4**

Consider a parallel-plate transmission line separated by a dielectric with conductance with G. Show that

$$G = \frac{\sigma a}{d}$$

where $\sigma$ is the conductivity of the dielectric

**Problem 8.5**

A transmission line is connected to a load impedance of $Z_L$. The characteristic impedance of the transmission line is 50 Ω, and the SWR is 4. The transmission line doesn't introduce any phase shift in the reflected wave.

What is the resistance to be connected to the transmission line in order to match the characteristic impedance of the line.

**Problem 8.6**

The characteristic impedance of a lossless line is 50 Ω. It is connected to a load of impedance 44 + i22. Find $Z_{in}$ at $l = \frac{\lambda}{8}$.

**Problem 8.7**

Consider a lossless transmission line. The distance at which the first voltage minimum occurs is 5 cm from the load. The distance at which the first l the load is 15 cm. Given SWR = 2. Find $\rho_L$.

**Problem 8.8**

Locate the following normalized impedance on the Smith chart.

(a) 0.4 + 0.5i (b) 0.3 − i0.4 (c) 0.3i (d) 0.5

**Problem 8.9**

Use the Smith chart to find the reflection coefficient corresponding to the normalized load impedance 0.5 − i2.

**Problem 8.10**

A transmission line with characteristic impedance $Z_o$ = 50 Ω has a load impedance of 75 + 100i

(a) Find the reflection coefficient at the load.

(b) Find the reflection coefficient at 0.15λ from the load.

**Problem 8.11**

The normalized impedance is given by 2 + i using Smith chart find SWR.

**Problem 8.12**

The load impedance connected to the transmission line is given by

$$Z_L = 47 - 17i$$

The characteristic impedance of the transmission line is 75 Ω. The transmission line operates at 21.6 MHz. The length of the transmission line is 50 m. Calculate the

(a) Input impedance.

(b) Input impedance at 2.8 m from the load.

**Problem 8.13**

If in Problem 8.11 the characteristic impedance of the transmission line is 50 Ω. Find the input admittance of the transmission line.

**Problem 8.14**

The characteristic impedance of a transmission line is 100 Ω. The length of the transmission line is 0.25λ. It is terminated with an load impedance of $Z_L$ = 140 + i130 Ω using Smith chart. Find the location of first voltage maximum and first voltage minimum from the load.

# Chapter 9
# Waveguides

## 9.1 Introduction

Waveguides as the name implies guide the electromagnetic waves from one point to the other. The electromagnetic wave is repeatedly reflected from opposite walls of the guide and is carried from one end to the other end in an zigzag path.

Waveguides are mostly used at microwave frequencies. From Table 8.1 and Eq. 7.123 the resistance per unit length of the transmission line is proportional to $\sqrt{\upsilon}$. Therefore the power loss in the transmission lines becomes unacceptable at higher frequencies. Because of the above said reason the transmission through transmission line becomes inefficient at microwave frequencies. At such high frequencies waveguides are used to transmit electromagnetic waves efficiently.

The most commonly used waveguides are rectangular waveguides which are shown in Fig. 9.1a.

## 9.2 Transverse Electromagnetic, Transverse Electric and Transverse Magnetic Waves

In Sect. 8.11 we discussed about the propagation of TEM waves in the parallel-plate transmission line. The waves propagate in the $z$-direction, while the electric field points in the $x$-direction, and the magnetic field points in the $y$-direction, or in other words there is no components of electric or magnetic fields in the direction of the wave propagation. These waves are called transverse electromagnetic waves, as they don't have electric and magnetic field components along the propagation direction.

© Springer Nature Singapore Pte Ltd. 2020
S. Balaji, *Electromagnetics Made Easy*,
https://doi.org/10.1007/978-981-15-2658-9_9

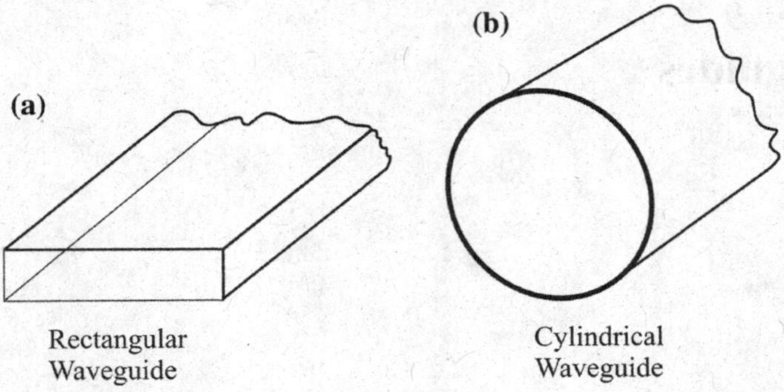


**Fig. 9.1** Types of waveguides

In the case of waveguides there are other possible wave propagation, namely

(a) transverse electric wave and
(b) transverse magnetic wave.

In the case of transverse electric (TE) waves the electric field is perpendicular to the direction of wave propagation, while the magnetic field has components in the direction of wave propagation. There is no component of electric field in the direction of wave propagation. The electromagnetic wave is referred to as transverse magnetic (TM) wave where the magnetic field is perpendicular to the direction of wave propagation, while the electric field has components in the direction of wave propagation. There is no component of magnetic field in the direction of wave propagation. The transverse electromagnetic wave (TEM) can propagate at any given frequencies. But the TE or TM wave can propagate in the waveguide when the frequency of the electromagnetic wave is higher than the certain frequency called cut-off frequency.

## 9.3 Wave Equation in Cartesian Coordinates

A waveguide consists of conductors separated by a dielectric. Therefore, the waveguide can be described by Eqs. 7.84a and 7.84b. For a perfect dielectric $\sigma = 0$. Hence wave equations and 7.84b can be written as

$$\nabla^2 \mathbf{E} = -\omega^2 \varepsilon \mu \, \mathbf{E} \tag{9.1}$$

$$\nabla^2 \mathbf{H} = -\omega^2 \varepsilon \mu \, \mathbf{H} \tag{9.2}$$

We assume wave propagation in $z$-direction and harmonic time dependency for the fields. We also assume each field varies as $e^{-\gamma z}$ where $\gamma$ the propagation constant is given by

$$\gamma = \alpha + i\beta \tag{9.3}$$

$\alpha$ is the attenuation constant, and $\beta$ is the phase constant.
The fields are given by

$$\begin{aligned}\mathbf{E}(x,y,z) &= \mathbf{E}_0(x,y)e^{-\gamma z}e^{i\omega t}\\ &= \mathbf{E}_0(x,y)e^{-\alpha z}e^{i(\omega t-\beta z)}\end{aligned} \tag{9.4}$$

$$\begin{aligned}\mathbf{H}(x,y,z) &= \mathbf{H}_0(x,y)e^{-\gamma z}e^{i\omega t}\\ &= \mathbf{H}_0(x,y)e^{-\alpha z}e^{i(\omega t-\beta z)}\end{aligned} \tag{9.5}$$

In component form the above equations can be written as

$$E_x\hat{\mathbf{i}} + E_y\hat{\mathbf{j}} + E_z\hat{\mathbf{k}} = \left(E_{ox}\hat{\mathbf{i}} + E_{oy}\hat{\mathbf{j}} + E_{oz}\hat{\mathbf{k}}\right)e^{-\gamma z}e^{i\omega t} \tag{9.6}$$

$$H_x\hat{\mathbf{i}} + H_y\hat{\mathbf{j}} + H_z\hat{\mathbf{k}} = \left(H_{ox}\hat{\mathbf{i}} + H_{oy}\hat{\mathbf{j}} + H_{oz}\hat{\mathbf{k}}\right)e^{-\gamma z}e^{i\omega t} \tag{9.7}$$

Substituting **E, H** from Eqs. 9.4 and 9.5 into wave Eqs. 9.1 and 9.2 writing the wave equations in Cartesian coordinates we get,

$$\left[\frac{\partial^2}{\partial x^2} + \frac{\partial^2}{\partial y^2}\right]\mathbf{E} = -(\omega^2\mu\varepsilon + \gamma^2)\mathbf{E} \tag{9.8}$$

$$\left[\frac{\partial^2}{\partial x^2} + \frac{\partial^2}{\partial y^2}\right]\mathbf{H} = -(\omega^2\mu\varepsilon + \gamma^2)\mathbf{H} \tag{9.9}$$

The above equations can be written in component from as

$$\left[\frac{\partial^2}{\partial x^2} + \frac{\partial^2}{\partial y^2}\right]E_{ox} = -(\omega^2\mu\varepsilon + \gamma^2)E_{ox} \tag{9.10}$$

$$\left[\frac{\partial^2}{\partial x^2} + \frac{\partial^2}{\partial y^2}\right]E_{oy} = -(\omega^2\mu\varepsilon + \gamma^2)E_{oy} \tag{9.11}$$

$$\left[\frac{\partial^2}{\partial x^2} + \frac{\partial^2}{\partial y^2}\right]E_{oz} = -(\omega^2\mu\varepsilon + \gamma^2)E_{oz} \tag{9.12}$$

$$\left[\frac{\partial^2}{\partial x^2} + \frac{\partial^2}{\partial y^2}\right] H_{ox} = -(\omega^2 \mu\varepsilon + \gamma^2) H_{ox} \tag{9.13}$$

$$\left[\frac{\partial^2}{\partial x^2} + \frac{\partial^2}{\partial y^2}\right] H_{oy} = -(\omega^2 \mu\varepsilon + \gamma^2) H_{oy} \tag{9.14}$$

$$\left[\frac{\partial^2}{\partial x^2} + \frac{\partial^2}{\partial y^2}\right] H_{oz} = -(\omega^2 \mu\varepsilon + \gamma^2) H_{oz} \tag{9.15}$$

All the above six equations can be solved to obtain the components and finally the electric and magnetic fields **E** and **H**. It is to be noted that the field components satisfy Maxwell's equations.

Therefore, if one field component is identified by solving one of the wave equations other field components can be identified by use of Maxwell's equations as follows

In a non-conducting source free medium with no charges and no currents and with $\sigma = 0$ Eqs. 6.164 and 6.167 can be written as

$$\nabla \cdot \mathbf{D} = 0 \tag{9.16}$$

$$\nabla \cdot \mathbf{B} = 0 \tag{9.17}$$

$$\nabla \times \mathbf{E} = \frac{-\partial \mathbf{B}}{\partial t} \tag{9.18}$$

$$\nabla \times \mathbf{H} = \frac{-\partial \mathbf{D}}{\partial t} \tag{9.19}$$

Using relations $\mathbf{D} = \varepsilon\mathbf{E}$, $\mathbf{B} = \mu\mathbf{H}$ and Eqs. 6.229, 9.16–9.19 can be written as

$$\nabla \cdot \mathbf{E} = 0 \tag{9.20}$$

$$\nabla \cdot \mathbf{H} = 0 \tag{9.21}$$

$$\nabla \times \mathbf{E} = -i\omega\mu\mathbf{H} \tag{9.22}$$

$$\nabla \times \mathbf{H} = i\omega\varepsilon\mathbf{E} \tag{9.23}$$

The Maxwell's equations $\nabla \times \mathbf{E} = -i\omega\mu\mathbf{H}$ can be written in component form as

$$\frac{\partial E_{oz}}{\partial y} + \gamma E_{oy} = i\omega\mu H_{ox} \tag{9.24}$$

$$\frac{\partial E_{oz}}{\partial x} + \gamma E_{ox} = -i\omega\mu H_{oy} \tag{9.25}$$

$$\frac{\partial E_{oy}}{\partial x} - \frac{\partial E_{ox}}{\partial y} = -i\omega\mu H_{oz} \tag{9.26}$$

Similarly $\nabla \times \mathbf{H} = i\omega\varepsilon\mathbf{E}$ can be written in component form as

$$\frac{\partial H_{oz}}{\partial y} + \gamma H_{oz} = i\omega\varepsilon E_{ox} \tag{9.27}$$

$$\frac{\partial H_{oz}}{\partial x} + \gamma H_{ox} = i\omega\varepsilon E_{oy} \tag{9.28}$$

$$\frac{\partial H_{oy}}{\partial x} - \frac{\partial H_{ox}}{\partial y} = i\omega\varepsilon E_{oz} \tag{9.29}$$

Using Eqs. 9.24–9.29 we can write

$$E_{ox} = \frac{-1}{\gamma^2 + \omega^2\mu\varepsilon}\left[i\omega\mu\frac{\partial H_{oz}}{\partial y} + \gamma\frac{\partial E_{oz}}{\partial x}\right] \tag{9.30}$$

$$E_{oy} = \frac{1}{\gamma^2 + \omega^2\mu\varepsilon}\left[i\omega\mu\frac{\partial H_{oz}}{\partial x} - \gamma\frac{\partial E_{oz}}{\partial x}\right] \tag{9.31}$$

$$H_{ox} = \frac{1}{\gamma^2 + \omega^2\mu\varepsilon}\left[i\omega\varepsilon\frac{\partial E_{oz}}{\partial y} - \gamma\frac{\partial H_{oz}}{\partial x}\right] \tag{9.32}$$

$$H_{oy} = \frac{-1}{\gamma^2 + \omega^2\mu\varepsilon}\left[i\omega\varepsilon\frac{\partial E_{oz}}{\partial x} + \gamma\frac{\partial H_{oz}}{\partial y}\right] \tag{9.33}$$

Therefore Eqs. 9.12 and 9.15 can be solved to obtain $E_{oz}$ and $H_{oz}$. Then $E_{oz}$ and $H_{oz}$ can be substituted in Eqs. 9.30–9.33 to obtain the remaining components of fields.

## 9.4 Parallel-Plate Waveguide

In this section we will discuss about wave propagation in parallel-plate waveguide. The propagation of TEM wave in a parallel-plate transmission line is discussed in Sect. 8.11. Apart from TEM waves, TE and TM waves can also propagate in the parallel-plate waveguide.

Assume that the plates in $y$-direction in Fig. 8.7 is of infinite extent. Hence there is no field variation along the $y$-direction. Therefore we can replace $\frac{\partial \mathbf{E}}{\partial y}$ or $\frac{\partial \mathbf{H}}{\partial y}$ by zero. Then Eqs. 9.30–9.33 can be written as

$$E_{ox} = \frac{-\gamma}{\gamma^2 + \omega^2 \mu\varepsilon} \frac{\partial E_{oz}}{\partial x} \tag{9.34a}$$

$$E_{oy} = \frac{i\omega\mu}{\gamma^2 + \omega^2 \mu\varepsilon} \frac{\partial H_{oz}}{\partial x} \tag{9.34b}$$

$$H_{ox} = \frac{-\gamma}{\gamma^2 + \omega^2 \mu\varepsilon} \frac{\partial H_{oz}}{\partial x} \tag{9.35}$$

$$H_{oy} = \frac{-i\omega\varepsilon}{\gamma^2 + \omega^2 \mu\varepsilon} \frac{\partial E_{oz}}{\partial_x} \tag{9.36}$$

(a) **Transverse Magnetic Mode**

For transverse magnetic mode we know that there is no magnetic field in the direction of wave propagation. Then $H_{oz} = 0$. The longitudinal electric field component is not equal to zero $E_{oz} \neq 0$.

As then from Eqs. 9.34b and 9.35 $E_{oy} = 0$, $H_{ox} = 0$

Let

$$p^2 = \gamma^2 + \omega^2 \mu\varepsilon \tag{9.37}$$

Noting down that $\frac{\partial \mathbf{E}}{\partial y}$ can be set to zero from Eq. 9.12

$$\left[\frac{\partial^2}{\partial x^2} + p^2\right] E_{oz} = 0 \tag{9.38}$$

The solution for the above equation is

$$E_{oz} = A \sin px + B \cos px \tag{9.39}$$

The boundary conditions should be applied to evaluate the constants in the above equation.

We know that inside the conductor the electric field is zero. From the boundary condition (6.181) $E_{1t} = E_{2t}$. Therefore in the parallel-plate waveguide at the conducting boundaries the tangential component of electric field is zero.

$E_{oz}$ being the tangential component (the other tangential component $E_{oy}$ is zero) the conductor in Fig. 8.7, $E_{oz}$ obeys the following boundary conditions.

(a) $E_{oz} = 0$ at $x = 0$
(b) $E_{oz} = 0$ at $x = 0$

Applying boundary condition a in Eq. 9.39 we obtain B = 0. Applying boundary condition b in Eq. 9.39 we obtain $\sin pd = 0 \Rightarrow pd = n\pi$.

$$\Rightarrow p = \frac{n\pi}{d} \tag{9.40}$$

where n = 0, 1, 2, 3...

Therefore

$$E_{oz} = A_n \sin\frac{n\pi}{d}x \tag{9.41}$$

The other components of the fields can be obtained as follows.

Substituting 9.41 in 9.36 and using Eqs. 9.37 and 9.40 we obtain

$$H_{oy} = \frac{i\omega\varepsilon A_n}{p}\cos\frac{n\pi x}{d} \tag{9.42}$$

Substituting 9.41 in 9.33 and using Eqs. 9.37 and 9.40 we obtain

$$E_{ox} = \frac{-\gamma}{p}A_n\cos\frac{n\pi x}{d} \tag{9.43}$$

The components $E_{oz}$, $H_{oy}$ and $E_{ox}$ are given by Eqs. 9.41, 9.42 and 9.43, and we have already seen $H_{oz}$, $E_{oy}$ and $H_{ox}$ are zero. Substituting the values of components in Eqs. 9.6 and 9.7 **E** and **H** can be obtained. **E** and **H** can be plotted in the parallel-plate waveguide for given n each n leading to pattern of fields. Each field pattern is called a mode and is designated as $TM_n$ mode. For n = 0 we observe that from Eq. 9.41 $E_{oz} = 0$. However for n = 0 $H_{oy}$ and $E_{ox}$ are nonzero, both being perpendicular to the direction of motion of wave. Hence for n = 0 the $TM_0$ mode corresponds to the TEM mode. Other modes are $TM_1$, $TM_2$. modes, etc.

**Transverse Electric (TE) mode**

For the case of transverse electric mode there is no component of electric field in the direction of wave propagation. Hence $E_{oz} = 0$. In this case the longitudinal component of magnetic field is nonvanishing $[H_{oz} \neq 0]$. As $E_{oz} = 0$ then from Eqs. 9.36 and 9.33 $H_{oy} = 0$ and $E_{ox} = 0$.

$$\text{Let } p^2 = \gamma^2 + \omega^2\mu\varepsilon \tag{9.44}$$

Noting down that $\frac{\partial \mathbf{H}}{\partial y}$ can be set to zero from Eq. 9.15

$$\left[\frac{\partial^2}{\partial y^2} + p^2\right]H_{oz} = 0 \tag{9.45}$$

The solution of the above equation is

$$H_{oz} = C \sin px + D \cos px \tag{9.46}$$

The boundary conditions should be applied to evaluate the constants. However the tangential component of magnetic field is not zero at the conducting boundaries. But the tangential components of electric field are zero at the boundaries. From Eqs. 9.34a, 9.44 and 9.46.

We can write

$$E_{oy} = \frac{i\omega\mu p}{p^2} C \cos px - \frac{i\omega\mu p}{p^2} D \sin px \tag{9.47}$$

$$E_{oy} = \frac{i\omega\mu}{p} C \cos px - \frac{i\omega\mu}{p} D \sin px \tag{9.48}$$

As $E_{oy}$ is the tangential component of the electric field to the conducting surfaces in Fig. 8.7. (The other tangential component $E_{oz}$ is zero.)

(c) $E_{oy} = 0$ at $x = 0$
(d) $E_{oy} = 0$ at $x = d$

Substituting boundary condition c in Eq. 9.47 gives C = 0,
Substituting boundary condition d in Eq. 9.47 gives $\sin pd = 0 \Rightarrow pd = n\pi$

$$\Rightarrow p = \frac{n\pi}{d} \tag{9.49}$$

Therefore from Eqs. 9.46, 9.48 and 9.49 we can write

$$H_{oz} = D_n \cos \frac{n\pi x}{d} \tag{9.50}$$

$$E_{oy} = -\frac{i\omega\mu}{p} D_n \sin \frac{n\pi x}{d} \tag{9.51}$$

From Eqs. 9.35, 9.44, 9.49 and 9.50

$$H_{ox} = \frac{\gamma}{p} D_n \sin \frac{n\pi x}{d} \tag{9.52}$$

$H_{ox}$, $E_{oy}$ and $H_{ox}$ are given by Eqs. 9.50, 9.51 and 9.52. We have already seen the remaining components $E_{oz}$, $H_{oy}$ and $E_{ox}$ are zero. The components can be substituted in Eqs. 9.6 and 9.7, and **E** and **H** can be obtained.

Similar to TM wave propagation for each value of n there are different field patterns within the parallel-plate waveguide each pattern leading to different $TE_n$ mode.

For n = 0 both $E_{oy}$, $H_{ox}$ become zero hence $TE_o$ mode is not possible in parallel-plate waveguide.

## Cut-off Frequency

We know that the propagation constant $\gamma$ is (Eq. 9.3)

$$\gamma = \alpha + i\beta \tag{9.53}$$

where $\alpha$ is the attenuation constant $\beta$ is the phase constant.

From Eqs. 9.37 and 9.44

$$p^2 = \gamma^2 + \omega^2\mu\varepsilon \tag{9.54}$$

$$\Rightarrow \gamma = \sqrt{p^2 - \omega^2\mu\varepsilon} \tag{9.55}$$

At high frequencies $\omega^2\mu\varepsilon \gg p^2$, then in Eq. 9.55 $\gamma$ is imaginary. Therefore from Eq. 9.53 $\gamma = i\beta$ and the em wave propagates in the parallel-plate waveguide with phase constant $\beta$. The condition $\omega^2\mu\varepsilon \gg p^2$ holds good at high frequencies. Let the frequency be reduced such that

$$\omega^2\mu\varepsilon \gg p^2 \tag{9.56}$$

Then the propagation constant $\gamma$ given by Eq. 9.55 becomes zero, and no wave propagates in the waveguide. The corresponding frequency when no wave propagates in the waveguide is called cut-off frequency and is denoted as $\upsilon_c$. From Eq. 9.56

$$\omega_C^2\mu\varepsilon = p^2 \tag{9.57}$$

$$\Rightarrow \upsilon_c = \frac{p}{2\pi\sqrt{\mu\varepsilon}} \tag{9.58}$$

when the frequency is reduced further then $\omega^2\mu\varepsilon \ll p^2$ and $\gamma$ in Eq. 9.55 becomes real. Then from Eq. 9.53 $\gamma = \alpha$ the attenuation constant. Equations 9.4 and 9.5 contain $e^{-\alpha z}$ term, the wave attenuates rapidly with $z$, and the em wave is evanescent.

Thus at low frequencies below cut-off frequencies $\upsilon_c$ there is no wave propagation in the waveguide.

At high frequencies greater then $\upsilon_c$ there is wave propagation in the waveguide with phase constant β. Substituting $p = \frac{n\pi}{d}$ in Eq. 9.58.

$$\upsilon_c = \frac{n}{2d\sqrt{\mu\varepsilon}} \tag{9.59}$$

From Eq. 9.59 it is observed that for each value of n there is a different cut-off frequency. Thus each $TE_n$, $TM_n$ mode has a different cut-off frequency for each n.

We have seen that the cut-off frequencies are the frequency which makes $\gamma = 0$. From Eq. 9.37 and 9.44 γ is identical for both $TE_n$ and $TM_n$ mode for given n is the parallel-plate waveguide.

The mode having the lowest cut-off frequency is called dominant mode of the waveguide. For TM wave propagation n = 0 is the lowest possible mode ($TM_o$ mode), which we have already seen corresponds to TEM wave. From Eq. 9.59 $\upsilon_c = 0$ for n = 0. Thus for a parallel-plate waveguide the dominant mode is TEM mode.

**Example 9.1**
Show that the propagation of TE and TM waves between the two plates in the parallel-plate capacitor is equivalent to superposition of the two plane waves propagation in the zigzag manner between the plates?

**Solution**
Consider TE wave propagation. From Eq. 9.49 and 9.51

$$E_{oy} = \frac{-i\omega\mu}{p} D_n \sin px \tag{9.60}$$

For TE wave propagation we have already seen that $E_{ox}$ and $E_{oz}$ are zero. Also for wave to propagate in the waveguide $\alpha = 0$.

In Eq. 9.6 omitting the time part then we can write

$$E_x\hat{\mathbf{i}} + E_y\hat{\mathbf{j}} + E_z\hat{\mathbf{k}} = E_{oy}e^{i\beta x}\hat{\mathbf{j}} \tag{9.61}$$

where we have used Eq. 9.3 in 9.6 with $\alpha = 0$.

Comparing the like terms

$$E_y = E_{oy}e^{-i\beta x} \tag{9.62}$$

Substituting the value of $E_{oy}$ from 9.60

$$E_y = \frac{-i\omega\mu}{p} D_n \sin px\; e^{-i\beta x} \tag{9.63}$$

The above equation can be written as

$$E_y = \frac{-\omega\mu}{2p} D_n \left[e^{ipx} - e^{-ipx}\right] e^{-i\beta z} \tag{9.64}$$

$$E_y = \frac{-\omega\mu}{2p} D_n \left[e^{ipx-i\beta z} - e^{-ipx-i\beta z}\right] \tag{9.65}$$

Now let us consider the TM case from Eq. 9.40 and 9.42

$$H_{oy} = \frac{-i\omega\varepsilon}{p} A_n \cos px \tag{9.66}$$

Following exactly the same procedure in TM case as in TE case we finally obtain

$$H_y = \frac{-i\omega\varepsilon}{2p} A_n \left[e^{ipx-i\beta z} + e^{-ipx-i\beta z}\right] \tag{9.67}$$

In Eqs. 9.65 and 9.67 consider the first term $e^{-ipx-i\beta z}$ which represents plane wave propagation. The term $e^{-i\beta z}$ represents the plane wave propagation in the positive $z$-direction, while $e^{-ipx}$ represents a plane wave propagation in the negative $x$-direction.

Consider the second term $e^{-ipx-i\beta z}$. The term $e^{-i\beta z}$ represents a plane wave propagation in the positive $z$-direction. While $e^{-ipx}$ represents a plane wave propagates in the positive $x$-direction. The above results are shown in Fig. 9.2a. The figure contains waves bouncing buck and forth between the parallel plates.

Thus the propagation of TE and TM waves between the parallel plates is equivalent to propagation of two plane waves.

However in Fig. 9.2a for TE and TM waves the direction of **E** and **H** will be different and is not shown in the figure.

In Fig. 9.2a the wave vector is given by

$$k = \omega\sqrt{\mu\varepsilon} \tag{9.68}$$

From Fig. 9.2b

$$\cos\theta = \frac{p}{k} \tag{9.69}$$

Substituting Eq. 9.48 in 9.69

$$\theta = \cos^{-1}\left(\frac{n\pi}{kd}\right) \tag{9.70}$$

From the above equation as n is discrete, $\theta$ is also discrete. Thus the plane waves of TE and TM modes in the parallel-plate waveguide can be guided only for discrete values of $\theta$.

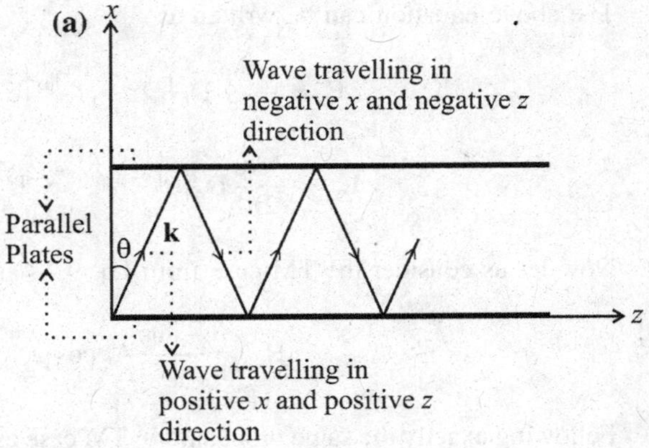


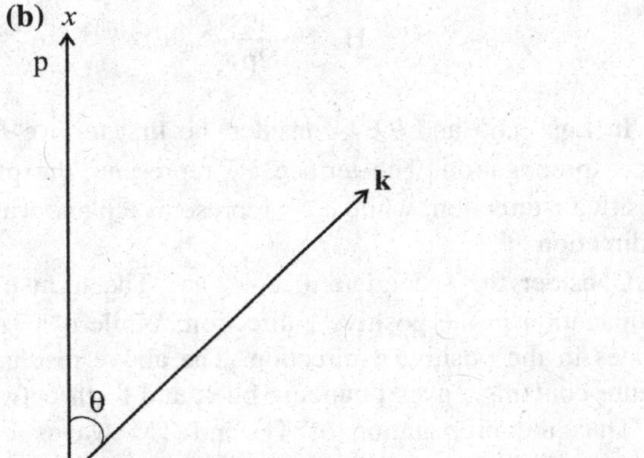


**Fig. 9.2** **a** An electromagnetic wave propagating in the parallel plate waveguide by oblique reflections at the conducting walls of the waveguide. **b** Wave vector of the electromagnetic wave propagating in the parallel plate waveguide

## 9.5 Propagation of TEM, TE and TM Waves in Parallel-Plate Waveguide

We have seen the propagation of TEM waves in parallel-plate transmission line in Sect. 8.11. At sufficiently low frequency TEM mode propagates in the transmission line along the $z$-direction with **k** pointing in the $z$-direction as shown in Fig. 8.8. The electric field points in the $x$-direction.

The magnetic field points in the $y$-direction. Thus the electric and magnetic fields are mutually perpendicular and are simultaneously perpendicular to the direction of wave propagation.

As the frequency is increased and becomes more than the cut-off frequency, TE and TM modes start propagating in the parallel-plate waveguide. The initial TEM mode may still be present between the parallel plates. In TEM mode the wave propagation is in $z$-direction with **k** pointing in $z$-direction. But as cut-off frequency is exceeded the wave propagation becomes zigzag as shown in Fig. 9.2a. The TEM mode doesn't have any cut-off frequency and propagates in all possible frequencies.

Thus a transmission line supports a TEM wave. On the other hand a waveguide supports TE and TM mode. Depending upon the design the waveguide may or may not support TEM waves.

**Example 9.2**

A parallel-plate waveguide is filled with a dielectric slab whose refractive index is 1.5, with d = 2 cm. Determine the maximum operating frequency such that only TEM mode propagates.

**Solution**

Setting $v = \frac{1}{\sqrt{\varepsilon\mu}}$ in Eq. 9.5

$$\upsilon_c = \frac{nv}{2d}$$

The velocity of the wave propagation in the medium is given by

$$v = \frac{c}{\text{refractive index}} = \frac{3 \times 10^8}{1.45}$$

$$\upsilon_c = 5n\,\text{GHz}$$

For n = 1, $\upsilon_c = 5$ GHz.

Therefore the operating frequency should be less than 5 GHz for TEM wave alone to propagate.

**Example 9.3**

A parallel-plate waveguide is filled with a dielectric with refractive index n = 1.45, and separation between the plates is d = 3 cm.

(a) Calculate the cut-off frequency in terms of n.
(b) What are the $TE_n$ and $TM_n$ modes which can propagate at 14 GHz.

**Solution**

(a) We know that the velocity of propagation is

$$v = \frac{c}{\text{refractive index}} = \frac{3 \times 10^8}{1.45}$$
$$v = 2 \times 10^8\ \text{m/s}.$$

Setting $v = \frac{1}{\sqrt{\varepsilon\mu}}$ in Eq. 9.59

$$\upsilon_c = \frac{nv}{2d}$$

$$\upsilon_c = \frac{n \times 2 \times 10^8}{2 \times 3 \times 10^{-2}}$$

$$\upsilon_c = 1 \times 10^{10}\,n\text{Hz}.$$

$$\Rightarrow \upsilon_c = 10n\,\text{GHz}.$$

(b) Given the operating frequency is $14 \times 10^9$ Hz or 14 GHz.
For $TE_1$, $TM_1$ mode.

$$\upsilon_c = 10\,\text{GHz}.$$

For $TE_2$, $TM_2$ modes

$$\upsilon_c = 20\,\text{GHz}.$$

Therefore $TE_1$ and $TM_1$ modes are alone possible because for those modes $\upsilon > \upsilon_c = 14\,\text{GHz} > 10\,\text{GHz}$.
The case n = 0 for $TM_o$ mode is TEM mode which propagates in possible frequencies. $n \geq 2$ modes will not propagate because for those modes $\upsilon < \upsilon_c$.

## 9.6 General Solution of Wave Equation

In the parallel-plate waveguide discussed in Sect. 9.3 we assumed that the plates are of infinite extent in the *y*-direction. The fields were independent of *y*, and hence the field variation along *y*-direction is not considered. But in general fields do vary with *y*-direction, and hence the field variation in *y*-direction needs to be considered.

From Eqs. 9.34a to 9.36 we observe that all the components depend on $E_{oz}$ and $H_{oz}$. Therefore once $E_{oz}$ and $H_{oz}$ components are identified remaining all other components can be obtained from Eqs. 9.34a to 9.36. Therefore first we will solve Eqs. 9.12 and 9.15 by separation of variables method to obtain $E_{oz}$ and $H_{oz}$.

The solution for Eq. 9.12 is

$$E_{oz} = X(x)Y(y) \tag{9.71}$$

with

$$p^2 = \gamma^2 + \omega^2\mu\varepsilon \tag{9.72}$$

Equations 9.12 can be written as

$$\frac{\partial^2 E_{oz}}{\partial x^2} + \frac{\partial^2 E_{oz}}{\partial y^2} = -p^2 E_{oz} \tag{9.73}$$

Substituting 9.71 in 9.73 and simplifying

$$\frac{1}{X}\frac{\partial^2 X}{\partial x^2} + \frac{1}{Y}\frac{\partial^2 Y}{\partial y^2} = -p^2 \tag{9.74}$$

As in Sect. 3.8 the left-hand side contains two terms each function of $x$ or $y$ only. The sum of the two terms is equal to a constant $-p^2$. Therefore each term should be a constant. Let us call these constants $-k_x^2, -k_y^2$. Thus

$$\frac{1}{X}\frac{\partial^2 X}{\partial x^2} = -k_x^2 \tag{9.75}$$

$$\frac{1}{Y}\frac{\partial^2 Y}{\partial y^2} = -k_y^2 \tag{9.76}$$

From Eqs. 9.74, 9.75 and 9.76 we can write

$$k_x^2 + k_y^2 = p^2 \tag{9.77}$$

Substituting the value of $p^2$ from 9.72 and rearranging.

$$\gamma = \sqrt{(k_x)^2 + (k_y)^2 - \omega^2 \mu \varepsilon} \tag{9.78}$$

The solutions for Eqs. 9.75 and 9.76 are

$$X(x) = A_1 \sin k_x x + A_2 \cos k_x x \tag{9.79}$$

$$Y(y) = A_3 \sin k_y y + A_4 \cos k_y y \tag{9.80}$$

Substituting Eqs. 9.79 and 9.80 in Eq.9.71

$$E_{oz}(x, y) = [A_1 \sin k_x x + A_2 \cos k_x x] \times [A_3 \sin k_y y + A_4 \cos k_y y] \tag{9.81}$$

where $A_1$, $A_2$, $A_3$, $A_4$ are constants. Using Eqs. 9.15 and following exactly the same procedure for $H_{oz}$ we obtain

$$H_{oz}(x, y) = [C_1 \sin k_x x + C_2 \cos k_x x] \times [C_3 \sin k_y y + C_4 \cos k_y y] \tag{9.82}$$

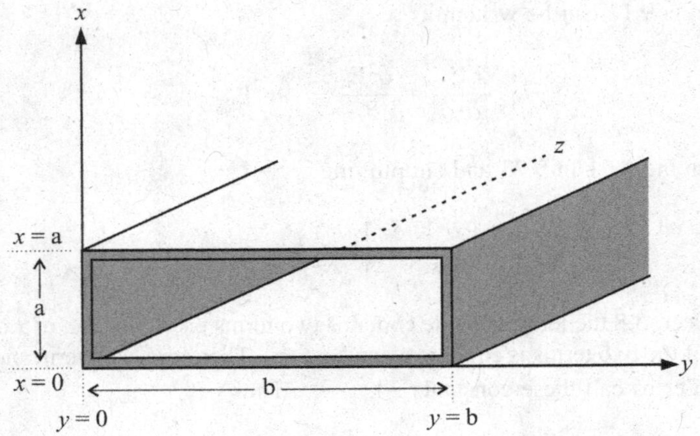


**Fig. 9.3** A rectangular waveguide

## 9.7 Rectangular Waveguides

In the previous sections we discussed in detail about parallel-plate waveguides in which we assumed the plates to be of infinite extent in the *y*-direction. However these plates are always finite, and electromagnetic energy leaks through the sides.

Therefore in all practical applications all the waveguides are closed structure the most familiar being hollow rectangular waveguides. The rectangular waveguide shown in Fig. 9.3 consists of perfectly conducting walls in which em waves propagate in the *z*-direction.

In contrast to parallel-plate waveguide, rectangular waveguides support only TM and TE modes but not TEM waves. As we will see similar to parallel-plate waveguide there is a cut-off frequency for wave propagation in rectangular waveguides

### TM Mode

For the case of TM mode we know that $H_{oz} = 0$ and $E_{oz} \neq 0$. In Fig. 9.3 $E_{oz}$ is the tangential component to the conducting plates, and it obeys the boundary condition 6.181–$E_{1t} = E_{2t}$. In Fig. 9.3 at the boundary between the conductor and interior of the waveguide the tangential component should be zero because inside the conductor we know that the electric field is zero.

Therefore the boundary conditions are

(a) For the conducting plate at $(x = 0, y, z)$ $E_{oz} = 0$.
(b) For the conducting plate at $(x = a, y, z)$ $E_{oz} = 0$.
(c) For the conducting plate at $(x, y = 0, z)$ $E_{oz} = 0$.
(d) For the conducting plate at $(x, y = b, z)$ $E_{oz} = 0$.

Applying boundary conditions a, c in Eq. 9.81 we obtain $A_2 = 0$ and $A_4 = 0$. Therefore Eq. 9.81 can be written as

$$E_{oz} = A_1 A_3 \sin k_x x \sin k_y y \tag{9.83}$$

$$\Rightarrow E_{oz} = A \sin k_x x \sin k_y y \tag{9.84}$$

where $A = A_1\, A_3$.

Applying boundary condition b in Eq. 9.84

$$\sin k_x a = 0 \tag{9.85}$$

$$\Rightarrow k_x = \frac{m\pi}{a} \text{ with } m = 0, 1, 2, 3 \tag{9.86}$$

Applying boundary condition d in Eq. 9.84 gives

$$\sin k_y b = 0 \tag{9.87}$$

$$\Rightarrow k_y = \frac{n\pi}{b} \text{ with } n = 0, 1, 2, 3\ldots \tag{9.88}$$

Substituting Eqs. 9.86, 9.88 in Eq. 9.84

$$E_{oz} = A \sin\left(\frac{m\pi x}{a}\right) \sin\left(\frac{n\pi y}{b}\right) \tag{9.89}$$

With $H_{oz} = 0$ and using Eq. 9.84 and 9.72 we can write Eqs. 9.30–9.33 as

$$E_{ox} = \frac{-\gamma}{p^2} k_x A \cos k_x x \sin k_y y \tag{9.90}$$

$$E_{oy} = \frac{-\gamma}{p^2} k_y A \sin k_x x \cos k_y y \tag{9.91}$$

$$H_{ox} = \frac{i\omega\varepsilon}{p^2} k_y A \sin k_x x \cos k_y y \tag{9.92}$$

$$H_{oy} = \frac{-i\omega\varepsilon}{p^2} k_x A \cos k_x x \sin k_y y \tag{9.93}$$

Substituting Eqs. 9.84, 9.90–9.93, $H_{oz} = 0$ in Eqs. 9.6, 9.7 **E** and **H** can be obtained.

Each combination of the integers m and n determines a possible solution or a mode denoted as $TM_{mn}$ mode.

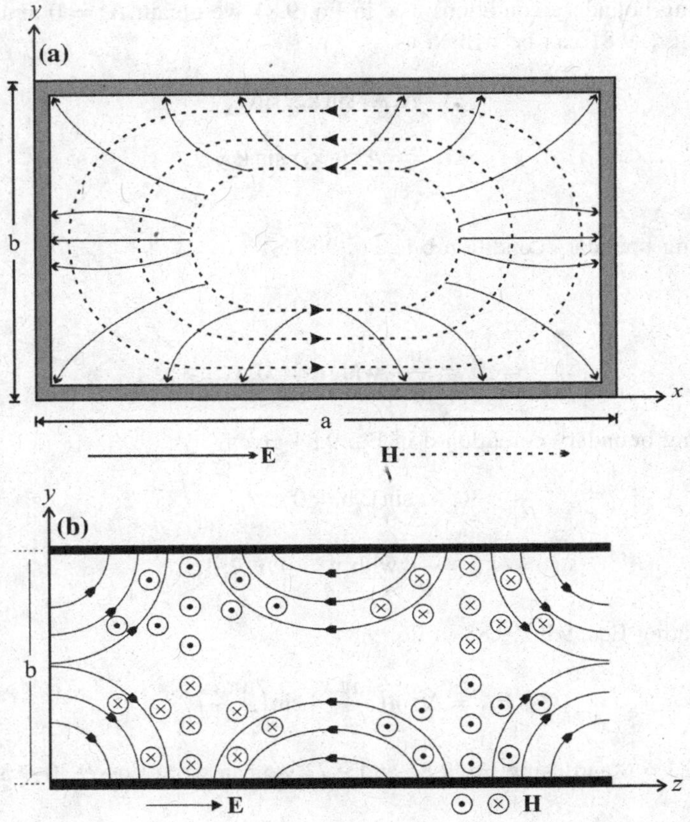


**Fig. 9.4** **a, b** Field configuration in a rectangular waveguide for $TM_{11}$ modes

The field distribution inside the guide depends on m, n values for $TM_{mn}$ mode, and the field in the waveguide is distinct for each mode.

If either m or n is zero then from Eq. 9.89 $E_{oz} = 0$. With $H_{oz} = 0$ all the field components are zero from Eqs. 9.90 to 9.93. Hence no field exists in the waveguide. Thus the lowest possible mode is $TM_{11}$ mode. The field configurations for the $TM_{11}$ mode are shown in Fig. 9.4.

**Transverse Electric Mode**

For the transverse electric mode we know that $E_{oz} = 0$. But the magnetic field component $H_{oz}$ is nonzero.

However as per boundary conditions 6.183 the tangential component of magnetic field is nonzero. But the tangential components of electric fields are zero at the conducting boundaries of the waveguide. Therefore the boundary conditions are

(e) In Fig. 9.3 at ($x$, $y$ = 0, $z$) the tangential component $E_{ox}$ is zero
(f) In Fig. 9.3 at ($x$, $y$ = b, $z$) the tangential components $E_{ox}$ is zero
(g) In Fig. 9.3 at ($x$ = 0, $y$, $z$) the tangential component $E_{oy}$ is zero
(h) In Fig. 9.3 at ($x$ = a, $y$, $z$) the tangential component $E_{oy}$ is zero.

With $E_{oz} = 0$ and Eqs. 9.72, 9.82 and 9.30

$$E_{ox} = \frac{-i\omega\mu}{p^2}\begin{bmatrix}(C_1 \sin k_x x + C_2 \cos k_x x) \\ \times (C_3 k_y \cos k_y y - C_4 k_y \sin k_y y)\end{bmatrix} \tag{9.94}$$

Applying the boundary condition e to Eq. 9.94 gives

$$C_3 = 0 \tag{9.95}$$

Noting down that $k_y$ in Eq. 9.94 has nonzero values, substituting boundary condition f in (9.94)

$$\sin k_y b = 0 \tag{9.96}$$

$$\Rightarrow k_y = \frac{n\pi}{b} \quad n = 0, 1, 2\ldots \tag{9.97}$$

Substituting Eqs. 9.95, 9.97 in Eq. 9.82

$$H_{oz} = [C_1 \sin k_x x + C_2 \cos k_x x]\left[C_4 \cos\left(\frac{n\pi y}{b}\right)\right] \tag{9.98}$$

With $E_{oz} = 0$ and from Eqs. 9.98, 9.31 and 9.72 we can write

$$E_{oy} = \frac{i\omega\mu}{p^2}\begin{bmatrix}(C_1 k_x \cos k_x x - C_2 k_x \sin k_x x) \\ \times \left(C_4 \cos \frac{n\pi y}{b}\right)\end{bmatrix} \tag{9.99}$$

Applying boundary condition g in Eq. 9.99 we obtain

$$C_1 = 0 \tag{9.100}$$

As $k_x$ can take nonzero values the boundary condition h with Eq. 9.99 gives

$$\sin k_x a = 0 \tag{9.101}$$

$$\text{or } k_x = \frac{m\pi}{a} \text{ with } m = 0, 1, 2\ldots \tag{9.102}$$

Substituting Eqs. 9.100 and 9.102 into 9.98

$$H_{oz} = C_2C_4 \cos\frac{m\pi x}{a}\cos\frac{n\pi y}{b} \quad (9.103)$$

With $C = C_2\ C_4$ we can write

$$H_{oz} = C\cos\left(\frac{m\pi x}{a}\right)\cos\left(\frac{n\pi y}{b}\right) \quad (9.104)$$

With $E_{oz} = 0$. the remaining field components can be obtained by substituting Eq. 9.104, 9.72, 9.97 and 9.102 in Eqs. 9.30–9.33.

$$E_{ox} = \frac{i\omega\mu}{p^2} C\ k_y \cos\ k_x x \sin k_y y \quad (9.105)$$

$$E_{oy} = \frac{-i\omega\mu}{p^2} C\ k_x \sin k_x x\ \cos k_y y \quad (9.106)$$

$$H_{ox} = \frac{\gamma}{p^2} C\ k_x \sin k_x x \cos k_y y \quad (9.107)$$

$$H_{oy} = \frac{\gamma}{p^2} C\ k_y \cos k_x x \sin k_y y \quad (9.108)$$

Substituting Eqs. 9.104, 9.105–9.108 and $E_{oz} = 0$ in Eqs. 9.6 and 9.7 **E** and **H** can be obtained.

When both m, n are zero we observe that except $H_{oz}$ all the field components are zero. Hence $TE_{00}$ mode is not possible in a rectangular waveguide. However for $TE_{10}$ mode $H_{ox}$, $H_{oz}$ and $E_{oy}$ components are nonzero, while remaining components are zero.

Similarly for $TE_{01}$ mode $H_{oy}$, $H_{oz}$, and $E_{ox}$ components are nonzero, while remaining components are zero.

Thus $TE_{10}$, $TE_{01}$ mode can exist in the waveguide.

As we will see the dominant mode for the rectangular waveguide is $TE_{10}$ mode for $a > b$. The field patterns for $TE_{10}$ mode are shown in Fig. 9.5.

**Cut-off frequency**

The propagation constant $\gamma$ is given by Eq. 9.3

$$\gamma = \alpha + i\beta \quad (9.109)$$

where $\alpha$ is the attenuation constant and $\beta$ is the phase constant.

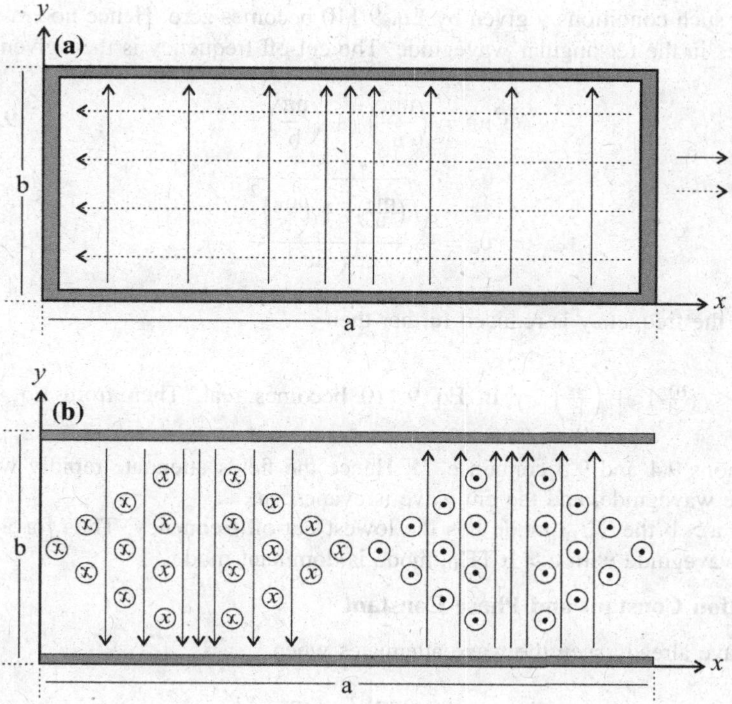


**Fig. 9.5** **a, b** Field configuration in a rectangular waveguide for $TE_{10}$ modes

From Eqs. 9.78, 9.86, 9.88, 9.97 and 9.102 we can write

$$\gamma = \sqrt{\left(\frac{m\pi}{a}\right)^2 + \left(\frac{n\pi}{b}\right)^2 - \omega^2\mu\varepsilon} \tag{9.110}$$

when the frequency is very high such that

$\omega^2\mu\varepsilon \gg \left(\frac{m\pi}{a}\right)^2 + \left(\frac{n\pi}{b}\right)^2$ then in Eq. 9.110 $\gamma$ is imaginary. Therefore from Eq. 9.109 $\gamma = i\beta$. Hence the em wave propagates in the parallel-plate waveguide with phase constant $\beta$.

Next as the frequency is reduced such that

$$\omega^2\mu\varepsilon = \left(\frac{m\pi}{a}\right)^2 + \left(\frac{n\pi}{b}\right)^2 \tag{9.111}$$

Under such conditions $\gamma$ given by Eq. 9.110 becomes zero. Hence no em wave propagates in the rectangular waveguide. The cut-off frequency is then given by

$$\omega_c^2 \mu\varepsilon = \left(\frac{m\pi}{a}\right)^2 + \left(\frac{n\pi}{b}\right)^2 \tag{9.112}$$

$$\Rightarrow \upsilon_c = \frac{\sqrt{\left(\frac{m\pi}{a}\right)^2 + \left(\frac{n\pi}{b}\right)^2}}{2\pi\sqrt{\mu\varepsilon}} \tag{9.113}$$

When the frequency is reduced further then

$\omega^2 \mu\varepsilon \ll \left(\frac{m\pi}{a}\right)^2 + \left(\frac{n\pi}{b}\right)^2$ $\gamma$ in Eq. 9.110 becomes real. Then from Eq. 9.109 $\gamma = \alpha$.

Equations 9.4 and 9.5 contain $e^{-\alpha z}$. Hence the fields attenuate rapidly with $z$, inside the waveguide, and the em wave is evanescent.

When $a > b$ the $TE_{10}$ mode has the lowest cut-off frequency. Thus for a rectangular waveguide with $a > b$ $TE_{10}$ mode is dominant mode.

**Attenuation Constant and Phase Constant**

As we have already seen the wave attenuates when

$$\omega^2 \mu\varepsilon \ll \left(\frac{m\pi}{a}\right)^2 + \left(\frac{n\pi}{b}\right)^2$$

Then the propagation constant from Eqs. 9.109 and 9.110 is

$$\gamma = \alpha = \sqrt{\left(\frac{m\pi}{a}\right)^2 + \left(\frac{n\pi}{b}\right)^2 - \omega^2 \mu\varepsilon} \tag{9.114}$$

Substituting Eqs. 9.112 in 9.114

$$\alpha = \sqrt{\omega_c^2 \mu\varepsilon - \omega^2 \mu\varepsilon} \tag{9.115}$$

$$\alpha = \omega\sqrt{\mu\varepsilon}\sqrt{\frac{\upsilon_c^2}{\upsilon^2} - 1} \tag{9.116}$$

with $v = \dfrac{1}{\sqrt{\mu\varepsilon}}$

$$\alpha = \frac{\omega}{v}\sqrt{\frac{\upsilon_c^2}{\upsilon^2} - 1} \tag{9.117}$$

The wave propagates in the waveguide when

$$\omega^2\mu\varepsilon \gg \left(\frac{m\pi}{a}\right)^2 + \left(\frac{n\pi}{b}\right)^2$$

Then the propagation constant from Eq. 9.109 and 9.110 is

$$\gamma = i\beta = i\sqrt{\omega^2\mu\varepsilon - \left(\frac{m\pi}{a}\right)^2 - \left(\frac{n\pi}{b}\right)^2} \tag{9.118}$$

Substituting Eq. 9.112 in 9.118

$$\beta = \sqrt{\omega^2\mu\varepsilon - \omega_c^2\mu\varepsilon} \tag{9.119}$$

$$\beta = \omega\sqrt{\mu\varepsilon}\sqrt{1 - \frac{\omega_c^2\mu\varepsilon}{\omega^2\mu\varepsilon}} \tag{9.120}$$

With $v = \frac{1}{\sqrt{\mu\varepsilon}}$

$$\Rightarrow \beta = \frac{\omega}{v}\sqrt{1 - \frac{\upsilon_c^2}{\upsilon^2}} \tag{9.121}$$

Similar calculations of $\alpha$ and $\beta$ can be made for parallel-plate waveguide. See Problem 9.1

**Wave Impedance**

We define the wave impedance as

$$Z = \frac{E_{ox}}{H_{oy}} = \frac{-E_{oy}}{H_{ox}} \tag{9.122}$$

We will calculate the wave impedance for TM and TE mode.

**TM Mode**

Substituting Eqs. 9.90–9.93 in 9.122

$$Z_{TM} = \frac{\gamma}{i\omega\varepsilon} \tag{9.123}$$

For $\upsilon < \upsilon_c$ the wave propagates in the waveguide with $\gamma = i\beta$. Hence Eq. 9.123 is then

$$Z_{TM} = \frac{\beta}{\omega\varepsilon} \tag{9.124}$$

Using Eq. 9.121 and noting down that $v = \frac{1}{\sqrt{\mu\varepsilon}}$ and $\eta = \sqrt{\frac{\mu}{\varepsilon}}$ where $\eta$ is the intrinsic impedance of the dielectric medium filling the waveguide. Equation 9.124 can be written as

$$Z_{TM} = \eta\sqrt{1 - \frac{\upsilon_c^2}{\upsilon^2}} \tag{9.125}$$

Thus for $\upsilon > \upsilon_c$ the wave propagates in the guide with impedance $Z_{TM}$ given by Eq. 9.125 which is purely resistive.

For $\upsilon > \upsilon_c$ we know that $\gamma = \alpha$ and Eq. 9.123 is

$$Z_{TM} = -i\frac{\alpha}{\omega\varepsilon} \tag{9.126}$$

where $\alpha$ is given by Eq. 9.117. Thus for $\upsilon < \upsilon_c$ the impedance is purely reactive, and there is no power flow in the guide.

**TE Waves**

Substituting Eqs. 9.105–9.108 in 9.122

$$Z_{TE} = \frac{i\omega\mu}{\gamma} \tag{9.127}$$

For $\upsilon > \upsilon_c$ we know that $\gamma = i\beta$ and hence Eq. 9.127

$$Z_{TE} = \frac{\omega\mu}{\beta} \tag{9.128}$$

Substituting Eq. 9.121 in 9.128 and with $\eta = \sqrt{\frac{\mu}{\varepsilon}}$

$$Z_{TE} = \frac{\eta}{\sqrt{1 - \frac{\upsilon_c^2}{\upsilon^2}}} \tag{9.129}$$

Thus for $\upsilon > \upsilon_c$ the wave propagates in the guide with impedance $Z_{TE}$ given by Eq. 9.129, and the impedance is purely resistive.

For $\upsilon < \upsilon_c$ we know that $\gamma = \alpha$ and Eq. 9.127 is

$$Z_{TE} = \frac{i\omega\mu}{\alpha} \tag{9.130}$$

where $\alpha$ is given by Eq. 9.117. Thus for $\upsilon < \upsilon_c$ the impedance is purely reactive, and there is no power flow in the medium.

**Example 9.4**

A rectangular waveguide is filled with a non-magnetic lossless dielectric whose relative permittivity is $\varepsilon_r = 1.96$. The dimensions of the waveguide are a = 1.6 cm and b = 1.1 cm. A $TM_{11}$ wave propagates in the waveguide with frequency $\upsilon = 16\,\text{GHz}$. Find

(a) cut-off frequency
(b) phase constant
(c) Is the $TM_{11}$ mode evanescent?
(d) If the wave is the propagating wave, calculate the phase velocity?

**Solution**

(a) From Eq. 9.113

$$\upsilon_c = \frac{1}{2\sqrt{\mu\varepsilon}}\sqrt{\left(\frac{m}{a}\right)^2 + \left(\frac{n}{b}\right)^2}$$

For a non-magnetic material $\mu = \mu_o$ and with $\mu = \mu_o\mu_r, \varepsilon = \varepsilon_o\varepsilon_r$ We have

$$\upsilon_c = \frac{1}{2\sqrt{\mu_o\varepsilon_o}\sqrt{\varepsilon_r}}\sqrt{\left(\frac{m}{a}\right)^2 + \left(\frac{n}{b}\right)^2}$$

with $\frac{1}{\sqrt{\mu_o\varepsilon_o}} = c$ the speed of light and $\sqrt{\varepsilon_r} = \sqrt{1.96} = 1.4$. For $TM_{11}$ we write

$$\upsilon_c = \frac{3 \times 10^8}{2 \times 1.4}\sqrt{\frac{1}{(1.6 \times 10^2)^2} + \frac{1}{(1.1 \times 10^2)^2}}$$

$$\upsilon_c = 11.7\,\text{GHz}$$

(b) We know that

$$\begin{aligned} k &= \omega\sqrt{\mu\varepsilon} \\ &= 2\pi\upsilon\sqrt{\mu\varepsilon} \\ &= \frac{2\pi \times 16 \times 10^9 \times 1.4}{3 \times 10^8} \\ \Rightarrow k &= 468\,\text{m}^{-1} \end{aligned}$$

From Eq. 9.121 we have

$$\beta = \frac{\omega}{v}\sqrt{1 - \frac{\upsilon_c^2}{\upsilon^2}}$$

As $v = \frac{1}{\mu\varepsilon}$ then we can write

$$\beta = \omega\sqrt{\mu\varepsilon}\sqrt{1 - \frac{\upsilon_c^2}{\upsilon^2}}$$

$$\beta = k\sqrt{1 - \frac{\upsilon_c^2}{\upsilon^2}}.$$

$$\Rightarrow \beta = 468\sqrt{1 - \frac{11.7^2}{16^2}}$$

$$\beta = 319\ \text{rad}\,\text{m}^{-1}$$

(c) As $\upsilon(= 16\,\text{GHz}) > \upsilon_c(11.7\,\text{GHz})$. Hence $TM_{11}$ wave propagates in the guide.

$$v_p = \frac{\omega}{\beta} = \frac{2\pi\upsilon}{\beta}$$

(d)
$$= \frac{2 \times 3.14 \times 16 \times 10^9}{319}$$

$$\Rightarrow v_p = 3.14 \times 10^8\ \text{m/sec.}$$

**Example 9.5**
Consider a rectangular waveguide which carries a $TE_{10}$ mode at a frequency of 12 GHz. The dimension of the waveguide is 3 cm × 2 cm. Write expressions for cut-off wavelength and guide wavelength and calculate cut-off wavelength for the above mentioned case.

**Solution**
With $v = \frac{1}{\mu\varepsilon}$ Eq. 9.113 can be written as

$$\upsilon_c = \frac{v}{2\pi}\sqrt{\left(\frac{m\pi}{a}\right)^2 + \left(\frac{n\pi}{b}\right)^2}$$

$$\Rightarrow \upsilon_c = \frac{v}{2}\sqrt{\left(\frac{m}{a}\right)^2 + \left(\frac{n}{b}\right)^2}$$

As $v = \upsilon_c\lambda_c$ we have

$$\lambda_c = \frac{2}{\sqrt{\left(\frac{m}{a}\right)^2 + \left(\frac{n}{b}\right)^2}} \tag{9.131}$$

For $TE_{10}$ mode

$$\lambda_c = 2a = 2 \times 3 = 6\,\text{cm}.$$

The wave propagates in the waveguide with a wavelength

$$\lambda_g = \frac{2\pi}{\beta}$$

This wavelength is called guide wavelength.
Substituting for β from Eq. 9.121

$$\lambda_g = \frac{2\pi \frac{V}{\omega}}{\sqrt{1 - \frac{\upsilon_c^2}{\upsilon^2}}}$$

$$\Rightarrow \lambda_g = \frac{\lambda}{\sqrt{1 - \frac{\upsilon_c^2}{\upsilon^2}}}$$

**Example 9.6**
Write the instantaneous field expression for TE mode in the rectangular waveguide for em wave propagation?

**Solution**
We know that for a em wave to propagate in the rectangular waveguide $\gamma = i\beta$. Hence to obtain the instantaneous field expression we add $e^{-\gamma z}e^{i\omega t} = e^{i(\omega t - \beta z)}$ term to all the field components as per Eqs. 9.4 and 9.5.

Therefore from Eqs. 9.104

$$H_{oz} = C\cos\left(\frac{m\pi x}{a}\right)\cos\left(\frac{n\pi y}{b}\right)e^{i(\omega t - \beta z)} \tag{9.133}$$

The above equation can be written as

$$H_{oz} = C\cos\left(\frac{m\pi x}{a}\right)\cos\left(\frac{n\pi y}{b}\right)[\cos(\omega t - \beta z) + i\sin(\omega t - \beta z)] \tag{9.134}$$

Taking the real part

$$H_{oz} = C\cos\left(\frac{m\pi x}{a}\right)\cos\left(\frac{n\pi y}{b}\right)\cos(\omega t - \beta z) \tag{9.135}$$

Noting down that $\gamma = i\beta$, $k_x = \frac{m\pi}{a}$ and $k_y = \frac{n\pi}{b}$Eqs. 9.105–9.108 can be written as

$$E_{ox} = \frac{i\omega\mu}{p^2}Ck_y\cos k_x x\sin k_y y[\cos(\omega t - \beta z) + i\sin(\omega t - \beta z)] \tag{9.136}$$

$$E_{oy} = \frac{-i\omega\mu}{p^2}Ck_x\sin k_x x\cos k_y y[\cos(\omega t - \beta z) + i\sin(\omega t - \beta z)] \tag{9.137}$$

$$H_{ox} = \frac{i\beta}{p^2} Ck_x \sin k_x x \cos k_y y[\cos(\omega t - \beta z) + i \sin(\omega t - \beta z)] \quad (9.138)$$

$$H_{oy} = \frac{i\beta}{p^2} Ck_y \cos k_x x \sin k_y y[\cos(\omega t - \beta z) + i \sin(\omega t - \beta z)] \quad (9.139)$$

Taking the real part's in Eqs. 9.136–9.139.

$$E_{ox} = \frac{-\omega\mu}{p^2} Ck_y \cos\frac{m\pi x}{a} \sin\frac{n\pi y}{b} \sin(\omega t - \beta z) \quad (9.140)$$

$$E_{oy} = \frac{\omega\mu}{p^2} Ck_x \sin\frac{m\pi x}{a} \cos\frac{n\pi y}{b} \sin(\omega t - \beta z) \quad (9.141)$$

$$H_{oy} = \frac{-\beta}{p^2} Ck_x \sin\frac{m\pi x}{a} \cos\frac{n\pi y}{b} \sin(\omega t - \beta z) \quad (9.142)$$

$$H_{oy} = \frac{-\beta}{p^2} Ck_y \cos\frac{m\pi x}{a} \sin\frac{n\pi y}{b} \sin(\omega t - \beta z) \quad (9.143)$$

**Example 9.7**
Consider a hollow air-filled waveguide. The cut-off frequency for $TE_{10}$ mode is 3 GHz and that of $TE_{01}$ mode is 5 GHz. Find

(a) Dimensions of the guide
(b) Cut-off frequency for $TE_{11}$ mode.

**Solution**

(a) From Eq. 9.113 for air-filled guide

$$\upsilon_c = \frac{1}{2\sqrt{\mu_0 \varepsilon_0}} \sqrt{\left(\frac{m}{a}\right)^2 + \left(\frac{n}{b}\right)^2}$$

$$\upsilon_c = \frac{c}{2} \sqrt{\left(\frac{m}{a}\right)^2 + \left(\frac{n}{b}\right)^2}$$

where $c = 3 \times 10^8$ m/s.
Therefore for $TE_{10}$ mode

$$(\upsilon_c)_{10} = \frac{c}{2a}$$

$$\Rightarrow a = \frac{c}{2(\upsilon_c)_{10}} = \frac{3 \times 10^8}{2 \times 3 \times 10^9}$$

Thus a = 5 cm.

For $TE_{01}$ mode

$$(\upsilon_c)_{01} = \frac{c}{2b}$$

$$\Rightarrow b = \frac{c}{2(\upsilon_c)_{01}} = \frac{3 \times 10^8}{2 \times 5 \times 10^9}$$

$$= 3\,\text{cm.}$$

(b) $$(\upsilon_c)_{11} = \frac{3 \times 10^8}{2} \sqrt{\left(\frac{1}{0.05}\right)^2 + \left(\frac{1}{0.03}\right)^2}$$

$$(\upsilon_c)_{11} = 5.8\,\text{GHz}$$

**Example 9.8**

**Write the instantaneous field expressions for TM mode in rectangular waveguide for em wave propagation**

**Solution**

**We know that for a em wave to propagate in the rectangular waveguide $\gamma = i\beta$. Hence to obtain instantaneous field expressions we add $e^{-\gamma z} e^{i\omega t} = e^{i(\omega t - \beta z)}$ term to all the field components as per Eq. 9.4 and 9.5.**

Therefore from Eq. 9.89

$$E_{oz} = A \sin\left(\frac{m\pi x}{a}\right) \sin\left(\frac{n\pi y}{b}\right) e^{i(\omega t - \beta z)} \tag{9.144}$$

$$E_{oz} = A \sin\left(\frac{m\pi x}{a}\right) \sin\left(\frac{n\pi y}{b}\right) [\cos(\omega t - \beta z) + i \sin(\omega t - \beta z)] \tag{9.145}$$

Taking the real part

$$E_{oz} = A \sin\left(\frac{m\pi x}{a}\right) \sin\left(\frac{n\pi y}{b}\right) \cos(\omega t - \beta z) \tag{9.146}$$

Following exactly similar procedure from Eqs. 9.91 to 9.93.

$$E_{ox} = \frac{-i\beta}{p^2} k_x A \cos k_x x \sin k_y y [\cos(\omega t - \beta z) + i \sin(\omega t - \beta z)] \tag{9.147}$$

$$\Rightarrow E_{ox} = \frac{\beta}{p^2} k_x A \cos k_x x \sin k_y y \sin(\omega t - \beta z) \tag{9.148}$$

$$E_{oy} = \frac{-\beta}{p^2} k_y A \sin k_x x \cos k_y y [\cos(\omega t - \beta z) + i \sin(\omega t - \beta z)] \tag{9.149}$$

$$E_{oy} = \frac{\beta}{p^2} k_y A \sin k_x x \cos k_y y \sin(\omega t - \beta z) \tag{9.150}$$

$$H_{ox} = \frac{i\omega\varepsilon}{p^2} k_y A \sin k_x x \cos k_y y [\cos(\omega t - \beta z) + i \sin(\omega t - \beta z)] \tag{9.151}$$

$$H_{ox} = \frac{-\omega\varepsilon}{p^2} k_y A \sin k_x x \cos k_y y \sin(\omega t - \beta z) \tag{9.152}$$

$$H_{oy} = \frac{-i\omega\varepsilon}{p^2} k_x A \cos k_x x \sin k_y y [\cos(\omega t - \beta z) + i \sin(\omega t - \beta z)] \tag{9.153}$$

$$H_{oy} = \frac{\omega\varepsilon}{p^2} k_x A \cos k_x x \sin k_y y \sin(\omega t - \beta z) \tag{9.154}$$

**Impossibility of a TEM Wave in a Hollow Waveguide**

A TEM wave doesn't have any field components in the propagation direction. Consider any hollow waveguide such as the rectangular waveguide in Fig. 9.3. As the wave is propagating in the *z*-direction there will be no electric field or magnetic fields in *z*-direction. Hence $E_z$ and $H_z$ are both zero.

We have already discussed magnetic field lines form closed loops. Such a loop should lie in *xy* plane in Fig. 9.3 because $H_z$ is zero for a TEM wave. By Eq. 6.171 the line integral of **H** around any closed loop should be equal to axial [*z*-axis] conduction and displacement currents passing through the loop.

As there is no inner conductor in a hollow waveguide there cannot be any axial conduction current. An axial displacement current requires on axial electric field along the *z*-direction. However for a TEM wave $E_z$ component is zero,·and hence there is no electric field along *z*-direction. The complete absence of axial conduction and displacement components indicates that there cannot be any closed loops of magnetic field in the *xy* plane inside the hollow conductor.

Therefore TEM wave cannot exist in a hollow or dielectric field waveguide.

## 9.8 Cavity Resonators

At high frequencies (> 300 MHz) circuits with lumped circuit elements whose dimensions are comparable with operating wavelength radiate energy. To circumvent the problem, at high frequencies cavity resonators are used instead of lumped circuit elements. Consider the rectangular waveguide shown in Fig. 9.3. If both open ends of the rectangular waveguide are closed then a closed cavity is formed, and the structure is referred to as cavity resonator. Similar to rectangular waveguide em wave propagates in the cavity but undergoes multiple reflections at the closed ends and forms standing wave patterns. The standing wave patterns which satisfy

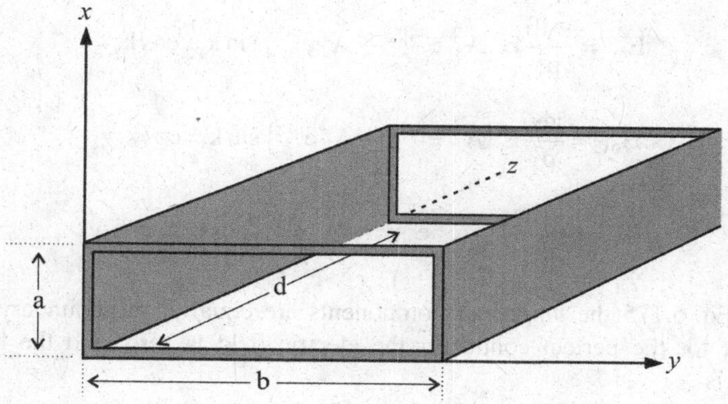


**Fig. 9.6** A cavity resonator

the boundary conditions at the six faces alone can exist in the cavity. A cavity resonator with sides a, b and d is shown in Fig. 9.6.

We have already seen that the wave propagating in $z$-direction can be described by $e^{-\gamma z}$ or $e^{-i\beta z}$.

In Fig. 9.6 the wave travelling in the positive $z$-direction suffers a reflection at $z = d$ and travels in the negative $z$-direction. The wave travelling in the negative $z$-direction can be described by $e^{i\beta z}$. Both TM and TE waves can exist in the cavity resonator, and we will discuss them in detail below.

### TM Mode

For the case of TM mode from Eq. 9.144 we can write

$$E_{oz} = A^{+} \sin k_x x \sin k_y y e^{-i\beta z} + A^{-} \sin k_x x \sin k_y y e^{i\beta z} \tag{9.155}$$

where we have eliminated the time part. The first term corresponds to the wave travelling in the $+z$-direction, while the second term corresponds to wave travelling in $-z$-direction, and we know that for TM wave

$$H_{oz} = 0 \tag{9.156}$$

Note down that $\gamma$ should be replaced by $\gamma = i\beta$ for wave travelling in the + $z$-direction and $-\gamma = -i\beta$ for wave travelling in the $-z$-direction.

Substituting Eqs. 9.155, 9.156 and 9.72 in Eqs. 9.30–9.33

$$E_{ox} = \frac{-i\beta}{p^2} k_x \left[A^{+} e^{-i\beta z} - A^{-} e^{-i\beta z}\right] \cos k_x x \sin k_y y \tag{9.157}$$

$$E_{oy} = \frac{-i\beta}{p^2} k_y \left[A^+ e^{-i\beta z} - A^- e^{i\beta z}\right] \sin k_x x \cos k_y y \tag{9.158}$$

$$H_{ox} = \frac{i\omega\varepsilon}{p^2} k_y \left[A^+ e^{-i\beta z} + A^- e^{i\beta z}\right] \sin k_x x \cos k_y y \tag{9.159}$$

$$H_{oy} = \frac{-i\omega\varepsilon}{p^2} k_x \left[A^+ e^{-i\beta z} + A^- e^{i\beta z}\right] \cos k_x x \sin k_y y \tag{9.160}$$

From Eq. 6.175 the tangential components are equal at the boundary. Noting down that for the perfect conductor the electric field is zero, and the boundary condition is

$$E_{ox} = E_{oy} = 0 \text{ at } z = 0 \text{ and } z = d \tag{9.161}$$

Applying the boundary conditions 9.161 to Eq. 9.157 at $z = 0$ gives

$$A^+ - A^- = 0 \tag{9.162}$$

$$\Rightarrow A^+ = A^- = A \tag{9.163}$$

Substituting Eq. 9.163 in 9.157

$$E_{ox} = \frac{-z\beta}{p^2} k_x A \cos k_x x \sin k_y y \sin \beta z \tag{9.164}$$

Applying the boundary condition 9.161 in 9.164 at $z = d$ gives

$$\sin \beta d = 0 \tag{9.165}$$

$$\Rightarrow \beta d = \ell\pi \text{ where } \ell = 0, 1, 2 \ldots \tag{9.166}$$

$$\Rightarrow \beta = \frac{\ell\pi}{d} \tag{9.167}$$

Thus the fields, with the aid of Eq. 9.163, can be finally written as (Eqs. 9.155, 9.156, 9.164, 9.158, 9.159 and 9.160).

$$E_{oz} = 2A \sin k_x x \sin k_y y \cos \beta z \tag{9.168}$$

$$E_{ox} = \frac{-2\beta}{p^2} k_x A \cos k_x x \sin k_y y \sin \beta z \tag{9.169}$$

$$E_{oy} = \frac{-2\beta}{p^2} k_y A \sin k_x x \cos k_y y \sin \beta z \tag{9.170}$$

$$H_{ox} = \frac{2i\omega\varepsilon}{p^2} k_y A \sin k_x x \cos k_y y \cos \beta z \tag{9.171}$$

$$H_{oy} = \frac{-2i\omega\varepsilon}{p^2} k_x A \cos k_x x \sin k_y y \cos \beta z \tag{9.172}$$

$$H_{oz} = 0 \tag{9.173}$$

where β is given by Eq. 9.167

From Eqs. 9.168 to 9.173 it is observed that the lowest-order TM mode for a rectangular cavity resonator is $m = 1$, $n = 1$ and $\ell = 0$.

**TE Mode**

For the case of TE mode from Eq. 9.133

$$H_{oz} = C^+ \cos k_x x \cos k_y y e^{-i\beta z} + C^- \cos k_x x \cos k_y y e^{i\beta z} \tag{9.174}$$

as usual we have eliminated the time part. While the first term corresponds to wave travelling in the +*z*-direction, the second term corresponds to the wave travelling in the −*z*-direction. We know that for a TE wave

$$E_{oz} = 0 \tag{9.175}$$

As already stated $\gamma$ should be replaced by $\gamma = i\beta$ for wave travelling in +*z*-direction and $-\gamma = -i\beta$ for the wave travelling in the −*z*-direction. Substituting Eqs. 9.174, 9.175 and 9.72 in Eqs. 9.30–9.33.

$$E_{ox} = \frac{i\omega\mu}{p^2} k_y \left[C^+ e^{-i\beta z} + C^- e^{i\beta z}\right] \cos k_x x \sin k_y y \tag{9.176}$$

$$E_{oy} = \frac{-i\omega\mu}{p^2} k_x \left[C^+ e^{-i\beta z} + C^- e^{i\beta z}\right] \sin k_x x \cos k_y y \tag{9.177}$$

$$H_{ox} = \frac{i\beta k_x}{p^2} \left[C^+ e^{-i\beta z} - C^- e^{i\beta z}\right] \sin k_x x \cos k_y y \tag{9.178}$$

$$H_{oy} = \frac{i\beta k_x}{p^2} \left[C^+ e^{-i\beta z} - C^- e^{i\beta z}\right] \cos k_y x \sin k_y y \tag{9.179}$$

Applying the boundary condition 9.161 at $z = 0$ to Eq. 9.176

$$C^+ + C^- = 0 \tag{9.180}$$

$$\Rightarrow C^+ = -C^- = C \tag{9.181}$$

Substituting Eq. 9.181 in 9.176

$$E_{ox} = \frac{-2\omega\mu k_y C}{p^2} \cos k_x x \sin k_y y \sin \beta z \tag{9.182}$$

Applying boundary condition 9.161 at $z = d$ to Eq. 9.182 gives

$$\beta d = \ell\pi \text{ where } \ell = 0, 1, 2 \ldots \tag{9.183}$$

$$\Rightarrow \beta = \frac{\ell\pi}{d} \tag{9.184}$$

Substituting Eqs. 9.181 in Eqs. 9.174, 9.176–9.179 we get

$$H_{oz} = -2iC \cos k_x x \cos k_y y \sin \beta z \tag{9.185}$$

$$E_{ox} = \frac{2\omega\mu k_y}{p^2} C \cos k_x x \sin k_y y \sin \beta z \tag{9.186}$$

$$E_{oy} = \frac{-2\omega\mu}{p^2} k_x C \sin k_x x \cos k_y y \sin \beta z \tag{9.187}$$

$$H_{ox} = \frac{2i\beta k_x C}{p^2} \sin k_x x \cos k_y y \cos \beta z \tag{9.188}$$

$$H_{oy} = \frac{2i\beta k_y C}{p^2} \cos k_x x \sin k_y y \cos \beta z \tag{9.189}$$

$$E_{oz} = 0 \tag{9.190}$$

From Eqs. 9.185 to 9.190 we observe that for electric and magnetic fields to exist either m or n can be zero and $\ell$ must be 1.

From Eqs. 9.110, 9.167, 9.184 with $\gamma = i\beta$ the resonant frequency for both TM and TE modes is given by

$$\upsilon_{mn\ell} = \frac{1}{2\sqrt{\mu\varepsilon}} \sqrt{\left(\frac{m}{a}\right)^2 + \left(\frac{n}{b}\right)^2 + \left(\frac{\ell}{d}\right)^2} \tag{9.191}$$

## 9.9 Quality Factor

A resonant cavity has walls made of conductor which has a finite conductivity and hence loses stored energy. The quality factor is the measure of loss of stored energy.

The energy is stored in the electric and magnetic fields in the cavity resonator. Let $W_e$ and $W_m$ be the energies stored in electric and magnet fields, respectively. Then the total energy stored is

$$W_T = W_e + W_m \tag{9.192}$$

The quality factor is defined as the ratio between time-averaged stored energy to the energy loss in one cycle

$$Q = \frac{2\pi \times \text{time} - \text{averaged stored energy}}{\text{Energy loss in one cycle}} \tag{9.193}$$

$$Q = \frac{2\pi W_T}{P_{loss} t} \tag{9.194}$$

$$Q = \frac{2\pi W_T \upsilon}{P_{loss}} \tag{9.195}$$

$$Q = \frac{\omega W_T}{P_{loss}} \tag{9.196}$$

where t is the time period, $\upsilon$ is the frequency, $\omega$ the angular frequency, and Q is the measure of the bandwidth of the cavity.

**Example 9.9**
Consider a air-filled cubic cavity.

(i) Identify the dominant modes for the cavity and calculate the resonant frequency.
(ii) For a cubical cavity of the side 3 cm calculate the resonant frequency.

**Solution**
As we have already seen the lowest possible mode for TM wave propagation in the cavity is $TM_{110}$ mode. For TE wave propagation the lowest possible mode is $TE_{011}$ or $TE_{101}$ mode.

Thus there are three lowest-order modes in the rectangular cavity: $TM_{110}$, $TE_{011}$ and $TE_{101}$ mode.

(i) For a cubical cavity with a = b = d from Eq. 9.191 we observe all the three modes ($TM_{110}$, $TE_{011}$ and $TE_{101}$) have the same resonant frequency.
$\upsilon_{110} = \frac{c}{\sqrt{2}a}$

Where $c = \frac{1}{\sqrt{\mu_o \varepsilon_o}}$ for air-filled cavity is the velocity of the em wave and is equal to $3 \times 10^8 \text{m/s}$. As all the three modes are having same frequency the modes are degenerate.

(ii) Given $a = 3\text{cm} = 3 \times 10^{-2}$ m

$$\upsilon_{110} = \frac{c}{\sqrt{2}a}$$

$$\upsilon_{110} = \frac{3 \times 10^8}{\sqrt{2}(3 \times 10^{-2})}$$

$$\upsilon_{110} = 7\,\text{GHz}$$

## Exercises

9.1 For the wave propagating in a parallel plate waveguide show that

(a) for the operating frequency $\upsilon < \upsilon_c$ the attenuation constant is given by

$$\alpha = p\sqrt{1 - \left(\frac{\upsilon}{\upsilon_c}\right)^2}$$

(b) for the operating frequency $\upsilon > \upsilon_c$ show that the phase constant is given by

$$\beta = \omega\sqrt{\mu\varepsilon}\sqrt{1 - \left(\frac{\upsilon_c}{\upsilon}\right)^2}$$

9.2 Consider the parallel-plate waveguide shown in Fig. 8.7. Show that when the frequency of the em wave propagating in the waveguide is greater than $\upsilon_c$ the wavelength of the em wave is given by

$$\lambda_g = \frac{2\pi}{\omega\sqrt{\mu\varepsilon}} \frac{1}{\sqrt{1 - \frac{\upsilon_c^2}{\upsilon^2}}}$$

9.3 Consider a parallel-plate waveguide with waves propagating in the *z*-direction. Show that the phase velocity $v_p$ and group velocity $v_g$ are given by

$$v_p = \frac{1}{\sqrt{\mu\varepsilon}\left[1 - \left(\frac{\upsilon_c}{\upsilon}\right)^2\right]^{1/2}},$$

$$v_g = \frac{1}{\sqrt{\mu\varepsilon}}\left[1 - \left(\frac{\upsilon_c}{\upsilon}\right)^2\right]^{1/2}$$

9.4 Calculate the group velocity of

(a) $TE_1$ and $TM_1$ modes
(b) TEM mode in Example 9.3.

9.5 Write the instantaneous field expressions for TM and TE modes in parallel-plate waveguide for em wave propagation.

9.6 Calculate the guide wavelength and wave impedance for the waveguide in Example 9.5.

9.7 For the rectangular waveguide shown in Fig. 9.3. calculate the average axial power flow in the *z*-direction for $TE_{10}$ mode.

9.8 Consider the $TE_{10}$ wave propagation in the rectangular waveguide. Show that the $TE_{10}$ em wave propagation is equivalent to superposition of two planes wave propagation in the guide in *z*-direction.

9.9 A hollow air-filled rectangular waveguide has a dimension of a = 5 cm and b = 2.5 cm.

(a) Calculate the cut-off frequency for $TE_{10}$ mode.
(b) The waveguide has a guide wavelength of 0.1 m. Calculate the frequency of operation.

9.10 For the waveguide given in Problem 9.9 find the frequency range over which only $TE_{10}$ propagates in the guide.

9.11 Show that the phase and group velocities for any mode are related by $v_p v_g = \frac{c^2}{n^2}$

9.12 For the Problem 9.10 calculate the phase velocity and group velocity.

9.13 Consider an air-filled rectangular cavity resonator. Find the dominant modes and the resonant frequency for

(i) a > b > d
(ii) a > d > b where a, b and d have usual meaning in Fig. 9.6.

9.14 An air-filled rectangular cavity with a = 4 cm, b = 1 cm and d = 3 cm operates in dominant mode. Use the results of Problem 9.13 to calculate the resonant frequency.

# Chapter 10
# Antennas

## 10.1 Introduction

In the previous chapters we have seen propagation of electromagnetic (em) waves in free space and various mediums, discussed in detail about guided wave propagation in transmission lines and wave guides. However em waves can be effectively radiated into free space from a source for various applications like broadcasting and the devices used to accomplish radiation of em waves in free space are called antennas.

An antenna is a device which converts a guided electromagnetic wave into em wave propagation in free space. As shown in Fig. 10.1a source generator feeds an transmission line and the wave propagates along the transmission line. The transmission line is connected to the antenna and the antenna radiates the em waves in free space. In this case the antenna functions as a transmitter. Antennas can also be used as receivers in which case it acts as a sensor for em wave.

Thus the antenna acts as a matching device between the transmission line or waveguide and the surrounding medium and can be used effectively to transmit or receive em energy.

## 10.2 Types of Antenna

There are number of different types of antenna. *Wire Antenna* consists of a long wire suspended above the ground and are used for transmitting and receiving purposes. Dipole antenna as the name implies consists of two poles as shown in Fig. 10.2a.

It's the simple form of antenna consisting of two straight wires lying along the same axis. Helical antenna as the name implies consists of a conducting wire wound in the form of the helix and is shown in Fig. 10.2b.

© Springer Nature Singapore Pte Ltd. 2020
S. Balaji, *Electromagnetics Made Easy*,
https://doi.org/10.1007/978-981-15-2658-9_10

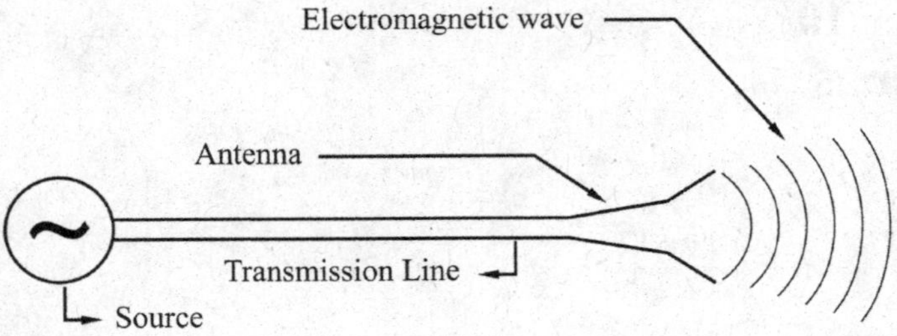


**Fig. 10.1** A source generator feeds the transmission line. The electromagnetic energy from the transmission line is coupled to the Antenna which radiates the energy into free space

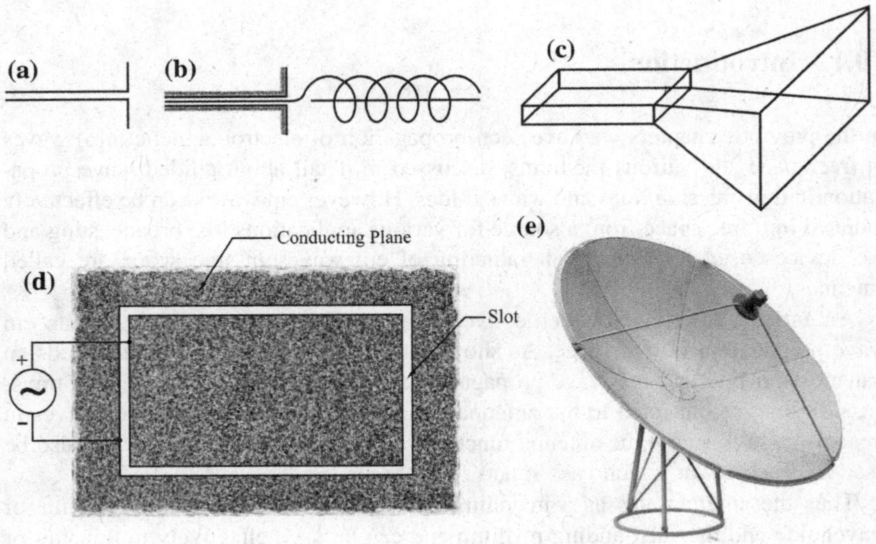


**Fig. 10.2** Types of Antennas a. Dipole Antenna b. Helical Antenna c. Horn Antenna d. Slot Antenna e. Parabolic Antenna

Horn type antenna shown in Fig. 10.2c is an example of aperture antenna. Horn antenna is basically a waveguide containing a large opening at one end.

Another type of antenna is a slot antenna which consists of slot cut in metal plate as shown in Fig. 10.2d.

A parabolic antenna as the name implies consists of a parabolic reflector as shown in Fig. 10.2e.

Because of dish like appearance it is also called dish antenna. The dish acts like a mirror for the em waves and reflects the waves to the required destination.

We will now proceed to determine the radiation fields of various basic types of Antennas like Hertzian dipole etc.

## 10.3 Hertzian Dipole

An infinitesimal current element $\mathrm{I}d\boldsymbol{l}'$ is an Hertzian dipole and is shown in Fig. 10.3. The dipole located at the origin O carries a uniform current I given by

$$\mathrm{I} = \mathrm{I_o} \cos \omega t \tag{10.1}$$

For current element $\mathrm{I}d\boldsymbol{l}'$ from Eq. 6.215 the vector potential can be written as

$$\mathbf{A}(\mathbf{r}, t) = \frac{\mu}{4\pi} \frac{\mathrm{I}(t_r)}{\mathscr{r}} d\boldsymbol{l}' \hat{\mathbf{k}} \tag{10.2}$$

where the dependance of I on $\mathbf{r}'$ is not explicitly stated.

For a small current element $\mathrm{I}d\boldsymbol{l}'$ we need not integrate in Eq. 10.2.

Substituting Eq. 6.213 in 10.2

$$\mathbf{A} = \frac{\mu}{4\pi} \frac{\mathrm{I}\left(t - \frac{\mathscr{r}}{v}\right) d\boldsymbol{l}'}{\mathscr{r}} \hat{\mathbf{k}} \tag{10.3}$$

where $\mathrm{I}\left(t - \frac{\mathscr{r}}{v}\right)$ corresponds to retarded currents and $v = \frac{1}{\sqrt{\varepsilon\mu}}$.

For the Hertzian dipole the current is pointing in the *z*-direction in Fig. 10.3. Hence **A** given by Eq. 10.3 also points in the *z*-direction. Therefore *z*-component of **A** alone is present in Eq. 10.3. $A_{zs}$ is the *z*-component of vector potential corresponding to point P is shown in Fig. 10.3. For convenience the component of vector potential in spherical polar coordinates $-A_{rs}$, $A_{\theta s}$ is shown in the figure.

**Fig. 10.3** .

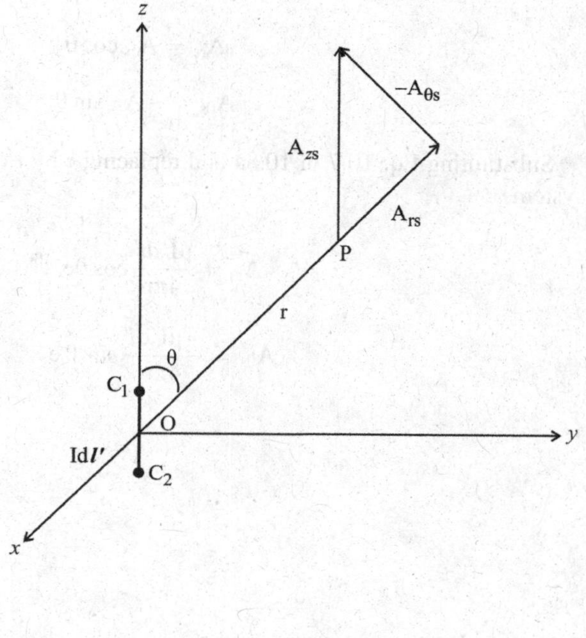

As $I = I_o \cos \omega t$

$$I\left(t - \frac{r}{v}\right) = I_o \cos\left[\omega\left(t - \frac{r}{v}\right)\right]. \tag{10.4a}$$

$$\Rightarrow I\left(t - \frac{r}{v}\right) = I_o \omega(\omega t - \beta z) \tag{10.4b}$$

$$\Rightarrow I\left(t - \frac{r}{v}\right) = \mathrm{Re}\left[I_o e^{i(\omega t - \beta_r)}\right] \tag{10.4c}$$

The phaser form of the current is

$$I_s = I_o\, e^{-i\beta r} \tag{10.5}$$

Therefore the phaser form of vector potential is

$$\begin{aligned} \mathbf{A_s} &= A_{zs}\hat{\mathbf{k}} \\ &= \frac{\mu I_o d\,l'}{4\pi r} e^{-i\beta r}\hat{\mathbf{k}} \end{aligned} \tag{10.6}$$

From the above equation

$$A_{zs} = \frac{\mu_o I\, d\,l'}{4\pi r} e^{-i\beta r} \tag{10.7}$$

Calculation of field components from $\mathbf{A_s}$ becomes easy in spherical polar coordinates. Let us express $\mathbf{A_s}$ in spherical polar coordinates.

From Fig. 10.3

$$A_{rs} = A_{zs} \cos\theta \tag{10.8a}$$

$$A_{\theta s} = -A_{zs} \sin\theta \tag{10.8b}$$

Substituting Eq. 10.7 in 10.8a and replacing $r$ by r of spherical polar coordinate system

$$A_{rs} = \frac{\mu I_o dl'}{4\pi r} \cos\theta e^{-i\beta r} \tag{10.9a}$$

$$A_{\theta s} = \frac{-\mu I_o dl'}{4\pi r} \sin\theta\, e^{-i\beta r} \tag{10.9b}$$

and there is no component of $\mathbf{A_s}$ in $\phi$ direction

$$A_{\phi s} = 0 \tag{10.10}$$

we know that

$$\mathbf{B_s} = \mu \mathbf{H_s} = \nabla \times \mathbf{H_s} \tag{10.11}$$

From Eqs. 10.9a and 10.10 we note that $A_{\phi s} = 0$ and $A_{rs}$, $A_{\theta s}$ are independent of $\phi$. Then from Eq. 1.118 we can write

$$\begin{aligned} \mathbf{H_s} &= H_{\phi s}\hat{\boldsymbol{\phi}} \\ &= \frac{1}{\mu r}\left[\frac{\partial}{\partial r}(rA_{\theta s}) - \frac{\partial A_{rs}}{\partial \theta}\right]\hat{\boldsymbol{\phi}} \end{aligned} \tag{10.12}$$

Thus the only surviving component of $\mathbf{H_s}$ is $H_{\phi S}$ with

$$H_{rs} = 0, H_{\theta s} = 0 \tag{10.13}$$

Substituting Eqs. 10.9a, 10.9b in 10.12

$$H_{\phi s} = \frac{I_0 d\,l'}{4\pi}\sin\theta\left[\frac{i\beta}{r} + \frac{1}{r^2}\right]e^{-i\beta r} \tag{10.14}$$

For a given medium with no currents $\mathbf{J_s} = 0$ and $\mathbf{D_s} = -\varepsilon\,\mathbf{E_s}$ we can write Eq. 6.223 as

$$\nabla \times \mathbf{H_s} = i\omega\varepsilon\,\mathbf{E_s} \tag{10.15}$$

Utilizing Eqs. 10.13, 10.14 and 1.1118 we can write

$$\nabla \times \mathbf{H_s} = \frac{1}{r\sin\theta}\frac{\partial}{\partial\theta}\left[\sin\theta\,H_{\phi s}\right]\hat{\mathbf{r}} - \frac{1}{r}\frac{\partial}{\partial r}\left[rH_{\phi s}\right]\hat{\boldsymbol{\theta}} \tag{10.16}$$

Substituting Eq. 10.14 in 10.16 and then utilizing Eq. 10.15 we get

$$E_{rs} = \frac{I_0 d\,l'}{2\pi}\eta\cos\theta\left[\frac{1}{r^2} - \frac{i}{\beta r^3}\right]e^{-i\beta r} \tag{10.17a}$$

$$E_{\theta s} = \frac{\eta I_0 d\,l'}{4\pi}\sin\theta\left[\frac{i\beta}{r} + \frac{1}{r^2} - \frac{i}{\beta r^3}\right]e^{-i\beta r} \tag{10.17b}$$

$$E_{\phi s} = 0 \tag{10.17c}$$

where $\eta = \frac{\beta}{\omega\varepsilon} = \sqrt{\frac{\mu}{\varepsilon}}$ is the intrinsic impedance of the medium.

### Near and Far Field Regions

In Eqs. 10.14 and 10.17a we see terms containing $\frac{1}{r}$, $\frac{1}{r^2}$ and $\frac{1}{r^3}$.

Based on the distance of the point P in Fig. 10.3 from the source we can classify near and far field regions.

When $r \ll \frac{\lambda}{2\pi}$ then the region is called near field region. As $r \ll \frac{\lambda}{2\pi}$ in Eq. 10.14 $\frac{1}{r^2}$ term dominates and in Eq. 10.17a $\frac{1}{r^3}$ term dominates. Also for smal r, $e^{-i\beta r}$ term can be approximated to unity.

Therefore in Eq. 10.14 $\frac{1}{r}$ term can be neglected

$$\mathbf{H}_{\phi s} = \frac{I_0 d\,l'}{4\pi r^2} \sin\theta \hat{\boldsymbol{\phi}} \tag{10.18}$$

As stated $\frac{1}{r}$, $\frac{1}{r^2}$ term in Eq. 10.17a can be neglected for near field regions. We can write the current $I_o = \frac{\partial q}{\partial t} = i\omega q$ with the aid of Eq. 6.229. Also with $\frac{\eta}{\beta} = \frac{1}{\omega\varepsilon}$ we can write Eq. 10.17a as

$$E_{rs} = \frac{q d\,l'}{4\pi\varepsilon} \frac{2\cos\theta}{r^3} \tag{10.19a}$$

$$E_{\theta s} = \frac{q\, d\,l'}{4\pi\varepsilon} \frac{\sin\theta}{r^3} \tag{10.19b}$$

The net electric field is then

$$\mathbf{E_s} = E_{rs}\hat{\mathbf{r}} + E_{\theta s}\hat{\boldsymbol{\theta}} \tag{10.20}$$

Substituting Eq. 10.19a in 10.20

$$\mathbf{E_s} = \frac{q d\,l'}{4\pi\varepsilon r^3}[2\cos\theta\hat{\mathbf{r}} + \sin\hat{\boldsymbol{\theta}}] \tag{10.21}$$

Comparing Eqs. 10.21 and 2.251 we observe that $\mathbf{E_s}$ is the field produced by static electric dipole. Hence the terms varying as $\frac{1}{r^3}$ are called as electrostatic field terms. Equation 10.18 gives the magnetic field of a short circuit filament as per Biot-Savart law. The $\frac{1}{r^2}$ term is called inductive field.

For $r \ll \frac{\lambda}{2\pi}$ the region is called far-field region.

In that case $\frac{1}{r^2}$, $\frac{1}{r^3}$ terms can be neglected.

Then from Eqs. 10.13, 10.14

$$\mathbf{H_s} = H_{\phi s}\hat{\boldsymbol{\phi}} = \frac{i\beta I_0 d\,l'}{4\pi r} \sin\theta\, e^{-i\beta r}\hat{\boldsymbol{\phi}} \tag{10.22}$$

From Eq. 10.17a

$$E_{rs} = 0 \tag{10.23a}$$

$$E_{\theta s} = \eta \frac{I_o dl'}{4\pi r} (i\beta) \sin\theta \, e^{-i\beta r} \tag{10.23b}$$

Substituting Eq. 10.22 in 10.23b

$$E_{\theta s} = \eta H_{\phi s} \tag{10.23c}$$

Substituting Eq. 10.23a in 10.20

$$\mathbf{E_s} = \eta H_{\phi s} \hat{\boldsymbol{\theta}} \tag{10.24}$$

From Eqs. 10.22 and 10.24 the magnetic field is pointing in the $\hat{\boldsymbol{\phi}}$ direction and electric field is pointing in $\hat{\boldsymbol{\theta}}$ direction. Hence electric and magnetic fields are perpendicular to each other. These fields extend far away from the antenna and are responsible for radiation flow from antenna. Hence far away fields are also known radiation fields.

**Power Flow and Radiation Resistance**

Let us calculate the time average power density for far field regions. From Eq. 7.75

$$<\mathcal{P}> = \frac{1}{2} \mathrm{Re}\left(\mathbf{E_s} \times \mathbf{H_s^*}\right) \tag{10.25}$$

Substituting Eqs. 10.22 and 10.24 in 10.25

$$<\mathcal{P}> = \frac{1}{2} \eta \left|H_{\phi s}\right|^2 (\hat{\boldsymbol{\theta}} \times \hat{\boldsymbol{\phi}}) \tag{10.26}$$

$$<\mathcal{P}> = \frac{\eta \left|H_{\phi s}\right|^2}{2} \hat{\mathbf{r}} \tag{10.27}$$

Substituting $H_{\phi s}$ from 10.22

$$<\mathcal{P}> = \frac{\beta^2 I_0^2 (d\,l')^2}{32\pi^2 r^2} \eta \sin^2\theta \, \hat{\mathbf{r}} \tag{10.28}$$

Considering a far away point located at r the total power passing through a closed spherical surface

$P_{rad}$ is

$$P_{rad} = \oint_s <\mathcal{P}> \cdot \mathbf{ds} \tag{10.29}$$

From Eq. 1.113

$$\mathbf{ds} = r^2 \sin\theta \, d\theta \, d\phi \, \hat{\mathbf{r}} \tag{10.30}$$

Substituting Eqs. 10.28, 10.30 in 10.29

$$P_{rad} = \frac{\beta^2 I_o^2 (dl')^2}{32\pi^2} \eta \int_0^{\pi} \sin^3\theta \, d\theta \int_0^{2\pi} d\phi \tag{10.31}$$

$$P_{rad} = \frac{\beta^2 \eta I_o^2 (dl')^2}{32\pi^2} \left(\frac{4}{3}\right)(2\pi) \tag{10.32}$$

$$P_{rad} = \frac{\beta^2 I_o^2 (dl')^2 \eta}{12\pi} \tag{10.33}$$

For free space $\eta = 120\pi$ and with $\beta = \frac{2\pi}{\lambda}$ Eq. 10.33 becomes

$$P_{rad} = 40\pi^2 \left[\frac{dl'}{\lambda}\right]^2 I_o^2 \tag{10.34}$$

As already noted the current is given by $I = I_o \cos \omega t$. We know that the power dissipated is given by

$$\begin{aligned} P_{rad} &= I_{rms}^2 R_{rad} \\ P_{rad} &= \frac{1}{2} I_o^2 R_{rad} \end{aligned} \tag{10.35}$$

Comparing Eqs. 10.34 and 10.35

$$R_{cad} = 80\pi^2 \left[\frac{dl'}{\lambda}\right]^2 \tag{10.36}$$

$R_{rad}$ is called radiation resistance. It is the characteristic quantity of the dipole described here. Radiation resistance signifies the power the dipole can radiate for a given current.

From Eq. 10.35 antenna's with large radiation resistance delivers large amount of power to space.

From Eq. 10.36 $R_{rad}$ is directly proportional to length of the antenna $dl$ and inversely proportional to $\lambda$. Therefore a large antenna has large radiation resistance and delivers more power to source for a given amount.

**Directional Characteristics**

The dipole discussed above radiates different amount of power in different directions.

Radiation intensity, directive gain …… etc., have been defined to measure non-isotropic radiation from the dipole.

Radiation intensity $U(\theta, \phi)$ is defined as

$$U(\theta, \phi) = r^2 \times \text{Time average Poynting vector at the point } (r, \theta, \phi) \tag{10.37}$$

$$U(\theta, \phi) = r^2 < \mathcal{P} > \tag{10.38}$$

The total time average power radiated is

$$P_{rad} = \oint < \mathcal{P} > \cdot ds \tag{10.39}$$

Noting down that in spherical polar coordinates $ds = r^2 \sin\theta \, d\theta \, d\phi$ the differential solid angle can be expressed as

$$d\Omega = \frac{ds}{r^2} = \sin\theta \, d\theta \, d\phi \tag{10.40}$$

Substituting Eqs. 10.38, 10.40 in 10.39.

$$P_{rad} = \oint U \, d\Omega \tag{10.41}$$

The directive gain of the antenna is defined as

$$G(\theta, \phi) = \frac{\text{Radiation intensity in direction } (\theta, \phi)}{\text{Average radation intensity}} \tag{10.42}$$

$$G(\theta, \phi) = \frac{U(\theta, \phi)}{P_{rad}/4\pi} = \frac{4\pi U(\theta, \phi)}{P_{rad}} \tag{10.43}$$

The directivity of the antenna is the maximum directive gain of the antenna. It is defined as the ratio between three maximum radiation intensity to the average radiation intensity

$$D = \frac{U_{max}}{U_{av}} = \frac{4\pi U_{max}}{P_{rad}} \tag{10.44}$$

### Power Gain and Efficiency

Equation 10.29 gives the radiated power and based on radiated power directive gain has been defined in Eq. 10.43.

However as radiation takes place in the antenna ohmic power loss $P_o$ occurs in the antenna and other supporting transmission structures. Thus it $P_{in}$ is the input power then,

$$P_{in} = P_{rad} + P_o \tag{10.45}$$

The new power gain $G_P$ of the antenna including the Ohmic power loss is defined as

$$G_P = \frac{4\pi U_{max}}{P_{in}} \tag{10.46}$$

The radiation efficiency is defined as the ratio between radiated power to the input power

$$\mathcal{E}_r = \frac{P_{rad}}{P_{in}} \tag{10.47}$$

As discussed previously, infinitesimal current element $I d\,l'$ is an Hertzian dipole. Such a current filament can be assumed to exist if we imagine two small fixed spherical conductors at $C_1$, $C_2$ in Fig. 10.3 separated by a distance $d\,l'$.

The two spherical conductors are separated by thin straight wire. The conductor at $C_1$ carries a time varying charge q(t) while the conductor at $C_z$ carries a time varying charge of −q(t) and a current I(t) flows between them; thereby the system forms an electric dipole.

**Example 10.1**
Find the radiation intensity of the Hertzian dipole.

**Solution**
Substituting Eq. 10.28 in 10.38

$$U = \frac{\beta^2 I_0^2 (d\,l')^2}{32\pi^2} \eta \sin^2\theta \tag{10.48}$$

**Example 10.2**
Find the directive gain and directivity of the Hertzian dipole?

**Solution**

From Eq. 10.43 the directive gain is

$$G(\theta,\phi) = \frac{4\pi, U(\theta,\phi)}{P_{rad}} \tag{10.49}$$

Substituting Eq. 10.48 and 10.33 in 10.49

$$G(\theta,\phi) = 4\pi\left[\frac{\beta^2 I_o^2 (d\,l')^2}{32\pi^2}\right]\eta\sin^2\theta\left[\frac{12\pi}{\beta^2 I_o^2 (d\,l')^2\eta}\right] \tag{10.50}$$

$$G(\theta,\phi) = 1.5\sin^2\theta \tag{10.51}$$

The directivity is the maximum value of $G(\theta,\phi)$, which occurs when $\theta = \pi/2$ in Eq. 10.51.

Hence

$$D = \text{Maximum of } G(\theta,\phi) = 1.5 \tag{10.52}$$

The directivity is usually expressed in decibels as $D = 10 \log_{10} 1.5 = 1.76$ dB.

**Example 10.3**

Plot the **E**-field pattern for the far field regions of the Hertzian dipole.

**Solution**

From Eq. 10.23b

$$E_{\theta s} = \frac{i\beta\eta I d\,l'}{4\pi r}\sin\theta\, e^{-i\beta r} \tag{10.53}$$

Equation 10.53 is maximum when $\theta = 90°$ and is zero when $\theta = 0°$ and 180°. We define the normalized field component as the ratio between the field at the given point to its maximum value.

$$\text{Normalised } E_{\theta s} = \frac{E_{\theta s}}{(E_{\theta s})_{max}} \tag{10.54}$$

Thus by substituting $E_{\theta s}$ and $(E_{\theta s})_{max}$

$$\text{Normalaised } E_{\theta s} = \sin\theta \tag{10.55}$$

The plot of normalised $E_{\theta s}$ is shown in Fig. 10.4.

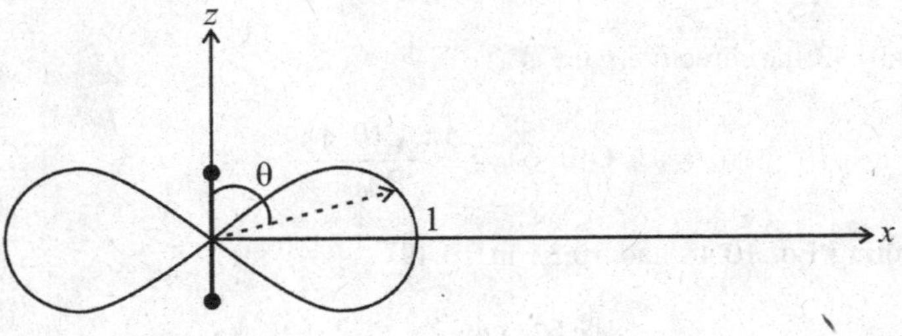


**Fig. 10.4** .

## 10.4 Magnetic Dipole

Consider two different variables being described by separate equations, the equations having the same mathematical form. As the mathematical equations are identical their solutions should also have identical form.

Hence if a set of solutions for one mathematical equation corresponding to one variable is obtained then solution for another set of mathematical equations corresponding to another variable can be obtained by swapping the symbols in the solutions. This is known as concept of duality.

Consider the Maxwell's equation 6.230–6.233 for source free regions i.e., $\rho_s = 0, \mathbf{J_s} = 0$ with $\mathbf{D_s} = \varepsilon\,\mathbf{E_s}, \mathbf{B_s} = \mu\mathbf{H_s}$ then we can write

$$\nabla \cdot \mathbf{E_s} = 0 \tag{10.56}$$

$$\nabla \cdot 0 \tag{10.57}$$

$$\nabla \times \mathbf{E_s} = -i\omega\mu\,\mathbf{H_s} \tag{10.58}$$

$$\nabla \times \mathbf{H_s} = i\omega\mu\,\mathbf{E_s} \tag{10.59}$$

As we swap $\mathbf{E_s}$ by $\mathbf{H_s}$, $\mathbf{H_s}$ by $-\mathbf{E_s}$, $\mu$ by $\varepsilon$ and $\varepsilon$ by $\mu$ we observe that Eqs. 10.56–10.59 are reproduced once again and obeys the concept of duality. By using the concept of duality the fields of the magnetic dipole radiating em energy can be obtained as described below.

Consider a magnetic dipole carrying a current $I = I_o \cos \omega t$ with radius R as shown in Fig. 10.5.

The magnetic moment of the dipole from Eq. 4.209 is

$$\mathbf{m} = I\left(\pi R^2\right)\hat{\mathbf{k}} \tag{10.60}$$

Now Eqs. 2.261 and 4.220 suggest that the electric field due to a electric dipole with electric dipole moment $\mathbf{p} = q\mathbf{d}$ has a similar form of the magnetic field due to a magnetic dipole with magnetic dipole moment $\mathbf{m} = I\pi R^2\hat{\mathbf{k}}$.

**Fig. 10.5** .

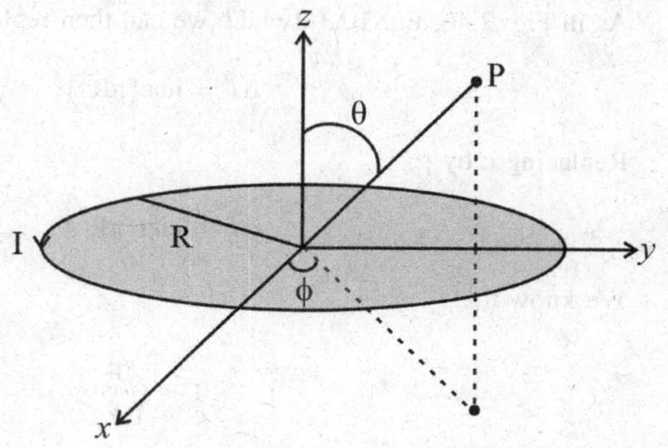


Equations 2.261 and 4.220 are reproduced below

$$\mathbf{E} = \frac{1}{4\pi\varepsilon r^3}[3(\mathbf{p}\cdot\hat{\mathbf{r}})\hat{\mathbf{r}} - \mathbf{p}] \tag{10.61}$$

$$\mathbf{H} = \frac{1}{4\pi r^3}[3(\mathbf{m}\cdot\hat{\mathbf{r}})\hat{\mathbf{r}} - \mathbf{m}] \tag{10.62}$$

From Eq. 6.229

$$I = \frac{\partial q}{\partial t} = i\omega q \Rightarrow q = \frac{I_o}{i\omega} \tag{10.63}$$

The electric dipole moment can be written with aid of Eq. 10.63 as

$$p = qd = \frac{Id}{i\omega} \tag{10.64}$$

Note down that **p** points in *z*-direction in Fig. 2.46.

Substituting for p in Eq. 10.61 and replacing d by $i\omega\varepsilon(\pi R^2)$ and **E** by **H** we observe that Eq. 10.61 is transformed into Eq. 10.62.

As noted in Sect. 10.3 an Hertzian dipole can be assumed to be an electric dipole. Therefore utilizing the concept of duality the fields obtained for the Hertzian dipole in Eqs. 10.13, 10.14, 10.17a can be used to determine the fields of the magnetic dipole.

The following replacements can be made.

$$\left.\begin{array}{l}\mathbf{E}\rightarrow\mathbf{H},\ \mu\rightarrow\varepsilon\\ \mathbf{H}\rightarrow-\mathbf{E},\ \varepsilon\rightarrow\mu\end{array}\right\} \tag{10.65}$$

As in Fig. 2.46, Eq. 10.3 $d = d\,l'$ we can then replace

$$d\,l' = i\omega\varepsilon\left(\pi R^2\right) \tag{10.66}$$

Replacing $\varepsilon$ by $\mu$

$$d\,l' = i\omega\mu\left(\pi R^2\right) \tag{10.67}$$

We know that

$$\eta = \sqrt{\frac{\mu}{\varepsilon}} \tag{10.68}$$

Replacing $\mu$ by $\varepsilon$, $\varepsilon$ by $\mu$

$$\eta = \sqrt{\frac{\mu}{\varepsilon}} \to \sqrt{\frac{\varepsilon}{\mu}} = \frac{1}{\eta} \tag{10.69}$$

Therefore $\eta$ should be replaced by $\frac{1}{\eta}$.
Executing the above replacements in Eqs. 10.13, 10.14 and 10.17a.

$$E_{rs} = 0, E_{\theta s} = 0 \tag{10.70}$$

$$E_{\phi s} = \frac{-i\omega\mu I_o\left(\pi R^2\right)}{4\pi}\sin\theta\left[\frac{i\beta}{r} + \frac{1}{r^2}\right]e^{-i\beta r} \tag{10.71}$$

$$H_{rs} = \frac{i\omega\mu I_o\left(\pi R^2\right)}{2\pi\eta}\cos\theta\left[\frac{1}{r^2} - \frac{i}{\beta r^3}\right]e^{-i\beta r} \tag{10.72}$$

$$H_{\theta s} = \frac{i\omega\mu I_o\left(\pi R^2\right)}{4\pi\eta}\sin\theta\left[\frac{i\beta}{r} + \frac{1}{r^2} - \frac{i}{\beta r^3}\right]e^{-i\beta r} \tag{10.73}$$

$$H_{\phi s} = 0 \tag{10.74}$$

Equations 10.70–10.74 give the fields produced by the magnetic dipole.

**Far Field Regions**

The fields in the far field regions are of interest, and we know from Eqs. 10.70 and 10.74.

$$E_{rs} = 0, E_{\theta s} = 0 \text{ and } H_{\phi s} = 0. \tag{10.75}$$

The remaining field components from Eqs. 10.22, 10.23a are (by using Eqs. 10.65, 10.67 and 10.69)

$$E_{\phi s} = \frac{\beta I_o \omega \mu \left[\pi R^2\right]}{4\pi r} \sin\theta\, e^{-i\beta r} \tag{10.76}$$

$$H_{rs} = 0 \tag{10.77}$$

$$H_{\theta s} = \frac{-E_{\phi s}}{\eta} \tag{10.78}$$

The net electric and magnetic fields are given by

$$\mathbf{E_s} = E_{\phi s}\hat{\boldsymbol{\phi}} \tag{10.79a}$$

$$\mathbf{H_s} = \mathbf{H_{\theta s}}\hat{\boldsymbol{\theta}} \tag{10.79b}$$

**Power Flow and Radiation Resistance**

As in Hertzian dipole case let us calculate $<\mathcal{P}>$, $P_{rad}$ for magnetic dipole.

From Eq. 10.25

$$<\mathcal{P}> = \frac{1}{2} R_e\left(\mathbf{E_s} \times \mathbf{H_s^*}\right) \tag{10.80}$$

Substituting Eq. 10.79a in 10.80

$$<\mathcal{P}> = \frac{1}{2} R_e E_{\phi s} H_{\theta s}^* (\hat{\boldsymbol{\phi}} \times \hat{\boldsymbol{\theta}}) \tag{10.81}$$

$$<\mathcal{P}> = \frac{1}{2} R_e E_{\phi s} H_{\theta s}^* (-\hat{\mathbf{r}}) \tag{10.82}$$

Substituting Eq. 10.78 in 10.82

$$<\mathcal{P}> = \frac{1}{2} \frac{\left|E_{\phi s}\right|^2}{\eta} \hat{\mathbf{r}} \tag{10.83}$$

Substituting Eq. 10.76 in 10.83

$$<\mathcal{P}> = \frac{1}{2\eta} \frac{\beta^2 I_0^2 \omega^2 \mu^2 \left[\pi R^2\right]^2}{16\pi^2 r^2} \sin^2\theta\, \hat{\mathbf{r}} \tag{10.84}$$

$P_{rad}$ can be calculated following the same procedure as in Hertzian dipole case. From Eq. 10.29

$$P_{rad} = \oint <\mathcal{P}> \cdot ds \tag{10.85}$$

Substituting Eqs. 10.84, 10.30 in 10.85

$$P_{rad} = \frac{1}{2\eta}\frac{\beta^2\omega^2\mu^2}{16\pi^2}\left[I_0\pi R^2\right]^2 \int_0^{\pi} \sin^3\theta d\theta \int_0^{2\pi} d\phi \tag{10.86}$$

$$P_{rad} = \frac{1}{2\eta}\frac{\beta^2\omega^2\mu^2}{16\pi^2}\left[I_0\pi R^2\right]^2\left(\frac{4}{3}\right)(2\pi) \tag{10.87}$$

Noting down that $\eta = \frac{\beta}{\omega\varepsilon} = \sqrt{\frac{\mu}{\varepsilon}}$
We can simplify Eq. 10.87 as

$$P_{rad} = \frac{4}{3}\eta\,\pi^3\left[\frac{I_0(\pi R^2)}{\lambda^2}\right]^2 \tag{10.88}$$

Comparing 10.88 with 10.35 we can write the radiation resistance of the magnetic dipole as

$$R_{rad} = \frac{8}{3}\eta\pi^3\left[\frac{\pi R^2}{\lambda^2}\right]^2 \tag{10.89}$$

The fields so obtained by concept of duality can be directly derived from magnetic vector potential **A** as we did for Hertzian dipole. But the derivation is lengthy.

**Example 10.4**
Find the directive gain and directivity of the magnetic dipole.

**Solution**
Substituting Eq. 10.38 in 10.43

$$G(\theta, \phi) = \frac{4\pi r^2}{P_{rad}} <\mathcal{P}> \tag{10.90}$$

Substituting Eqs. 10.84 and 10.87 in 10.90

$$G(\theta, \phi) = 1.5\sin^2\theta \tag{10.91}$$

The directivity is the maximum value of $G(\theta, \phi)$ which occurs when $\theta = \pi/2$ in Eq. 10.91

Hence

$$D = \text{Maximum of } G(\theta, \phi)$$
$$D = 1.5 \tag{10.92}$$

The directivity is usually expressed in decibels as

$$D = 10\,\log_{10} 1.5 = 1.76\,\text{dB}.$$

## 10.5 Half Wave Dipole Antenna

Half wave dipole antenna as the name implies is $\lambda/2$ long where $\lambda$ is the wavelength. The half wave dipole antenna is shown in Fig. 10.6a. A voltage source is connected to the mid point of the antenna via a transmission line. Because of this reason the half-wave dipole antenna is also called a center-fed antenna. In order to find the fields of the half wave dipole antenna we need to know the current distribution in the antenna.

However there is no way to know the current distribution in the antenna except at the ends where the current is zero.

Considering the dipole antenna to be a open circuited transmission line we can assume the current distribution to be sinusoidal. Hence

$$I = I_o \cos \beta z \tag{10.93}$$

Let us now calculate the fields of an half-wave dipole. In Fig. 10.6b the half-wave dipole is divided into number of Hertzian dipoles.

One such Hertzian dipole of length dz is shown in figure. The Hertzian dipole lies at a distance of r′ from a far away point P.

For the faraway point the field due to the Hertzian dipole at point P from Eq. 10.23b is given by

$$dE_{\theta s} = \frac{iIn\beta dz}{4\pi r'} \sin\theta\, e^{-i\beta r'} \tag{10.94}$$

The field due to the entire half wave dipole antenna is obtained by integrating the fields from all Hertzian dipoles

$$E_{\theta s} = \int_{z=-\lambda/4}^{z=\lambda/4} dE_{\theta s} \tag{10.95}$$

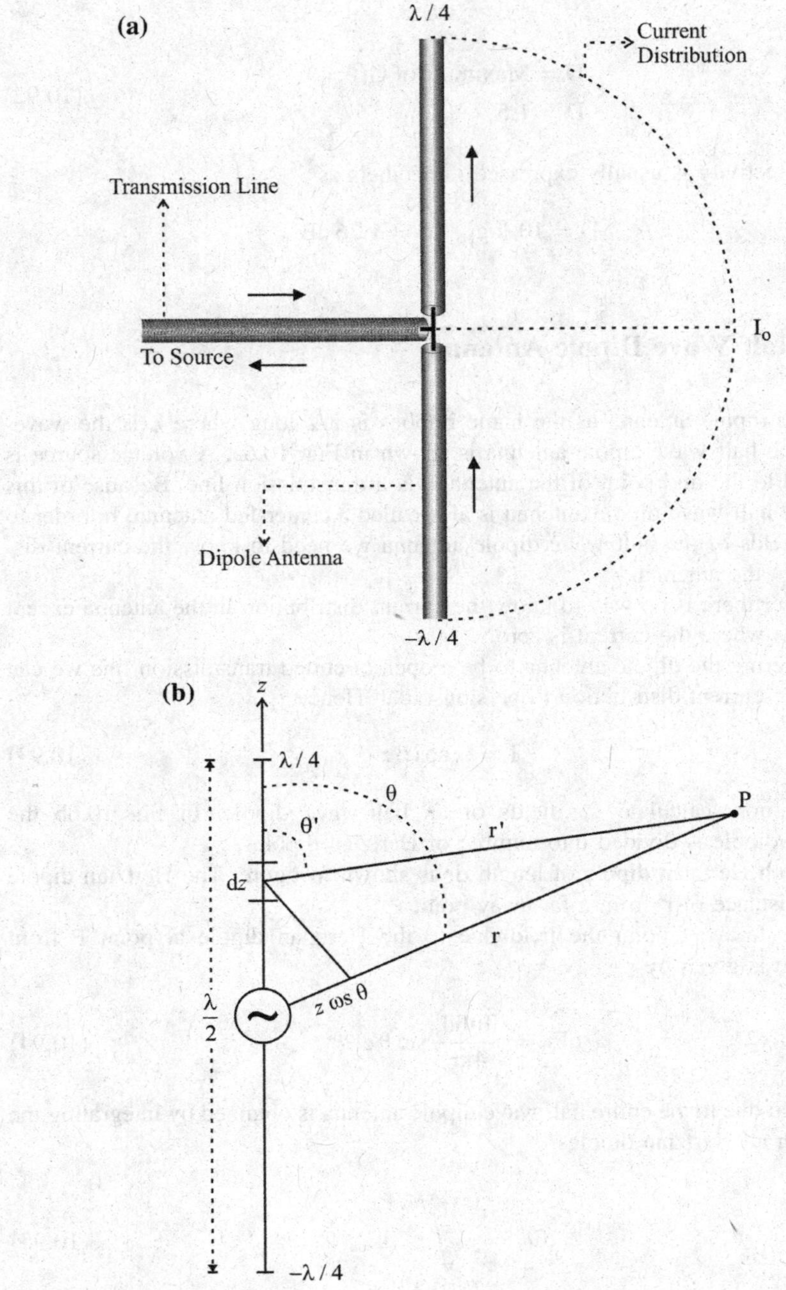
(a)
λ / 4
Current Distribution
Transmission Line
To Source
I$_o$
Dipole Antenna
−λ / 4
(b)
z
λ / 4
θ
P
θ'
r'
dz
r
z ωs θ
λ/2
~
−λ / 4

Fig. 10.6 .

From Fig. 10.6b

$$r' = r - z\cos\theta \tag{10.96}$$

For a faraway point in Eq. 10.94 $r' = r - z\,\omega s\theta \approx r$ in the denominator as the approximation involves little change in value of $dE_{\theta s}$, if we substitute $r' \approx r$ in the denominator. In the exponent $r' = r - z\cos\theta$ is retained as such because any approximation in the exponent will be significant and lead to erroneous values of $E_{\theta s}$.

Therefore Eq. 10.94 can be written as

$$dE_{\theta s} = \frac{iI\eta\,\beta dz}{4\pi r}\sin\theta e^{-i\beta r}e^{i\beta z\cos\theta} \tag{10.97}$$

Substituting Eqs. 10.93, 10.97 in 10.95

$$E_{\theta s} = \frac{iI_o\eta\beta}{4\pi r}\sin\theta e^{-i\beta r}\int_{-\lambda/4}^{\lambda/4}\cos\beta z\, e^{i\beta z\cos\theta}dz \tag{10.98}$$

Let us evaluate the integral separately.

$$\int_{-\lambda/4}^{\lambda/4}\cos\beta z\, e^{i\beta z\cos\theta}dz \tag{10.99}$$

Noting down that

$$\int e^{az}\cos bz dz = \frac{e^{az}[a\cos bz + b\sin bz]}{a^2 + b^2}$$

With $a = i\beta\cos\theta$, $b = \beta$ the integral in Eq. 10.99 becomes

$$= \left\{\frac{e^{i(z\beta)\cos\theta}[i\beta\cos\theta\cos\ \beta z + \beta\sin\ \beta z]}{\beta^2 - \beta^2\cos^2\theta}\right\}_{-\lambda/4}^{\lambda/4} \tag{10.100}$$

Setting $z = \frac{\lambda}{4}, \frac{-\lambda}{4}$ we can write

$$\cos\ \beta z = \cos\left(\frac{2\pi}{\lambda}\right)\left(\frac{\lambda}{4}\right) = 0$$

$$\cos\ \beta z = \cos\left(\frac{2\pi}{\lambda}\right)\left(\frac{-\lambda}{4}\right) = 0$$

$$\sin\beta z = \sin\left(\frac{2\pi}{\lambda}\right)\left(\frac{\lambda}{4}\right) = 1$$

$$\sin\beta z = \sin\left(\frac{2\pi}{\lambda}\right)\left(\frac{-\lambda}{4}\right) = -1.$$

$$z\beta = \frac{\lambda}{4}\left(\frac{2\pi}{\lambda}\right) = \frac{\lambda}{2}, z\beta = \left(\frac{-\lambda}{4}\right)\left(\frac{2\pi}{\lambda}\right) = \frac{-\pi}{2}$$

Therefore applying the limits in Eq. 10.100

$$= \frac{e^{i\pi/2(\cos\theta)}[0+\beta] - e^{-i\pi/2\cos\theta}[0-\beta]}{\beta^2\sin^2\theta} \tag{10.101}$$

$$= \frac{\beta\left[e^{i\pi/2\cos\theta} + e^{-i\pi/2\cos\theta}\right]}{\beta^2\sin^2\theta} \tag{10.102}$$

As $e^{ix} + e^{-ix} = 2\cos x$ we have

$$= \frac{2\cos\left(\frac{\pi}{2}\cos\theta\right)}{\beta\sin^2\theta} \tag{10.103}$$

Therefore

$$\int\limits_{-\lambda/4}^{\lambda/4} \cos\beta z e^{i\beta z\cos\theta} dz = \frac{2\cos\left(\frac{\pi}{2}\cos\theta\right)}{\beta\sin^2\theta} \tag{10.104}$$

Substituting Eq. 10.104 in 10.98

$$E_{\theta s} = \frac{iI_o\eta\beta}{4\pi r}\sin\theta e^{-i\beta z}\left[\frac{2\cos\left(\frac{\pi}{2}\cos\theta\right)}{\beta\sin^2\theta}\right] \tag{10.105}$$

Simplifying

$$E_{\theta s} = \frac{iI_o\eta}{2\pi r}e^{-i\beta r}\left[\frac{\cos\left(\frac{\pi}{2}\cos\theta\right)}{\sin\theta}\right] \tag{10.106}$$

Equation 10.106 gives the electric field for far field region of the half wave dipole antenna.

The magnetic field is given by Eq. 10.24

$$H_{\phi s} = \frac{E_{\theta s}}{\eta} \tag{10.107}$$

The fields are given by

$$\mathbf{E_s} = E_{\theta s}\hat{\boldsymbol{\theta}} \tag{10.108}$$

$$\mathbf{H_s} = H_{\phi s}\hat{\boldsymbol{\phi}} \tag{10.109}$$

**Power Flow and Radiation Resistance**

From Eq. 10.27

$$<\boldsymbol{\mathcal{P}}> = \frac{1}{2}\eta\left|H_{\phi s}\right|^2\hat{\mathbf{r}}$$

Substituting Eqs. 10.106, 10.107 in 10.109

$$<\boldsymbol{\mathcal{P}}> = \frac{\eta I_o^2}{8\pi^2 r^2}\left[\frac{\cos^2\left(\frac{\pi}{2}\cos\theta\right)}{\sin^2\theta}\right]\hat{\mathbf{r}} \tag{10.110}$$

The total power radiated by Eq. 10.29 is

$$P_{rad} = \int <\boldsymbol{\mathcal{P}}> \cdot\, d\mathbf{s} \tag{10.111}$$

Substituting Eqs. 10.110 in 10.111 and with

$$ds = r^2 \sin\theta\, d\theta\, d\phi$$

$$P_{rad} = \int_0^{\pi} \frac{\eta\, I_o^2 \cos^2\left(\frac{\pi}{2}\cos\theta\right)}{8\pi^2 r^2 \sin^2\theta} r^2 \sin\theta\, d\theta \int_0^{2\pi} d\phi \tag{10.112}$$

$$P_{rad} = \frac{\eta I_o^2}{4\pi}\int_0^{\pi} \frac{\cos^2\left(\frac{\pi}{2}\cos\theta\right)}{\sin^2\theta} d\theta \tag{10.113}$$

The integral in Eq. 10.113 can be evaluated numerically to give 1.22 so that

$$P_{rad} = \frac{1.22}{4\pi}\eta I_o^2 \tag{10.114}$$

Comparing Eq. 10.114 with 10.35

$$\frac{1.22}{4\pi}\eta I_o^2 = \frac{1}{2} I_o^2 R_{rad} \tag{10.115}$$

$$\Rightarrow R_{rad} = \frac{1.22}{2\pi}\eta \tag{10.116}$$

For free space $\eta = 377\,\Omega$.
Then

$$R_{rad} = 73.14\,\Omega \tag{10.117}$$

High radiation resistance of the half-wave dipole antenna suggest that it can deliver greater amounts of power in space as compared to Hertzian dipole.

**Example 10.5**
Find the directive gain and directivity of the half-wave dipole antenna.

**Solution**
From Eq. 10.90

$$G(\theta, \phi) = \frac{4\pi r^2 < \mathcal{P} >}{P_{rad}} \tag{10.118}$$

Substituting Eqs. 10.110, 10.114 in 10.118

$$G(\theta, \phi) = \frac{\frac{4\pi\eta I_o^2}{8\pi^2}\left|\frac{\cos^2\left(\frac{\pi}{2}\cos\theta\right)}{\sin^2\theta}\right|}{\frac{1.22}{4\pi}\eta I_0^2} \tag{10.119}$$

$$G(\theta, \phi) = 1.64\left[\frac{\cos^2\left(\frac{\pi}{2}\cos\theta\right)}{\sin^2\theta}\right] \tag{10.120}$$

The directivity is the maximum value of $G(\theta, \phi)$ which occurs when $\theta = \pi/2$ in Eq. 10.120.
Hence

$$\begin{aligned} D &= \text{Maximum of } G(\theta, \phi) \\ D &= 1.64 \end{aligned} \tag{10.121}$$

In decibels

$$D = 10 \log_{10} 1.64 = 2.15$$

## 10.6 Antenna Arrays

The antenna's so far discussed Hertzian dipole, magnetic dipole and half-wave dipole have wide spread radiation pattern.

In number of applications we require directed power—power radiated by antenna to be focussed in a particular direction. This objective cannot be achieved by a single antenna. However number of antennas can be arranged in space and inter connected such that desired directional radiation characteristics can be obtained. Such a collection of antenna's are called antenna array.

Consider two antenna's placed at a distance d apart along the $x$-axis as shown in Fig. 10.7.

One of the antenna marked antenna 0 is located at the origin. The other antenna marked antenna I is located at a distance d from the origin along the $x$-axis. A faraway point P is located at a distance of r and $r_1$ from antenna 0 and antenna I

**Fig. 10.7** .

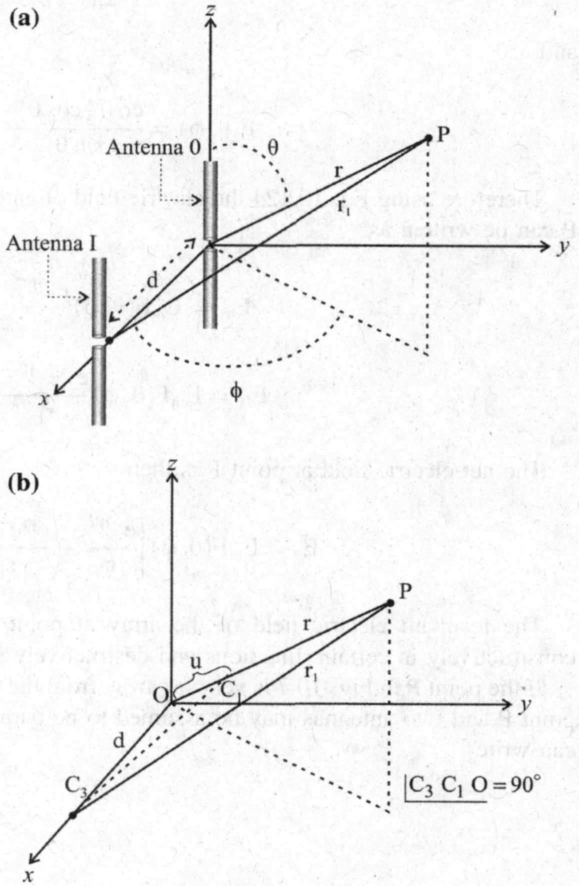

respectively. The current in both the antenna is same in magnitude but the phase of current in antenna I leads that of antenna 0 by an angle of $\alpha$ so that

$$\text{In antenna } 0\; I = I_o \tag{10.122}$$

$$\text{In antenna } I\; I = I_o e^{i\alpha} \tag{10.123}$$

Let us express the electric field radiated by each antenna as

$$E_{\theta s} = E_m \frac{e^{-i\beta r}}{r} F(\theta, \phi) \tag{10.124}$$

where $F(\theta, \phi)$ is the field pattern and $E_m$ is the maximum value of electric field. For example for the half-wave dipole from Eq. 10.106.

$$E_m = \frac{i I_o \eta}{2\pi} \tag{10.125}$$

and

$$F(\theta, \phi) = \frac{\cos\left(\frac{\pi}{2}\cos\theta\right)}{\sin\theta} \tag{10.126}$$

Therefore using Eq. 10.124 the electric field of antenna 0 and antenna I at point P can be written as

$$E_{os} = E_m F(\theta, \phi) \frac{e^{-i\beta r}}{r} \tag{10.127}$$

$$E_{Is} = E_m F(\theta, \phi) \frac{e^{i\alpha} e^{-i\beta r_1}}{r_1} \tag{10.128}$$

The net electric field at point P is then

$$E_s = E_m F(\theta, \phi) \left[\frac{e^{-i\beta r}}{r} + \frac{e^{i\alpha} e^{-i\beta r_1}}{r_1}\right] \tag{10.129}$$

The resultant electric field of the array at point P can be made to interfere constructively in certain directions and destructively in certain directions.

If the point P in Fig. 10.7 is very far away from the antenna then the lines joining point P and two antennas may be assumed to be parallel. Then from Fig. 10.7 we can write

$$\frac{1}{r_1} \simeq \frac{1}{r} \tag{10.130}$$

and

$$r_1 \approx r - d \sin\theta \cos\phi \tag{10.131}$$

In the denominator of Eqs. 10.129, 10.130 can be used at a faraway point. However for the exponent in Eqs. 10.129, 10.131 must be used because using $r \approx r_1$ in the exponent leads to large variation and hence erroneous results, even though difference between r and $r_1$ is small.

Substituting Eqs. 10.130, 10.131 in 10.129

$$E_s = E_m \frac{F(\theta, \phi)}{r} e^{-i\beta r}\left[1 + e^{i\beta d \sin\theta \cos\phi} e^{i\alpha}\right] \tag{10.132}$$

$$E_s = E_m \frac{F(\theta, \phi)}{r} e^{-i\beta r} e^{i\Psi/2}\left(2 \cos\frac{\Psi}{2}\right) \tag{10.133}$$

where

$$\Psi = \beta d \sin\theta \cos\phi + \alpha \tag{10.134}$$

The magnitude of the electric field is

$$\begin{aligned} |\mathbf{E_s}| &= \sqrt{E_s E_s^*} \\ &= \frac{2E_m}{r} |F(\theta)|\,|\cos\Psi/2| \end{aligned} \tag{10.135}$$

Here $|F(\theta, \phi)|$ is called the element factor. $|\cos(\Psi/2)|$ is called the normalised array factor. Thus from Eq. 10.135 (Fig. 10.8).

We observe that the total field pattern of an array is equal to element pattern times the array factor. This is known as principle of pattern multiplication. The element factor depends on the radiation pattern of individual radiation elements. The array factor depends on the nature of the entire array.

**Example 10.6**

Compute the normalised array factor and find the values of the electric field for the H-plane if two dipoles are placed at $d = \lambda/2$ in Fig. 10.7a.

The currents in both the dipoles are in phase.

**Fig. 10.8** .

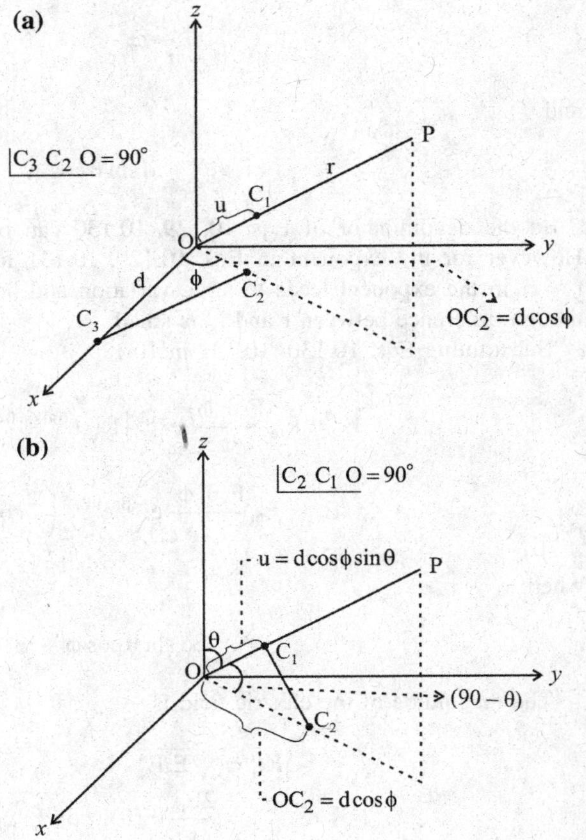


**Solution**

Because the currents are in phase $\alpha = 0$. We have

$$\beta d = \frac{2\pi}{\lambda} \cdot \frac{\lambda}{2} = \pi$$

Then from Eq. 10.134

$$\Psi = \pi \cos \phi$$

For the H-plane $\theta = \pi/2$ and hence from Eq. 10.126 $F(\theta, \phi) = 1$.
Therefore the field in Eq. 10.135 depends on $|\cos \Psi/2|$. Now the array factor

$$|\cos \Psi/2| = \left|\omega s\left(\frac{\pi}{2} \cos \phi\right)\right| \qquad (10.136)$$

In Fig. 10.7a as $\phi$ becomes zero and $\pi$, the array factor and hence the field becomes zero. Thus along the $x$-axis the field is zero in Fig. 10.7a. The array factor and hence the field has a maximum at $\pm\pi/2$ along the $y$-axis. This type of array is called broad side arrays.

Another type of array is end-fire array and is discussed in Problem 10.8.

## 10.7 Receiving Antenna and Friis Equation

In the previous sections we discussed about transmitting antenna which radiates power to the surroundings. Antennas can be used to received em energy and such antenna's are called receiving antennas.

In Fig. 1.20a the surface abcd is held perpendicular to the incoming vector and in Fig. 1.20b the surface abcd is inclined at an angle to the incoming vector. As discussed in Sect. 1.12 when the surface is held perpendicular to the incoming vector the entire area is shown to the incoming vector. When the surface is inclined (Fig. 1.20b) the "Effective Area" shown by the surface to the incoming vector decreases.

In the case of receiving antennas—if the incoming em wave is perpendicular to the antenna then as in Fig. 1.20a the entire area of the antenna is "visible" to the em wave. However the incoming em wave is not always perpendicular to the antenna and is mostly inclined at a angle to the antenna surface as shown in Fig. 10.9. In such situations the effective area shown by the receiving antenna to the incoming em wave will be different.

We define the effective area as the ratio between average power received or intercepted by the receiving antenna to the average power density of the incident wave.

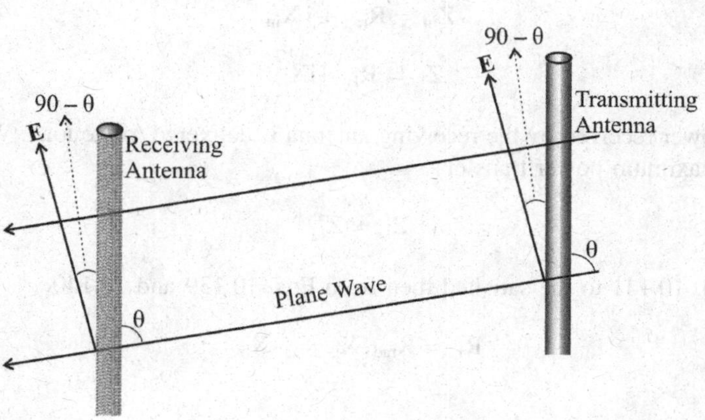


**Fig. 10.9** .

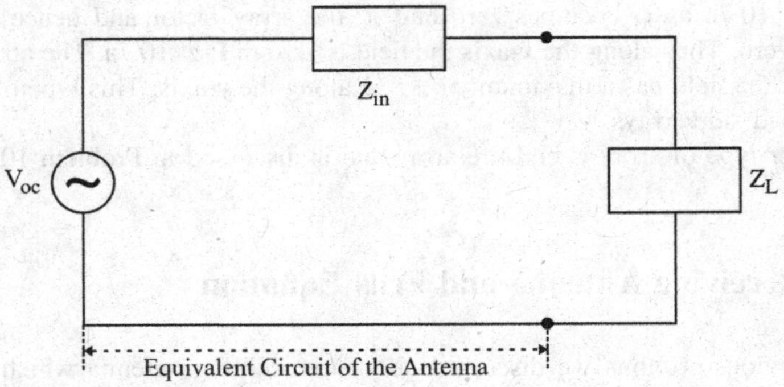


**Fig. 10.10** .

$$A_{er} = \frac{P_{int}}{<\mathcal{P}>} \tag{10.137}$$

The receiving antenna can be modeled using therein equivalent circuit as shown in Fig. 10.10.

The em wave induces voltage in the antenna terminals and it is denoted as open circuit voltage in the Fig. 10.10. In Fig. 10.9 the tangential component of the electric field E sin θ is responsible for the voltage induced in the receiving antenna and hence we can write

$$V_{oc} = E\,l \sin\theta \tag{10.138}$$

where $l$ is the length of the receiving antenna. The antenna impedance and the impedance of the external load connected to the antenna is denoted as $Z_{in}$ and $Z_L$ respectively. The impedances are given by

$$Z_{in} = R_{rad} + i\,X_{in} \tag{10.139}$$

$$Z_L = R_L + iX_L \tag{10.140}$$

The power received by the receiving antenna is delivered to the load. We know that for maximum power transfer

$$Z_L = Z_{in}^* \tag{10.141}$$

For Eq. 10.141 to the satisfied then from Eqs. 10.139 and 10.140.

$$R_L = R_{rad}, X_L = -X_{in} \tag{10.142}$$

With Eq. 10.142 the circuit in Fig. 10.10 reduces to a source $V_{oc}$ connected to a resistance 2 $R_{rad}$. The time averaged power delivered to the load is therefore

$$P_{int} = \frac{1}{2}\left[\frac{|V_{oc}|}{2R_{rad}}\right]^2 R_{rad} \tag{10.143}$$

$$P_{int} = \frac{|V_{oc}|^2}{8\,R_{rad}} \tag{10.144}$$

The time average power at the antenna [Eq. 10.83] is

$$<\mathcal{P}> = \frac{E^2}{2\eta} \tag{10.145}$$

Substituting Eq. 10.138 in 10.144

$$P_{int} = \frac{1}{8R_{rad}} E^2 l^2 \sin^2\theta \tag{10.146}$$

Substituting Eqs. 10.145, 10.146 in 10.137

$$A_{er} = \frac{\eta}{4R_{rad}} l^2 \sin^2\theta \tag{10.147}$$

Substituting Eq. 10.36 in 10.147 (with $d\,l' = l$ Eq. 10.36) and for free space $\eta = 120\pi$ we have

$$A_{er} = \frac{\lambda^2}{4\pi}\left(1.5\,\sin^2\theta\right) \tag{10.148}$$

Substituting Eq. 10.51 in 10.149

$$A_{er} = \frac{\lambda^2}{4\pi} G_r(\theta, \phi) \tag{10.149}$$

$G_r(\theta, \phi)$ is the directive gain of the Hertzian dipole. The effective area is independent of the length of the antenna in Eq. 10.147. Hence $A_{er}$ delivered in Eq. 10.147 is valid for all antennas.

From Eqs. 10.38 and 10.43

$$<\mathcal{P}> = \frac{P_{rad} G_t(\theta, \phi)}{4\pi r^2} \tag{10.150}$$

Here $G_t(\theta, \phi)$ is the directive gain of the transmitting antenna. From Eq. 10.137

$$P_{int} = <\mathcal{P}> A_{er} \quad (10.151)$$

Substituting Eqs. 10.149, 10.150 in 10.151

$$P_{int} = P_{rad} G_t G_r \left[\frac{\lambda}{4\pi r}\right]^2 \quad (10.152)$$

Equation 10.152 is known as Friis transmission formula.

It gives the relationship between the power intercepted by the receiving antenna to the power transmitted by the transmitting antenna. Equation 10.152 applies to the cases where $r >> \lambda$ or in other words both transmitting antenna and receiving antenna should be in far field of each other.

For a given em wave of wavelength $\lambda$, $\frac{A_{er}}{G_r(\theta,\phi)}$ in Eq. 10.149 is a constant. Therefore we can write similar equation for transmitting antenna.

$$A_{et} = \frac{\lambda^2}{4\pi} G_t(\theta, \phi) \quad (10.153)$$

**Example 10.7**
The maximum effective area of a receiving antenna is 10 $m^2$. Find the wavelength of the incoming em wave if the receiving antenna is a half wave dipole.

**Solution**

(a) From Eq. 10.149

$$(A_e)_{max} = \frac{\lambda^2}{4\pi}[G_r(\theta, \phi)]_{max}$$

From Eq. 10.121

$$D = [G_r(\theta, \phi)]_{max} = 1.64$$

Hence

$$(A_e)_{max} = \frac{\lambda^2}{4\pi}[1.64]$$

$$\lambda^2 = \frac{4\pi}{1.64}(A_e)_{max}$$

$$\Rightarrow \lambda^2 = \frac{4\pi}{1.64}[10]$$

$$\Rightarrow \lambda = 8.75\,\text{m}.$$

## 10.8 The Radar Equation

Radar which means radio detection and ranging is an electromagnetic system which is used to detect objects. The transmitter transmits a em signal which gets reflected by the object. The reflected signal is detected by the signal and is used to analyze the position of the object. A typical radar system is shown in Fig. 10.11. Generally the transmitting antenna itself acts as a receiving antenna. This is accomplished with the help of send-receive (S/R) switch as shown in Fig. 10.11.

The power density $<\mathcal{P}>$ at the target which is located at a distance r from the antenna in Fig. 10.11 is given by Eq. 10.150.

$$<\mathcal{P}> = \frac{P_{rad} G_t(\theta, \phi)}{4\pi r^2} \tag{10.154}$$

where $P_{rad}$ is the transmitted power $G_t(\theta, \phi)$ is the directive gain of the antenna in the direction of the target. The incident power is scattered isotropically by the target in all directions. Let $\sigma$ be the backscattering or radar cross-section of the target. Then the power that is scattered in all directions is

$$\sigma <\mathcal{P}> \tag{10.155}$$

Therefore the power density reflected back to the antenna is

$$P_{ref} = \frac{\sigma <\mathcal{P}>}{4\pi r^2} \tag{10.156}$$

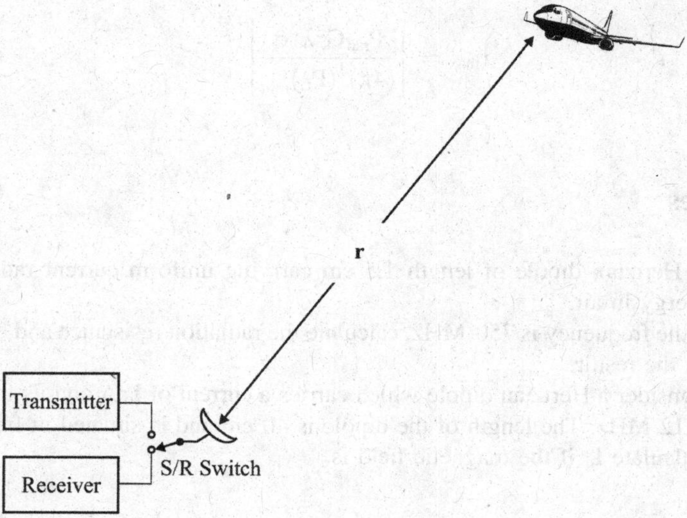


**Fig. 10.11**

The power received by the antenna is

$$P_L = P_{ref} A_e \tag{10.157}$$

Substituting Eq. 10.156 in 10.157

$$P_L = \frac{\sigma < \mathcal{P} >}{4\pi r^2} A_e \tag{10.158}$$

Substituting Eq. 10.154 in 10.158

$$P_L = \frac{P_{rad}}{(4\pi r^2)^2} \sigma A_e G_t(\theta, \phi) \tag{10.159}$$

Substituting Eq. 10.153 in 10.159

$$P_L = \frac{P_{rad}}{(4\pi)^3} \left(\frac{\lambda}{r^2}\right)^2 \sigma [G_t(\theta, \phi)]^2 \tag{10.160}$$

The above equation is called radar equation. The transmitted power varies as $\frac{1}{r^2}$ and hence the reflected power varies as $\frac{1}{r^4}$.

The detector has to detect the signal in the presence of noise. Therefore a minimum amount of reflected power is necessary for the signal to be detected. If the target is faraway, the reflected signal is weak. Therefore there is a "maximum range" beyond which the radar cannot detect the target.

If $(P_L)_{min}$ is the detectable power then from Eq. 10.160 the maximum range $r_{max}$ is

$$r_{max} = \left[\frac{P_{rad} G_t^2 \lambda^2 \sigma}{(4\pi)^3 (P_L)_{min}}\right]^{1/4} \tag{10.161}$$

## Exercises

10.1 A Hertzian dipole of length 1.5 cm carrying uniform current radiates em energy in air.
If the frequency is 750 MHz, calculate the radiation resistance and comment on the result.

10.2 Consider a Hertzian dipole which carries a current of $I_o$ A and its frequency is 12 MHz. The length of the dipole is 40 cm and is situated in free space. Calculate $I_o$ if the magnetic field is

$$H = \frac{0.16i}{r} \sin\theta e^{-i\beta r}$$

10.3 Calculate the total power radiated in Problem 10.2.

10.4 In Problem 10.2 calculate the magnitudes of **E** and **H** at (300 m, 90°, 0°).

10.5 A magnetic dipole of radius 7 cm carries a current of $I = 90\cos(\omega t - \beta z)$A, where $\omega = 336$ M rad/s. The dipole radiates in free space. Write expressions for electric and magnetic fields.

10.6 A magnetic dipole of radius $R = \frac{\lambda}{30}$ radiates in free space. Calculate the radiation resistance and power radiated by the loop, if it carries a peak current of $I_o = 50$ mA.

10.7 A half-wave dipole antenna is radiating in free space. The magnetic field at the distance of 10 km from the dipole is $3 \times 10^{-4}$A/m at $\theta = \pi/2$. The dipole antenna is operating at 90 MHz.

(a) Determine the length of the antenna.
(b) The total power radiated by the antenna.

10.8 Compute the normalised array factor and field for H-plane field pattern if two dipoles are placed at $d = \lambda/4$ in Fig. 10.7a. Given $\alpha = -\pi/2$.

10.9 Find the electric field strength of the em wave if the power intercepted by the receiving antenna in Example 10.7 is $60 \times 10^{-9}$ W. The antenna radiates in free space.

10.10 The wavelength of an em wave radiated in free space by an half wave dipole antenna is 5 m. An Hertzian dipole acting as receiving antenna is placed at a distance of 50 km from the transmitter-the half wave dipole antenna such that $\theta = 90°$. The power received by the receiving antenna is 1 μW. Calculate the power transmitted by the transmitter.

10.11 A radar system uses a single antenna for transmitting and receiving signals whose gain is 30 dB and the transmitted power of the radar system is 4.5 MW. The minimum detectable signal of the radar system is 0.7 mW. The radar locates the target whose cross-section is 12 $m^2$. Find the distance from the target to the antenna if the radar operates at 3 GHz.

10.12 A radar system is used to track an object lying at a distance of 100 km and having cross-section of 10 $m^2$. The antenna whose effective area is 11 $m^2$ radiates a power of 200 KW. The radar operates at 3 GHz. Calculate the power received by the antenna.

# Correction to: Electromagnetics Made Easy

## Correction to:
## S. Balaji, *Electromagnetics Made Easy*, https://doi.org/10.1007/978-981-15-2658-9

The original version of the book was published without incorporating the belated corrections provided by the author.

In general, except for $\nabla^2$, since $\nabla$ is a vector quantity and $\nabla^2$ is a scalar quantity, all del $\nabla$ symbols throughout the book should be in bold font.

Example: As an example note down the following equations

a. $\nabla u$ should be read as $\boldsymbol{\nabla} u$ meaning del in the above equation is an vector quantity.

b. $\nabla . \mathbf{A}$ should be read as $\boldsymbol{\nabla} . \mathbf{A}$ meaning del in the above equation is an vector quantity.

c. $\nabla \times \mathbf{A}$ should be read as $\boldsymbol{\nabla} \times \mathbf{A}$ meaning del in the above equation is an vector quantity.

d. $\nabla \times \nabla \times \mathbf{A}$ should be read as $\boldsymbol{\nabla} \times \boldsymbol{\nabla} \times \mathbf{A}$ meaning del in the above equation is an vector quantity.

In general whenever del appears in gradient, divergence and curl it should be read with bold face meaning it is a vector quantity.

On the other hand $\nabla^2 V$ should be read as such $\nabla^2 V$ without any bold face as $\nabla^2$ is a scalar quantity.

---

The online version of this book can be found at
https://doi.org/10.1007/978-981-15-2658-9

© Springer Nature Singapore Pte Ltd. 2020
S. Balaji, *Electromagnetics Made Easy*,
https://doi.org/10.1007/978-981-15-2658-9_11

In Chapter 1, the figure order has been changed to 1.4, 1.5 and 1.6 in page number 7, Equation "1.79" should be read as

$$= \lim_{\Delta s \to 0} \frac{\oint_{PQRS} \mathbf{A} \cdot d\boldsymbol{l}}{\Delta s} = \left[ \frac{\partial A_z}{\partial y} - \frac{\partial A_y}{\partial z} \right] \quad (1.79)$$

in the Section 1.15 and Example "1.10" should be read as

(i) For face ABCD, $x = 0$ $\mathbf{ds} = -dy\, dz\, \hat{\mathbf{i}}$ as outward drawn normal to ABCD is along $x$-axis

$$\iint_{\substack{ABCD\\x=0}} \mathbf{A} \cdot dy\, dz\, (-\hat{\mathbf{i}})$$

$$= \iint_{\substack{ABCD\\x=0}} \left( x\hat{\mathbf{i}} + y\hat{\mathbf{j}} + z\hat{\mathbf{k}} \right) \cdot dy\, dz\, (-\hat{\mathbf{i}})$$

$$= \iint_{\substack{ABCD\\x=0}} (-x\, dy\, dz) = 0 \text{ as } x = 0$$

(ii) For the face PQRS, $x = 1$ and $\mathbf{ds} = dy\, dz\, \hat{\mathbf{i}}$ . Hence

$$\iint_{\substack{PQRS\\x=1}} \mathbf{A} \cdot dy\, dz\, \hat{\mathbf{i}}$$

$$= \iint_{\substack{PQRS\\x=1}} \left( x\hat{\mathbf{i}} + y\hat{\mathbf{j}} + z\hat{\mathbf{k}} \right) \cdot dy\, dz\, \hat{\mathbf{i}}$$

$$\iint_{\substack{PQRS\\x=1}} x\, dy\, dz = 1$$

(iii) For the face SADP, $y = 0$ and $\mathbf{ds} = -\mathrm{d}x\,\mathrm{d}z\,\hat{\mathbf{j}}$ as the outward drown normal to SADP is along $y$-axis

$$= \iint\limits_{\substack{\text{SADP}\\ y=0}} \mathbf{A} \cdot \mathrm{d}x\,\mathrm{d}z\,(-\hat{\mathbf{j}})$$

$$= \iint\limits_{\substack{\text{SADP}\\ y=0}} \left(x\hat{\mathbf{i}}+y\hat{\mathbf{j}}+z\hat{\mathbf{k}}\right) \cdot (-\mathrm{d}x\,\mathrm{d}z)\,\hat{\mathbf{j}}$$

$$= \iint\limits_{\substack{\text{SADP}\\ y=0}} y(-\mathrm{d}x\,\mathrm{d}z) = 0 \text{ as } y = 0$$

(iv) For the face RBCQ, $y = 1$ and $\mathbf{ds} = \mathrm{d}x\,\mathrm{d}z\,\hat{\mathbf{j}}$

$$= \iint\limits_{\substack{\text{RBCQ}\\ y=1}} \mathbf{A} \cdot (\mathrm{d}x\,\mathrm{d}z)\hat{\mathbf{j}}$$

$$= \iint\limits_{\substack{\text{RBCQ}\\ y=1}} \left(x\hat{\mathbf{i}}+y\hat{\mathbf{j}}+z\hat{\mathbf{k}}\right) \cdot (\mathrm{d}x\,\mathrm{d}z)\hat{\mathbf{j}}$$

$$= \iint\limits_{\substack{\text{RBCQ}\\ y=1}} y\,\mathrm{d}x\,\mathrm{d}z = 1$$

(v) For the face PQCD, $z = 0$ and $\mathbf{ds} = \mathrm{d}x\,\mathrm{d}y\,(-\hat{\mathbf{k}})$ as the outward drawn normal is along $z$-axis

$$= \iint\limits_{\substack{\text{PQCD}\\ z=0}} \mathbf{A} \cdot (\mathrm{d}x\,\mathrm{d}y)(-\hat{\mathbf{k}})$$

$$= \iint\limits_{\substack{\text{PQCD}\\ z=0}} \left(x\hat{\mathbf{i}}+y\hat{\mathbf{j}}+z\hat{\mathbf{k}}\right) \cdot (-\mathrm{d}x\,\mathrm{d}y)\,\hat{\mathbf{k}}$$

$$= \iint\limits_{\substack{\text{PQCD}\\ z=0}} (-z\,\mathrm{d}x\,\mathrm{d}y) = 0 \text{ as } z = 0.$$

(vi) For the face SRBA, $z = 1$ and $d\mathbf{s} = dx\,dy\,\hat{\mathbf{k}}$

$$= \iint_{\substack{SRBA\\z=1}} \mathbf{A}\cdot dx\,dy\,\hat{\mathbf{k}}$$

$$= \iint_{\substack{SRBA\\z=1}} \left(x\hat{\mathbf{i}}+y\hat{\mathbf{j}}+z\hat{\mathbf{k}}\right)\cdot dx\,dy\,\hat{\mathbf{k}}$$

$$= \iint_{\substack{SRBA\\z=1}} z\,dx\,dy = 1$$

in Section 1.16.

In Chapter 4, Equation "4.113" should be read as

$$\mathbf{A} = \frac{\mu_o}{4\pi}\int_{S'} \frac{\mathbf{K}ds'}{\imath} \tag{4.113}$$

and "4.159" should be read as

$$\frac{N}{L}dx \tag{4.159}$$

In Chapter 5, The equation under the text "Integrating over a given volume" in Example 5.2 should be read as

$$\int_{\tau} \nabla\cdot B d\tau = 0$$

In Chapter 6, Equation below "6.12" should be read as

$$\text{As} \oint \mathbf{E}\cdot d\boldsymbol{l} = 0$$

In Chapter 10, Equation "10.57" should be read as

$$\boldsymbol{\nabla}\cdot\mathbf{H}_s = 0 \tag{10.57}$$